Sustainable and Resilient Engineering

Sustainable and Resilient Engineering

Drivers, Metrics, Tools, and Applications

Second Edition

Krishna R. Reddy, Ph.D., P.E., F.ASCE, ENV SP
Professor of Civil and Environmental Engineering
University of Illinois, Chicago, Illinois, USA

Claudio Cameselle, Ph.D.
Professor of Chemical Engineering
University of Vigo, Spain

Jeffrey A. Adams, Ph.D., P.E., ENV SP
Principal Consultant
ENGEO Incorporated, San Ramon, California, USA

WILEY

Published by John Wiley & Sons, Inc., Hoboken, New Jersey.
Published simultaneously in Canada.

Library of Congress Cataloging-in-Publication Data applied for

Hardback ISBN: 9781394267682
ePDF ISBN: 9781394267705
ePUB ISBN: 9781394267699

Cover Design: Wiley
Cover Image: © Artur Debat/Getty Images

Set in 9.5/12.5pt STIXTwoText by Straive, Pondicherry, India

This book is dedicated to our families for their unwavering support, love, and sacrifice:

- *Krishna R. Reddy – Hema (wife), Nishith (son), and Navya (daughter)*
- *Claudio Cameselle – Claudio and Rosa (parents), Susana (wife), and Lucia and Guilherme (children)*
- *Jeffrey A. Adams – Amy (wife) and Lizzie (daughter)*

Contents

Preface for the Second Edition *xix*
Preface for the First Edition *xxi*
About the Companion Website *xxv*

Section I Drivers, Environmental, Economic and Social Impacts, and Resiliency *1*

1 Emerging Challenges, Sustainability, Resiliency, and Sustainable and Resilient Engineering *3*
1.1 Introduction *3*
1.2 Emerging Challenges *4*
1.2.1 Increased Consumption and Depletion of Natural Resources *4*
1.2.1.1 Easter Island Example *4*
1.2.1.2 Metallic Ores Consumption Example *5*
1.2.2 Growing Environmental Pollution *6*
1.2.3 Increasing Population *7*
1.2.4 Increasing Waste Generation *9*
1.2.5 Increasing Greenhouse Gas Emissions *11*
1.2.5.1 Climate Change and Extreme Weather Impacts: Hurricane Sandy Example *15*
1.2.6 Decline of Ecosystems *16*
1.2.7 Loss of Biodiversity *17*
1.2.8 Social Injustice *17*
1.2.9 Urban Sprawl *19*
1.3 The Master Equation or IPAT Equation *20*
1.4 What Is Sustainability? *21*
1.4.1 United Nations Sustainable Development Goals (SDGs) *23*
1.4.2 What Is Sustainable Engineering? *27*
1.5 What Is Resiliency? *33*
1.5.1 What Is Resilient Engineering? *33*
1.5.2 Difference Between Resiliency and Sustainability *34*
1.6 Integrated Sustainable and Resilient Engineering *35*
1.7 Summary *35*
1.8 Questions *36*
1.9 Problems *37*
References *43*

2 Environmental Concerns *47*
2.1 Introduction *47*
2.2 Global Warming and Climate Change *48*
2.2.1 Intergovernmental Panel on Climate Change (IPCC) Assessment Report *56*
2.3 Desertification *57*
2.4 Deforestation *59*
2.5 Loss of Habitat and Biodiversity *60*
2.6 Ozone Layer Depletion *61*
2.7 Air Pollution *63*
2.8 Smog *65*
2.9 Acid Rain *65*
2.10 Water Usage and Pollution *68*
2.11 Eutrophication *74*
2.12 Salinity *74*
2.13 Wastes and Disposal *75*
2.14 Land Contamination *83*
2.15 Visibility *84*
2.16 Odors *84*
2.17 Aesthetic Degradation *85*
2.18 Land Use Patterns *85*
2.19 Thermal Pollution *85*
2.20 Noise Pollution *86*
2.21 Summary *86*
2.22 Questions *87*
2.23 Problems *88*
References *92*

3 Social, Economic, and Legal Issues *97*
3.1 Introduction *97*
3.2 Social Issues *98*
3.2.1 Society *98*
3.2.2 Developed and Developing Societies *99*
3.2.3 Social Sustainability Concept *100*
3.2.4 Social Indicators *100*
3.2.5 Social Impact and Social Vulnerability Assessment Framework *102*
3.2.6 Social Sustainability Implementation *107*
3.2.7 Equity *107*
3.2.8 Environmental Justice *107*
3.3 Economic Issues *110*
3.3.1 Economic Assessment Framework *111*
3.3.2 Life Cycle Costing *113*
3.3.3 True-cost Accounting *114*
3.4 Legal and Governance Issues *115*
3.5 Environmental, Social and Governance (ESG) Reporting *116*
3.6 Summary *117*
3.7 Questions *118*
3.8 Problems *118*
References *122*

4 Availability and Depletion of Natural Resources *125*
4.1 Introduction *125*
4.2 Types and Availability of Resources *126*
4.2.1 Fossil Fuels *126*
4.2.2 Radioactive Fuels *132*
4.2.3 Mineral Resources *132*
4.2.4 Water Resources *134*
4.2.5 Other Elemental Cycles *136*
4.3 Resource Depletion *141*
4.3.1 Causes of Resource Depletion *141*
4.3.2 Effects of Resource Depletion *142*
4.3.3 Overshooting *146*
4.3.4 Urban Metabolism *147*
4.4 Summary *147*
4.5 Questions *148*
4.6 Problems *149*
References *153*

5 Disaster Resiliency *157*
5.1 Introduction *157*
5.2 Climate Change and Extreme Events *158*
5.2.1 Rising Temperatures *158*
5.2.2 Sea Level Rise *160*
5.2.3 Extreme Weather Events *160*
5.2.4 Droughts and Floods *161*
5.2.5 Wildfires *162*
5.3 Impacts of Extreme Events *163*
5.3.1 Hurricanes, Storms, and Floods *163*
5.3.2 Wildfires by Drought and Record Heat *166*
5.4 What Is Resiliency? *166*
5.5 Initiatives and Policies on Resiliency *172*
5.6 Resilient Design of Infrastructure *176*
5.7 Resiliency Assessment Framework *178*
5.7.1 Technical Resiliency *180*
5.7.2 Cascading Impacts of Resiliency *182*
5.8 Resilient Infrastructure Examples *184*
5.8.1 San Francisco Firehouse Resilient Design *184*
5.8.2 San Francisco Resilient CSD Design *184*
5.8.3 Resilient Environmental Remediation *188*
5.9 Managed Retreat and Cost-Benefits *194*
5.10 Soft Infrastructure *194*
5.11 Challenges *195*
5.12 Summary *195*
5.13 Questions *196*
5.14 Problems *196*
References *199*

Section II Sustainability and Resiliency Metrics and Assessment Tools *203*

6 Sustainability and Resiliency Indicators, Metrics, and Assessment Tools *205*
6.1 Introduction *205*
6.2 Attributes of Sustainability and Resiliency Indicators *206*
6.3 Sustainability Indicators *206*
6.4 Sustainability Metrics *217*
6.5 Sustainability Assessment Tools *219*
6.6 Resiliency Indicators *221*
6.7 Resiliency Metrics *226*
6.8 Resiliency Assessment Tools *228*
6.9 Integrated Sustainability and Resilience Assessment Tools *230*
6.10 Summary *230*
6.11 Questions *231*
6.12 Problems *232*
References *233*

7 Material Flow Analysis and Material Budget *235*
7.1 Introduction *235*
7.2 Budget of Natural Resources *236*
7.3 Constructing a Budget *238*
7.4 Material Flow Analysis *238*
7.5 Material Flow Analysis: Wastes *242*
7.6 National Material Account *245*
7.7 Summary *250*
7.8 Questions *250*
7.9 Problems *251*
References *253*

8 Carbon Footprint Analysis *255*
8.1 Introduction *255*
8.2 Global Warming Potential and Carbon Footprint *256*
8.3 Measuring Carbon Footprint *257*
8.3.1 Define the Scope of Your Inventory *257*
8.3.2 Measure Emissions and Establish a Baseline *258*
8.3.3 Develop Targets and Strategies to Reduce Emissions *258*
8.3.4 Offset Unavoidable Emissions *258*
8.3.5 Independent Verification *261*
8.4 Standards for Calculating the Carbon Footprint *261*
8.5 GHG Inventory: Developments in the United States *262*
8.6 USEPA: Greenhouse Gas Reporting Program *262*
8.7 Tools for GHG Inventory *262*
8.8 Carbon Footprint Case Study *264*
8.9 Programs to Mitigate GHG Emissions *270*
8.10 Climate Action Plans *272*
8.11 Carbon Markets: Credits/Offsets *274*
8.12 Summary *274*
8.13 Questions *275*

8.14 Problems *275*
References *276*

9 Life Cycle Assessment *279*
9.1 Introduction *279*
9.2 Life Cycle Assessment *280*
9.2.1 Definition and Objective *280*
9.2.2 Procedure *281*
9.2.3 History *282*
9.3 LCA Methodology *284*
9.3.1 Goal and Scope Definition *285*
9.3.2 Life Cycle Inventory (LCI) *286*
9.3.3 Life Cycle Impact Assessment (LCIA) *289*
9.3.4 Interpretation *294*
9.4 LCA Tools and Applications *296*
9.5 Summary *299*
9.6 Questions *300*
9.7 Problems *301*
References *303*

10 Streamlined Life Cycle Assessment *305*
10.1 Introduction *305*
10.2 Streamlined LCA (SLCA) *306*
10.3 Expanded SLCA *310*
10.4 Simple Example of SLCA *312*
10.5 Applications of SLCA *318*
10.6 Summary *324*
10.7 Questions *324*
10.8 Problems *324*
References *326*

11 Economic Input–Output Life Cycle Assessment *327*
11.1 Introduction *327*
11.2 EIO Model *328*
11.3 EIO-LCA *330*
11.4 EIO-LCA Model Results *331*
11.4.1 Interpretation of Results *331*
11.4.2 Uncertainty *331*
11.4.3 Other Issues and Considerations *332*
11.5 Projects Using the EIO-LCA Model *332*
11.6 Conventional LCA versus EIO-LCA *335*
11.7 EIO versus Physical Input–Output (PIO) Analysis *337*
11.8 Summary *339*
11.9 Questions *340*
11.10 Problems *340*
References *341*

12 Environmental Health Risk Assessment *343*
12.1 Introduction *343*
12.2 Emergence of the Risk Era *344*
12.3 Risk Assessment and Management *344*
12.3.1 Hazard Identification *345*
12.3.2 Dose–Response Assessment *346*
12.3.3 Exposure Assessment *347*
12.3.4 Risk Characterization *349*
12.4 Ecological Risk Assessment *350*
12.5 Summary *352*
12.6 Questions *352*
12.7 Problems *353*
References *354*

13 Emerging Sustainability and Resiliency Assessment Tools *355*
13.1 Introduction *355*
13.2 Environmental Assessment Tools/Indicators *355*
13.3 Economic Assessment Tools *358*
13.3.1 Life-Cycle Costing *358*
13.3.2 Cost–Benefit Analysis *359*
13.4 Ecosystem Services Valuation Tools *360*
13.5 Environmental Justice Tools *361*
13.6 Integrated Sustainability Assessment Tools *362*
13.7 Integrated Sustainability and Resilience Assessment Tools *364*
13.8 Summary *367*
13.9 Questions *367*
13.10 Problems *368*
References *370*

Section III Sustainable and Resilient Engineering Practices *373*

14 Emerging Technologies for Sustainable and Resilient Engineering *375*
14.1 Introduction *375*
14.2 Emerging Smart Technologies *376*
14.2.1 Big Data and Artificial Intelligence *376*
14.2.2 Sensors and Remote-Sensing Technologies (Satellites, UAVs, GIS) *378*
14.2.3 3D Printing and Robotics *380*
14.2.4 Digital Twinning *380*
14.2.5 Internet of Things *382*
14.3 Engineered Materials for Circular Economy *383*
14.3.1 Recycled Waste Materials *383*
14.3.2 Nanomaterials *383*
14.4 Resource Conservation in Energy Sector *384*
14.4.1 Alternative Energy Sources *384*
14.4.2 Innovative Energy Storage Technologies *386*

14.5 Nature-Based Solutions *387*
14.5.1 Bio-Based Materials *387*
14.5.2 Biomimicry *388*
14.5.3 Green Infrastructure *389*
14.5.4 Synthetic Biology *389*
14.6 Summary *389*
14.7 Questions *390*
14.8 Problems *391*
References *391*

15 Sustainable and Resilient Energy Engineering *395*
15.1 Introduction *395*
15.2 Environmental Impacts of Energy Generation *396*
15.2.1 Air Emissions *397*
15.2.2 Solid Waste Generation *400*
15.2.3 Water Resource Use *401*
15.2.4 Land Resource Use *401*
15.3 Nuclear Energy *401*
15.4 Strategies for Clean Energy *403*
15.5 Renewable Energy *405*
15.5.1 Solar Energy *405*
15.5.2 Wind Energy *406*
15.5.3 Water Energy *409*
15.5.4 Geothermal Energy *410*
15.5.5 Biomass Energy *413*
15.6 Economic Considerations *416*
15.7 Sustainability and Resiliency Considerations *418*
15.8 Summary *418*
15.9 Questions *420*
15.10 Problems *420*
References *422*

16 Sustainable and Resilient Materials, Waste Management, and Circular Economy *425*
16.1 Introduction *425*
16.2 Sustainable Materials *426*
16.2.1 Eco-design and Green Materials *426*
16.2.2 Environmental Product Declarations (EPDs) *427*
16.2.3 Extended Producer Responsibility *427*
16.3 Waste Generation and Problems *428*
16.3.1 Types of Waste *429*
16.3.2 Nonhazardous Waste *429*
16.3.3 Hazardous Waste *429*
16.3.4 Effects and Impacts of Waste *430*
16.4 Waste Management *430*
16.4.1 Pollution Prevention *431*
16.4.2 Green Chemistry *432*

16.4.3 Waste Minimization *433*
16.4.4 Reuse/Recycling *434*
16.4.5 Energy Recovery *435*
16.4.6 Landfilling *437*
16.5 Integrated Waste Management *437*
16.6 Sustainable Waste Management *440*
16.7 Circular Economy *442*
16.8 Resiliency Considerations *443*
16.8.1 Resilient Materials *443*
16.8.2 Resilient Waste Management *443*
16.8.3 Resilient Circular Economy *444*
16.9 Summary *444*
16.10 Questions *445*
16.11 Problems *446*
References *447*

17 Sustainable and Resilient Buildings *451*
17.1 Introduction *451*
17.2 Green Building History *452*
17.3 Why Build Green? *453*
17.4 Green Building Concepts *454*
17.4.1 Embodied Carbon and Carbon Neutral Approaches *454*
17.5 Components of Green Building *455*
17.6 Green Building Rating – LEED *457*
17.6.1 LEED Case Study: HBS Tata Hall Executive Education Center *463*
17.7 Sustainable Buildings *466*
17.8 Resiliency Considerations *466*
17.9 Summary *468*
17.10 Questions *468*
17.11 Problems *469*
References *470*

18 Sustainable and Resilient Civil Infrastructure *473*
18.1 Introduction *473*
18.2 Principles of Sustainable Infrastructure *474*
18.3 Civil Infrastructure *475*
18.4 Envision™: Sustainability Rating of Civil Infrastructure *476*
18.4.1 Envision™ Case Study: Berryessa Transit Center *477*
18.5 Sustainable Infrastructure Practices *481*
18.6 Built-Infrastructure Projects: Examples *484*
18.6.1 Transportation Infrastructure Projects *484*
18.6.1.1 Sustainable Airports *484*
18.6.2 Water and Wastewater Treatment Projects *485*
18.6.2.1 Sustainable Water Treatment *485*
18.6.2.2 Reuse of Treated Wastewater *486*

18.6.3 Stormwater Treatment/Green Infrastructure Projects *486*
18.6.3.1 Green Roofs *487*
18.6.3.2 Permeable Pavements *487*
18.6.3.3 Rainwater Harvesting *488*
18.6.3.4 Rain Gardens and Planter Boxes *491*
18.6.3.5 Bioswales *491*
18.6.3.6 Constructed Wetlands and Tree Canopies *491*
18.7 Resiliency Considerations *494*
18.8 Summary *494*
18.9 Questions *495*
18.10 Problems *495*
References *496*

19 Sustainable and Resilient Land Management *499*
19.1 Introduction *499*
19.2 Land Management Problems *501*
19.3 Contaminated Land Remediation Approach *503*
19.4 Green and Sustainable Remediation Technologies *504*
19.5 Sustainable Remediation Framework *509*
19.6 Sustainable Remediation Indicators, Metrics, and Tools *512*
19.7 Case Studies *513*
19.8 Land Remediation Challenges and Opportunities *514*
19.9 Resiliency Considerations *516*
19.10 Summary *516*
19.11 Questions *517*
19.12 Problems *518*
References *519*

20 Climate Geoengineering *521*
20.1 Introduction *521*
20.2 Climate Geoengineering *526*
20.3 Carbon Dioxide Removal (CDR) Methods *527*
20.3.1 Subsurface Sequestration *527*
20.3.2 Surface Sequestration *528*
20.3.3 Marine Organism Sequestration *529*
20.3.4 Direct Engineered Capture *530*
20.4 Solar Radiation Management (SRM) Methods *531*
20.4.1 Sulfur Injection *532*
20.4.2 Reflectors and Mirrors *534*
20.5 Applicability of CDR and SRM *535*
20.6 Climate Geoengineering – A Theoretical Framework *536*
20.7 Risks and Challenges *536*
20.8 Summary *538*
20.9 Questions *539*
20.10 Problems *539*
References *540*

Section IV Sustainable and Resilient Engineering Applications *543*

21 Sustainable and Resiliency Assessment in Engineering Projects *545*
21.1 Introduction *545*
21.2 Integrated Sustainability and Resiliency Assessment Framework *546*
21.2.1 Resiliency Assessment *546*
21.2.2 Sustainability Assessment *549*
21.2.2.1 Environmental Sustainability *550*
21.2.2.2 Social Sustainability *550*
21.2.2.3 Economic Sustainability *551*
21.2.3 Integrated Resiliency and Sustainability Assessment *551*
21.3 Resiliency and Sustainability Assessment of Water Disinfection Technologies at Stickney Water Reclamation Plant *553*
21.3.1 Problem Statement *553*
21.3.2 Project Background *554*
21.3.2.1 Technical Design of Alternatives *555*
21.3.3 Integrated Resiliency and Sustainability Assessment *558*
21.3.3.1 Resiliency Assessment *558*
21.3.3.2 Sustainability Assessment *567*
21.3.3.3 Integrated Resilient Sustainability Index *574*
21.3.4 Conclusions and Recommendations *574*
21.4 Resiliency and Sustainability Assessment to Design and Remediate a Contaminated Site *576*
21.4.1 Problem Statement *576*
21.4.2 Project Background *577*
21.4.2.1 Technical Design of Alternatives *577*
21.4.3 Integrated Resiliency and Sustainability Assessment *581*
21.4.3.1 Resiliency Assessment *581*
21.4.3.2 Sustainability Assessment *583*
21.4.3.3 Integrated Resilient Sustainability Index *591*
21.4.4 Lessons Learned *594*
21.4.5 Conclusions *594*
21.5 Resiliency and Sustainability Assessment of Different Alternatives for Typical Highway Pavement *594*
21.5.1 Problem Statement *594*
21.5.2 Project Background *595*
21.5.2.1 Technical Design of Alternatives *595*
21.5.3 Integrated Resiliency and Sustainability Assessment *597*
21.5.3.1 Resiliency Assessment *597*
21.5.3.2 Sustainability Assessment *598*
21.5.3.3 Integrated Resilient Sustainability Index *603*
21.5.4 Conclusions *604*
21.6 Resiliency and Sustainability Assessment of Community Gardens Versus Hydroponics: Chicago's Sustainable Food Systems *605*
21.6.1 Problem Statement *605*
21.6.2 Project Background *605*
21.6.2.1 Technical Design of Alternatives *605*

21.6.3 Integrated Resiliency and Sustainability Assessment *606*
21.6.3.1 Resiliency Assessment *606*
21.6.3.2 Sustainability Assessment *607*
21.6.3.3 Integrated Resilient Sustainability Index *615*
21.6.4 Lessons Learned *615*
21.6.5 Conclusions *616*
21.7 Resiliency and Sustainability Evaluation of Green Infrastructure in a Chicago Neighborhood *616*
21.7.1 Problem Statement *616*
21.7.2 Project Background *617*
21.7.2.1 Technical Design *618*
21.7.3 Integrated Resiliency and Sustainability Assessment *620*
21.7.3.1 Resiliency Assessment *620*
21.7.3.2 Sustainability Assessment *621*
21.7.3.3 Integrated Resilient Sustainability Index *626*
21.7.4 Lessons Learned *626*
21.7.5 Conclusions *627*
References *628*

Index *631*

Preface for the Second Edition

Sustainability concepts and sustainable engineering practices have been rapidly evolving over the past several years, and their importance has been made amply clear in the first edition of this book. Simultaneously, resiliency and climate-resilient engineering have gained immense significance as communities and governments consider adaptation strategies to counter the effects of climate change, emphasizing the need for their integration into engineering practices. Recognizing the critical importance of combining resiliency with sustainability principles, the second edition of this book reflects this paradigm shift by addressing both aspects with equal prominence. To better represent this integrated approach, the title of the book has been updated to "Sustainable and Resilient Engineering."

Resiliency considerations are emphasized throughout this book. Chapter 5, titled "Disaster Resiliency," has been updated to reflect the growing importance of addressing climate change and extreme weather impacts on engineering projects. A new chapter, Chapter 14 – Emerging Technologies for Sustainable and Resilient Engineering, has been added to emphasize the role of advanced technologies that may be applied to achieve sustainability and resiliency in engineering projects. Chapter 17, which focuses on sustainable and resilient buildings, has been revised to include case studies illustrating the application of the LEED rating system, while Chapter 18, which focuses on sustainable and resilient civil infrastructure, has been revised to provide an example of application of the ENVISION rating system. Additionally, the case studies of sustainable engineering presented in the first edition have been replaced with a new set of examples that integrate both resiliency and sustainability, showcasing methods for quantifying these aspects in engineering projects.

Further, relevant information has been updated in each chapter to reflect current innovations, advancements, and the latest topics of the state of the practice. Additionally, select chapters include example problems to enhance understanding of resiliency and sustainability concepts. To further support learning, a comprehensive set of problems has been provided at the end of each chapter. These problems may be assigned as homework in relevant courses or used for practicing the application of the concepts discussed.

The second edition strives to capture the current state of knowledge in sustainable and resilient engineering practices, aiming to serve as a valuable resource for students, researchers, professionals, policymakers, and other stakeholders, including communities.

The authors would like to express their gratitude to the graduate students at the University of Illinois Chicago – Jagadeesh Kumar Janga, Banuchandra Nagaraja, and Gaurav Verma – for their invaluable assistance. Special thanks go to Jagadeesh Kumar Janga for his significant contributions toward the revisions of the book chapters. The case studies included in this edition are derived from class projects completed by students in the Sustainable Engineering graduate course at the University of Illinois Chicago, whose contributions are sincerely acknowledged.

The authors extend their heartfelt appreciation to their families, colleagues, and friends for their unwavering support throughout the revision of this book.

- Krishna Reddy wishes to express his heartfelt love and gratitude to his wife, Hema; son, Nishith; and daughter, Navya. He also extends his sincere appreciation to his colleagues, as well as his former and current graduate students at the University of Illinois Chicago, for their invaluable encouragement and support. Additionally, he is deeply grateful for the motivation and love he has received from his professional friends worldwide.
- Claudio Cameselle would like to express his heartfelt love and gratitude to his wife, Susana, daughter, Lucía, and son, Guilherme. He also extends his sincere appreciation to his sister, Rosa María; brother-in-law, Jose Manuel; and nephew, Adrian. A very special and loving thanks goes to his father, Claudio, and to his late mother, Rosa, who is always in his thoughts. Additionally, he is deeply grateful to his colleagues at the University of Vigo, whose support has been invaluable in writing this book. He also wishes to thank his international collaborators for their shared work and experiences.
- Jeff Adams would like to express his heartfelt love and gratitude to his wife, Amy; daughter, Lizzie; and his gratitude for shared ideas, experiences, and camaraderie to his colleagues at ENGEO Incorporated and the faculty, students, and staff at the Paul H. O'Neill School of Public and Environmental Affairs, Indiana University Bloomington. He would also like to thank his professional and academic mentors, friends, and family in the United States and worldwide for their support and inspiration.

Finally, the authors would like to express their gratitude to Summers Scholl, Executive Editor of Physical Sciences at Wiley, and Satvinder Kaur, Managing Editor at Wiley, for their invaluable encouragement and support in ensuring the timely completion of this second edition.

Krishna R. Reddy, Chicago, USA
Claudio Cameselle, Vigo, Spain
Jeffrey A. Adams, San Ramon, California, USA
15 December 2024

Preface for the First Edition

The concept of *sustainability* is embedded everywhere. Over the past couple of decades, an unprecedented awareness has arisen among private citizens, governmental bodies, and the business world that human activities continue to impart an enormous effect on the natural environment. Human activities are not only affecting the natural environment but they are also affecting human health, well-being, and mankind as a whole. Among the most striking environmental impacts are the rapid depletion of natural resources, pollution of environment (air, soil, and water), and global warming and climate change. As the natural environment is impacted, strains are placed on society and the economy.

Consider the manifestation of sea-level rise resulting from climate change. With increases in mean sea level, coastline habitats can become inundated, and salt water intrudes into estuary environments, greatly disrupting flora and fauna in these ecosystems. The built environment is also significantly affected; ports, roads, bridges, and other shoreline structures and improvements are diminished in serviceability or lose function altogether. In many cases, these result in adverse economic effects, either through loss of jobs, associated profits and tax revenues, or require large expenditures to maintain operation of these systems. The lost economic value can have a direct adverse effect on society, leading to increased unemployment, decreased fiscal resources for ancillary social services, and an overall diminished quality of life.

In meeting these increasing challenges faced by humankind and planet Earth, there is an increased focus among engineers, technical and design professionals, government regulators, the business communities, and the general public to seek alternatives of development that meet the needs of the present without compromising the ability of future generations to meet their own needs. This precise definition of sustainability is one that serves as the guiding premise of this book. Furthermore, as we look for sustainable solutions to these global challenges, the aforementioned stakeholders increasingly look for solutions that maximize the benefit when considered among environmental, economic, and societal dimensions. The maximization of these benefits or the minimization of the adverse impacts in these dimensions, commonly described as the "triple bottom line," is also a recurring theme of this book. The dimensions or parameters of the triple bottom line may be further described as the following:

- *Environmental* – diversity and interdependence within living systems, the goods and services produced by the world's ecosystems, and the impacts of human-generated wastes (the planet)
- *Economic* – flow of financial capital and the facilitation of commerce (prosperity)
- *Social* – interactions between institutions/firms and people, functions expressive of human values, aspirations, well-being, and ethical issues (people)

Sustainable engineering involves the application of sustainability principles and concepts to design environmentally friendly, economically viable, and socially equitable projects, products, and systems. It offers a means to develop designs that may positively affect social, economic, and environmental considerations across the entire life cycle of a project, product, or system. Currently, a common goal of sustainable engineering is to engineer systems that are flexible enough to adapt to a range of dynamic, external stressors imposed by the natural environment while prudently utilizing resources and minimizing waste generation and emission.

This book is organized into four sections. Section I focuses on key drivers that identify the need to incorporate sustainability principles into engineering practices. The major concerns and challenges to minimize the impacts on the environment, economy, and society – the three dimensions of the triple bottom line – are discussed. This section also highlights the importance of resiliency in engineering design and the key approaches to incorporate it in conjunction with sustainability in engineering project designs. Section II provides a comprehensive presentation of metrics and tools for the sustainability assessment of alternative designs for a project, product, or activity. Detailed explanations on several topics, including material flow analyses and material budgets, carbon footprint analysis, life-cycle assessment and streamlined life-cycle assessment, economic input and output models, environmental health-risk assessments, and other emerging assessment tools are provided in this section. Section III details several sustainable engineering practices, including topics such as sustainable energy engineering, sustainable waste management, green and sustainable buildings and civil infrastructure, green and sustainable remediation, and climate geoengineering. Section IV provides a range of case studies demonstrating the application of sustainability assessment tools to assess sustainability in engineering project designs. These are grouped into several categories, including environmental and chemical engineering projects, civil and materials engineering projects, and infrastructure engineering projects. Each chapter concludes with questions that may be used for review, contemplation, or as coursework exercises.

This book is primarily intended as a textbook for graduate students interested in sustainability and sustainable engineering. However, the structure and the contents of the book make it a valuable reference source for engineers, technicians, administrators, and members of the public interested in the topic of sustainability. The authors have collectively amassed decades of research, teaching, and consulting experience and have strived to write this book so that it can be valuable to practitioners and academic professionals alike.

The authors offer their deepest gratitude to a number of individuals and organizations who have provided assistance, support, patience, brainstorming, and inspiration toward the completion of this book. Many students who have taken the Sustainable Engineering graduate course offered by Krishna Reddy at the University of Illinois at Chicago (UIC) provided motivation to undertake this book project, added to the content, and developed the example projects presented in Chapters 20–22. Their hard work is gratefully acknowledged. The authors are grateful to Girish Kumar, a doctoral student at UIC, for his immensely valuable assistance in preparing, reviewing, and editing the chapters. The authors are also thankful to other graduate students at UIC, particularly Jyoti Chetri, who also reviewed the chapters. Claudio Cameselle is thankful for the Fulbright Fellowship that allowed him to visit UIC and undertake this book project. The support of researchers and staffs at the University of Vigo, especially Susana Gouveia, has been vital in the completion of this book. Jeff Adams is immensely thankful to ENGEO Incorporated, particularly Uri Eliahu, Shawn Munger, Joe Tootle, Scott Johns, and Divya Bhargava, all of whom have thoughtfully provided love and encouragement. Finally, the authors thank Bob Esposito, Michael Leventhal, and the entire Wiley editorial group (Beryl Mesiadhas and Grace Paulin) for their cooperation during this book project. We hope that this book will help students and professionals alike to learn and advance the concepts of sustainability and develop sustainable engineering projects.

Krishna R. Reddy, Chicago, USA
Claudio Cameselle, Vigo, Spain
Jeffrey A. Adams, San Ramon, USA
25 January 2019

About the Companion Website

This book is accompanied by instructor website. The URL for the site is

www.wiley.com/go/reddy/sustainableresilientengineering2e

The Instructor's website includes:

- PowerPoint Slides
- Solutions to Chapter Problems

Section I

Drivers, Environmental, Economic and Social Impacts, and Resiliency

1

Emerging Challenges, Sustainability, Resiliency, and Sustainable and Resilient Engineering

Chapter Goal and Objectives

This chapter aims to introduce the drivers and concepts of sustainability and resilience, highlighting the importance of integrating them into engineering practice. The specific objectives are to (i) provide examples of emerging challenges that necessitate a focus on sustainability and resilience, (ii) explain the definitions of sustainability and resilience and clarify the differences between these two concepts, and (iii) emphasize the importance of sustainable and resilient engineering in promoting sustainable development.

1.1 Introduction

The concepts of *resiliency* and *sustainability* are everywhere. In recent years, ever-growing numbers of people around the world have become more aware of strains placed on the Earth. These strains have been manifested in a variety of ways – accelerated exploitation of natural resources, increased waste generation, pollution of air, soil, and water, and climate change. Not only have private citizens taken notice, but governments and the business world have also taken steps to address sustainability. Numerous intra- and intergovernmental initiatives and agreements have been developed to address the strains on the environment and to identify measures that encourage more sustainable practices. Businesses, too, have realized that sustainability is a good practice for a variety of reasons. New systems and products have been developed that are more protective and less wasteful of resources, and the pursuit of the "triple bottom line" (TBL) of sustainability has been increasingly applied in new projects and products. The TBL is the reference framework in sustainability that accounts for financial as well as social and environmental metrics.

Similarly, the increasing impacts of climate change, along with other uncertainties such as economic instability and social unrest, make it essential for the world to embrace the concept of resilience. In engineering, technical robustness against both expected and unexpected shocks has historically been the primary focus of design and implementation. This approach can be seen as a form of resiliency consideration. However, growing concerns over climate change and other uncertainties necessitate the engineering community to consider other dimensions of resiliency, such as environmental, economic, and social impacts of a failure. Despite the importance of these dimensions of resiliency, they are often overlooked during engineering decision-making processes.

Sustainable and Resilient Engineering: Drivers, Metrics, Tools, and Applications, Second Edition. Krishna R. Reddy, Claudio Cameselle, and Jeffrey A. Adams.

Companion website: www.wiley.com/go/reddy/sustainableresilientengineering2e

Several key questions have emerged that necessitate contemplation. What emerging challenges are forcing us to think about sustainability and resiliency? What is sustainability? What is resiliency? How do we take action to further these concepts? What are sustainable engineering (SE), resilient engineering (RE), and integrated sustainable and resilient engineering, and what role can they play in sustainable development? Of equal importance, how do we determine success in pursuit of these initiatives, and how do we measure our progress toward these goals? These evolving and increasingly significant concepts are the focus of this book.

This chapter describes the broader emerging challenges that are forcing us to think about sustainability and resiliency. Next, the general definitions and interpretations of the meaning of sustainability and resiliency are presented. Finally, sustainable and resilient engineering and its role in achieving sustainable development are described.

1.2 Emerging Challenges

Before we delve into the concepts, applications, methods, and measures related to sustainability and resiliency, let us examine several acute problems and related examples that are faced worldwide and are increasingly having a measurable, detrimental effect on the planet.

1.2.1 Increased Consumption and Depletion of Natural Resources

A key consideration of sustainability focuses on our ability to preserve resources for future generations. This is extremely important, as many essential resources (e.g. precious metals, fossil fuels) are non-renewable and are limited in quantity. For many of these resources, we are on a current trajectory of utilization/exploitation in which near-total depletion of economically viable reserves is a very realistic possibility. The alarming rates of consumption of a number of resources not only spell trouble for the availability of these resources for future generations but also cause unintended secondary, yet catastrophic, side effects on the environment.

1.2.1.1 Easter Island Example

A classic example of the catastrophic consequences of the primary and secondary effects of natural resource depletion is the collapse of a civilization on *Easter Island*. Located in the southeastern Pacific Ocean and arguably the most remote habitable region on the planet, Easter Island gained its name from the sighting/discovery of the island on Easter Sunday, 1722, by Dutch sailors. It was subsequently annexed by Chile in 1888. Large stone statuary called moai, created by early Rapa Nui peoples, were important monuments to a sophisticated culture and civilization that had once flourished on Easter Island but had devolved into a small, primitive culture at the time of European discovery (DiNapoli et al. 2021).

Although Easter Island is subject to a cold and dry climate, it was at one time heavily forested with palms, conifers, and sandalwood. The first Polynesians arrived at Easter Island in the fifth century and numbered no more than 20 or 30. The harsh climate and nutrient poor soils restricted agricultural activity to the cultivation of sweet potatoes. Nevertheless, a sophisticated and advanced society flourished among the Rapa Nui. To allow for agricultural activity, much of the land was deforested. Trees were also harvested to provide structural materials for housing and boat fabrication and for use as fuel. However, a significant number of trees were also harvested to create a track system to maneuver the large moai from quarry locations to sites where they were erected. As the population grew upward of 7000 persons, these resources were further utilized to meet increasing demand.

By the 1600s, the entire island had been deforested. The lack of timber resources eliminated the ability to construct fishing boats and wooden structures. The elimination of bark materials prevented the fabrication of cloth materials. Furthermore, the deforestation greatly accelerated soil erosion, and the agricultural capacity of the already nutrient-poor soils was again severely reduced. Elaborate rituals centered around the moai statuary diminished, placing even greater strain on the social fabric of the declining society, including the breakdown of social and religious conventions. Because boats could no longer be fabricated without timber, the Rapa Nui were trapped on the remote island. Eventually, they were forced to resort to primitive cultural practices where available shelter had been reduced to available caves. Ongoing turbulent conditions fueled conflict/warfare, slavery, and even reports of cannibalism.

Ultimately, the once great civilization had collapsed. At the time of European discovery, the population had declined precipitously. The collapse of island's agricultural capacity and activity resulted in widespread, ongoing starvation. Subsequent contacts from seafaring groups, such as whalers, introduced sexually transmitted diseases and smallpox. Peruvian slave parties also captured numerous Rapa Nui for use in the slave trade. By 1877, only 111 Rapa Nui remained on Easter Island, and at the time of Chilean annexation, the Chilean government confined the remaining inhabitants to one village.

Rapa Nui remain on Easter Island to this day. Archaeologists brought attention to the island in the mid-twentieth century, which in turn has stimulated tourism and led to the restoration of some of the moai statuary. Conditions have improved for the Rapa Nui, although unemployment remains high and alcoholism and related social strains are quite prevalent. Additionally, they are still dependent on imported food.

Nevertheless, the story of the Rapa Nui has been repeated with other people – the Mayan and Inca people of Central and South America as well as the ancient Greeks and Romans. All of these people offer a cautionary tale – when a society disregards the health of its environment, places excessive strains on vital resources, such as soil and water, can lead to a collapse of agricultural activity and other aspects of economy and culture. When basic necessities such as food, clothing, and shelter become scarce, a disparity between "haves" and "have-nots" is often exacerbated, leading to mistrust and resentment between classes. As the problem grows, conflict is inevitable, and collapse of the underlying civilization will occur.

1.2.1.2 Metallic Ores Consumption Example

Several examples can be presented with respect to unsustainable utilization of natural resources. Let us take the example of the usage of metallic ores, and as an example, let us examine the use of zinc (Graedel and Allenby 2010). Consider zinc use over a sustainability design period of 50 years and a global population of 7.5 billion people. Estimated global zinc reserves consist of 330×10^{12} g (330 Tg) of zinc. Considering a 50-year period (after which all resources will be depleted), 6.6 Tg may be used per year. Assuming an even allocation among 7.5 billion people, each person would be entitled to 0.9 kg/year. Allowing for recycling (assuming a 30% increase in zinc supply), the available per capita allotment may be increased to 1.2 kg/person/year.

Consider these industrial/consumer uses of zinc: zinc alloys – 38% (0.5 kg/person/year); construction coatings – 31% (0.4 kg/person/year); zinc chemicals – 15% (0.15 kg/person/year); vehicle plating – 15% (0.15 kg/person/year); cast, rolled zinc – 4% (0.05 kg/person/year).

However, current automobiles contain approximately 32 kg of zinc (Ortego et al. 2020). Assuming a 1.2 kg/year allocation, this would result in a new car purchase once every 27 years – only if no other uses of zinc were allocated. When compared to the uses listed above, zinc would need to be removed from all other product streams and uses to meet automobile use demand at this rate.

Of course, it is a valid argument that not every person on the planet will consume an equal amount of zinc, allowing for some to consume vastly more than others in this example. Nevertheless, when finite resources are available for use, it is necessary to consider the limits of the resource. A famous example takes this further. Imagine a resource that is finite and is open to use to a finite number of users. Even in this case, the users may be motivated to use more of the resource because the side effects do not offset the gain. This famous example is detailed in a seminal paper, *the Tragedy of the Commons* (Hardin 1968, 2009).

Here is the scenario – 20 farmers use a communal pasture to graze 10 cows each or 200 cows. Each farmer is utilizing 1/20 of the resource – or 5%. Now, let us say one farmer adds an 11th cow. He is increasing his benefit by 10%, but the consequences of the additional cow (let us assume the additional loading begins a process of overgrazing) are borne by all equally, so he suffers 5% of the consequences. He has just increased his marginal benefit of adding one cow by this difference. The farmer has an incentive to act in this manner, but so do all of the farmers. It is reasonable to expect that other farmers will also choose to add cows, increasing their respective marginal benefit, but all of the farmers ultimately will suffer the consequences of overuse together.

The *Tragedy of the Commons* remains a controversial study, and many have pointed out shortcomings or exceptions to the model, but it is a simple illustration that individual parties often have incentives to exploit a resource, even if the consequences will eventually spoil that resource. In other words, it remains difficult to get multiple parties to act together to preserve resources. This is vividly illustrated in Figure 1.1, which shows the effect of overfishing on the population of sardines near the Pacific coast of North America. Sardine populations have dramatically declined since their peak in 2006. This led to a fishing ban enacted to protect the collapsing population. The fishing ban is not expected to be lifted in the short or mid-term, until the sardine population reaches acceptable levels to support fishing again.

1.2.2 Growing Environmental Pollution

The growth in consumption of natural resources has led to increasingly problematic side effects – one in particular is increased environmental pollution. Air, soil, and water are being

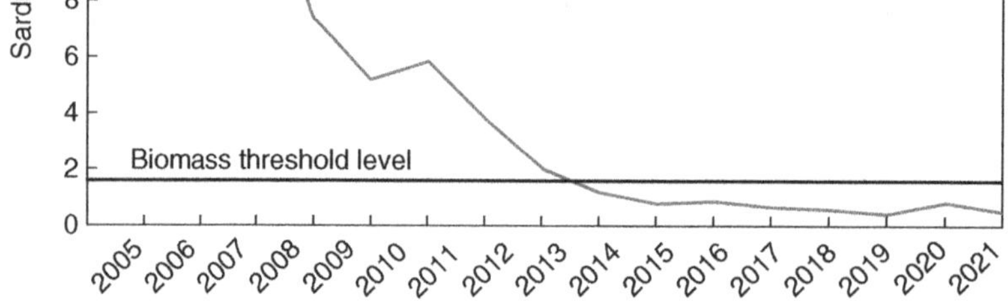

Figure 1.1 Sardine population declining up to below the threshold level. *Source:* Kuriyama et al. (2022)/ The National Oceanic and Atmospheric Administration (NOAA)/Public domain.

increasingly polluted from human development and activities. Control and remediation of the effects of pollution are crucial in order to protect the environment and public health, and a movement toward sustainability is not possible without control or prevention.

Water pollution can have many effects. Pollutant loading can directly affect domestic water sources, such as surface water or groundwater resources. Sediment runoff can affect water flow in navigable waterways. In addition to direct deposition of toxic materials to surface water, pollution sources can also have a long-term detrimental effect on surface water bodies (Schweitzer and Noblet 2018).

Soil impacts are also increasing. In the form of mining, waste products are often landfilled at or near the extraction point. Noxious by-product chemicals capable of leaching are often disposed of as well, and when mixed with tailings and rainwater, the resulting toxic "soup" can significantly pollute soil as well as groundwater. Not only can mining practices affect soil, but also other practices such as deforestation, overgrazing, timber harvesting, and agricultural management can lead to releases of potentially toxic materials or otherwise exacerbate environmental impacts such as erosion of topsoil. Further, landfilling practices used in waste management can also affect soil. Older landfills often do not have protective measures in place, such as liners or leachate collection systems, and authorized or unauthorized hazardous waste disposal facilities accept acutely toxic or recalcitrant compounds, such as aromatic and aliphatic hydrocarbons, chlorinated solvents, PCBs, pesticides, and heavy metals, which can be released into the environment (Li et al. 2019).

In addition to water and soil, air is also increasingly polluted by human activities. Combustion of wood and fossil fuels leads to emissions of gases and particulate matter, many of which can affect health or even climate. Evaporated volatile organic compounds (VOCs), carbon monoxide, heavy metals, and particulates can lead to harmful impacts on public health, including elevated asthma or cancer rates. Some emission constituents are subject to secondary reactions in the atmosphere, leading to increased smog, near-surface ozone, or depletion of higher atmospheric ozone (Mannucci et al. 2015).

1.2.3 Increasing Population

Much of the strain placed on the planet is associated with ever-increasing uses of natural resources, the unintended side effects, and pollution that results from their use of land, air, and water. However, even if measures were taken to curtail these uses and mitigate the effects of pollution, a third problem leading to strains that may not be as easy to curtail, is the rapid increase in the total human population as well as increasing acceleration in population growth.

The statistics are staggering – over 8 billion humans are currently alive on Earth. This number has increased by approximately 2 billion in only 25 years and has doubled from approximately 4 billion in the past 50 years (Figure 1.2). Currently, the world population grows by 75 million on an annual basis. If current growth rate predictions prove to be true, the world population will exceed 9.1 billion by the year 2050. Figure 1.2 demonstrates the rapid acceleration in population growth over the past millennium. It also demonstrates how population totals may grow, level off, or contract by the year 2200 assuming worldwide average female fertility rates over this period of time (Wakabayashi 2020).

With additional people added at such an accelerating pace, goods and resources become scarcer. Of course, the growing population will need to use resources for food, clothing, and shelter, as well as participate in a meaningful way in local, national, and/or international economies. Yet, challenges are already faced by a significant portion of the world's population in even trying to meet the most basic human needs. Approximately 985 million people experience extreme poverty (defined as income of less than or equal to US $1 per day). Over 800 million people face severe

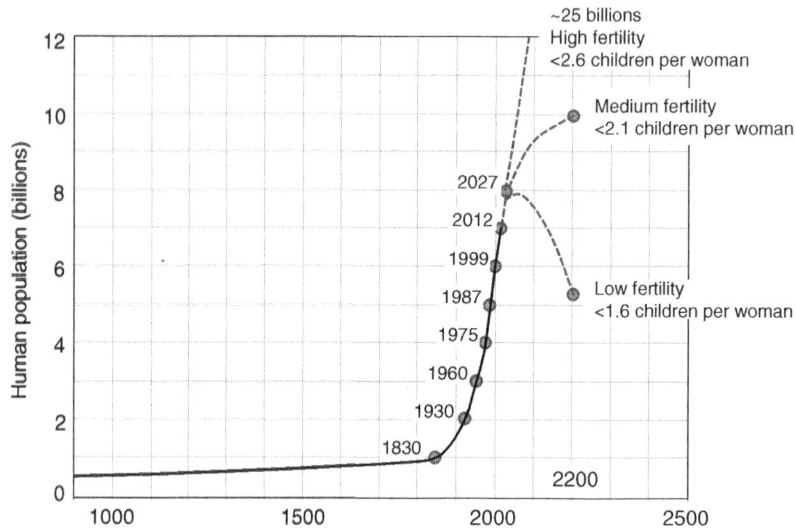

Figure 1.2 World's population explosion and projection to the year 2200. *Source:* Data from United Nations.

malnutrition. A startling number of preschool children, over 6 million, die from hunger/malnutrition-related maladies per year (Dasandi 2014).

With the rapid increase in world population, similar appreciable growth in world economic output has occurred. Global economic production has doubled in the past 25 years, a rate that even exceeds the rapid rate in population growth. This development has led to a great benefit for many millions of people, as the growth has indeed lifted millions out of poverty. Gross domestic products in numerous developing countries have grown significantly over this time period, and scores of nations throughout the world have grown into world economic powers from modest economic positions within a generation. Nevertheless, economic development in many of these countries has proven to be uneven at best; large inequalities in wealth have developed within these countries as well as compared to the positions of larger, more developed countries, such as those in Europe or North America. As a result, real income in many developing countries is actually falling. Countries with high birth rates are especially susceptible to this growing inequity; stabilization of birth rates will be essential in assuring more equitable economic growth for a given population in the future.

Example 1.1

The city of Beijing has a recorded population of 21.7 million people in the year 2017. The annual population growth rate has been recorded at 2.1% per year. China's average birthrate has declined swiftly over the past five years, from 12.4 births for every 1000 citizens in 2017 to 7.52 births for every 1000 citizens in 2021. At the same time, the average death rate in China was recorded at a constant value of 7 deaths for every 1000 citizens over the five-year period. Assuming that Beijing's birth and death rate are consistent with that of China's average, calculate the following:

a) Assuming the decrease in birth rates is linear between 2017 and 2021, determine the population increase due to newborns each year in Beijing.
b) Assuming birth, death, and migrant influx/outflux as the only reasons for population change, calculate the population growth due to migrant influx each year, comment on the observed trends.

Solution:

a) Considering a linear decrease in birth rates between 2017 and 2021, the birth rates for the years 2017, 2018, 2019, 2020, and 2021 are 12.4, 11.18, 9.96, 8.74, and 7.52 births for every 1000 citizens, respectively.
Population increase = (birth rate × population in the corresponding **year)**
Population increase In 2017 = $21.7 \times (12.4/1000) = 0.269$ million people
Population increase in 2018 = $[21.7 \times (1 + 2.1\%)] \times (11.18/1000) = 0.2477$ million people
Population increase in 2019 = $[22.1557 \times (1 + 2.1\%)] \times (9.96/1000) = 0.2253$ million people
Population increase in 2020 = $[22.62 \times (1 + 2.1\%)] \times (8.74/1000) = 0.2018$ million people
Population increase in 2021 = $[23.096 \times (1 + 2.1\%)] \times (7.52/1000) = 0.1774$ million people

b) **Natural growth = growth due to birth-decrease due to death**
Natural growth in 2017 = $0.269 - 21.7 \times (7/1000) = 0.11718$ million people
Natural growth in 2018 = $0.24771 - 22.1557 \times (7/1000) = 0.09262$ million people
Natural growth in 2019 = $0.2253 - 22.62 \times (7/1000) = 0.06696$ million people
Natural growth in 2020 = $0.2018 - 23.096 \times (7/1000) = 0.0402$ million people
Natural growth in 2021 = $0.17733 - 23.581 \times (7/1000) = 0.01226$ million people

Migrant growth = total growth-natural growth
Migrant growth in 2017 = $21.25 \times 2.1\% - 0.11718 = 0.329$ million people
Migrant growth in 2018 = $21.7 \times 2.1\% - 0.09262 = 0.36$ million people
Migrant growth in 2019 = $22.1557 \times 2.1\% - 0.06696 = 0.4$ million people
Migrant growth in 2020 = $22.62 \times 2.1\% - 0.04 = 0.43$ million people
Migrant growth in 2021 = $23.096 \times 2.1\% - 0.01226 = 0.47$ million people

This trend suggests that despite the birth rate decrease, the overall population of Beijing continued to grow steadily, primarily driven by migrants.

1.2.4 Increasing Waste Generation

As the population continues to grow at an accelerating rate and is coupled with increasing use of natural resources, an additional by-product of these factors is also placing strain on the environment of the planet – increased waste generation. Enormous quantities of waste are being generated every year, and this growth continues unabated. It is especially troubling and detrimental to the environment because much of this waste generation growth is occurring in the developing world, where waste management and disposal practices are very basic and not protective of the environment. Much of the developed world has placed an increased focus on waste minimization and diversion, such as reuse and/or recycling; however, large quantities of waste materials cannot be practicably repurposed and therefore require disposal or incineration. As a result, the detrimental environmental effects of both proper and improper waste management continue to grow, further spoiling the environment. Collectively, countries around the world must find ways to reduce this increasing trend of waste generation, as these practices cannot be considered sustainable.

The framework to reduce waste generation and the pollution effects that lead to fouling of the environment is straightforward, and it is a necessary undertaking such that we do not jeopardize the quality of life for present or future generations. Ultimately, we need to identify waste products that have the greatest impact on the environment. Once these materials have been identified, the activities acting as a source of waste materials can be identified, and steps can be taken to reduce or minimize the activity through alterations in generation practices or by identifying more protective alternative activities. Measures may also be taken to prevent the introduction of unavoidable waste products into the environment in such a way that environmental impacts may occur. Further, pollution impacts that have already occurred may be identified and remedied through a variety of traditional or innovative environmental remediation techniques. Ultimately, in order to achieve sustainability, civilization will need to identify technologically feasible means to transition from polluting activities to environmentally protective activities.

Before techniques and alternatives to reduce waste generation can be identified, it is important to discuss types of waste. In 1976, the Resource Conservation and Recovery Act (RCRA) was enacted into law in the United States. While this was not the first attempt to regulate waste disposal practices, it was the first comprehensive legal framework that reached nearly all facets of waste generation, storage, and disposal for a wide range of waste classifications. Generally speaking, solid waste is generated in a variety of residential, commercial, and industrial settings. Much of this material is not considered hazardous and thus falls into a general category of municipal solid waste (MSW). Figure 1.3 shows the MSW production by type in the United States. Much of this material can be recycled and reused or repurposed. Figure 1.4 shows the end fate of the MSW waste stream in the United States. Notable progress in recycling programs has greatly expanded the percentage of waste that is recycled. However, two-thirds of MSW are still not being repurposed and are either being landfilled or incinerated.

As an example of MSW management, consider landfills. Landfills have been used for countless generations in waste management; however, only in the past century have advances been made to help mitigate the environmental side effects of landfilling. Sanitary landfills have been constructed to manage odors, protect underlying soil and groundwater, and facilitate the collection and management of generated landfill gas and leachate. Leachate is a liquid by-product of landfill operation. It consists of infiltrated water (often from rainfall) as well as liquid and "leached" solid wastes from waste materials. Leachate often contains elevated concentrations of a variety of organic and

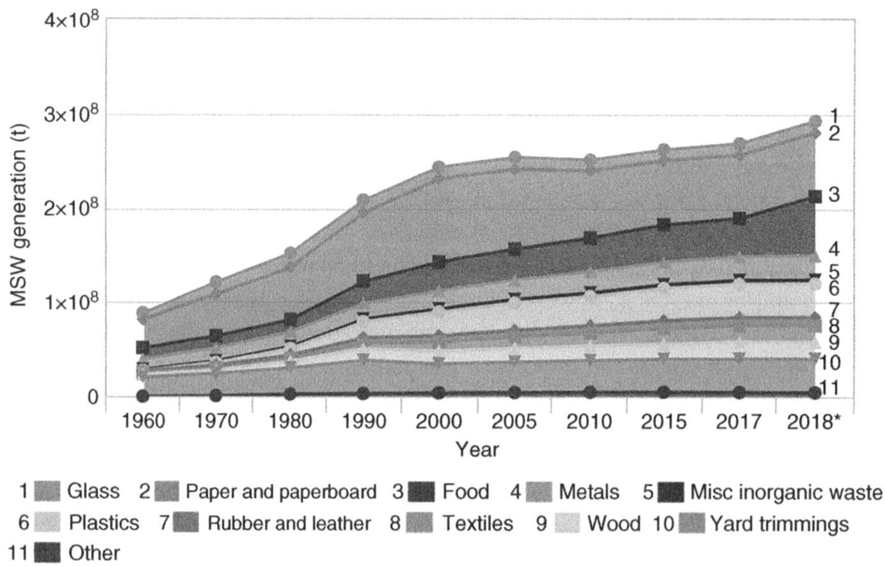

Figure 1.3 Total municipal solid waste (MSW) generated in the United States. *Source:* USEPA (2023)/United States Environmental Protection Agency.

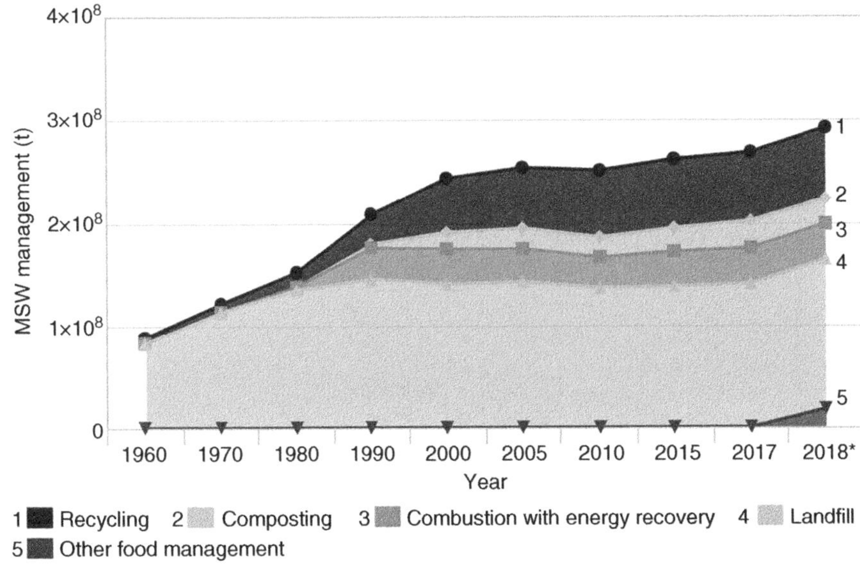

Figure 1.4 Municipal solid waste (MSW) recovery by material and MSW management in the United States. *Source:* USEPA (2023)/United States Environmental Protection Agency.

inorganic chemical compounds, heavy metals, and other potentially deleterious materials. If not controlled properly, it can enter groundwater, where it can adversely affect aquifer sources (Ozbay et al. 2021). Landfill-induced groundwater contamination is a serious problem throughout all 50 states, not only from recently constructed landfills but more often from old uncontrolled or abandoned landfill sites.

Landfills also generate copious amounts of gases that need to be controlled. Buried wastes commonly undergo anaerobic decomposition, resulting in the production of biogas. This biogas primarily contains methane, carbon dioxide, and hydrogen-containing compounds. If not controlled, this can be a significant health and explosion hazard, both at the landfill site and off-site due to horizontal and vertical migration into nearby structures. However, when properly captured and treated, biogas is a valuable resource that can be used as a fuel for power generation. As an example, in 2008, commercial landfill gas produced electricity and natural gas for 1.4 million homes (Sadasivam and Reddy 2014). Because it burns relatively clean, its capture and use can displace the use of dirtier fossil fuels, leading to a relative reduction in greenhouse gas emissions and air pollutants.

1.2.5 Increasing Greenhouse Gas Emissions

Increased resource generation and utilization leads not only to increased solid/liquid waste generation but also to increased gas emissions. Gas emissions, particularly those from agriculture and combustion processes in the industrial, transportation, and energy generation sectors, can lead to serious air pollution consequences, but they can also facilitate climate change. Although climate change occurs due to a variety of natural physical processes, overwhelming evidence suggests that anthropogenic sources also contribute to climate change. The magnitude of this contribution and the resulting effects of climate change are still being debated, but very little scientific basis exists that objectively refutes contributions to these effects from anthropogenic sources.

Much of the anthropogenic portion comes from the emission of greenhouse gases. Greenhouse gases, such as carbon dioxide, methane, nitrous oxide, and water vapor, as well as a range of fluorocarbons, act to trap heat from the earth's surface that would normally radiate to space harmlessly (Figure 1.5). This effect, similar to the trapping of heat generated by sunlight within a greenhouse (commonly known as the Greenhouse Effect), warms the lower atmosphere. Higher temperatures result in increased generation of water vapor, which can act to trap more heat. However, higher temperatures also accelerate plant metabolism, which leads to utilization (and removal) of carbon dioxide from the atmosphere. These phenomena are still being studied (Rae et al. 2021).

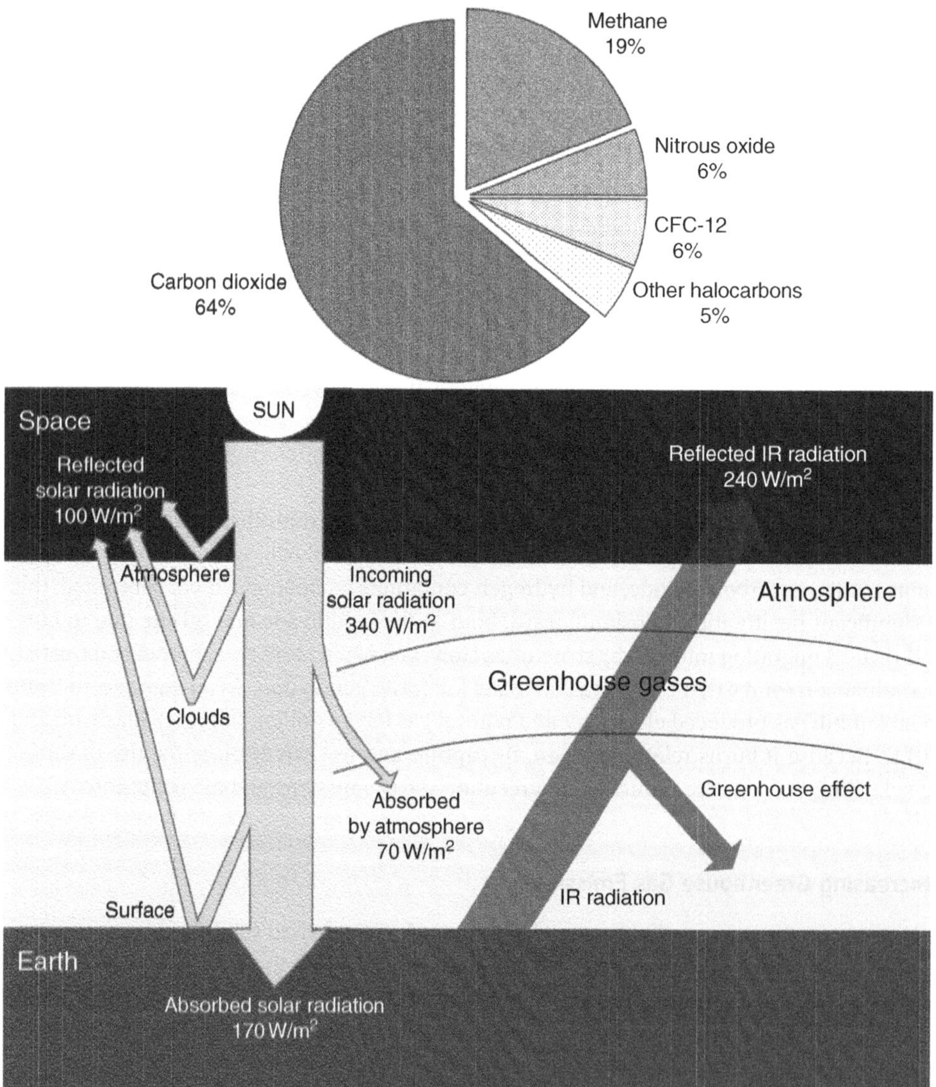

Figure 1.5 Greenhouse gas effect on the atmosphere and the estimated relative contribution of main greenhouse gases.

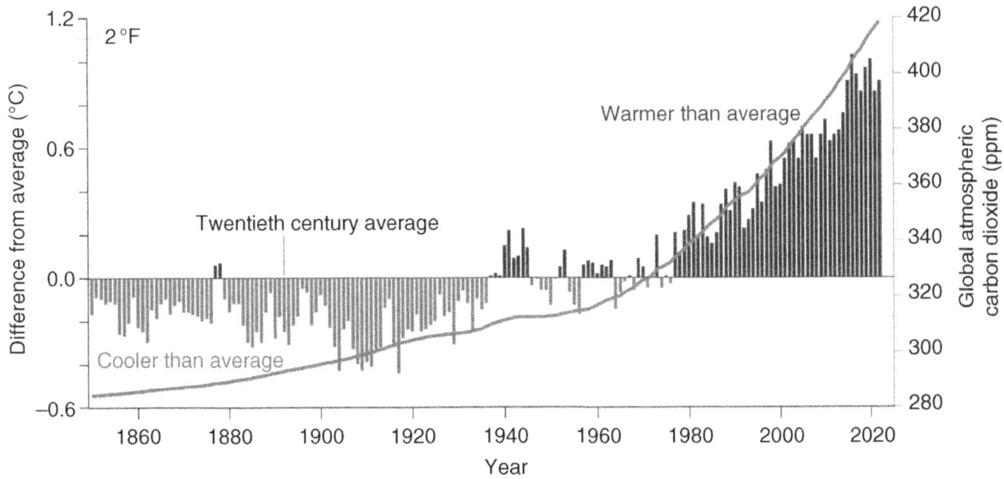

Figure 1.6 Average global temperature change (bars) and atmospheric carbon dioxide (line). *Source:* NOAA (2023)/Public domain.

Nevertheless, substantial increases in greenhouse gases have been generated during the industrial era, contributing to a rapid increase in global temperatures. Figure 1.6 depicts the correlation between global temperature and atmospheric carbon dioxide concentrations over the past 180 years.

Predictions of further impact are difficult to make and offer wide ranges due to variations in mathematical models. By 2100, global temperatures are predicted to increase between 2.5 and 10.4 °F (1.4–5.8 °C), with temperatures expected to increase even more in the United States (IPCC 2023). Increased global temperatures can lead to drastic shifts in weather patterns, with more precipitation in some areas, less in others, and more intense storm events. These effects can drastically affect agricultural activities and induce strains on ecosystems and biodiversity. Furthermore, increased temperatures are expected to accelerate the melting of polar ice and alpine glaciers, leading to increases in sea elevation, which in turn can inundate coastal areas.

As a short example of the current impact and sustainability of greenhouse gas emissions, consider carbon dioxide and a sustainability period of 50 years as well as a representative world population of 7.5 billion people. If global emissions were limited to a doubling of atmospheric carbon dioxide concentrations, this would be 7–8 pg (picograms) of carbon ($7–8 \times 10^{-15}$ kg). Based on an allocation to 7.5 billion people, this would result in 1 Mg (megagram) of carbon per year per person. Assuming a 40/25/35% breakdown of transportation, energy, and general society emissions, this would translate to 400, 250, and 350 kg of carbon dioxide per year, respectively.

With respect to generation, a typical automobile emits 62 g of carbon/km, and coal-fired power generates 89 kg of carbon/GJ. This translates to 6450 km/year of automobile transport and 2.8 GJ/year/person in terms of allowable carbon emissions. However, in Germany, the current per capita distance of auto travel is 15 000 km/year, and per capita use of energy is 270 GJ/year (EUROSTAT 2024a, 2024b, 2024c). The current rates are way above an allocated level of emissions that would only lead to a doubling of carbon atmospheric concentrations.

If these levels of consumption and emission generation continue, current predictions for these inputs are for a 3–5 °C increase in global temperature. To put it in perspective, a 5 °C difference in global temperature represents the difference between a warm period and an ice age. As indicated above, such a temperature will have a significant effect on sea level, agricultural practices, and numerous other activities that can adversely affect coastal and inland populations alike.

Example 1.2

The United States has been monitoring its greenhouse gas (GHG) emissions over the years. In 2021, the total GHG emissions were recorded at 6340 million metric tons of CO_2-equivalent. Due to the new environmental policies taken up by the government to reduce GHG emissions, the annual GHG emissions are expected to decrease by 2% each year.

Questions:

a) Calculate the total GHG emissions for the US for the year 2025, assuming the annual decrease rate remains constant.
b) Determine the total decrease in GHG emissions from 2020 to 2025.
c) Compare the total GHG emissions for 2025 with and without the implementation of the new environmental policies. How much emissions reduction is achieved due to the policy change?

Solution:

a) Assuming an annual decrease of 2%, starting in the following year of 2022, we can calculate the estimated total GHG emissions for the year 2025 with the following formula:

$$E_{2025} = E_{2021}(1-r)^t$$

where r is the targeted annual decrease of 2% and t is the time period between 2021 and 2025 of four years. Calculating for E2025, we have:

$$E_{2025} = 6340(1-0.02)^4 = 5848 \text{ million metric tons } CO_2e$$

If the US maintains their new environmental policies, it is estimated that the total GHG emissions for 2025 is 5848 million tons of CO_2e.

b) The US Environmental Protection Agency (EPA) generates the annual report Inventory of US Greenhouse Gas Emissions and Sinks. The 2023 report (which inventories 1990–2021) notes that in 2020, the US generated 6026 million tons of CO_2e. Therefore, the total decrease in GHG emissions from 2020 to 2025 is:

$$6026 - 5848 = 168 \text{ million metric tons } CO_2e$$

c) Using the recorded GHG emissions in the years 2020 and 2021, we can calculate the annual growth rate of GHG emissions, r', and prior to the enactment of new environmental policies:

$$E_{2021} = E_{2020}(1-r')^t$$
$$r' = \frac{E_{2021}}{E_{2020}} - 1 = \frac{6340}{6026} - 1 = 0.052 \ (i.e., 5.2\%)$$

We can now calculate estimated GHG emissions in the year 2025, assuming the growth rate stays similar without implementation of the new environmental policies, and compare it to the emissions for 2025 with enactment of the proposed environmental policies that were calculated earlier.

Emissions without enactment of environmental policies:

$$6340 \times (1 + 0.052)^4 = 7768 \text{ million metric tons of } CO_2\, e$$

Emission reduction achieved due to this policy change is therefore:

$$7768 - 5848 = 1920 \text{ million metric tons } CO_2e.$$

1.2.5.1 Climate Change and Extreme Weather Impacts: Hurricane Sandy Example

On the evening of October 29, 2012, Hurricane Sandy unleashed its full force upon New York City, bringing with it a storm surge that the city's infrastructure was ill-prepared to handle. At Consolidated Edison's 13th Street substation in lower Manhattan, the waters began to rise ominously. Though the substation had been engineered to withstand water levels up to 12.5 ft, the storm surge from Sandy reached up to 14 ft. As the water overwhelmed the system, the transformer exploded, plunging lower Manhattan into darkness. Power outages spread to surrounding neighborhoods as officials pre-emptively disconnected areas to prevent overloading the already-strained power grid. Critical facilities across the city, including hospitals, had to be evacuated due to the lack of electricity. At New York University's Medical Center, two backup generators failed, leaving the facility without power.

Hurricane Sandy was not only notable for the destruction it caused but also for its unprecedented scale. It was the largest Atlantic hurricane ever recorded in terms of diameter, with tropical storm-force winds extending across a staggering 1150 miles (Gutner 2012). The storm wreaked havoc across multiple countries, causing nearly $70 billion USD in damages, an amount that, when adjusted for inflation, equals approximately $93 billion in 2023. Tragically, it also claimed the lives of 254 people across eight countries, from the Caribbean to Canada.

Sandy's destruction stands as a stark reminder of the growing threat posed by climate change. As the planet continues to warm, the frequency and intensity of extreme weather events are on the rise. These events, which used to be rare occurrences, are increasing in frequency and intensity. Globally, dozens of extreme weather events are reported every year, with increasingly severe impacts on communities and infrastructure. In the United States alone, since 1980, there have been 395 weather and climate disasters in which damages exceeded $1 billion, resulting in a staggering total loss of $2.77 trillion and claiming 16 449 lives (NOAA 2024). The increasing frequency of such events in recent years, as depicted in Figure 1.7, raises critical concerns about the safety, reliability, and future resilience of public infrastructure systems.

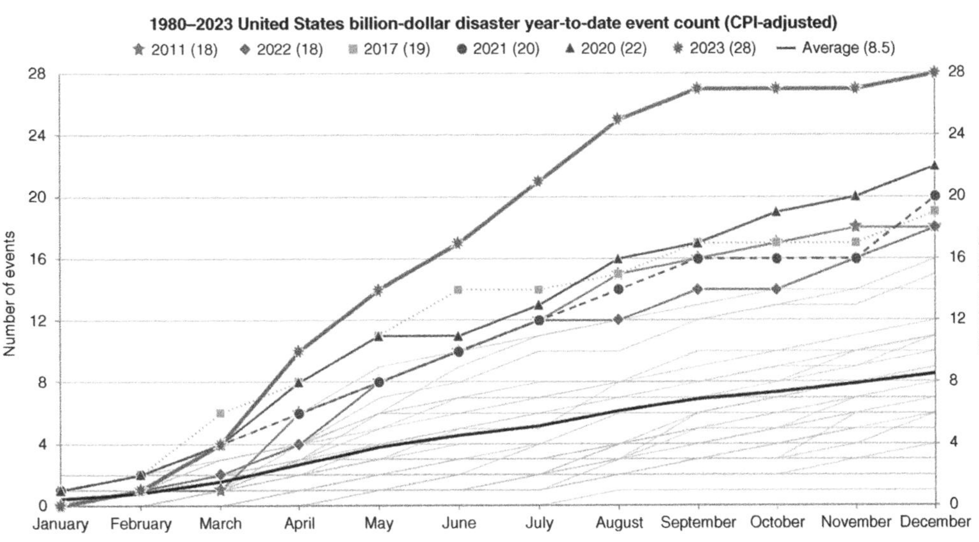

Figure 1.7 Month-by-month accumulation of billion-dollar disasters for each year on record. The colored lines represent the top six years for most billion-dollar disasters. All other years are colored light gray. *Source:* NOAA (2024)/The National Oceanic and Atmospheric Administration (NOAA)/Public domain.

These alarming statistics are not unique to the United States; nations across the world face similar, if not worse, consequences from climate-induced disasters. Traditional engineering design standards, which were developed under more predictable climate patterns, are now being revised to address the heightened risks posed by climate change. It is no longer sufficient to design infrastructure merely to meet historical weather conditions. Instead, engineers have to prioritize the resilience of systems to ensure that they can withstand and adapt to the unpredictable, often severe impacts of climate change. This concept is increasingly referred to as climate resiliency – essential for safeguarding public infrastructure and mitigating the effects of future climate disasters.

1.2.6 Decline of Ecosystems

The ongoing harm and destruction of natural environments resulting from waste generation and emissions lead to an additional problem – the decline of ecosystems. Healthy ecosystems are critical to human civilization – of primary importance; they support human life and economies by providing a supply of materials to be used in goods and services. However, these resources are not being managed in a healthy or sustainable manner. Human populations and consumption, commercial activity, and industrial activity are leading to the depletion and contamination of groundwater, topsoil, forests, and fisheries.

The global economy is highly dependent on natural, renewable resources that are provided by the wide range of ecosystems on the planet. The primary, direct products available in these ecosystems include fresh water, food, fuel, wood/timber, leather, wool, furs, and other products. Secondary and/or refined/synthesized products include raw materials for fabrics, oils, chemicals, and many others. Furthermore, the utilization of these materials in industries such as agriculture, forestry, and fishing is responsible for 50% of jobs worldwide (Vallecillo et al. 2019).

Ecosystems are also a primary source for service-based economic roles. Some services that derive directly from ecosystem management include waste management, climate regulation, erosion control, water treatment, and vector management. These combined goods and services form "ecosystem capital" – economic development, income, and overall human well-being are strongly based on the effective management of this capital. These goods and services are available and feasible as long as the ecosystems from which they are derived are healthy and protected. Therefore, the ecosystem capital in a particular nation and its income-generating capacity represent a critical component of a particular nation's wealth.

As the new millennium dawned, an assessment was performed to evaluate overall ecosystem health. The assessment focused on the services that are provided by ecosystems and provided an agenda for ecosystem restoration. Further, the *Millennium Ecosystem Assessment* gathered information on the world's ecosystems. This study, which included the work of 1360 scientists in 95 countries, gathered, analyzed, and synthesized ecosystem-related information to focus on the links between ecosystem goods/services and the well-being of human civilization (Marzec 2018).

The conclusions of the *Millennium Ecosystem Assessment* were sobering: it found that humans have altered the world's ecosystems more rapidly and profoundly over the past 50 years than at any time in the human history. Further, over 60% of ecosystem goods and services are being degraded or not being utilized in a sustainable manner. In the absence of a reversal of these trends, there will soon be critical consequences to human civilization. Ideally, the *Assessment* will serve as a knowledge base for sound decisions and management that can be undertaken by policymakers and managers in both the private and public sectors.

1.2.7 Loss of Biodiversity

With the ongoing decline of ecosystems comes another related problem – the loss of biodiversity. Biodiversity is generally defined as the variability among living organisms and the greater ecological complexes to which they belong. Millions of species exist on the planet, and many have not been studied or assessed. Of an estimated 5–30 million organisms, only approximately 2 million have been described. Many of these have been adversely affected. Since 1970, vertebrate species have reportedly declined by 27% (Butchart et al. 2010).

Stresses and impacts on biodiversity come from several causes. Many are linked to strains on host ecosystems, including pollution, introduction of invasive species that "crowd out" native species or are predatory on native species, and conversion of natural lands into developed land use, such as agriculture, timber harvesting, commercial/industrial uses, or residential uses. Additionally, species may be exploited for commercial values, either legally or illegally. For instance, many species continue to be hunted, killed, and marketed even after national or international prohibitions have been enacted.

The loss of biodiversity can have substantial negative effects on the environment and human civilization. Diverse species play a critical role in agriculture, the overall food chain, as well as in advances in pharmaceuticals and medicine. Further, a robust, diverse range of flora and fauna are critical in an ecosystem's self-healing mechanisms after major natural or artificial disruptions, such as hurricanes, forest fires, or flooding.

As a counterpoint to overexploitation, the protection of biodiversity can provide short-term and long-term economic benefits. Land and resource stewardship performed in a sustainable manner can have significant economic benefits. These resources can be a source of economic wealth through their incorporation into goods and services, especially among the poor. Further, many argue there are aesthetic and, even more powerfully, moral obligations to protect biodiversity and the range of flora and fauna present on the planet (Vellend 2017).

1.2.8 Social Injustice

Effective stewardship of resources, ecosystems, and their side effects on biodiversity requires some relative form of intervention, both economically and politically. Yet the means and degree of intervention are continuously subject to debate, both locally and globally. Figure 1.8 shows a matrix outlining political and economic systems. On the vertical axis is a depiction of political systems – moving upward along the axis indicates an increasing libertarian view, which espouses that justice is based on the equality of opportunity toward inputs or events. Moving downward along the axis indicates a more egalitarian view, which defines justice as the equality of an outcome based on inputs or events. The horizontal axis depicts different types of economic values. To the left indicates a greater degree of communitarianism, in which welfare is optimized by the absorption of individual economic activity into the community. A position to the right of the access indicates a greater focus on corporatism, in which welfare is optimized by the free economic activity of individuals.

With respect to many other nations, the economic and political system of the United States has long been positioned toward the upper right-hand corner of the matrix – indicating a greater focus on libertarian political and corporate economic policies. At first glance, some would argue that this is at the opposite end of the matrix from where a sustainable-inducing position would be located. However, there are many points to refute this. First, many of the worst environmental atrocities and catastrophes have occurred in nations of totalitarian government regimes and communal economic

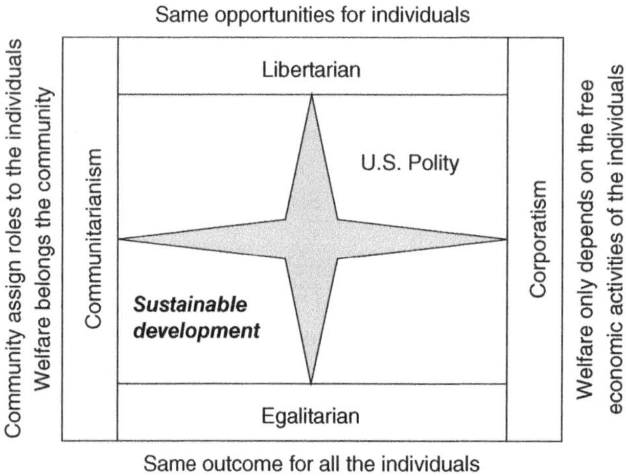

Figure 1.8 Relations between political and economic systems.

principles – no individual or group could challenge the will of these governments and a lack of economic competition diminished innovations or better alternatives to practices that led to these outcomes. Second, less regulation and more enterprise can lead to market-based solutions to address sustainability. When financial gain is to be made, many economic participants will pursue these gains. The goal should be to find a way to align sustainability goals and financial goals. This is increasingly being applied with the "triple bottom line" – where emphasis is placed on financial, social, and environmental metrics for a particular economic activity.

When it comes to environmental and resource stewardship, it is also important to facilitate just relationships within both economic and political systems. An increasing focus is being placed on environmental justice – a concept that addresses the increased notoriety of environmental racism. The concept of environmental racism is better defined in terms of socioeconomic status, although sadly, this range in status often follows along racial or ethnic lines. Environmental racism postulates that waste sites and harmful industries are often placed in areas of lower economic status residents – which often include substantial portions of minority or historically disadvantaged residents. This is further exacerbated by the perception of greater infrastructure expenditures and facility improvements, including institutional buildings and facilities, roads, and utility conveyance projects, in areas of greater political activity or economic clout, which often includes greater percentages of white residents. Whether the racial component in specific cases is coincidental or a real influencing factor is irrelevant, the perception is that these decisions are made along racial lines, and this leads to greater perceptions of environmental racism.

The concept of environmental justice has received greater focus in recent years throughout the developed world. Citizens of all racial/ethnic and economic backgrounds have coalesced into groups to pursue environmental justice issues. Citizen and "watchdog" groups increasingly bring effective action toward environmental justice goals and to monitor progress toward effective, desirable outcomes (Figueroa 2022).

Increased focus on social and environmental justice is also critical in the developing world. In many instances, this has not occurred in areas with rapid economic growth, leading to an alarming disparity between the wealthy and poor. The economic developments in developing countries are often driven by natural resource utilization or low-tech industries. In this process, people often lose their land or their ability to utilize it, leading to strains on their individual economic prospects. Decreased access to food, shelter, and other basic needs in turn often leads to increased health problems. These factors tend to repeat, and more people are trapped in poverty. In fact, over a billion people face extreme poverty from economic and environmental injustice that stems from this pattern. This is exacerbated in developing countries, as they often have weak or immature governing institutions. Corruption between politicians, wealthy individuals, and other officials who are intermediaries between these groups leads to the inappropriate diversion or outright theft of vast sums of money. Reforms, strengthened institutions, and accountability for governments, corporations, and their respective individuals are necessary to reverse these trends.

Wealthy, industrialized nations also have a role to play in combating these forms of injustice. Many industrialized nations or large developing countries engage in unfair trade practices that prevent others from participating fairly. Tax and tariff systems and quotas often restrict imports that may emanate from other developing countries. Some countries (especially those with heavy mass production-focused manufacturers) will flood markets with products at depressed prices (often for less than the cost to manufacture the products) with the aim to damage competition from other competing countries. These practices destroy competitors and create too many barriers in developing countries from competing in a fair environment. In addition to providing forms of political, advisory, and economic aid to developing countries, wealthy, developed nations can play a role in eliminating these unfair practices. By forming and enforcing free trade agreements as well as educating the marketplace, these unethical and often illegal destructive trade practices can be limited when meaningful sanctions are put in place toward offending nations.

1.2.9 Urban Sprawl

Governments and their institutions not only have a role in facilitating economic and social justice but they also play a greater role in advancing humans to preserve their built environment. Urban sprawl is causing an adverse impact on the environment in many developed and developing worlds. Urban sprawl is difficult to define, but it can be described as the pursuit of a built environment that focuses more on low-density residential areas, retail areas (such as low-rise shopping malls), industrial facilities, and other facilities. These features often are polynuclear in nature (areas of intense activities are often situated at relatively large distances from each other) and need to be connected by transportation linkages, such as multilane highways (Artmann et al. 2019).

Urban sprawl leads to an ever-expanding boundary for cities into the surrounding countryside. In areas without or with weak planning codes/regulations or traditions, sprawl can occur without due consideration of side effects or a defined limit of where growth will be fostered or controlled. Expansion of the built environment with respect to land uses (residential, commercial, and industrial) and transportation linkages often leads to the destruction of open areas and natural lands. Explosive development around traditional city centers often happened when cheap land was available for development without any forethought to planning principles. Eventually, this development extended out toward existing entities (towns and villages) that had their own jurisdictional authority or, in cases where planning regulations existed, their own land use plans and codes; in many cases, keeping residential and commercial uses separate in manners that make sense within the specific jurisdiction but are incongruous when considered from a regional standpoint. In the absence of land use regulatory frameworks or where conflicting approaches existed, growth has often occurred without undertaking a regional approach to planning or infrastructure development, leading to inefficiency, as well as areas overserved or underserved by infrastructure.

Uncontrolled urban sprawl can lead to significant side effects that can affect sustainability principles. A polynuclear land use approach emanating from unchecked sprawl that requires relatively intensive transportation systems drives up the energy requirements for transportation. Increased needs from transportation affect the environment – approximately 28% of greenhouse gas emissions are related to transportation-based activity (Rode et al. 2017). Much of this is due to intensive automobile use – cars are responsible for approximately 80% of these emissions – and numerous cities in the United States (many lacking a robust mass transit infrastructure) cannot meet air quality standards.

Greater sprawl takes up a greater footprint of the landscape. Impervious materials – paving for roads and parking, roofs, etc. – increasingly replace natural pervious ground surfaces, leading to increased water runoff, erosion, and flooding. This runoff also picks up pollutants – sediments, oils, and agricultural chemicals – leading to a general degradation in water quality.

Increased land development resulting from low-density sprawl also takes already productive land out of use. Sprawl leads to the removal of an estimated 1.3 million acres of productive agricultural land per year. As a result, there is a greater reliance on food products from increasingly distant locales – local sources have been replaced by sources often hundreds or thousands of miles away. This, in turn, leads to greater emissions from the requisite transportation activity to deliver these products. Further, 1.8 million acres per year of natural lands are lost due to development. This can affect the habitats of species, either through direct reduction or fragmentation of wild areas that can affect species which need large areas to thrive (Lawler et al. 2014).

Sprawl also can generate several secondary effects that can affect the quality of life of people and communities as a whole. A greater reliance on automobile trips in lower-density areas results in higher mileage trips, leading to more energy consumption and wear and tear on roads. A greater number of trips per capita also lead to a greater absolute number of accidents, injuries, and fatalities. When infrastructure cannot keep up with population growth or when constraints are in place (e.g. existing land uses or natural landforms like water bodies or mountains) to limit expansion of existing systems, traffic congestion is inevitable, leading to lost productivity, increased transit times, and a greater waste of energy to power idling vehicles. Building additional roadways to increase transportation capacity requires greater expenditure, which may lead to an increased rate of taxes. The greater trip distances that result from sprawl lead to greater reliance on automobile transportation and a like reduction in physically active transportation, such as biking or walking. As a result, reduced physical activity, increased obesity, and related health effects tend to rise.

1.3 The Master Equation or IPAT Equation

To assess the combined impacts of resource use, environmental impact, and population, the "*Master Equation*" may be used to assess the overall impact of these inputs. The equation is defined as follows:

$$\begin{aligned}\text{Overall impact} &= \text{Population} \times (\text{Resource use/person}) \\ &\quad \times (\text{Environmental impact/unit of resource use})\end{aligned} \tag{1.1}$$

This equation is also known as the IPAT equation (Chertow 2000) first proposed by Ehrlich and Holdren (1971), where I = Impact, P = Population, A = Affluence, and T = Technology. As for measurements of inputs, resource use can be set to gross domestic power (GDP) and converted to a per capita value to account for a person-by-person basis. As can be seen, with increased population and resource use (which are relatively easy to measure), a reduction in environmental impact per unit of resource use is necessary and must be of sufficient magnitude if overall impact is to be minimized.

Example 1.3
Given:

- A country has a population of 50 million people.
- The affluence level, measured as GDP per capita, is \$30 000.
- The technology factor, which represents the environmental impact per unit of GDP, is 0.25 units of impact per \$1000 of GDP.

Calculate the total environmental impact of the country using the IPAT equation.

Solution:

$$\begin{aligned}\text{Total environmental impact} &= \text{population} \times \text{affluence (GDP per capita)} \times \text{technology factor}\\ &= 50\ \text{M people} \times \$30\,000/\text{person} \times (0.25\ \text{units}/\$1000)\\ &= 375\ \text{M units}\end{aligned}$$

1.4 What Is Sustainability?

Several problems faced on a global scale that could severely affect the environment and ultimately human civilizations have been presented in preceding sections. There is a growing desire to pursue sustainable solutions to address these growing problems, but what exactly is sustainability? There is no universally accepted definition, but several common definitions convey similar concepts. The most commonly used definition comes from one presented by the 1987 United Nations World Commission on Environment and Development (commonly known as the Brundtland Commission, named after the commission chair, Gro Harlem Brundtland, former Prime Minister of Norway). In the Commission's report, *Our Common Future*, sustainability was defined as "Development that meets the needs of the present without compromising the ability of future generations to meet their own needs" (UN 1987). This definition is succinct and powerful in conveying the need to act responsibly so that the current and future generations can have a high quality of life while protecting the planet. This definition, however, does not inform on how this can be accomplished. The term 'sustainability' has since been defined differently by different organizations. However, it is essential to note that any definition of sustainability is incomplete without considering all the TBL aspects: environment, society, and the economy.

The United Nations (UN) adopted the 2030 Agenda (Transforming our World: the 2030 Agenda for Sustainable Development) in September 2015 with effective date 1 January 2016 (UN 2015). The agenda is a commitment to achieve sustainable development by 2030 world-wide. It consists of 17 global goals (known as Sustainability Development Goals or simply SDGs) with associated 169 specific targets. The SDGs cover multiple social, economic, and environmental development issues, including no poverty, quality education, clean water and sanitation, affordable and clean energy, sustainable cities and communities, and others. These goals are vital and ambitious but necessary to shift the world onto a sustainable and resilient path.

The concept of sustainability and sustainable development supersedes the concept of "environmentalism" that has been in existence for several years. In addition to protection of the environment, sustainability incorporates the concept of equity – the needs of the present shall be met, but of equal importance is the preservation of resources for equally deserving future generations. The incorporation of sustainability means different things to different people, too. Economists typically consider sustainability and its aspects with respect to growth, efficiency, and resource use. Sociologists focus on human needs, including equity, empowerment, social cohesion, and cultural identity. Environmentalists tend to focus on the preservation of the integrity of the environment, living within the limits, means, and carrying capacity of the planet, and exploring ways to minimize environmental impacts while remediating existing impacts from past and present pollution occurrences.

The pursuit and achievement of sustainable solutions presents several challenges, not the least of which is that development and economic activity often have goals that, at first glance, can be in direct conflict with environmental stewardship or resource preservation. However, sustainable solutions are increasingly desired, and the concept of the TBL has been a valuable framework in which to consider the relative efficacy of these solutions. The TBL (Figure 1.9) consists of three measurement schemes to consider these solutions (Halpern et al. 2013):

- *Environmental* – Diversity and interdependence within living systems, the goods and services produced by the world's ecosystems, and the impacts of human-generated wastes (the planet).
- *Economic* – Flow of financial capital and the facilitation of commerce (prosperity).
- *Social* – The interactions between institutions/firms and people, functions expressive of human values, aspirations, and well-being, and ethical issues (people).

Sustainable solutions are those that maximize the relative return to these three categories and their representative metrics. More sustainable solutions lead to a greater intersection of these metrics.

A collective transition to more sustainable practices may be hard to imagine, but it is necessary. A stable human population must recognize that the earth has a finite set of resources as well as a finite ability to accept and absorb waste products. If sustainable practices are not incorporated,

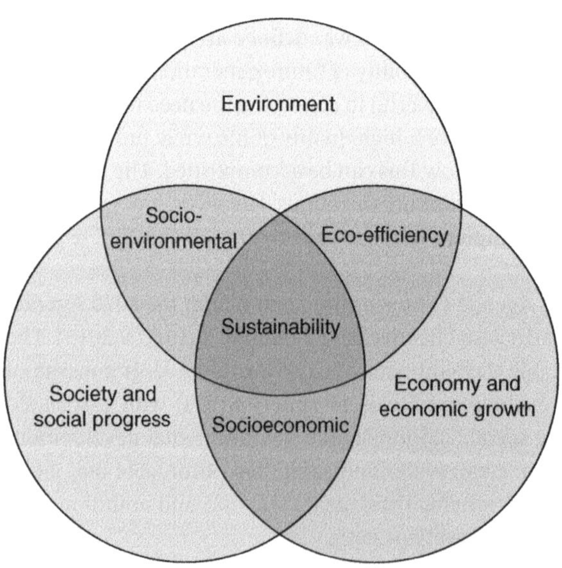

Figure 1.9 The triple bottom line: environment, economic, and social aspects, in sustainability.

catastrophic consequences may result. A pursuit of greater sustainability requires care and dedication to the preservation of the natural world and a commitment to justice and equity among various sections of people on the planet.

Several actions or characteristics are vital for the transformation to more sustainable practices. First, population growth must slow down to a more stable, sustained pace. Political and social institutions need to increasingly embrace environmental stewardship and approaches to reduce poverty and inequity. Resource utilization needs to progress from a sole focus on short-term economic benefit and output to one that incorporates a long-term view with due consideration for the protection of ecosystems and natural resources. Technology needs to play a role in the reduction of greenhouse gases and transition from pollution and waste-intensive processes to more environmentally protective activities. Finally, land use and planning should increasingly focus on community development practices that discourage sprawl-causing development and encourage higher-density uses, including more livable cities.

In summary, the sustainability concept, if not the specific definitions, is relatively easy to comprehend. However, the ability to measure sustainability remains a work in progress. Nevertheless, efforts have been made, and sustainability measuring tools are beginning to evolve. As an example, in 2007, the United Nations Commission on Sustainable Development listed 96 indicators under 14 themes to measure sustainability or sustainability in a particular country. Some examples are presented in Table 1.1. However, sustainability at project level requires a new set of sustainability indicators, which are addressed in Section II of this book.

1.4.1 United Nations Sustainable Development Goals (SDGs)

The 2030 Agenda for Sustainable Development, adopted by all United Nations Member States in 2015, serves as a universal blueprint for fostering peace and prosperity for both people and the planet, both now and in the years to come. Central to this agenda are the 17 Sustainable Development Goals (SDGs), representing a compelling call to action for all nations, whether developed or developing, to collaborate in a worldwide partnership. These goals are:

- *Goal 1.* End poverty in all its forms everywhere (No Poverty)
- *Goal 2.* End hunger, achieve food security and improved nutrition, and promote sustainable agriculture (Zero Hunger)
- *Goal 3.* Ensure healthy lives and promote well-being for all at all ages (Good Health and Well-being)
- *Goal 4.* Ensure inclusive and equitable quality education and promote lifelong learning opportunities for all (Quality Education)
- *Goal 5.* Achieve gender equality and empower all women and girls (Gender Equality)
- *Goal 6.* Ensure availability and sustainable management of water and sanitation for all (Clean Water and Sanitation)
- *Goal 7.* Ensure access to affordable, reliable, sustainable, and modern energy for all (Affordable and Clean Energy)
- *Goal 8.* Promote sustained, inclusive and sustainable economic growth, full and productive employment, and decent work for all (Decent Work and Economic Growth)
- *Goal 9.* Build resilient infrastructure, promote inclusive and sustainable industrialization, and foster innovation (Industry, Innovation, and Infrastructure)
- *Goal 10.* Reduce inequality within and among countries (Reduced Inequalities)
- *Goal 11.* Make cities and human settlements inclusive, safe, resilient, and sustainable (Sustainable Cities and Communities)

Table 1.1 Indicators for measuring sustainability.

Theme	Subtheme	Indicator (example)
Poverty	Income poverty	Proportion of population living below national poverty line
	Income inequality	Ratio of share in national income of highest to lowest quintile
	Sanitation	Proportion of population using an improved sanitation facility
	Drinking water	Proportion of population using an improved water source
	Access to energy	Share of households without electricity or other modern energy services
	Living conditions	Proportion of urban population living in slums
Governance	Corruption	Percentage of population having paid bribes
	Crime	Number of intentional homicides per 100 000 population
Health	Mortality	Under-five mortality rate
	Health care delivery	Percentage of population with access to primary health-care facilities
	Nutritional status	Nutritional status of children
	Health status and risks	Morbidity of major diseases such as HIV/AIDS, malaria, tuberculosis
Education	Education level	Net enrolment rate in primary education
	Literacy	Adult literacy rate
Demographics	Population	Population growth rate
	Tourism	Ratio of local residents to tourists in major tourist regions and destinations
Natural hazards	Vulnerability to natural hazards	Percentage of population living in hazard-prone areas
	Disaster preparedness and response	Human and economic loss due to natural disasters
Atmosphere	Climate change	Carbon dioxide emissions
	Ozone layer depletion	Consumption of ozone depleting substances
	Air quality	Ambient concentration of air pollutants in urban areas
Land	Land use and status	Land degradation
	Desertification	Land affected by desertification
	Agriculture	Arable and permanent cropland area
	Forests	Proportion of land area covered by forests
Oceans, seas, and coasts	Coastal zone	Percentage of total population living in coastal areas
	Fisheries	Proportion of fish stocks within safe biological limits
	Marine environment	Proportion of marine area protected

Table 1.1 (Continued)

Theme	Subtheme	Indicator (example)
Freshwater	Water quantity	Water use intensity by economic activity
	Water quality	Presence of fecal coliforms in freshwater
Biodiversity	Ecosystem	Proportion of terrestrial area protected, total and by ecological region
	Species	Change in threat status of species
Economic development	Macroeconomic performance	Gross domestic product (GDP) per capita
	Sustainable public finance	Debt to GNI (Gross National Income) ratio
	Employment	Labor productivity and unit labor costs
	Information and communication technologies	Internet users per 100 population
	Research and development	Gross domestic expenditure on R&D as a percentage of GDP
	Tourism	Tourism contribution to GDP
Global economic partnership	Trade	Current account deficit as percentage of GDP
	External financing	Net Official Development Assistance (ODA) given or received as a percentage of GNI
Consumption and production patterns	Material consumption	Material intensity of the economy
	Energy use	Annual energy consumption, total and by main user category
	Waste generation and management	Generation of hazardous waste
	Transportation	Modal split of passenger transportation

Source: UN (2007)/United Nations.

- *Goal 12.* Ensure sustainable consumption and production patterns (Responsible Consumption and Production)
- *Goal 13.* Take urgent action to combat climate change and its impacts (Climate Action)
- *Goal 14.* Conserve and sustainably use the oceans, seas, and marine resources for sustainable development (Life Below Water)
- *Goal 15.* Protect, restore, and promote sustainable use of terrestrial ecosystems, sustainably manage forests, combat desertification, and halt and reverse land degradation and halt biodiversity loss (Life on Land)
- *Goal 16.* Promote peaceful and inclusive societies for sustainable development, provide access to justice for all and build effective, accountable and inclusive institutions at all levels (Peace, Justice and Strong Institutions)
- *Goal 17.* Strengthen the means of implementation and revitalize the Global Partnership for Sustainable Development (Partnerships for the Goals).

These 17 Goals specify a total of 169 targets, which in turn are monitored using 247 Indicators. Figure 1.10 illustrates the most common graphical representation of these SDGs. These goals acknowledge the necessity of simultaneously addressing poverty and other forms of deprivation, alongside implementing strategies that enhance health and education, diminish inequality, and stimulate economic growth. This approach is also integral to addressing climate change and conserving our oceans and forests (https://sdgs.un.org/goals).

The Division for Sustainable Development Goals (DSDG) within the United Nations Department of Economic and Social Affairs (UNDESA) offers essential support and builds capacity for the SDGs and their associated themes, such as water, energy, climate, oceans, urbanization, transport, science, and technology.

Annually, the UN Secretary-General delivers a report on the progress of the SDGs. This report is prepared in collaboration with the UN System, drawing on the global indicator framework. It utilizes data from national statistical systems and regional information. Details about SDG indicators and reports can be found at https://unstats.un.org/sdgs. Furthermore, the Global Sustainable Development Report is published every four years to support the quadrennial review of the SDGs at the General Assembly. Authored by an Independent Group of Scientists selected by the Secretary-General, the most recent edition was *The Sustainable Development Goals Report* released in 2024 (https://unstats.un.org/sdgs/report/2024/).

The SDGs are relevant to every country and sector, including cities, businesses, schools, and organizations, all of which are encouraged to take action (Universality). These goals are interconnected within a system, meaning we cannot focus on achieving just one; we need to work toward all of

Figure 1.10 United Nations Sustainable Development Goals (UN SDGs). *Source:* United Nations Development Programme.

them (Integration). Realizing these goals requires significant, fundamental changes in our lifestyle on Earth (Transformation).

The ASCE Policy statement 517 endorses the civil engineering components of the United Nations' SDGs. ASCE collaborates with both domestic and international organizations to involve engineers in helping disadvantaged communities. This involves building capacity and creating sustainable, appropriate solutions to combat poverty. There is a growing demand for resources and services like energy, food, land, water, transportation, materials, waste disposal, construction, health care, environmental cleanup, telecommunications, and infrastructure. Engineers play a vital role in meeting these needs across different scales, especially in the developing world, from small remote communities to large urban areas. It is essential to educate a new generation of engineers who are equipped to tackle the challenges in the developing world and cater to the needs of disadvantaged communities. Engineers are key to building a more sustainable, stable, and secure world.

The latest report on the SDGs published in 2024 provides a status update on the progress toward achieving the 17 Sustainable Development Goals. The individual data by country and SDG is available in the report. Figure 1.11 shows the results by groups of countries, categorized by regions or economic levels. As seen in Figure 1.11, the highest index corresponds to OECD countries and high-income countries. The lowest index values correspond to Sub-Saharan Africa, Oceania, and low-income countries. The detailed analysis of the individual results for each SDG shows that only a few of these goals have been achieved in specific countries or regions, while other goals are facing significant challenges in many regions of the world. Notably, SDGs 2, 14, 15, and 16 show serious difficulties in being achieved globally and by region according to current data, indicating that additional efforts are still needed to change the trend and ensure these SDGs can be met by the 2030, target set by the United Nations in the design of this program.

1.4.2 What Is Sustainable Engineering?

With a definition or at least a general idea of sustainability in place, it is time to discuss how sustainability principles and goals can be applied to the world around us. Of course, for centuries, engineers have applied scientific principles to devise solutions in nearly all aspects of our lives, including water, sanitation, mobility, energy, information management and propagation, food, health care, shelter, and communications. For much of this time, engineers have developed solutions with an implicit notion of limitless resources and little concern with respect to the management of waste or by-products or at least solutions independent of these factors. Additionally, engineers have commonly sought solutions that have not fully considered the social implications of their solutions.

With an understanding that, in fact, resources, waste management, and social implications are indeed critical factors, the concepts of sustainable engineering have begun to emerge. Sustainable engineering is the development of engineering solutions to advance human life to maximize benefits and minimize adverse impacts on the environment, the economy, and society ("the triple bottom line") throughout the life cycle of a project. Sustainable engineering can be applied regardless of value chain position or the magnitude of a project – it is scalable and applied across molecular, product, process, and system design.

The goals of sustainable engineering are shared among many specific disciplines. As presented in the American Academy of Environmental Engineers (AAEE) Environmental Engineering Body of Knowledge (2008), a joint declaration was issued with the American Association of Engineering

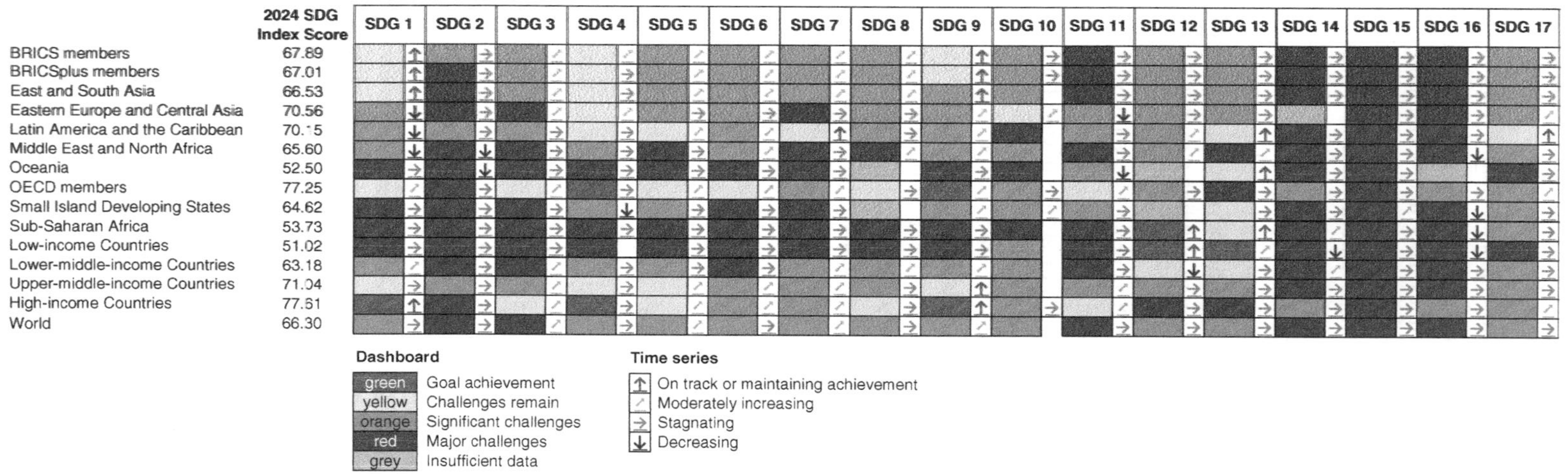

Figure 1.11 Performance scores and trends of individual Sustainable Development Goals (SDGs) in 2024 for different groups of countries categorized by regions or economic levels.

Societies (AAES), American Institute of Chemical Engineers (AIChE), the American Society of Mechanical Engineers (ASME), and the National Academy of Engineering (NAE) that stated the following: “Creating a sustainable world that provides a safe, secure, healthy life for all the people is a priority for the US engineering community. Engineers must deliver technically viable, commercially feasible, and environmentally and socially sustainable solutions.” The American Society of Civil Engineers (ASCE) addresses sustainable development in its Policy statement 418 by stating, “Sustainability is a set of environmental, social, and economic conditions (aka “The Triple Bottom Line”) in which all of society has the capacity and opportunity to maintain and improve its quality of life indefinitely without degrading the quantity, quality, or the availability of environmental, social, and economic resources” (ASCE 2023a).

Lately, the concept of green engineering (GE) has begun to emerge, but there are important, clear distinctions between green engineering and sustainable engineering. Generally stated, green engineering is the development of engineering solutions that are protective of human and environmental health throughout the life cycle of a project. Sustainable engineering is broader than green engineering in that it aims to develop engineering solutions that are not only protective of human and environmental health but are also economically and socially responsible throughout the life cycle of a project. In other words, green engineering is a subset of sustainable engineering with a focus on minimizing environmental impacts. Table 1.2 and Figure 1.12 summarize key principles of green engineering and sustainable engineering, respectively.

Table 1.2 Principles of green engineering.

Inherent rather than circumstantial	Designers need to strive to ensure that all materials and energy inputs and outputs are as inherently nonhazardous as possible
Prevention instead of treatment	It is better to prevent waste than to treat or clean up waste after it is formed
Design for separation	Separation and purification operations should be designed to minimize energy consumption and materials use
Maximize efficiency	Products, processes, and systems should be designed to maximize mass, energy, space, and time efficiency
Output-pulled versus input-pushed	Products, processes, and systems should be “output pulled” rather than “input pushed” through the use of energy and materials
Conserve complexity	Embedded entropy and complexity must be viewed as an investment when making design choices on recycle, reuse, or beneficial disposition
Durability rather than immortality	Targeted durability, not immortality, should be a design goal
Meet need, minimize excess	Design for unnecessary capacity or capability (e.g. “one size fits all”) solutions should be considered a design flaw
Minimize material diversity	Material diversity in multicomponent products should be minimized to promote disassembly and value retention
Integrate material and energy flows	Design of products, processes, and systems must include integration and interconnectivity with available energy and materials flows
Design for commercial “afterlife”	Products, processes, and systems should be designed for performance in a commercial “afterlife”
Renewable rather than depleting	Material and energy inputs should be renewable rather than depleting

Source: Anastas and Zimmerman (2003). Reproduced with permission of American Chemical Society.

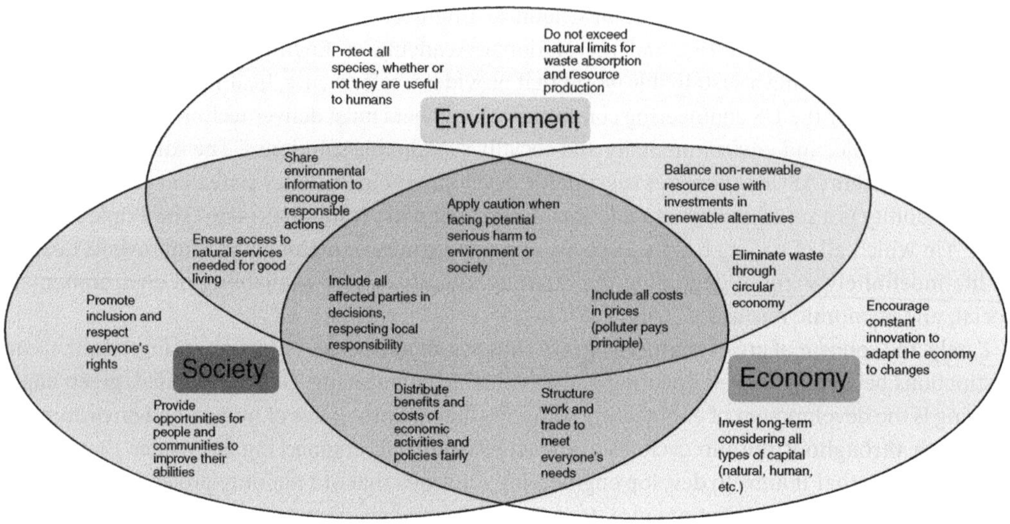

Figure 1.12 Classification of sustainable engineering principles across environmental, social, and economic criteria.

Of equal importance to environmentally protective green concepts and financial considerations are several socially focused dimensions. In helping to maximize the benefit within the context of these dimensions, it is important to identify to the extent possible the social issues and values that a particular activity or project will affect, and it is equally important to consider both the positive and the negative social consequences. For many reasons, perhaps due to practice, cultural, or social traditions, it is important that the social context even be considered, as it has not been incorporated in the past for countless projects or activities. To the extent possible, it is important to provide a culture/environment around a project or activity such that dialogue with respect to socially related aspects can occur. In such an environment, social values that may not have been seriously considered or even identified can be integrated into the project-related, decision-making process. Once this culture and dialogue are in place, decisions with respect to important social dimensions may be thoughtfully considered. Some important dimensions that may be considered include effects on population factors, such as birth rate, literacy, life span, cultural acceptance/openness (art, religion, etc.), and gender roles and degree of equity.

Aspects of a project can be designed or re-designed to maximize social or cultural benefits to the maximum extent feasible, while at the same time maximizing relative benefit with respect to traditional financially driven metrics, including costs, rate of return, and consumer demand. Of course, this is an iterative process, and it is rare, if not impossible, to maximize return or benefit to all three triple bottom line aspects. However, with iterations and thoughtful consideration of these three frames of reference, the goal is to maximize the relative benefit of each frame, incorporate into design, execution, and management, and carry out the action to attain the desired result.

Although the framework and the nomenclature of sustainable engineering are relatively new, some concepts have been around for a long time, and there has been a steady progression for decades toward its current state. Figure 1.13 shows how this trend has evolved over the past 40–50 years. For instance, prior to 1970, very little thought was given to social or environmental aspects. However, with significant environmental issues and related government regulation

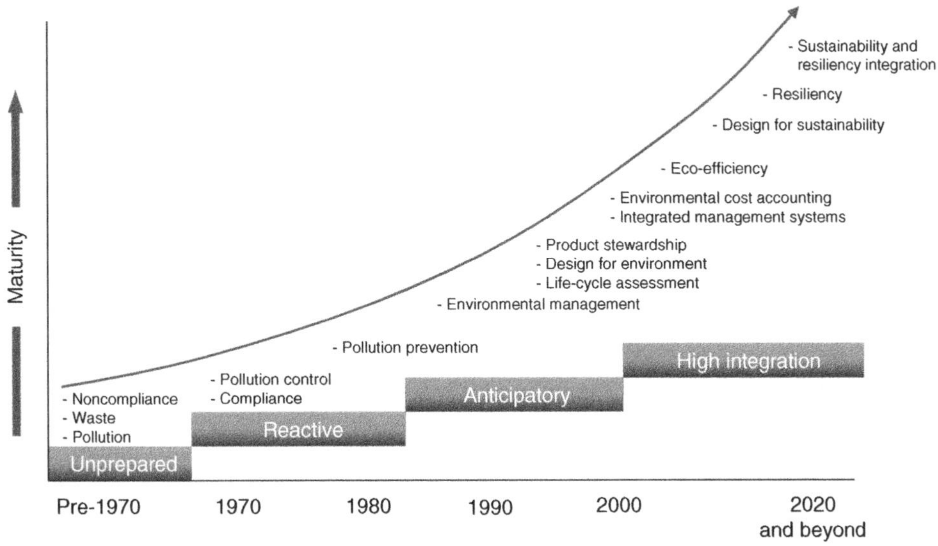

Figure 1.13 Evolution of concepts in sustainable and resilient engineering.

increasingly becoming part of the public consciousness in the 1970s, projects began to react and comply, then anticipate, and later even guide evolving environmental attitudes. As this evolution progressed, increased attention was paid to socially responsible and culturally aware practices, with an increased incorporation into design phases of projects and activities.

As we have introduced and discussed the aspects of sustainable engineering, there are several emerging general strategies or goals that are increasingly being considered core to sustainable engineering. Ultimately, there is a goal to pursue a full integration of engineered systems within the overall natural system of which it is a part. Additionally, it is important to seek flexible, scalable solutions that may be adapted from a micro- to macrolevel or a local-to-global scale. Further, thoughtful designs are based on a system approach that incorporates green chemistry, processes, and material use with comprehensive considerations of life cycle impacts with respect to energy and material inputs as well as emissions and other waste outputs. Finally, these designs should consider socially responsible outcomes as well as traditional economic metrics, including revenue, profit, profit margin, return on investment (ROI), and return on equity (ROE), among many others.

As mentioned, sustainable engineering operates at various scales. From a local standpoint, concepts and principles may be incorporated by individual firms as they attempt to design projects or products that are protective of the environment, aim to reduce pollution or waste generation, and utilize resources efficiently. On the next larger scale, firms or other entities may cooperate on larger initiatives, such as intra or intersector initiatives, standards of practice, joint ventures, collaborative studies, or facility development. Finally, on a regional or global scale, larger multinational firms, entire industries, or government entities may participate in even large initiatives, such as budget and cycle studies, materials and energy flow studies, research initiatives, and modeling scenarios that can ultimately lead to practice regulations.

An example of this scalability comes from the automotive industry. Imagine a single firm that fabricates a component of an automobile, such as axles. At the firm level, they can find sustainable designs or implement systems that eliminate waste or add efficiency to the design. At the next scale, they can participate with their client (one of the large automobile companies, such as General

Motors or Toyota) as well as with the company's other subcontractors to develop eco-friendly integrated systems for overall design or for manufacturing principles. At the next larger scale, they may work within the industry to find larger-scale sustainable solutions, such as those that may affect the design of roads and infrastructure, refining of traditional motor fuels and transitions to alternative fuels. Finally, at the global scale, they may work with several intra-industry groups as well as those from other industries or from governments to look for other initiatives that directly look to solve society's needs, such as alternative transportation modes, or land use that may optimize trip length and density.

A concrete example of this concept is the design and manufacture of Sertraline by Pfizer, a large pharmaceutical company (USEPA 2002; Nameroff et al. 2004). Sertraline is the active ingredient in Zoloft, a market-leading drug for the treatment of depression. Hundreds of millions of prescriptions have been written for Zoloft in the United States alone. By applying some of the sustainable engineering principles described in this chapter, Pfizer was able to eliminate the use of 140 metric tons per year of titanium tetrachloride and eliminate the generation of 440 metric tons of titanium dioxide waste per year. In addition, 150 metric tons of hydrochloric acid waste and 100 metric tons of sodium hydroxide needed for production were eliminated on a yearly basis. Other raw materials were also eliminated. These reductions saved Pfizer substantial money with respect to raw materials purchases and waste management. Corresponding resources were preserved, and Pfizer's financial metrics improved.

When designing products or projects that aim to be sustainable, it is important to measure the degree of sustainability involved in the fabrication, use, and disposal of the product. Not only should this study span the project or product itself, but it should also consider all inputs and activities associated with the final disposition of the product or project. Such a study may include a life cycle assessment (LCA). An LCA includes an inventory of inputs (e.g. raw materials and energy) and outputs (e.g. products, emissions, and solid waste) of the product over its entire life. The inventories that are assembled are converted into categories of impact; some of these may include emissions-related effects on climate change, acidification, aquatic toxicity, or human health. The LCA can further assign a quantitative weighting system of the individual categories to compute a single impact number.

LCAs can be incredibly powerful and useful in assessing sustainability. They consider the entire cycle of a project or product and related effects. It allows for the consideration and incorporation of several value-based assessment criteria, as well as latitude with respect to the weighting of these criteria. Their use also promotes comparison between project/product alternatives as well as individual components. Unfortunately, there are some limitations associated with LCAs. They are costly and time-consuming to perform. There is an inherent imprecision of some data associated with the attempt to quantify qualitative aspects or virtues. Additionally, it is difficult to incorporate and/or measure all relevant data, and some impacts may be inadvertently overlooked. Nevertheless, they are useful tools in working to achieve more sustainable designs with respect to material and energy inputs as well as waste management/reduction.

One of the concepts, whether with respect to the degree of sustainability, the TBL, or applications of LCAs, is the effect of globalization. Increasingly, virtually all aspects of economic activity are moving toward a global approach and participation. Very little economic activity of any importance occurs without some form of global impact. The global aspects of economic activity need to be carefully considered when assessing sustainability. Trade policies and regulations, the interaction of design and manufacturing in distant locations, transportation, and the flow of inputs, outputs, and waste by-products (such as emissions) are important factors to consider when assessing sustainable alternatives.

1.5 What Is Resiliency?

The impacts of climate change, such as extreme weather events discussed in the previous section, along with other shocks and stressors that may affect a system throughout its life cycle, motivate the adoption of resilience in the design and implementation of engineering systems. The term resilience originates from the Latin word "*resilire*", meaning "to jump back" or "to recoil." This traditional concept of resilience, emphasizing recovery from stressors, has been recognized for decades. The term is used across various domains, including psychology, biology, ecology, sociology, and engineering. Since the scope of this book is limited to engineering systems' resilience, our discussion will focus accordingly.

While resilience is interpreted differently across fields, most definitions share a consensus (Hosseini et al. 2016). Many intersect with concepts like robustness, fault tolerance, flexibility, survivability, and agility. In general, resilience is understood as the ability of a system to sustain its operations despite sudden shocks, stresses, changing conditions, or disturbances. In the modern-day engineering practice, resiliency encompasses four main concepts: rebound capability, robustness, flexibility, and sustained adaptability (Woods 2015). Rebound capability refers to a system's ability to recover from catastrophic failure or performance deterioration due to stress. Robustness denotes the system's capacity to absorb recurring disturbances. Flexibility contrasts brittleness, referring to a system's ability to avoid sudden failure in the face of unexpected or unprecedented challenges. In the context of climate change, adaptability – the system's capacity to adjust to continuous changes in conditions – has become increasingly relevant.

Another dimension of resilience concerns its scope, whether referring to the technical resilience of a single project or the resilience of a socio-technical system as a whole. A holistic perspective of resilience reflects the ability of communities to cope with external stresses resulting from social, political, or environmental changes. This view – that resilience is a socio-technical system's potential to achieve desirable performance – implies that resilience itself is not the end goal but rather a formal concept tied to the overarching aims of engineering, such as ecological safety and human well-being (Doorn et al. 2019).

1.5.1 What Is Resilient Engineering?

Resilient engineering focuses on developing solutions that can withstand and recover from unexpected shocks or stressors. For example, climate-resilient engineering involves designing and constructing projects that address climate-related impacts such as sea level rise, heavy rainfall, droughts, storms, and cyclones. Traditionally, engineering aims to design systems with a certain level of resilience so that they can endure expected stressors or weather events over the project's life cycle. However, this conventional understanding of resilience is becoming outdated as challenges continue to evolve.

Modern, resilient engineering requires building systems that can maintain functionality despite sudden shocks, stresses, changing conditions, or disturbances. Climate effects, like natural disasters, cause variations in system loads and often influence a project's overall performance. Designing a technically resilient system can lead to overconsumption of environmental, human, and economic resources, making such designs unsustainable in the long term. On the other hand, compromising technical performance can result in catastrophic system failures during extreme events, causing environmental, social, and economic losses. Therefore, there is a trade-off between these two aspects, with the TBL elements impacted either way. It is crucial to balance technical resilience with the TBL aspects to ensure sustained resiliency.

Many professional engineering organizations are actively promoting resilient engineering. For example, the American Society of Civil Engineers (ASCE) is encouraging resilient engineering through its newly released toolkit, which centralizes resources to help civil engineers and community leaders adopt modern building codes and standards. The toolkit highlights hazard-specific standards for climate-resilient engineering like ASCE 7-22: "Minimum design loads and associated criteria for buildings and other structures" (ASCE 2021). The toolkit also features the Envision program, which supports sustainable and resilient infrastructure design. ASCE's efforts also focus on educating policymakers and local communities to build infrastructure capable of withstanding climate-related and other hazards. Similarly, to ensure the resilient design of power grid systems, the ASME made several recommendations: accelerating the replacement of aging infrastructure, shortening supply chains to reduce replacement times for key components, hardening the power supply at the national level against threats from weather and cyber/physical attacks, deploying new technologies, encouraging diverse clean energy technologies, supporting deployment of energy storage, and updating the operating systems for the electric grid. Similarly, recommendations were made by AIChE for water resiliency and the Interstate Technology Regulatory Council (ITRC) for resilient remediation of contaminated sites.

1.5.2 Difference Between Resiliency and Sustainability

The concepts of sustainability and resilience share many similarities and are often used interchangeably across various applications, though their meanings and purposes are distinct. It is important to note that both sustainability and resilience serve as descriptors of systems, rather than being independent entities. These systems can range from the global economy to an individual's physical or mental well-being. Both concepts emphasize the state of a system over time, focusing on its persistence under normal conditions and its ability to withstand disruptions. This common focus on system survivability leads to the use of similar research methodologies, such as similar time scales (life cycle). Sustainability and resilience are both influenced by global political trends, with frameworks and agendas promoting both concepts in areas such as urbanization. The overlap in their goals has led to the creation of integrated sustainability and resilience departments within government agencies, universities, and nonprofit organizations.

However, despite their similarities, sustainability and resilience are two distinct entities. They operate on different spatial and temporal scales, with sustainability often focusing on broader spatial areas and longer timeframes than resilience. "Resiliency refers to the ability of a system to withstand and recover from unexpected shocks or stressors, such as extreme weather events resulting from climate change. Sustainability represents a holistic approach that encompasses environmental, economic, and social considerations. It aims to continuously improve the quality of life and infrastructure without compromising the needs of future generations" (Reddy et al. 2024). Particularly with respect to climate change, resilience is the ability to adapt, withstand, and recover from climate impacts, whereas sustainability is an approach adopted to reduce the rate and extent of climate change. In some cases, resilience may be achieved at one scale or timeframe but compromised at another, which is not the case with sustainability.

A lack of understanding of the similarities and differences between sustainability and resiliency can create challenges in implementation, leading to missed opportunities for synergy and the potential for conflicting objectives. Hence, it is important to clearly understand the differences and similarities between the concepts of resilient and sustainable engineering.

1.6 Integrated Sustainable and Resilient Engineering

Both sustainability and resilience of engineered systems are important for human existence and survival; these terms cannot be viewed separately because having resilient systems that are not sustainable may exacerbate the climate change and other impacts making the built environment non-resilient in the future. Conversely, sustainable systems that are not resilient to extreme events defy the primary purpose of building such systems, thus making them inherently unsustainable. Integrated sustainable and resilient engineering refers to the practice of developing solutions that not only endure extreme shocks and stressors but do so sustainably by minimizing resource consumption and environmental impact as well as maximizing the environmental and social benefits. In other words, the solution ensures resilience while minimizing the impacts on the environment, economy, and society.

In order to properly integrate these concepts, one needs clearly to understand such similarities and differences and try to maintain a balance between the conflicting objectives of sustainability and resiliency. Traditional engineering might be inadequate in such cases, and this paradigm shift to integrated resilient and sustainable engineering necessitates innovative engineering practices. Examples of such innovations in the field of infrastructure include the adoption of green infrastructure instead of gray infrastructure for urban stormwater management, and the adoption of nature-based solutions such as floodable parks instead of concrete seawalls, among others (Reddy et al. 2024). The paradigm shift toward integrated sustainable and resilient engineering represents a vital step for future engineering practices, prioritizing a balanced approach to infrastructure development that can effectively respond to climate change and other global challenges. Some professional engineering organizations such as ASCE and ITRC have also recognized the importance of integrating these concepts and stipulated guidelines on integrating these aspects into the design and implementation of engineering projects (ASCE 2023b, 2023c; ITRC 2021).

1.7 Summary

When defining sustainability, it is important to remember that it is a concept that transcends and supersedes familiar topics associated with environmentalism. Sustainability entails that the human development or activity currently should not jeopardize the needs and the development of future generations. Sustainability is measured across several factors, and the concept of the TBL assesses sustainability within a framework that considers the overall effects of a project or product with respect to the environment, economics, and the society. Sustainable practices, especially to be applied in the future, must be applied in such a way that several emerging challenges (as described in this chapter) are addressed and incorporated.

Sustainable engineering is the means by which sustainable concepts can be applied to systems, projects, and products. Sustainable engineering is defined as a tool for engineering responsibly and professionally by integrating environmental, social, and economic considerations during the life cycle of a project. Finally, the strategies of sustainable engineering aim to engineer systems that can be integrated into natural systems, are flexible enough to adapt to a range of scales from a local to global basis, incorporate design based on a systems approach, and incorporate judicious resource use while minimizing waste generation.

Resilience, on the other hand, differs from sustainability in that it focuses on the ability of systems to withstand and recover from shocks or stressors, such as extreme weather events. A resilient system may not always be sustainable, as it could require significant resource use that contradicts sustainability goals. Conversely, a sustainable system may not be resilient if it cannot endure unexpected challenges. Both concepts must work together to create effective long-term solutions. Integrated sustainable and resilient engineering combines both principles, promoting innovative solutions that balance resource efficiency with the ability to endure unexpected shocks and stressors. Achieving this requires a paradigm shift in traditional engineering, embracing innovative engineering practices to ensure long-term viability and adaptability in engineered systems.

1.8 Questions

1.1 Describe briefly the case of Easter Island, Chile. Based on this case, explain how and why sustainability is important.

1.2 Explain environmental impacts and other emerging challenges in the context of sustainability.

1.3 Outline three strategies you would recommend to lessen/abate waste generation.

1.4 Discuss how impacts on rainforest ecosystems could affect human populations that live in other areas.

1.5 Explain the "Master Equation." Discuss the validity of this equation.

1.6 Explain the "Tragedy of Commons." How is it relevant in a sustainability context?

1.7 Define "Sustainability" and "Triple Bottom Line." How does it fit into the current regulatory framework?

1.8 Name three specific metrics that can be used to assess the degree of sustainability from within the environmental dimension.

1.9 Name three specific metrics that can be used to assess the degree of sustainability from within the economic dimension.

1.10 Name three specific metrics that can be used to assess the degree of sustainability from within the social dimension.

1.11 Explain a simple approach to quantify sustainability. Give some examples.

1.12 How can one evaluate sustainability under complex and dynamic conditions?

1.13 Explain the "grand objectives" of sustainability. How can they be related to environmental concerns and targeted activities and specific actions?

1.14 What is sustainability (or sustainable) engineering? Why is it becoming more critical in current engineering practice?

1.15 What are the general strategies for sustainable engineering? Explain them.

1.16 Discuss a consumer product of your choosing, listing five raw material inputs and five emissions/outputs that could be included in an LCA analysis.

1.17 Explain the concept of resiliency as defined in various domains.

1.18 Why is resiliency of engineered systems important?

1.19 List potential similarities and differences between sustainability and resiliency.

1.20 What are the main objectives of integrating resiliency and sustainability?

1.21 Explain integrated resilient and sustainable engineering concept with examples.

1.9 Problems

1.1 The country of Indonesia has a forest that covers almost 95 million ha (hectares). The forest is shrinking due to increased logging activities, forest clearance for palm oil and pulpwood in addition to mining activities. The current rate of deforestation is 650 000 hectares per year. However, the government plans to implement conservation measures that will reduce the rate of deforestation by 2% each year. Indonesia is one of the major sources of greenhouse gas emissions resulting from land use. In the last 2 decades, the country has lost forest area approximately 27 million ha, equivalent to 20 Gt of CO_2-equivalent emissions.

a) Calculate the area of the forest that will be left after five years if the conservation measures are implemented.
b) Determine the total area of the forest that will be lost due to deforestation over these five years.
c) If the government introduces a reforestation program that plants trees over 150 000 hectares each year starting from the first year, what will be the net change in the forest area after 5 years?
d) Analyze how the situation would differ if the rate of deforestation did not decrease each year. What would be the forest area after five years in this scenario?
e) Calculate the annual amount of CO_2-equivalent emissions that would result from Indonesia's land use activities after 10 years, considering that the target rates of deforestation and reforestation are met by the government, and compare it with the case if the rate of deforestation continues at current rate without any reforestation. Comment on the results.

1.2 The city of Delhi has been experiencing increasing levels of air pollution due to vehicle emissions (38.8%), road dust (18.1%), power plants (10.9%), construction activities (8.4%), residential (6.3%), landfill fires (4.6%), open burning (3.8%), industries (3.4%), agricultural burning (1.3%), and other sources (4.3%). The percentage contributions given are contributors to daily PM2.5 concentration levels. (PM2.5 refers to fine particulate matter that is 2.5 μm or smaller in diameter).

In an effort to reduce air pollution, the city implements a policy that reduces the on-road traffic to half of what it is now. The air quality index (AQI) due to a pollutant (I_p) is determined using the following equation:

$$I_p = \frac{I_{Hi} - I_{Lo}}{BP_{HI} - BP_{Lo}} (C_p - BP_{Lo}) + I_{Lo}.$$

where I_p = the index for pollutant p
C_p = the truncated concentration of pollutant p
BP_{Hi} = the concentration breakpoint that is greater than or equal to C_p
BP_{Lo} = the concentration breakpoint that is less than or equal to C_p
I_{Hi} = the AQI value corresponding to BP_{Hi}
I_{Lo} = the AQI value corresponding to BP_{Lo}

The values corresponding to I_{Hi}, I_{Lo}, BP_{Hi}, BP_{Lo} for PM2.5 are given in the following table.

I_{Hi}	I_{Lo}	BP_{Hi} (μg/m^3)	BP_{Lo} (μg/m^3)	Category
0	50	0.0	12.0	Good
51	100	12.1	35.4	Moderate
101	150	35.5	55.4	Unhealthy for sensitive groups
151	200	55.5	150.4	Unhealthy
201	300	150.5	250.4	Very unhealthy
301	400	250.5	350.4	Hazardous
401	500	350.5	500.4	Hazardous

On a given day, the truncated concentration of PM2.5 (C_p) was recorded to be 226 μg/m^3.

a) Determine the PM2.5 concentrations resulting from different sources on the given day.
b) Calculate the AQI corresponding to existing PM2.5 concentrations on the given day.
c) If the city's goal to reduce traffic by half was a success, what would the reduced PM2.5 concentration and the corresponding AQI be?
d) If the government suspends agricultural burning, construction activities, and open burning activities owing to the worsening air quality, what would the reduced PM2.5 concentration and the corresponding AQI be?

Note: Refer to EPA's Technical Assistance Document for the Reporting of Daily Air Quality – the Air Quality Index for examples on calculating the AQI.

USEPA (2018). Technical Assistance Document for the Reporting of Daily Air Quality – the Air Quality Index (AQI). US Environmental Protection Agency. Available at: https://document.airnow.gov/technical-assistance-document-for-the-reporting-of-daily-air-quailty.pdf (accessed 10 December 2024).

1.3 The concentration levels of different pollutants are recorded as follows in Delhi, India on a given day:

Pollutant	Concentration ($\mu g/m^3$)
PM2.5	154.5
PM10	257
O_3	9.1
NO_2	13.3
SO_2	1.3
CO	900

Calculate the air quality index corresponding to each pollutant, and determine the overall air quality index (AQI) of Delhi on the given day.

Note: Refer to EPA's Technical Assistance Document for the Reporting of Daily Air Quality – the Air Quality Index for examples on calculating the AQI.

1.4 The city of New York generates an average of 1.6 kg of municipal solid waste (MSW) per person per day. The current population of the city is 8.65 million people, growing at an average 0.5% per year. Percentage of waste managed though different methods and the corresponding GHG emissions for New York are given below:

Management technique	Fraction of MSW	Greenhouse gas emissions
Recycled	19%	100 kg CO_2e per tonne of recycled waste
Incinerated	15%	415 kg CO_2e per tonne of MSW incinerated
Landfilled	66%	781.05 kg CO_2e per tonne of MSW landfilled

Average composition of MSW collected in New York is as follows:

Organic	47%
Paper	23%
Plastic	14%
Construction debris	7%
Metals	4%
Glass	3%
Appliances and electronics	1%
Household hazardous waste	0%
Miscellaneous	1%

a) Calculate the total amount of waste generated by the city in one year at the current rate.
b) Determine the total amount of waste that will be generated in the next year, considering the 5% increase in waste generation per person.

c) Calculate the greenhouse gas emissions from each management method and the total greenhouse gas emissions from MSW management in New York city at current generation rates.
d) Identify the recyclables in the MSW matrix, and what is their fraction of the total MSW generated.
e) NYC aims to reduce the landfilling rate of MSW by 75%. Assuming that the city meets its target of reducing the landfilling rate and all the recyclables in the MSW matrix are recycled, determine the percentage of GHG emissions reduced.

1.5 The city of Beijing generates an average of 0.22 kg of municipal solid waste (MSW) per person per day. The current population of the city is 21.5 million people, growing at an average 2% per year. Percentage of waste managed though different methods and the corresponding GHG emissions for Beijing are given below:

Management technique	Fraction of MSW	Greenhouse gas emissions
Recycled	20%	100 kg CO_2e per tonne of recycled waste
Composted	15%	400 kg CO_2e per tonne of MSW composted
Incinerated	15%	415 kg CO_2e per tonne of MSW incinerated
Landfilled	50%	781.05 kg CO_2e per tonne of MSW landfilled

Average composition of MSW collected in Beijing is as follows:

Waste component	Composition (%)
Kitchen waste	71.71
Paper (non-recyclable/recyclable)	6.30/5.87
Plastics (non-recyclable/recyclable)	2.75/7.09
Metals	0.15
Glass	3.92
Textile	0.60
Others	1.20

a) Calculate the total amount of waste generated by the city in one year at the current rate.
b) Determine the total amount of waste that will be generated in the next year, considering the 2.5% increase in waste generation per person.
c) Calculate the greenhouse gas emissions from each management method and the total greenhouse gas emissions from MSW management in Beijing city at current generation rates
d) Identify the recyclables in the MSW matrix, and what is their fraction of the total MSW generated
e) Beijing aims to increase the recycling and composting rate of MSW. Assuming that the city can increase the recycling rate by 10% in the next one year, and composting rate by 5%, determine the percentage of GHG emissions reduced due to the diversion of these waste streams from landfilling.

1.6 Scientists have discovered and identified 8.7 million different species of animals and plants on the Earth until now. However, due to deforestation alone Earth is losing approximately 50 000 species each year. Annual deforestation rate stands as 25 million ha currently. Assume the global deforestation rate reduces by 2% each year for the next 10 years due to various national and international policies.

a) Calculate the number of species that the Earth is expected to lose due to deforestation in the next 10 years.
b) Determine the total number of species remaining on the Earth after 20 years, considering the annual reduction in the loss is only due to deforestation.
c) If the policies are more successful and reduces the deforestation rate by 3.5% each year instead of 2%, how many species will be lost in the next 10 years?
d) Apart from deforestation, list other possible reasons for loss of biodiversity, and briefly explain some conservation efforts that can be undertaken to avoid the loss of biodiversity.

1.7 In India, the income distribution is highly skewed. A latest survey shows that the richest 10% of the population earns 57% of the total national income, while the poorest 50% earns only 13% of the total income; the income distribution is as shown in the table below:

Population (%)	National income (%)
Top 10	57
Next 40	28
Next 10	3
Bottom 40	10

The per capita national income of India is $2400 USD, and the population of India is approximately 1.4 billion.

a) Calculate the total income earned by the richest 10% of the population.
b) Determine the total income earned by the poorest 40% of the population.
c) If the government of India decides to implement a policy to redistribute 5% of the total income from the richest 10% to the poorest 40%, calculate the new total income for both groups after redistribution.
d) Calculate the Gini coefficient of income inequality based on the given data. Compare it with the Gini coefficients of other countries and comment on the results.
e) Discuss the potential impacts of this income redistribution on reducing income inequality. What are some other measures that could be taken to address social injustice in India?

Note:

- Search on the internet for information about Gini coefficients of different nations.
- Method to calculate Gini coefficients can be obtained at Hasell (2023).

Hasell, J. (2023). Measuring inequality: what is the Gini coefficient?". Available at: OurWorldinData.org. Retrieved from: 'https://ourworldindata.org/what-is-the-gini-coefficient' (accessed 10 December 2024).

1.8 The city of Bangalore, India, has been expanding rapidly due to urban sprawl. In 2000, the city had a built-up area of 200 km^2 and had a population of 5 561 000. By 2015, the city's built-up area had increased to 275 km^2, while the population grew to 10 000 000. The city is planning further expansion, and the projected growth rate of the built-up area is 3% per year, while the population is expected to grow at a rate of 3% per year.

a) Determine the projected built-up area of Bangalore in 2025.
b) Calculate the projected population of Bangalore in 2025.
c) If the increase in built-up area leads to GHG emissions of 75 kg CO_2e/m^2 area, what would be the increase in GHG emissions after 10 years due to the increase in the built-up area?
d) If the per capita waste generation rate of Bangalore stands at 500 g/day, what would be the total waste generated in the year 2025?
e) Discuss the potential impacts of urban sprawl on Bangalore's infrastructure, environment, and quality of life. What strategies could be employed to manage urban sprawl effectively?

1.9 Given:

- Country A: Population = 10 million, Affluence (GDP per capita) = \$20 000, Technology factor = 0.2 units/\$1000 GDP.
- Country B: Population = 20 million, Affluence (GDP per capita) = \$15 000, Technology factor = 0.3 units/\$1000 GDP.

Using IPAT equation,

a) Calculate the total GDP for each country.
b) Calculate the total environmental impact for each country.
c) Compare the impacts of Country A and Country B.

1.10 The data regarding car ownership in the metropolitan cities of Chicago and New York are given below:

City	Total number of households	Number of cars owned per household	Manufacturing emissions (kg CO_2e/car)
Chicago	1.08M	1.12	5000
New York	3.2M	0.63	5000

Assuming the all the cars have similar rate of emissions per manufactured car, calculate the following using the IPAT equation.

a) Calculate the total GHG emissions due to manufacturing the existing cars in Chicago and New York. Comment on the results.
b) Assuming that each car travels 37 miles a day on an average and each car emits 400 g CO_2e emissions per mile, calculate the daily greenhouse gas emissions from Chicago and New York. Compare and comment on the results.

References

Anastas, P.T. and Zimmerman, J.B. (2003). Design through the twelve principles of green engineering. *Environmental Science & Technology* 37 (5): 94A–101A.

Artmann, M., Inostroza, L., and Fan, P. (2019). Urban sprawl, compact urban development and green cities. How much do we know, how much do we agree? *Ecological Indicators* 96: 3–9.

ASCE (2023a). Policy statement 418 - The role of the civil engineer in sustainable development. American Society of Civil Engineers. https://www.asce.org/advocacy/policy-statements/ps418---the-role-of-the-civil-engineer-in-sustainable-development (accessed 9 September 2024).

ASCE (2023b). *Standard Practice for Sustainable Infrastructure*. Reston, VA: ASCE https://doi.org/10.1061/9780784415733.

ASCE (2023c). *Integrating Resilience and Sustainability into Civil Engineering Projects* (ed. C. Field and C. Zawislak). Reston, VA: ASCE https://doi.org/10.1061/9780784484814.

Butchart, S.H., Walpole, M., Collen, B. et al. (2010). Global biodiversity: indicators of recent declines. *Science* 328 (5982): 1164–1168.

Chertow, M.R. (2000). The IPAT equation and its variants. *Journal of Industrial Ecology* 4 (4): 13–29.

Dasandi, N. (2014). International inequality and world poverty: a quantitative structural analysis. *New Political Economy* 19 (2): 201–226.

DiNapoli, R.J., Lipo, C.P., and Hunt, T.L. (2021). Triumph of the commons: sustainable community practices on Rapa Nui (Easter Island). *Sustainability* 2021 (13): 12118.

Doorn, N., Gardoni, P., and Murphy, C. (2019). A multidisciplinary definition and evaluation of resilience: the role of social justice in defining resilience. *Sustainable and Resilient Infrastructure* 4 (3): 112–123. https://doi.org/10.1080/23789689.2018.1428162.

Ehrlich, P.R. and Holdren, J.P. (1971). Impact of population growth. *Science: American Association for the Advancement of Science* 171 (3977): 1212–1217.

EUROSTAT (2024a). Transport. Eurostat Statistics Explained. https://ec.europa.eu/eurostat/statistics-explained/index.php?title=Transport (accessed 10 October 2024).

EUROSTAT (2024b). Environment. Eurostat Statistics Explained. https://ec.europa.eu/eurostat/statistics-explained/index.php?title=Environment (accessed 10 October 2024).

EUROSTAT (2024c). Energy. Eurostat Statistics Explained. https://ec.europa.eu/eurostat/statistics-explained/index.php?title=Energy (accessed 10 October 2024).

Figueroa, R.M. (2022). Environmental justice. In: *The Routledge Companion to Environmental Ethics* (ed. B. Hale, A. Light, and L. Lawhon), 767–782. Routledge.

Graedel, T.E. and Allenby, B.R. (2010). *Industrial Ecology and Sustainable Engineering*. Hoboken, NJ: Prentice Hall.

Gutner, T. (2012). *Hurricane Sandy grows to largest Atlantic tropical storm ever*. CBS News WBZ-TV Meteorology Update. https://www.cbsnews.com/boston/news/hurricane-sandy-grows-to-largest-atlantic-tropical-storm-ever/ (accessed 8 September 2024).

Halpern, B.S., Klein, C.J., Brown, C.J. et al. (2013). Achieving the triple bottom line in the face of inherent trade-offs among social equity, economic return, and conservation. *Proceedings of the National Academy of Sciences* 110 (15): 6229–6234.

Hardin, G. (1968). The tragedy of the commons. *Science* 162 (3859): 1243–1248.

Hardin, G. (2009). The tragedy of the commons∗. *Journal of Natural Resource Policy Research* 1 (3): 243–253.

Hosseini, S., Barker, K., and Ramirez-Marquez, J.E. (2016). A review of definitions and measures of system resilience. *Reliability Engineering & System Safety* 145: 47–61.

IPCC (2023). Climate change 2023: synthesis report. In: *Contribution of Working Groups I, II and III to the Sixth Assessment Report of the Intergovernmental Panel on Climate Change* (ed. Core Writing Team, H. Lee, and J. Romero), 184. Geneva, Switzerland: IPCC https://doi.org/10.59327/IPCC/AR6-9789291691647.

ITRC (Interstate Technology & Regulatory Council) (2021). *Sustainable Resilient Remediation SRR-1. SRR Team*. Washington, DC: ITRC www.itrcweb.org (accessed 10 April 2024).

Kuriyama, Peter T.; Hill, Kevin T.; Zwolinski, Juan P. (2022). Update assessment of the Pacific sardine resource in 2022 for U.S. management in 2022–2023. Southwest Fisheries Science Center, US. Report: NOAA-TM-NMFS-SWFSC; 662. https://doi.org/10.25923/57mx-dj29.

Lawler, J.J., Lewis, D.J., Nelson, E. et al. (2014). Projected land-use change impacts on ecosystem services in the United States. *Proceedings of the National Academy of Sciences* 111 (20): 7492–7497.

Li, C., Zhou, K., Qin, W. et al. (2019). A review on heavy metals contamination in soil: effects, sources, and remediation techniques. *Soil and Sediment Contamination: An International Journal* 28 (4): 380–394.

Mannucci, P.M., Harari, S., Martinelli, I., and Franchini, M. (2015). Effects on health of air pollution: a narrative review. *Internal and Emergency Medicine* 10: 657–662.

Marzec, R.P. (2018). Securing the future in the anthropocene: a critical analysis of the millennium ecosystem assessment scenarios. *Elementa: Science of the Anthropocene* 6: 42.

Nameroff, T.J., Garant, R.J., and Albert, M.B. (2004). Adoption of green chemistry: an analysis based on US patents. *Research Policy* 33 (6–7): 959–974. https://doi.org/10.1016/j.respol.2004.03.001.

NOAA (2023). If carbon dioxide hits a new high every year, why isn't every year hotter than the last? https://www.climate.gov/news-features/climate-qa/if-carbon-dioxide-hits-new-high-every-year-why-isn%E2%80%99t-every-year-hotter-last (accessed 20 February 2024).

NOAA (2024). U.S. billion-dollar weather and climate disasters. https://www.ncei.noaa.gov/access/billions/ (accessed 8 September 2024).

Ortego, A., Calvo, G., Valero, A. et al. (2020). Assessment of strategic raw materials in the automobile sector. *Resources, Conservation and Recycling* 161: 104968.

Ozbay, G., Jones, M., Gadde, M. et al. (2021). Design and operation of effective landfills with minimal effects on the environment and human health. *Journal of Environmental and Public Health* 2021: 6921607.

Rae, J.W., Zhang, Y.G., Liu, X. et al. (2021). Atmospheric CO_2 over the past 66 million years from marine archives. *Annual Review of Earth and Planetary Sciences* 49: 609–641.

Reddy, K.R., Janga, J.K., and Kumar, G. (2024). Sustainability and resilience: a new paradigm in geotechnical and geoenvironmental engineering. *Indian Geotechnical Journal* 1–22. https://doi.org/10.1007/s40098-024-00899-5.

Rode, P., Floater, G., Thomopoulos, N. et al. (2017). Accessibility in cities: transport and urban form. In: *Disrupting Mobility*, 239–273. Springer International Publishing.

Sadasivam, B.Y. and Reddy, K.R. (2014). Landfill methane oxidation in soil and bio-based cover systems: a review. *Reviews in Environmental Science and Biotechnology* 13 (1): 79–107. https://doi.org/10.1007/s11157-013-9325-z.

Schweitzer, L. and Noblet, J. (2018). Chapter 3.6: Water contamination and pollution. In: *Green Chemistry* (ed. B. Török and T. Dransfield), 261–290. Elsevier https://doi.org/10.1016/B978-0-12-809270-5.00011-X.

UN (1987). *Report of the World Commission on Environment and Development: Our Common Future*. Oslo, Norway: United Nations General Assembly, Development and International Co-operation Environment.

UN (2007). *Indicators of Sustainable Development: Guidelines and Methodologies*, 3e. New York: Economic and Social Affairs. United Nations. ISBN: 978-92-1-104577-2.

UN (2015). *Transforming our World: The 2030 Agenda for Sustainable Development*. United Nations https://sustainabledevelopment.un.org/post2015/transformingourworld/publication (accessed 24 January 2019).

USEPA (2002). *Presidential Green Chemistry Challenge Award Recipients*. Document Number EPA-744-K-02-002. Washington, DC: EPA.

USEPA (2023). National overview: facts and figures on materials, wastes and recycling. https://www.epa.gov/facts-and-figures-about-materials-waste-and-recycling/national-overview-facts-and-figures-materials (accessed 13 February 2024).

Vallecillo, S., La Notte, A., Zulian, G. et al. (2019). Ecosystem services accounts: valuing the actual flow of nature-based recreation from ecosystems to people. *Ecological Modelling* 392: 196–211.

Vellend, M. (2017). The biodiversity conservation paradox. *American Scientist* 105 (2): 94.

Wakabayashi, Y. (2020). Population dynamics and the environment. In: *Human Geoscience. Advances in Geological Science* (ed. Y. Himiyama, K. Satake, and T. Oki). Singapore: Springer https://doi.org/10.1007/978-981-32-9224-6_18.

Woods, D.D. (2015). Four concepts for resilience and the implications for the future of resilience engineering. *Reliability Engineering & System Safety* 141: 5–9.

2

Environmental Concerns

Chapter Goal and Objectives
Sustainability is underpinned by the concept of the triple bottom line, which considers the interconnected dimensions of environmental, economic, and social factors in decision-making processes. For a comprehensive approach to sustainable practices, it is crucial to understand and address a broad spectrum of environmental issues that may arise at different scales. This chapter seeks to provide an overview of the key environmental factors that must be considered when assessing sustainability in engineering and development projects. The specific objectives of this chapter are to (i) identify and categorize the major environmental issues, ranging from global to local concerns; (ii) highlight how these issues can impact not only the immediate project environment and surrounding communities but also extend their influence to the global population and the planet at large; and (iii) explore how these environmental challenges intersect with broader economic and social implications, reinforcing the need for holistic solutions. By addressing these multi-scale environmental concerns, the chapter will provide a background for integrating environmental considerations into the project's sustainability, ensuring a balance between environmental protection, economic viability, and social equity.

2.1 Introduction

When considering a range of projects, which may include a tangible product, a real estate development, or a productive service/activity, it is often beneficial to assess and quantify the environmental impacts associated with a range of alternative projects/designs for the purposes of a comparative assessment. In order to measure the degree of sustainability of a project, all potential environmental impacts, or as many as practicable, should be considered and incorporated into an environmental impact assessment, especially when those impacts are quantifiable. These may include both direct and indirect environmental impacts. However, it can often be difficult to quantify some impacts resulting from pre-project issues, such as raw materials' extraction/fabrication, project activities, outputs (e.g. emissions and financial implications), and post-use issues (e.g. waste management and disposition). In this regard, a life cycle environmental impact assessment will identify and allow for the selection of the project/design that minimizes overall environmental

Sustainable and Resilient Engineering: Drivers, Metrics, Tools, and Applications, Second Edition. Krishna R. Reddy, Claudio Cameselle, and Jeffrey A. Adams.

Companion website: www.wiley.com/go/reddy/sustainableresilientengineering2e

impact. Of course, in addition to environmental impacts, various economic and social implications should be considered to select a project as the most sustainable option.

As discussed in Chapter 1, environmental impacts constitute one of the dimensions (pillars) of sustainability. To assess environmental concerns, it is useful to utilize a reference framework, such as "the triple bottom line." As previously discussed, the triple bottom-line sustainability framework is measured using three criteria – environmental, economic, and social.

Environmental concerns are present during all phases of a project, and they occur over a range of scope and magnitudes. These concerns may occur on a local scale, a regional scale, or a global scale. It is important to remember that these concerns are not universal – many potential environmental concerns that warrant attention may not occur for a specific project. Additionally, virtually every activity will have one or more environmental impacts. Several aspects of a project may result in side effects that pertain to a specific or a range of environmental concerns. However, it is also important to point out that many of these may be disregarded, depending on their magnitude or when considered with the benefits and drawbacks of moving forward with a specific project. The key point is to determine the degree of potential impact that may occur and determine the degree to which human health or the environment may be affected. Separating important effects from "background noise" is a key skill in assessing the sustainability of a project, and it is important that these environmental concerns be assessed on a qualitative or quantitative basis.

To perform these assessments, it is critical to understand a range of environmental concerns that can occur from the projects or activities under assessment. This chapter provides background on various environmental issues and impacts related to sustainability. This background is essential to serve as a context for sustainable engineering practices.

2.2 Global Warming and Climate Change

There is an overwhelming scientific consensus that the Earth is warming and undergoing climate change. Although climate change results from a variety of natural physical processes, and global climate appears to follow a natural oscillatory cycle, the scientific evidence suggests that anthropogenic sources, such as greenhouse gases (e.g. methane and carbon dioxide), are contributing to the climate change. The magnitude of this contribution and the resulting effects on climate change in the biosphere is still being debated, but there is a consensus among the scientific community that these anthropogenic sources are contributing in a meaningful way to the climate change. Global warming and climate change can have a range of adverse effects, including increased extreme weather events, effects on natural habitats and species, and adverse effects on the ability of ecosystems to naturally adapt and cope with extreme weather and geophysical events. Collectively, much of the anthropogenic contribution to global warming and climate change is occurring through the "greenhouse effect."

The greenhouse effect is a beneficial process that helps to maintain the Earth's average temperature around 15 °C/59 °F, making life on the Earth possible. Without the greenhouse effect, the average temperature in the planet would be −18 °C/0 °F, a 33 °C/59 °F difference (Archer 2016). This effect is not exclusive to the Earth; the temperature of other terrestrial planets is also influenced by the greenhouse effect (Figure 2.1). Simply stated, the greenhouse effect occurs when the energy from the sun, in the form of electromagnetic radiation, heats the surface of the Earth, which in turn regulates Earth's weather and climate. Much of this energy radiates back into space. As this radiated energy travels through the atmosphere, several "greenhouse gases" (mainly water vapor and carbon dioxide) trap some of this energy, much in the same way a greenhouse traps heat. The trapped energy serves to warm the Earth. Figures 2.2 and 2.3 depict the mechanisms of the greenhouse effect.

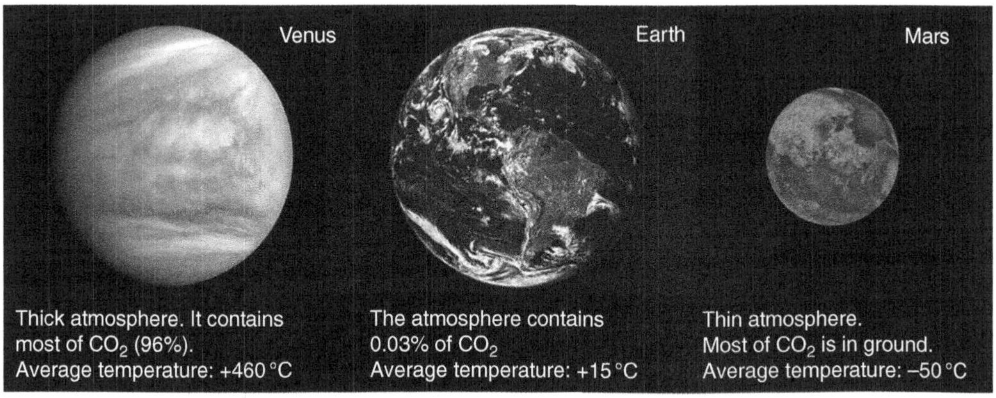

Figure 2.1 Greenhouse effect on three planets (Venus, Earth and Mars).

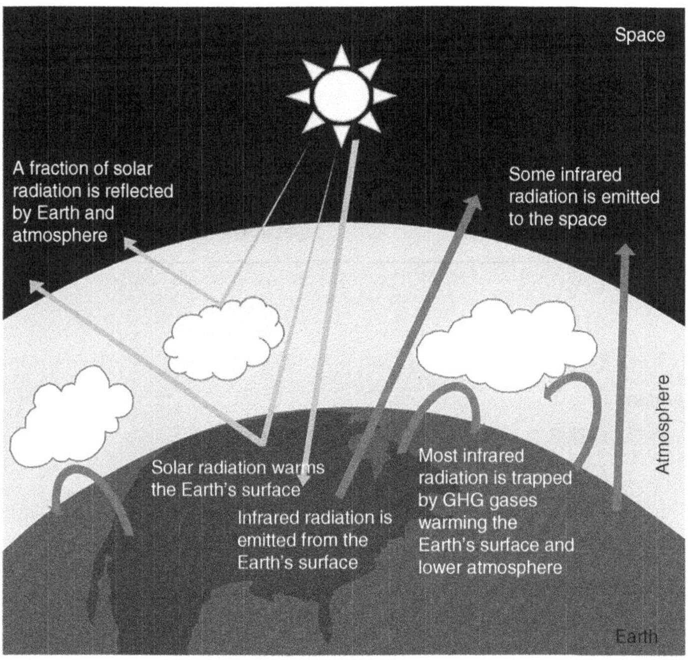

Figure 2.2 Greenhouse effect on planet Earth.

Some of the greenhouse gases, such as carbon dioxide, methane, nitrous oxide, and water vapor, are generated by both natural and anthropogenic processes. Other gases, such as fluorocarbons, are solely generated from anthropogenic processes. Specific gases exhibit a relative "strength" or contribution to the greenhouse effect. The global warming potential (GWP) is an index that assesses the heat-trapping ability of a similar mass of a particular gas. This dimensionless value is expressed as a factor of carbon dioxide's heat-trapping ability (carbon dioxide's GWP is standardized to 1). Table 2.1 presents the GWP for several greenhouse gases based on the Intergovernmental Panel on Climate Change (IPCC) Sixth Assessment Report (SAR) (IPCC 2023).

Anthropogenic contributions of greenhouse gases come from a variety of sources. Carbon dioxide emanates from the combustion of fossil fuels, mainly for energy production and transportation. From a mass balance perspective, 1 kg of fuel-based carbon becomes 3 kg of atmospheric carbon

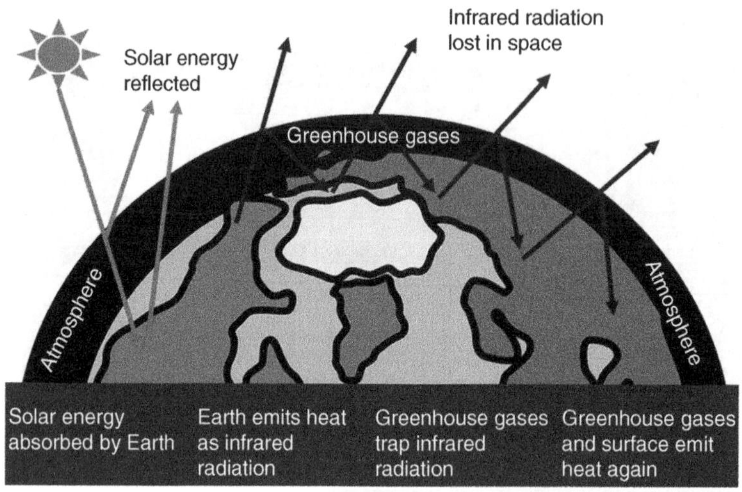

Figure 2.3 Greenhouse gases trap the energy from the Sun heating the Earth's surface.

Table 2.1 Global Warming Potential (GWP) for some greenhouse gases in the atmosphere.

Industrial designation/common name of the greenhouse gas	Chemical formula	GWP based on IPCC Sixth assessment report 2023
Carbon dioxide	CO_2	1
Methane	CH_4	27.9
Nitrous oxide	N_2O	273
Substances controlled by Montreal protocol		
CFC-11	CCl_3F	5560
CFC-12	CCl_2F_2	11 200
CFC-13	$CClF_3$	16 200
CFC-113	CCl_2FCClF_2	6520
CFC-114	$CClF_2CClF_2$	9430
CFC-115	$CClF_2CF_3$	9600
Halon-1301	$CBrF_3$	7200
Halon-1211	$CBrClF_2$	1930
Halon-2402	$CBrF_2CBrF_2$	2170
Carbon tetrachloride	CCl_4	2200
Methyl bromide	CH_3Br	2.43
Methyl chloroform	CH_3CCl_3	161
HCFC-21	$CHCl_2F$	160
HCFC-22	$CHCLF_2$	1960
HCFC-123	$CHCl_2CF_3$	90.4
HCFC-124	$CHClFCF_3$	597
HCFC-141b	CH_3CCl_2F	860

Table 2.1 (Continued)

Industrial designation/common name of the greenhouse gas	Chemical formula	GWP based on IPCC Sixth assessment report 2023
HCFC-142b	CH_3CClF_2	2300
HCFC-225ca	$CHCl_2CF_2CF_3$	137
HCFC-225cb	$CHClFCF_2CClF_2$	568
Hydrofluorocarbons (HFCs)		
HFC-23	CHF_3	14 600
HFC-32	CH_2F_2	771
HFC-41	CH_3F_2	135
HFC-125	CHF_2CF_3	3740
HFC-134	CHF_2CHF_2	1260
HFC-134a	CH_2FCF_3	1530
HFC-143	CH_2FCHF_2	364
HFC-143a	CH_3CF_3	5810
HFC-152	CH_2FCH_2F	21.5
HFC-152a	CH_3CHF_2	164
HFC-161	CH_3CH_2F	4.84
HFC-227ea	CF_3CHFCF_3	3600
HFC-236cb	$CH_2FCF_2CF_3$	1350
HFC-236ea	CHF_2CHFCF_3	1500
HFC-236fa	$CF_3CH_2CF_3$	8690
HFC-245ca	$CH_2FCF_2CHF_2$	787
HFC-245fa	$CHF_2CH_2CF_3$	962
HFC-365mfc	$CH_3CF_2CH_2CF_3$	914
HFC-43-10mee	$CF_3CHFCHFCF_2CF_3$	1600
Perfluorinated compounds		
Sulfur hexafluoride	SF_6 23	25 200
Nitrogen trifluoride	NF_3	17 400
PFC-14	CF_4	7380
PFC-116	C_2F_6	12 400
PFC-218	C_3F_8	9290
PFC-318	c-C_4F_8	10 200
PFC-31-10	C_4F_{10}	10 000
PFC-41-12	C_5F_{12}	9220
PFC-51-14	C_6F_{14}	8620
PCF-91-18	$C_{10}F_{18}$	7480
Trifluoromethyl sulfur pentafluoride	SF_5CF_3	18 500
Perfluorocyclopropane	c-C_3F_6	9200

(*Continued*)

Table 2.1 (Continued)

Industrial designation/common name of the greenhouse gas	**Chemical formula**	**GWP based on IPCC Sixth assessment report 2023**
Fluorinated ethers		
HFE-125	CHF_2OCF_3	14 300
HFE-134	CHF_2OCHF_2	6630
HFE-143a	CH_3OCF_3	616
HCFE-235da2	$CHF_2OCHClCF_3$	539
HFE-245cb2	$CH_3OCF_2CF_3$	747
HFE-245fa2	$CHF_2OCH_2CF_3$	878
HFE-347mcc3	$CH_3OCF_2CF_2CF_3$	576
HFE-347pcf2	$CHF_2CF_2OCH_2CF_3$	980
HFE-356pcc3	$CH_3OCF_2CF_2CHF_2$	277
HFE-449sl	$C_4F_9OCH_3$	460
HFE-569sf2	$C_4F_9OC_2H_5$	60.6
HFE-43-10pccc124 (H-Galden 1040×)	$CHF_2OCF_2OC_2F_4OCHF_2$	3220
HFE-236ca12 (HG-10)	$CHF_2OCF_2OCHF_2$	6060
HFE-338pcc13 (HG-01)	$CHF_2OCF_2CF_2OCHF_2$	3320
HFE-227ea	$CF_3CHFOCF_3$	3600
HFE-236ea2	$CHF_2OCHFCF_3$	2590
HFE-236fa	$CF_3CH_2OCF_3$	1100
HFE-245fa1	$CHF_2CH_2OCF_3$	934
HFE 263mf	$CF_3CH_2OCH_3$	2.06
HFE-329mcc2	$CHF_2CF_2OCF_2CF_3$	3770
HFE-338mcf2	$CF_3CH_2OCF_2CF_3$	1040
HFE-347mcf2	$CHF_2CH_2OCF_2CF_3$	963
HFE-356mec3	$CH_3OCF_2CHFCF_3$	264
HFE-356pcf2	$CHF_2CH_2OCF_2CHF_2$	831
HFE-356pcf3	$CHF_2OCH_2CF_2CHF_2$	484
HFE 365mcf3	$CF_3CF_2CH_2OCH_3$	1.6
HFE-374pc2	$CHF_2CF_2OCH_2CH_3$	12.5
Perfluoropolyethers		
PFPMIE	$CF_3OCFCF_3CF_2OCF_2OCF_3$	10 300
Hydrocarbons and other compounds – direct effects		
Chloroform	$CHCl_3$	20.6
Methylene chloride	CH_2Cl_2	11.2
Methyl chloride	CH_3Cl	5.54
Halon-1201	$CHBrF_2$	380

Source: Adapted from IPCC (2023).

dioxide upon combustion. Increased carbon dioxide also results from deforestation since plants are a sink for carbon dioxide, and the removal of plants via mass deforestation reduces this carbon dioxide absorption ability. Methane, with its GWP of 27.9 indicating a much stronger greenhouse effect than carbon dioxide, is emitted from landfills, livestock and farming activities, natural gas pipeline losses, and land use/wetlands alterations, considerably adding to the greenhouse emissions into the atmosphere.

It is important to note that the Earth has a natural absorption capacity for carbon dioxide and other greenhouse gases; Figure 2.4 shows this relationship, and Figure 2.5 depicts the planetary carbon cycle. Nevertheless, a substantial amount of greenhouse gases have been generated during

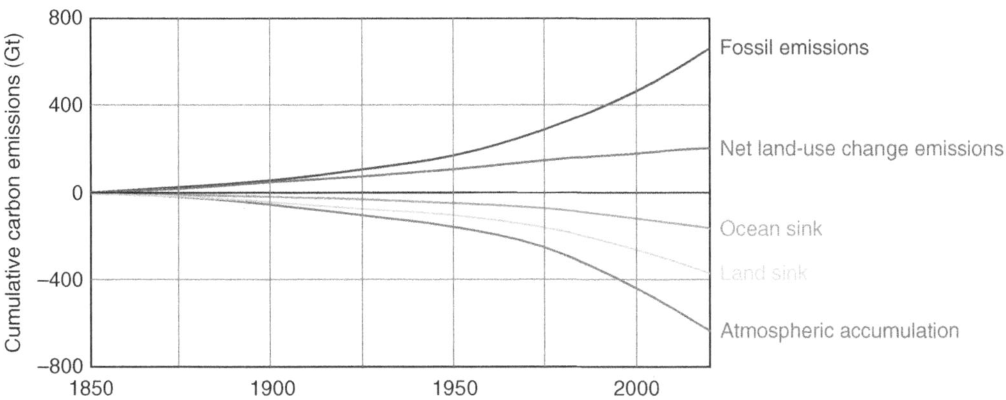

Figure 2.4 Cumulative carbon emissions and Earth's capacity for CO_2 absorption. *Source:* NOAA's Earth System Research Laboratory.

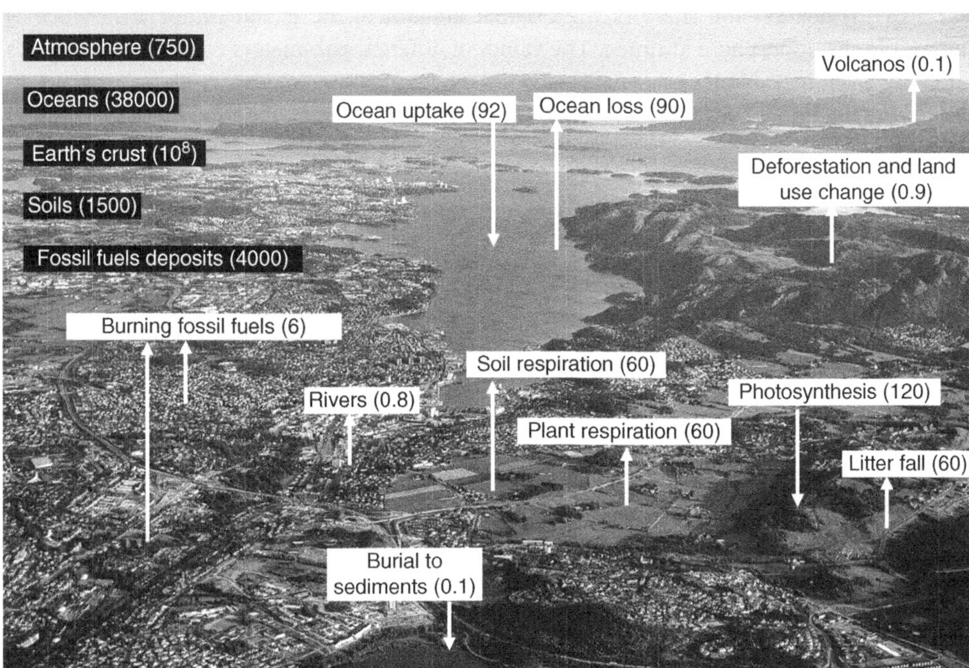

Figure 2.5 Global carbon cycle. Pools in black boxes (Pg). Fluxes in white arrows (Pg/year).

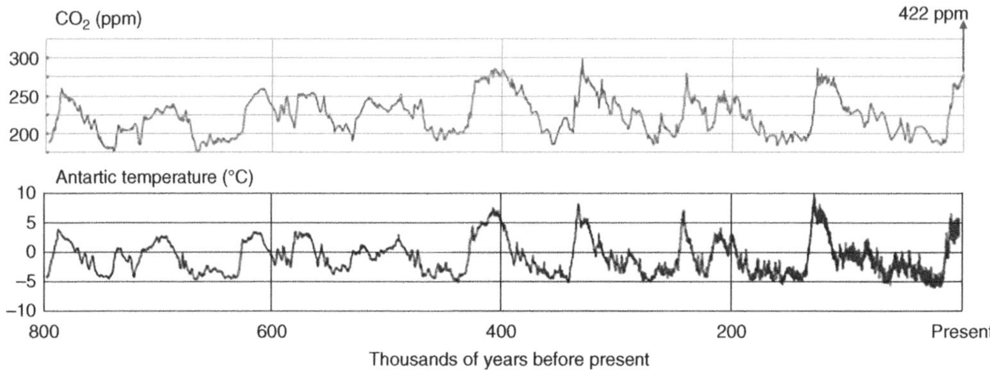

Figure 2.6 Influence of atmospheric CO_2 concentration in the average temperature of Earth over thousands of years. *Source:* Adapted from Lüthi et al. (2008) and Jouzel et al. (2007).

the industrial era, contributing to a measured rapid increase in global temperatures. Figure 2.6 depicts the average temperature on the Earth over thousands of years, and it highlights the increase of temperature because of the massive CO_2 emissions to the atmosphere dating from the onset of the industrial revolution.

Example 2.1

If the following relationship holds for the time decay of any given quantity of carbon dioxide in the atmosphere,

$$RCO_2(t) = a_0 + \sum_{i=1}^{3} a_i e^{-\frac{t}{\tau_i}}$$

Here, $RCO_2(t)$ denotes the share of the original amount of the gas at time $t = 0$, which is still present in the atmosphere at time t. The values of different parameters corresponding to this equation are given in Table E2.1.

Considering the initial concentration of CO_2 at $t = 0$ to be 416 ppm, determine the following:

- What is the remaining amount of CO_2 (ppm) in the atmosphere, if no additional emissions are assumed, after 10/25/50/100/500/1000 years?
- Plot the function of $RCO_2(t)$ for $t = 0$ to 1000 years.
- Determine the half-life of CO_2 in the atmosphere, and the average lifetime of CO_2 in the atmosphere.
- If CO_2 emissions are increasing at 2.4 ppm/year, plot the concentrations of CO_2 with time for 100 years. (Don't forget to consider the decay rates of newly emitted CO_2.)

Table E2.1 Numerical values for the parameters in the equation for RCO_2.

Variable/order	0	1	2	3
a	0.2173	0.2240	0.2824	0.2763
τ	–	394.4	36.54	4.304

Sources: Numbers and equations from "Anthropogenic and Natural Radiative Forcing," Table 8.A.1 in part 1 of the 5th IPCC Assessment report as well as "Anthropogenic and Natural Radiative Forcing, Supplementary Material," Section 8.SM.11.1: Equations for the Global Warming Potential of part 1 of the 5th IPCC Assessment report.

Solution:

Considering the parameters provided in Table E2.1, and using the decay relationship given in the question, the fraction of CO_2 originally emitted can be calculated:

Sample Calculation: for $t = 45$ years,

$$RCO_2 = 0.2173 + 0.2243 \times e^{-\frac{45}{394.4}} + 0.2824 \times e^{-\frac{45}{36.54}} + 0.2763 \times e^{-\frac{45}{4.304}}$$

$$RCO_2 = 0.499839 \cong 0.5$$

Half of the initial concentration of CO_2 remains in the atmosphere after 45 years, hence it can be called the half-life of the CO_2 in the atmosphere. Similarly, calculations can be performed for other time periods, the values of these calculations are provided below:

Time (yrs)	10	25	50	100	500	1000
RCO_2	0.68	0.57	0.49	0.41	0.28	0.24

Based on the RCO_2 values obtained above, the CO_2 concentrations in the atmosphere can be calculated as follows:

After 10 years $= 0.68 \times 416 = 282.88$ ppm
After 25 years $= 0.57 \times 416 = 237.12$ ppm
After 50 years $= 0.49 \times 416 = 203.84$ ppm
After 100 years $= 0.41 \times 416 = 170.56$ ppm
After 500 years $= 0.28 \times 416 = 116.48$ ppm
After 1000 years $= 0.24 \times 416 = 99.84$ ppm

The plot of RCO_2 vs time until $t = 1000$ years is given below:

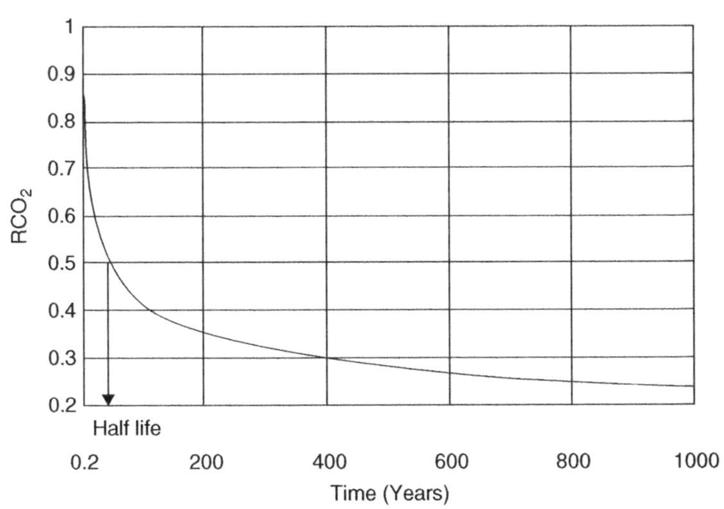

The half-life of CO_2 based on the given decay relationship was calculated to be 45 years.

If the CO_2 concentrations are increasing at 2.4 ppm/year:

Sample calculation for $t = 5$ years:

The RCO_2 values based on the given decay relationship for $t = 1, 2, 3, 4$, and 5 years are 0.935, 0.881, 0.838, 0.802, and 0.771 respectively.

Considering additional emissions of 2.4 ppm/year, concentration of CO_2 after five years can be calculated as: $416 \times 0.771 + 2.4 \times 0.802 + 2.4 \times 0.838 + 2.4 \times 0.881 + 2.4 \times 0.935 + 2.4 = 331.43$ ppm

Similar calculations can be performed for different time periods to plot the curve of changes in CO_2 concentration with time, below is the plot for t = 100 years

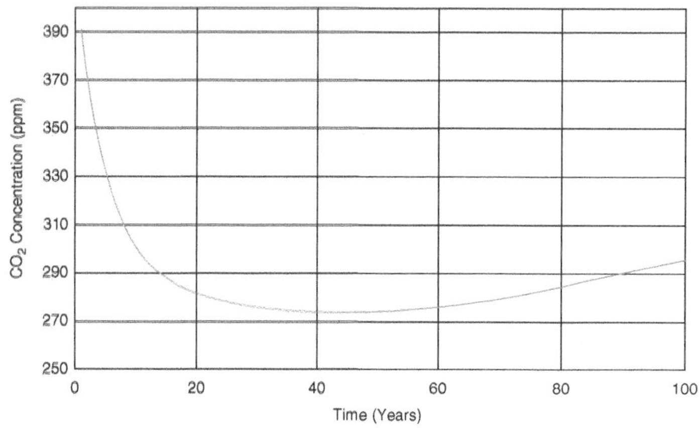

As we can see in the Figure, with annual emissions of 2.4 ppm/year, the CO_2 concentrations in the atmosphere will decrease initially, but as the time progresses the accumulation of new emissions overtakes the decay rate and the atmospheric concentrations of CO_2 will start increasing.

2.2.1 Intergovernmental Panel on Climate Change (IPCC) Assessment Report

Many resources have been devoted to the ongoing study of greenhouse gases and resulting climate change. The IPCC was established by the United Nations in 1988. The IPCC has provided assessments and reports on global climate change dating to 1990, when the first report was published. In 2023, the SAR was issued (IPCC 2023). Utilizing inputs from thousands of scientific experts and hundreds of authors, the SAR provided convincing evidence that emission-related human activity is already having a severe impact on the global climate and sea level. Specifically, the IPCC SAR concluded that "An increasing body of observations gives a collective picture of a warming world and other changes in the climate system." Also, "There is new and stronger evidence that most of the warming observed over the past 50 years is attributable to human activities."

Predictions of future impacts are difficult to make and involve wide ranges due to variations in mathematical models. Figure 2.7 depicts two models and their predicted fossil fuel emissions for the next 50 years. Regarding the effects of these continued emissions, by 2100, global temperatures are predicted to increase between 2.5 and 10.4 °F, with temperatures expected to increase even more in the United States (NCA 2014).

Increased global temperatures can lead to drastic shifts in weather patterns, with intense precipitation and storm events in some areas and less in others (Figure 2.8). These events can drastically affect agricultural activities and induce strains on ecosystems and biodiversity. Furthermore, increased temperatures accelerate the melting of polar ice and alpine glaciers, leading to increases in sea levels, which in turn can inundate coastal areas. These climatic effects may have drastic impacts on several economic and resource sectors, including agriculture, water resources, energy, transportation, public health, and infrastructure. Specific climate effects include the following:

- *Precipitation* – Many areas will experience increased precipitation, including more days with precipitation, more intense events, and more storm events.

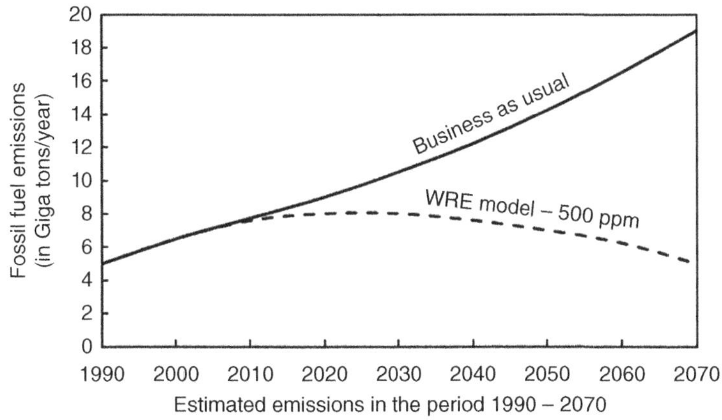

Figure 2.7 Emissions of greenhouse gases due to the utilization of fossil fuels. (WRE: Wigley, Richels, and Edmonds model for "stabilization at 500 ppm.") *Source:* Adapted from Pacala and Socolow (2004).

- *Snow cover* – With increases in temperature, snowpack and related water-holding capacity decreases. Snow seasons are shorter, precipitation events consist of rain as opposed to snow, and runoff patterns and timings are altered.
- *Tropical storms* – An increased number of more intense tropical storms are expected to make landfall, as depicted in Figure 2.8c. The Atlantic and Gulf Coasts as well as Hawaii are susceptible to hurricanes. Additionally, one-fourth of US crude oil production is situated in the hurricane-prone Gulf of Mexico (EIA 2017a).
- *Flooding/drought* – Altered precipitation patterns will trigger more frequent and severe flooding in some regions and droughts in others. Approximately, 3800 towns and cities of over 2500 inhabitants are located in floodplains in the United States (Encyclopedia of Arkansas 2023), and increases in flooding will lead to increased damage to infrastructure and property, which is already averaging approximately $6 billion in annual losses.

2.3 Desertification

Dry lands – defined as areas with arid, semiarid, and dry subhumid climates – constitute approximately 40% of the Earth's land surface. These regions play a vital role for civilization in that they provide much of the world's arable land and rangeland for livestock. However, nearly 70% of these regions are under stress through land degradation, or desertification, which reduces the productive use of these regions and ultimately decreases food security. The degree of land degradation falls into one of the following four categories (Lal et al. 2003):

- *Light* – Lands that are mildly salinized or eroded, rangelands that contain approximately 70% native vegetation.
- *Moderate* – Lands that have been moderately eroded or salinized, soil chemical and physical integrity has been compromised, and rangelands that have 30–70% native vegetation.
- *Severe* – Lands that include frequent gullies or hollows, poor crop production, and rangelands that have less than 30% native vegetation.
- *Extreme* – No crop growth occurs, and restoration is impossible; damage may have been caused by extreme erosion or salinization.

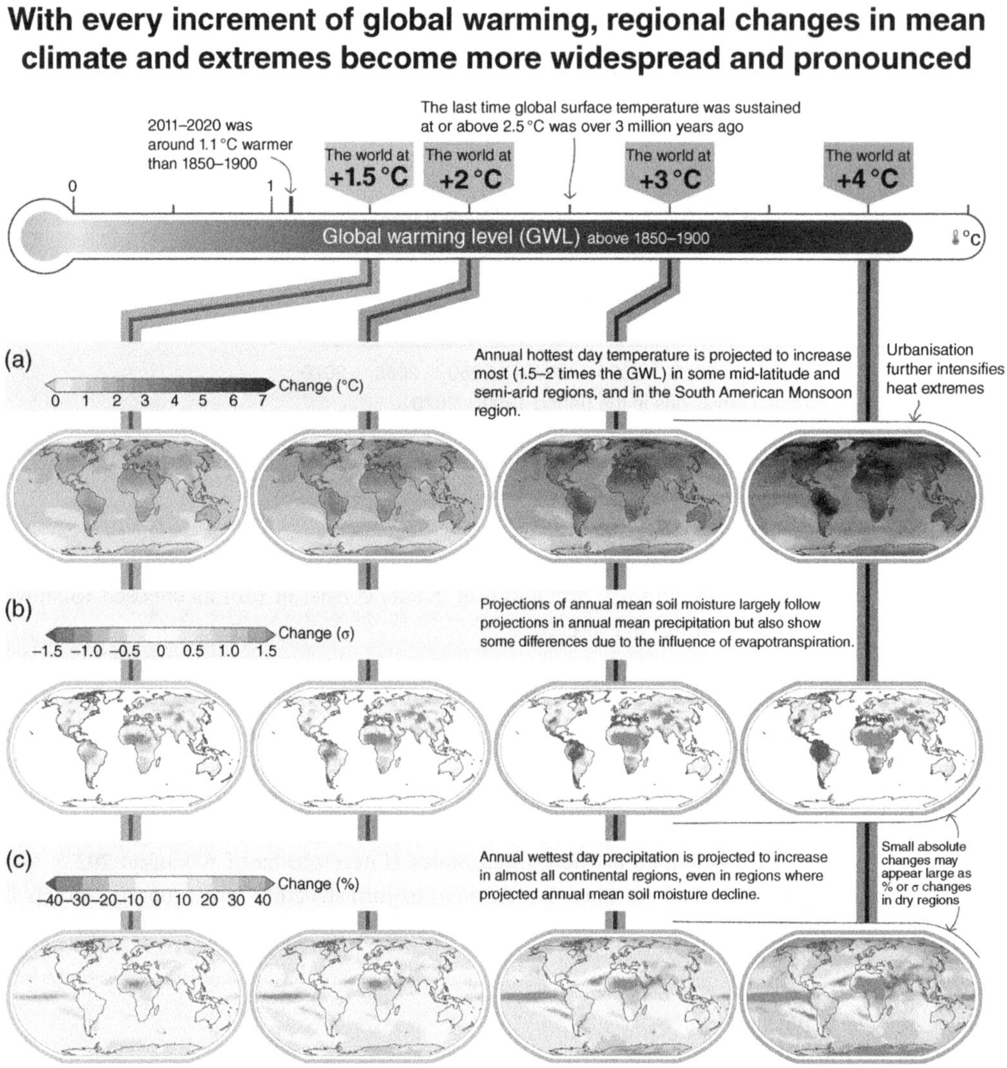

Figure 2.8 Climate change (a) annual hottest-day temperature change, (b) annual mean total column soil moisture change, (c) annual wettest-day precipitation change. *Source:* IPCC (2023)/IPCC.

Land degradation occurs because of the overexploitation of land resources or productive capacity. Overcultivation, deforestation, and overgrazing are some of the common causes of land degradation. Soils susceptible to land degradation in dry lands are particularly vulnerable to erosion, chemical and physical deterioration of the soil structure.

Unfortunately, more than 250 million people are directly affected by the effects of land degradation, and 1 billion people in more than 100 countries are at risk. Many of these regions include some of the world's poorest and most economically disadvantaged people. For instance, in Africa, land degradation is a constant threat to physical and economic survival. Recurrent drought conditions exacerbate ongoing soil degradation problems, which in turn magnify the effects of subsequent droughts, both of which magnify the effects and extent of famines (Reynolds et al. 2007).

Land degradation has many consequences for the environment and human health. The ability of the land and its resilience to climatic disturbances is reduced. Soil productivity is also reduced, resulting in reduced food production. Vegetation cover is reduced, often resulting in the replacement of edible plants with nonedible plants. Further, the detrimental effects of reduced vegetation cover increase susceptibility to erosion, which often results in flooding, increased surface runoff, reduced water quality, sedimentation of water bodies, and siltation of reservoirs and navigable channels. Health problems may also be exacerbated; wind-blown dust can result in chronic or acute respiratory problems, eye infections, allergies, and mental stress.

The UN Convention to Combat Desertification (UNCCD) was enacted by the United Nations in 1996 to address the issue of land degradation (UNCCD 2017). The convention established at the national level addresses the underlying causes of desertification and drought. The convention also seeks to identify and implement preventative measures that help to stop and even to invert the general trend of the degradation of productive soil and the desertification of land.

2.4 Deforestation

Deforestation is a form of land degradation that is occurring globally at an alarming rate. During the 1990s, more than 110 million hectares (or 11 million hectares/year) of forested lands were lost. Much of this loss occurred in developing countries. Unfortunately, about 45% of the Earth's remaining tropical rain forests are now in a degraded state (Thiele 2016), and deforestation continues to occur mainly in tropical and subtropical forests.

When reviewing these statistics, it would seem logical that the cause of this deforestation is unsustainable timber harvesting. However, this is not the case – the primary causes of deforestation are related to attempts to create agricultural land by clearing forests as well as harvesting to gain wood for fuel. Approximately, half of the wood harvested in the world is used as fuel or converted to charcoal, mostly in developing countries. On the other hand, harvested timber in developed countries is primarily used for industrial uses and as construction materials.

Ironically, deforestation to create agricultural land in developing countries is often counter-productive. For instance, soils in tropical or coniferous forests are relatively nutrient-poor and often do not provide prime agricultural land when cleared. The original forests often provide better economic capacity as well as better overall food security than the resulting clear land. Deforestation also reduces biodiversity and eliminates habitat for fauna that often could serve as a sustainable economic resource.

The effect of deforestation goes beyond the destruction of wildlife habitat. When trees are removed, the water absorption and retention capacity of soils are often reduced, contributing to increased surface water runoff and reduced water infiltration. In turn, groundwater recharge is reduced, and aquifer depletion may result. Because groundwater is the primary freshwater source for one-third of the world's population, this reduction in capacity can have disastrous consequences for agriculture, economy, and human well-being (Thiele 2016). Increased surface runoff also increases land degradation effects, including erosion of soils, which further degrades the agricultural capacity of these lands leading to decreased water quality and increased sedimentation in nearby waterways. This, in turn, has a significant secondary effect on the host ecosystem as well as economic prospects, since aquatic life can be adversely affected, leading to reduced hatcheries and reefs that are supportive of aquatic life and related commerce. Further, when water infiltration is reduced, areas of standing water may increase. In tropical areas, this can lead to increased malaria transmission, which annually claims the lives of 750 000 children under the age of 5 years (WHO 2023). As with land degradation, strategies to combat deforestation have been enacted. Efforts have been taken on an international level with the formation of the Intergovernmental Panel on Forests (IPF) and its successor, the Intergovernmental Forum on Forests (IFF).

2.5 Loss of Habitat and Biodiversity

The previous section has outlined several primary and secondary effects resulting from deforestation. Land degradation and deforestation not only result in the loss of economically valuable natural resources within an ecosystem but also have a significant adverse effect on other plant and animal species. With the elimination of forests, supportive habitats for a rich diversity of plant and animal species can be irrevocably harmed. As an example, current data suggests that the Earth's primary tropical forests can be essentially eliminated within four to five decades. It is well known that these forests support the widest range of species diversity on the planet. With increased habitat reduction or outright destruction, recent studies have indicated that rates of extinction will continue to accelerate in the future (Ceballos et al. 2015). As these forests continue to be eliminated, it is estimated that 0.2–0.3% of all species within these forests become extinct on an annual basis. These percentages are believed to translate to the loss of 4000–6000 species annually, which is approximately 10 000 times the natural rate of species extinction (Rain Forest Maker 2024).

These reductions in biodiversity are occurring because of the increased conversion of open space to agricultural land or urban use, overharvesting of plants and animals, draining and filling of wetlands, overapplication of agricultural chemicals, overfishing, and air pollution. In 2005, the Millennium Ecosystem Assessment was released, which highlighted a substantial and irreversible loss of biodiversity on the Earth. The assessment concluded that approximately 10–30% of bird, mammal, and amphibian species are threatened with extinction due to human activities (MEA 2005). Broken down by species classification, this includes the following:

- One out of eight bird species.
- One out of four mammal species.
- One out of four conifer species.
- One out of three amphibians.
- Six out of seven marine turtles.

As a species becomes endangered, its breeding capacity is greatly diminished, and its available gene pool is significantly reduced. This, in turn, reduces its capacity to adapt to environmental changes. If left unchecked, this cycle continues until the species becomes extinct. When a species is lost completely, its genetic material is lost as well. The elimination of the species from an ecosystem can have a significant effect, as its elimination from the food chain/food web can significantly upset the balance of other species, either due to its elimination as a predator or prey. Aggressive or invasive species can quickly evolve into a dominant role as these natural checks and balances are removed. This can result in a chain reaction of effects that can be of a global magnitude.

While the reduction of biodiversity has an obvious impact on ecosystem and environmental health, it also results in significant adverse impacts on human health. Disruptions to the food chain can directly or indirectly affect plant-based and animal-based food sources. Additionally, potential or existing sources for medicines and other pharmaceutical products can be lessened or eliminated. In response, communities and governments worldwide have increasingly focused on the protection of habitats and biodiversity by focusing on protection of air and water, maintenance of soil fertility, surface water management, waste treatment, and waste reduction.

The benefits of biodiversity are many. With respect to the ecosystems, healthy biodiversity and habitat are directly linked to water resource and soil protection, pollution breakdown and absorption, contributions to climate stability, and the ability of an ecosystem to recover from natural disasters or other rare/unforeseen events. Biodiversity also leads directly to beneficial economic output in the forms of food, medicinal sources, timber, ornamental plants, and a diverse gene pool that can benefit wild and domesticated species. Finally, these factors also contribute positively to culture and society, with increased opportunities for research and education, recreation and tourism, and reinforcement of cultural values.

Biodiversity is becoming increasingly important at an intergovernmental level. The United Nations Convention on Biological Diversity (UNCBD) was established in 1992 and has been ratified by more than 175 countries (CBD 2014). The convention set forth three main goals: (i) the conservation of biodiversity, (ii) the sustainable use of its components, and (iii) the fair and equitable sharing of the benefits from the use of genetic resources. The convention also established the Cartagena Protocol in 2000, with the goal of protecting the planet's species and ecosystems from potential risks and modified organisms and of establishing an advanced informed agreement procedure. Thus, countries are provided with the necessary information to make informed decisions before agreeing to the import of these organisms. The protocol was a landmark agreement because it was the first use of a formal "precautionary approach" as set in the 1992 Rio Declaration on Environment and Development (Rio Summit) (UNCED 1992).

2.6 Ozone Layer Depletion

As discussed in Section 2.7, the presence of ozone near the ground surface is a form of air pollution. The presence of ozone in the stratosphere, however, protects lifeforms from the deleterious effects of ultraviolet (UV) light. UV light, adjacent to the visible light spectrum, is invisible to the naked human eye and is highly energetic, especially UV-C light. Stratospheric ozone blocks the penetration of UV-C and reduces the incidence of UV-B by 95% and UV-A by 5%, respectively.

Ozone is formed in the stratosphere when UV light splits O_2 molecules, resulting in free oxygen (O). The free oxygen may quickly react with O_2 molecules to form O_3 (ozone) molecules. Figure 2.9 depicts the occurrence of ozone in the atmosphere as a function of altitude above the Earth's surface.

Ozone levels fluctuate with the seasons. In the tropics, they remain relatively stable throughout the year, with slight increases around mid-year due to changes in atmospheric circulation. In both the Northern and Southern Hemispheres, ozone levels rise during spring as a result of poleward transport from the tropics and then, decrease due to sunlight-driven chemical reactions that lead to the thinning of the ozone layer. This effect is especially pronounced in the Southern Hemisphere, where it contributes to the formation of the Antarctic ozone hole. This "thinning" is further exacerbated by the presence of chlorofluorocarbons (CFCs) in the atmosphere. Rowland and Molina (1974) concluded that CFCs could damage ozone due to the release of chlorine atoms. CFCs are degraded by UV light, resulting in the generation of free chlorine. The free chlorine, in turn, reacts with ozone to form chlorine monoxide and molecular oxygen. In 1985, noticeable thinning (a 50% reduction) of ozone was detected over the South Pole. Cold temperatures during the Antarctic

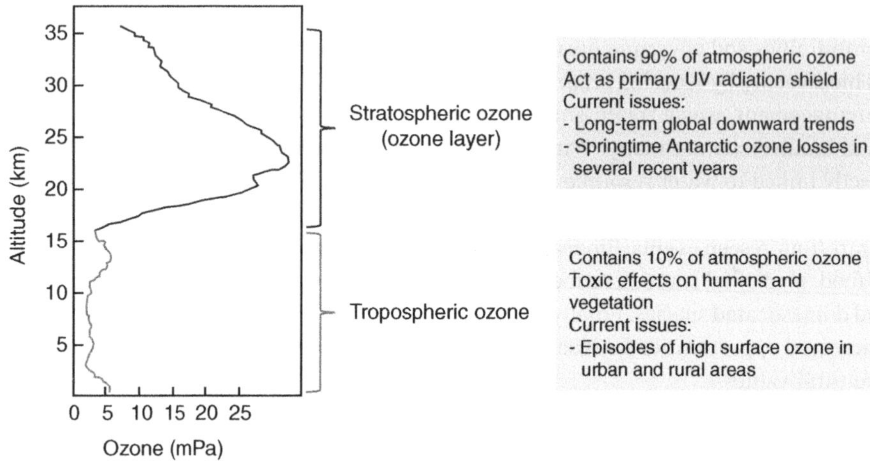

Figure 2.9 Occurrence of ozone in the atmosphere. *Source:* The National Oceanic and Atmospheric Administration (NOAA)/http://www.ozonelayer.noaa.gov/science/basics.htm/ Public domain.

winter and the vortex in the atmosphere induced by the Earth's rotation facilitate the accumulation of chlorine over Antarctica. Fortunately, summer air replenishes ozone, somewhat reversing the cycle.

Nevertheless, the Antarctic ozone hole has grown to the size of North America, and temperate regions in the southern hemisphere are being affected.

Several human health impacts have measurably increased with impacts on the ozone layer. Some of these include the increased occurrence of melanoma and nonmelanoma skin cancers, acceleration of eye cataracts, and reduced immune system effectiveness. As an example, two out of three people in Queensland, Australia are expected to develop skin cancer due to exposure to UV light not blocked by the absent ozone. Additionally, increased UV-B exposure can affect plant growth and crop yields as well as damage aquatic ecosystems.

Because of their stability, halogenated hydrocarbon compounds have been often used as coolant compounds in refrigerators, air conditioners, and heat pumps. They have also been used in the manufacture of plastic foams. Due to their nonreactive nature, they were also used as propellants in aerosol applications. Because of their destructive effect on stratospheric ozone, CFC use in aerosols was banned in 1978. While manufacturers of these products switched to alternative compounds, CFC use has continued in other products. In 1987, UN member countries met in Montreal, leading to the formation of the Montreal Protocol. A total of 194 nations agreed to scale back the use of CFCs by 50% by 2000 (UNEP 2017). Subsequent amendments moved the total phase out of CFCs to 1996. Additionally, other similar compounds have dangerous effects on the ozone layer. For instance, bromine compounds, commonly used in soil fumigants and pesticides, are 60 times as potent as chlorine with respect to ozone depletion.

Although CFCs are still in use to some extent, some substitutes have been developed. For instance, hydrofluorocarbons (HFCs) do not contain chlorine, and fluorine does not pose a threat to the ozone layer. These measures have positive effects; although ODS concentrations in the atmosphere peaked in 1994, the concentrations have been gradually decreasing due to worldwide efforts to curb the use of CFCs, and it is expected that pre-1980 levels will be reached by 2050 (Ennis 2017). However, the risks to the ozone layer and associated health impacts from ozone layer depletion persist.

2.7 Air Pollution

The atmosphere contains a range of naturally occurring gases – nitrogen, oxygen, carbon dioxide, argon, and water vapor – plus numerous trace gases, including ozone, helium, hydrogen, and nitrogen oxides. Naturally occurring aerosols – microscopic liquid or soil particles resulting from the entry of dust, pollen, sea salts, and other sources – are also present in the atmosphere. However, numerous pollutants, many of these from anthropogenic sources, have been emitted to the atmosphere, resulting in a range of potential harmful effects.

Air pollutants can be classified as primary or secondary pollutants. Primary pollutants are the direct by-products of combustion or evaporation emitted to the atmosphere; some include volatile organic compounds (VOCs), particulates, carbon monoxide, nitrogen oxides (NO_x), sulfur dioxide, and lead. Secondary pollutants are the result of the reactions of primary pollutants with other compounds within the atmosphere. Some of these include ozone, pero-oxyacetyl, nitrates, sulfuric acids, and nitric acids.

The presence of air pollutants can result in a range of adverse effects to human health. Acutely elevated concentrations of air pollutants can be lethal to people suffering from cardiopulmonary diseases. Even moderate levels can affect cardiac functions, affecting those with heart disease. Several common pollutants can also have chronic effects. For example, sulfur dioxide and ozone can lead to irritation and/or damage to the lungs and/or breathing passages. Particulate matter can lead to respiratory and cardiovascular diseases. Nitrogen oxides can impair lung function and affect the immune system. Carbon monoxide reduces the oxygen-carrying capacity of blood. Acute concentrations of carbon monoxide can be fatal, but even chronic lower concentrations can lead to heart disease.

Some air pollutants are also carcinogenic. Benzene, a common VOC present in motor fuels, solvents, other industrial chemicals, explosives, and smoke, is clearly correlated to several types of cancers, including leukemia, blood disorders, and immune system disorders. Some diesel fuel-related emissions are also considered likely carcinogens.

Lead is an example of an anthropogenic air pollutant with significant human health consequences. Lead was historically used as a common additive to gasoline for anti-knock purposes. With its presence in leaded gasoline as a primary source, potential exposure pathways include the inhalation of lead-containing emissions. Additionally, subsequent deposition of airborne lead (aerially deposited lead) resulting from combustion provides an exposure pathway through ingestion or dermal contact. Lead poisoning can have serious health consequences, including learning disabilities in children and hypertension in adults. Although lead was gradually phased out of gasoline in the 1970s and 1980s, elevated lead concentrations in blood persisted in children and adults through the 1980s. Fortunately, the EPA mandated the elimination of leaded gasoline by 1996, leading to a dramatic reduction of lead concentrations in the environment (Dapul and Laraque 2014).

Air pollutants can directly affect vegetation. Large-scale sulfur dioxide emissions have been linked to the widespread destruction of vegetated areas. Because of its ability to affect carbon and nutrient plant metabolism, ground ozone emissions can reduce or destroy orchards, cereal crops, or forests. Areas in California and Appalachia that are susceptible to increased ozone concentrations have experienced direct tree damage or secondary damage as weakened trees become more susceptible to pests and wildfires.

Common air pollutants can also have a detrimental effect on structures and the built environment. Particulate matter can discolor building façades and windows. Ozone emissions react with rubber, leading to degradation and deterioration. Sulfur oxides and nitrogen oxides corrode metals and deteriorate stonework and masonry. While many materials can be cleaned or repaired, the effects on many historic statues and architecture/artworks are beyond repair.

With the increased deterioration of air quality following the end of World War II, by the 1960s, it was becoming more obvious that increased emissions were having a deleterious effect on the atmosphere. Increased public awareness ultimately led to the governmental regulation of air pollution. In 1970, Congress passed the Clean Air Act. This landmark legislation (amended on several occasions) serves as the foundation of US anti-air pollution efforts. Ambient standards were set for criteria pollutants, and control measures were established. Some criteria pollutants include sulfur dioxide, carbon monoxide, nitrogen oxides, ozone, and lead. The National Ambient Air Quality Standards (NAAQS) set primary standards for short-term and long-term conditions such that these pollutants would not harm human health. Table 2.2 provides a summary of these standards.

The Clean Air Act requires the EPA to review pollutants every five years, and adjustments are made with advances in scientific knowledge. Additionally, pollutants have been added to the regulatory environment. For instance, in 1980, National Emission Standards for Hazardous Air Pollutants (NESHAPs) were established, requiring tracking and regulation of 187 toxic pollutants. The National Emission Inventory was established, which consists of a database of criteria and pollutants that are regulated.

Table 2.2 EPA has set National Ambient Air Quality Standards for six principal pollutants.

Pollutant		Primary/secondary	Averaging time	Level	Form
Carbon monoxide (CO)		Primary	8 h	9 ppm	Not to be exceeded more than once per year
			1 h	35 ppm	
Lead (Pb)		Primary and secondary	3-month rolling average	0.15 μg/m^3	Not to be exceeded
Nitrogen dioxide (NO_2)		Primary	1 h	100 ppb	98th percentile of 1-h daily maximum concentrations, averaged over 3 yr
		Primary and secondary	1 yr	53 ppb	Annual mean
Ozone (O_3)		Primary and secondary	8 h	0.070 ppm	Annual fourth-highest daily maximum 8-h concentration, averaged over 3 yr
Particle pollution (PM)	$PM_{2.5}$	Primary	1 yr	12.0 μg/m^3	Annual mean, averaged over 3 yr
		Secondary	1 yr	15.0 μg/m^3	Annual mean, averaged over 3 yr
		Primary and secondary	24 h	35 μg/m^3	98th percentile, averaged over 3 yr
	PM_{10}	Primary and secondary	24 h	150 μg/m^3	Not to be exceeded more than once per year on average over 3 yr
Sulfur dioxide (SO_2)		Primary	1 h	75 ppb	99th percentile of 1-h daily maximum concentrations, averaged over 3 yr
		Secondary	3 h	0.5 ppm	Not to be exceeded more than once per year

Source: USEPA (2017a)/US Environmental Protection Agency/Public domain.

The Clean Air Act has been very useful in addressing air pollution. The regulatory philosophy is to keep criteria pollutant emissions below primary standard levels, leading to an overall improvement in public health. Emitting industries are required to implement controls to address their polluting emissions. Additionally, regions in violation are required to meet compliance. However, enforcement can be difficult, as cities and states are often unwilling to enforce or control emissions to achieve compliance with the air quality standards.

2.8 Smog

Several of the air pollutants discussed in Section 2.7 may form smog. Smog often takes one of two common forms. The first form is industrial smog, which is a combination of smoke and fog. Industrial smog consists of a grayish mix of soot, sulfur compounds, and water vapor. Commonly the result of coal burning, it is most prevalent in cold climates with industrialized areas. London was historically infamous for "London Fog" (actually industrial smog), but it is now common in China, India, Korea, and Eastern European countries (Elsom 2014). The second common form of smog, photochemical smog, is primarily associated with vehicle exhaust in warm, sunny areas. Pollutants, such as VOCs and nitrogen oxides, interact with sunlight in warm climates to form a brown, irritating haze. The creation of photochemical smog from automobile exhaust, as identified by Arie Jan Haagen-Smit of the California Institute of Technology, is presented in Eq. (2.1):

$$\text{NMHC (nonmethane hydrocarbons)} + \text{NO} + h\upsilon \rightarrow \text{NO}_2 + \text{other products} \tag{2.1}$$

Nitrogen dioxide further reacts as shown in Eqs. (2.2) and (2.3):

$$\text{NO}_2 + h\upsilon\ (\upsilon < 410\,\text{nm}) \rightarrow \text{NO} + \text{O}^{\bullet} \tag{2.2}$$

$$\text{O}_2 + \text{O}^{\bullet} \rightarrow \text{O}_3 \tag{2.3}$$

The free oxygen reacts with molecular oxygen to produce ground ozone. Exposure to 0.1–1 ppm of ground ozone is an irritant as described above, leading to headaches, burning eyes, and irritation of breathing passages.

As demonstrated by the reactions outlined above, photochemical smog can be controlled by limiting the concentrations of NMHCs (nonmethane hydrocarbons) and nitrogen oxides in the atmosphere. One successful method has been the widespread adoption of automobile catalytic converters. These devices, introduced in the United States in the 1970s, are designed to capture NO, CO, and unburned hydrocarbons from exhaust.

Several large metropolitan areas, including Los Angeles, Beijing, New Delhi, and Mexico City, have been afflicted with chronic smog issues for decades. On some occasions, conditions can exist where acute smog can form, resulting in significant health effects. A famous example occurred in Donora, Pennsylvania, in October 1948. A temperature inversion trapped industrial smog from coal-fired and sulfur-emitting industries. The smog outbreak lasted for several days. Within a month, at least 70 people died, and an estimated 6000 people were struck with a variety of illnesses. While local industries denied responsibility, public outrage at this event as well as others eventually led to governmental intervention through legislation.

2.9 Acid Rain

Air pollution can also have secondary effects on public health and the environment. With the emission of sulfur dioxide and nitrogen oxides, sulfuric and nitric acids can form due to interaction with atmospheric moisture. This, in turn, can result in the deposition of these acids via acid rain, a phenomenon that affects broad areas of industrialized regions of the world (Kulshrestha 2020).

Rainfall is normally slightly acidic (on the order of pH = 5.6). However, acid rain (defined as precipitation with pH less than 5.5) can be 10–1000 times more acidic. Rain and snow in eastern North America have been measured with pH ranging from 3.0 to 4.6. Mountain forests near Los Angeles have exhibited measurable pH as low as 2.8.

Several natural sources of acid deposition occur. An estimated 50–70 million tons of sulfur dioxide are generated per year from volcanoes, sea spray, and microbial processes. Between 30 and 40 million tons of nitrogen oxides are generated per year from lightning, natural combustion of biomass, and microbial processes. However, significant anthropogenic sources also exist. Between 100 and 130 million tons of sulfur dioxide are emitted primarily from the combustion of coal. Between 60 and 70 million tons of nitrogen oxides are emitted yearly from vehicular fuel combustion sources. Much of these emissions occur in industrialized areas, but downwind areas are also affected. Fortunately, rates of emission are decreasing due to increased regulations. Although emissions have increased fourfold since 1900, deposition has decreased 33–35% in the past 15 years, as new technologies have been implemented and obsolete coal-burning plants have been decommissioned. Nevertheless, these compounds are still generated, and acid rain generation still continues.

The deposition of these acids can have long-term impacts. For instance, consequences are still observed from emission/deposition that occurred decades ago. Acid precipitation can result in the leaching of toxic heavy metals from soil and rock, where they can be mobilized and become available for biological uptake in the environment. Acid deposition can also have a detrimental effect on aquatic environments, as lowered pH can adversely affect enzymes, hormones, and proteins of organisms. Because natural river and lake environments commonly have pH ranging from 6 to 8, deviations can impact the regulatory systems of organisms. Future generations may also be affected; small changes in pH can adversely affect egg and sperm cells.

It is important to note that natural systems have the ability to buffer, or neutralize acid deposition, and not all areas are harmed by deposition. Limestone as well as derived soils have an appreciable buffering capacity. Additionally, the Clean Air Act and related regulations have led to reductions in atmospheric NO_x and SO_x concentrations. Specifically, Title IV of the Clean Air Act has led to significant reductions, which have occurred faster than expected. This approach has included mandates to introduce emissions-curbing technologies (e.g. scrubbers and use of low-sulfur coal) as well as free-market incentives that utilize emissions allowances. These market-based approaches have fostered technological innovations and provided incentives to reduce emissions while saving costs related to emissions allowances.

Example 2.2 Acid Rain Impact from Industrial Emissions

In an industrial area, a factory releases sulfur dioxide (SO_2) and nitrogen dioxide (NO_2) into the atmosphere. The release amounts are given in kilograms per hour:

Compound	Release amount (kg/h)
Sulfur dioxide (SO_2)	100
Nitrogen dioxide (NO_2)	80

The local precipitation pattern indicates an average rainfall of 5 mm per hour. Assume that the emitted gases readily react with rainwater, forming sulfuric acid (H_2SO_4) from SO_2 and nitric acid (HNO_3) from NO_x.

- Determine the amount of sulfuric acid (H_2SO_4) and nitric acid (HNO_3) formed per hour, considering the chemical reactions.

- Calculate the pH of the resulting rainwater after the formation of sulfuric acid and nitric acid. Assume that both H_2SO_4 and HNO_3 are fully dissociated in the rainwater.
- Explain the potential impact of acid rain on local ecosystems, infrastructure, and water bodies.

Solutions:

a) Assuming that water and oxygen are in excess, the chemical reactions for the formation of sulfuric acid (H_2SO_4) and nitric acid (HNO_3) are as follows:

$$2SO_{2(g)} + O_{2(g)} + 2H_2O_{(l)} \rightarrow 2H_2SO_{4(aq)}$$
$$3NO_{2(g)} + H_2O_{(l)} \rightarrow 2HNO_{3(aq)} + NO_{(g)}$$

We can then use stoichiometry to obtain the molar mass per hour of sulfuric acid from sulfur dioxide.

$$\text{mol/h SO}_2 = \frac{\text{Mass of SO}_2\text{/h}}{\text{Molar mass of SO}_2} = \frac{100{,}000\text{ g/h}}{64\text{ g/mol}} = 1{,}562.5\text{ mol/h}$$
$$\text{mol/h H}_2\text{SO}_4 = \text{mol/h SO}_2 = 1{,}562.5\text{ mol/h}$$

which, we can then convert to kg/h of H_2SO_4.

$$= 1{,}562.5\text{ mol/h} \times 98\text{ g/mol} \times 0.001\text{ kg/g} = 153\text{ kg/h}$$

Calculations for the amount of nitric acid per hour can be similarly calculated.

$$\text{mol/h NO}_2 = \frac{\text{Mass of NO}_2\text{/h}}{\text{Molar mass of NO}_2} = \frac{80{,}000\text{ g/h}}{46\text{ g/mol}} = 1{,}739.13\text{ mol/h}$$
$$\text{mol/h HNO}_3 = \frac{3}{2}\text{mol/h NO}_2 = 2{,}608.7\text{ mol/h}$$
$$\text{kg/h HNO}_3 = 2{,}608.7\text{ mol/h} \times 63\text{ g/mol} \times 0.001\text{ kg/g} = 164\text{ kg/h}$$

b) To obtain the pH of the resulting rainwater, we must first calculate the volume of water (in liters) precipitated in the industrial area. Assuming that the industrial area occupies 0.40 km^2 ($\approx$ 100 acres) area, the volume of water per hour precipitated is

$$\text{l/h precipitation} = 5\text{ mm/h} \times 0.40\text{ km}^2 \times 1000\text{ l/m}^3 = 2{,}000{,}000\text{ l/h}$$

We can then use this volume of precipitation to obtain the molarity (M) of the total.
H^+ ions in precipitation. In aqueous form, sulfuric acid and nitric acid ionize to

$$H_2SO_4 \rightarrow 2H^+ + SO_4^{2-}$$

$$HNO_3 \rightarrow H^+ + NO_3^-$$

Assuming complete ionization, the total amount of H^+ in the solution is

H_+ (from acid dissolution) $= 2 \times 1{,}562.5 + 2{,}608.7 = 5{,}733.7$ mol
H_+ (present in rain water) $= 10^{-5.6}$ mol/l $\times$ 2,000,000 l $= 5$ mol (assuming the pH of rain water to be 5.6)

As observed, the molar concentration of H^+ in rain is minimal compared to the number of moles of H^+ that can result from the dissolution of emitted gases.

$$\text{Molar concentration of H}_+\text{ in the rain water is: } \frac{(5733.7 + 5)\text{ mol}}{2{,}000{,}000\text{ l}} = 0.00287\text{ M}$$

Finally, assuming that both H_2SO_4 and HNO_3 are fully dissolved, we can calculate the pH of the resulting rainwater with the formula $pH = -\log [H^+]$

$$pH = -\log(0.00287) = 2.54$$

Note: The purpose of this exercise is to demonstrate how elevated levels of these gases in the atmosphere can affect the pH levels of rain. However, in reality, full dissolution of sulfur and nitrogen oxides does not occur. In general, partial dissolution results in a decreased pH, typically around 4, for acid rain.

c) Acid rain, originating from increased sulfur dioxide and nitrogen oxides released into the atmosphere through human activities such as burning fossil fuels and industrial processes, poses a significant and diverse threat to the environment. In local ecosystems, it induces soil acidification, adversely affecting plant growth and nutrient availability, while also damaging leaves and needles, leading to a reduction in biodiversity. Aquatic environments face challenges as acid rain lowers water pH, putting fish and amphibians at risk and intensifying the leaching of toxic metals from the soil. Infrastructure, including buildings and monuments, confronts corrosion issues as acid rain erodes limestone, marble, and calcareous structures, jeopardizing both aesthetics and structural integrity. Metal structures like bridges and pipelines undergo accelerated corrosion, resulting in increased maintenance costs and a shortened lifespan. Acidification of surface water bodies, such as lakes and rivers, disrupts aquatic life, while groundwater quality is compromised by the leaching of toxic elements into aquifers, posing threats to drinking water supplies and human health. Given these extensive impacts, urgent measures, including the reduction of sulfur dioxide and nitrogen oxide emissions through air quality regulations and cleaner technologies, are crucial to mitigating the consequences of acid rain on local ecosystems, infrastructure, and water bodies.

2.10 Water Usage and Pollution

Access to clean, fresh water sources is essential to human life. Protection of these resources is equally important. Only 3% of the Earth's water is fresh; the remainder is saline and unusable without some type of desalination measure. Of the available fresh water, about 30% comes from groundwater sources. Nearly 69% of fresh water is trapped in ice caps and glaciers. Only approximately 0.3% exists as surface water. Further, approximately 87% of surface fresh water is present in lakes, 11% exists in fresh wetlands and swamps, and approximately 2% is present in river environments (Figure 2.10) (Ritchie and Roser 2023).

Human fresh water usage has increased over the past century, as demonstrated in Figure 2.11. Fresh water is used for a variety of key human activities, including agriculture, industrial processes, and domestic uses. Consequently, sources of fresh water are under stress, and such stress has been growing due to the increasing demands for water in developed and less-developed countries (Figure 2.12). For instance, drought-induced surface water depletion results in habitat loss, decreased aquifer recharge, impacts on municipal water supplies, and concentration of pollutants. Groundwater depletion due to unsustainable pumping rates and/or reduced recharge can affect private or public water systems, the drying of wells, deterioration of water quality, and land subsidence. Groundwater depletion is especially problematic in semiarid and arid regions of the Midwestern and western United States, where the lowering of localized and regional water tables has been especially acute.

Water pollution can take many forms, and it can have significant environmental effects. The sources of pollution can be point-source and nonpoint source. Point-source pollution comes from a single location or system, such as industrial facilities, wastewater treatment facilities, and mines

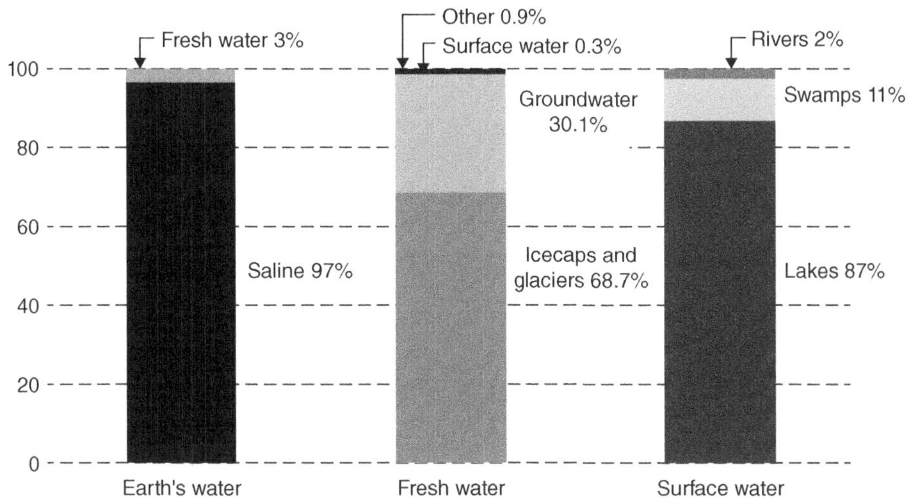

Figure 2.10 Distribution of water on the planet Earth.

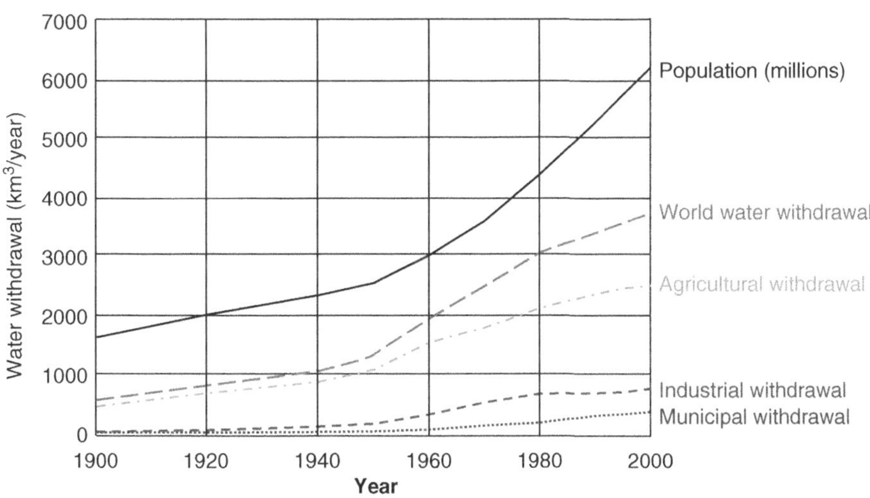

Figure 2.11 Freshwater use and population growing. *Source:* Adapted from Shiklomanov (1998).

or other points of resource extraction/refinement. Because of their specific location, point sources are easy to identify and, therefore, are relatively straightforward to monitor and regulate. Nonpoint sources consist of sources that do not have a specific point of discharge; rather, they often consist of discharge resulting from areal or regional runoff. Some examples include agricultural runoff, stormwater runoff from vegetated or paved surfaces, and runoff from construction sites. Nonpoint sources rarely feature the acute loading or toxicity of point sources; however, the volume from non-point sources can be substantial, and these discharges may be difficult to control and regulate. When dealing with point sources, it can be efficient to treat water before releasing it into the environment, whereas nonpoint sources are most efficiently managed by reducing the runoff by some means.

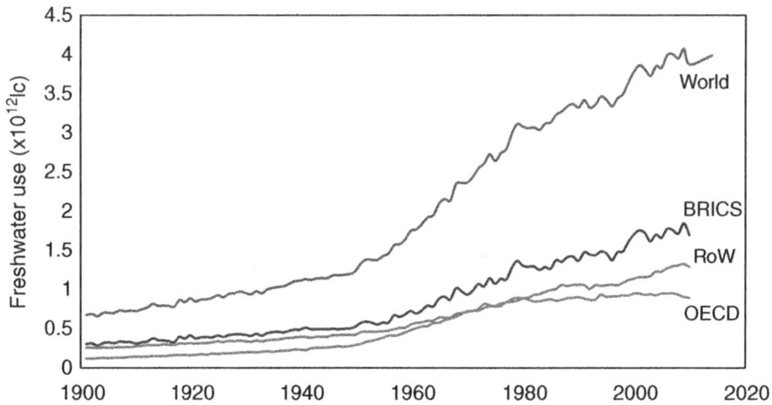

Figure 2.12 Freshwater use. BRICS: Brazil, Russia, India, China, and South Africa. OCDE: Organization for Economic Cooperation and Development. ROW: rest of the world.

Pathogens are one class of water pollutants. Bacteria, viruses, and parasites often found in human and animal excrement can carry disease. Infections from pathogens can result in widespread symptoms and health problems. Public health departments play an active role in minimizing exposure to pathogens, using measures such as purification and disinfection of public water supplies, sanitary collection and treatment of wastes, regulations pertaining to sanitary food preparation practices, and domestic and personal hygiene practices.

Organic wastes are a second type of water pollutants. Organic matter can come from human or animal wastes and by-products, remnant food materials, and decaying plant matter. With the exception of plastic and other select synthetic materials, much of this matter is biodegradable. Bacteria and other organisms will utilize and consume these materials, but oxygen is utilized as well. Biological oxygen demand (BOD) is a quantitative measure of the relative amount of organic matter present in a given water body. Because the saturation limit of dissolved oxygen (DO) is relatively low (8–10 ppm), the presence and subsequent utilization of excessive organic matter can lead to rapid depletion of dissolved oxygen.

Chemical compounds constitute a third category of water pollutants. These can be classified as inorganic chemicals, which include heavy metals, acids, or other compounds (including salts). Likewise, organic chemicals pose a threat, including industrial chemicals, such as solvents, polychlorinated biphenyls, detergents, or pesticides. Many chemicals can be toxic to aquatic life at very low concentrations; furthermore, these compounds, if considered "bioaccumulative," can become concentrated and pose a threat to organisms in a more advanced position on the food chain. Additionally, higher concentrations can alter water chemistry, leading to secondary effects on aquatic life.

An acute example of chemical pollution is oil spills. These typically occur in marine or near-shore environments due to accidental releases from oil platforms, tankers, or pipelines. The spills may involve crude oil or lighter-phase refined products, such as gasoline or diesel fuel, as well as remnant heavier-phase products, such as bunker oil. Cleanup and recovery can be difficult, expensive, and time-consuming; these efforts can be affected by temperature, weather conditions, currents, and shore topography/conditions. Spills can take months or years to clean up, and as a result, drastic effects to aquatic life are quite common. Oil spills can be equally lethal to birds, especially seabirds, which are closer to the source of oil. Oil causes matting and separation of bird's feathers and exposure of their delicate skin to extreme temperatures. This not only affects their ability to fly but can also be toxic when ingested during preening. Oil spills can also have a range of physiologic and toxic effects to fish and mammals.

Excessive nutrient loading, which commonly includes nitrogen, phosphorus, and potassium, can have significant effects. While these inorganic nutrients are essential for plant life, excessive nutrients in runoff, either from point sources such as wastewater plants with insufficient treatment) or nonpoint sources, including agricultural runoff, lawns, and golf courses, can lead to algal blooms. Specifically, when excess nitrogen is introduced into a water body, growth of phytoplankton accelerates. Excess phytoplankton can lead to excessive growth of zooplankton as well. The consumption of phytoplankton by zooplankton, as well as the decay of these microorganisms, can lead to an acceleration of oxygen consumption, depleting oxygen concentrations in water bodies. The resulting condition, hypoxia, can severely affect the balance of life in water bodies, and if unchecked, "dead zones" may develop. Dead zones have occurred in water bodies on a seasonal or ongoing basis.

A well-known example of dead-zone formation has occurred in the Gulf of Mexico near the Mississippi River Delta. The Mississippi River collects 40% of the continental United States watershed runoff. Much of this watershed is situated in agricultural-intensive areas, and as a result, the runoff often contains agricultural chemicals (including fertilizers) and biomass from animal wastes (Beaty 2015). Much of the runoff water and sediments are devoid of oxygen, and the Louisiana Gulf Coast has become susceptible to dead-zone formation during the summer months. Because this area includes an abundant fishing ground valued at approximately $2.8 billion, the formation of dead zones can have a significant economic effect (de Mutsert et al. 2016).

To combat the potential for hypoxia occurrences, in 1998, Congress passed the 1998 Harmful Algal Bloom and Hypoxia Research and Control Act (NCOOS 2017). Further, following confirmation of the relationship between dead-zone occurrences and elevated nitrogen concentrations, an interagency taskforce in 2000 identified options to reduce the occurrence and size of dead zones. Options for controlling this include reduced fertilizer applications and means to restore natural denitrification processes. The latest actions recommend a 45% reduction in nitrogen and phosphorus runoff, but an action plan to reduce the hypoxic area by 50% by 2015 has not led to a reduction in the size of the dead zone (USEPA 2015).

Water pollution has affected many countries and civilizations over the generations. As an example, untreated chemicals and sewage were commonly dumped into waterways during the early stages of the Industrial Revolution. Not surprisingly, public health issues commonly arose as drinking and domestic water supplies were affected. Once sewage-borne pathogens were demonstrated to pose significant health risks, cities began to install rudimentary wastewater disposal systems. Unfortunately, these systems often only served to concentrate wastes, and many receiving bodies evolved into little more than cesspools. Although the public began to take notice, early regulations, such as the Federal Water Pollution Control Act of 1948, did little to combat growing water pollution. Following World War II and a rapid acceleration of industrial and chemical facility prowess, water pollution conditions greatly worsened. Many water bodies throughout the United States were greatly impacted. An acute example, and one that galvanized public outrage, occurred in 1969 when chemical pollution within the Cuyahoga River in Cleveland caught on fire (an event that had occurred several times in the past).

Partially inspired by this event, and with public outcry increasing, Congress enacted the Clean Water Act in 1972. By charging the US Environmental Protection Agency (USEPA) with restoring and maintaining waters from a chemical, physical, and biological standpoint, the Clean Water Act has proven to be very effective.

Several other regulations and criteria have either been attributable to or have evolved since the passage of the Clean Water Act. A total of 167 substances have been identified as criteria pollutants based on some characteristics (e.g. toxicity and pH), and maximum allowable concentrations have been assigned and enforced. The concentrations may include criteria maximum concentration, defined as the highest single or "acute" concentration beyond which impacts may occur, or as criteria continuous concentration, the highest sustained or "chronic" concentration beyond which impacts may occur (USEPA 2012).

These criteria have been used to develop other standards. For instance, the Safe Drinking Water Act (SDWA) set forth maximum contaminant levels (MCLs) for 94 substances that may affect drinking water USEPA (2017b). Another example is the National Pollution Discharge Elimination System (NPDES), which addresses point-source pollution (USEPA 2017c). Permits are required for wastewater and industrial discharges. Total maximum daily loads (TMDLs) are promulgated to address mainly nonpoint sources (USEPA 2017d).

Example 2.3

Untreated sewage from a small city of population 20 000 was discharged into a nearby lake. Two rivers flow into and out of the river so that the volume of the lake remains constant, as shown in the figure below.

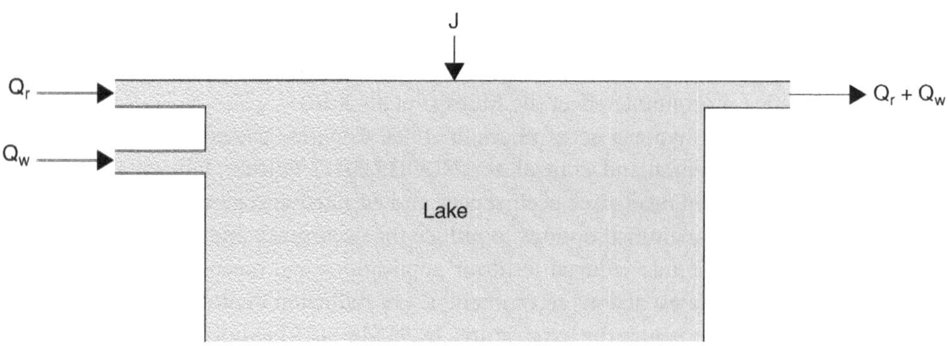

Assume that the lake water is kept well mixed because of strong winds. The following data is given:

- Lake: Volume = 100,000 m^3, Average depth = 10 m, Temperature = 21 °C.
- Incoming river: Flow rate = 0.5 m^3/s, Dissolved oxygen = 8.5 mg/l.
- Oxygen transfer from atmosphere to lake $J = K\,(DO_{sat} - DO)$:
  $K = 3 \times 10^{-5}$ cm/s, $DO_{sat} = 8.9$ mg/l.
  (K – Mass transfer coefficient, DO_{sat} – Saturation concentration of dissolved oxygen in water exposed to air, DO – Dissolved oxygen in lake water).
- Sewage: Flow rate = 0.08 m^3/s, BOD content = 120 mg/l, BOD decay rate = 0.2 d^{-1}, Dissolved oxygen = 8 mg/l.

Use first-order decay model for biochemical oxygen demand (BOD). Determine the dissolved oxygen content in the lake water (mg/l) under steady-state conditions. Ignore the effect of aquatic life on DO.

Note: Refer to any environmental engineering textbook for examples on similar problems.

Solutions:
For the given problem, we have to solve the mass balance for the DO in the given lake.

- Volume of the lake = 100 000 m^3
- Flow rate of the incoming river = $Q_{inflow} = 0.5\,m^3/s$.
- Flow rate of the sewage = $Q_{sewage} = 0.08\,m^3/s$.
- Flow rate of the outgoing river = $Q_{outflow} = 0.58\,m^3/s$ (given in the problem, the volume of the lake is constant, $Q_{inflow} = Q_{outflow}$).
- Dissolved oxygen in the incoming river = $DO_{inflow} = 8.5\,mg/l = 8500\,mg/m^3$.
- Dissolved oxygen in the sewage = $DO_{sewage} = 8\,mg/l = 8000\,mg/m^3$.
- BOD content in the sewage = $BOD_{sewage} = 120\,mg/l = 120\,000\,mg/m^3$.
- BOD decay rate = $k' = 0.2\,day^{-1}$
- Volume of the lake = $0.5M\,m^3$.
- Surface area of the lake = volume/average depth = $0.5M\,m^3/10\,m = 50\,000\,m^2$.
- Atmospheric oxygen transfer coefficient = $k = 3 \times 10^{-5}\,cm/s = 3 \times 10^{-7}\,m/s$
- Saturation concentration of dissolved oxygen = $DO_{sat} = 8.9\,mg/l = 8900\,mg/m^3$

Calculate the Hydraulic Residence Time (HRT) i.e., average duration water stays in the lake:

$$HRT = \frac{V}{Q_{outflow}} = \frac{10^5}{0.58} = 1.72 \times 10^5\,s \times \frac{1}{86400} day/s \cong 2\,days$$

Initial steady-state BOD (BOD_0) in the lake:

$$BOD_0 = BOD_{sewage} \times \frac{Q_{sewage}}{Q_{inflow} + Q_{sewage}}$$

$$BOD_0 = 120 \times \frac{0.08}{0.58} = 16.55\frac{mg}{l} = 16\,550\,mg/m^3$$

To maintain mass balance of oxygen in the lake, the following equivalence can be established under steady-state conditions:

$$\left[\left(Q_{inflow} \times DO_{inflow}\right) + \left(Q_{sewage} \times DO_{sewage}\right) + J\right] \times HRT = BOD_0 \times \left(1 - e^{-k \times HRT}\right)$$
$$\times V + Q_{outflow} \times DO \times HRT$$

We can calculate DO, i.e., dissolved oxygen in the lake at steady state (same as the DO in the outflow river) using the above equivalency. To find J, atmospheric transfer of oxygen, we can use the coefficient given in the problem:

$$J = k \times Area\ of\ lake \times (DO_{sat} - DO)$$
$$= 3 \times 10^{-7}\frac{m}{s} \times \frac{5 \times 10^5 m^3}{10\,m} \times (8900 - DO) = (133.5 - 0.015DO)mg/s$$

Now, solving the oxygen mass balance presented before:

$$(0.50 \times 8500 + 0.08 \times 8000 + 133.5 - 0.015\,DO) \times 1.72 \times 10^5 = 16\,550 \times \left(1 - e^{-0.2 \times 2}\right)$$
$$\times 100000 + 100000\,DO$$

$$DO = 3104.13\,mg/m^3 = 3.14\,mg/l$$

Therefore, because of the sewage discharge, the DO levels in the lake have drastically reduced to 3.14 mg/l

2.11 Eutrophication

Hypoxia, as described before, is a serious form of water pollution. It is also one manifestation of eutrophication and is defined as an ecosystem response to the addition of natural or artificial substances, such as nutrients, to an aquatic system (van Beusekom 2018). The additions may come from substances associated with anthropogenic activities, such as wastewater or fertilizer-laden runoff, due to runoff from urban or agricultural activities. It also can occur as a natural process, called natural eutrophication, due to the accumulation of materials in a depositional environment or from ephemeral flows. Natural eutrophication is a natural succession process within water bodies, such as lakes and ponds.

Typically, eutrophication results in the excessive growth of plants. Although this leads to an increase in oxygen levels, subsequent decay of dead plant matter results in decreases in oxygen levels. Additionally, the stimulation of phytoplankton and zooplankton results in a rapid reduction in water aesthetics and water quality. An increase in these organisms leads to an increase in turbidity, which in turn decreases the depth of sunlight penetration into a given water body. As a result, benthic (bottom-rooted) plants as well as submerged aquatic vegetation cannot receive the necessary sunlight to survive. Therefore, oligotrophic lakes, which are home for a wide range of aquatic life and exhibit several attractive qualities with respect to aesthetics, can be overtaken by phytoplankton and zooplankton. With the growth and decay cycles of these materials, oxygen levels decrease, turbidity increases, and plants and animals are "choked out" as well as poisoned by secreted toxins. A eutrophic condition occurs in the lake, in which it becomes an unappealing body of algae and devoid of other forms of life.

Eutrophication can be slowed or reversed through a range of intervening activities. Active treatments can be applied to specific bodies of water. Herbicides can be applied to control phytoplankton or submerged aquatic vegetation to manageable levels. Of course, these can be toxic to other life forms, and the decay of the resulting dead matter will also consume oxygen. Plant matter may be harvested to maintain a balance, or aerators may be installed to increase dissolved oxygen levels. Water levels may also be drawn down to control plant matter. However, these are temporary measures, as plants will inevitably grow back.

Eutrophication may be controlled on a larger scale by limiting the amount of nutrients that can migrate into water bodies. The NPDES permit process has proven to be effective in reducing point-source pollution. Additionally, controls placed on some materials, such as phosphates, as well as monitoring of nitrogen and phosphorus, have proven effective in limiting nutrient runoff. The TMDLs described in earlier sections have been established to limit pollutant loading into water bodies. Best management practices (BMPs), often consisting of policies or intervention technologies, can also be applied to intercept, treat, and minimize loading from runoff.

With an effective intervention of various authorities, the effects of eutrophication on water bodies can be limited or recovery may be induced. Lake Washington, situated to the east of Seattle, had become an unpleasant water body covered with blue-green algae during the 1940s and 1950s due to the discharge of 20 million gallons per day of treated wastewater. Following the efforts of concerned citizens, wastewater flow was diverted away from the lake, and gradually, the lake was fully recovered by 1975 (Cooke et al. 2016). Citizens all around the United States have acted to protect similar water bodies.

2.12 Salinity

Salinity is defined as the degree of dissolved salt (such as sodium chloride, magnesium and calcium sulfates, and bicarbonates) present in water or soil (Ivushkin et al. 2019). It is often reported in units of parts per thousand (grams per kilogram) as well as milligrams per liter (parts per million). Plants (called halophytes) and microbes such as bacteria (called halophiles) can adapt to live in saline conditions. Organisms that can tolerate a wide range of salinity conditions are referred to as euryhalines.

The degree of salinity is believed to contribute to global concentrations of carbon dioxide. Carbon dioxide is less soluble in highly saline water. Additionally, saline intrusion is a critical problem in estuary. As global sea levels rise, more saline waters flow back into estuaries, which can affect the salinity of surface water and groundwater that reside in these locations. This, in turn, can greatly affect aquatic life, especially those not adaptable to saline conditions.

2.13 Wastes and Disposal

Waste generation and disposal is a constant concern with respect to the environment. The proper handling and disposal of waste, as well as its potential negative effects on the environment, is highly dependent on the specific type of waste. Several categories of waste exist based on their specific origin and characteristics, including municipal solid waste (MSW), hazardous waste, e-waste, medical waste, and radioactive waste.

MSW consists of the wastes generated by households and small businesses. It differs from hazardous waste and nonhazardous industrial waste (e.g. construction/demolition waste, agricultural waste, mining waste, and sewage sludge) in that these materials are regulated more stringently. MSW generation in the United States is generally increasing, both on a total basis and a per capita basis. For instance, in 1960, the average person generated 2.7 lb of waste per day. By 2007, this had increased to 4.6 lb of waste per day (USEPA 2016). These wastes vary in composition by region and season. Figure 2.13 shows the relative contribution of materials to the MSW waste stream. An

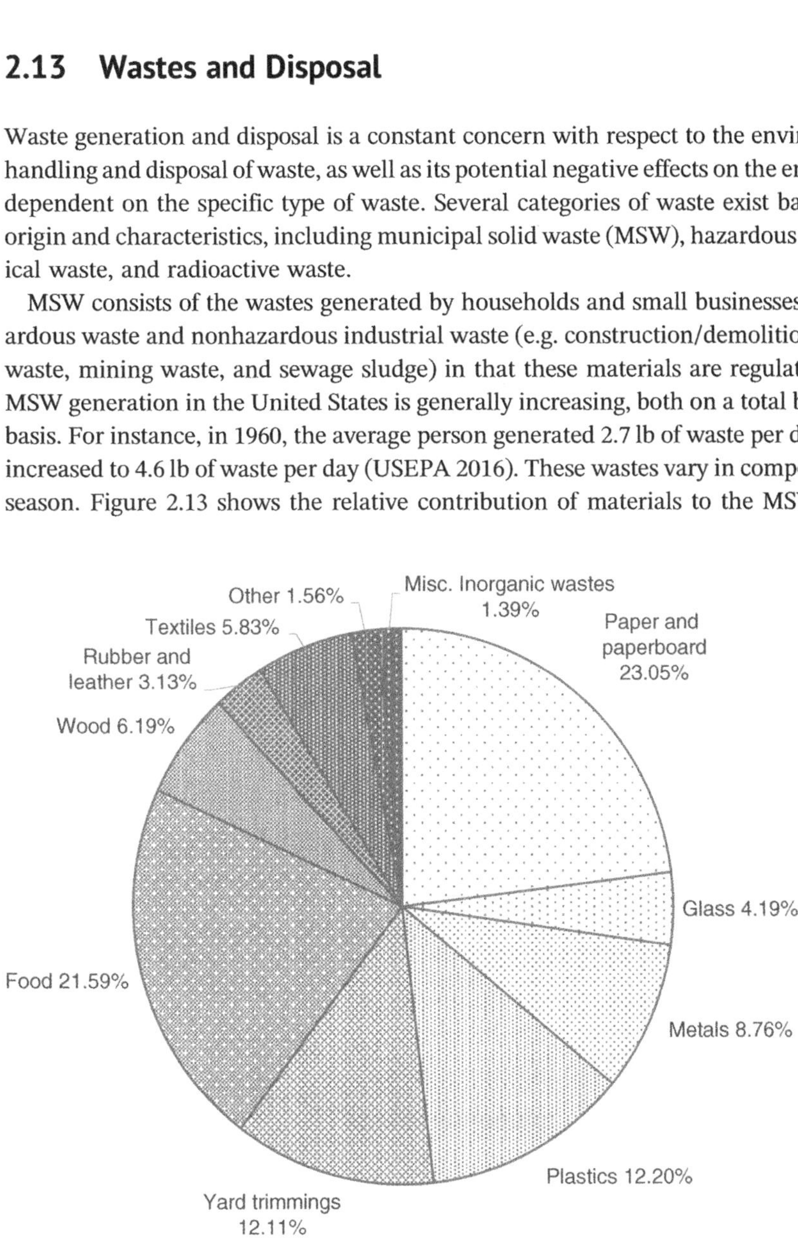

Figure 2.13 Municipal solid waste (MSW) composition in the United States in 2018. *Source:* Adapted from USEPA (2023).

increasing portion of this waste stream is diverted to recycling applications. The remainder of the waste stream is landfilled or incinerated. Unfortunately, for several reasons, it is getting progressively more difficult to safely manage and dispose of these wastes in an appropriate manner.

MSW is collected by local government agencies and/or their agents. In some cases, local governments directly collect and transport wastes using their own workers and equipment. In other cases, they outsource the work to private firms. One of the three fee models may be used to pay for waste collection – taxes, direct service billing to households/businesses, or "pay as you go" tipping fees.

In the past, waste disposal was commonly performed in an uncontrolled or unregulated manner with little regard for its effects on the environment. MSW was commonly burned or buried in uncontrolled dumps. Uncontrolled burning contributed to air pollution, and open dumps had serious effects on soil, air, and water quality. Because of growing public pressure, these uncontrolled methods were phased out, with engineered landfills becoming a more preferred solution.

Landfilling of wastes has been a common waste disposal method for generations. In the past, uncontrolled dumping was commonly performed outside of populated areas within natural depressions (wetlands, gullies, and abandoned quarries), where land was cheap and where nuisance conditions (odors, vectors, etc.) would not affect nearby residents. Techniques were gradually developed to deal with these and other serious environmental side effects. As an example, soil cover was placed on waste dumps daily to help suppress odors and for vector control. As landfilling practices continued to evolve, engineered covers and liners were developed to control infiltration as well as exfiltration of liquids. Gas capture systems were also developed to control nuisance and/or dangerous gases developed within the landfill. Figure 2.14 depicts a typical engineered landfill cross-section.

Many of the engineering advances in landfill design, construction, and operation came about because of the side effects and "by-products" that are generated during the life of a landfill. One by-product is leachate generation. Leachate is a liquid formed within landfills from infiltrated rainwater combined with waste-derived liquids present within the landfill. Because it is commonly acidic, it can mobilize pollutants from solid waste materials. Leachate commonly includes a wide range of pollutants, including heavy metals, organic chemicals, and organic matter. If left uncontrolled, leachate can enter and migrate into groundwater, polluting domestic water sources. Leachate-based groundwater pollution is a common problem throughout the United States and is

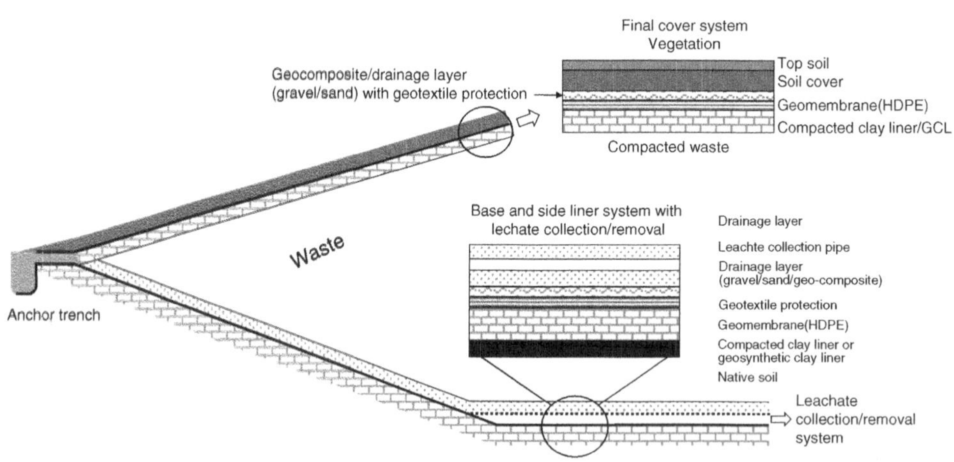

Figure 2.14 Typical municipal solid waste landfill.

commonly associated with old uncontrolled and/or abandoned landfill/dump sites without leachate collection systems or liners in place. Fortunately, leachate collection systems, consisting of permeable drainage layers, polymer liners, and drainage tiles, can effectively be used to prevent unwanted leachate migration. Monitoring wells are also typically installed to confirm that groundwater has not been polluted.

In addition to noxious liquid generation, the decomposition of refuse within landfills results in the generation of prodigious amounts of biogas, which primarily consists of methane, carbon dioxide, and other nonmethanogenic organic compounds. Biogas is highly flammable and mobile – if not controlled; it can migrate and enter structures, presenting an explosion hazard. Additionally, it can be harmful to plant life and can kill vegetation if it reaches the surface. Fortunately, biogas is also potentially valuable – if captured and purified, can be used as a fuel. As an example, in 2008, commercial landfill gases were used to provide natural gas and electricity for 1.4 million homes (DOE 1999). Because this landfill operation by-product can be used to offset the use of fossil fuels, resources may be preserved, and greenhouse gas emissions may be reduced.

Biogas is generated during the decomposition of wastes. However, the process of decomposition of waste is not always uniform, some wastes are readily degradable, some may take a longer time, and some may not be degradable at all. For example, plastics resist decomposition, and even biodegradable polymers (such as those derived from cornstarch, cellulose, lactic acid, or soybeans) can take decades to decompose. Even readily biodegradable materials may take years or decades to degrade. Decades-old newspapers are commonly exhumed in readable condition from landfills. In fact, newspapers are used to date the age of decades-old trash disposal when exhumed for engineering or forensic purposes.

Because of uneven rates of decomposition, settlement rates associated with waste degradation can be problematic. Settlements are typically large, and differential settlements can occur due to the differing rates of decomposition. This can place significant stresses on landfill covers and monitoring/collection equipment, although design allowances are typically included in these devices. Because of the settlement and compressibility of waste, buildings are not commonly sited on landfill deposits. However, closed landfills have been converted to sites for recreational uses, including golf courses or playgrounds.

With advances in technologies, landfills have evolved into well-engineered systems. Ideally, landfills are constructed on areas at higher elevations, stable ground, groundwater tables at deeper depths, and away from airports (given their propensity to attract birds, which can have a detrimental effect on aircraft). However, because of both the real and perceived environmental effects associated with landfills, it has been increasingly difficult to approve new landfills or expansions of existing facilities. In fact, between 1988 and 2007, the number of active landfills dropped from approximately 8000 to 1754 facilities (Vaughn 2011). Although the USEPA does not believe that there is a capacity issue, it presumably will get harder to approve facilities in the future as anti-growth "NIMBY" ("Not In My Backyard") or "BANANA" ("Build Absolutely Nothing Anywhere Near Anything") sentiments persist near suitable landfill locations. Other alternatives will need to be considered if landfill capacity cannot meet the demand. Additionally, with increasing costs, illegal dumping is also a growing concern.

One alternative to this is outsourcing, i.e. the transfer of wastes across state or national boundaries. Currently, 11 states export over one million tons of waste per year, while 13 states import over one million tons of waste per year. This, understandably, can create economic benefits for the importing states but vociferous opposition and resentment toward the exporting states.

Incineration is another option increasingly being considered. A total of 89 facilities in the United States burn 32 million tons of waste per year (Vaughn 2011). This process is considered as a waste reduction and not a disposal method. Although combustion reduces waste weight by 70% and volume by 90%, it results in the generation of ash, which is again a waste product. In some cases, fly ash and/or bottom ash are incorporated into building materials, including concrete.

Combustion does offer several additional benefits. First, neither households nor small businesses (the generators of MSW) need to alter waste generation activities. Second, combustion can be incorporated into waste-to-energy (WTE) facilities. Figure 2.15 depicts the operation of a hypothetical WTE facility. Two-thirds of such facilities can capture this energy; MSW combustion releases 35% as much energy when burned, and currently, such facilities provide electricity for 2.3 million homes, which in turn eliminates the need for 9.4 billion gallons of diesel per year (EIA 2017b). Further, to achieve more efficient performance, many facilities separate recyclable materials before burning, enhancing resource recovery.

Of course, there are drawbacks associated with combustion. Air pollution is a constant concern, and even small volumes of potentially hazardous household materials can lead to toxic emissions as well as the release of nuisance-causing odorous compounds. Facilities are difficult to site and expensive to build as well. Finally, to be economically feasible with respect to WTE, these facilities need a steady stream of MSW to generate energy. To provide an example, consider an MSW facility receiving 3000 tons MSW a day from a service area of one million people. Assuming typical efficiency levels, over the course of the year, 80% of 1 million tons of MSW is burned, producing electricity for 65 000 homes. Additionally, with 12% recovery due to recycling, 40 000 tons of metals are recycled. Of course, the concerns associated with such facilities are cited as drawbacks, including emissions, odors, and traffic.

In the past decades, federal regulations have been enacted to enforce local control or consolidate such control at the federal level. For instance, in 1965, the Solid Waste Disposal Act gave the Bureau

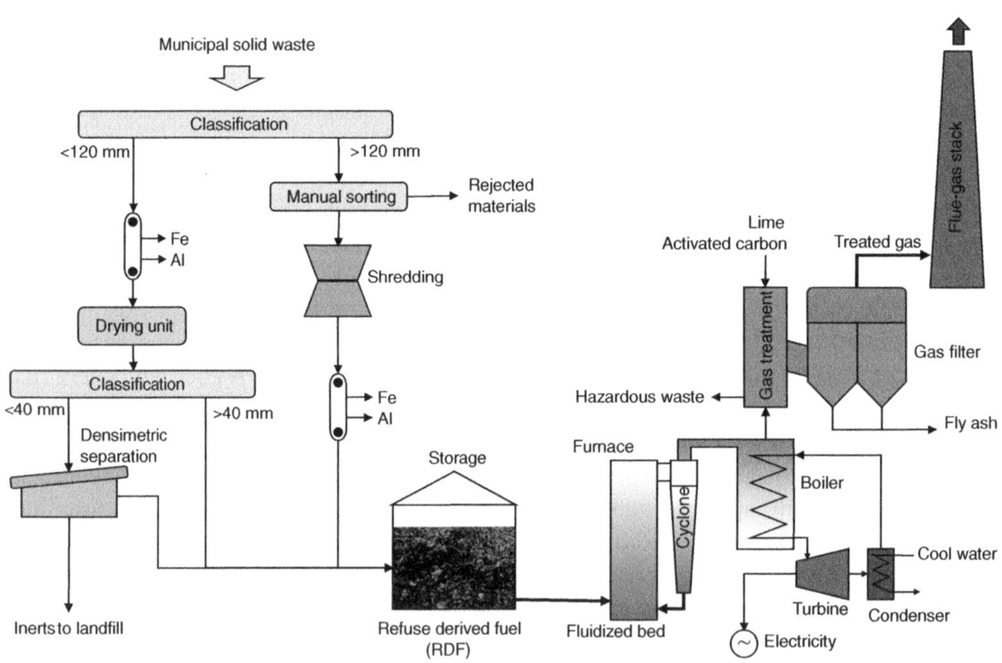

Figure 2.15 Waste to energy (WTE) process. Wastes are classified and shredded before incinerated. Heat is used to produce electric energy.

of Solid Waste Management jurisdiction over MSW, although it was limited to financial and technical issues. The Resource Conservation and Recovery Act (RCRA) was passed in 1976 to manage and regulate both hazardous and nonhazardous wastes as well as underground storage tanks (USTs). RCRA also laid emphasis on the recovery and reuse of materials through recycling.

While several alternatives exist for handling MSW, an integrated approach of source reduction, WTE, recycling, compositing, and "traditional" landfilling will likely be needed to address the growing challenges associated with MSW. Several local, state, and federal programs have been enacted nationwide to encourage a smarter, integrated approach to MSW.

In contrast to nonhazardous wastes, hazardous wastes exhibit, by definition, at least one characteristic that warrants special care (USEPA 2009a):

- *Ignitability* – Fire-related properties, including those materials spontaneously combustible or with flash points less than 140 °F.
- *Corrosivity* – Materials with pH less than 2 or greater than 12.5, including those capable of corroding metallic containers.
- *Reactivity* – Materials susceptible to unstable conditions such as explosions, toxic fumes, gases, or vapors when heated, compressed, or mixed with water under normal conditions.
- *Toxicity* – Substances that can induce harmful or fatal effects if ingested, absorbed, or inhaled.

Electronic waste, also known as e-waste, refers to discarded electronics such as televisions, computerized and peripherals, telephones, batteries, and audio and video equipment. It is estimated that 3 million tons of e-waste is generated in the United States per year, including 27 million televisions, 205 million computer products, and 140 million cellular phones (Liu et al. 2023). Less than 15–20% of these materials are refurbished, even though they often include valuable resources as well as potentially hazardous materials. Some of these materials include heavy metals, organic compounds, and flame retardants. The improper disposal of these materials is a growing concern to the environment as well as the waste of potentially valuable materials. Recently, programs have been developed to encourage special waste diversion, source reduction, and recovery of these materials.

Medical waste is also a specific environmental concern. Because of its inherent dangers, it is regulated by the United States Environmental Protection Agency (USEPA), Occupational Safety and Health Administration (OSHA), Centers for Disease Control and Prevention (CDC), and Agency for Toxic Substances and Disease Registry (ATSDR). Following well-publicized events in the 1980s when medical waste washed up on the New Jersey shore, Congress enacted the Medical Waste Tracking Act (MWTA) to evaluate risks from disposal. Seven wastes under MWTA include microbiological wastes, human blood products, pathological human wastes, contaminated animal wastes, isolation wastes, contaminated sharps, and uncontaminated sharps. Additionally, the USEPA considered including wastes that had been in contact with infectious agents or blood, including sponges, dressings, surgical gloves, drapes, slides, and laboratory coats.

Radioactive waste, unlike nonhazardous and hazardous waste, is not regulated by RCRA. Instead, it is under the jurisdiction of the Nuclear Regulatory Commission (USNRC 2017). Radioactive wastes are characterized into four categories: high-level wastes, transuranic waste, low-level waste, and mill tailings. The health and environmental dangers due to radiation may persist for hundreds of thousands of years since various radioactive wastes decay at different rates.

The Clean Air and Clean Water Acts prevented the disposal of hazardous wastes in the air or water; unfortunately, this induced unregulated land disposal. It also led to three common disposal methods.

Deep well injection consisted of injection of materials (commonly pesticides, explosives, fuels, and volatile organics) into deep sealed boreholes thousands of feet below the surface. These materials react with natural deposits, rendering them less hazardous. The USEPA oversees this practice through the Underground Injection Control Program. A total of 121 wells operate in the United States; most are located in the Gulf Coast region (USEPA 2017e).

A second method consists of surface impoundments. These consist of depressions (such as ponds) in which materials are placed to induce the settlement of solids and evaporation of water, consolidating the hazardous material. Assuming that the bottom is well sealed and loading is equal to the rate of evaporation, these facilities can receive wastes indefinitely or until solids fill the impoundment or are removed. However, surface water infiltration, such as that generated by stormwater capture, can affect operation. In 2007, it was reported that 781 million pounds of toxics were released to 18 000 surface impoundments at 7500 facilities. Two-thirds of the impoundments contained materials deemed carcinogenic or posing other health risks. Many of the impoundments were within close proximity to groundwater, and many were not lined. Furthermore, many were located near population centers; over 20 million people live within 1.2 miles of an impoundment. Further, 24% were considered a threat to the environment (Lim et al. 2010).

Hazardous materials may also be placed in landfills. The landfills meet specific RCRA criteria: they typically include sophisticated leachate removal and monitoring systems, engineered caps, and surveillance and monitoring programs. Concentrated liquids or solids are placed in drums. Best-demonstrated available technologies (BDATs) are also used to reduce mobility or toxicity, including stabilization, incineration, and chemical oxidation. Currently, only 23 landfills receive off-site hazardous waste (403 million pounds in 2007).

Because of their acute nature, it is essential that hazardous wastes are properly handled. RCRA Subtitle C was developed to manage hazardous wastes, including a regulatory framework for the generation, transportation, storage, and disposal of hazardous waste as well as technical standards for the design and operation of treatment, storage, and disposal facilities (TSDFs). Additionally, hazardous wastes are subjected to "cradle-to-grave" tracking, requiring detailed records for generation, transport, and disposal. The copies of these records must go to the EPA as well.

Although RCRA covered nonhazardous and hazardous waste management for ongoing facilities, it did not address abandoned sites. Several well-publicized incidents highlighted cases where hazardous wastes were negligently handled. One instance is known as Love Canal. In the Niagara Falls, New York area, development occurred over the previously abandoned Love Canal. The resulting development and infrastructure construction pierced the clay-lined canal. Over time, noxious odors were observed, and significant acute and chronic health problems were reported by the citizens. Eventually, follow-up testing and analysis determined the presence of widespread soil and groundwater contamination, and the United States Federal government paid for the relocation of hundreds from the Love Canal area. A second site was the Valley of the Drums. This site, a 23-acre toxic waste site in Bullitt County, Kentucky, was named after the waste-containing drums strewn across the area. In 1966, the site caught fire and burned for over a week (Reddy 2014). Ultimately, it caught the attention of state officials, leading to its fencing off and capping in the 1980s.

As a result, in 1980, the Comprehensive Environmental Response, Compensation, and Liabilities Act (CERCLA) or "Superfund" was passed to address the cleanup of these hazardous sites, specifically addressing funding, liability, and prioritization of hazardous and/or abandoned waste sites. A $1.6 billion fund was created from taxes levied on chemical and petroleum industries to finance the cleanup of hazardous waste sites and litigation brought against potentially responsible parties (PRPs). Additional funds ($8.5 billion) were appropriated in 1986 with the passage of the Superfund Amendments and Reauthorization Act (SARA). A $500 million fund was also appropriated for the remediation of leaking USTs (LUST). Additionally, community right-to-know provisions were adopted. Controversially, SARA established provisions for cleanup-related legal and financial liability. Disclosure requirements related to annual releases of hazardous substances were also included. Explicit liability provisions directed at current landowners and related innocent landowner provisions became a paramount concern for all entities associated with land transactions.

Both CERCLA and SARA significantly underestimated the potential costs and timing associated with environmental cleanups. When CERCLA was first enacted, approximately 36 000 contaminated sites were identified; of these, 1200 were placed on the national priority list (NPL) (OSWER 2010). In October 2024, 1340 sites remained on the NPL, and 458 sites had been closed (USEPA 2024). In FY 2023, the EPA's efforts led to the closure of 84 Superfund sites. The total cost associated with these closures was approximately \$1.1 billion. On average, the cost per site closure was about \$13.1 million. The benefits for communities around these sites were substantial, with remediated properties being returned to productive use, enhancing public health and environmental quality. These closures also positively impacted nearly 900 000 people living within a one-mile radius of the sites, with a significant focus on addressing environmental justice (EJ) concerns in disadvantaged areas. This has helped improve living conditions and reduce health risks in vulnerable communities.

Because of fears associated with CERCLA liability, impacted properties with significant reuse potential remained idle and contaminated for long periods of time. Many of these sites became known as Brownfields. A Brownfield is an abandoned, idled, or underutilized industrial or commercial site where expansion or redevelopment is complicated by actual or perceived environmental contamination. The real or perceived contamination can range from minor aesthetic issues to significant subsurface contamination. In the early 1990s, the federal government took action to provide inducements to encourage Brownfield redevelopment. The USEPA also took measures to clarify liability provisions as well as provide for indemnity for prospective purchasers. In 2002, the United States Congress passed the Small Business Liability Relief and Brownfields Revitalization Act to provide certain relief for small businesses from liability imposed under the CERCLA, promote the cleanup and reuse of brownfields, provide financial assistance for brownfields revitalization, and enhance State response programs (USEPA 2009b).

In 2005, the USEPA established the All Appropriate Inquiries (AAI) requirements, which became law on 1 November 2006 (USEPA 2017f). The purpose of AAI was to establish liability protection under CERCLA for innocent landowners, contiguous property owners, or *bona fide* prospective purchasers provided that certain criteria were met. AAI has been a very important milestone in encouraging land acquisition and development. By establishing a framework, prospective land purchasers have a discrete set of actions they must perform to avoid open-ended liability and costs. In this manner, they can help eliminate the unknowns associated with a potential redevelopment project, which facilitates a return to productive use for many impacted properties.

Finally, another source of soil contamination is related to the USTs, which are common at service stations, corporation yards, and a variety of commercial and industrial settings. The underground placement of tanks reduces the danger of explosion and fire, however, because tanks eventually deteriorate, USTs are a very common source for soil and groundwater contamination. Leaks, especially in older tanks, can result in substantial contamination before they are detected. Subtitle I of RCRA regulates USTs and includes provisions for notification, methods for release detection, design and construction standards, and reporting, recordkeeping, and financial responsibility. Corrective actions pertaining to releases from USTs are also regulated under Subtitle I. USTs containing hazardous wastes are regulated under Subtitle C. Furthermore, a LUST trust fund receives 0.1 cents per gallon of fuel purchased to finance federal oversight activities.

Example 2.4

Let us say a community of population 50 000 plans to employ an integrated waste management system to manage their daily amount of MSW generated. Details of the MSW generated and their composition are as follows:

- MSW generated – 2.5 kg/person/day

Component	Fraction
Kitchen waste	45%
Yard waste	10%
Recyclables	20%
Others	25%

Details of the MSW management plants are as follows:

- Landfill area: 6 ha, average maximum depth to be filled – 20 m.
- MSW composting plant capacity: 50 tons/day.
- Recycling facility capacity: 25 tons/day.

Calculate the following:

- Amount of waste generated per day by the community.
- Amount of waste landfilled by the community per day, assuming that the composting and recycling facilities run at capacity.
- Composition of waste landfilled assuming both composting and recycling facilities run at capacity.
- Amount of waste diverted by employing composting and recycling facilities.
- Assuming that the waste is compacted at 700 kg/m^3 density, determine the useful life of the landfill site.

Solutions:

- **Amount of waste generated per day by the community**
  = Waste produced per person × population of the community
  = 2.5 kg/person/day × 50 000 people = 125 000 kg/day
- **Amount of waste landfilled by the community per day, assuming that the composting and recycling facilities run at capacity**
  = Total waste – (sum of recycled waste and composted waste)
  = 125 000kg/day – (50 000 kg/day+ 25 000 kg/day) = 50 000 kg/day
- **Composition of waste landfilled, assuming both composting and recycling facilities run at capacity**
  Assuming kitchen waste is going to compost: total waste = 45% of 125 000
  = 0.45 × 125 000 kg/day = 56 250 kg/day
  Capacity of composting plant = 50 000 kg/day
  Difference = 56 250 kg/day – 50 000 kg/day = 6250 kg/day or 5%

  Quantity of recycled waste generated per day = 20% of 125 000 kg/day
  = 0.2 × 125 000 kg/day = 25 000 kg/day

  The recyclable waste has zero waste.

Therefore, the composition of the landfill:

Yard waste (10%) + other waste (25%) + part of kitchen waste (5%).

- **Amount of waste diverted by employing composting and recycling facilities**
  = Capacity of composting plant + capacity of recycling plant
  = 50 000 kg/day + 25 000 kg/day = 75 000 kg/day
- Assuming that the waste is compacted at 700 kg/m^3 density, determine the useful life of the landfill site
  Volume of the landfill = 6 ha, i.e. 60 000 $m^2 \times 20$ m = 1 200 000 m^3
  Density of landfill = 700 kg/m^3
  Total mass capacity of landfill = 700 $kg/m^3 \times$ 1 200 000 m^3 = 8.4×10^8 kg
  Total amount of waste landfilled per day = 50 000 kg/day
  Therefore, useful life of landfill = 8.4×10^8 kg / 50 000 kg/day
  = 16,800 days $\approx$ 46 years

2.14 Land Contamination

Hazardous waste sites, those with contamination from a range of toxic materials, are a serious threat to public health. Both active and inactive sites pose a concern. However, active sites can be monitored, and site activities can be regulated or changed to be more protective. More-invasive intervention techniques can be implemented at inactive sites, such as the capping of large-scale cleanups. Inactive sites were often operated without protective measures in place; for instance, old landfills and dumps were often constructed without liners, which could lead to more widespread soil and groundwater contamination.

Toxic chemicals fall into several categories. Some can be easily degraded and assimilated into the environment, where they may be diluted from stabilized to safe levels; however, others, such as heavy metals and synthetic organics, are resistant to degradation and persist in the environment. Many heavy metals and products (e.g. lead, mercury, arsenic, cadmium, tin, chromium (VI), zinc, copper, and products containing these substances) are persistent, toxic compounds. They have been commonly used in many industrial applications, including metalwork, plating, batteries, and electronics. They were also once commonly used in paints, glazing, inks, and dyes. Some can be soluble in water, and if absorbed, they are commonly bioaccumulative within the body. They can interfere with enzyme function, birth defects, learning disabilities, and severe mental illness.

Synthetic and derived organic compounds can also be toxic to human health and the environment. These compounds are often used as the basis for plastics, fibers, coatings, solvents, pesticides, and preservatives. They are resistant to degradation, allowing them to persist in toxic form within the environment. Upon absorption into the human body, these compounds can interfere with enzyme action. Acute exposure can lead to poisoning or death, and chronic exposure can result in mutagenic, carcinogenic, and teratogenic effects. Several compounds pose acute dangers to health and the environment. Some of these, collectively labeled as persistent organic pollutants (POPs), include halogenated hydrocarbons (those containing chlorine, bromine, fluorine, and iodine), and chlorinated hydrocarbons, which include plastics, pesticides, solvents, and insulation. In 2004, the Stockholm Convention banned or restricted the use of 12 of these compounds (nicknamed the "dirty dozen") because of their toxic, carcinogenic, or endocrine disruption effects at low concentrations (Batool et al. 2016). Additional compounds (including the "nasty nine") have since been added (UNEP 2008).

A very common compound which is used in a variety of industrial and consumer applications but exhibits a significant impact on health and the environment is tetrachloroethylene or perchloroethylene (PCE, or "PERC"). PCE is a colorless, nonflammable, halogenated hydrocarbon used as a dry-cleaning solvent, and industrial solvent, and home products. As a dense, nonaqueous phase liquid (DNAPL), it is heavier than water; when it is exposed to the environment, it is mobile, sinks through groundwater, and is highly resistant to degradation. Acute exposure can lead to dizziness, fatigue, and headaches, and because it has proven to be carcinogenic to rats and mice, it is suspected as a human carcinogen. Dry cleaning accounts for 10% of the United States' annual PCE consumption of 370 million pounds per year, and the remainder is primarily used to make CFCs (ICIS 2017). The use of PCE is being phased out; several European Union are considering a ban on its use, and California has mandated its ban in dry cleaning and within related equipment by 1 January 2023 (CARB 2007). Other common organic compounds that have been increasingly linked to health effects include phthalates (used to soften plastics) and perchlorate (used in rocket fuels and other flammable products). These compounds have entered and impacted the environment (soil sediments, water) as well as affected human health due to their presence in a range of products.

2.15 Visibility

Visibility is defined as the greatest distance over which one can see and identify familiar objects with the naked eye (Hyslop 2009). Visibility is based on two factors – first is the perception abilities of the individual, and second is related to the degree to which light coming from the object is absorbed or scattered. Two principal causes of decreased visibility include the emission of small particles on the same order of visible light, about 0.2–0.7 μm, as well as the emission of reactive gases that are subsequently converted into small particles. Visibility impacts can affect urban or rural areas. The EPA passed the Regional Haze Rule in 1999 with the goal of improving visibility and reducing emissions in parks and wilderness areas.

2.16 Odors

Odors, resulting from both organic and inorganic compounds, can have different degrees of unpleasantness. Regardless of the aesthetics and thresholds, some odors can be indicative of a potential health hazard; others may result in a perceived danger or affect quality of life. From a public health standpoint, odors may have a direct adverse effect on health or may induce other symptoms associated with illness (Piccardo et al. 2022).

Odors can emanate from a variety of sources. Some industrial sources include wood treatment facilities, paper mills, landfills, refineries, solvent processing facilities, and wastewater treatment plants. Nonindustrial sources may include pesticide and fertilizer applications, feedlots, and facilities with numerous idling trucks. Odors from these sources elicit far more community complaints than odorless air pollutants.

In some cases, odors do not represent overexposure to a specific compound. In other cases, such as VOC exposures, odors can be indicative of environmental quality problems as well as a sign of unhealthy overexposure. The resulting health effects depend on several factors, including sensitivity and response to odors on an individual basis. Health effects may include irritation and adverse effects on mucous membranes, air passages, blood vessels, heart, stomach and intestines, brain, and psychological well-being.

2.17 Aesthetic Degradation

Aesthetic degradation is any reduction in environmental quality that people interpret as a reduction in attractiveness (Verburg 2014). This may include the siting of unsightly industrial or institutional facilities or other cases where the built environment affects the natural environment. These may also include visual degradation of soil, air, and water. Specific sources can include the placement of debris, unnatural colors or textures, turbidity or haziness to water/air, or scum and floating materials on water. With respect to water, natural, or anthropogenic substances can degrade water quality, such as water color, odor, or clarity. The monitoring of aesthetics can be relatively inexpensive. It is primarily based on subjective criteria, as visual appreciation will vary from person to person. It can also be a source of controversy, as the range of personal senses of aesthetics can make it difficult to reach a consensus on these matters.

2.18 Land Use Patterns

Land use pertains to how the physical environment is utilized for resources and/or converted to the built environment. It can include specific resource utilization, such as mining, forestry, or agricultural uses, or it can pertain to consumption, such as residential, commercial, industrial, or recreational activities. Economic and social factors are paramount to land use decisions and patterns. As the built environment expands, there is often stress or conflict with the natural environment. The sprawl of metropolitan areas can often lead to the loss of natural vegetation, open space, wetlands, and other habitats. Agricultural lands are also often lost to growing communities. As these land use conversions continue, more public attention is drawn to the environmental impacts that result from a range of side effects, including loss of vegetation and habitat, traffic, and utilization of other resources.

2.19 Thermal Pollution

Thermal pollution is the degradation of the environment due to processes that change ambient temperature. These effects can have significant consequences and are commonly observed due to the discharge of heated or cooled water resulting from human activity. A typical occurrence is the release of cooling water from power plants, refineries, paper mills, chemical plants, and steel mills. These cooling waters are often warmer than the receiving water, and the resulting elevation in temperature can affect the ecosystem. Thermal effects can also occur from urban runoff and stormwater runoff that may be generated from changes in vegetation. As impervious surfaces replace natural vegetation, runoff temperature may be elevated as compared to a natural condition, leading to similar effects as observed within cooling water discharge. When warmer water is discharged, dissolved oxygen rates decrease because oxygen solubility decreases. Also, the increased temperature can enhance metabolism in organisms, leading to greater utilization of oxygen and decreased dissolved oxygen levels, ultimately resulting in eutrophic conditions.

Thermal pollution can also be directly toxic to aquatic life since enzyme systems can be suddenly disrupted or shocked. Organisms can be killed by sudden thermal changes that are beyond the tolerance limits of their metabolic systems. Ironically, in some instances, thermal pollution can be viewed as beneficial (thermal enrichment). Heated water can result in longer commercial fishing seasons and the reduction of winter ice cover in cold areas.

Thermal pollution effects can be controlled with several strategies. Heat transfer can be safely induced using cooling towers, cooling ponds, or cogeneration applications where the heat may be safely extracted or dissipated. Cooling water can also be retained and recirculated through the heat-generating system, thereby reducing direct discharge into vulnerable water bodies. Stormwater runoff can be treated by increasing residence times using bioretention and infiltration systems, thereby allowing a gradual reduction in temperature.

2.20 Noise Pollution

Noise pollution is excessive environmental noise that can disrupt the quality of life. Noise pollution may emanate from human, animal, or machine-related sources. Typical sources of noise may include construction or transportation activities as well as industrial activities. It may also result from undesirable activities and practices as well as inappropriate urban planning in which incompatible activities are placed near one another. Noise pollution can have physical health effects, such as hearing loss, cardiovascular effects, such as hypertension, as well as psychological effects, and diminished well-being. Animals can be affected, as noise can diminish or eliminate natural predator or prey behaviors.

Federal regulations have been enacted to abate noise pollution. Title IV of the Clean Air Act focuses on noise pollution (USEPA 2018). Additionally, the Noise Control Act of 1972 and the Quiet Communities Act of 1978 provide a regulatory framework for noise pollution reduction. These regulations as well as other mandated or industry-led activities have identified methods to reduce noise, such as design improvement in roads and vehicles, quieter jet engines, and refinements in industrial equipment design.

2.21 Summary

Environmental concerns of varying scope and magnitude may occur during all phases of a project. These concerns may occur on a local scale, a regional scale, or a global scale, but regardless of the scope and scale, virtually every activity will have one or more environmental impacts.

It is also important to point out that environmental impacts can have a significant detrimental effect on human health and the ecosystem. Separating important effects from "background noise" is a key skill in assessing the sustainable impacts of a project, and it is important that these environmental concerns are assessed on a qualitative or quantitative basis.

When assessing a project or an activity, it is essential to understand a range of environmental concerns that can occur. This chapter summarizes a wide range of environmental concerns that can directly affect human health or the greater environment. When assessing the sustainability of a project, all potential environmental impacts, or as many as practicable, should be considered and incorporated into the assessment, especially the impacts that are quantifiable. These may include both direct and indirect environmental impacts. It is also often beneficial to assess and quantify the environmental impacts associated with a range of alternative projects/designs for the purposes of a comparative assessment. Such an assessment will identify and allow for the selection of the project/design that causes the least overall environmental impact.

2.22 Questions

2.1 What are the environmental concerns from a sustainability perspective? Explain in detail.

2.2 What are the global-scale environmental concerns? Explain them briefly.

2.3 What are the regional-scale environmental concerns? Explain them briefly.

2.4 What are the local-scale environmental concerns? Explain them briefly.

2.5 What are greenhouse gases? How are they helpful in maintaining favorable temperatures on the Earth for the survival of most of the living organisms? What are the consequences of high amounts of greenhouse gases in the atmosphere?

2.6 Explain global climate change and list the main greenhouse gases of anthropogenic origin. What are the sources and impacts?

2.7 Describe how the GWP assesses the relative strength of greenhouse gases?

2.8 Name some key desert areas, and identify two large metropolitan areas near each location?

2.9 Why is deforestation to create agricultural land often counter-productive?

2.10 Discuss in detail why losses in genetic diversity can be harmful to ecosystems?

2.11 Describe the difference between stratospheric ozone and ground-level ozone.

2.12 Which are the compounds that abet ozone layer depletion? Describe in detail the process of how CFCs result in ozone-layer depletion.

2.13 Explain the physiological and psychological effects of lead poisoning.

2.14 Discuss meteorological and topographical conditions that may exacerbate industrial or photochemical smog.

2.15 What are the two types of smog? Explain briefly.

2.16 What is acid rain and how does it occur? Explain briefly.

2.17 What is eutrophication? Explain briefly.

2.18 Provide specific examples of point source and nonpoint source water pollution.

2.19 Explain how climate change and water cycle are related.

2.20 Explain how climate change and water hazards are related.

2.21 Explain how climate change and water quality are related.

2.22 Discuss the different types of solid wastes and characteristics of each.

2.23 What are the characteristics of hazardous wastes? Explain briefly.

2.24 What is land degradation and mention the different categories of land degradation.

2.25 What are the different sources of odor in the environment, and how does it impact human health?

2.26 What is thermal pollution and what are its effects on the environment? Explain briefly.

2.27 List examples of health effects caused by noise pollution.

2.23 Problems

2.1 When speaking about the global warming effects of greenhouse gases, it is important to know how to understand the concentration of such gases in the atmosphere. The average concentrations of following greenhouse gases were reported in the atmosphere as per the latest reports: CO_2 – 416 ppm, CH_4 – 2 ppm, N_2O – 335 ppb increasing at 2.4 ppm/year, 0.01 ppm/year, and 0.7 ppb/year, respectively.

Calculate the following:

- How much are the concentrations of each gas in percentage (%) in the atmosphere?
- How much volume of CO_2, CH_4, and N_2O are contained in a cubic meter (m^3) of air?
- How large are the mass shares of CO_2, CH_4, and N_2O in the atmosphere (% and ppm), assuming that all the gases and the atmospheric air are ideal gases?

2.2 For the gas concentrations obtained in Problem 2.1, determine the GWP of a kg of individual substance ($W/m^2/kg$). Use the concentration-specific atmospheric radiative forcing ($W/m^2/ppb$) from the 6th IPCC assessment report.

2.3 Determine the GWP of all the three gases mentioned above, over a time horizon of 25 and 100 years.

Data: Obtain the specific radiative forcing values for CH_4, CO_2, and N_2O from the IPCC 6th Assessment Report on Climate Change. The effect of a compound on climate is not limited to its direct radiative forcing. Compounds can perturb the carbon cycle affecting atmospheric CO_2 concentrations. Chemical reactions from emitted compounds can produce or destroy other greenhouse gases or aerosols, obtain the correction factors owing to this compounding effect for different gases from here (https://www.industrialecology.uni-freiburg.de/Content/IEooc_Background2_Exercise2_GWP.pdf)

Note: Go through different tasks of this exercise carefully, to understand how to solve the above problems (https://www.industrialecology.uni-freiburg.de/Content/IEooc_Background2_Exercise2_GWP.pdf)

University of Freiburg (2024). Industrial Ecology Open Online Course (IEooc) Part I: Background Background 2: Climate, sustainability, and the contribution of industrial ecology. Available at: https://www.industrialecology.uni-freiburg.de/Content/IEooc_Background2_Exercise2_GWP.pdf

2.4 If the following relationship holds for the time decay of any given quantity of methane in the atmosphere:

$$R_{CH_4}(t) = e^{-\frac{t}{\tau_{CH_4}}}$$

Consider the initial concentration of CH_4 to be 2 ppm and the constant, $\tau_{CH_4} = 12.4$ yrs, determine the following:

- Considering no additional emissions of methane into the atmosphere, what will be the concentration of CH_4 in the atmosphere after the time τ_{CH_4} has passed?
- What is the half-life (time-after which half the CH_4 will remain) of CH_4 in atmosphere?
- Plot the above function $R_{CH_4}(t)$, for time $t = 0$ to 100 years.
- What is the average lifetime of CH_4 emitted to the atmosphere?
- Now consider, that the CH_4 emission rates are 0.01 ppm/year, determine the concentration of CH_4 in the atmosphere after 25 years, and plot the concentration of CH_4 with time. (Don't forget to consider the decay rate of additionally emitted CH_4).

2.5 What is the temperature inversion phenomenon? Describe with the help of a schematic. If the variation of temperature with height on a given day is recorded as follows:

Altitude (m)	0	50	100	150	200	250	300	350
Temperature (°C)	15	14	13	16	15	14	13	12

- Plot a graph for altitude versus air temperatures.
- Determine the lapse rate between 250 and 350 m.
- Do you find any temperature inversion in this profile? If so, where?
- Determine whether fumigation would occur if smoke at 20 °C were discharged from a chimney at altitudes 100 and 250 m.

2.6 Let us say that the oxygen depletion potential (ODP) of chlorofluorocarbon-11 (CFC-11) is 1.0. If a refrigerator leaks 2 kg of CFC-11 into the atmosphere, calculate the equivalent amount of ozone depletion compared to releasing the same mass of pure chlorine gas (Cl_2) into the atmosphere. The molecular weight of CFC-11 is 137.37 g/mol, and that of chlorine is 35.45 g/mol. In addition, clearly explain the main reasons for ozone depletion. List out some compounds or gases that lead to the depletion of the ozone layer and explain with the help of chemical reactions. Please cite any sources for your information.

2.7 In a coal-fired power plant, the daily combustion of 1000 tons of coal, of which 2% is sulfur, leads to the complete burning of sulfur, resulting in emissions into the atmosphere from a stack with an effective height of 100 m. Given a wind speed of 6 m/s, calculate:

a) The Downwind Ground-Level Sulfur Dioxide (SO_2) Concentration:
   - Determine the concentration of SO_2 along the plume centerline at ground level 10 km downwind from the power plant.

b) The Maximum Ground-Level SO_2 Concentration for Class B Stability:
   - Calculate the maximum ground-level concentration of SO_2 under class B stability conditions, using a conservative wind speed value.

c) The Downwind Distance of Maximum Ground-Level SO_2 for Class B Stability:
   - Determine the distance downwind at which the maximum ground-level concentration of SO_2 occurs under class B stability conditions.

2.8 A wastewater treatment plant discharges effluent into a nearby water body. The concentrations of nutrients in the effluent and the corresponding eutrophication potential factors are given below:

Compound	**Concentration**	**Eutrophication potential (kg PO_4^{3-}-eq)**
Ammonium nitrogen (NH_4-N)	10 mg/l	0.35
Nitrate nitrogen (NO_3-N)	5 mg/l	0.1
Orthophosphate phosphorus (PO_4-P)	8 mg/l	1.0

Tasks:

- Calculate the eutrophication potential of the wastewater discharge for one year based on the given concentrations, if the plant discharges 700 MGD (million gallons per day).
- If the wastewater treatment plant implements a technology to reduce ammonium nitrogen concentrations by 50%, recalculate the new eutrophication potential and discuss the potential improvement in water quality.
- Explain the potential environmental impact of the discharged contaminants on the receiving water body in terms of eutrophication, considering both the original and modified scenarios.

2.9 At a distance of 55 m from the center of a roadway, the sound levels were recorded to be 70 dBA. Estimate the sound level at a distance of 15 m from the center of the road. If the average speed of traffic triples but the traffic volume remains the same, estimate the sound level at a distance of 15 m from the center of the road.

2.10 A water sample was analyzed for dissolved, suspended, and volatile solids. A 100 ml water sample was first filtered through a clean glass-fiber filter, following which the filter with the residue was ignited at 550 °C. The following observations were made:

- Weight of filter = 550 mg.
- Weight of residue + filter = 585 mg.
- Weight of residue + filter after ignition = 564 mg.

50 ml of the filtered water sample was placed in an empty evaporating dish (weight – 38.00 g), and after evaporation, the weight of residue + dish was found to be 38.04 g.

- Determine the total suspended solids, total dissolved solids, and the total volatile solids in the sample.
- Compare the values with federal regulations for the corresponding limits in drinking water, and comment on the observed results.

2.11 A multiple-fermentation test was performed on the water sample and the results are as shown below:

Dilution series	Results after incubation
10 ml	2 positive 1 negative
1 ml	3 positive
0.1 ml	1 positive 2 negative

Determine the Most Probable Number (MPN) value of the sample. State whether this water is suitable for drinking or not?

2.12 The following salinity values were found in some well-known water bodies worldwide

Water body	Salinity (%)
Dead sea	33.0
Mediterranean sea	3.8
Red sea	3.6
Baltic sea	1.0

Consider the following salt composition in all the water bodies:

- Sodium chloride – 77.7%
- Magnesium chloride – 10.9%
- Magnesium sulfate – 4.7%
- Calcium sulfate – 3.6%
- Potassium sulfate – 2.5%

Tasks:

- Determine the mass of each salt in 1 l on the water from different water bodies.
- Represent the salinity values of different water bodies in ppm.
- Discuss the impacts of water salinity on CO_2 solubility, marine and aquatic life, and interactions between seawater and fresh surface water sources.
- Comment on the salinity values of these water bodies.
- Is there a recommended value of salinity in water bodies considering, sustainability of marine ecosystems, and the dissolution of CO_2?

2.13 Given below is the composition of landfilled municipal soil waste in two different communities.

Component	Community 1	Community 2
Paper	4.6	19.0
Yard waste	5.0	6.3
Wood	1.8	29.6
Food	61.9	2.3
Textiles	3.1	1.1
Metal	1.1	1.5
Glass	2.1	10.6
Plastic	9.8	2.4
Rubber	1.3	–
Miscellaneous, inert	6.8	33.2
Soil-like particles, inert	2.5	–

Determine the following:

- Find a source for typical moisture contents of different components and determine the average moisture content of the MSW mix.
- If the MSW mixes are to be filled in a landfill at a bulk compacted density of 900 kg/m^3 determine the porosity and degree of saturation of the waste. (Obtain the specific gravities of different components from any online sources, or assume appropriately).
- Carefully examine the compositions of the landfilled waste. What does it tell you about the MSW management strategies of the respective communities?

References

Archer, D. (2016). *The Long Thaw: How Humans are Changing the Next 100,000 Years of Earth's Climate.* Princeton University Press.

Batool, S., Ab Rashid, S., Maah, M.J. et al. (2016). Geographical distribution of persistent organic pollutants in the environment: a review. *Journal of Environmental Biology* 37 (5): 1125.

Beaty, T.A. (2015). Life on the Mississippi: reducing the harmful effects of agricultural runoff in the Mississippi River Basin. *Ohio Northern University Law Review* 41: 819–965.

van Beusekom, J. E. (2018). Eutrophication. Handbook on Marine Environment Protection: Science, Impacts and Sustainable Management, 429–445.

California Air Resources Board (CARB) (2007). Final Regulation Order: Airborne Toxic Control Measure for Emissions of Perchloroethylene Associated with Dry Cleaning Operations and Requirements for Manufacturers and Distributors of Perchloroethylene. https://ww2.arb.ca.gov/sites/default/files/classic/toxics/dryclean/approvedregorder-1.pdf (accessed 20 March 2025).

CBD (2014). *Global Biodiversity Outlook 4*, 155. Montréal: Secretariat of the Convention on Biological Diversity. ISBN: ISBN-92-9225-540-1.

Ceballos, G., Ehrlich, P.R., Barnosky, A.D. et al. (2015). Accelerated modern human-induced species losses: entering the sixth mass extinction. *Science Advances* 1 (5): e1400253.

Cooke, G.D., Welch, E.B., Peterson, S., and Nichols, S.A. (2016). *Restoration and Management of Lakes and Reservoirs.* CRC Press.

Dapul, H. and Laraque, D. (2014). Lead poisoning in children. *Advances in Pediatrics* 61 (1): 313–333.
DOE (1999). Natural Gas 1998: Issues and Trends. Energy Information Administration. Office of Oil and Gas. U.S. Department of Energy. *Report no. DOE/EIA-0560(98).* Washington, DC.
EIA (2017a). Petroleum and other liquids: crude oil production. US Energy Information Administration. Independent Statistics and Analysis. https://www.eia.gov/dnav/pet/pet_crd_crpdn_adc_mbblpd_a.htm (accessed 16 October 2018).
EIA (2017b). Landfill gas and biogas. US Energy Information Administration. Independent Statistics and Analysis. https://www.eia.gov/energyexplained (accessed 10 April 2017).
Elsom, D. (2014). *Smog Alert: Managing Urban Air Quality.* Routledge.
Encyclopedia of Arkansas (2023). Floods. https://encyclopediaofarkansas.net/entries/floods-5148/ (accessed 6 October 2024).
Ennis, C.A. (2017). Scientific Assessment of Ozone Depletion: 2006. *World Meteorological Organization Global Ozone Research and Monitoring Project—Report No. 50.* Geneva, Switzerland.
Hyslop, N.P. (2009). Impaired visibility: the air pollution people see. *Atmospheric Environment* 43 (1): 182–195. https://doi.org/10.1016/j.atmosenv.2008.09.067.
ICIS (2017). Perchloroethylene uses and market data. ICIS web site. https://www.icis.com/resources/news/2007/11/06/9076131/perchloroethylene-uses-and-market-data (accessed 25 July 2017).
IPCC (2023). *Climate Change 2023: Synthesis Report. Contribution of Working Groups I, II and III to the Sixth Assessment Report of the Intergovernmental Panel on Climate Change* (eds. Core Writing Team, H. Lee, and J. Romero), 184 pp. Geneva, Switzerland: IPCC. https://doi.org/10.59327/IPCC/AR6-9789291691647.
Ivushkin, K., Bartholomeus, H., Bregt, A.K. et al. (2019). Global mapping of soil salinity change. *Remote Sensing of Environment* 231: 111260.
Jouzel, J., Masson-Delmotte, V., Cattani, O. et al. (2007). Orbital and millennial Antarctic climate variability over the past 800,000 years. *Science* 317 (5839): 793–796.
Kulshrestha, U. (2020). Acid rain. In: *Managing Air Quality and Energy Systems* (ed. B.D. Faith), 707–727. CRC Press.
Lal, R., Iivari, T., and Kimble, J.M. (2003). *Soil Degradation in the United States: Extent, Severity, and Trends.* CRC Press.
Lim, S.R., Lam, C.W., and Schoenung, J.M. (2010). Quantity-based and toxicity-based evaluation of the US Toxics Release Inventory. *Journal of Hazardous Materials* 178 (1-3): 49–56.
Liu, K., Tan, Q., Yu, J., and Wang, M. (2023). A global perspective on e-waste recycling. *Circular Economy* 2 (1): 100028.
Lüthi, D., Le Floch, M., Bereiter, B. et al. (2008). High-resolution carbon dioxide concentration record 650,000–800,000 years before present. *Nature* 453 (7193): 379–382.
MEA (2005). *Millennium Ecosystem Assessment – Information to Conserve Ecosystems and Enhance Human-Well Being.* Washington, DC: New Island.
de Mutsert, K., Steenbeek, J., Lewis, K. et al. (2016). Exploring effects of hypoxia on fish and fisheries in the northern Gulf of Mexico using a dynamic spatially explicit ecosystem model. *Ecological Modelling* 331: 142–150.
NCA (2014). *The U.S. Global Change Research Program's Third Assessment Report.* National Climate Assessment. U.S. Global Change Research Program. ISBN: 9780160924026. nca2014.globalchange.gov.
NCOOS (2017). Harmful algal bloom and hypoxia research and control act. National Centers for Coastal Ocean Science. https://coastalscience.noaa.gov/research/habs/habhrca (accessed 10 April 2017).
OSWER (2010). Fiscal Year 2010 End of Year Report. Office of Solid Waste and Emergency Response. Washington, DC: United States Environmental Protection Agency.
Pacala, S. and Socolow, R. (2004). Stabilization wedges: solving the climate problem for the next 50 years with current technologies. *Science* 305 (5686): 968–972.

Piccardo, M.T., Geretto, M., Pulliero, A., and Izzotti, A. (2022). Odor emissions: a public health concern for health risk perception. *Environmental Research* 204: 112121.

Rain Forest Maker (2024). Extinction of Rainforest Species Slows Future Growth. https://www.rainforestmaker.org/extinction-of-rainforest-species-slows-future-growth.html (accessed 6 October 2024).

Reddy, K.R. (2014). Evolution of geoenvironmental engineering. *Environmental Geotechnics* 1 (3): 136–141.

Reynolds, J.F., Smith, D.M.S., Lambin, E.F. et al. (2007). Global desertification: building a science for dryland development. *Science* 316 (5826): 847–851.

Ritchie, H. and Roser, M. (2023). *Water Use and Stress*. Our World in Data.

Rowland, F.S. and Molina, M.J. (1974). Stratospheric ozone destruction catalyzed by chlorine atoms from photolysis of chlorofluoromethanes. *Transactions of the American Geophysical Union* 55 (1974): 1153–1153.

Shiklomanov, I.A. (1998). *World Water Resources. A New Appraisal and Assessment for the 21st Century. Paris: United Nations, Educational, Scientific and Cultural Organization (UNESCO).*

Thiele, L.P. (2016). *Sustainability*. Wiley.

UNCCD (2017). United Nations convention to combat desertification. http://www2.unccd.int (accessed 7 April 2017).

UNCED (1992). Rio declaration on environment and development. Rio de Janeiro, Brazil: United Nations Conference on Environment and Development.

UNEP (2008). Stockholm Convention. Protecting human health and environment from persistent organic pollutants. http://chm.pops.int/TheConvention/ThePOPs (accessed 10 April 2017).

UNEP (2017). The Montreal protocol on substances that deplete the ozone layer. Ozone Secretariat: United Nations Environment Programme. http://ozone.unep.org (accessed 9 April 2017).

USEPA (2009a). Hazardous waste characteristics. *US Environmental Protection Agency*. Report. 30 pages.

USEPA (2009b). Brownfields fact sheet – EPA brownfields grants, CERCLA liability, and all appropriate inquiries. EPA 560-F-09-026, Washington, DC.

USEPA (2012). Water Quality Standards Handbook. Chapter 3: Water Quality Criteria. *Report No. EPA-823-B-12-002*.

USEPA (2015). Mississippi River/Gulf of Mexico watershed nutrient task force. US Environmental Protection Agency. *HTF 2015 Report to Congress*.

USEPA (2016). Municipal solid waste generation, recycling, and disposal in the United States: facts and figures. https://archive.epa.gov/epawaste/nonhaz/municipal/web/html/msw99.html (accessed 9 April 2017).

USEPA (2017a). National ambient air quality standards (NAAQS). US Environmental Protection Agency. https://www.epa.gov/criteria-air-pollutants (accessed 10 April 2017).

USEPA (2017b). Safe drinking water act (SDWA). https://www.epa.gov/sdwa (accessed 10 April 2017).

USEPA (2017c). National pollutant discharge elimination system (NPDES). https://www.epa.gov/npdes (accessed 10 April 2017).

USEPA (2017d). Program overview: total maximum daily loads (TMDL). https://www.epa.gov/tmdl/program-overview-total-maximum-daily-loads-tmdl (accessed 10 April 2017).

USEPA (2017e). Underground injection control regulations and safe drinking water act provisions. Environmental Protection Agency. https://www.epa.gov/uic (accessed 7 March 2025).

USEPA (2017f). Brownfields all appropriate inquiries. US Environmental Protection Agency. https://www.epa.gov/brownfields/brownfields-all-appropriate-inquiries (accessed 10 April 2017).

USEPA (2018). Clean air act title IV – noise pollution. US Environmental Protection Agency. https://www.epa.gov/clean-air-act-overview/clean-air-act-title-iv-noise-pollution (accessed 25 July 2018).

USEPA (2023). National overview: facts and figures on materials, wastes and recycling. https://www.epa.gov/facts-and-figures-about-materials-waste-and-recycling/national-overview-facts-and-figures-materials (accessed 13 February 2024).

USEPA (2024). Superfund enforcement FY 2023 annual results. US Environmental Protection Agency. https://www.epa.gov/enforcement/superfund-enforcement-fy-2023-annual-results (accessed 6 October 2024).

USNRC (2017). Radioactive waste. United States Nuclear Regulatory Commission. https://www.nrc.gov/waste.html (accessed 10 April 2017).

Vaughn, J. (2011). *Environmental Politics: Domestic and Global Dimensions*, 6e. Boston, MA: Wadsworth, Cengage Learning.

Verburg, P.H. (2014). The representation of human-environment interactions in land change research and modelling. In: *Understanding Society and Natural Resources*, Chapter 8 (ed. M.J. Manfredo, J.J. Vaske, and E.A. Duke), 161–177. Dordrecht: Springer. ISBN: 978-94-017-8958-5.

WHO (2023). Malaria. World Health Organization. https://www.who.int/news-room/fact-sheets/detail/malaria. (accessed 10 October 2024).

3

Social, Economic, and Legal Issues

Chapter Goal and Objectives
This chapter aims to enhance engineering professionals' understanding of the social, economic, and legal aspects of sustainability by examining how social systems interact with environmental and economic dimensions to promote equitable and just development. The specific objectives are to (i) explore challenges in both developed and developing societies, highlight the importance of social indicators for measuring well-being and equity, and introduce frameworks for assessing social impact and vulnerability; (ii) cover economic aspects, focusing on the economic assessment framework, Life Cycle Costing, and True-cost Accounting to evaluate long-term sustainability; and (iii) address key legal and governance issues, and the growing role of Environmental, Social, and Governance (ESG) reporting in shaping sustainable business practices.

3.1 Introduction

As presented in Chapter 1, the concepts of sustainability and sustainable/resilient development transcend the idea of "environmentalism." In addition to environmental protection and stewardship, the concept of sustainability incorporates the notion of equity, including social, economic, and legal aspects. From an economic standpoint, sustainability is usually framed with respect to growth, efficiency, and resource use, whereas the social framework often focuses on human needs, including social equity, empowerment, social cohesion, and cultural identity.

The global concept of sustainability defines sustainable solutions as those that maximize the positive impacts of the three dimensions of the triple bottom line (TBL) – environmental, social, and economic. The concept of the TBL has continued to evolve into a valuable framework for the assessment of a project or an activity. More sustainable solutions lead to a greater intersection of these metrics, and as sustainable solutions are increasingly desired, it is important to understand the social, economic, and legal issues that become relevant when considering a project or an activity. The pursuit of resilient infrastructure incorporates an increasing focus on the social vulnerability of residents, neighborhoods, and populations. In this chapter, the social and economic issues associated with sustainability are presented and discussed. Furthermore, the importance of legal issues cannot be underestimated due to their influence on economic activities and social organization. A brief outline of these issues is presented in this chapter.

Sustainable and Resilient Engineering: Drivers, Metrics, Tools, and Applications, Second Edition. Krishna R. Reddy, Claudio Cameselle, and Jeffrey A. Adams.

Companion website: www.wiley.com/go/reddy/sustainableresilientengineering2e

3.2 Social Issues

Regardless of the precise definition that one may choose or prefer for sustainability, each definition, in some way, includes the necessity to consider the needs of future generations. Further, while considering the needs of future generations, it is important to consider the effects of actions on local society as well as how these actions may affect people across the globe. The effects of these actions, both positive and negative, are reflected in the social aspects of the triple bottom line.

Let us now examine the social dimension of the triple bottom line.

3.2.1 Society

Society is a social system that reproduces the biological and cultural aspects of humankind within a territory. In addition to the reproduction of a human population, societies aim to create biological and physical entities that provide utility, are assigned some form of ownership, and may be produced or reproduced. Crop cultivation and the raising of livestock, man-made infrastructure, and industrial output are three examples of these entities.

Considered on a larger scale, we can divide social systems into these entities (stocks) and actions undertaken by these entities or stimuli imparted to these entities (denoted as flows in Figure 3.1). Some examples of stocks include humans themselves, the output of human-directed activity (crops/livestock, industrial output, infrastructure), and physical territory. Flows include reproduction, migration, labor activity, energy inputs and outputs, material inputs and outputs, appropriation of net primary productivity (NPP), and agricultural harvesting/culling.

The structure that a particular society creates in order to govern itself (i.e. creates rules and norms, enacts and enforces rules and norms, and interprets rules and norms) is an important aspect of a society. Assuming a particular society is able to impose its collective free will and choose a system of its own particular liking (something that, unfortunately, the majority of peoples over the millennia have not been able to do), the respective system is often a reflection of the importance of a range of attributes emphasized by that society. As discussed in Chapter 1, different nation-states throughout history have placed a wide range of emphasis on economic systems, from centrally planned to free markets; the type or inspiration of a legal framework, often based on a particular religion, a series of cultural norms, or historical precedent; and the degree central government strength/control, from libertarian to authoritarian approaches. These different economic and

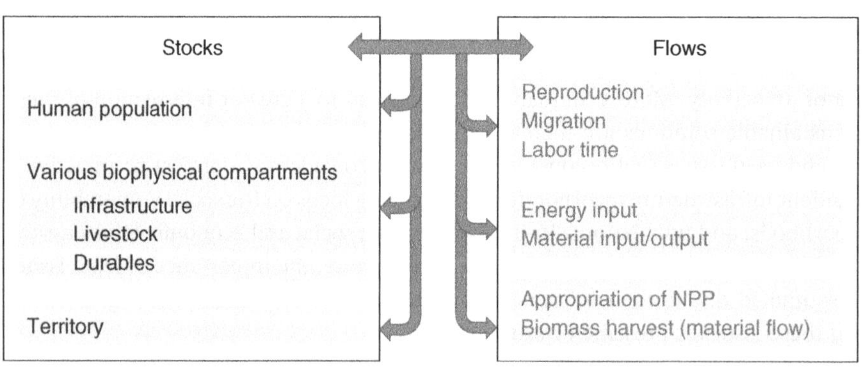

Figure 3.1 Relevant biophysical dimensions of social systems.

governance models have an enormous influence on society's development and structure, as well as on the links and relationships among society members.

3.2.2 Developed and Developing Societies

The characteristics of a society (i.e. organization, equality, and liberty) may vary depending on the wealth and standard of living of each country. The access to resources and opportunities, public services, and well-being can be very different depending on the level of development of the country and the available resources and services for its citizens. Within the frame of reference of social sustainability, a compelling aspect is how societies evolve and deteriorate under stress. In general, wealthier societies are more stable societies, whereas poorer societies struggle to provide the necessary inputs to ensure a stable society. Further, the degree of stability may be influenced by the range of wealth distribution equity or inequity.

In discussing these differences, it is important to distinguish between more (or most) developed countries (MDCs) versus less (or least) developed countries (LDCs). An MDC is typically associated with highly developed economies, advanced technological infrastructure, and relatively high standards of living. LDCs, on the other hand, are typically associated with low standards of living, undeveloped or underdeveloped economies, and underdeveloped industrial bases.

LDCs often serve as lower-cost sources of labor for international companies (often based in MDCs) and attract lower-value industries because of their underdeveloped economies and relatively less skilled workforce. Unfortunately, these industries are often significant contributors to environmental pollution. LDCs harbor an increasing percentage of the world's industrial facilities, leading to increased pollution (Wang et al. 2021). An example of this is the textile industry in Tirupur, India. This region had traditionally served as a center of textile fabrication, which included 4000 independent small business units (Arumugam et al. 2020). Until the 1980s, many of these businesses manufactured white cotton undershirts ("banians") commonly worn in India. However, a shift in production occurred in response to lucrative international markets associated with colored T-shirts. The focus of the textile industry in Tirupur shifted, and industry participants began to incorporate significant dying operations. This, combined with an appreciable increase in emissions associated with the increase of production to meet the demand of the external market, resulted in a significant increase in pollution related to this industry. Salt used in the dying process resulted in the generation of highly saline wastewater laden with excess dyes and other chemicals. The disposal of this wastewater further impacted water sources and the environment in general. Specifically, the uncontrolled discharge of wastewater contaminated the groundwater and made it unusable for the local community for domestic use.

When analyzing sustainable engineering-related issues associated with LDCs, it is important to remember that the relative degree of achieving sustainable outcomes will be largely based on the activities undertaken in the developing world. LDCs hold large percentages of non-renewable resources but often consume relatively small amounts. However, these resources are increasingly utilized, especially in growing industries located in LDCs.

Fortunately, LDCs do hold significant advantages when compared to the humble industrialization beginnings of most MDCs. For instance, LDCs are starting at a far more advanced technological "starting point" as compared to where many MDCs started. LDCs can learn from the development experiences, both good and bad, exhibited by MDCs in the past. For instance, with respect to energy utilization and infrastructure, LDCs may opt to avoid the incorporation and utilization of fossil fuels, instead choosing to utilize low-emission or emission-free energy sources, such as wind or solar energy sources. Since LDCs have not developed an underlying infrastructure to an appreciable extent, they have the opportunity to "do it right the first time" and avoid the past mistakes or regrets experienced by MDCs (Michaelowa and Sippel 2022).

3.2.3 Social Sustainability Concept

Sustainability is associated with the preservation of resources for the benefit of future generations; thus, it is common to refer to sustainability primarily in terms of natural resources, their economic and environmental benefits and drawbacks, and the impact of present actions on future generations. However, there is also an increased focus on the concept of social resources as well as access to these within the same generation or across generational lines. Some of these social resources include culture and cultural resources, human rights, labor rights, and corporate governance. The concept of available access to these within and among generations is referred to as "social sustainability" (Ruggerio 2021). With respect to human development, the environment and natural resources may be used to achieve better standards of living. Income and social resources are also important factors in achieving quality of life and well-being.

As the key issues that affect society and social sustainability can vary significantly, there is no straightforward definition of social sustainability. The Western Australia Council of Social Services (WACOSS) defines social sustainability as the following (McKenzie 2004):

> *Social sustainability occurs when the formal and informal processes, systems, structures, and relationships actively support the capacity of current and future generations to create healthy and livable communities. Socially sustainable communities are equitable, diverse, connected and democratic and provide a good quality of life.*

WACOSS has outlined several dimensions associated with social sustainability. These include the following (WACOSS 2013):

- *Equity* – The community provides equitable opportunities and outcomes for all its members, especially the poor and the vulnerable.
- *Diversity* – The community promotes and encourages diversity.
- *Social cohesion/interconnection* – The community provides processes, systems, and structures that promote connectedness within and outside of the community at the formal, informal, and institutional level.
- *Quality of life* – The community ensures that basic needs are met and fosters a good quality of life for all its members (e.g. health, housing, education, employment, and safety).
- *Democracy and governance* – The community provides democratic processes and open and accountable governance structures.
- *Maturity* – The individual accepts responsibility for consistent growth and improvement through broader social attributes (e.g. communication styles, behavioral patterns, indirect education, and philosophical explorations).

3.2.4 Social Indicators

The evolution of a society toward more sustainable conditions, or the level of social sustainability, can be measured through several important social issues that can be considered as the indicators of the status of a society. These issues or indicators can give a quantitative or qualitative measure of the detrimental effect on society and the ability of societies to achieve social well-being (Table 3.1).

Table 3.1 Global social issues that affect the sustainability of societies.

Poverty
Exploitation
Deteriorated physical and mental health
Lack of essential services
Civil unrest
Crime
Depression

Poverty is defined as the lack of sufficient wealth to meet basic human needs. It is caused by a lack of essential resources, lack of education, political and societal oppression, and similar factors. Often, poverty results in physical and mental health deterioration because of a lack of resources to acquire proper or enough nutrition and health resources. Ultimately, poverty can result in death, either directly or indirectly, and can exacerbate civil unrest and crime.

In its most basic definition, exploitation is the unfair treatment of people, often to achieve a physical, economic, or social benefit. It occurs when social, financial, or physical power is used to force less empowered members of society to serve the needs or desires of more powerful members of society to the detriment of themselves. Often, it exacerbates the differences between these classes, as the exploiting classes amass more wealth and power, and the exploited classes become weaker and poorer. It is often manifested by the physical exploitation of people as labor or through financial exploitation. If left unchecked or unabated, it can lead to civil unrest.

Another key issue is mental/physical health. When a society is in a stressed or depressed condition, it can lead to widespread deterioration of individual mental and physical health. Some of the causes include inadequate nutrition, exercise, and/or recreation, increased pollution, physical strain and injury, stress, lack of mental stimulation, and lack of purpose or meaningful contribution opportunities. Both physical health and mental health are directly connected to the physical environment, and effects on the physical environment can transcend and affect the overall health. When mental and physical health is adversely affected, it can lead to reduced productivity, a strong effect on quality of life, and a strong enhancement of mortality factors.

One manifestation of diminished mental health is depression. This is typically associated with feelings of unhappiness, hopelessness, powerlessness, and lack of self-worth. It is often rooted in poor treatment, poor health, circumstances of poverty, exploitation, and isolation. It can result in reduced societal contributions and diminished quality of life.

Mental and physical health can also drastically suffer when there is a lack of essential services or a lack of access to such services. In many cases, this access is adversely affected when there are insufficient systems in place to provide these services. This diminished access can result from poverty, ineffective or corrupt governance, lack of community empathy or responsibility, crime, natural disasters, or other natural or man-made barriers. If the basic needs of the community are not met, significant stress will be placed on society, which could lead to reduced economic output, diminished physical and mental health, and an acceleration of crime, poverty, and other ill effects on the society.

When poverty, corruption, and a lack of essential services exist, it can lead to civil unrest. These reactions are manifested in rebellion and mob/mass action of citizens against the systems and organizations that are in place. Civil unrest is often rooted in feelings of resentment toward poor management, neglect, and exploitation, often at the hands of government institutions. Civil unrest can lead to disruptions of already stressed services and infrastructure and can lead to significant changes or replacement (sometimes for the better; other times for the worse) of the organizations that are perceived to be inadequately meeting the needs of the society.

3.2.5 Social Impact and Social Vulnerability Assessment Framework

Projects and other interventions, usually associated with infrastructure and industrial activities, may have an important influence on the surrounding communities. A Social Impact Assessment (SIA) may analyze and measure the impact of such interventions in the community. The International Association of Impact Assessment (IAIA) states that a SIA includes the processes of analyzing, monitoring, and managing the intended and unintended social consequences, both positive and negative, of planned interventions (policies, programs, plans, and projects) and any social change processes invoked by those interventions. Its primary purpose is to bring about a more sustainable and equitable biophysical and human environment.

The SIA is a method to evaluate the impact of a project or activity in the social community, in a way similar to the Environmental Impact Assessment (EIA). The EIA was first developed in the United States in 1970s, and the SIA has been developed to give a response to the increasing demands of society about the impact of different interventions in social and biophysical media (Bryan 2021).

SIAs are performed by professionals with diverse backgrounds, including expertise in sociology, anthropology, geography, development studies, planning, and other fields. They are often carried out with EIAs, but they have not been as widely adopted, and often play a minor role in EIAs. Nevertheless, the present tendency is to incorporate the SIA with the same weight of EIA in an Environmental and Social Impact Assessment (ESIA) (Bryan 2021). Several frameworks have been developed on a national basis; we will take a brief look at the framework commonly used in the United States.

The US framework includes several key aspects as follows:

- Achieve an extensive understanding of local and regional settings to be affected by the action or policy
  - Identify interested/affected stakeholders and other parties
  - Develop baseline profiles of local and regional communities
- Focus on key elements of the human environment
  - Identify the key social and cultural issues that could be affected or related to the action/policy
  - Select social and cultural variables that measure and explain the issues identified
- Identify research methods, assumptions, and significance
  - Research methods should be holistic in scope
  - Research methods must describe cumulative social effects
  - Ensure methods and assumptions are transparent and replicable
  - Select appropriate forms and levels of data collection and analysis
- Provide quality information for use in decision-making
  - Collect sufficient and useful quantitative and qualitative data
  - Ensure methods are scientifically robust
  - Ensure the integrity of the collected data
- Ensure that any environmental justice issues are fully described and analyzed
  - Ensure methods consider underrepresented and vulnerable stakeholders and populations
  - Consider the distribution of all impacts to different social groups

- Undertake evaluation, monitoring, and mitigation
  - Establish mechanisms for evaluation and monitoring of the action, policy, or program
  - Provide an effective mitigation plan where needed
  - Identify data gaps and plan for filling these needs

Figure 3.2 and Table 3.2 illustrate a flowchart and a matrix, respectively, associated with this SIA framework. Examples of key SIA indicators are as follows:

- *Population characteristics* – Present and expected population, ethnic/racial diversity, demographic mix, and seasonal variations.
- *Community and institutional structures* – Local government and connections to the larger system, patterns of employment and economic diversification, voluntary organizations, and religious and other interest organizations.
- *Political and social resources* – Distribution of power/authority, public participation, income/wealth distribution, and legal and civil rights.
- *Individual and family* – Health, education, safety, and friendship networks.
- *Community resources* – Natural resources/land use, physical environment, recreation, availability of housing and services, viability of community life, and historical/cultural resources.

When compiling and quantifying these indicators, a ranking system may be applied to normalize the quantification process. Scaled rankings may be applied to both positive and negative impacts; these are often based on the number of people or percentage of the population that may be affected.

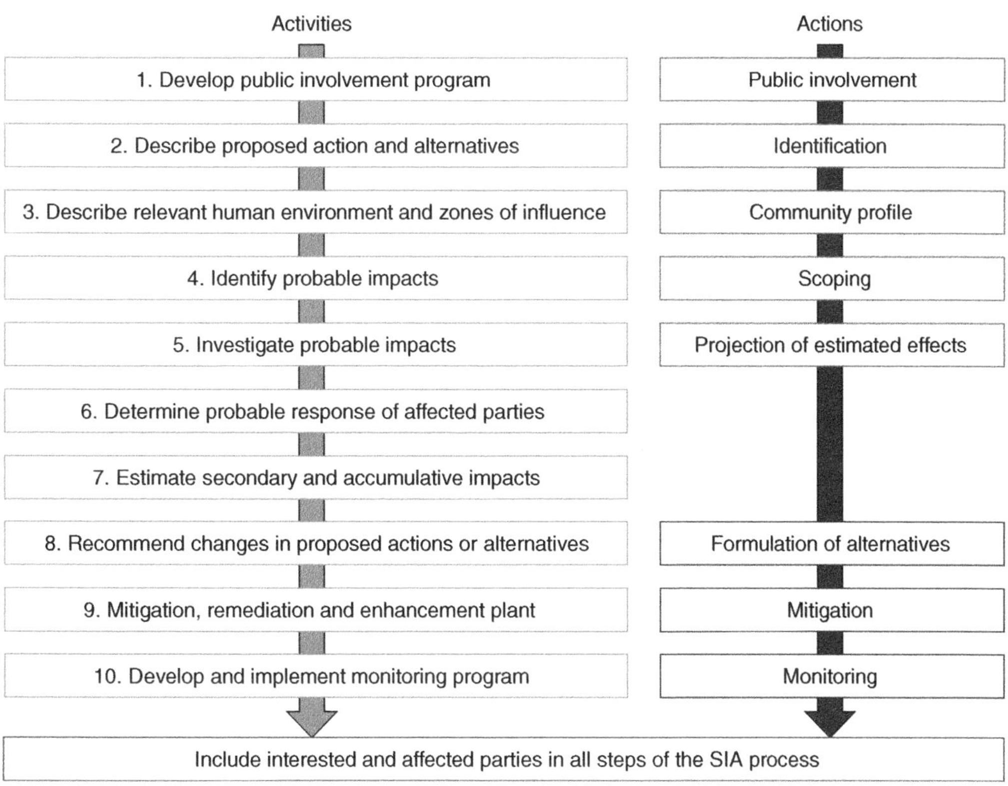

Figure 3.2 Flowchart for social impact assessment process.

Table 3.2 Matrix of social impact assessment variables.

Social impact assessment variables	General planning, policy development, and preliminary assessment	Detailed planning, funding, and impact assessment	Construction implement	Operation/ maintenance	Decommission/ abandonment
Population change					
Population size, density, and change					
Ethnic and racial composition and change					
Relocating people					
Inflow/outflow of temporaries					
Presence of seasonal residents					
Community and institutional structures					
Voluntary associations					
Interest group activity					
Size and structure of local government					
Historical experience with change					
Employment/ income characteristics					
Employment equity of disadvantaged groups					
Local/regional/ national linkages					
Industrial/ commercial diversity					
Presence of planning and zoning					
Political and social resources					
Distribution of power and authority					

Table 3.2 (Continued)

Social impact assessment variables	General planning, policy development, and preliminary assessment	Detailed planning, funding, and impact assessment	Construction implement	Operation/ maintenance	Decommission/ abandonment
Conflict newcomers and old-timers					
Identification of stakeholders					
Interested and affected parties					
Leadership capability and characteristics					
Interorganizational cooperation					
Community and family changes					
Perception of risk, health and safety					
Displacement/ relocation concerns					
Trust in political and social institutions					
Residential stability					
Density and acquaintanceships					
Attitudes toward proposed action					
Family and friendship networks					
Concerns about social well-being					
Community resources					
Change in community infrastructure					
Indigenous populations					
Changing land use pattern					
Effects on cultural, historical, sacred and archaeological resources					

The rankings can be applied to individual subgroups or the community, and the scores may be summed as appropriate.

The quantification and measurement of positive and negative attributes remain to be a work in progress; however, it is important to incorporate it on an objective basis to the maximum degree practicable. Moreover, when negative, substantive impacts should be identified, and mitigation efforts should be made to minimize or avoid the impact.

It is important to note that quantification is not a goal in and of itself – it is meant to be a process where a comparison and assessment can be made to allow for informed decisions about project design, implementation, and mitigation as necessary.

The quantification of the variables identified in the SIA can be based on a ranking system considering the amount or fraction of people that it is affected, positively or negatively, by a specific action or activity. Thus, for example, when evaluating the social costs and benefits of a specific project or action, a value from −4 to 0 or 0 to 4 can be assigned to the variables based on the number of people affected by that action (Table 3.3).

The social impacts can be assessed and quantified for individuals or small groups and the whole community separately. Thus, the SIA may identify what groups or individuals are more affected by an action or a project and, in turn, analyze its influence on the whole community. The social benefits for the individuals and the community can be added, and then the costs can be subtracted to get an overall score for each action, project, or alternative. Thus, the SIA may decide between different alternatives based on the benefits/costs for the community, or the SIA may even be used to propose modifications to a project or an action to reduce those negative impacts on the community or individuals. In the case of a social impact that is irreversible or of substantial significance, no matter how small the affected group is, those negative impacts must be highlighted in the final assessment, and such impacts should be avoided with alternative solutions or modifications to the project. The main limitation of this quantification and assessment method is associated with the subjectivity of many social variables or impacts that may change the final scoring of the SIA over a wide range. Regardless, it is necessary to investigate social sustainability for the completeness of a TBL assessment.

Table 3.3 Ranking system for quantification of variables in SIA based on the fraction of people affected.

To assess social costs: rank −4–0	
0	No or minimal disadvantage to individuals or communities
−1	Significant disadvantage to a small group of individuals or minimal disadvantage to <5% of population
−2	Disadvantage to 5–20% of population
−3	Disadvantage to 20–50% of population
−4	Disadvantage to >50% of population
To assess social benefits: rank 0–4	
0	Minimal or no advantage to individuals or communities
1	Significant advantage to a small group of individuals or minimal advantage to <5% of population
2	Advantage to 5–20% of population
3	Advantage to 20–50% of population
4	Advantage to >50% of population

3.2.6 Social Sustainability Implementation

Social sustainability attributes are necessarily implemented at the project/activity level. In designing and implementing projects, it is important to incorporate and respond to the input of the stakeholders in the local community, consider local opinions and contexts in design aspects, and facilitate designs that promote social inclusion, cohesion, and accountability. Citizen participation, engagement, and empowerment are key facets since the populace is always a key stakeholder, and acceptance of a project requires their emotional investment. To achieve this, it is important to perform adequate social analyses, which in turn allows for the identification of social opportunities as well as mitigation of social impacts and risks. When designing projects, it is important to understand the social environment as much as the physical environment, and pursue the goals of social sustainability along with economic and environmental sustainability.

3.2.7 Equity

When considering the social dimensions of sustainability, stakeholders rarely seek or affect equal outcomes. Expecting equality would violate the other dimensions of sustainability, namely financial and environmental. Disparities and variations have and will exist with respect to economic participation and outcomes for participants, and natural environmental as well as anthropogenic activities and related stressors will occur. Just as the Pacific coastline is present in California and not in Iowa, anthropogenic activities and uses that impact the environment have to go somewhere, even as sustainability practitioners act to minimize deleterious effects.

While we cannot induce or expect equal outcomes, we can act to bring about *equitable* outcomes. Equality is defined as having the same resources or opportunities. Of course, there are expectations of equality – equal rights to vote, job opportunities for qualified individuals, etc. Yet, we focus on equity or equitable outcomes. Equity is, conversely, focused on outcomes. Equity is defined as a fair or impartial circumstance – while the same resources or opportunities may not be provided, we act to provide the correct resources or opportunities for a fair or equal outcome (Milken Institute School of Public Health 2020).

3.2.8 Environmental Justice

For decades, across communities and nations around the world, people with lower incomes and power, often within disadvantaged communities measured across any number of dimensions, have been subjected to an outsized burden of environmental stressors. Whether due to natural phenomena (i.e. low-lying areas subject to flood risk) or anthropogenic activities with noted deleterious side effects (e.g. transportation corridors, such as railroads or highways, industrial uses, such as refineries and factories, or agricultural uses, such as concentrated animal feed operations (CAFOs)), the intentional siting of these facilities, or the development of lower-income communities subjected to these stressors is a common situation. It is reasonable to pose a "chicken or the egg" question – do disadvantaged communities develop in less desirable places because of low land values, or do less desirable land uses emerge near disadvantaged communities due to favorable economics, less political resistance, or racism? The answer to both causes is "yes."

When we frame this undue burden on disadvantaged communities, whether due to environmental racism or other causes, there is a growing movement in the United States to abate these occurrences and/or undo the decisions and the resulting effects of the past. This has generated a growing interest among populations, policymakers, businesses, governments, and local and national stakeholders in environmental justice. Environmental justice (EJ) is the fair treatment and meaningful involvement of all people regardless of race, color, national origin, or income with respect to the development, implementation and enforcement of environmental laws, regulations and policies (USEPA 2024).

Environmental justice is typically organized around three dimensions (Carley and Konisky 2020).

Distributional justice focuses on the distribution of both burden and benefit across population groups, with a stated objective of ensuring populations have equal access to benefits while specific groups do not bear an outsized share of burden.

Procedural justice aims to provide access to those who want to participate in decision-making processes, providing a procedural process that is fair, equitable, and inclusive to all who choose to participate.

Recognition justice focuses on the understanding and acknowledgement of historic and ongoing inequalities borne by populations, prescribing approaches that seek to reconcile these inequalities (Figure 3.3).

Some scholars describe a fourth pillar, *restorative justice*, in which advocates support government action or other interventions to either avoid distributional, recognitional, or procedural injustices or to alleviate or correct past injustices.

As an example of environmental distributional injustice, consider the effect of CO_2 emissions and increasing global temperatures. High-income countries have historically generated the majority of emissions, but atmospheric CO_2 concentrations are essentially constant around the world. As a result, everyone experiences the same rise in CO_2 atmospheric concentrations, and the most severe physical manifestations of climate change, such as sea level rise, flooding, extreme precipitation, and droughts, may be experienced in areas that contribute relatively little to total emissions and atmospheric CO_2 concentrations. Figure 3.4a shows that developed countries are responsible for

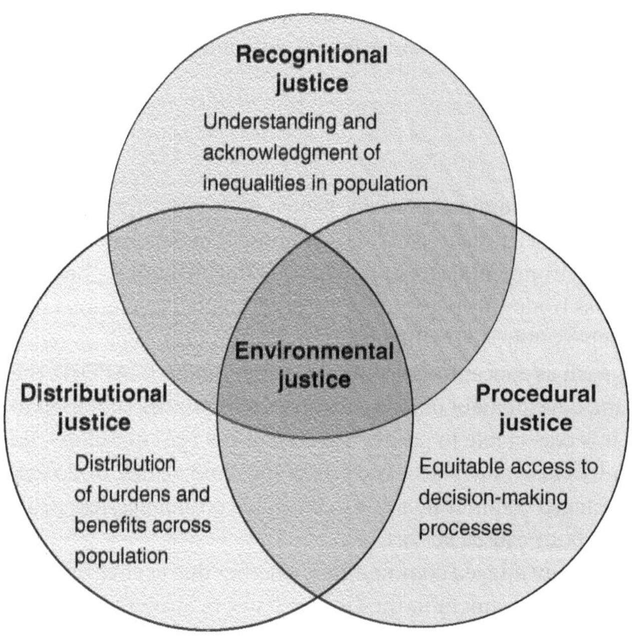

Figure 3.3 The three dimensions of environmental justice.

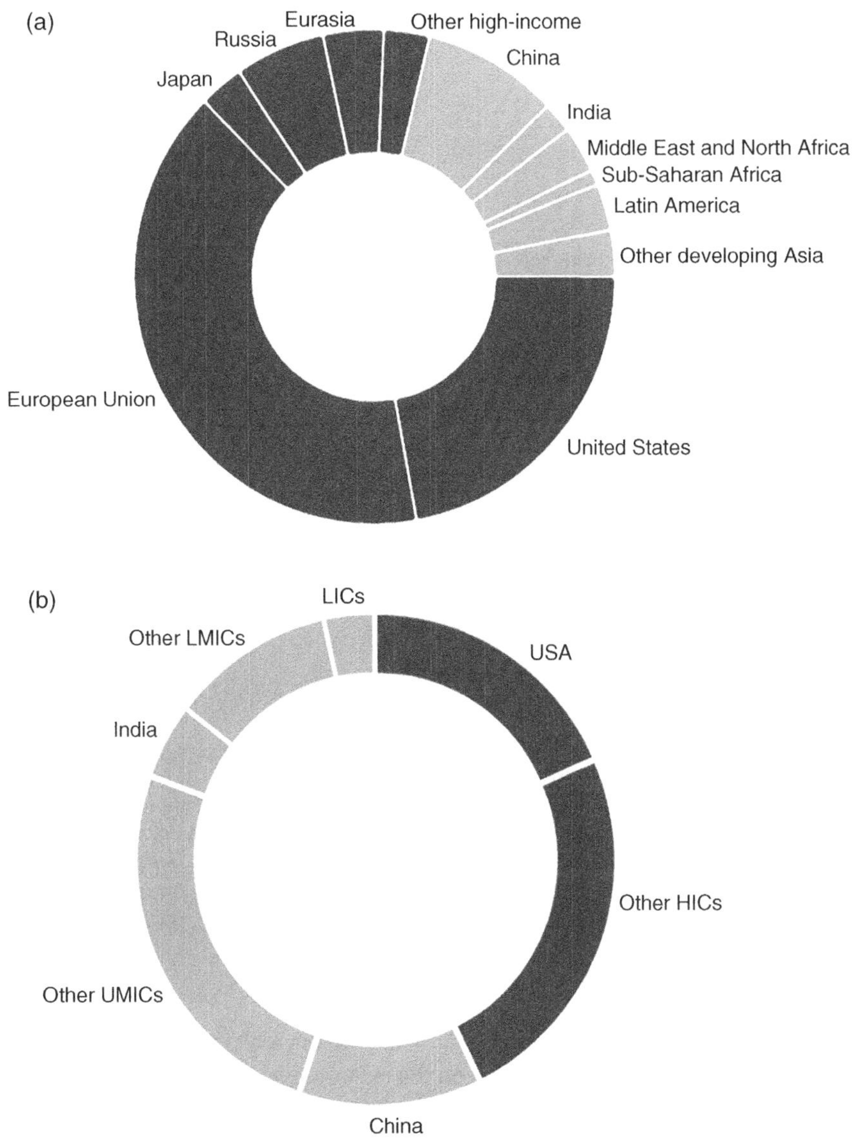

Figure 3.4 Who caused climate change historically? (a) GHG emissions 1850–2011, (b) GHG emissions 1850–2022. *Source:* CGD 2024.

79% of the GHG emissions while other emerging economies and less developed countries contributed to a minor extent. As a result, many emerging-economy and lower-income countries experience climate injustice due to greater physical sensitivity to the impacts of the climate change as compared to high-income countries, even though the former are not responsible for the legacy GHG emissions (Figure 3.5). This situation has been evolving over time and now developed countries contributed only 43% to the GHG emissions due to the increasing emissions of emerging economies (Figure 3.4b).

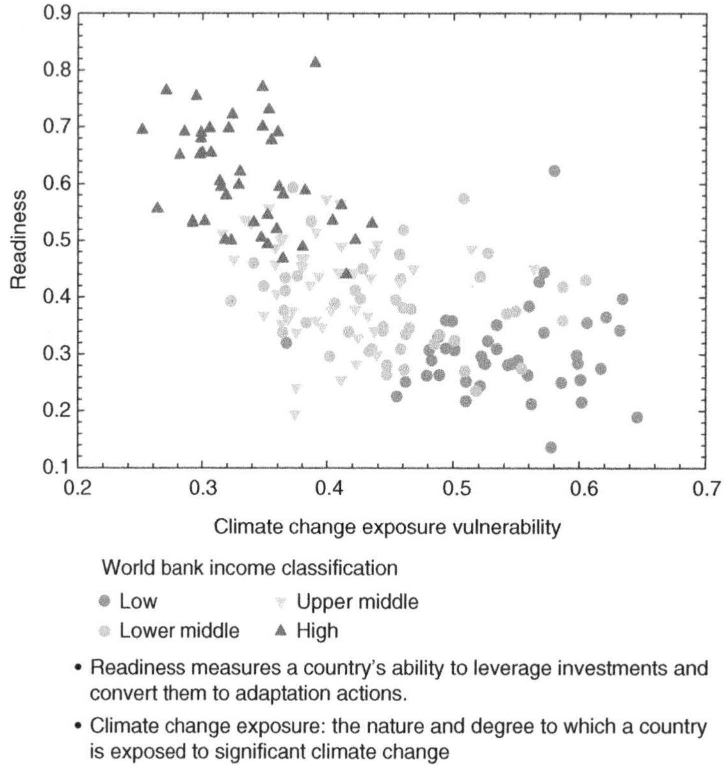

Figure 3.5 Unequal costs of climate change. Poorer countries face greater risks for climate change and are less able to adapt. *Source:* Adapted from Notre Dame Global Adaptation Initiative (ND-GAIN) University of Notre Dame, Scores for 2022.

3.3 Economic Issues

Many economic activities, such as mining, industry, and transportation, rely upon the utilization of natural resources, and they have an enormous impact on the environment and on the future availability of those resources. Thus, any discussion about sustainability must include an economic context. When considering the economic aspects of sustainability, it is important to consider a multidimensional economic analysis of a particular activity or set of activities that accounts for ecological and environmental effects. Additionally, social, cultural, and public health-related aspects should also be incorporated.

There are different ways in which the economic context can be defined. One model offers two distinctive economic disciplines: the *physical economy*, defined as activities in which tangible natural resources act as a limit to activity, and the *money economy*, in which activities are defined in monetary means and are not constrained by resources or physical laws. Figure 3.6 depicts a schematic of capital flow within an economy, in which investments are made in productive capacity, which in turn yield beneficial output. However, the important question to consider is how to measure economic output from a sustainability perspective? There are traditional methods – cost–benefit analyses, either considering a payback method or using a discount rate. However, it is difficult to incorporate these approaches in environmental or social issues. Alternatively, other

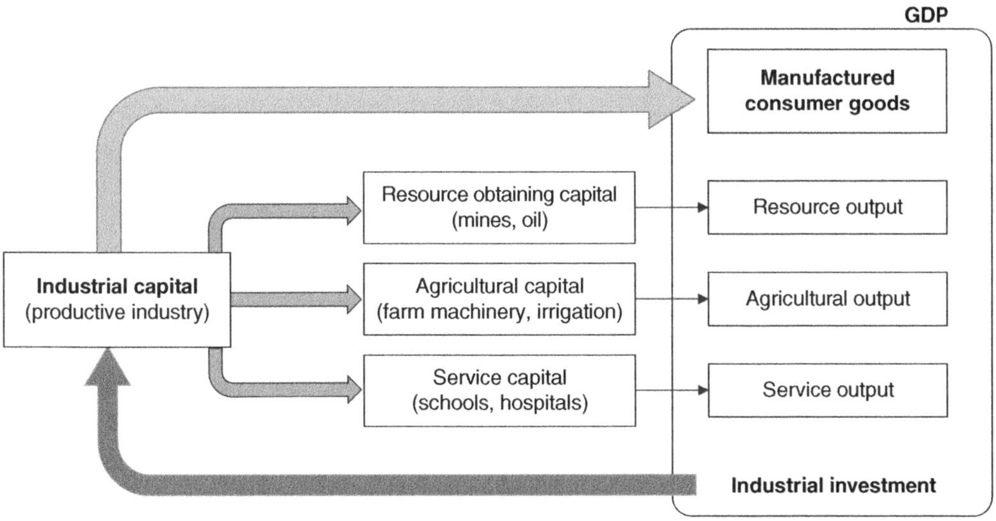

Figure 3.6 Capital flow in the productive economy.

methods, such as the human capital method, the cost of illness method, the preventive/mitigative expenditure method, the wage differential method, the contingent valuation method, and the surrogate actions method, may be able to incorporate different human effects with varying degrees of precision (Dixon et al. 2013). While all of these methods try to assign dollar values to the aforementioned issues, they can be complicated, have inherent drawbacks, and should be applied with caution.

3.3.1 Economic Assessment Framework

To provide a robust assessment framework for environmental issues, "green accounting" and "environmental accounting" methods have been developed. The purpose is to accurately estimate environmental costs associated with economic activity. Some additional factors may include the cost of raw materials, manufacturing costs, costs associated with resource use, waste and scrap generation, and product design. Environmental costs are assigned to the associated activity or activities with the purpose of managing these costs using traditional cost-benefit analyses.

A multi-tier environmental costing framework can be used – it consists of five tiers associated with the costs and benefits of economic activity (Langfield-Smith et al. 2015).

- Tier I costs are those that are normally captured in traditional economic engineering applications and include the following: capital equipment, materials, labor, supplies, utilities, structures, and salvage value.
- Tier II costs incorporate environmental costs that are commonly applied to overhead, including the following: off-site waste management charges; waste treatment equipment and operating expenses; permit costs; sampling and analysis of waste streams; waste/emissions inventory assessments; documentation and inspection of waste storage and handling; emergency response plans; documentation of chemical usage; development, implementation, and reporting associated with pollution prevention plans and related activities.
- Tier III incorporate environmental liability costs, including the following: compliance obligations; remediation obligations; fines and penalties; obligations to compensate private parties

for personal injury, property damage, and economic loss; punitive damages; and natural resource damages. Tier III costs depend on the probability that an event will occur, the costs associated with the event, and when the event occurs.

- Tier IV consists of costs and benefits (internal to a company) associated with improved environmental performance.
- Tier V consists of costs and benefits (external to a company) associated with improved environmental performance.

Example 3.1

A small manufacturing company is considering purchasing a new piece of machinery to increase its production capacity. The management is conducting a cost-benefit analysis to decide whether the investment is viable.

Costs:

- The new machine costs $150,000.
- Installation costs are estimated to be $20,000.
- Operating costs are expected to increase by $5,000 per year due to the new machine.
- The machine has a useful life of 10 years, after which it will have no salvage value.

Benefits:

- The machine is expected to increase annual revenue by $40,000 due to increased production capacity.
- Efficiency savings from the new machine are estimated to reduce other production costs by $10,000 per year.

Additional Information:

- The company uses a discount rate of 8% for its investments.
- Ignore taxes and depreciation for this analysis.

Questions:

1) Calculate the Net Present Value (NPV) of the investment.
2) Determine the Internal Rate of Return (IRR) and compare it with the company's discount rate.
3) What is the Payback Period for the investment?

Solution:

1) Net Present Value (NPV)

The formula for NPV is:

$$NPV = \sum_{1}^{t} \frac{Cash\ Flow_t}{(1+r)^t} - Initial\ Investment$$

where:

t is the year (1–10)
r is the discount rate (8% or 0.08)
Cash Flow = Revenue Increase + Efficiency Savings − Operating Cost Increase
Initial Investment = Machine Cost + Installation Cost = $150,000 + $20,000 = $170,000

Annual Net Cash Flow Calculation:
\$40,000 (increased revenue) + \$10,000 (efficiency savings) – \$5,000 (additional operating cost) =\$45,000
NPV Calculation:

$$NPV = \sum_{t=1}^{10} \frac{45{,}000}{(1 + 0.08)^t} - 170{,}000$$

Calculating this sum, the Net Present Value (NPV) of the investment is \$131,954
Since the NPV is positive, the investment is financially viable.

2) Internal Rate of Return (IRR)

The IRR is the discount rate that makes the NPV equal to zero:

$$0 = \sum_{t=1}^{10} \frac{45{,}000}{(1 + IRR)^t} - 170{,}000$$

Using iteration, we solve for IRR, and the result is IRR = 23.18%.
Since the internal rate of return is significantly greater than the company's discount rate, the investment is financially attractive.

3) Payback Period

The Payback Period is how long it takes for the cumulative cash flows to recover the initial investment.
Annual Net Cash Flow is \$45,000
The Initial Investment is \$170,000
Payback Period = Initial Investment/Annual Net Cash Flow = \$170,000/\$45,000 = 3.78 years

3.3.2 Life Cycle Costing

Life cycle analysis (LCA), which will be presented in greater detail in Chapter 9, is a rapidly evolving framework in which the costs and benefits associated with an economic activity are assessed over the full life of a product or an activity. According to the International Organization for Standardization (ISO), LCA is the "compilation and evaluation of inputs, outputs, and the potential environmental impacts of a product system throughout its life cycle" (ISO 2006). Further, economic and social aspects are, typically, outside the scope of the LCA. Other tools may be combined with LCA for more extensive assessments.

Two extensions of LCA include life cycle costing (LCC) to account for the economic dimension, and accounting for the work environment and employment to incorporate the social dimension (Toniolo et al. 2020). For LCC, the costs induced by a product, both indirect and direct, in public and private activities, are accounted. It is important to determine proper cost categories, cost bearers, cost models, and cost aggregations.

Cost categories are grouped according to economics (budget costs, market costs, social costs, and alternative costs); life cycle stages (research and development, primary production, manufacturing, use, and disposal); and activity types (design, transport, sales, and manufacturing). Cost bearers include producers, supply chain participants/activities, owners, users, regional/national society, and global society.

Several traditional cost models may be used. Some include net present value (NPV), average yearly cost, steady-state cost, annuity, payback time, and cost–benefit ratio. The discount rate used in these analyses is also key – it may be statutorily assigned, market-based (CPI, CCI, Libor), or a specific company's internal preference. Additionally, the appropriate time horizon needs to be determined.

3.3.3 True-cost Accounting

Another approach gaining popularity is true-cost accounting (TCA), a method that seeks to calculate the social, environmental, and economic costs and benefits of a product over its lifespan instead of focusing solely on the cost of its initial purchase (Baker et al. 2020). TCA tries to accomplish the following:

- *Goal definition and scoping* – Identifies and defines the project and purpose of the assessment.
- *Streamlining the analysis* – Refines the first step by connecting the objectives and other elements to sustainability metrics and impact categories.
- *Identify potential risks* – Evaluates the relative importance of the impact categories and the current feasibility of expressing the costs for each attribute of the project.

To perform the TCA, the Tier I through V costs are assigned, and an impact assessment is conducted to determine the largest contributory costs for each category and the information is then used for the decision-making.

Example 3.2
The following table gives the economic values of ecosystem services.

	($/ha/yr)				
Item	**Total**	**Supply service**	**Adjustment service**	**Habitat service**	**Cultural service**
Open ocean	9	0	7	2	1
Coral reef	206,873	20,078	186,795	0	0
Coast	77,907	1,453	76,144	164	146
Coastal wetland	960	0	960	0	0
Inland wetland	282	167	115	0	0
Rivers and lakes	812	3	129	681	0
Tropical rainforest	29	0	12	17	0
Temperate forest	1,281	3	1,277	0	0
Forest land	5,066	25	130	1,005	3,907

a) Select a country of your choice and check for the total amount of each type of ecosystem existing in these countries, and calculate the total annual economic value of the ecosystem services in that country.
b) Assume that a local government decides to implement a scheme where industries that destroy ecosystems are expected to pay 10× the annual monetary value of the ecosystem service destroyed. Now consider an industrial activity led to the destruction of 1000 acres of forestland and 250 acres of temperate forest. Estimate the amount of money the company has to pay to the government.

Solution:

a) Australia was selected here for the analysis.

Item	Total economic value ($/ha/yr)	Total area of item in Australia (million ha)	Total ecosystem service value of Australia ($ million/yr)
Open ocean	$9	810	7 290
Coral reef	$206 873	2.9	599 932
Coast	$77 907	25.76	2 006 884
Coastal wetland	$960	3	2 880
Inland wetland	$282	33.5	9 447
Rivers and lakes	$812	118	95 816
Tropical rainforest	$29	6.7	194
Temperate forest	$1 281	19	24 339
Forest Land	$5 066	134	678 844
		Total value	***3 425 626***

b) Estimated fee for ecosystem service destroyed

	Ecosystem destroyed		**Total Economic Value**		
	(acres)	**(ha)**	**($/ha/year)**	**Multiplier**	**Fee**
Temperate forest	250	101.17	1281	10	$1 296 006
Forest land	1000	404.69	5066	10	$20 501 375
Total fee:					$21 797 380

3.4 Legal and Governance Issues

Table 3.4 shows a matrix that highlights the evolution of environmental regulations. In the past, evolutions in the framework were directed at remediation – the handling and remedy of waste and the polluted environment. Today, framework changes are primarily focused on compliance – the regulation and mitigation of effects associated with emissions. In the future, the framework will likely again shift focus toward more sustainable economic output, with a greater focus on life cycles.

The interaction of national and international frameworks poses an interesting dilemma. A national statutory framework is primarily limited within a nation's borders. However, environmental issues – such as acid precipitation, GHG emissions, ozone depletion, and watershed pollution – often extend across borders and have continental or global effects. The scale of these

Table 3.4 Evolution of the environmental regulations.

Time	Activity	Focus	Geographical temporary scale	Regulation	Leader
Past	Remediation	Waste	Local Immediate	Command and control	USA
Present	Compliance	Emissions	Point source Immediate	Command and control, end of pipe	More developed countries
Future	IE/DfE	Products and services life cycle	Regional and global systems All time scales	Establish boundary conditions	EU, Netherlands and Germany

DfE: Design for Environment; IE: Industrial Ecology.

effects and the political bodies and related legal frameworks do not match; one nation's laws cannot solely be applied to battle environmental issues of such extent. Therefore, it is critical to develop international frameworks or political bodies that can act on a global basis.

Legal frameworks should be cooperative in nature. They should be established to pursue both intragenerational and intergenerational equity. They should also focus less on sanctions and punitive activity and find a means to facilitate cooperation through incentives or rewards for positive actions. Additionally, they should be established where legal jurisdiction can match the reach of the particular environmental issue. To put it another way, legal frameworks should attempt to invoke the following (Rogers et al. 2008):

- *Motivation for initiators* – Quality of living standards should be, in the very least, maintained at current levels, but behaviors that lead to an improvement should be encouraged.
- *Intergenerational equity* – The condition of the environment should be preserved "as-is" or better for future generations.
- *Intragenerational equity* – Efforts should be made to provide a decent standard of living for all through "humane globalization."

3.5 Environmental, Social and Governance (ESG) Reporting

ESG – environmental, social, and governance – has become a well-known buzzword in the business world. Specifically, it represents non-financial risk and opportunity dimensions that affect business operations, performance, and perceptions. Specifically, it moves beyond strictly financial performance – it is an approach or framework that companies pursue to incorporate and measure against sustainability's triple bottom line. ESG has evolved into the framework to assess and measure these issues (UNPD 2023).

As commonly used, ESG frameworks that businesses can use to measure their performance criteria along these sustainability-related dimensions. These may include waste minimization and carbon emissions mitigation efforts, water and materials use, transparency in management activities, corporate philanthropic efforts, equity in hiring practices and governing board/body representation, or customer data privacy protections (Gartner 2024). There are a seemingly endless set of metrics that can be applied to environmental, social, and governance criteria; organizations will typically select those that are meaningful to the operation and mission of the organization as well as those that are salient to a range of stakeholders. Numerous standards have been developed and applied for ESG reporting.

The following is a summary of four notable reporting frameworks (TechTarget 2024).

- *International financial reporting standards (IFRS) sustainability disclosure standards* – Dating to 2022 and formed by the merger of two "parent" reporting frameworks, IFRS creates a set of disclosure standards that can be used globally to report ESG data to investors. One set focuses on disclosures of sustainability-related financial information, and the second set focuses on climate-related risks and opportunities.
- *Global reporting initiative (GRI) standards* – These standards "enable any organization – large or small, private or public – to understand and report on their impacts on the economy, environment and people in a comparable and credible way, thereby increasing transparency on their contribution to sustainable development." The framework includes sets of universal, sector-specific, and topic-based sustainability reporting standards and may be used by investors, government policymakers and other stakeholders.
- *Carbon disclosure project (CDP)* – This framework allows companies to report on business risks and opportunities related to climate change, water security, and deforestation, which in turn earns a public-facing "grade." Reportedly, CDP is used by 18 700 companies worldwide as well as city governments.
- *Task force on climate-related financial disclosures (TCFD)* – This framework was established to develop recommendations for disclosures of financial risks. The reporting is commonly used by investors, lenders, and insurance underwriters.

3.6 Summary

Although it is difficult to separate environmental issues and concerns from considerations of sustainability, it is equally important to consider issues and concerns associated within social, economic, and legal frameworks. An understanding of these concepts is essential when considering and assessing projects or activities from a TBL framework. Sustainable solutions should strive to maximize the overall TBL and their representative metrics. More sustainable solutions lead to a greater intersection of these metrics.

When considering social metrics, projects and activities should be assessed on how they may exacerbate or ameliorate poverty, exploitation, physical and mental health, social services, civil unrest, or crime. Economic metrics are of critical importance as both for-profit and not-for-profit entities seek to maximize rates of return on investment measured in numerous ways, including cost–benefit analyses, discounted rate models, or simple payback models. Additionally, accounting and financial models are evolving in which direct or indirect costs are assigned to a range of sustainability-related factors. Legal frameworks are also important, as a keen understanding of national legal models are key to any successful venture within that specific country. However, international and intergovernmental legal frameworks will likely become more essential in the future to deal with environmental and sustainability issues that span across national jurisdictions and shared borders.

The concepts associated with the social, economic, and legal issues presented in this chapter are essential in providing an appropriate context of sustainability presented and applied throughout this book.

3.7 Questions

3.1 Mention some of the examples for SIA indicators.

3.2 Select a regional society with which you are familiar and discuss indicators of social sustainability associated with it.

3.3 What are the economic issues related to sustainability perspective?

3.4 Discuss the difference between the physical economy and the money economy.

3.5 What is environmental accounting? How do you evaluate environmental cost?

3.6 Explain briefly the multitier environmental costing framework.

3.7 Explain life cycle costing.

3.8 Discuss the differences between NPV and payback financial assessments.

3.9 Define CCI, CPI, and LIBOR. How may these be used in financial assessments?

3.10 Explain true cost accounting.

3.11 Select a product or a project, and list economic metrics in each of the five Tiers of environmental costing that may be applied.

3.12 What are legal issues related to sustainability?

3.13 List examples where extranational legal issues between nations may arise with respect to environmental issues.

3.8 Problems

3.1 Let us say that the costs of a pollution control program are estimated to be 100$ per year. The expected benefits of the program are $25 a year initially after implementation of the program, and these benefits keep doubling for every five years of implementation until they reach $200 a year. The benefits of implementing the program after this stay constant at $200 a year at current prices. Calculate the net benefits from implementing the program after 50 years considering a discount rate of 4%.

3.2 The following table gives the social costs of greenhouse gas emissions as estimated by the USEPA for the years 2020 to 2080.

	SC-GHG and near-term Ramsey discount rate								
	$SC\text{-}CO_2$ (2020 dollars per metric ton of CO_2)			$SC\text{-}CH_4$ (2020 dollars per metric ton of CH_4)			$SC\text{-}N_2O$ (2020 dollars per metric ton of N_2O)		
	Near-term rate			Near-term rate			Near-term rate		
Emission Year	2.5%	2.0%	1.5%	2.5%	2.0%	1.5%	2.5%	2.0%	1.5%
2020	120	190	340	1300	1600	2300	35 000	54 000	87 000
2030	140	230	380	1900	2400	3200	45 000	66 000	100 000
2040	170	270	430	2700	3300	4200	55 000	79 000	120 000
2050	200	310	480	3500	4200	5300	66 000	93 000	140 000
2060	230	350	530	4300	5100	6300	76 000	110 000	150 000
2070	260	380	570	5000	5900	7200	85 000	120 000	170 000
2080	280	410	600	5800	6800	8200	95 000	130 000	180 000

a) Considering a near-term rate of 2.0%, calculate the total social costs of greenhouse gases emitted by the US in the year 2020.

b) Given below are the annual emission rates of greenhouse gases from an industry.

Greenhouse gas	Emissions (ton/year)
CO_2	1000
CH_4	5
N_2O	1

Scenario 1: Assuming that the industry continues to emit the same levels of greenhouse gases each year for the next 50 years, determine the total social cost of CO_2 (Consider a near-term interest rate of 2.5%).

Scenario 2: Now considering that the company plans for a phase-wise shift to greener practices to reduce its greenhouse gas emissions, which can result in a reduction of greenhouse gas emissions at 2.5% each year for each gas until the total greenhouse gas emissions from the company reach 50% of its current levels. Calculate the total social cost of greenhouse gas emissions by the company by the year 2070. (Consider a near-term interest rate of 2.5%)

Compare both scenarios and comment on the results.

3.3 The following are the net benefits from two projects for the next two years. Calculate the overall net benefit for all three years included at current prices considering discount rates of 5% and 11%.

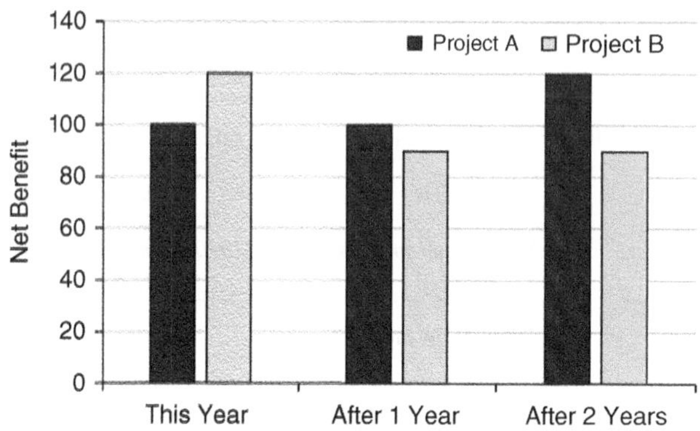

3.4 Assume a food packaging company is considering replacing its conventional plastic packaging with a more sustainable biodegradable alternative. The management wants to evaluate this decision from different financial and environmental perspectives using CBA, LCC, and TCA.

Costs and Savings:

- Initial investment in new packaging machinery: $500 000
- Cost per unit of conventional packaging: $0.10
- Cost per unit of biodegradable packaging: $0.15
- Annual production: 5 million units
- Expected life span of the machinery: 10 years
- Savings from reduced environmental fines and penalties: $50 000 per year
- Additional annual marketing benefits due to improved brand image: $100 000
- Salvage value of the machinery: $50 000 after 10 years
- Environmental and Social Considerations:
- Reduction in carbon emissions by switching to biodegradable packaging: 1000 tons per year. The cost of carbon offset is $25 per ton.
- Reduction in waste management costs due to biodegradable materials: $30 000 per year
- Increase in water usage by the new process: 10 000 gallons per year. Local water cost: $0.03 per gallon.
- The social cost of increased labor for the new packaging process: $20 000 per year.

Questions:

a) Calculate the net benefit over 10 years, considering initial investment, operational costs, savings from environmental fines, and additional marketing benefits.
b) Determine the total life cycle cost over 10 years, including initial investment, operational costs, and savings from waste management.
c) Estimate the true cost per unit by incorporating environmental and social costs/benefits into the operational cost.

Note: Calculate all costs at current prices considering a discount rate of 5%.

3.5 Here are some representative numbers illustrating the costs and benefits of an initiative aimed at pollution control. The existing monthly emissions stand at 10 tons. Determine the emission threshold at which overall benefits are highest. Demonstrate that this threshold differs from the emission level associated with the highest benefit–cost ratio. Discuss the possible reasons behind this inconsistency.

Emissions (ton/month)	**10**	**9**	**8**	**7**	**6**	**5**	**4**	**3**	**2**	**1**	**0**
Benefits ($ mil)	0	4	8	18	32	44	54	62	68	72	74
Costs ($ mil)	0	2	4	6	9	14	21	36	48	64	86

- Determine the emission threshold at which overall benefits are highest. Demonstrate that this threshold differs from the emission level associated with the highest benefit–cost ratio. Discuss the possible reasons behind this inconsistency.
- Now assume that as a result of the initiative the emissions stand at 4 tons/month for 10 consecutive years. Calculate the net benefits resulting from the initiative at current prices considering a discount rate of 6%.
- Let us say the initiative worked partly and the emissions stand at 7 tons/month for 8 consecutive years starting from 2024. Assuming half of these emissions are CO_2 emissions calculate the social cost of carbon.

Note: The social costs of carbon given by USEPA can be found at USEPA (2023). Report on the Social Cost of Greenhouse Gases: Estimates Incorporating Recent Scientific Advances. US Environmental Protection Agency. Report number EPA-HQ-OAR-2021-0317. Available at: https://www.epa.gov/system/files/documents/2023-12/epa_scghg_2023_report_final.pdf. (accessed 8 November 2024).

3.6 In a community co-habitated by two consumer groups, namely workers and retirees, a polluting factory causes PM 2.5 emissions. The total damage inflicted on workers is represented by $3p^2$, where p denotes the pollution level (in kg PM2.5 emissions; PM2.5 refers to fine particulate matter that is 2.5 μm or smaller in diameter), while retirees experience a total damage of 3 times the damage experienced by the workers group. According to an assessment conducted, the factory accrues savings of $40p\text{-}5p^2$ through polluting at level p.

- Determine the combined (encompassing both consumer types) marginal damage for the public nuisance. Calculate the value of marginal damage at a pollution level of 4 units.
- Create graphs for the marginal savings and aggregate marginal damage, with pollution depicted on the horizontal axis.
- In the absence of regulation or negotiation, calculate the factory's pollution level. What constitutes the society's optimal pollution level?
- Commencing from the uncontrolled pollution level determined in the above question, ascertain the marginal willingness to pay for pollution abatement (A) for each consumer group. Find the aggregate marginal willingness to pay for abatement.
- Once again, starting from the uncontrolled pollution level, identify the firm's marginal cost of pollution abatement. Determine the optimal level of A.

- Now, find any environmental economics related textbook, read about concepts of marginal costs, marginal savings, marginal damage, and pollution abatement costs and savings.

Note: Please find solutions to similar problems at Standorf (2007). Problem set 2 suggested solutions. Available at: https://web.stanford.edu/class/econ155/coursework/CourseMaterials/ProblemSets/PS2Solutions.pdf (accessed 8 November 2024).

References

Arumugam, K., Karthika, T., Elangovan, K., and Kumar, A.R. (2020). Assessment of groundwater pollution due to textile industrial activities in and around Tirupur Region, Tamil Nadu, India. *Nature Environment and Pollution Technology* 19 (3): 1203–1209.

Baker, L., Castilleja, G., De Groot Ruiz, A., and Jones, A. (2020). Prospects for the true cost accounting of food systems. *Nature Food* 1 (12): 765–767.

Bryan, H. (2021). The assessment of social impacts. In: *Natural Resource Management* (ed. A.W. Ewert), 145–166. Routledge.

Carley, S. and Konisky, D.M. (2020). The justice and equity implications of the clean energy transition. *Nature Energy* 5 (8): 569–577.

CGD (2024). Climate and development in three charts: an update. Center for Global Development. https://www.cgdev.org/blog/climate-and-development-three-charts-update (accessed 13 December 2024).

Dixon, J., Scura, L., Carpenter, R., and Sherman, P. (2013). *Economic Analysis of Environmental Impacts.* Routledge.

Gartner (2024). Environmental, Social and Governance (ESG). https://www.gartner.com/en/finance/glossary/environmental-social-and-governance-esg- (accessed 31 June 2024).

ISO (2006). *14040: 2006. Environmental Management-Life Cycle Assessment-Principles and Framework.* European Committee for Standardization.

Langfield-Smith, K., Thorne, H., Smith, D.A., and Hilton, R.W. (2015). *Management Accounting: Information for Creating and Managing Value*, 7e. North Ryde: NSW McGraw-Hill Education Australia Pty Ltd.

McKenzie, S. (2004). *Social Sustainability: Towards Some Definitions*, 31. Hawke Research Institute, University of South Australia Magill.

Michaelowa, A. and Sippel, M. (2022). How can global carbon markets promote low-carbon cities in developing countries? Lessons learnt from the clean development mechanism. In: *Planning Climate Smart and Wise Cities: A Multidisciplinary Approach* (ed. K.-G. Kim and M. Thioye), 291–315. Springer Nature.

Milken Institute School of Public Health (2020). Equity vs. equality: what's the difference? Milken Institute School of Public Health. The George Washington University. https://onlinepublichealth.gwu.edu/resources/equity-vs-equality (accessed 9 July 2025).

Rogers, P.P., Jalal, K.F., and Boyd, J.A. (2008). *An Introduction to Sustainable Development.* London: Glen Educational Foundation, Inc. ISBN: 978-1-84407-520-6 (hardback).

Ruggerio, C.A. (2021). Sustainability and sustainable development: a review of principles and definitions. *Science of the Total Environment* 786: 147481.

TechTarget (2024). 10 top ESG reporting frameworks explained and compared. https://www.techtarget.com/sustainability/feature/Top-ESG-reporting-frameworks-explained-and-compared (accessed 8 November 2024).

Toniolo, S., Tosato, R.C., Gambaro, F., and Ren, J. (2020). Life cycle thinking tools: life cycle assessment, life cycle costing and social life cycle assessment. In: *Life Cycle Sustainability Assessment for Decision-Making* (ed. J. Ren and S. Toniolo), 39–56. Elsevier.
UNPD (2023). The short guide to ESG. United Nations Developed Program. https://www.undp.org/serbia/publications/short-guide-esg (accessed 31 June 2024).
USEPA (2024). Learn about environmental justice. United States Environmental Protection Agency. https://www.epa.gov/environmentaljustice/learn-about-environmental-justice (accessed 30 July 2025).
WACOSS (2013). Sustainable community services. WA State Election 2013 Party Policies. wacoss.org.au/StateElection2013/PartyPolicies/SustainableCommunityServices.aspx.
Wang, S., He, Y., and Song, M. (2021). Global value chains, technological progress, and environmental pollution: inequality towards developing countries. *Journal of Environmental Management* 277: 110999.

4

Availability and Depletion of Natural Resources

Chapter Goal and Objectives

This chapter aims to explore the exploitation and depletion of natural resources, focusing on various types such as fossil fuels, radioactive fuels, mineral resources, rare earth elements, lithium, and water. The specific objectives of this chapter are to (i) highlight the growing demand for these resources, the challenges posed by their limited supply, and the environmental, economic, and geopolitical impacts of depletion; (ii) present the causes of resource depletion, including overconsumption, population growth, and unsustainable extraction practices, while discussing their effects on ecosystems, economies, and societies; (iii) examine the phenomenon of overshooting, where resource use surpasses the Earth's regenerative capacity; and (iv) introduce the concept of urban metabolism, emphasizing how cities drive resource consumption and waste generation, and the need for more sustainable, circular urban systems to mitigate the long-term impacts of resource depletion.

4.1 Introduction

Since the primary focus of sustainability principles is the preservation of natural resources for future generations, it is critical to consider the prudent utilization of natural resources. When we discuss the availability and utilization of resources, it is important to make a distinction between renewable and nonrenewable resources. A nonrenewable resource cannot be reproduced or regenerated to any appreciable extent within the timeframe of human life, and any material rate of consumption of a nonrenewable resource will result in its eventual depletion. In contrast, renewable resources can be replaced by natural processes and are a result of natural forces/phenomena that are persistent and virtually inexhaustible in the environment. This may also include recyclable materials, which when used for a certain period, can be reformulated and/or recaptured for future beneficial use. Examples include timber, solar power, wind power, wave/tidal power, and geothermal energy.

In order to achieve sustainability, the long-term protection and preservation of natural resources is critical. In particular, it is important to preserve nonrenewable resources, as they are finite. This chapter presents information on the availability, use, and depletion of natural resources, which are essential aspects of sustainability assessments.

Sustainable and Resilient Engineering: Drivers, Metrics, Tools, and Applications, Second Edition. Krishna R. Reddy, Claudio Cameselle, and Jeffrey A. Adams.

Companion website: www.wiley.com/go/reddy/sustainableresilientengineering2e

4.2 Types and Availability of Resources

4.2.1 Fossil Fuels

Many developed nations are dependent on the use of nonrenewable fossil fuels for energy. These fuel sources are a result of the decomposition of plant and animal matter over the course of millions of years due to environmental changes. Primary nonrenewable sources include coal, petroleum, and natural gas. Oil shales and tar sands are regarded as secondary sources. These fuel sources are relatively easily refined and used in a variety of energy-consuming applications but result in the generation of high amounts of carbon dioxide emissions.

Coal is considered the most abundant fossil fuel, with an estimated reserve of one trillion metric tons (USEIA 2024a). Coal is most abundant in Asia, Eastern Europe, and the United States. As of 2005, China was the global leader in extraction (44%), followed by the United States (20%), India (8%), and Australia (7%). It is commonly burned as a primary source of heat, and it is used to convert water to steam for electrical generation.

An important concept regarding coal use (or use of another fuel source) is "peak coal." It is defined as the point in time at which a maximum global production peak is reached, after which the rate of production will enter a period of terminal decline. It may be measured in two ways – by mass of coal used or by energy output. For coal, the energy output per mass has dropped since 2000, indicating that the energetic peak will come sooner than the mass peak. Estimates for the mass peak vary widely, but research by the University of Newcastle (Australia) (ENN 2010) estimates that the peak may occur between now and 2048.

Petroleum, or crude oil, is a second primary fossil fuel. It is versatile and may be easily refined into a range of products, including motor fuels, material feedstock, lubricants, and heating oils. More than 50% of proven reserves are present in the Middle East (USEIA 2024b). However, resources are limited, and unless consumption is controlled, resources could be depleted within 30 years. From a US peak oil perspective, domestic oil production peaked around 1970. Further, global production has stabilized close to 2005 highs (approximately 74 million barrels per day) (IEA 2024). Figures 4.1 and 4.2 depict actual and predicted sources of petroleum.

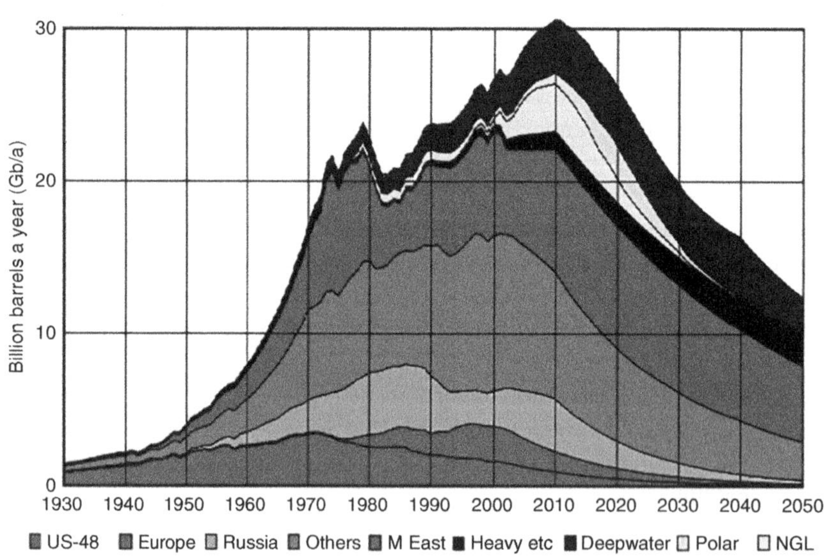

Figure 4.1 Production of oil and future predictions. *Source:* Gralo/Wikimedia Common/Public domain.

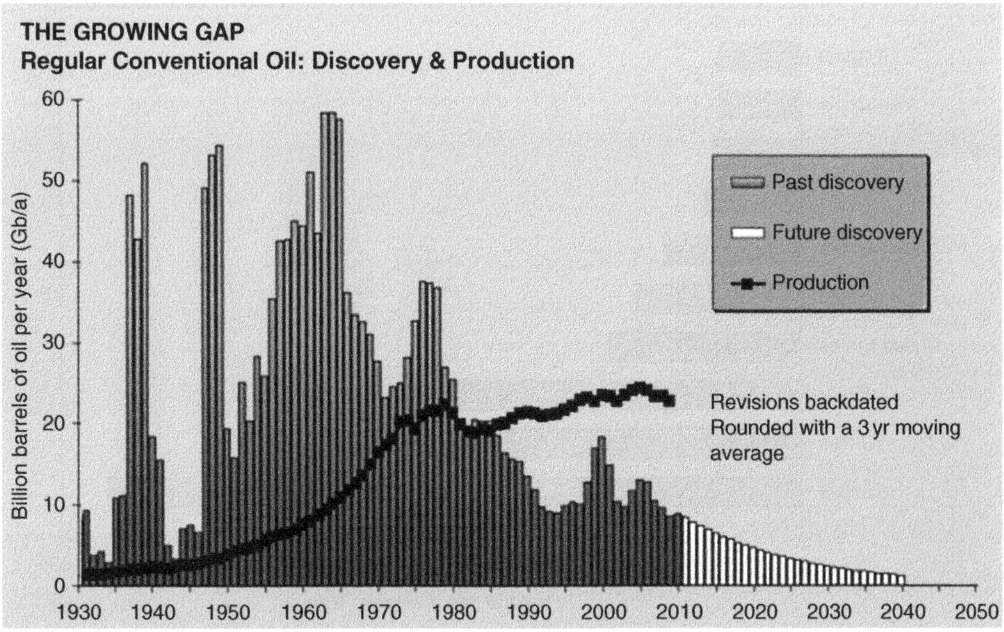

Figure 4.2 Oil production and reserves. *Source:* Post Carbon Institute/http://www.resilience.org/primer/.

Oil can be difficult to locate and extract. Current technology can extract about 40% of oil from most wells (USDOE 2024), and much of the oil that has been considered relatively easy to withdraw from the Earth has been extracted and utilized. However, new extraction technologies are allowing for the extraction of additional reserves, and oil fields long considered exploited are again being returned to an economically productive state. Further, higher prices for crude oil can render oil reserves that require more difficult and costly extraction to become economically feasible. Alternative technologies for increased oil extraction rates are horizontal drilling and fracking. Horizontal drilling involves drilling a well horizontally after reaching a certain depth, allowing access to larger reserves. Fracking involves injecting high-pressure fluid and sand into the well to crack the rock and extract trapped oil and gas. While these technologies show higher oil production rates and recovery factors, they have also raised environmental concerns due to potential groundwater contamination, air pollution, and induced seismic activity. The percentage of oil and gas extracted through horizontal drilling and fracking varies by region and reservoir characteristics, but it has become a dominant method in many areas, contributing substantially to oil production (USEIA 2018).

Natural gas is a mixture of gaseous hydrocarbons (methane, ethane, propane, and butane) where methane is the most abundant. It can be refined and reconstituted, and it is cheap. It can be stored and transported more easily than coal or oil and is relatively environmentally friendly. Nearly one-quarter of world energy is derived from natural gas, and its consumption has nearly doubled in the past 30 years (USEIA 2024c), with significant growth in consumption exhibited by developing countries. Figures 4.3 and 4.4 depict natural gas reserves and production.

Oil shales and tar sands are used less frequently but are considered a secondary source of fossil fuel. Oil shale contains kerogen, a waxy form of carbon that can be vaporized at very high temperatures and condensed into shale oil. Tar sand contains low-quality thick crude oil. Both oil shales and tar sands require significant energy inputs for extraction; however, when global petroleum prices are high, their extraction and use become cost-effective.

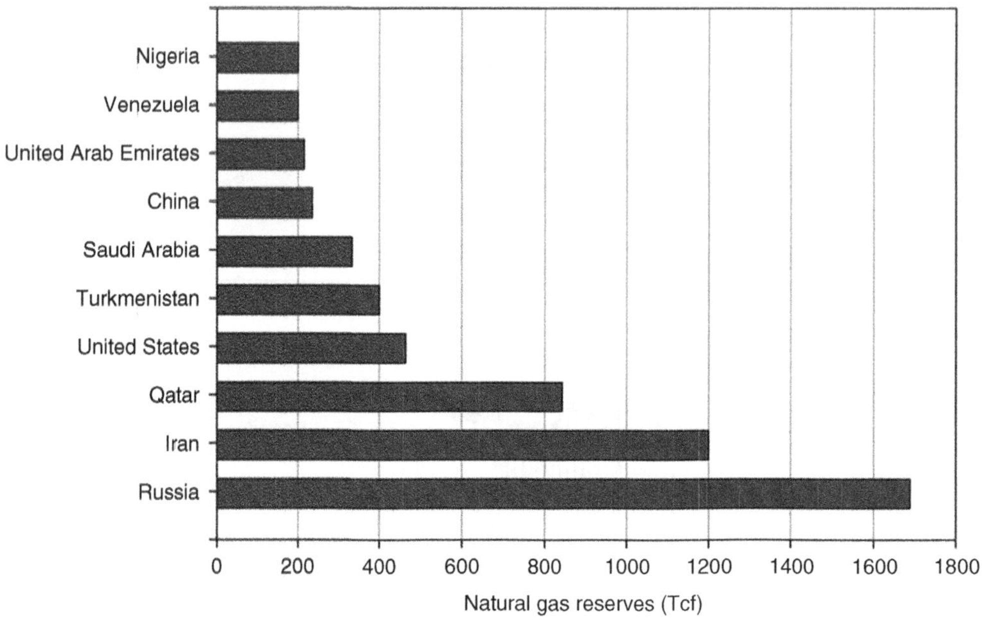

Figure 4.3 Natural gas reserves (Tcf: trillion cubic feet). *Source:* U.S. Energy Information Administration/ https://www.eia.gov/naturalgas/(accessed 31 May 2024)/Public domain.

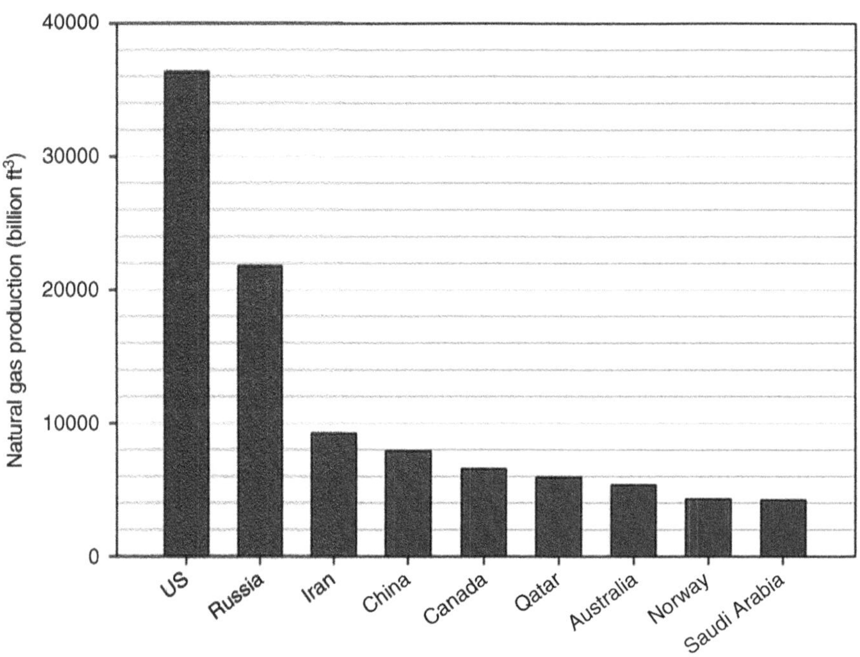

Figure 4.4 Natural gas production in 2014. *Source:* U.S. Energy Information Administration/https://www.eia.gov/naturalgas/(accessed 31 May 2024)/ Public domain.

Example 4.1

Two countries have different levels of natural gas reserves, extraction rates, consumption rates, and potential for future discovery. The data provided for the year 2023 is as follows:

Country	Reserves (trillion ft^3)	Annual extraction rate (trillion ft^3)	Annual consumption rate (trillion ft^3)	Potential for future discovery (% increase in reserves year over year)
1	100	2 (+2%)	1.5 (+3%)	10.0
2	200	3 (−1.5%)	2.0 (+1%)	12.5

Note: The figures in brackets are the annual rates of change in the corresponding values.

Considering that the annual rates of change remain constant for the next 10 years, and the potential for discovery reduces by 20% each year for all the countries, determine the following:

a) Calculate the remaining reserves for each country at the end of 2033, considering both the extraction rates and the potential for future discoveries.
b) Determine the year in which each country will exhaust its natural gas reserves if the current extraction and consumption rates continue unchanged.
c) If each country aims to extend the availability of its natural gas reserves by 50% beyond the exhaustion year calculated in question b, by what percentage should each country reduce its annual extraction rate, assuming no changes in consumption rates and no further discoveries?

Solution:

a) Calculate the remaining reserves for each country at the end of 2033

Country 1:
Total quantity of natural gas extracted for the period of 2024–2033 is given by:

$$2 \times (1.02)^1 + 2 \times (1.02)^2 + 2 \times (1.02)^3 + \ldots + 2 \times (1.02)^{10}$$

$$= 2 \times \left(1.02\frac{(1.02)^{10} - 1}{1.02 - 1}\right) \left(\because x + x^2 + x^3 + \ldots + x^n = x\left(\frac{x^n - 1}{x - 1}\right)\right)$$

$$= 22.34 \ trillion \ ft^3$$

New reserves discovered from 2024 to 2033:

Given that potential for future discovery in 2023 is 10%, so new reserves found in 2024 can be calculated as $100 \times 0.1 = 10$ trillion ft^3.

For the year 2025, the potential for new reserves reduces by 20% of the previous years' discovery rate reducing the new reserves found to 8% in 2025. Hence, the new reserves discovered can be calculated as $(100+10) \times 0.08 = 8.8$ trillion ft^3, increasing the total extractable reserves to 118.8 trillion ft^3

Similarly, the total natural reserves can be calculated after each year, as shown in the Table below:

Year	New reserves discovered (%)	Total natural reserves	Extraction rate	Cumulative extraction	Total remaining natural reserves
2024	10.00	110.00	2.04	2.04	107.96
2025	8.00	118.80	2.08	4.12	114.68
2026	6.40	126.40	2.12	6.24	120.16
2027	5.12	132.88	2.16	8.41	124.47
2028	4.10	138.32	2.21	10.62	127.70
2029	3.28	142.85	2.25	12.87	129.98
2030	2.62	146.59	2.30	15.17	131.43
2031	2.10	149.67	2.34	17.51	132.16
2032	1.68	152.18	2.39	19.90	132.28
2033	1.34	154.22	2.44	22.34	131.89

Note: All quantities are in trillion ft^3.

The maximum amount of natural reserves were calculated to be 162.7 trillion ft^3 owing to negligible rate of new reserve discovery after the year 2060.

$$\textit{Initial reserves} + \textit{new reserves discovered} - \textit{total extraction}$$

$$= 100 + 54.22 - 22.34 = 131.89 \textit{ trillion } ft^3$$

Total quantity of natural gas consumed between 2024 and 2033:

$$1.5 \times \left(1.03\frac{(1.03)^{10}-1}{1.03-1}\right) = 17.71 \textit{ trillion } ft^3$$

Country 2:

For Country 2, the extraction rate is targeted to reduce by 1.5% each year unlike Country 1, where the extraction rate is expected to continuously increase. Now considering this, and setting up a calculation table similar to that of the previous problem, the following data can be obtained for Country 2:

Year	New reserves discovered (%)	Total natural reserves	Extraction rate	Consumption rate	Cumulative extraction	Total remaining natural reserves
2024	12.50	225.00	2.96	2.02	2.96	222.05
2025	10.00	247.50	2.91	2.04	5.87	241.63
2026	8.00	267.30	2.87	2.06	8.73	258.57
2027	6.40	284.41	2.82	2.08	11.56	272.85
2028	5.12	298.97	2.78	2.10	14.34	284.63
2029	4.10	311.21	2.74	2.12	17.08	294.14
2030	3.28	321.41	2.70	2.14	19.78	301.64
2031	2.62	329.84	2.66	2.17	22.44	307.40
2032	2.10	336.76	2.62	2.19	25.05	311.70
2033	1.68	342.41	2.58	2.21	27.63	314.77

Note: All quantities are in trillion ft^3.

The maximum amount of natural reserves were calculated to be 366 trillion ft^3 owing to negligible rate of new reserve discovery after the year 2060.

Total remaining natural reserve at the end of 2033 are 314.77 trillion ft^3.

b) Determine the year in which each country will exhaust its natural gas reserves.

Country 1:
Let the number of years from 2023 that the countries will exhaust their reserves be n.

For Country 1 to exhaust its reserves, the cumulative extraction should exceed total natural reserves:

$$(Total\ Extraction) = (Total\ Available\ Reserves)$$

$$2 \times \left(1.02\frac{(1.02)^n - 1}{1.02 - 1}\right) = 162.7$$

By trial and error, we can find $n = 48$; hence, Country 1 will exhaust its reserves by the year 2071. Alternatively, the same can be found by expanding the calculation table that we have setup earlier for Country 1.

Country 2:
Although Country 2 targets to reduce the extraction rate by 1.5% each year, the extraction rate cannot be less than the consumption rate. Hence, we assume that the extraction rates reduce by 1.5% each year until the extraction rate equals the consumption rate and then is equal to the consumption rate. The extraction rate equals the consumption rate in 2039. Using the calculation table setup before, the reserves in Country 2 are expected to exhaust in 2125.

c) If each country aims to extend the availability of its natural gas reserves by 50% beyond the exhaustion year calculated in question b, by what percentage should each country reduce its annual extraction rate, assuming no changes in consumption rates and no further discoveries?

Country 1:
To increase the availability by 50% beyond the exhaustion year, $n = 48 \times 1.5 = 72$ (that corresponds with the year 2095), let us say that the extraction rate is reduced by a factor of x, x can then be calculated using

$$2 \times \left(1.02\frac{(1.02)^{72} - 1}{1.02 - 1}\right) \times \frac{1}{x} = 162.7$$

$$x \approx 2$$

Therefore, to extend the availability by 50%, Country 1 should reduce the extraction rate by a factor of 2.

Country 2:
Country 2's reserves are calculated to be exhausted by 2125, which is 102 years from 2023. Now to extend this by 50% means that the reserves have to last until $2176 = (2023 + 102 \times 1.5)$ and $n = 153$ years.

For Country 2, since the extraction rate reduces until it equals the consumption rate, and then equals the consumption rate. However, to simplify the problem, we will assume that the extraction rate = consumption rate for the entire 153 years (since extraction rate > consumption rate only till 2039 as per our earlier calculations). Now to calculate x, we can follow the same procedure as before consider annual consumption rate as well as the rate of change of annual consumption rate

$$2 \times \left(1.01 \frac{(1.01)^{153} - 1}{1.01 - 1}\right) \times \frac{1}{x} = 366$$

$$x \approx 2$$

Therefore to extend the availability of reserves by 50% beyond the exhaustion year we calculated previously, Country 2 should also reduce the extraction/consumption rate by a factor of 2.

4.2.2 Radioactive Fuels

Nuclear power relies upon the fission of fuel atoms (typically uranium) to produce heat, which in turn generates steam for electricity production. Uranium can also be bred into other compounds, such as plutonium, allowing for more efficient fuel use. Uranium may also be refined for use in nuclear explosives/weapons. Nuclear power accounts for approximately 6% of the world's energy and 13–14% of the world's electricity (NEI 2017). It is a clean source of energy since no emissions are generated; however, spent fuel is acutely dangerous for a long period of time. The nuclear power industry produces a relatively small amount of waste, with 97% of it classified as low- or intermediate-level waste (LLW or ILW); the rest is high-level waste (HLW), whose radioactivity decreases significantly over time, dropping to one-thousandth after 40 years of storage in nuclear facilities. A typical large reactor generates about 25–30 tonnes of used fuel annually, and globally around 400 000 tonnes of used fuel have been produced, with one-third being reprocessed and sent back to nuclear plants (WNA 2018).

When considering the peak use of uranium, there is debate as to know if the peak has occurred. Some argue that the peak happened in the 1980s, whereas others claim that a second peak could happen around 2035 (WNA 2016). World demand is expected to continue increasing. It is expected that between 80 and 100 kilotons of uranium will be needed per year by 2025. Uranium is present in many rock formations as well as in seawater. Like many other metals, it is found in relatively low concentrations, which limits its economical extraction. Typically, anything containing over 0.075% uranium is considered ore and can be mined.

4.2.3 Mineral Resources

A mineral resource is "a concentration or occurrence of material of intrinsic economic interest in or on the Earth's crust in such form, quantity, or quality that there are reasonable prospects for eventual economic extraction." These may be further divided into three categories: inferred, indicated, and measured, depending on the degree of confidence in their amount, quality, and concentration. Figure 4.5 depicts the elemental resources in the Earth's crust.

Inferred mineral resources are those whose concentration, quality, and amount can be estimated with only a low level of confidence. Information derived from geologic evidence (e.g. trenches, outcrops) is used to classify a possible mineral resource in this group. Indicated mineral resources are those whose characteristics – tonnage, grade, and density – can be estimated to a reasonable level of confidence based on sampling. If the sampling goes further, an "indicated resource" may be classified as a "measured mineral resource" when the estimate regarding content, density, tonnage, and quality can be made with a high degree of confidence. Mineral resources may also be classified as "reserve" when they are known to be economically feasible for extraction. Probable reserves are those with similar characteristics to a proven reserve but with a lower level of confidence (USGS 2024).

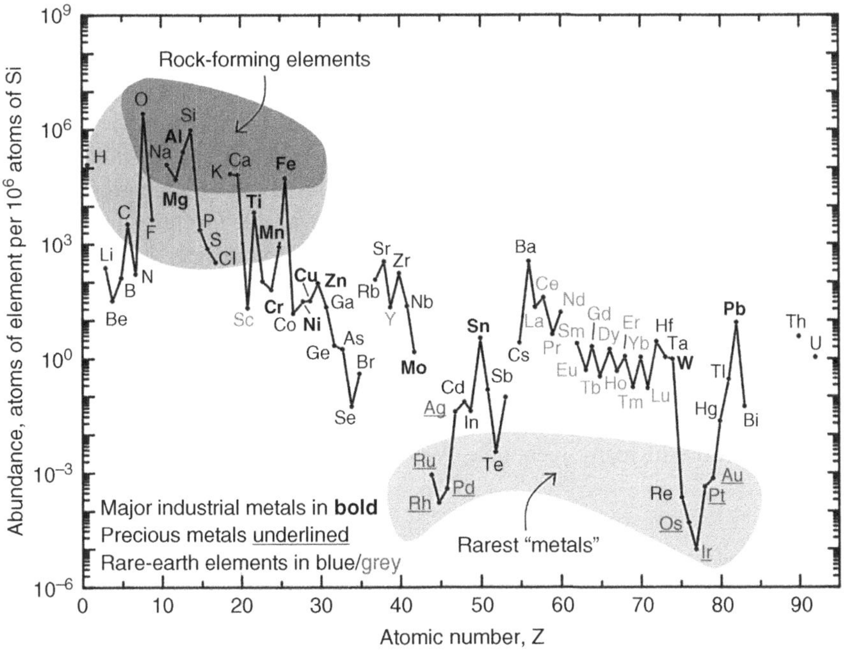

Figure 4.5 Elements abundance in the Earth's crust. *Source:* Haxel et al. (2002)/U.S. Geological Survey/Public domain.

Metal ores include various minerals such as oxides, sulfides, silicates, or native metals. They can be found in the Earth's crust at various concentrations. The rocks and minerals in the deposits must first be processed to separate the metal ore from the waste rock. Then, the metal is extracted from the ore by reducing the metal to its elemental state.

With increased demands for the mineral of interest, there is an increased demand for the parent ore. With increased demand comes a greater value and more exploitation. However, this results in higher extraction and a decrease in ore supply as reserves are depleted. Additionally, greater mining efforts often result in detrimental environmental effects.

One example of a mineral resource is copper. Copper, due to its ability to be recycled, has been reused many times over. An estimated 80% of all copper ever mined is still available for use (REDELCA 2023). Chief copper producers include Chile, the United States, and Peru. Copper is a very important industrial metal and is commonly used in power and transmission cables, electrical and automotive equipment, ammunition, and jewelry. Copper has been in use for over 10 000 years, but 95% in use has been extracted since 1900 (NSE 2024). As developing countries demand more copper, supplies are becoming more strained, leading to increased prices and increased copper theft.

In 2023, global copper production reached approximately 22 million metric tons, with consumption growing due to the transition to renewable energy and electric vehicles (USGS 2024). Copper demand is expected to continue increasing in the coming years, with a projected production of 24.5 million metric tons for 2024. However, starting in 2027, many current mines could be depleted, which, combined with growing demand, could lead to a supply deficit towards the end of the decade. According to projections, copper production could face a significant deficit by 2030, when annual demand could reach 36.6 million metric tons, while projected supply would be only 30.1 million metric tons (IWCC 2023).

4.2.4 Water Resources

Water is a vital natural resource. Over 70% of the Earth's surface is covered by water. Of course, the majority of this is salt water (97.5%). Table 4.1 presents the major salt compounds present in seawater, which can be effectively extracted and used for useful purposes. Only 2.5% of the water on Earth is usable fresh water. Unfortunately, of this fresh water, nearly 70% is trapped in icecaps in Antarctica and Greenland; much of the remainder is present as soil moisture or deep groundwater not accessible for human use. Ultimately, the percentage of water available for human use is only about 0.007%. This water is in lakes and reservoirs, rivers and shallow groundwater. The water consumed from natural resources is replenished by precipitations (rainfall and snowfall), and these sources are therefore considered renewable. Figure 4.6 depicts the water cycle responsible for movement and replenishment of water.

Freshwater is essential for several uses, including domestic use (approximately 8%), irrigation (70%), and industrial use (22%). Most supplies are derived from surface runoff, although groundwater is also a primary source in many locations. Water use for these sources has increased

Table 4.1 Composition of major constituents in seawater (Salinity: 35).

Constituent	Concentration (mol/kg-soln)
Na^+	0.46911
Mg^{2+}	0.05283
Ca^{2+}	0.01036
K^+	0.01021
Cl^-	0.54922
SO_4^{2-}	0.02824

Source: USDOE (1994)/US Department of Energy/Public domain.

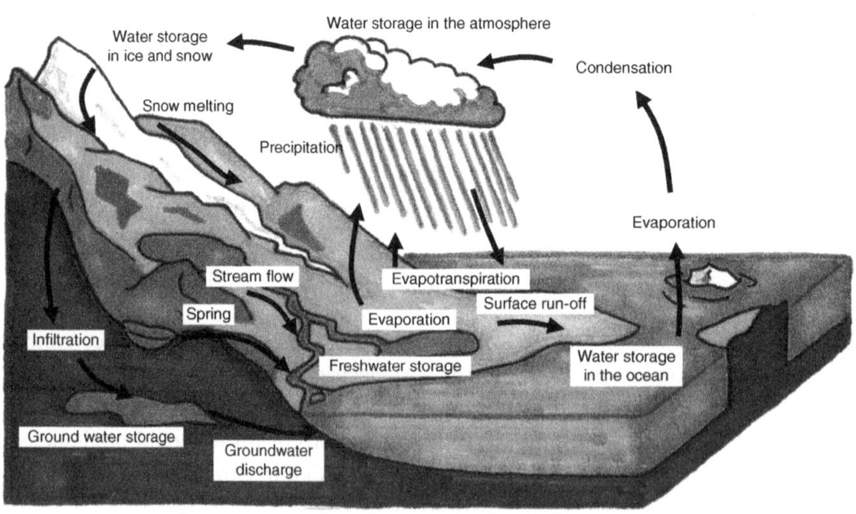

Figure 4.6 Natural cycle of water.

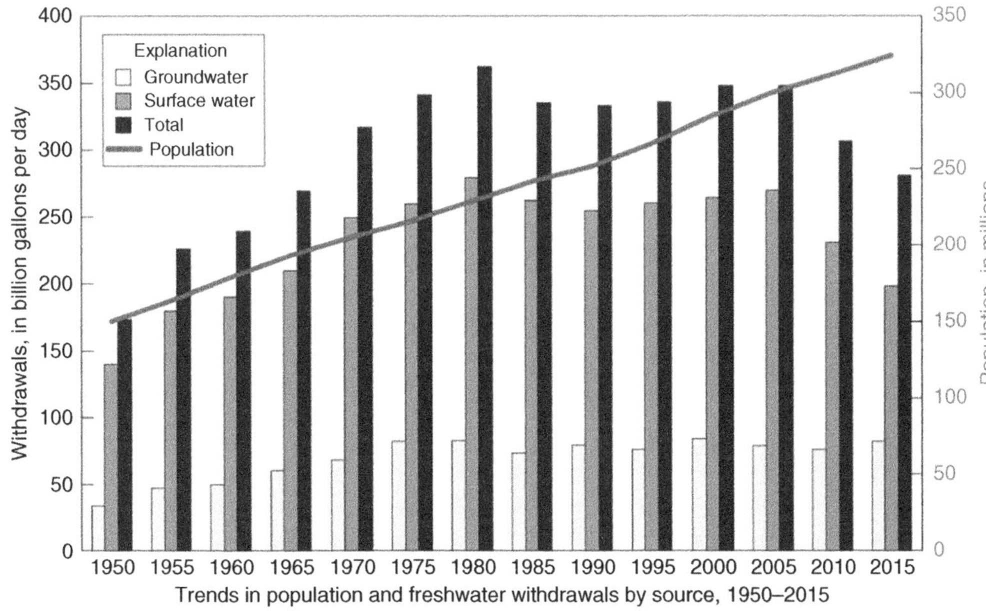

Figure 4.7 Increasing demand of water resources by population in the USA. *Source:* U.S. Geological Survey/ https://water.usgs.gov/edu/wateruse-trends.html./ Public domain.

dramatically over recent decades, as depicted in Figure 4.7. With increased population projected in the future, the demand for water will only increase (Bunsen et al. 2021). Unfortunately, the supply of usable freshwater is decreasing (due to impacts from contamination). As an example, 2 million tons of wastewater and other effluents drain into the world's water. Further, in developing countries, 70% of untreated industrial wastes are dumped into surface waters, where they contribute to contamination of the useable water supply (Weerasooriya et al. 2021). Therefore, water scarcity will only increase in the future, compounded by the fact that water use has been growing at more than twice the growth of population over the past century. Further, much of the increase in freshwater withdrawals is occurring and will continue to occur in developing countries for the next decade. Currently, 30% of accessible freshwater meets this need; however, it is projected that by 2025 this will increase to 70% (WWAP 2018).

Freshwater supplies are essential for health, economic activity, and overall quality of life. Unfortunately, approximately one of every six people in the world lacks basic water services, and over 20% of people do not have access to sanitation services (WHO 2024). As a result, millions of people die of water-borne diseases every year. Water is also the basis of international conflict. It is essential to provide access to clean freshwater universally for adequate health, welfare, and security of people worldwide. To accomplish this, it is important to provide freshwater sources for domestic use while finding efficient use and allocation for irrigation and industrial purposes.

With respect to water availability, it is helpful to assess the "water footprint" (WFN 2024). The water footprint assesses the amount of water necessary in all the aspects of production, delivery, use, and disposal of goods and services (Figure 4.8). The water footprint illustrates how daily consumption/utilization and the problems of water depletion and pollution exist in the respective regions where consumption/utilization occurs. The water footprint of nearly every type of good or service can be comparatively assessed.

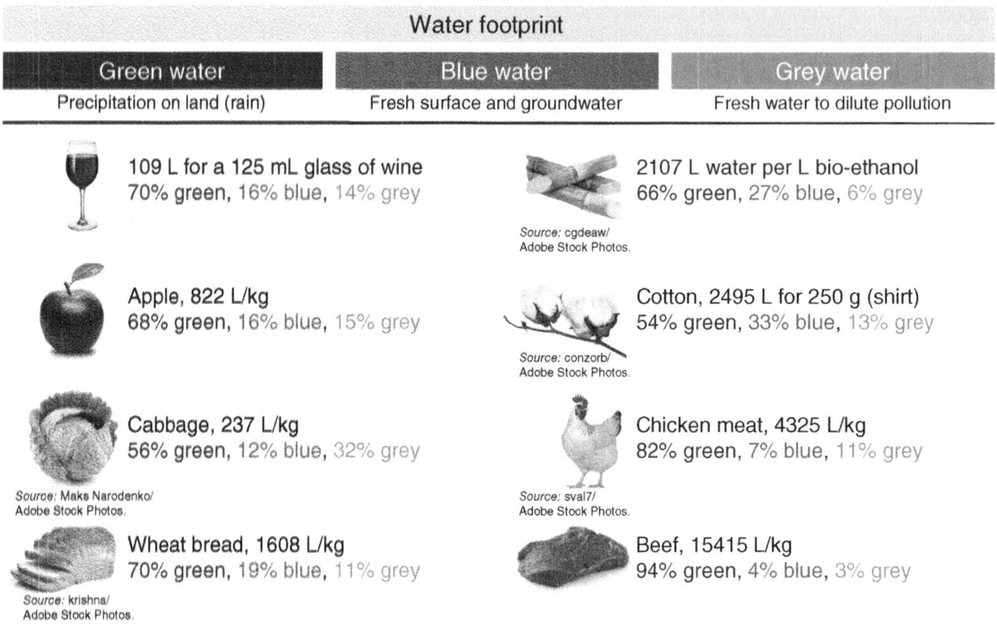

Figure 4.8 Water footprint for selected products.

4.2.5 Other Elemental Cycles

About 15–20 elements are commonly utilized and deposited or redistributed throughout the global ecosystem in a manner and timeframe where the cycle can be meaningfully analyzed. From a biological standpoint, the four most important elements and related cycles include carbon, nitrogen, sulfur, and phosphorus (Stevenson and Cole 1999).

Figure 4.9 presents a schematic of the global carbon cycle with the sinks and sources (as well as their relative magnitudes). Carbon is provided to the atmosphere through decomposition and respiration in the form of carbon dioxide (Eq. (4.1)):

$$C_6H_{12}O_6 + 6O_2 = 6CO_2 + 6H_2O + \text{energy} \tag{4.1}$$

Carbon is brought back to the land and oceans via photosynthesis (Eq. (4.2)):

$$6CO_2 + 6H_2O + \text{sunlight} \xrightarrow{\text{chlorophyll}} C_6H_{12}O_6 + 6O_2 \tag{4.2}$$

Approximately 6000 Pg of carbon are present within reserves of fossil fuels (Friedlingstein et al. 2022). Considering approximately 5 Pg/year of carbon is liberated from the combustion of fossil fuels on an annual basis, this results in approximately 1200 years' worth of total fossil fuel reserves.

Sulfur is present in the lithosphere, hydrosphere, and atmosphere (Brimblecombe 2015). In the lithosphere, sulfur is present in elemental form as well as in the mineral form (e.g. gypsum). Approximately 7800 trillion tons are present in sedimentary rocks, while 2375 trillion tons are present in oceanic rocks; approximately 1280 trillion tons of dissolved sulfate ions are present in the hydrosphere, while 6 million tons of sulfur (mostly carbonyl sulfide) is present in the atmosphere (Wang et al. 2021).

Figure 4.10 presents the sulfur cycle. Prior to industrial activity, approximately 55 million tons of reduced sulfide or dusts and 20 million tons of SO_2 from volcanic eruptions would be transported annually between the atmosphere and land, with an additional 140 million tons transported as

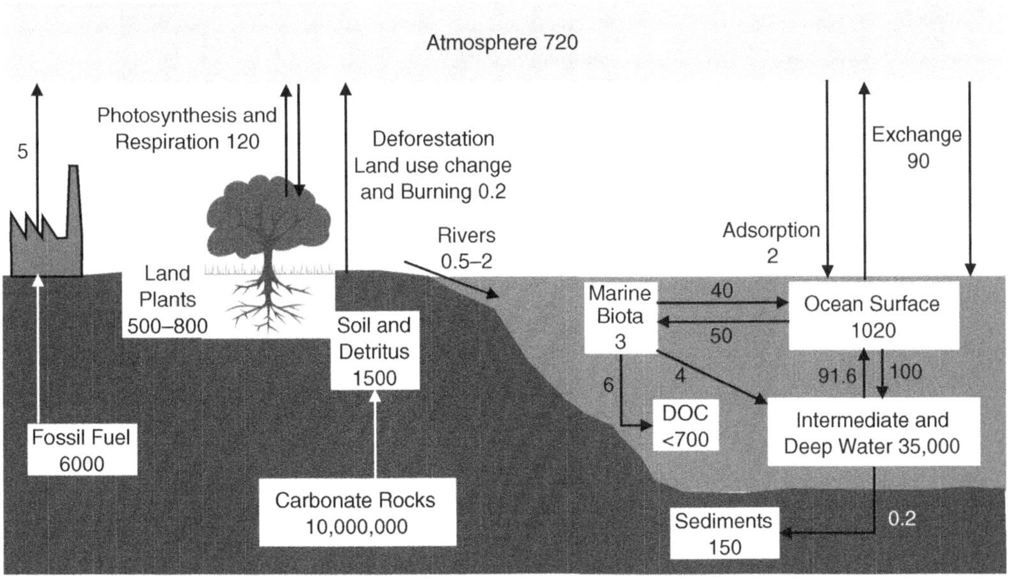

Figure 4.9 Global carbon cycle. (Reservoirs in Pg and fluxes in Pg/yr. 1 Pg = 10^{15} g).

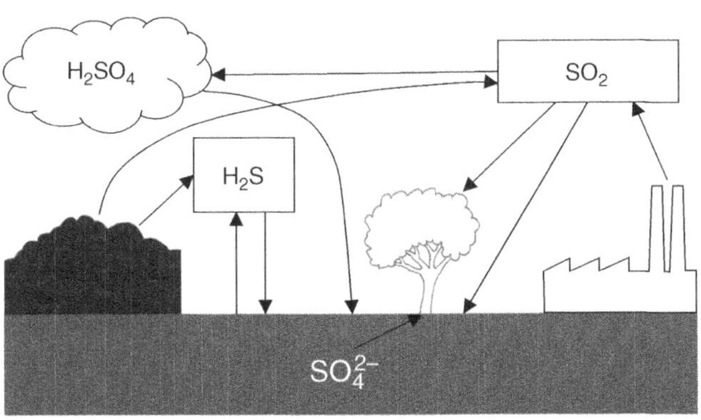

Figure 4.10 Global Sulfur Cycle.

ocean spray between the atmosphere and the oceans. However, global human activities now release nearly 150 million tons annually: 93 million tons of SO_2, 10 million additional tons of sulfur-containing dust, 29 million tons via fertilizers, and 13 million tons via wastewater.

Nitrogen is the most prevalent gas in the atmosphere, constituting approximately 78% of atmospheric gases. It is also ubiquitous in the lithosphere. Figure 4.11 shows the sources and sinks of nitrogen in the environment, and Figure 4.12 illustrates the natural nitrogen budget and global nitrogen cycle, respectively. Nitrogen is a fundamental element for plant growth, and it is intensively used in agriculture to increase land productivity. As a consequence, there are increasing amounts of nitrogen in water that lead to eutrophication of the rivers, lakes, and sea estuaries (Zhang et al. 2020).

Phosphorus in rocks and sediments is estimated to be 19 billion tons and 8.4×10^5 billion tons, respectively, with an estimated additional 100 billion tons present in soil, peat, and litter

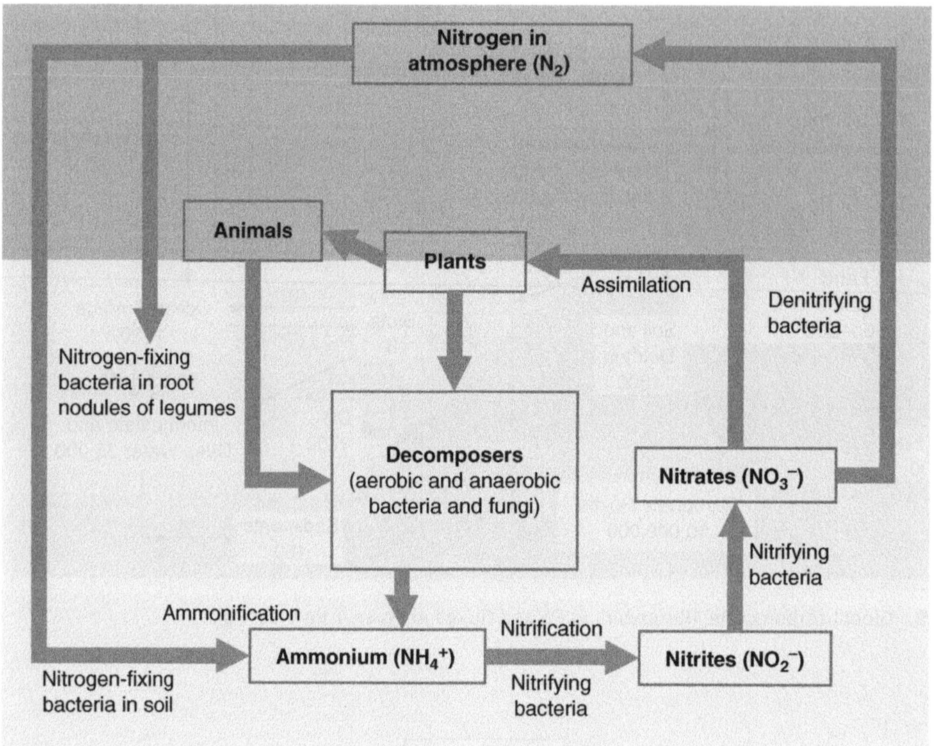

Figure 4.11 Sources and sinks of nitrogen in the environment.

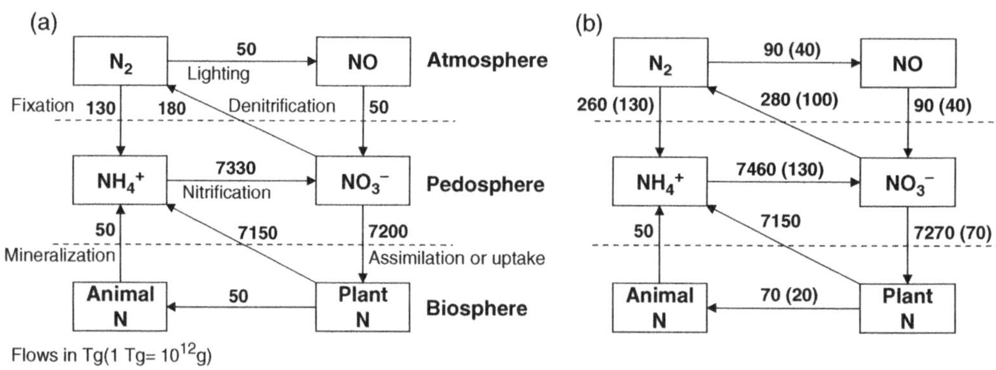

Figure 4.12 (a) Natural nitrogen budget and (b) global nitrogen cycle.

(Ruttenberg 2001). The major anthropogenic source of phosphorus is fertilizers, constituting approximately 14 million tons annually. Erosion, deposition, and dust emissions transport phosphorus to the hydrosphere and atmosphere. Figures 4.13 and 4.14 depicts the phosphorus cycle, and their flows and inventory. Similar to nitrogen, phosphorus is also used in agriculture and is also responsible for the eutrophication of waterbodies.

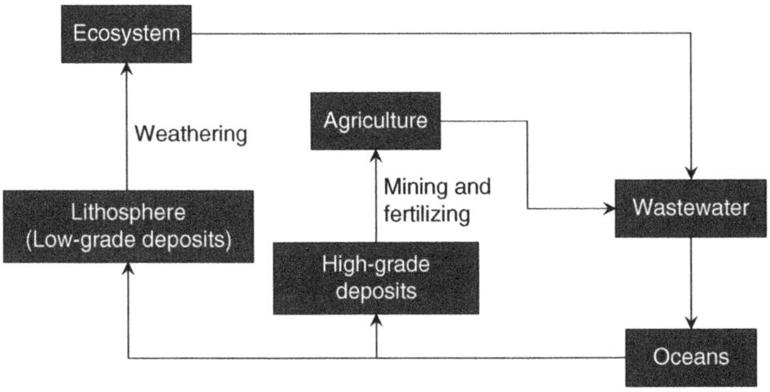

Figure 4.13 Global Phosphorus Cycle.

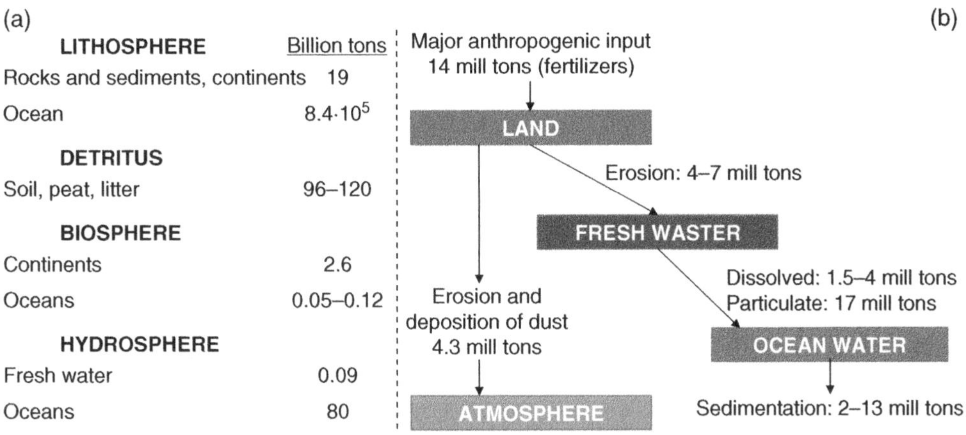

Figure 4.14 The phosphorus cycle: (a) global inventory and (b) present flow of P (tons per year).

Example 4.2

Consider a simplified model of the global phosphorus cycle that involves the following key components:

Weathering of Rocks – This is the primary source of phosphorus, from where it is released into the soil and water bodies. Assume the annual phosphorus release due to rock weathering is estimated to be 20 Tg (teragrams, where 1 Tg = 10^{12} g).

Agricultural Runoff – Human activities, particularly the use of phosphate fertilizers in agriculture, contribute significantly to phosphorus runoff into nearby water bodies. Assume that agricultural runoff adds an additional 18 Tg of phosphorus to aquatic systems annually.

Ocean Sedimentation – A portion of the phosphorus in water bodies is eventually deposited into ocean sediments. Assume that the annual rate of phosphorus sedimentation in the ocean is 10 Tg.

Uptake by Terrestrial Plants – Plants absorb phosphorus from the soil for their growth. Assume that the annual uptake of phosphorus by terrestrial plants is 30 Tg.

Decomposition and Return to Soil – When plants and animals die, the phosphorus contained in their biomass is returned to the soil through decomposition. Assume that this process annually returns 28 Tg of phosphorus to the soil.

Given these components, calculate the following:

a) What is the annual net change in phosphorus in terrestrial systems (soil + plants)?
b) What is the annual net change in phosphorus in aquatic systems?
c) Assuming there is a balance in the global phosphorus cycle, how much phosphorus is transported from aquatic systems to terrestrial systems annually through geological processes not listed above (e.g. uplift and erosion of ocean sediments)?

Solutions:
Given data, sources of phosphorus (P):

P released into soil + water bodies (weathering of rocks) = 20 Tg
P released into water bodies + aquatic systems (human activities, agricultural runoff) = 18 Tg
P sedimented into oceans = 10 Tg
Annual uptake of P by terrestrial plants = 30 Tg
P released into soil (decomposition and return) = 28 Tg

a) Annual Net Change in Phosphorus in Terrestrial Systems:

The inputs into the terrestrial system are:

Weathering of rocks: 20 Tg
Decomposition and return to oil: 28 Tg

The output from the terrestrial system is the uptake by plants, which is 30 Tg.
Therefore, the net change in phosphorus in terrestrial systems can be calculated as:

Net change = Inputs − Output
Net change = (20 + 28) Tg−30 Tg
Net change = 48 Tg−30 Tg
Net change = 18 Tg

So, the annual net change in phosphorus in terrestrial systems is 18 Tg.

b) Annual Net Change in Phosphorus in Aquatic Systems:

The inputs into the aquatic system are:

Agricultural Runoff: 18 Tg

The output from the aquatic system is ocean sedimentation, which is 10 Tg.
The ocean sedimentation represents phosphorus that settles down to the ocean floor, effectively removing it from the aquatic system. Therefore, in the context of this calculation, it serves as an output from the aquatic system.
Therefore, the net change in phosphorus in aquatic systems can be calculated as:

Net change = Inputs − Output
Net change = 18 Tg−10 Tg
Net change = 8 Tg

So, the annual net change in phosphorus in aquatic systems is 8 Tg.

c) Phosphorus Transported from Aquatic Systems to Terrestrial Systems:

To determine the phosphorus transported from aquatic systems to terrestrial systems, we rely on the assumption that there is a balance in the global phosphorus cycle. (The assumption of a balanced global phosphorus cycle is based on the principle of conservation of mass).

If there is a net outflow of phosphorus from aquatic systems (such as through sedimentation or other processes), it implies that an equivalent amount of phosphorus must be entering terrestrial systems to maintain balance.

This assumption is based on the idea that the phosphorus cycle operates as a closed system on a global scale, i.e, any phosphorus leaving aquatic systems must be balanced by inputs into terrestrial systems from geological processes not listed above.

Total phosphorus input from the listed processes is from weathering of rocks, agricultural runoff, and decomposition = (20 + 18 + 28) Tg = 66 Tg.

Total output of phosphorus from sedimentation and plants uptake = (10 + 30) Tg = 40 Tg.

The difference in the phosphorus input and output of 26 Tg is being transferred from aquatic to terrestrial systems through unlisted processes.

4.3 Resource Depletion

A sustained economic development based on the consumption of nonrenewable resources is impossible within a closed system, such as Earth. The concept of "resource depletion" refers to the exhaustion of raw materials in a region beyond the rate of restoration. Resource depletion is most commonly caused by various human activities such as farming, fishing, mining exploitation, and fossil fuels consumption. It is surprising how little attention is directed toward resource depletion while much attention is focused on climate change. Interestingly, these two challenges are linked because many measures to combat climate change will also reduce resource utilization.

4.3.1 Causes of Resource Depletion

The causes of resource depletion are many, including the following:

- Overconsumption/unnecessary use of resources
- Nonequitable distribution of resources
- Overpopulation
- Slash and burn agricultural practices
- Technological development
- Erosion
- Habitat and biodiversity destruction
- Irrigation
- Mining/extraction
- Aquifer depletion
- Inadequate forestry stewardship
- Pollution or contamination of resources.

"The Limits to Growth" was a relevant study published in 1972. This study alerted the governments, corporations, and society that human activities would start to exceed the capacity of the planet if the humanity did not change their rate of resource consumption (Meadows et al. 1972). The authors opined that within a generation, various precious metals and oil would likely be scarce and much more expensive due to the increasing demand and relatively lower availability. Since that study was published, resource use and waste generation, instead of decreasing, have continued to grow at an accelerated pace due to population growth and the development of the economies of many countries. Thus, researchers estimate that humanity's ecological footprint is at least 20% beyond what the Earth can sustain (Bastianoni et al. 2020). In a 2011 report, The United Nations Environment Programme estimated that the demand of minerals, ores, fossil fuels, and biomass will be 140 billion tons by 2050 (UNEP 2011). This is three times the current consumption, and much more than the Earth can supply. To combat this, there needs to be a collective commitment to improve recycling and repurposing of materials as well as investments into renewable energy.

4.3.2 Effects of Resource Depletion

Utilization and future resource depletion is illustrated in Figures 4.15–4.18. Figures 4.15 and 4.16 show the utilization of metals in the twentieth century as well as the history of Australian ores, plotted as the percent of metal mined in rock. The trend of mining poorer and poorer metal ores is obvious. Figures 4.17 and 4.18 demonstrate the energy required to produce or extract a range of materials. The energy required to extract and process copper ore at different ore grades is roughly linear on a log–log scale. It indicates that mining of an ore with a grade ten times poorer than that being mined today results in energy requirements roughly ten times as much as those required today, mostly to separate the ore mineral grains from their parent rock. Hence, energy limitations are readily imaginable if we continue extracting copper from an ore of rapidly diminishing quality. Finally, Figure 4.19 depicts the environmental disturbance and toxicity associated with specific mining activities. The toxicity of metals differs greatly; however, this chart (Figure 4.19) indicates a much higher risk associated with the mining of arsenic, cadmium, or mercury compared with copper, iron, or zinc.

Ultimately, resources are being used at increasing rates, which in turn requires increased energy for extraction and transportation. The uncontrolled mining and acquisition of these materials can have significant adverse effects on the environment. Oddly, materials continue to be utilized in an increasingly unsustainable manner as compared to other alternatives. The continued scarcity of these materials and the prices associated with them will be subjected to amplified pressures from increased demand and decreased supply.

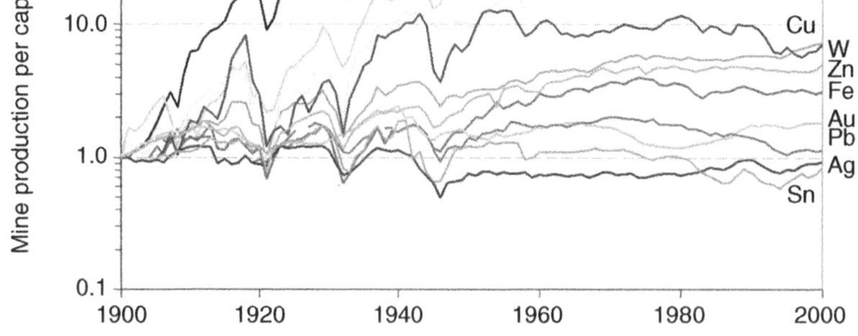

Figure 4.15 Metal utilization in the twentieth century.

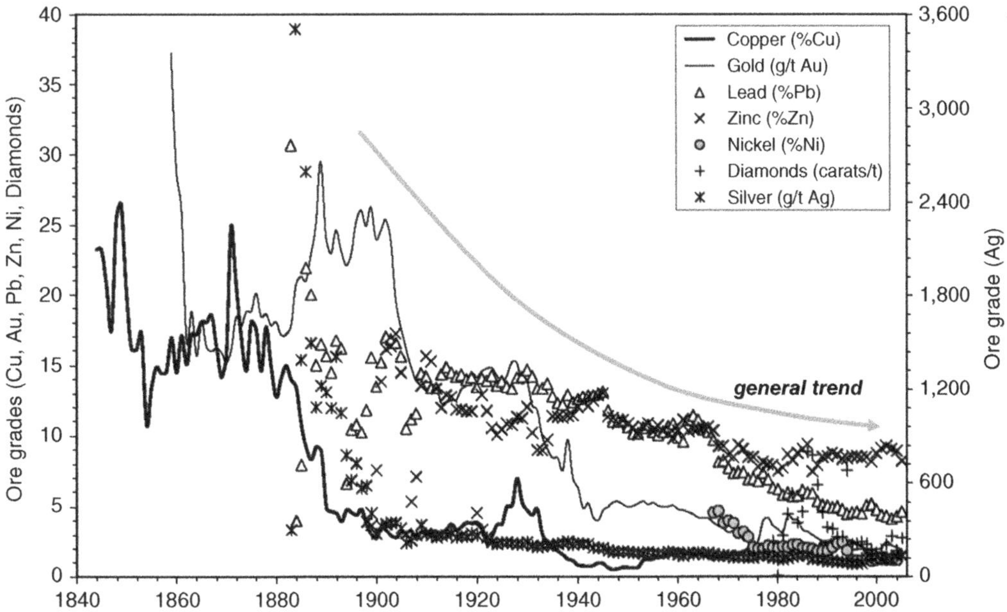

Figure 4.16 Consumption of the Australian ores.

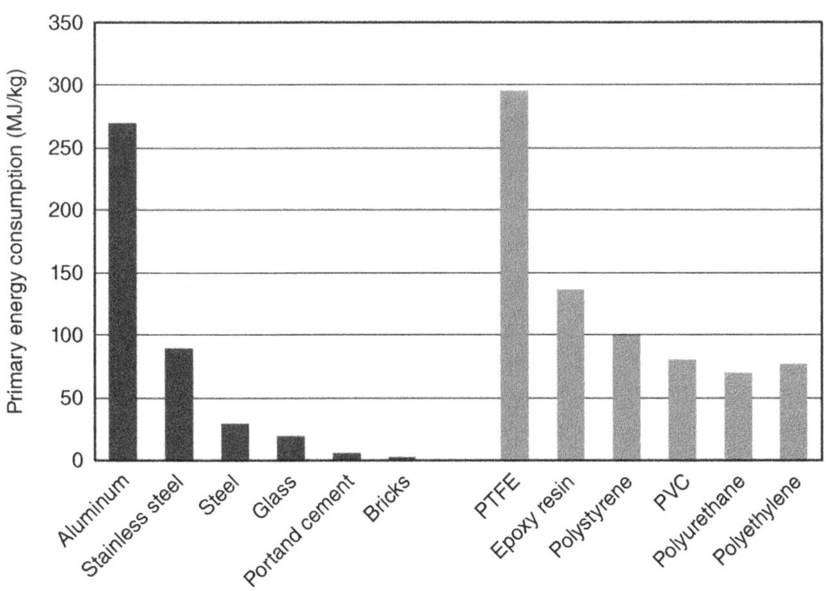

Figure 4.17 Energy requirements per unit of material. *Source:* Schuckert et al. (1997)/with permission of Munster Technology University.

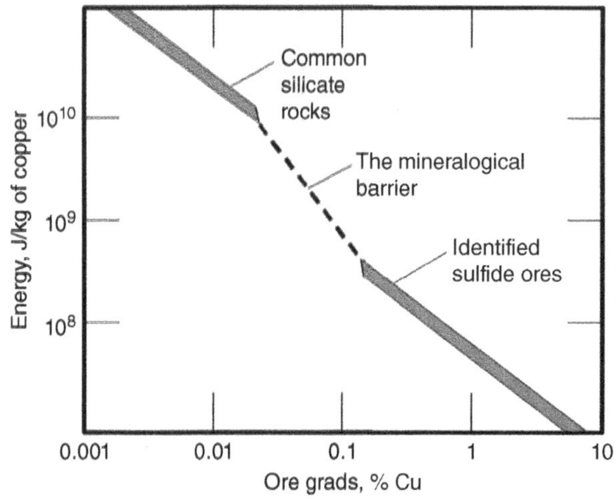

Figure 4.18 Energy required for copper mining as a function of ore grade. *Source:* With Gordon et al. (1987)/Reproduced with permission of Harvard University Press.

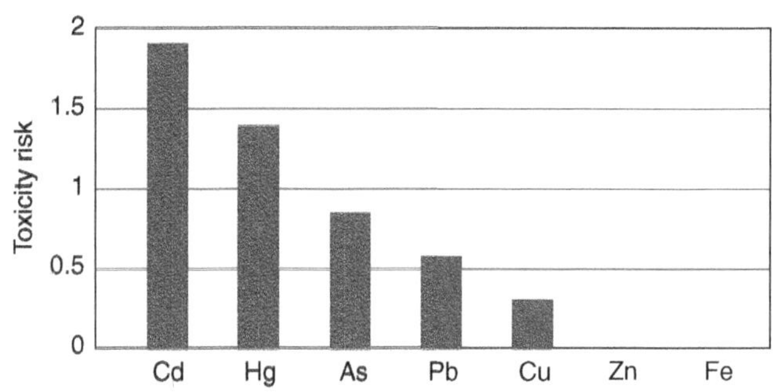

Figure 4.19 Toxicity risk potential of several common metals.

Example 4.3

A mining company is evaluating a section of a gold deposit to decide whether it is economically viable to develop a new open-pit mine. The section under consideration is a rectangular block measuring 500 m in length, 200 m in width, and 150 m in depth. The gold distribution within this block is known to be normally distributed due to extensive sampling and analysis. Estimate the recoverable reserves of gold in this block using a cut-off grade strategy and determine if the section is economically viable for mining operations. Assume the mean gold grade of the block is 2 g/t (grams per ton) with a standard deviation of 0.5 g/t.

- Calculate the cut-off grade for gold mining in this deposit, considering the gold price is \$1500 per ounce, mining cost is \$25 per ton of ore, processing cost is \$15 per ton of ore, and the mine's operational cost is \$10 million per year (1 troy ounce = 31.1035 g).
- Determine the tonnage of ore above the cut-off grade by integrating the grade-tonnage curve derived from the block's gold grade distribution.
- Estimate the total amount of recoverable gold in ounces above the cut-off grade.
- Calculate the gross revenue from the recoverable gold.

- Deduct the total mining, processing, and operational costs from the gross revenue to determine the net profit or loss.
- Assume a mining efficiency of 90% and a metallurgical recovery rate of 95%.

Solution:

For the purpose of this problem, the following symbols will be used, μ – mean gold grade, σ – standard deviation of gold grade, x – gold grade at any given point

Calculate the cut-off grade:

Cut-off grade is defined as the grade above which the mining remains economically viable, i.e., cost of mining and processing > selling price of gold.

Cost of mining and processing = \$25/t + \$15/t = \$40/t

Selling price of 1 g of gold = \$1500/ounce × (1/(31.1 ounce/g)) = \$48.23/g

If the cut-off grade is x (g/t), then the revenue can be calculated = x (g/t) × \$48.23/g × 0.9× 0.95 = \$41.23x (0.9 and 0.95 are mining efficiency and metallurgical recovery rates assumed)

For the mining to be economically viable, at the least revenue ≥ cost → 41.23x ≥ 40 → x ≥ 0.97

Therefore, the cut-off grade is 0.97 g/t

Ore-tonnage above the cut-off grade:

Total ore-tonnage = (500 × 200 × 150) m^3 × 2000 kg/m^3 = 33.07 Mt (assuming bulk density of ore to be 2000 kg/m^3)

Given the gold grade distribution is normal, the total ore-tonnage above the cut-off grade (G_{cut}) can be calculated as = P (G > G_{cut}) × 33.07 Mt

To calculate P (G > G_{cut}) for a normal distribution, we can calculate the

$$\text{z-value, } z = \frac{x - mean}{standard\ deviation} = \frac{0.97 - 2}{0.5} \approx -2$$

For a z-value of -2, P (G > G_{cut}) = 97.72% (can be inferred from any normal distribution cumulative distribution function tables)

Therefore, the ore-tonnage above the cut-off grade = 0.9772 × 33.07 Mt = 32.32 Mt

Total recoverable gold:

To calculate this, we need to integrate the cut-off grade distribution above cut-off grade with the corresponding tonnage = $32.32\,Mt \times \int_{0.97}^{\infty} x\,P(x)dx$, where x is the gold grade.

$$\text{For a normal distribution, } P(x) = \frac{1}{\sigma\sqrt{2\pi}} e^{-\frac{1}{2}\left(\frac{x-\mu}{\sigma}\right)^2} = 0.8e^{-8-2x^2+8x}$$

Therefore, the recoverable gold can be calculated as $32.32\,Mt \times \int_{0.97}^{\infty} x \times 0.8 \times e^{-8-2x^2+8x}\,dx$

Solving this integral directly is difficult; hence, a numerical solution was obtained and the total recoverable gold was calculated to be 32.32 Mt × 1.98 g/t = 64.06 Mg.

Considering the mining efficiency and metallurgical recovery rate, the actual total recoverable gold is 54.78 Mg

Net Profit or Loss Calculations:

Gross revenue from the actual total recoverable gold = 54.78 Mg × \$48.23/g = \$2.642 billion

Mining costs = \$25/t × 32.32 Mt = \$0.808 billion

Processing costs = \$15/t × 32.32 Mt = \$0.485 billion

Considering a typical annual mining rate of 10 Mt/year for open-pit gold mines

$$\text{Annual operational costs} = \$10 \text{ million per year} \times \frac{32.32\,Mt}{10\,Mt/year} = \$32.32 \text{ million}$$

Total Profit = Gross revenue – Mining costs – Processing costs – Operational costs

Therefore, the total profit for the section under consideration accounts to \$1.317 billion

4.3.3 Overshooting

Occasionally, resource depletion (or at least a significant nonsustainable period of consumption) can happen without intention or notice. This concept is known as overshooting. This may result from rapid population and/or economic growth coupled with a rapid increase in resource use. Unfortunately, there can be a delay in the perception of the overuse and the responses that would act to keep the system within its limits (EEA 2020).

When overshooting occurs, resource stocks collapse, and waste and pollution can rapidly accumulate. As a result, economic and human investment is necessarily diverted from other functions to deal with the waste (e.g. wastewater treatment, disposal). Although some positive attributes can occur, such as the invention and adoption of technologies that lessen the need for a given resource, waste accumulation can overcome natural absorption processes, and the additional energy and resources expended to develop the resource can add to the waste generation.

Overshooting can result from unfavorable consumption patterns. With respect to consumption, three general categories have been developed. *Background consumption* is the first category, which is defined as the consumption needed by an organism for survival and reproduction. *Overconsumption*, the second category, is defined as the level of consumption that undermines a species' own life support system and abilities through which individuals and collective groups can make choices. *Misconsumption,* the third category, undermines an individual's own well-being, even if there are no aggregate effects on the population or the species.

Consumption can also result in high rates of change with respect to resource use. For instance, iron and steel use/production in China has experienced a dramatic change. In 2002, China added more steel production capacity than all of the capacity that existed in the United States at that time (Figure 4.20). This behavior can be observed in other lesser-developed countries (LDCs), where it is common for resources to be extracted for consumption by others (Figure 4.21).

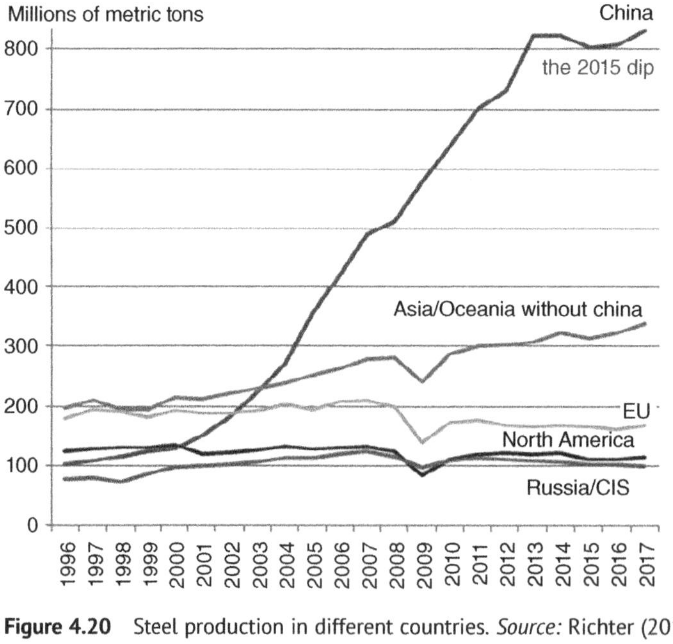

Figure 4.20 Steel production in different countries. *Source:* Richter (2018)/Wolf Street Corp.

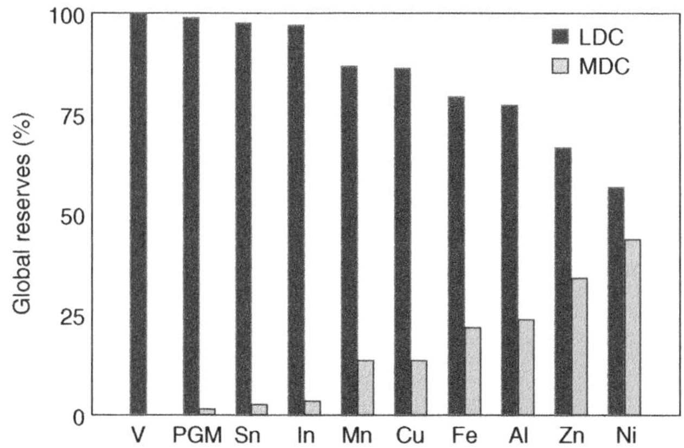

Figure 4.21 Global Reserves in LDCs and Moderately Developed Country of selected materials. *Source:* USGS (2008)/Mineral Commodity Summaries/US Geological Survey/Public domain.

4.3.4 Urban Metabolism

Consumption behavior may be modeled similar to a metabolic process. This concept, referred to as "urban metabolism," may be defined in terms of stocks and flows, with the contributing urban activities. The urban stocks that occur via urban metabolism are of importance due to the quantity associated with these stocks. The nature of the stocks consists of significant amounts of recyclable materials, which can serve as a valuable resource in themselves. Additionally, the quantity of waste flows associated with urban metabolism is important to consider for the design of adequate waste and wastewater handling facilities. The location of these stocks and flows are also important considerations in devising systems and processes to utilize stocks and handle flows. Most urban areas continue to show per capita increases in the use of water, energy, and materials, as well as the generation of wastes and emissions. Further, urban growth can affect microclimate and regional ecosystems due to resource use and urban heat island effects.

To date, urban metabolism studies are few and highly specialized and specific. Comparisons of cities and urbanized areas are not yet possible. To facilitate comparisons that could affect changes or mitigation activities, consistent metrics are necessary to compare performance and monitor progress.

4.4 Summary

Resources may be broadly classified into two types: renewable or nonrenewable. The greater attention is on nonrenewable resources, as overutilization of these resources can result in their depletion, which conflicts with the principles of sustainability.

Whether used as energy (fossil fuels), raw materials for manufacturing (petroleum, metals), or other essential processes (fresh water), if not preserved and protected, or at least used at sustainable levels, these resources may become overexploited, which may lessen or eliminate their ability to be utilized by future generations. Additionally, use and liberation of other key elements, such as carbon, sulfur, nitrogen, and phosphorus, can have detrimental effects to the environment when overexploited and redeposited in the environment at rates or volumes greater than what can be handled by the natural absorption capacity of the environment.

Overutilization of these resources can result in different complications. In the case of a general resource depletion, resources are used at a rate that exceeds the rate of generation. If this continues unabated, the reserves of the resource will become depleted. In some cases, "overshooting" may occur; this phenomenon can result in a series of events like a rapid collapse of resource availability.

In order to achieve sustainability, the long-term protection and preservation of natural resources is critical. In particular, given their finite nature, it is of utmost importance to preserve the nonrenewable resources currently available.

4.5 Questions

4.1 Name a few nonrenewable and renewable sources of energy.

4.2 Discuss the concept of "peak coal." How can estimates of "peak coal" vary by the metric used for estimation?

4.3 Explain how crude oil reserves and oil shale/tar sand exploitation is expected to have a more detrimental effect when considering sustainability over time.

4.4 What is the most common nuclear fuel used in nuclear energy generating facilities?

4.5 Discuss benefits and drawbacks associated with radioactive fuel use.

4.6 Explain why extraction of "virgin" copper reserves would be expected to become more expensive with time.

4.7 Select a consumer product and discuss its water cycle with respect to manufacturing.

4.8 Discuss adverse side effects associated with the mining and refining of heavy metals.

4.9 List specific examples of how agricultural practices could have an effect on the environment with respect to sustainability.

4.10 What are the major causes for resource depletion?

4.11 What are the effects of resource depletion? Explain briefly.

4.12 What are the issues related to resource depletion from sustainability perspective? How would you assess resource availability?

4.13 Explain how embedded energy, local supply, and environmental impacts associated with recovery may be considered in resource material acquisition.

4.14 Explain how alternate material substitutions can impact resources and environment.

4.15 What is overshooting? Explain briefly.

4.16 Describe how overshooting could have a sudden and detrimental effect on human societal units.

4.17 Discuss the concept of urban metabolism.

4.18 How could cities improve metrics related to urban metabolism, and why would cities have a metabolic advantage over less densely populated areas, such as rural communities?

4.6 Problems

4.1 As of 2023, the world's proven oil reserves are estimated to be around 1.65 trillion barrels. Based on an annual consumption rate of 35 billion barrels, calculate:

- How many years will it take to deplete these reserves, assuming the annual consumption remains constant?
- If global oil consumption increases by 2% annually, how many years will it take to deplete the reserves?
- Research and discuss potential implications of oil depletion for different sectors (e.g. transportation, energy production).

4.2 India relies significantly on coal for its energy needs. The total recoverable coal reserves in the country are estimated to be 388 billion metric tons as of 2021, and the extraction capacity is only 740 million metric tons per year. The current annual coal consumption for energy needs is around 876 million metric tons. The electricity consumed in India for the year 2021 is about 1 Gton growing rapidly at 8% per year.

- Determine the percentage of electricity needs fulfilled by coal-powered energy in the year 2021?
- What percentage of the coal used is being imported?
- Assuming that the coal production capacity increases at an average of 15% year on year for the foreseeable future, determine the year by which India will become self-sufficient in terms of coal requirements (considering the share of energy needs fulfilled by coal remain constant).
- Find information online regarding capacities of other sources of energy for India and their annual growth rates, and consider the facts presented in this problem. Determine how long would the coal reserves last in India, assuming that the country does not export any of the coal produced?
- Comment on the sustainability of the current state of energy use by a country like India drawing inferences from other countries and sustainable development goals.

Note: Make suitable assumptions for any missing data.

4.3 Four countries have different levels of natural gas reserves, extraction rates, consumption rates, and potential for future discovery. The data provided for the year 2023 is as follows:

Country	Reserves (trillion ft^3)	Annual extraction rate (trillion ft^3)	Annual consumption rate (trillion ft^3)	Potential for future discovery (% increase in reserves year over year)
1	100	2 (+2%)	1.5 (+3%)	10.0
2	150	4 (+3.5%)	3.0 (+2%)	15.0
3	200	3 (−1.5%)	2.0 (+1%)	12.5
4	250	6 (+4.0%)	4.5 (+2.5%)	0

Note: The figures in brackets are the annual rates of change in the corresponding values.

Considering that the annual rates of change remains constant for the next 10 years, and the potential for discovery reduces by 20% each year for all the countries, determine the following:

Calculate the remaining reserves for each country at the end of 2033, considering both the extraction rates and the potential for future discoveries (plot the extraction and consumption rate with time).

- Determine the year in which each country will exhaust its natural gas reserves if the current extraction and consumption rates continue unchanged.
- If each country aims to extend the availability of its natural gas reserves by 50% beyond the exhaustion year calculated in question 2, by what percentage should each country reduce its annual extraction rate, assuming no changes in consumption rates and no further discoveries?
- Now try to find similar data for real countries (at least 2) from online sources, and solve for all the above questions? (you can assume any data that you cannot find)

4.4 The global nitrogen cycle is crucial for maintaining the balance of nitrogen in terrestrial and aquatic ecosystems. Human activities, particularly the industrial synthesis of ammonia for fertilizers and the cultivation of nitrogen-fixing crops, have significantly altered the nitrogen fluxes in the environment. Consider the following data to answer the questions that follow:

- Natural nitrogen fixation (by lightning and biological fixation in soils and water) globally adds approximately 100–200 Tg (teragrams) of nitrogen to terrestrial ecosystems annually.
- Industrial nitrogen fixation for fertilizers and other chemicals adds about 120 Tg of nitrogen per year.
- The cultivation of nitrogen-fixing crops contributes an additional 40 Tg of nitrogen per year to terrestrial ecosystems.
- The estimated total denitrification process, which converts nitrates back to N_2 gas, removes about 210 Tg of nitrogen from terrestrial and aquatic ecosystems annually.

Questions:

- Calculate the total annual addition of nitrogen to terrestrial ecosystems from natural and human sources.

- Determine the net increase or decrease in nitrogen added to terrestrial ecosystems when considering the denitrification process.
- Discuss the potential ecological impacts of the net increase or decrease in nitrogen on terrestrial and aquatic ecosystems.
- Discuss strategies that could be implemented to mitigate the negative impacts of increased nitrogen in ecosystems.

4.5 Consider the following data for a region of area 10 000 m^2. Precipitation: 1200 mm/year, evapotranspiration: 800 mm/year, surface runoff: 200 mm/year, storage capacity from rainwater harvesting: 1000 m^3, outflow to river system (O): 150 mm/year. Using the given data calculate the following:

- Groundwater Recharge:
  - Determine the annual groundwater recharge for the region. Use the water budget equation considering the region is in a steady state.
- Water Budget Analysis:
  - Analyze the contribution of each component (precipitation, evapotranspiration, surface runoff, groundwater recharge, and outflow) to the region's water budget. Present in terms of percentages relative to the total precipitation.
- Assume a climate change scenario where the annual precipitation is expected to decrease by 10%, and evapotranspiration is expected to increase by 5% due to higher average temperatures:
  - Recalculate the groundwater recharge under this new scenario.
  - Discuss the potential implications of these changes on the regional water budget and water resource management strategies that could be adopted to mitigate adverse impacts.

4.6 As of 2023, the estimated global copper reserves are 830 million metric tons. The annual copper production is approximately 20 million metric tons, primarily used in electrical wiring, roofing and plumbing, and industrial machinery. Due to technological advancements and the increasing popularity of electric vehicles (which use about four times as much copper as conventional cars), the demand for copper is expected to rise by 3% annually. However, new copper discoveries and mining technologies are improving the efficiency of copper extraction, leading to a 1% increase in the total recoverable copper reserves each year.

- Calculate the total global copper consumption over the next 10 years, assuming the demand increases by 3% annually.
- Determine the total global copper reserves at the end of 10 years, taking into account the 1% annual increase in recoverable reserves.
- Based on the consumption and new reserve estimates, calculate how many years it will take for the current copper reserves to be depleted, assuming no significant changes in production efficiency or discovery rates beyond the 10-year period.
- Discuss the implications of these findings for global resource management and the importance of recycling and finding alternative materials.

4.7 A simplified representation of the global carbon cycle considering the inputs from fossil fuel burning in addition to the natural reservoirs is presented in the figure below:

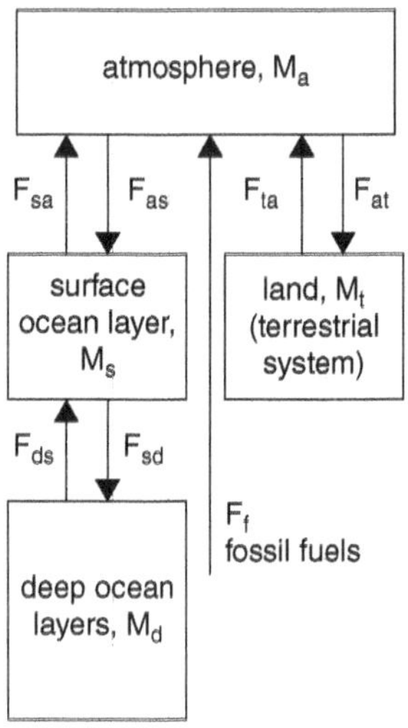

The fluxes of carbon into or out of each reservoir is represented by the following equations:

$$F_{as} = (0.143)M_a \tag{4.1}$$

$$F_{sa} = \left(10^{-25}\right)M_s^{9.0} \tag{4.2}$$

$$F_{at} = (16.2)M_a^{0.2} \tag{4.3}$$

$$F_{ta} = (0.0200)M_t \tag{4.4}$$

$$F_{ds} = (0.00129)M_d \tag{4.5}$$

$$F_{sd} = (0.450)M_s \tag{4.6}$$

where the units for masses in these equations is in Pg (Petagram = 10^{15} g).

In an unsteady state, say for example, when the inputs from fossil fuels is non-zero, the mass-balance equation for the rate of change of mass of carbon in a reservoir is given by

$$\frac{dM_a}{dt} = \left(10^{-25}\right)M_s^{9.0} + (0.0200)M_t - (0.143)M_a - (16.2)M_a^{0.2} + F_f \tag{4.7}$$

Similar equations can be employed for the other three environments (M_s, M_t, and M_d).

a) Considering that the carbon released from fossil fuels reduces to zero with immediate effect, and the total mass of carbon in all the four environments added together is 51 900 Pg. Determine the values of mass of carbon in each environment i.e. M_a, M_t, M_s, and M_d.
b) Considering that the carbon released due to fossil fuel burning reduces from the current level of 4 Pg/year to zero over the next 80 years, and the carbon cycle reaches a steady state after 80 years. Determine the masses of carbon in each environment after 80 years. Also, determine what fraction of the carbon produced from fossil fuel burning resides in the atmosphere after this period.

c) For this part, you have to write down mass balance equations similar to that of Eq. (4.7) for all four environments. Now considering an unsteady state, and an initial mass of M_a, M_t, M_s, and M_d to be 800, 4000, 1500, and 40 000, respectively, solve for the following problems:
d) Considering the carbon input from fossil fuel burning stays at a constant value of 4 Pg/year for the next 80 year period. Solve for the amount of carbon in each reservoir for the next 80 years taking fixed time interval of 5 years. i.e. calculate the mass of carbon in each environment for every 5 years for the next 80 years. Present the results in tabular form as well as on a graph.
e) Now, considering the input mass of carbon from fossil fuel burning reduces from current level of 4 Pg/year to zero over the next 80 years as in part-b, generate similar results as the previous problem and present your solution in tabular and graphical form.

Note: To understand further about this problem, read pages 1 through 7 carefully in this article: pratclif.com (2021). Introduction to biogeochemical cycles: material balances for carbon. Available at: http://pratclif.com/climatechange/Carbon_Cycle.pdf (accessed 24 March 2025).

References

Bastianoni, S., Niccolucci, V., Neri, E. et al. (2020). Sustainable development: ecological footprint in accounting. In: *Managing Human and Social Systems*, 301–320. CRC press.

Brimblecombe, P. (2015). BIOGEOCHEMICAL CYCLES. Sulfur Cycle. In: *Encyclopedia of Atmospheric Sciences (Second Edition)* (ed. G.R. North, J. Pyle, and F. Zhang), 187–193. Academic Press. ISBN 9780123822253. https://doi.org/10.1016/B978-0-12-382225-3.00015-3.

Bunsen, J., Berger, M., and Finkbeiner, M. (2021). Planetary boundaries for water–A review. *Ecological Indicators* 121: 107022.

USDOE (1994). *Handbook of Methods for the Analysis of the Various Parameters of the Carbon Dioxide System in Sea Water; Version 2* (eds. A.G. Dickson and C. Goyet). ORNL/CDIAC-74. https://www.nodc.noaa.gov/ocads/oceans/DOE_94.pdf.

EEA (2020). Is Europe living within the limits of our planet? European Environment Agency. EEA Report 01/2020. Luxembourg: Publications Office of the European Union. ISBN 978-92-9482-215-6. doi:10.2800/890673.

ENN (2010). Peak Oil, Then Coal. Environmental News Network. https://www.enn.com/articles/42071-peak-oil,-then-coal (accessed 7 October 2024).

Friedlingstein, P., O'Sullivan, M., Jones, M.W. et al. (2022). Global Carbon Budget 2021. *Earth System Science Data* 14 (4): 1917–2005. https://doi.org/10.5194/essd-14-1917-2022.

Gordon, R.B., Koopmans, T.C., Nordhaus, W.D., and Skinner, B.J. (1987). *Toward a New Iron Age?: Quantitative Modeling of Resource Exhaustion*. Harvard University Press.

Haxel, G.B., Hedrick, J.B., Orris, G.J., Stauffer, P.H., and Hendley, J.W. (2002). Rare earth elements: critical resources for high technology. United States Geological Survey (USGS). Fact Sheet 087-02. https://doi.org/10.3133/fs08702.

IEA (2024). Oil Market Report - December 2023. International Energy Agency. https://www.iea.org/reports/oil-market-report-december-2023 (accessed 7 October 2024).

IWCC (2023) IWCC Short-Term Forecasts for Copper. International Wrought Copper Council. https://www.coppercouncil.org/iwcc-press-release-short-term-forecasts-for-copper-oct-2023 (accessed 7 October 2024).

Meadows, D.H., Meadows, D.H., Randers, J., and Behrens, W.W. III (1972). *The Limits to Growth: A Report to the Club of Rome (1972).* New York: Universe Books.

NEI (2017). Nuclear statistics. Nuclear Energy Institute. https://www.nei.org (accessed 11 April 2017).

NSE (2024). NSE COPPER DERIVATIVES Performance Review 2023–24. https://nsearchives.nseindia.com/web/sites/default/files/inline-files/Performance%20review%20for%20FY23-24%20-%20Copper.pdf (accessed 7 October 2024).

REDELCA (2023). Copper. https://redelca.com/en/copper/ (accessed 7 October 2024).

Richter, W. (2018). Who Dominates Global Steel Production & Trade? Wolf Street. https://wolfstreet.com/2018/06/01/who-dominates-the-steel-trade-that-trump-just-hit-with-tariffs/ (accessed 31 May 2024).

Ruttenberg, K.C. (2001). Phosphorus cycle. In: *Encyclopedia of Ocean Sciences* (ed. J.H. Steele), 2149–2162. Academic Press. ISBN 9780122274305. https://doi.org/10.1006/rwos.2001.0277.

Schuckert, M., Beddies, H., Florin, H. et al. (1997). Quality requirements for LCA of total automobiles and its effects on inventory analysis. In: *Proceedings of the Third International Conference on Ecomaterials*, 325–329. Tokyo: Society of Non-Traditional Technology.

Stevenson, F.J. and Cole, M.A. (1999). *Cycles of Soils: Carbon, Nitrogen, Phosphorus, Sulfur, Micronutrients.* Wiley.

UNEP (2011). *Decoupling natural resource use and environmental impacts from economic growth, A Report of the Working Group on Decoupling to the International Resource Panel.* Fischer-Kowalski, M., Swilling, M., von Weizsäcker, E.U., Ren, Y., Moriguchi, Y., Crane, W., Krausmann, F., Eisenmenger, N., Giljum, S., Hennicke, P., Romero Lankao, P., Siriban Manalang, A., Sewerin, S.

USDOE (2024). Enhanced Oil Recovery. US Department of Energy. Office of Fossil Energy and Carbon Management. https://www.energy.gov/fecm/enhanced-oil-recovery (accessed 7 October 2024).

USEIA (2018). Hydraulically fractured horizontal wells account for most new oil and natural gas wells. U.S. Energy Information Administration. https://www.eia.gov/todayinenergy/detail.php?id=34732 (accessed 7 October 2024).

USEIA (2024a). Coal. U.S. Energy Information Administration. https://www.eia.gov/coal/ (accessed 31 May 2024).

USEIA (2024b). Petroleum & other liquids. Energy Information Administration. https://www.eia.gov/petroleum/ (accessed 31 May 2024).

USEIA (2024c). Natural gas. U.S. Energy Information Administration. https://www.eia.gov/naturalgas/ (accessed 31 May 2024).

USGS (2008). Mineral Commodity Summaries, 2009. US Geological Survey. Government Printing Office.

USGS (2024). National Minerals Information Center. U.S. Geological Survey. https://www.usgs.gov/centers/national-minerals-information-center (accessed 31 May 2024).

Wang, Z., Liu, Y., Liu, S. et al. (2021). Advances in the biogeochemical sulfur cycle in natural ecosystems. *Science of the Total Environment* 754: 142075. https://doi.org/10.1016/j.scitotenv.2020.142075.

Weerasooriya, R.R., Liyanage, L.P.K., Rathnappriya, R.H.K. et al. (2021). Industrial water conservation by water footprint and sustainable development goals: a review. *Environment, Development and Sustainability* 23 (9): 12661–12709.

WFN (2024). What is a water footprint. Water footprint Network. https://www.waterfootprint.org/water-footprint-2/ (accessed 7 October 2024).

WHO (2024). Water supply, sanitation and hygiene monitoring. World Health Organization and UNICEF Joint Monitoring Programme. https://www.who.int/teams/environment-climate-change-and-health/water-sanitation-and-health/monitoring-and-evidence/wash-monitoring (accessed 7 October 2024).

WNA (2016). Uranium mining overview. World Nuclear Association. http://www.world-nuclear.org (accessed 11 April 2017).

WNA (2018). Radioactive Waste – Myths and Realities. World Nuclear Association. https://world-nuclear.org/information-library/nuclear-fuel-cycle/nuclear-waste/radioactive-wastes-myths-and-realities (accessed 7 October 2024).

WWAP (2018). The United Nations World Water Development Report 2018: Nature-Based Solutions for Water. United Nations World Water Assessment Programme, UN-Water. Paris, UNESCO. ISBN: 978-92-3-100264-9.

Zhang, X., Ward, B.B., and Sigman, D.M. (2020). Global nitrogen cycle: critical enzymes, organisms, and processes for nitrogen budgets and dynamics. *Chemical reviews* 120 (12): 5308–5351.

5

Disaster Resiliency

Chapter Goal and Objectives

In previous chapters, we have discussed the importance of sustainability and resilience, defined relevant terms and concepts, and delved deeper into the drivers of sustainability. In this chapter, we will discuss the concepts of resilience in further detail and present how we can quantify the disaster resiliency of engineering systems. The specific objectives of this chapter are to (i) understand the increasing impacts of climate change, (ii) discuss concepts of resiliency and review policies in various professional organizations related to resiliency, and (iii) explore frameworks for the resilient design of infrastructure, presented with case studies and real-world examples of resilient infrastructure.

5.1 Introduction

Resiliency has increasingly gained salience in different fields, such as ecology, climate science, community, cyberspace, energy, and infrastructure. Correspondingly, related terms, including ecological resiliency, climate resiliency, cyber resiliency, energy resiliency, and infrastructure resiliency, have become popular, depending on the field of consideration. In general, resiliency is defined as the ability of a system to recover from a shock or disturbance and adapt to such changes that may occur in future. Therefore, ecological resiliency refers to the capacity of an ecosystem to recover from perturbations; climate resiliency means the ability of systems to recover from climate change-induced extreme events; cyber resiliency means the ability of a computer network to maintain its service during influences from external stressors, such as a security breach; energy resiliency means the ability to generate and supply energy even in adverse conditions; and infrastructure resiliency refers to engineered infrastructure that can withstand, respond to, or adapt to a vast range of disruptive events. The concept of resilience during and after extreme and/or unexpected events was originally introduced by Holling (1973) and Timmerman (1981) for ecological systems and human communities.

Disaster resiliency is another term commonly used with reference to built or engineered infrastructure (especially civil infrastructure) to survive impacts of extreme events associated with climate change. In this case, resiliency refers to the capability of an engineered system to withstand and recover from climate impacts, such as sea level rise, extreme flooding, extreme drought,

Sustainable and Resilient Engineering: Drivers, Metrics, Tools, and Applications, Second Edition. Krishna R. Reddy, Claudio Cameselle, and Jeffrey A. Adams.

Companion website: www.wiley.com/go/reddy/sustainableresilientengineering2e

wildfires, and other stressors. Unfortunately, the frequency and magnitude of these climate impacts are not accurately predicted or forecasted; hence, engineers are challenged to design infrastructure systems for nonstationary design criteria developed from observed impacts. Further, system designs need to balance the ability to withstand frequent, low-magnitude stressors with the ability to also withstand low-frequency, large-magnitude events; these goals are often not aligned and can often be at odds with each other (Anderies et al. 2019; Deslatte et al. 2023). Nevertheless, the ideal goal of designing resilient engineered systems is to have them undergo less damage during extreme events, and then recover to full function quickly.

Engineers have a moral obligation to consider resiliency in engineering designs to prevent loss of property and life while also protecting the environment. The expectation for these designs is that they will be capable of functioning to some degree following exposure to a stressor event, even if the event creates conditions well beyond their design limits. Further, there is an increasing focus on considering "safe-to-fail" infrastructure in place of increasingly robust "fail safe" infrastructure as climate change is increasingly introducing unknowns to upper limits of "credible" weather events that need to be incorporated for designs. Further, because of difficulties in quantifying resiliency in engineering projects due to the uncertainties involved, it can be challenging to design and implement resilient solutions.

This chapter presents stressors associated with climate change and extreme events, the concept of resiliency, a discussion of a range of initiatives developed by different entities to address resiliency, resiliency quantification and resilient design frameworks, examples of resilient designs, and finally, challenges associated with resilient engineering designs.

5.2 Climate Change and Extreme Events

In Chapter 2, greenhouse gases, global warming, and climate change were discussed. Scientific data from the Intergovernmental Panel on Climate Change (IPCC), National Oceanic Atmospheric Administration (NOAA), United States Environmental Protection Agency (USEPA), United States Global Change Research Program (USGCRP), and Climate Impacts Group (CIG) Scientific Studies indicate that certain climate trends are occurring on global, national, and local scales. In May 2018, for the first time in the recorded history, the average monthly level of CO_2 in the atmosphere exceeded 421 parts per million (ppm), according to the NOAA Global Monitoring Laboratory (NOAA 2024, accessed 2 April 2024). Further, since the CO_2 levels tend to be higher in indoor environments within urbanized areas partially because of combustion for cooking and heating, numerous urban populations are at risk from physiological effects of CO_2, including impaired cognition skills. The increase in atmospheric CO_2 concentrations also contributes to global warming that causes extreme events such as sea level rise, heat waves, flooding, drought, wildfires, and hurricanes. Recently, increases in CO_2 levels have been accelerating when compared to historic trends, with some experts predicting that CO_2 levels could reach 950 ppm by the end of the twenty-first century (IPCC 2021).

5.2.1 Rising Temperatures

If the higher estimates of atmospheric CO_2 levels occur, reaching up to 950 ppm by 2100, average global temperatures could increase as much as 6 °C. Whether temperatures reach this extreme projection or rise a fraction thereof, increased average temperatures can play havoc with a wide range of natural processes and affect both ecological and human well-being. As annual air temperatures are projected to increase, average temperatures by the 2050s will be higher than the warmest temperatures in the previous 100 years. These changing temperature patterns contribute to more severe drought and potentially lower groundwater tables, more frequent and intense heat waves in summer, less frequent and more intense cold events in winter, reduced amounts of snowpack and associated altered runoff patterns described above, and increased frequency and intensity of summer wildfires.

First, increased heat and humidity will act to increase the wet bulb temperature, defined as the temperature of adiabatic saturation, or the lowest temperature that air can be cooled to by the evaporation of water (National Weather Service 2024, accessed 1 October 2024). The wet bulb temperature is a key metric related to how the human body can dissipate heat. Above a certain temperature, the human body cannot acclimate to ambient temperatures and cannot dissipate heat fast enough to maintain a healthy function. At this point, the human body's function begins to break down, and vital organs are susceptible to damage and failure, endangering the lives of individuals exposed to this heat. Several global locations are susceptible to increases in wet bulb temperature, above which habitation will be difficult and dangerous without interventions like air conditioning. As a result, mass migrations of populations to more hospitable climates may occur, which in turn could stress the recipient areas and create unrest.

Not only could an increase in ambient temperatures lead to migration and unrest, but stresses to agriculture could also occur with increased temperatures. As average temperatures rise, annual and perennial crops will be adversely affected. Countless plant species require strict temperature ranges for a variety of biological processes necessary for growth, production, and propagation – even minor disruptions or imbalances can have adverse effects on a crop's production and resulting harvest – or even imperil the plants themselves. Some people may argue that even if this occurs and plants become difficult, if not impossible, to continue to cultivate in their current regions, they may be replaced by new crops – the "orange groves in Minnesota" argument. This is faulty logic for a number of reasons. First, crops require a range of suitable physical attributes to thrive, including appropriate temperature ranges, ideal precipitation, both in cumulative amounts and matched with the plant's growing cycle, duration and intensity of sunlight during the growing season, wind, fauna acclimated for pollination, and amenable soils. All of these are disrupted when temperatures increase, and farmers consider growing crops that previously had been grown for centuries in warmer climates. Additionally, both farmers and the societies they serve are reliant on the crops that have been grown and harvested in those respective regions. Farmers may not have the know-how or financial resources to make drastic changes in the crops they grow, and societies may have difficulty adapting to new products brought to market. Put another way, hypothetically speaking, corn/soybean farmers in Iowa are knowledgeable and resourceful for corn and soybean farming, but will they know how to grow sorghum or wheat effectively, and will they have the money to swap out their farming equipment to handle the new crops? Will there be a market in Iowa and elsewhere for sorghum?

With increased temperatures, more winter precipitation will fall as rain instead of snow, and snowmelt is expected to begin earlier in the spring. These increased temperature-induced changes are expected to result in a shorter snow season on average and earlier peak streamflow in rivers with a significant snowmelt component. These changing precipitation patterns contribute to flow changes in major snowmelt-influenced rivers with higher flows in winter and lower flows in summer, more frequent and severe river flooding, increased landslide risk due to saturation of soil, and increased erosion and riverine sediment transport in fall, winter, and spring (NOAA 2017; CIG 2018).

5.2.2 Sea Level Rise

Even with a drastic reduction in greenhouse gas emissions, sea level rise will continue through the end of the twenty-first century because of the latency of response of the oceans to temperature conditions at the Earth's surface. When assuming a high gas emissions scenario, these factors are combined with glacial and ice sheet melting, resulting in an estimated 4 ft of sea level rise on a global scale by 2100 (NOAA 2017). Even under a low greenhouse gas emissions scenario, thermal expansion of the oceans and melting of small mountain glaciers is projected to result in approximately 1 ft of absolute sea level rise by 2100 on a global scale. When combining rising sea levels with high tides, storm surges, and subsidence, a range of adverse physical effects may occur, including (i) an increase in the elevation, depth, or extent of inundation along the marine and coastal shorelines; (ii) an amplification of the inland reach of high tides, resulting in increased flooding further inland of the coastline, especially when compounded by severe storm events; (iii) movement of the saltwater intrusion further upstream in tidally influenced rivers; (iv) saltwater intrusion into groundwater; and (v) increased risks of landslide and rates of erosion along coastal bluffs.

All of these effects will have dire consequences on coastal ecosystems as well as coastal communities. Especially in areas of expansive coastal plains, populations will be susceptible to inundation or "sunny day flooding", resulting from the combination of elevated seas and tidal action. Over 900 million people (approximately 11% of the global population) across the world live in low-lying areas (less than 10 m above sea level) in coastal zones, which, as seas rise, will require massive adaptation projects to hold back rising waters or will have to emigrate to higher regions *en masse* as the sea reclaims low-lying coastal areas (IPCC 2019). Saltwater intrusion is an especially pernicious phenomenon – domestic water sources, including fresh surface water or groundwater – can be fouled and rendered unusable for consumption if salinity concentrations rise too much, causing dependent populations significant stress in trying to identify and utilize other water sources, which in arid areas may already be stressed or overdrafted. Coastal/estuary habitats can also be quite fragile – taking root in brackish waters, these resources and the plant/animal life that thrive within them can also be imperiled if seawater overpowers fresh water sources and alters the salinity.

5.2.3 Extreme Weather Events

The Global Change Research Act of 1990 requires the US Global Change Research Program (USGCRP) to develop national climate assessments every four years or less. These assessments are intended to evaluate the state of the science and the broad range of impacts of climate change in the United States. The draft Fourth National Climate Assessment (NCA4) report reviewed a wide range of topics of high importance more broadly, extending from human health and community well-being to the built environment, to businesses and economies, and to ecosystems and natural resources. The climate change trends are expected to accelerate in the decades ahead, contributing to the increased extreme events occurring individually or in some cases in combination. The notable extreme events to be considered include sea level rise, coastal flooding, increased flooding due to tidal influences, increased frequency and intensity of precipitation or drought, snowpack melting, earlier peak spring stream flow, shrinking glaciers, increased average annual air temperature, increased frequency of nighttime heat waves, longer frost-free seasons, increased wildfires, and ocean acidification. The IPCC projections indicate that climate change will notably impact the production of crops and fisheries, and several animal and plant species are projected to be endangered. Moreover, infrastructure damage and associated consequences, such as loss of life and property, can have an enormous impact on the environment, economy, and society.

5.2.4 Droughts and Floods

Climate change is significantly increasing the likelihood of more severe droughts and/or floods across various regions of the United States and the world. Rising global temperatures increase the moisture the atmosphere can hold, resulting in more storms and heavy rains but paradoxically also more intense dry spells as more water evaporates from the land and global weather patterns change. In particular, areas with arid climate like the US Southwest face heightened risks, as droughts are projected to become more frequent, intense, and prolonged. Warmer temperatures, driven by climate change, accelerate evaporation, depleting surface water and drying soils and vegetation. These conditions make already low precipitation levels more severe than they would be in cooler climates. Additionally, climate change is altering the timing and form of precipitation. For example, in the Northern Hemisphere, warmer winters are causing more precipitation to fall as rain rather than snow, leading to reduced snowpack. This presents challenges for water management systems that depend on spring snowmelt as well as for ecosystems that rely on cold water from snowmelt, such as those supporting salmon populations.

The Southwestern United States has already experienced a significant decline in annual precipitation since the early twentieth century, a trend that is expected to persist. Climate models suggest that while wet areas may become wetter, dry regions like the subtropics will likely get drier. In some regions, droughts may enter a vicious cycle, where dry soils and reduced vegetation absorb more heat, reinforcing high-pressure systems that further limit rainfall. Recent droughts in the United States, including the 2011–2019 California drought and the widespread 2020 drought, have been among the most severe in decades. These droughts, exacerbated by extreme heat waves, have withered vegetation and fueled record-setting wildfires, particularly in the West. The persistence of extreme drought, especially in the Pacific Northwest, highlights the urgency of addressing climate-induced water scarcity and its devastating impacts on ecosystems and human communities.

In addition to droughts, climate change is also exacerbating flood risks in both coastal and inland regions, primarily through changes in precipitation patterns, sea level rise, and intensification of storms. Warmer atmospheric conditions result in higher rates of evaporation, leading to more moisture in the atmosphere. This causes more intense rainfall events, overwhelming drainage systems and leading to flash floods. As storms become more intense and frequent, even regions that were historically less prone to floods are now experiencing heightened risks. Urbanization adds to the problem, with impervious surfaces preventing natural water absorption, worsening flood severity. Beyond physical damage to properties and infrastructure, flooding poses serious health risks, including water contamination and the spread of waterborne diseases.

By the 2080s, extreme precipitation events (24-hour rain events) are projected to increase in intensity by 22% on average, with a corresponding increase in the frequency (compared to the 1980s). This would be an increase to about eight days per year, on average, as opposed to just two days per year in the past. The increasing frequency and intensity of atmospheric rivers – those moist airflows that extend from the tropical Pacific to the West Coast of North America during winter – are expected to carry more moisture in the future, resulting in more frequent, more intense rainfall events (OCCRI 2017; CIG 2018).

Furthermore, climate change is altering snow patterns in ways that contribute to both flooding and water scarcity. As global temperatures rise, more precipitation falls as rain rather than snow, especially in regions that traditionally rely on snowpack, such as the Western United States and parts of Europe. This shift not only reduces the accumulation of snow during winter months but also leads to earlier snowmelt in the spring. When snow melts rapidly, it can overwhelm rivers and reservoirs, increasing the risk of spring floods. These floods are often more intense because the water is released over a shorter period, straining water management systems. Moreover, the reduction in snowpack has long-term implications for water availability. In many regions, snow acts as a natural reservoir, gradually releasing water into rivers and streams during warmer months. As snow levels decline and melt earlier, less water is available during the critical dry season. Coastal floods due to sea level rise, as discussed earlier, are another means of flooding due to climate change. Hence, with climate change intensifying the frequency and severity of floods, it is essential for communities to invest in resilient infrastructure, effective flood management, and sustainable urban planning to mitigate future risks.

5.2.5 Wildfires

Climate change is intensifying wildfires by creating hot, dry conditions that make fires more likely to ignite and spread. As global temperatures rise, the size, frequency, and severity of wildfires are expected to increase. Coincidentally, the rise in rates of wildfires releases CO_2 emissions into the atmosphere from combustion, further contributing to climate change as shown in Figure 5.1.

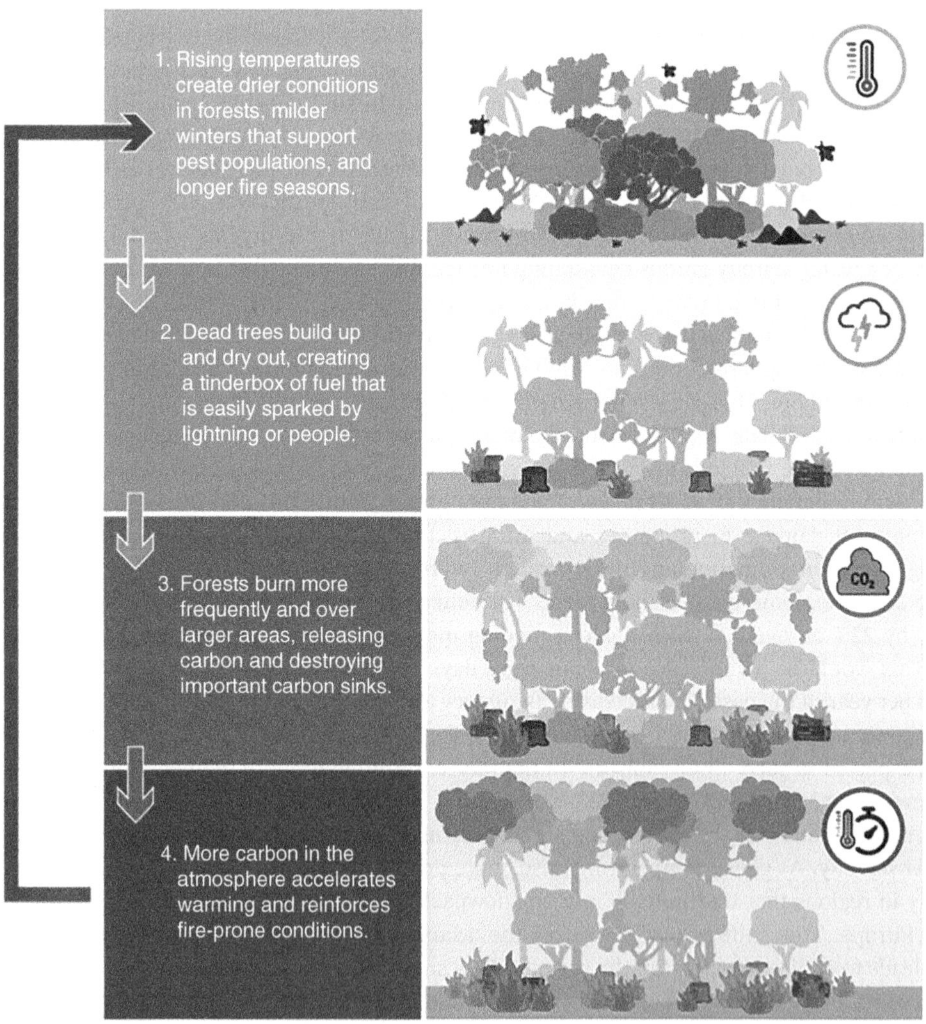

Figure 5.1 How rising temperatures impact forest fires. *Source:* Ruiz (2020).

Wildfires can start naturally, such as through lightning strikes, or from human activities like campfires, discarded cigarettes, or downed power lines. However, climate change exacerbates the situation by increasing severe heat and drought, which dry out vegetation and turn forests and grasslands into kindling. While fire serves a natural ecological role, the combination of a warming planet and poor forest management practices is contributing to the rise of more destructive wildfires.

Between 2001 and 2023, the area burned by forest fires increased by approximately 5.4% per year. This surge now leads to nearly 6 million more hectares of tree cover loss annually compared to 2001 – an area roughly the size of Croatia. Fires have also become a larger contributor to global tree cover loss, rising from 20% in 2001 to around 33% today, surpassing other drivers like mining and forestry (MacCarthy et al. 2024). The rising severity and frequency of wildfires are putting enormous strain on nearby communities and infrastructure, leading to damaged habitats, degraded air quality with associated health impacts, and contamination of drinking water supplies. As climate change continues to exacerbate these trends, it is crucial to make communities more resilient to wildfire impacts, using strategies such as using fire-resistant building materials and ensuring adequate spacing between buildings to create barriers that help limit the spread of wildfires.

5.3 Impacts of Extreme Events

It is essential that climate change and related impacts be dealt with in a comprehensive manner. To mitigate the effects associated with the climate change, it is desirable, and likely more efficient, to address the root causes of climate change. IPCC links anthropogenic greenhouse gas emissions to the key drivers that include population size, economic activity, lifestyle, energy use, land use patterns, technology, and climate policy. The 2015 Paris Agreement under the UN Framework Convention on Climate Change (UNFCCC) targets voluntary mitigation efforts by signatory countries; however, the implementation falls well short of criteria that can have a material impact on climate change (UNFCCC 2023, accessed 16 September 2024). Since the occurrence of extreme events and associated impacts on people and infrastructure have been quite evident, it is necessary to invest in adaptation to climate change. Designing built infrastructure that is adaptable to changing climate and capable of withstanding extreme impacts is necessary. Several notable recent natural disasters that overwhelmed existing infrastructure are described below.

5.3.1 Hurricanes, Storms, and Floods

Extreme weather events, such as hurricanes and severe storms, are occurring with greater frequency and intensity due to global warming and climate change. As temperatures increase, atmospheric moisture levels increase, creating more instability and energetic storm activity. These divergent weather patterns from historically normal weather and climatic activity highlight the need for mitigation of GHG emissions and resulting climate impacts.

Hurricane Sandy (also known as Superstorm Sandy) made landfall in October 2012. It was the deadliest and most destructive storm system ever to hit the New York metropolitan area. Its effects resulted in nearly $70 billion in damage due to the flooding of streets, tunnels, and subways, as well as the widespread loss of power. The storm affected basic infrastructure in Manhattan, destroyed numerous homes and businesses, and resulted in at least 53 fatalities. The extensive damage highlighted the inadequacy of current infrastructure in the New York metropolitan area, home to millions of people and a commercial center of primary global importance, to withstand and recover from such weather events.

Hurricane Harvey, the first Category 4 (and major) hurricane to make landfall along the middle Texas coast since Celia in 1970, struck on August 2017. The hurricane brought peak wind gusts of about 145 mph (233.3 km/h) and storm surges of 12 ft, inundating low-lying coastal areas in South and Southeast Texas. It dumped an estimated 27 trillion gallons of rain over the state, with some parts of Southeast Texas receiving as much as 51.88 inches of record-breaking rainfall. Unlike most hurricanes, instead of moving further inland and away from the coast, Hurricane Harvey stalled over Southeast Texas for days, resulting in excessive rainfall and catastrophic, deadly flash and river flooding. It became the wettest Atlantic hurricane ever recorded. In addition to the severe flooding, the peak wind gusts caused widespread structural damage along the coastal areas, and the added combination of storm surge flooding and flash flooding from the rainfall led to an estimated $125 billion in losses, making Hurricane Harvey the second-costliest hurricane in US history, after adjusting for inflation, second only to Hurricane Sandy.

Hurricane Helene originated as a broad low-pressure system in the western Caribbean Sea and intensified as it moved north through the Gulf of Mexico, eventually traveling inland and causing devastation along its 500-mile path from Florida to the Southern Appalachian Mountains. Over a two-day period, Helene transformed from scattered thunderstorms and rain into a Category 4 hurricane fueled by unusually warm waters in the Caribbean Sea and Gulf of Mexico. These conditions allowed the hurricane to intensify rapidly and store large amounts of moisture, which it later released as heavy precipitation inland.

Helene made landfall on 26 September 2024, in Florida's Big Bay region near the city of Perry as a Category 4 hurricane, with sustained wind speeds of 140 mph. This marked the first hurricane to strike the Big Bay area since 1851. At 400 miles in diameter, Helene was massive, comparable to Hurricane Katrina, and among the largest hurricanes to make landfall in the continental United States.

Upon landfall, Helene generated a storm surge of 15 ft in the Big Bay beachfront area. Even in Tampa Bay, located about 110 miles south of the storm's center, the storm surge exceeded 6 ft, highlighting the hurricane's widespread impact. Helene traveled hundreds of miles inland, affecting millions across six different states. The hurricane severely affected cities across Florida, Georgia, South Carolina, North Carolina, Tennessee, and Virginia, bringing intense rainfall of 12–15 inches – equivalent to four months' worth of rain in just two days – accompanied by heavy winds.

Typically, fast-moving hurricanes present less rain risk inland, but Helene was a major exception. The mountainous terrain of the Southern Appalachians enhanced rainfall through orographic precipitation, a process in which the storm was forced upward over mountains, causing it to cool and condense, thereby dropping additional precipitation. Streams and rivers swelled with runoff, as many areas had already been saturated from prior storms a few days before the hurricane's impact, leaving them especially vulnerable to Helene's flooding. The hurricane was estimated to have dropped about 40 trillion gallons of water across these states. Flooding in many areas extended beyond traditional 100-year flood zones – areas expected to flood only once every century – and reached several locations for the first time in over a century.

Helene's storm surge drove debris and beach sand inland, affecting coastal residences. Heavy rains and flooding left over four million people without electricity, while the strong winds uprooted trees and damaged infrastructure, closing hundreds of roads as debris accumulated and isolated communities. Asheville, North Carolina, a city previously considered safe from flooding due to its high elevation, was hit particularly hard, becoming largely submerged and cut off. In Afton, Tennessee, Helene washed out Kinser Bridge, which previously had stood over 60 ft above the river over which it traversed. Helene is regarded as the deadliest hurricane to strike the mainland United States since Hurricane Katrina in 2005, with a death toll of over 230 people. The initial estimates of $200 billion in damages would make Helene one of the costliest, if not the costliest, hurricanes in US history.

Hurricane Milton struck Florida's Gulf Coast area on 9 October 2024, making landfall near Siesta Key as a Category 3 storm with sustained winds of 120 mph. This impact came less than two weeks after the state endured significant damage from Hurricane Helene. Initially projected to make landfall in Tampa Bay as a Category 5 hurricane, Milton veered south and weakened slightly due to wind shear as it approached the coast. Upon landfall, Milton generated a storm surge of 6 ft, lower than the 15 ft expected if it had maintained Category 5 status. The hurricane continued moving east across Florida, emerging into the Atlantic Ocean after crossing the state. Rainfall totals of 10–15 inches led to widespread localized flooding, compounding the impacts on affected communities. Though Milton was less powerful than initially anticipated, it caused widespread devastation across Florida, with a reported death toll of 24 people. Milton spawned an unusually large number of tornadoes along its path, causing further destruction even as the hurricane weakened to Category 1. These tornadoes were responsible for a significant portion of the death toll.

As a hurricane makes landfall, friction between the land and the lower atmosphere slows down surface winds, while upper-level winds retain their speed. This difference in wind speed with height is known as vertical wind shear, which can contribute to the formation of tornadoes. The magnitude of wind shear is influenced by the heat and humidity of the air in the lower atmosphere; warm, moist air increases atmospheric instability, making conditions more conducive to tornado formation. Recent studies suggest that hurricanes in a warmer climate could produce a greater number of tornadoes compared to colder weather, as warmer temperatures increase the energy available for hurricanes.

Dubai Flooding. On 16 April 2024, Dubai saw 10 inches of rain – equivalent to a two-year rainfall – in less than 24 hours. This deluge caused severe health issues and infrastructure damage as a significant power outage was experienced in the city, disrupting flights as runways were inundated. The storm initially struck Oman two days earlier, causing over 20 fatalities. The National Guard's Search and Rescue Center conducted over 100 operations, providing critical rescues and medical evacuations as unexpected flooding shocked residents. The high sea temperatures around Dubai and the air temperatures increased potential evaporation rates, and the warmer air held more moisture, allowing bigger dumps of rainfall. Despite being built on sand, Dubai's massive grey infrastructure essentially hindered rainwater from draining into the sand, intensifying the flooding. Initial panic led to speculations of cloud seeding by the state being responsible for the flooding, which the meteorological agencies later discredited. The rainfall was likely due to the weather system exacerbated by climate change, with atmospheric temperature shifts and elevated sea temperatures contributing to the extreme rain. The effects of El Niño were noted, potentially amplifying rainfall by up to 40%. These rains marked the largest ever recorded in the United Arab Emirates since the records began in 1949, underscoring an urgent need for improved infrastructure and disaster preparedness to face increasingly severe weather patterns.

The **DANA** (Isolated Depression at High Levels) that affected **Valencia (Spain)** in October 2024 spurred heavy rains and flooding, causing dramatic impacts on the population and infrastructure. DANA was created by a mass of cold air at high altitudes that encountered warm and humid air from the Mediterranean Sea, which created atmospheric conditions conducive for heavy storm activity. With rainfall density approaching nearly 500 liters per square meter in some areas, severe flooding events resulted, especially in the Huerta Sur region, where many reported fatalities occurred. The consequences were devastating, with more than 200 fatalities and the destruction of hundreds of homes, businesses, and supporting infrastructure. With the commencement of preliminary recovery efforts, authorities estimated that the costs for recovery and repair will exceed 10 billion Euros.

With climate change, temperatures recorded in the Mediterranean Sea continue to increase; this increase in sea temperature increases the moisture in the air, which acts to fuel more intense and more frequent events. In the summer of 2024, the temperature of the Mediterranean Sea reached record levels, with an average surface temperature of 28.47 °C, and in October 2024, recorded temperatures were 2.1 °C above historical averages. In autumn, when the moist, warm air meets masses of cold air flowing from the north, the conditions for DANAs are created.

The worst monsoon rains in South Asia occurred during August 2017 (around the same time Hurricane Harvey inundated Houston). More than one-third of Nepal had reported flooding or mudslides from heavy rainfall, decimating crops across the country. In Bangladesh, record rainfall pushed rivers to some of the highest levels ever and submerged one-third of the country, damaging enormous areas of farmland. On India's West Coast, Mumbai received extreme rainfall, up to 322 mm (12.7 inches) in 24 hours, causing widespread flooding. Many other cities in the region saw extreme 24-hour rainfall rates that caused widespread and deadly flooding.

5.3.2 Wildfires by Drought and Record Heat

Wildfires occurred in Chile's central provinces during November 2016 to January 2017, with 119 active fires and forest fires, affecting an area of 1.35 million acres. At least 100 vineyards were damaged or destroyed. A total of 11 deaths were confirmed. Years of drought, combined with an extended period of extreme heat and dry weather, helped to fuel the historic wildfires. Over several preceding years, annual precipitation deficits ranged from 30% to 70%, resulting in acute drought and leading to long-lasting and widespread water shortage.

Multiple studies have identified human-caused climate change as a primary adverse factor on recent precipitation trends in Chile. Recent trends have highlighted increasing temperatures in central Chile, especially for the upper elevation stations, where precipitation amounts have also been decreasing. Higher temperatures lead to greater water loss from snow-covered areas (sublimation), crops and natural vegetation (evapotranspiration), and lakes and reservoirs (evaporation), which contributes to water deficits and increases wildfire risk. From 7 to 31 January 2025, a series of 14 destructive wildfires swept through Southern California, severely impacting the Los Angeles metropolitan area and San Diego County. Fueled by prolonged drought, extremely low humidity, dense vegetation growth from the prior wet season, and Santa Ana winds reaching up to 100 mph (160 km/h), the fires killed at least 30 people, displaced over 200,000 residents, destroyed more than 18,000 structures, and scorched over 57,000 acres of land. The Palisades Fire in Pacific Palisades and the Eaton Fire in Altadena accounted for most of the destruction and were among the most damaging wildfires in state history. Researchers have linked the severity of wildfires to climate change, with attribution studies showing that warming-driven shifts in global temperature and precipitation trends have increased the likelihood of such extreme fire conditions, underscoring the growing challenges of wildfire management in a warming climate.

5.4 What Is Resiliency?

The term *resiliency* is derived from the Latin word *resilio*, which means to jump back. Resiliency is defined differently in different fields such as ecology, social science, economy, and engineering, with relevant meanings and implications:

- *Ecological resilience* is defined as a measure of the persistence of systems and of their ability to absorb change and disturbance and still maintain the same relationships between populations or state variables. Stability represents the ability of a system to return to an equilibrium state after a temporary disturbance; the more rapidly it returns to equilibrium and the less it fluctuates, the more stable it would be (Holling 1973).

- *Social resilience* is defined as the ability of groups or societies to cope with external stresses and disturbances because of social, political, and environmental changes (Adger 2000).
- *Economic resilience* is defined as the inherent ability and adaptive response that enables individual business firms, local and regional economies, and national economies to avoid maximum potential losses (Rose and Liao 2005).
- *Engineering resilience* is defined as the capability of a system to maintain its functionality with minimum possible degradation in the face of internal and external changes. In particular, engineered infrastructure (e.g. water supply, electricity, and transportation) is recognized as a key element for communities' resilience in the context of disasters.

The general concept of resilience is depicted in Figure 5.2, and some common elements of resiliency include the system design that performs its function safely under normal conditions, the vulnerability potential for an extreme event to occur, the system's absorption of the event and resistance to damage, maintaining function at some desired level, and recovering to achieve target level of functionality after the event.

In general, disaster resilience is specifically used to refer to minimizing any reduction in quality of life (e.g. loss of life, injuries, disruption of services, and economic losses) due to a disaster. Specifically, disaster resilience is characterized by (i) reduced failure probabilities – the reduced likelihood of damage and failures to critical infrastructure, systems, and components; (ii) reduced consequences from failures – in terms of injuries, lives lost, damage, and negative economic and social impacts; and (iii) reduced time to recovery – the time required to restore a specific system or set of systems to normal or pre-disaster level of functionality. Based on these characteristics, resilience can be enhanced by reducing the likelihood of failure of critical infrastructure (thereby, reducing their cascading impacts) and speeding up the time it takes to make a full recovery.

In engineering, resilience is often understood as the technical robustness of systems in the face of disasters. However, resilience encompasses more than just this one dimension. For instance, when engineered infrastructure fails due to a disaster, the consequences extend beyond technical failure;

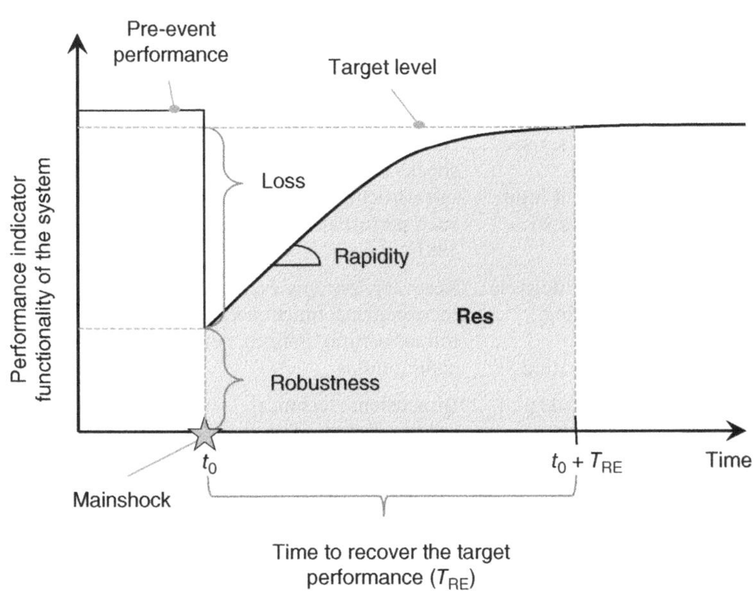

Figure 5.2 Concept of resilience. *Source:* Cimellaro et al. (2016)/with permission of American Society of Civil Engineers.

they affect the services the infrastructure system provides to society and can have environmental and economic impacts resulting from failure and subsequent recovery efforts. Similar to sustainability, resilience also includes multiple dimensions: technical, environmental, social, and economic, as defined by Reddy et al. (2024). To account for these dimensions, they define resilience as *"the capacity of a system to prepare for and overcome unexpected crises and adapt to turbulent changes, with minimal impacts on the environment, society, and the economy."*

Sometimes resilience is considered as one of the indicators of sustainability. However, the correlation between these two is more complicated. As stated in Chapter 1, the definition of sustainability given by the Brundtland Commission, formally known as the World Commission on Environment and Development (Brundtland 1987), is *the development that meet the needs of the present without compromising the ability of future generations to meet their own needs*. ASCE defines sustainability as *a set of economic, environmental and social conditions (aka 'the triple bottom line') in which all of society has the capacity and opportunity to maintain and improve its quality of life indefinitely without degrading the quantity, quality or the availability of economic, environmental and social resources* (ASCE 2018a). From a practical point of view, the triple bottom line impacts should be minimized to design sustainable systems (see, for example, Chapters 20–22). ASCE defines resilience as *"the capability to mitigate against significant all-hazards risks and incidents and to expeditiously recover and reconstitute critical services with minimum damage to public safety and health, the economy, and national security"* (ASCE 2018b). It is possible to have a sustainable design that can reduce resource and energy consumption, optimize waste management, and be economically efficient, but not necessarily maintain operating conditions in case of extreme events, making them not resilient. However, such systems are not truly sustainable. Resilience is essential to be sustainable and should be taken into account simultaneously over the life cycle of projects. Table 5.1 summarizes the similarities and differences between sustainability and resilience.

Although both sustainability and resilience are essential for future infrastructure, they might work against each other in some cases. Sustainable systems may not be resilient. Despite the holistic

Table 5.1 Similarities and differences among sustainability and resilience.

Categories	Sustainability	Resilience	Matching
Common definition	*Sustainable development* is development that meets the needs of the present without compromising the ability of future generations to meet their own needs (Brundtland 1987)	*Resilience* is the ability of human communities to withstand external shocks or perturbations to their infrastructure and to recover from such perturbations (Timmerman 1981)	o
Keywords often used for description in papers	Holistic, green, life cycle, life-cycle assessment, life-cycle costing, social costing, sustainable development, indicators, rating	Recovery, extreme events, disaster management, functionality, infrastructure, lifelines, networks, communities	+
Dimensions/ pillars	Dimensions: economic, ecological, social Strategies: efficiency, sufficiency, consistency	Dimensions: technical, organizational, social, economic Means: redundancy, resourcefulness	++
Target	Reduction of impacts and resource consumption in the three dimensions, inter- and intra-generational fairness	Goals: robustness, rapidity results: higher reliability, fast recovery, lower risk	+

Table 5.1 (Continued)

Categories	Sustainability	Resilience	Matching
Measuring labels/ medals	Labels: Platinum, gold, silver, certified, quantitative measures are constantly developed	A medal-based system is used in practice, but quantitative measures are becoming more popular among researchers	+
Quantification	Mostly based on indices summarizing different quantitative and qualitative indicators; result is a score	Often quantified by the index in Eq. (5.1) (Cimellaro et al. 2010a; Bocchini and Frangopol 2011)	o
Spatial scale	Most advanced on building level, activities on network level seldom included	Mainly community and network level (but also individual structure level)	o/+
Important instruments/ calculation methods in science	LCA, LCC, external costs, multicriteria decision-making, energy modeling, building information modeling	LCC, user cost analysis, multicriteria decision making, extreme events simulation	++
Important instruments/ calculation methods in practice	Decision supporting assessment schemes, mostly based on checklists compared with a reference structure	Classification of critical infrastructure in three levels Prioritization of interventions based on previous experience	+
Current relevance in science for infrastructure assessment	Interest growing since 2005, continued recent growth in interest because of several research projects started in the past several years	Interest growing since 2000, very significant momentum in the last three years; very topical these days	++
Current relevance in practice for infrastructure assessment	Medium relevance for newly built structures; several new systems recognizable that will increase relevance within the next five years. Work for the implementation in national and international standards underway	Approach very different from science; some associations are promoting the use of resilience criteria in the building codes, but not even a common vocabulary has been established	++

Note: ++ = perfect matching; + = good matching; o = no significant matching.

$$R = \frac{\int_{t_0}^{t_0 + t_h} Q(t)\mathrm{d}t}{t_h} \tag{5.1}$$

where R is the resilience index, t_h is the time horizon investigated, Q is the percentage “functionality” (or “quality” or “serviceability” of the system), and t is time.
Source: From Bocchini et al. (2014).

aspirations of both sustainability and resilience, it can be argued that systems that are resilient may not be sustainable, systems that are not resilient are inherently unsustainable, and systems that are sustainable might not be resilient. For example, dense and compact cities (e.g. featuring tall buildings, public transportation) are considered sustainable as they can reduce energy consumption, but such dense urban systems can make cities more vulnerable to extreme events (nonresilient). So, defining a limit for population density in a city might be the solution to have cities that are both sustainable and resilient. To the contrary, resilient systems may not be sustainable. For example, after Superstorm Sandy hit New York, sea gates were proposed at the narrow section of the New York Harbor entrance to avoid surge flooding and achieve resiliency. However, such a gate system would require long-time maintenance and could also adversely affect the surrounding ecology; hence, this system is not considered sustainable, even though it is considered resilient.

Example 5.1

Given the design factor of safety of a levee is 1.75 at the normal flood level. A sudden surge in the flood level, as an aftermath to an intense rainfall, reduces the factor of safety to 1.25. After the flood level recedes to normal flood level in five days, the factor of safety increases back to the initial level. The rate of increase in the factor of safety is assumed to be linear, as shown in the figure below. Identify different terms, as shown in the figure below, and calculate the performance loss.

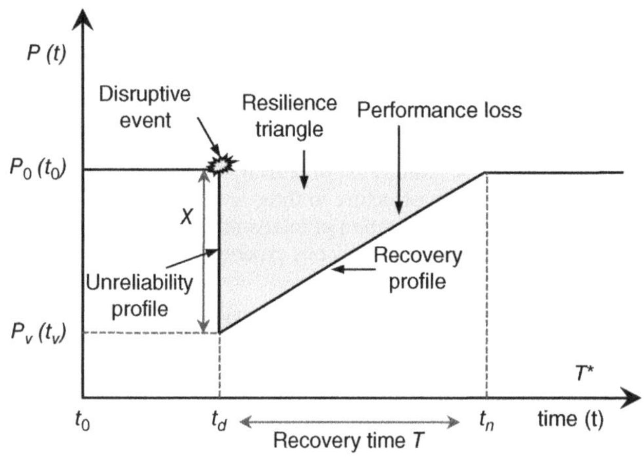

Considering $P_0\ (t_0)$ stays consistent for the design life of the levee, calculate the resilience of the levee considering 20 such disruptive events per year during the lifetime of the levee. The design life of the levee is 50 years. The resilience for one disruptive event is given by the following equation:

$$\Psi = P_0(t_0) - \frac{XT}{2}\left(\frac{1}{T^*}\right)$$

Solution:

Terms:

$P(t)$ = performance quality of a system which varies with time (t); loss of safety factor.

$P_0(t_0)$ = base safety factor; 1.75.

$P_v(t_v)$ = performance loss, disrupted safety factor, 1.25 at time of event (increases as it recovers).

t_d = time that disruption began.

t_n = time that recovery completes.

$T = (t_n - t_d)$ = length of disruption and recovery.

XT/2 = impacted area (shaded triangle area)

T^* = long time interval (in this case, the design life of the levee).

X = lost performance ($P_0(t_0) - P_v(t_v)$)

Given:

Safety factor initial = $P_0(t_0) = 1.75$
During the surge = $P_v(t_v) = 1.25$
Time taken for the flood level to be normal (T) = 5 days
Events per year = 20
Design life years = 50
Total events over lifetime (n); 20 events/year × 50 years = 1000 events.

Performance loss for one event, $X = P_0(t_0) - P_v(T_v) = 1.75–1.25 = 0.5$

$$\Psi = P_0(t_0) - n \times \frac{XT}{2}\left(\frac{1}{T^*}\right)$$

Resilience = 1.75 − 1000 × [(0.5 × 5)/2] × (1/ (50*365)) = 1.68

Example 5.2
Consider a transportation network designed to handle 40 000 vehicles per hour. An unprecedented flooding event after 5 months of construction has reduced the capacity of the network gradually to 10 000 vehicles per hour in 2 days. If the disrupted network stays at this state for 2 days and gradually increases back to pre-flood levels in 2.5 days, identify different terms, as indicated in the figure below, and calculate the performance loss due to the flooding.

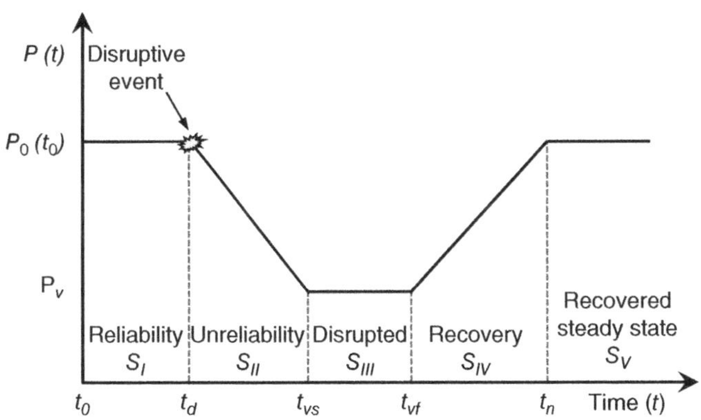

Note: Performance loss is the difference between the performance in case of an undisrupted state and the performance in case the disruption occurs.

Solution:
System States

- *Reliability state (S_I)* – The base line or original state of a system prior to a disruptive event at time t_d, operating normally at a performance value of P_0.
- *Unreliability state (S_{II})* – The system is in a vulnerable state following the disruptive event at time t_d. System performance gradually degrades to P_v at time t_{vs}.
- *Disrupted state (S_{III})* – The system operates in a state of degraded performance P_v until internal or external recovery effort can be made. The duration of this state is from time t_{vs} to t_{vf}.
- *Recovery state (S_{III})* – The system performs recovery efforts to restore its performance function. This restoration begins at t_{vf} and ends at t_n.

- *Recovered steady state (S_{IV})* – The system reaches a recovered steady state with a performance value equal to the system prior to the disruptive event.

Performance (P) and Time (t) Values

- P_0 – The performance of the system before the disruption. From the problem statement: 40 000 veh/h.
- P_v – The degraded performance of the system following the disruptive event and the deterioration occurring in the unreliability state, beginning at time t_{vs}. From the problem statement: 10 000 veh/h.
- t_d – Time of the disruption event. From the problem statement: 5 months
- t_{vs} – Beginning of the disrupted state following the initial disruptive event and unreliability event. Interpreted from the problem statement: 5 months and 2 days (or 48 hours)
- t_{vf} – End of the disrupted state and beginning of recovery efforts. Interpreted from the problem statement: 5 months and 4 days (96 hours)
- t_n – Time for the system to reach the recovered steady state. Interpreted from the figure: 5 months and 6.5 days (156 hours).

Assuming that time values are used in reference to one another, values of t can be interpreted relative to one another, such that the additional 5 months of t_d can be ignored in calculations.

Performance Loss
Performance loss can be quantified as the difference in performance of a system operating in a reliability state, minus the lost performance due to a disruption event. Graphically this can be interpreted as the region enclosed below by a constant performance curve of P_0 and above the disrupted performance curve. While more complex scenarios would involve integral or numerical analysis, we can evaluate the performance loss of the system geometrically as the area of a trapezoid:

$$\text{Performance loss} = \frac{(t_{vf} - t_{vs}) + (t_n - t_d)}{2}(P_0 - P_v)$$

Performance loss is thus quantified as loss of vehicles that could have been operative in the system:

$$\therefore \text{Performance loss} = \frac{2 + (2 + 2 + 2.5)}{2}(40\,000 - 10\,000)\text{veh/h} \times 24\,\text{h/d} = 3\,060\,000\,\text{vehicles}$$

The road network could not accommodate almost 3 million vehicles because of the flooding.

5.5 Initiatives and Policies on Resiliency

Many organizations and authorities have recognized the value of resiliency and developed proposals, policies, standards, or guidelines to support resiliency in general and for resilient infrastructure design in particular. Some of the selected efforts are briefly presented below.

US Presidential Policy Directives (PPDs) – The directives (PPD-8 and PPD-21) have defined resilience for communities to withstand and adapt to changing conditions (PPD 2011, 2013). Other organizations, such as The Infrastructure Security Partnership (TISP) and the Department of Homeland Security (DHS), have also suggested definitions for resilience for the functionality of built systems. According to PPD21 (Critical Infrastructure Security and Resilience), the term resilience refers to the ability to prepare for and adapt to changing conditions and withstand and recover rapidly from disruptions. Resilience includes the ability to withstand and recover from deliberate attacks, accidents, or naturally occurring threats or incidents. The PPD-8 calls for the DHS to "coordinate a comprehensive campaign to build and sustain national preparedness, including public outreach and community-based and private-sector programs to enhance national resilience..." The efforts by the US government to enhance the resilience of the national infrastructure was one of the stimuli that led to the identification of principles that could be used by designers to create infrastructure systems that would enable these systems to avoid, withstand, and recover from a broad range of threats.

Other US Federal Government Initiatives – Executive Order (E.O.) no. 14008 released in January 2021 directed federal agencies to pursue actions that will help us avoid and adapt to impacts from the most catastrophic events caused by a changing climate. Among other efforts, the order commits the United States to move quickly to build resilience against the impacts of climate change. Further, E.O. No. 14022 released in the June of 2022 calls on the Federal agencies to deploy Nature-Based Solutions to tackle climate change and enhance resilience. In June of 2022, the White House, along with the National Climate Task Force, introduced the "National Initiative to Advance Building Codes" to extend the climate protections and energy efficiency of consensus-based building codes across the country. This initiative will ensure that federally funded or financed new construction and substantially rehabilitated buildings and residential properties are constructed to be resilient to the increased frequency and intensity of climate-induced extreme weather events.

Rockefeller Foundation's 100 Resilient Cities Initiative – The Rockefeller Foundation has worked since December 2013 in helping cities around the world to become more resilient to physical, social, and economic challenges. The 100 Resilient Cities project aims to help individual cities to become more resilient and simultaneously to facilitate the building of a global practice of resilience among governments, NGOs, the private sector, and individuals. It defined urban resilience as the capacity of individuals, communities, institutions, businesses, and systems within the city to survive, adapt, and grow with no matter of chronic stresses and acute shocks they experience. The Rockefeller framework is built on four essential dimensions of urban resilience: health and well-being, economy and society, infrastructure and environment, and leadership and strategy. Each of the dimensions is divided into three more detailed drivers, as shown in Figure 5.3.

The United Nations (UN) – The UN defines resilience as follows: "... ability of a system, community or society exposed to hazards to resist, absorb, accommodate to and recover from the effects of a hazard in a timely and efficient manner, including through the preservation and restoration of its essential basic structures and functions." The UN Office for Disaster Risk Reduction (UNDRR) began in March 2015 to implement a new ISO standard for resilient and sustainable cities: ISO 371207 (ISO 2018). The United Nations Office for Disaster Risk Reduction at the request of the UN General Assembly adopted The Sendai Framework for Disaster Risk Reduction 2015–2030 at the Third World Conference in Sendai, Japan, on 18 March 2015. The Sendai Framework is the successor instrument to the Hyogo Framework for Action (HFA) 2005–2015: Building the Resilience of Nations and Communities to Disasters. The Sendai Framework also articulates the following: the need for improved understanding of disaster risk in all its dimensions of exposure, vulnerability and hazard characteristics; the strengthening of disaster risk governance, including national platforms; accountability for disaster risk management; preparedness to "Build Back Better"; recognition of stakeholders and their roles; mobilization of risk-sensitive investment to avoid the creation of new risk; resilience of health infrastructure, cultural heritage and

Figure 5.3 The framework of resilience design. *Source:* The Rockefeller Foundation (2015).

work-places; strengthening of international cooperation and global partnership, and risk-informed donor policies and programs, including financial support and loans from international financial institutions (UNDRR 2015).

The Sendai Framework aims to achieve substantial reduction of disaster risk and losses in lives, livelihoods, and health and in the economic, physical, social, cultural, and environmental assets of persons, businesses, communities, and countries. To pursue and achieve this goal, it is essential to enhance the implementation capacity and capability of developing countries, least developed countries, small island developing states, landlocked developing countries, and African countries, as well as middle-income countries facing specific challenges, including the mobilization of support through international cooperation for the provision of means of implementation in accordance with their national priorities.

In order to support the assessment of progress toward the above-mentioned goals and expected outcomes, seven global targets have been agreed under the Sendai Framework listed as follows:

1) Substantially reduce global disaster mortality by 2030, aiming to lower the average per 100 000 global mortality rates in the decade 2020–2030 compared to the period 2005–2015.
2) Substantially reduce the number of affected people globally by 2030, aiming to lower the average global figure per 100 000 in the decade 2020–2030 compared to the period 2005–2015.
3) Reduce direct disaster economic loss in relation to global gross domestic product (GDP) by 2030.
4) Substantially reduce disaster damage to critical infrastructure and disruption of basic services, among them health and educational facilities, including through developing their resilience by 2030.
5) Substantially increase the number of countries with national and local disaster risk reduction strategies by 2020.
6) Substantially enhance international cooperation to developing countries through adequate and sustainable support to complement their national actions for implementation of the present Framework by 2030.
7) Substantially increase the availability of and access to multihazard early warning systems and disaster risk information and assessments to people by 2030.

Intergovernmental Panel on Climate Change (IPCC) – The Fifth Assessment Report of the IPCC (IPCC 2015) defines resilience as "the capacity of social, economic, and environmental systems to cope with a hazardous event or trend or disturbance, responding or reorganizing in ways that maintain their essential function, identity, and structure, while also maintaining the capacity for adaptation, learning, and transformation." The report recommends creating a climate-resilient pathway as a measure to respond to climate change risks that may be concentrated in urban areas. This recommendation implies that, in terms of sustainable development, we must change our decision-making and behavior pertaining to the economy, society, technology, and politics. The Sixth Assessment Report of the IPCC (IPCC 2023) defines climate-resilient development as development that integrates climate adaptation and greenhouse gas mitigation to advance sustainable development for all. It emphasizes that such development is enabled by increased international cooperation, improved access to financial resources (particularly for vulnerable regions), and inclusive governance. The report repeatedly warns that every increment of further warming constrains climate-resilient development, and the window of opportunity for a liveable and sustainable future is rapidly closing. Immediate action is thus necessary to secure a climate-resilient future for humanity.

US Department of Defense (DOD) – The DOD considers four key properties for a resilient system: repel/resist/absorb, recover, adapt, and broad utility, as depicted in Figure 5.4. In this figure, the circles represent greater ability at the edges than in the center for each resilience property; the intersection of two or more circles can be viewed as providing a reasonably effective resilience solution. The intersection (i.e. nexus) of all four properties defines the space of resilient solutions with each solution in the space being proximal to an optimal solution for a particular property.

American Society of Civil Engineers (ASCE) – ASCE Policy Statement 493 supports additional basic and applied research efforts and the development of national standards in support of design and construction initiatives that increase the reliability, safety, security, and survivability of the nation's vast infrastructure (e.g. water, energy, utilities, buildings, and transportation) against natural and man-made disasters. Another, ASCE policy statement 500, supports initiatives that increase the resilience of infrastructure against man-made and natural hazards through education, research, planning, design, construction, operation, and maintenance. ASCE recognizes the importance of the development of performance criteria and uniform national standards that

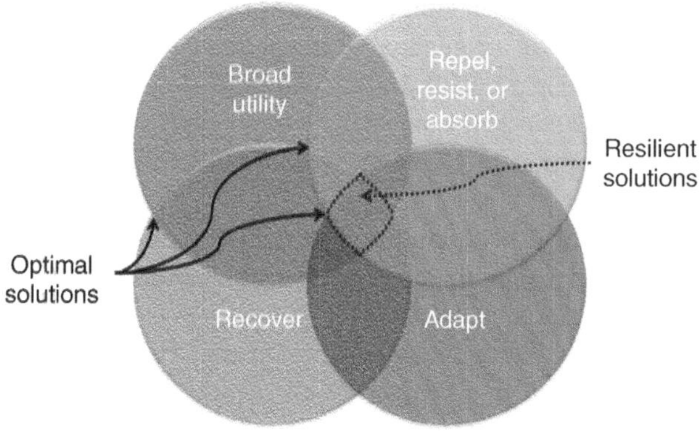

Figure 5.4 Four key properties of resilient DoD systems.

address interdependencies and establish minimum performance goals for infrastructure. An all-hazard, comprehensive risk assessment is recommended to be used for planning and design of infrastructure that considers event likelihood and consequence, mitigation strategies, monitoring, and recovery/return service. Designers and planners are encouraged to incorporate system resiliency (the ability of a project or system to withstand and recover to full functionality from extreme events quickly and efficiently) into the decision-making process. The increasing frequency and intensity of natural disasters as well as deliberate destructive events, combined with continually increasing population densities, reliance on technology, and interdependencies, are considered to be vulnerabilities in the nation's infrastructure. An important component of resilience is understanding the impact of the loss of infrastructure and the timeline and cost to restore its function following an extreme event. ASCE also established an infrastructure resilience division (IRD) to develop a unified approach to advancing the concepts of resilience within lifeline and infrastructure systems. More recently, ASCE has released a toolkit – Pathways to Resilient Communities – which centralizes resources to help civil engineers and community leaders adopt modern building codes and standards (ASCE 2023). The toolkit highlights hazard-specific standards for climate resilient engineering like ASCE 7-22: "Minimum design loads and associated criteria for buildings and other structures" (ASCE 2021). ASCE 7-22 was given an update to its flood mitigation chapter, which will improve the expected performance of buildings and structures located in flood hazard areas by using the 500-year flood as its threshold, up from the 100-year threshold to account for climate change. The toolkit also features the Envision™ program, which supports sustainable and resilient infrastructure design. ASCE's efforts also focus on educating policymakers and local communities to build infrastructure capable of withstanding climate-related and other hazards.

5.6 Resilient Design of Infrastructure

Disasters such as Superstorm Sandy in New York and Hurricane Harvey have revealed that civil infrastructure (e.g. buildings, water, and transportation) that normally functions as designed in high-frequency, low-magnitude events is highly vulnerable to extreme events. Therefore, infrastructure that can adapt to climate change impacts is critical for the economic security, the health and safety of people, and the health of our environment. As stated before, climate impacts could include sea level rise and coastal inundation, hurricane surge, riverine flooding and extreme rain events, landslides and erosions, wildfires, and drought.

Although resilience is an attractive concept, incorporating resilience in engineering practice for infrastructure design can be challenging. For resilient infrastructure design, first and foremost, a vulnerability assessment should be conducted in order to (i) identify climate change impacts that have the greatest potential to adversely affect infrastructure; (ii) understand the scope of infrastructure vulnerability to these impacts identifying types and instances of vulnerable infrastructure, where they are located, determining which specific types of infrastructure exhibit a high potential to be affected by climate change impacts, and what those vulnerabilities may be; (iii) develop a process for the project engineers to conduct a more detailed and site-specific vulnerability assessment; and (iv) inform the development of an adaptation design and operational strategy to increase resilience of infrastructure.

The prediction of climate change and associated impacts, including extreme events, has been a daunting task. Therefore, vulnerability assessments have been generally based on the analysis of historical observations assuming that the underlying environmental processes are stationary. However, recently observed data shows that the environmental processes are not stationary, and the past variability does not reflect future variability. This means that historical data may not be a good predictor of future variability, especially decades into the future when infrastructure designed today may well still be expected to be operable and in service.

Infrastructure along coastal areas or areas further inland may be vulnerable to high flooding hazards due to a number of regional and site-specific factors, such as proximity to tidally influenced freshwater rivers and tributaries affected by sea level rise; the extent of developed areas with non-permeable surfaces; the relative portion of localized low-lying land; the type and adequacy of flood controls; and the percentage of land within 100- or 500-year floodplains. For projects identified as vulnerable to climatic impacts, further evaluation of the risk should be done based on project-specific information.

Different standards are being used to address specific hazards (e.g. BRRT, FORTIFIED, and REDi), and many tools or assessment methods are being developed to measure project-scale resilience (e.g. Envision™, BRLA, PEER, SITES, LEED Pilot Credits, RELi, ANCR, and ICCR). Financial models are used to calculate full benefit–cost ratios to solidify the economic case for incorporating resiliency in design.

Many entities have initiated efforts to develop guidelines to incorporate the effects of extreme events such as sea level rise and flooding into infrastructure design. For example, the city of San Francisco (2015) has developed guidance for incorporating sea level rise into capital planning (http://onesanfrancisco.org/sea-level-rise-guidance). Similarly, the State of Washington (2017) developed guidance on incorporating climate impacts on contaminated site remediation. USEPA has developed guidance also on how to address climate impacts in selecting a remedial strategy or technology based on site-specific conditions.

Many analytical or decision tools are being developed for evaluating climate change vulnerabilities and impacts. For example, the following tools focus on the vulnerability of water resources (including stormwater, wastewater, and drinking water), sea level rise, and flooding:

- Sea Level Change Curve Calculator Tool (USACE 2014) is available to calculate three local sea level change scenarios (low, intermediate, and high curves) based on historic rates of sea level change, the modified National Research Council's projections, Intergovernmental Panel on Climate Change projections, and/or vertical land movement.
- National Stormwater Calculator with Climate Assessment Tool (USEPA 2017) was developed by USEPA to evaluate future climate vulnerabilities applicable to stormwater, based on national

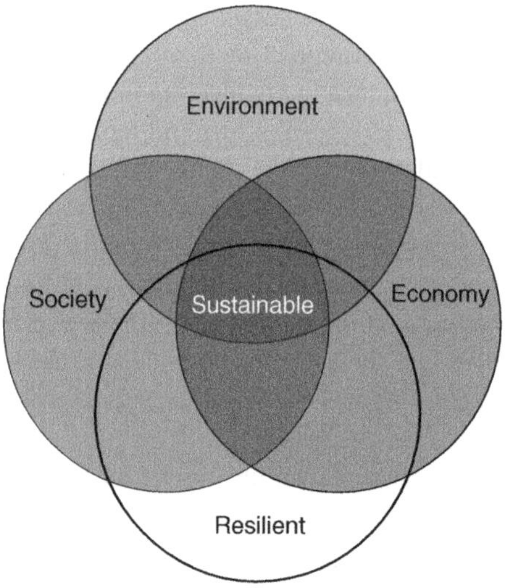

Figure 5.5 Integrated resilient and sustainable framework.

estimates of annual rainfall and frequency of runoff. Local soil conditions, land cover, and historic rainfall are also considered.

- Climate Resilience Evaluation and Awareness Tool (CREAT) 3.0 (USEPA 2016) is a valuable tool to assist site managers in understanding and assessing potential climate change impacts and risks to drinking water, wastewater, and stormwater systems.
- NOAA Sea Level Rise Viewer developed by the NOAA Office for Coastal Management (2018a) is a web mapping tool to visualize community-level impacts from coastal flooding or sea level rise (up to 10 ft above average high tides).
- NOAA Surging Seas is developed by the NOAA Office for Coastal Management (2018b). This web-based tool allows users to see areas potentially affected by sea level rise and storm surge, down to the neighborhood scale, and with risk timelines. The tool also provides population statistics and summaries of homes and land affected by sea level rise and storm surge.

Ideally, our target should be not only to design resilient infrastructure but also sustainable infrastructure. Hence, resilient assessments and sustainability assessments should be integrated and optimized resilient sustainable designs or solutions be developed. Such optimized approach is depicted in Figure 5.5. However, much more work is needed to establish integrated resilient and sustainable assessment frameworks and assessment/design methodologies.

5.7 Resiliency Assessment Framework

As mentioned before, the implications of resiliency of engineered systems extend beyond the traditional robustness of the system. An over-resilient design can impose environmental, social, and economic costs of the additional resource use while an under-designed system will have corresponding environmental, social, and economic implications of failure. There is a necessity to balance these aspects and ensure a resilient design that not only improves the resiliency of the engineered system to the disaster but also minimizes the corresponding triple bottom line implications. Engineers often face difficulties in taking into account these several dimensions into account for resilient designs mainly because of the difficulties in assessing these various dimensions of resilience. A framework to this effect which helps in resilient design of engineered systems considering not only the technical robustness but also the cascading impacts on environment, society, and the economy is developed by Reddy et al. (2024) as part of a broader integrated resilience and sustainability assessment framework. A schematic of resilience assessment framework is presented in Figure 5.6.

The first step of this framework is to identify the climate hazards of concern using various tools, as described in the previous section. Further, resiliency of engineered systems to the identified climate

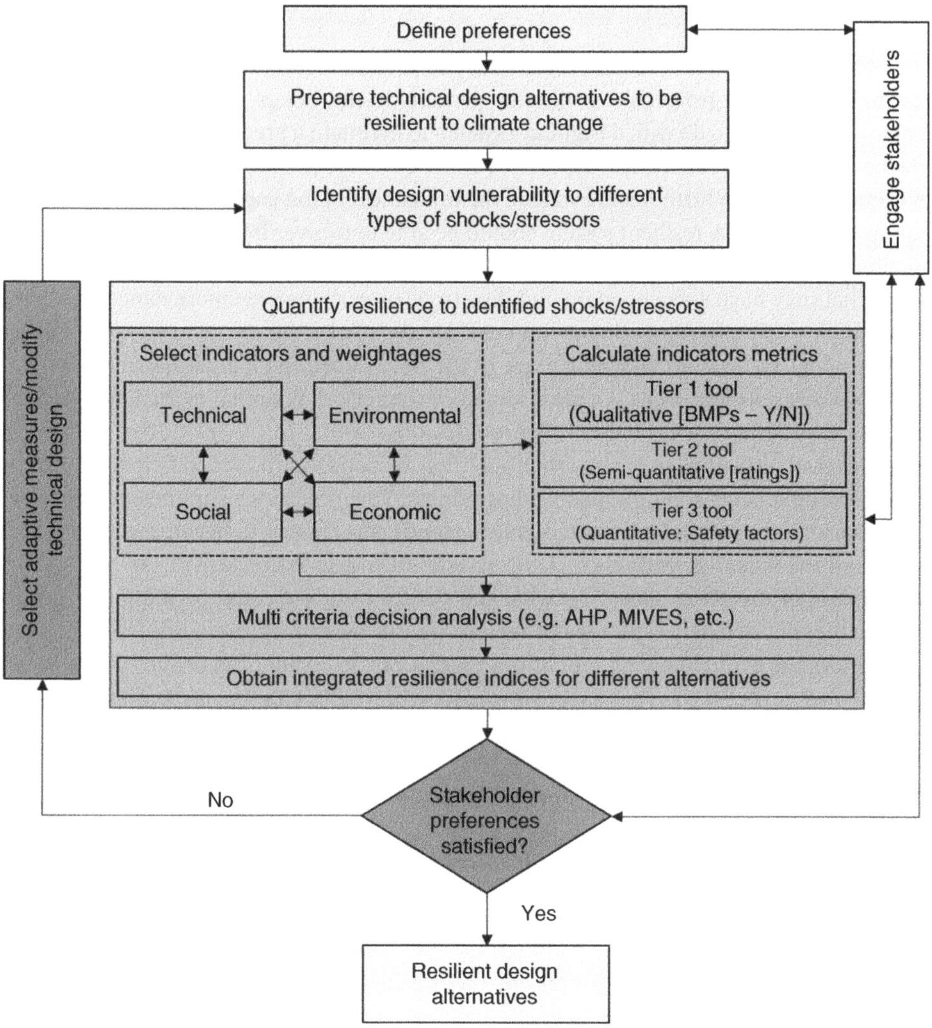

Figure 5.6 Resilience assessment framework. *Source:* Reddy et al. (2024)/with permission of Springer Nature.

hazards can be assessed by identifying and quantifying relevant indicators. To describe briefly, in order to quantify the four dimensions of resilience for a given engineering system – technical, environmental, social, and economic – one has to identify relevant indicators for resiliency through literature reviews and extensive brainstorming. Each of these indicators can be quantified for various designs of viable alternatives. As per the framework, depending on the quantification method, these indicators are divided into three tiers: Tier 1 being qualitative indicators (subjective: binary ratings or yes/no), Tier 2 being semi-quantitative indicators (quantifies using expert or stakeholder ratings on pre-defined rating scales), and Tier 3 being quantitative indicators (where the value of the indicator is determined quantitatively using an existing methodology/technique). This tiered approach allows for inclusion of all relevant indicators while providing the flexibility to use different types of variables for quantification. Further, post-quantification of the indicators, the values can be normalized and aggregated to obtain an overall resiliency index.

5.7.1 Technical Resiliency

It should be noted that technical resilience is the ability to provide required capability in the face of adversity, which is different from risk. While risk pertains to the loss of value due to uncertain future events, resilience has to do with designing a system to maintain a predesignated level of capability following a disturbance. Therefore, design for resiliency will require the engineer to take existing design frameworks a step further and evaluate the robustness of the engineered systems during and after an extreme event. A resilient system should be able to recover from damage rapidly and regain functionality if lost but, at the same time, needs to adapt and be prepared for the next extreme event that may be just as uncertain and disastrous or perhaps even more dangerous. Ideally, bouncing back or simply recovering is not considered sufficient to characterize a system as resilient. The resilient system can take advantage of the newly acquired knowledge and improve its response to the next extreme event. An ideal resilience framework is similar to the four-element community resilience framework shown in Figure 5.7.

In order to increase resilience, the concepts of monitor, respond, learn, and anticipate are considered. A monitoring scheme should support preparedness. The monitoring parameters should be carefully selected, with the selected parameters able to identify the positive or negative performance of the system and the actual performance of the system during an event, which can be used to address the impacts of the event as well as what the system response could be if another event occurs. By using the collected data, a post-event analysis may be performed, providing lessons about the system's performance and allowing monitoring/operations to be adjusted or the system to be modified/redesigned as necessary. The major complexity with resilient design is the unpredictable

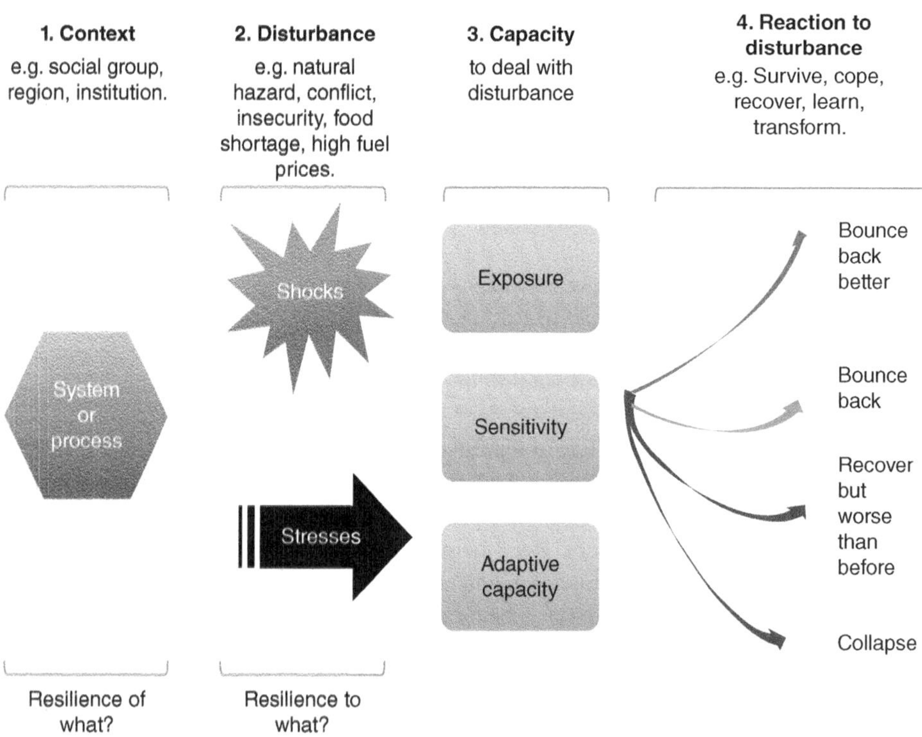

Figure 5.7 Resilience framework. *Source:* Combaz (2014)/with permission of Elsevier.

nature of extreme events; therefore, one should anticipate the occurrence of potential extreme events and prepare for them in light of project-specific demands and constraints and changing operating conditions.

The effectiveness of a resilient system depends upon its ability to anticipate, absorb, adapt to, and/or rapidly recover from a potentially disruptive event. Resiliency performance may be measured by metrics based on the data collected from previous shock situations (i.e. an extreme event). The data on the speed and time of recovery of the system (S_0, S_d, S_f, t_e, t_d, t_s, and t_f), as shown in Figure 5.8, may be determined and used as resilience metrics. These metrics can serve as the indicators for quantifying the technical resiliency of an engineered system, as defined in the framework introduced earlier.

Resilience of infrastructure has received greater attention after Hurricanes Katrina, Irene, and Sandy crippled domestic infrastructure, which stranded many people without basic infrastructure services or access to emergency facilities, and delayed evacuation and recovery processes. Over the past two decades, several studies (e.g. Rinaldi et al. 2001; O'Rourke 2007) have investigated resilience determinants in infrastructure systems to provide a theoretical framework for identifying components and characteristics of resilient infrastructure systems. Some researchers have conceptualized resilience using four properties (known as 4Rs of resilience): robustness, redundancy, resourcefulness, and rapidity (Bruneau et al. 2003). All these methodologies aid in quantifying

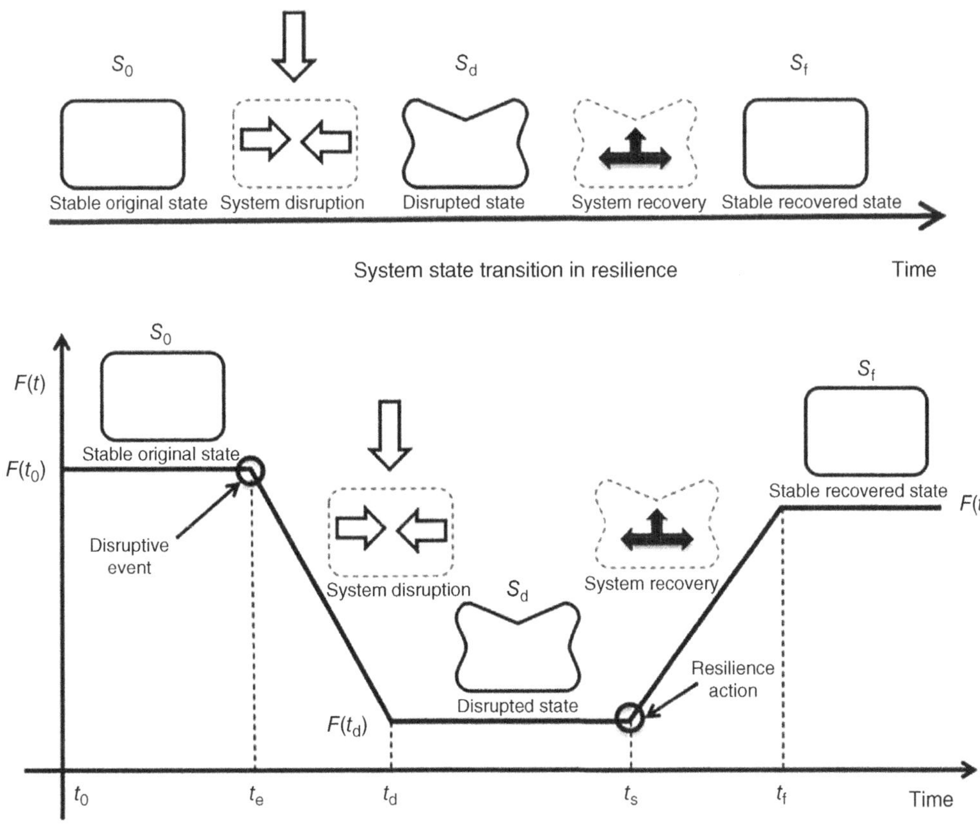

Figure 5.8 Delivery function transition in resilience. *Source:* Henry and Ramirez-Marquez (2012)/Reproduced with permission of Elsevier.

the technical resiliency indicators that can later be integrated with the cascading impacts of resiliency to develop a unified resiliency index.

5.7.2 Cascading Impacts of Resiliency

Most of the frameworks developed for resiliency of engineering systems primarily focus on technical functionality of the system and ignore the cascading impacts on environmental, social, and economic dimensions. By ignoring these dimensions, in the event of the compromise or failure of the system, the deleterious impacts can cascade along these dimensions, greatly impacting a community or wider society. Performing analyses to quantify these different dimensions from the planning phase can be challenging due to the absence of relevant frameworks and quantification methods. The framework introduced in this section facilitates analysis of these cascading impacts, providing an opportunity to design a system to minimize these impacts. First, indicators for these different dimensions need to be identified to accurately incorporate the cascading impacts of resiliency on triple bottom line dimensions. Example indicators for a levee system are provided below in Table 5.2. Further, inter-disciplinary collaborations are necessary to develop quantification approaches for different indicators and standardize the methodology for resiliency quantification and resilient designs. In the absence of such robust quantification techniques, rating systems or simple yes/no-based indicators, like those used in the Envision™, can be used to quantify these indicators. This allows engineers and designers to account for the resilience of engineered systems comprehensively by considering all dimensions.

Table 5.2 Example indicators to assess the resiliency of levee systems.

Requirement	Criteria	Indicators
Technical resiliency	Seepage	Performance deterioration (exit gradient)
		Time for deterioration
		Time in deteriorated state
		Performance recovery
		Time for recovery
		Functional loss of time
	Slope stability	Performance deterioration (FOS)
		Time for deterioration
		Time in deteriorated state
		Performance recovery
		Time for recovery
		Functional loss of time
Cascading environmental impacts	Air	Greenhouse gas emissions due to remediation and reconstruction
		Soil erosion leading to particulate matter suspending in the air
	Water	Disturbance to aquatic life
		Debris from failure, carried into water

Table 5.2 (Continued)

Requirement	Criteria	Indicators
	Human health	Possible water contamination
		Unsanitary conditions and waste accumulation
	Land	Soil erosion
		Change in land use pattern and associated effects
Cascading social impacts	Socio-individual	Sustained injuries due to failure
		Loss of life
	Socio-community	Disruption to transportation and emergency services
		Damages to public infrastructure
	Socio-economic	Public property damages and associated costs
		Impact on local businesses
	Socio-environmental	Habitat destruction
		Contamination of waterways
Cascading economic impacts	Direct costs	Cost of reconstruction
		Maintenance costs
	Indirect costs	Insurance premiums
		Restricted access
	Direct benefits	Long-term cost savings
		Water management
	Indirect benefits	Sustained local water source
		Recreational activities for the public

Recently, several researchers have emphasized the need to integrate both resiliency and sustainability in engineered systems, proposing unified approaches to address these aspects simultaneously and quantitatively throughout the life cycle. One such approach, proposed by Bocchini et al. (2014) for civil infrastructure, proposed a unified approach to address resilient and sustainable civil infrastructure (e.g. bridges, transportation) simultaneously and quantitatively over the infrastructure life cycle. The approach is based on a well-established framework of risk assessment (probability of occurrence and risk). The impact of infrastructure and its service states on society in normal operational conditions are assessed by sustainability analysis based on triple bottom line dimensions. The impacts on infrastructure after extreme events are assessed by resilient analysis based on the ability to deliver a certain service level after the event and then quickly recover to the desired functionality. These aspects are weighted by the associated probabilities of occurrence and combined in a global impact assessment. Another framework, developed by Reddy et al. (2024), allows for integrated, indicator-based quantification of resilience and sustainability beginning in the planning phase, ensuring that these considerations are incorporated from the outset. This framework operates incrementally, first ensuring that all considered alternatives are resilient to stressors and shocks, after which the resilient alternatives are subjected to a sustainability assessment to select the most sustainable option, ensuring both resiliency and sustainability in an integrated manner.

5.8 Resilient Infrastructure Examples

Several examples are presented in this section to understand how the efforts are being currently made addressing climate impacts in designing resilient engineering systems.

5.8.1 San Francisco Firehouse Resilient Design

San Francisco Public Works (SFPW) investigated floating pier (float) and fixed pier options and chose a float, in part to accommodate sea level rise. Swinerton Builders/Power Engineering Construction Joint Venture was awarded the project by SFPW. Liftech Consultants Inc. is designing the pier, access ramp, float, and piling; Shah Kawasaki Architects is designing the building (Cardno 2021).

Fire Station 35 is an Essential Facility with 50-year design life requiring resilience to 2070. Design sea level rise has been estimated at 2.8 ft by 2070, which is consistent with the "Guidance for Incorporating Sea Level Rise into Capital Planning in San Francisco: Assessing Vulnerability and Risk to Support Adaptation" dated 14 December 2015.

Figure 5.9a shows the project details. The facility has resilience for estimated sea level rise to the year 2100, including a 1% storm (base flood elevation, or [BFE]) + wave runup and setup + sea level rise. Local conditions, as shown in Figure 5.9b, are considered in the final design. The project design components included float guide pile height, access ramp and gangway with hinges at the landside connection and wheel supports at the float, and utilities with flexible joints at moving interfaces. The final design of the floating pier resilient to sea level rise is shown in Figure 5.9c.

Some items considered while designing for sea level rise are as follows:

- Long-term regional plan for addressing sea level rise.
- The vicinity of site and how it will be affected by sea level rise; if only focused on the specific site, then solutions may be less effective (e.g. building a barrier around the site that rising water will eventually flow around).
- How all the elements of the project need to adapt (e.g. float is moving, so structures and utilities connected to float from landside need to accommodate that movement).
- Certain adaptations may be more favorable than others (e.g. float is more aesthetically appealing than a fixed pier several feet above surroundings).
- Required resiliency of the project given its exposure and sensitivity (e.g. fire station on the water is an essential facility with high exposure and sensitivity to sea level rise, so it should have high resiliency).
- Preferred solution may be more expensive given above or other considerations, but cost estimates should be made on a long-term basis, which could include future renovations if initial design does not accommodate long-term uncertainty.

5.8.2 San Francisco Resilient CSD Design

San Francisco is implementing climate adaptation strategies in their Sewer System Improvement Program (SSIP). The SSIP is a programmatic approach to strategize, plan, and implement a portfolio of capital improvements in San Francisco's combined sewer discharge (CSD) system over a 20-year timeframe. In order to make this infrastructure resilient to sea level rise, the following design criteria are used:

- **New infrastructure** must accommodate expected sea level rise within the service life of the asset (i.e. 6 in by 2030, 11 in by 2050, 36 in by 2100) and be consistent with the City's Guidance for Incorporating Sea Level Rise into Capital Planning.

(a)

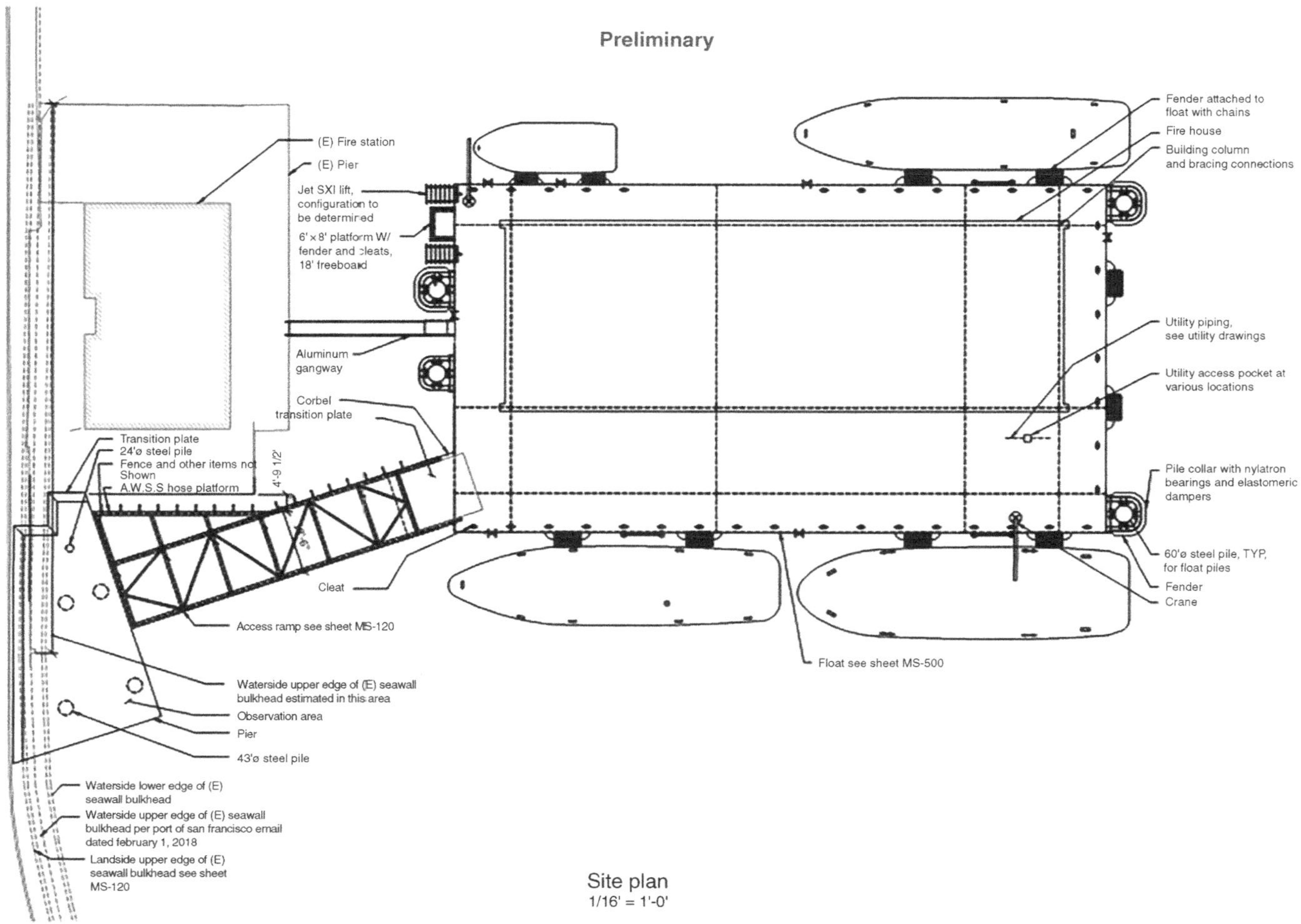

Figure 5.9 (a) Project details. (b) Water levels considered for design. (c) Floating fire house designed to be resilient to sea level rise. *Source:* City of San Francisco (2015).

(b)

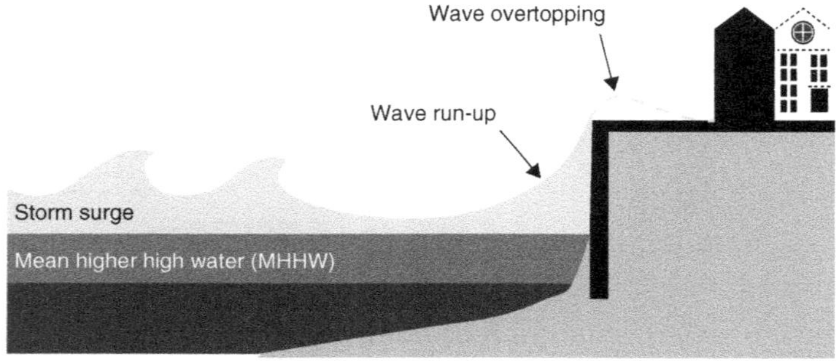

Factors that influence local water level conditions in addition to sea level rise

Factors affecting water level	Typical range CCSF pacific shoreline (a)	Typical range CCSF bay shoreline (a)	Period of influence	Frequency
Tides	5 to 7 ft	5 to 7 ft	Hours	Twice daily
Storm Surge	0.5 to 3 ft	0.5 to 3 ft	Days	Several times a year
Storm Waves	10 to 30 ft	1 to 4 ft	Hours	Several times a year
El Niños (within the ENSO cycle)	0.5 to 3 ft	0.5 to 3 ft	Months to Years	Every 2 to 7 years

Sources:

(a) Typical ranges for tides, storm surge, and storm waves for the CCSF Pacific Coast: BakerAECOM 2012. Intermediate Data Submittal #1. Scoping and Data Review. San Francisco Country, California. California Coastal Analysis and Mapping Project / Open Pacific Coast Study. Submitted to FEMA Region IX. February 2012.

(b) Typical ranges for tides, storm surge, and storm waves for the CCSF Bay shoreline: DHI. 2010. Regional Coastal Hazard Modeling Study for North and Central Bay. Submitted to FEMA Region IX. October 2011.

Figure 5.9 (Continued)

(c)

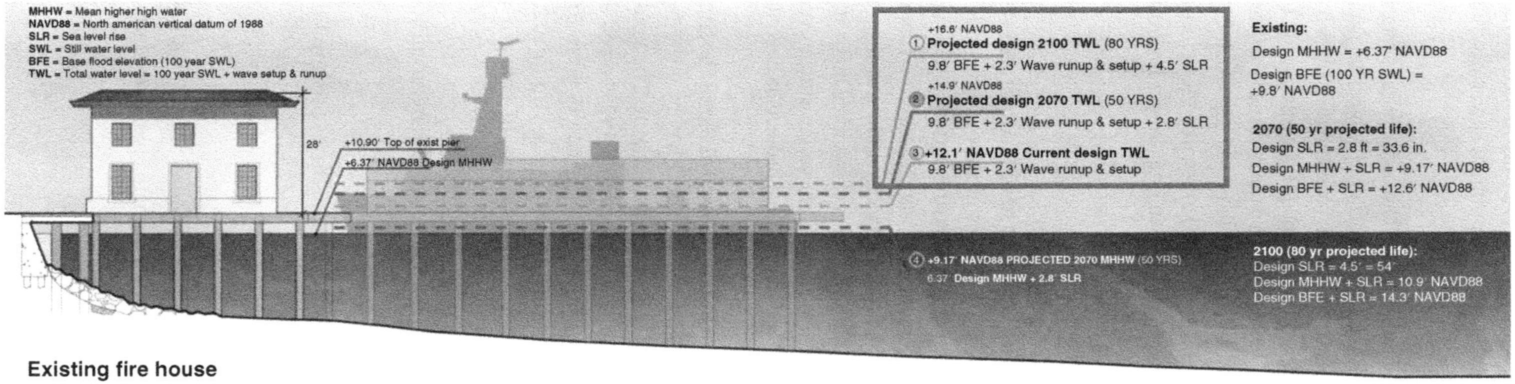

Existing fire house

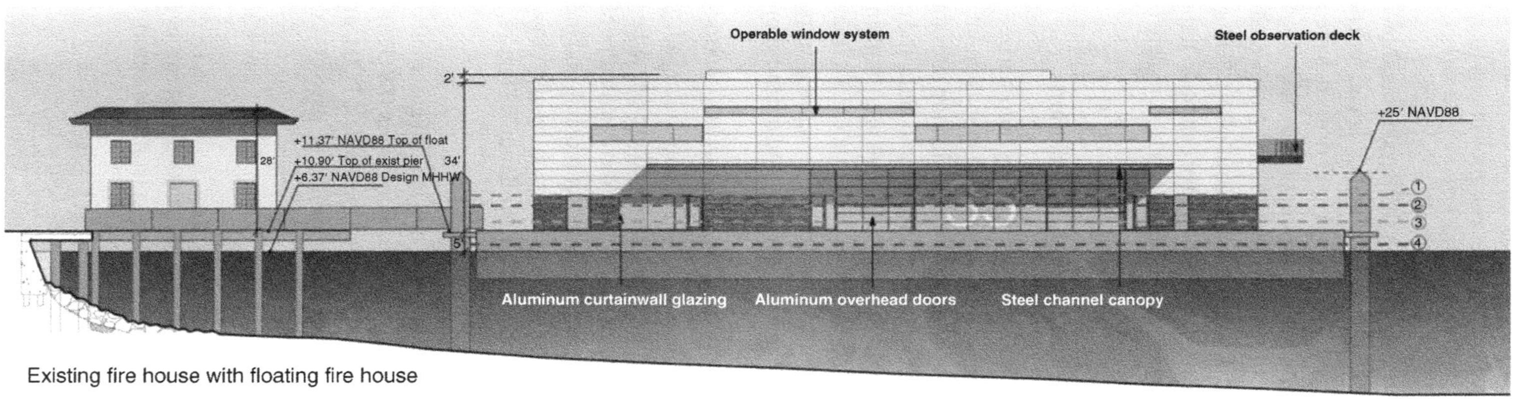

Existing fire house with floating fire house

Figure 5.9 (Continued)

- **Existing infrastructure** that is impacted by sea level rise, within the service life of the asset, will be modified based on sea level rise projections (e.g. pump stations, combined sewer discharge [CSD] structures).

There are 29 CSD structures on the Bayside, as shown in Figure 5.10a. The conceptual design of CSD is presented in Figure 5.10b, and how it can function during wet weather, dry weather, and a combination of dry weather and sea level rise is depicted in Figure 5.10b–d, respectively. Tides have a diurnal pattern with short durations of high tides (20–120 minutes). A tidal inflow calculation tool was developed to estimate inflows. Modeled projected tide levels (2-year storm surge + sea level rise) are used to estimate inflow volume at each CSD location. In addition to projecting when CSDs may need backflow prevention, timing of condition improvements was also estimated using annual inspection data. The capital strategy aims to complete both types of work in one project for implementation efficiency.

5.8.3 Resilient Environmental Remediation

Washington State Department of Ecology summarized cleanup sites that have incorporated resilience into the cleanup design or adaptive management plans. They discuss incorporating sea level rise into the remedial design at landfill cleanup sites located along the marine waterfront, at contaminated soil and groundwater sites located along the marine waterfront, and at sediment cleanup sites in a marine embayment. They have also implemented adaptive management during post-cleanup construction at a sediment site to address damage from multiple severe storm events within three days of each other. In addition, they have incorporated protection from severe storm events into the remedial design for a sediment and upland shoreline site. One of these projects is described briefly below.

The March Point Landfill site (also referred to as Whitmarsh Landfill) is located along the shoreline of Padilla Bay in Anacortes. It is situated along a lagoon connected to Padilla Bay by a 100-foot-wide channel under the Burlington Northern Santa Fe (BNSF) railroad embankment. The landfill operated for more than 20 years as an unregulated public dump and then around 1973 served as a county disposal area for household, commercial, and industrial wastes. From the late 1980s to 2011, a sawmill was operated at the site, which resulted in accumulations of wood waste up to 10 ft in thickness over large portions of the landfill (Figure 5.11a). The contaminants of concern are numerous:

- *Soil* – solid waste, wood waste, and landfill gas (methane); metals; total petroleum hydrocarbons; benzene; semi-volatile organic compounds (SVOCs); and PCBs.
- *Groundwater* – SVOCs, benzene, pesticides, PCBs, and metals.
- *Surface water* – metals, one pesticide, PCBs, SVOCs, and benzene seeping from the landfill to surface water.

The cleanup remedy consists of:

- A design to address impacts from 100-year storm events and sea level rise of 7.6 ft above mean higher high water (MHHW) due to tsunamis (short-term impact) or climate change (long-term impact) (Figure 5.11b–d).

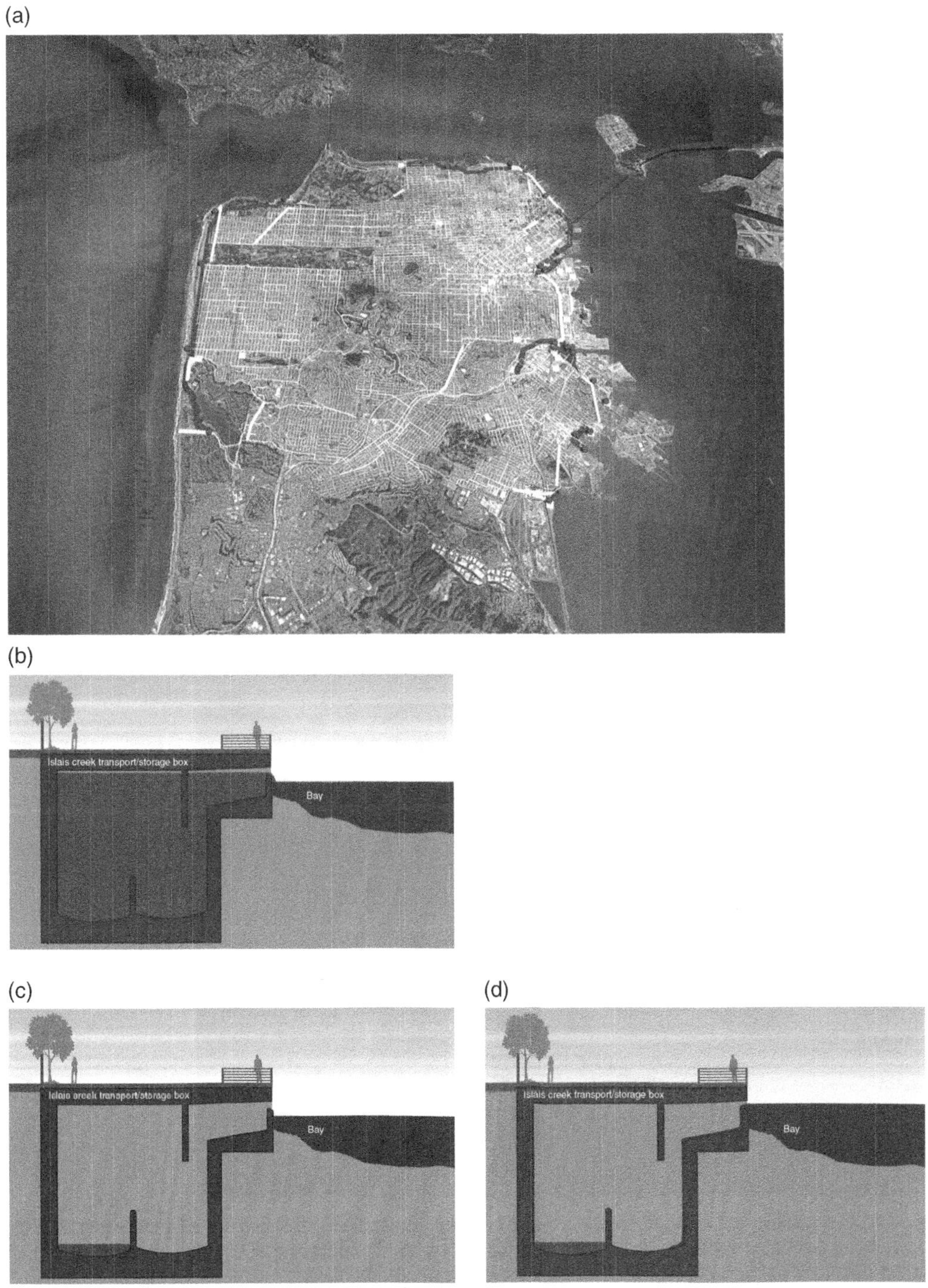

Figure 5.10 (a) Locations of combined sewer discharge (CSD) structures on the bayside. (b) Conceptual rendering of the CSD structure in wet weather. (c) Conceptual rendering of the CSD structure in dry weather. (d) Conceptual rendering of the CSD structure in dry weather-impacted by sea level rise. *Source:* City of San Francisco (2015).

- Moving solid waste from the edges of the landfill inward.
- Grading the waste to a mound.
- Installing a passive landfill gas collection system to vent gases into the atmosphere.
- Installing an engineered cap over the landfill with standard geosynthetic clay laminated liner (GCLL) above 16 ft mean lower low water (MLLW) to minimize or eliminate infiltration into the landfill.
- Placement of an enhanced geosynthetic clay liner over the landfill along the shoreline up to 16 ft MLLW to minimize or eliminate discharge of groundwater to surface water (Figure 5.11b–d).
- Constructing a perimeter access road around the landfill.
- Requiring 30 years of operation and maintenance.
- A tidal study and geotechnical and hydrogeological evaluations showed that (i) the shallow nature of Padilla Bay, (ii) the BNSF rail embankment, (iii) nearby hillside, and (iv) Highway 20 protected the lagoon and shoreline from increased wave heights that might result from sea level rise, as well as wave and current actions during 100-year storm events.

(a)

Figure 5.11 (a) March Point Landfill before cleanup showing the lagoon. (b) March Point Landfill plan view of the extent of GCL liner and waste. (c) March Point Landfill details of enhanced GCL liner. (d) March Point Landfill details of enhanced GCL liner. *Source:* State of Washington (2017).

(b)

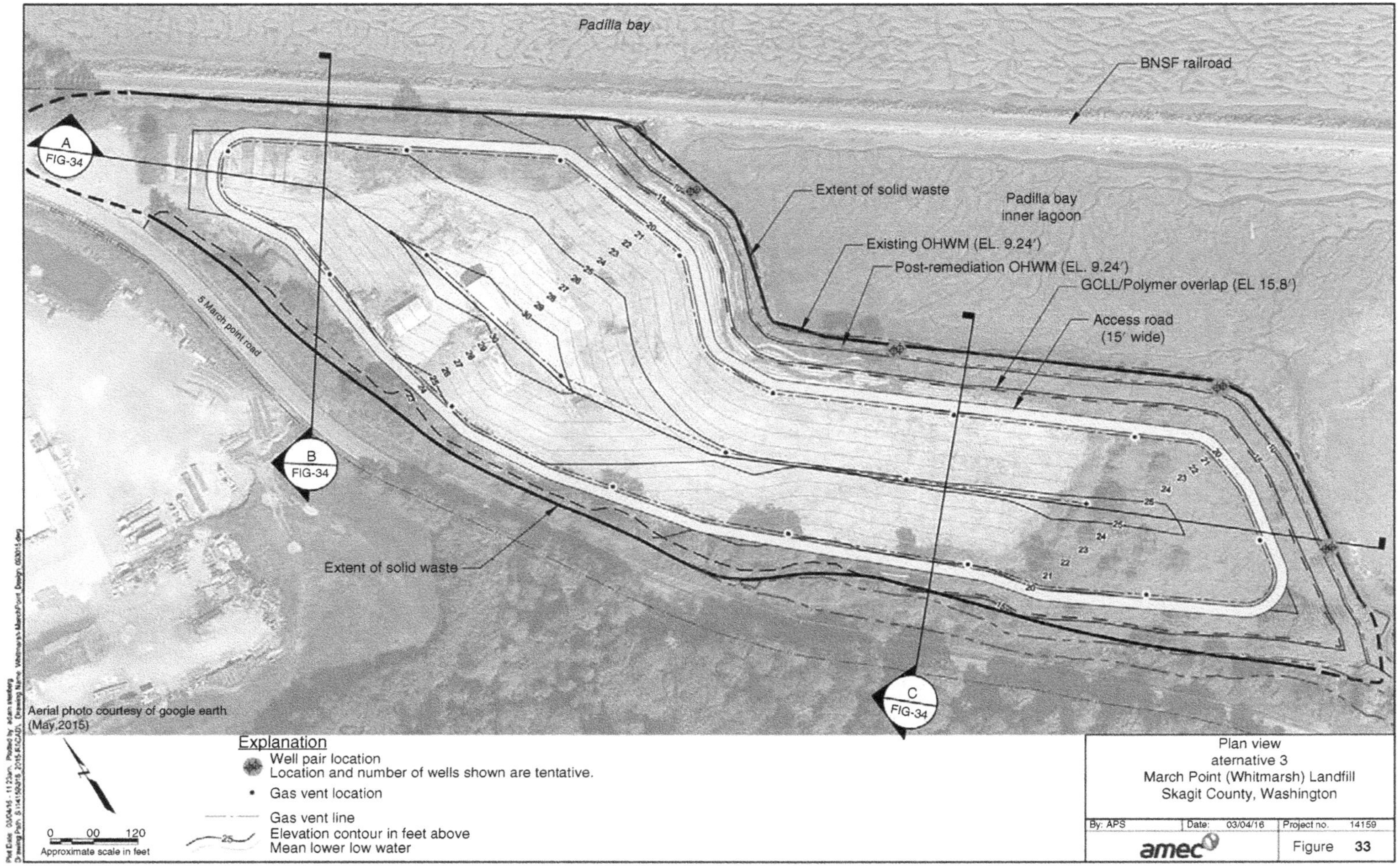

Figure 5.11 (Continued)

(c)

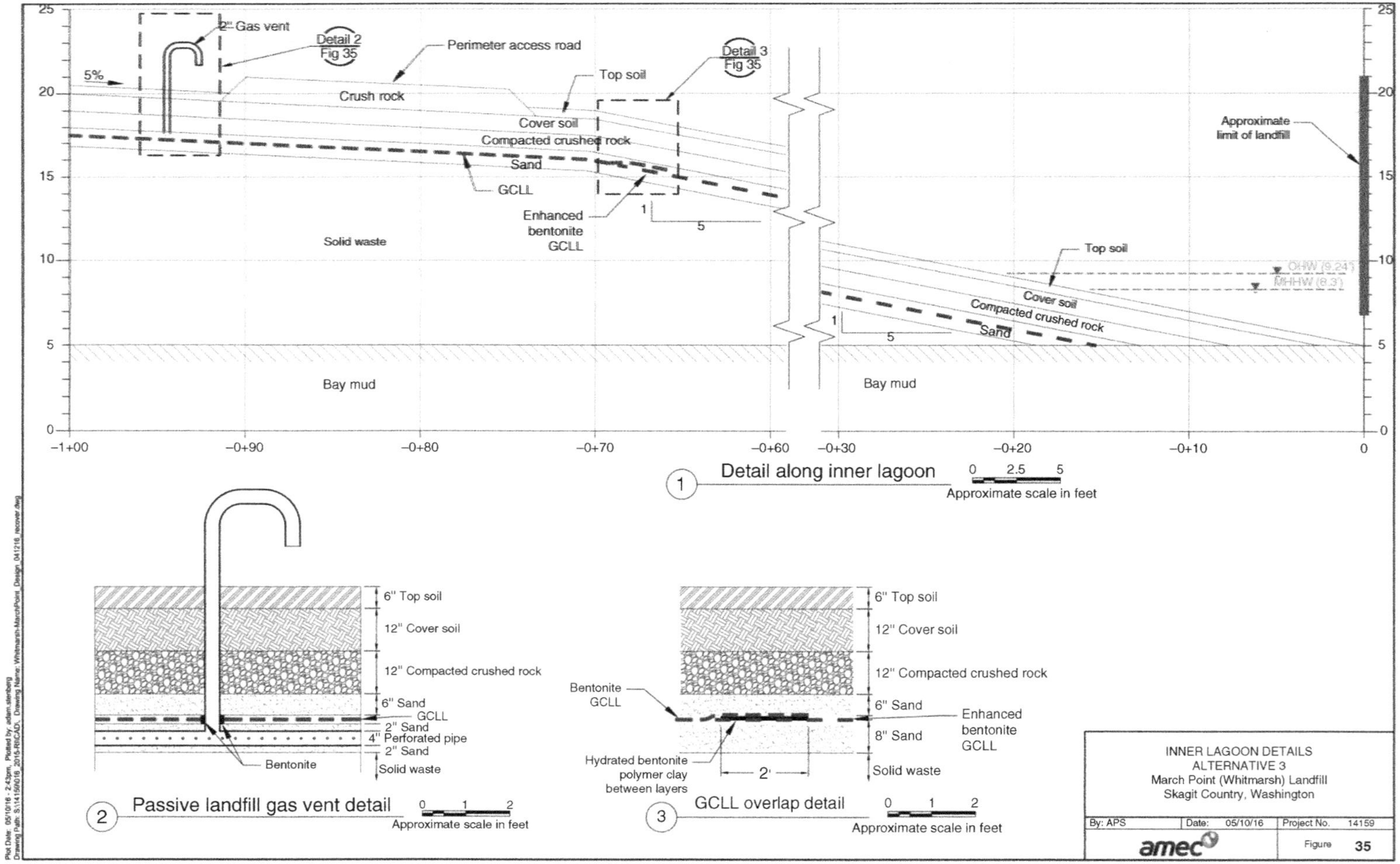

Figure 5.11 (Continued)

(d)

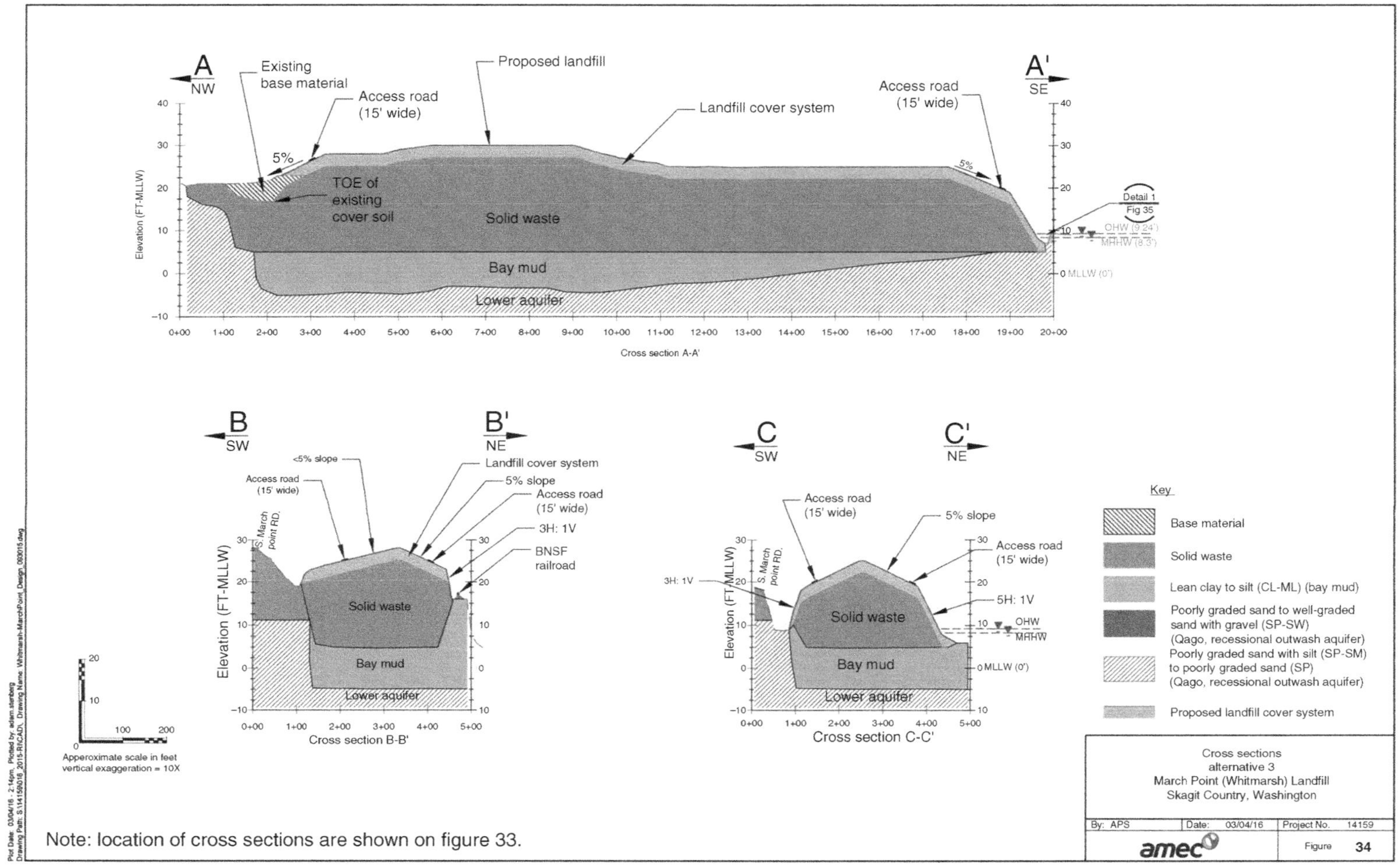

Figure 5.11 (Continued)

5.9 Managed Retreat and Cost-Benefits

Much of disaster resiliency is focused on devising and designing infrastructure systems that are robust enough to withstand stressors as well as exhibiting the ability to remain serviceable and/or recover in the event that a natural disaster strikes. In some instances, however, it is impractical or economically infeasible to design and construct/rehabilitate infrastructure to resist all credible and foreseen natural phenomena. The unit cost may be too high on an absolute basis, or in many cases, the service provided by the infrastructure, such as a transportation link, protection, etc., is not great enough to justify the infrastructure improvement.

As an example, consider coastal protection. Associated infrastructure may consist of a seawall, rockery/rip-rap berms/dunes, or other "hard edges", with the purpose of preventing coastal flooding/inundation resulting from high tides, wave action, or storm surge from large storm events. With sea levels rising due to climate change, more coastal expanses and adjacent low-lying areas are becoming increasingly susceptible to encroaching seawater, yet some uncertainty exists for how high walls may need to be erected or extended to offer protection. Yet, with high unit costs, incremental increases in wall thicknesses or height can render projects to be cost-prohibitive. Further, some relatively high-value land uses, such as ports, tourist facilities, and central business districts, may justify high protective infrastructure expenditures, other areas may not have such inherent land value, and protective infrastructure of any degree may become cost-prohibitive.

In such cases, managed retreat – allowing rising seas to "reclaim" shoreline – may be an appropriate or economically wise option. With managed retreat, select low-lying coastal areas, especially those susceptible to erosion losses due to wave action, are allowed to become inundated. These areas often cycle between inundation and emergence during daily tidal activity. Such an approach can offer advantages. First, these lands can support the development of important habitats, such as salt marshes or other coastal wetland features. They also provide a natural protective means to land further inland bay acting as a dissipative buffer and run-up area for wave action, tidal action, and/or storm surge. On the other hand, there are some disadvantages. First, land use decisions need to be made, often controversial decisions, about what lands may be designated for managed retreat. Invariably, this will include the loss of existing infrastructure, such as coastal roadways, or private landowner developments and improvements. If private lands are to be acquired for managed retreat, landowners must be duly compensated for their lands, and the economic losses of supporting public infrastructure should be carefully weighed against the benefits of a managed retreat approach for such land.

5.10 Soft Infrastructure

Well-designed, robust, and properly maintained infrastructure is essential toward achieving societal resiliency in the event of foreseen or unforeseen disasters. Of course, a competent institutional structure (i.e., local, state, and/or federal legislative and executive governments, regulatory agencies, utilities) is also an essential component toward the goal of resiliency. It is these institutions, after all, that are charged with the design, construction, maintenance, and operation of infrastructure systems, as well as the task of response to disasters that may affect the integrity and function of infrastructure systems.

It is helpful to think of these institutions as a type of infrastructure system itself. Physical systems whether "gray infrastructure"– roads, bridges, pipelines, structures, airports, river/seaports, power generation plants, etc., – or "green infrastructure" – stormwater systems such as bioretention, bioswales, floodways, coastal inland wetlands, etc., – may all be classified as "hard infrastructure". These are the systems that are designed, monitored, constructed, operated, and maintained – actively or passively – that are necessary for modern society to function. In addition, and of significant importance, "soft infrastructure" consists of the institutions, organizations, statutes, regulations, policies, revenue streams (i.e. tax and fee structures), etc., that are necessary to support hard infrastructure. We suggest that the reader consult the research of Anderies, Deslatte, DeCaro, and others for a detailed overview and research on the importance of design and operation of soft infrastructure.

5.11 Challenges

It is projected that the United States and many other countries worldwide sustain greater economic hardship from climate impacts such as hurricanes, rising sea levels, extreme flooding, and high temperatures. Many entities are planning to address resiliency in infrastructure projects. Many challenges exist to achieve improved resilience, including those listed below:

- Uncertainty in prediction of magnitude and frequency of extreme events. Thus, there is an inevitable need for adaptability and acceptance of some level of uncertainty.
- Climate change is a global problem and mitigation/adaptation strategies and development of standards require coordination of multiple nations and multiple jurisdictions.
- There is no acceptable way to quantify resilience; hence, it is not possible to assess alternatives to increase robustness to extreme events. Comprehensive frameworks such as the one discussed in this chapter can help.
- There is a lack of a census-based standard strategy to assess vulnerability to extreme events and resilient design.
- There are no mandates or incentives to design for resiliency, resulting in lack of investments. Life cycle costs and benefits should be used to justify pursuing resilient designs.
- There is a lack of educational programs on resiliency, especially in areas of the world where resilient systems are utmost needed. Lack of transfer of technology and knowledge may be impediments to developing or implementing innovative resilient systems.
- Resilient thinking should be highlighted as a moral obligation of an engineer to ensure the safety of the public.
- Lack of communication among the stakeholders (design engineers, policymakers, regulators, citizens, and investors) to build consensus for the need for design for resiliency and additional investments.

5.12 Summary

Resilience is a term common to many fields of study. Resilience helps systems to withstand, respond, and/or adapt to a vast range of disruptive events by preserving and even enhancing critical functionality. Greenhouse gases and climate change are found to be responsible for the growing concern of extreme events such as sea level rise, storm surge, extreme precipitation and flooding, and extreme heat. Superstorm Sandy, Hurricane Harvey, and other extreme events have highlighted the need for resilient civil infrastructure. Many entities developed policies or guidance on pursuing resilient systems. The framework for incorporating resiliency in infrastructure design is not well established, but a vulnerability assessment is needed and adaptive resilient systems need to be designed following similar frameworks, as described in this chapter. The effectiveness of a resilient infrastructure design depends upon its ability to anticipate, absorb, adapt to, and/or rapidly recover from a potential extreme event. Despite the growing importance of resilient infrastructure, there is a lack of guidance for engineers regarding how to specify, design, monitor, and assess the resilient civil infrastructure. In addition, educational and technology transfer programs as well as incentives to pursue resilience are needed.

5.13 Questions

5.1 What is meant by resiliency? What are the different fields that deal with resiliency and in what ways? Explain briefly.

5.2 Mention some of the disastrous climatic impacts.

5.3 Describe how changes in precipitation patterns and snowfall could affect runoff and flooding patterns.

5.4 What is disaster resilience and what are its characteristics?

5.5 Discuss in detail an infrastructure project with which you are familiar and identify ways to enhance its design to improve its resiliency.

5.6 List in detail some important indicators or inputs of a resiliency assessment tool that interests you.

5.7 What are the different policies and initiatives that were taken in regard to resilience? Explain briefly.

5.8 Mention some of the tools developed for evaluating climate change vulnerabilities and impacts.

5.9 What are the important considerations while designing for resilience against sea level rise?

5.10 What are the major challenges in adaptation and implementation of resiliency in engineering and infrastructure design?

5.14 Problems

5.1 Five dimensions of resilience are defined as shown below:

Resilience dimension	Description	Equation
(1) Recovery	Time required to return to the acceptable performance range	$t_n - t_d$
(2) Impact	The severity impact on performance	$P_0(t_0) - P_v(t_v)$
(3) Performance loss	Total performance loss is the area above the curve $P(t)$ between t_d and t_n	$(t_n - t_d) \times P_0(t_0) - \int_{t_d}^{t_n} f(t)dt$

Resilience dimension	Description	Equation
(4) Profile length $f(t)$	The length of the recovery profile as it reaches the acceptable performance level	$\int_{t_d}^{t_n} \sqrt{1+\left(\frac{dp}{dt}\right)^2}\, dt$
(5) Weighted-sum $g(t)$	A time-dependent deviation weighted sum to capture the speed and shape of the recovery response	$\sum_{i=1}^{n} a_i[g(a_i) - p(a_i)]$

If the recovery function is given by a quadrant of the circle, and $t_n - t_d = 50$ days, formulate the recovery function ($f(t)$). Given the pre-disruption performance is 100% and post-disruption performance has deteriorated to 50%, calculate all the five dimensions of the resilience, as defined above:

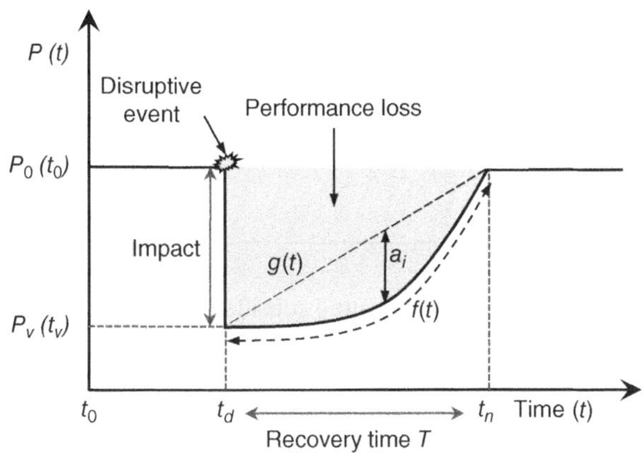

If the resilience of the system is given by the following equation:

$$\psi = w_1(\text{recovery}) + w_2(\text{impact}) + w_3(\text{performance loss}) + w_1(\text{profile length}) + w_5(\text{weighted} - \text{sum})$$

Calculate the resilience of the system considering equal weightages to all the dimensions of the resilience.

5.2 An engineering system has been subjected to a sudden disruption which reduced its functionality to 50%. The time for natural recovery ($t_n - t_d$) is 12 days and the natural recovery could bring back the performance to only 70%. The recovery function for natural recovery is represented by the dashed line in the figure below. Adaptation measures can be undertaken to bring back the performance to a 100% in the same time, the recovery profile in this scenario is represented by the solid line. Calculate the performance loss in case of both natural recovery and adaptive recovery for the given time period.

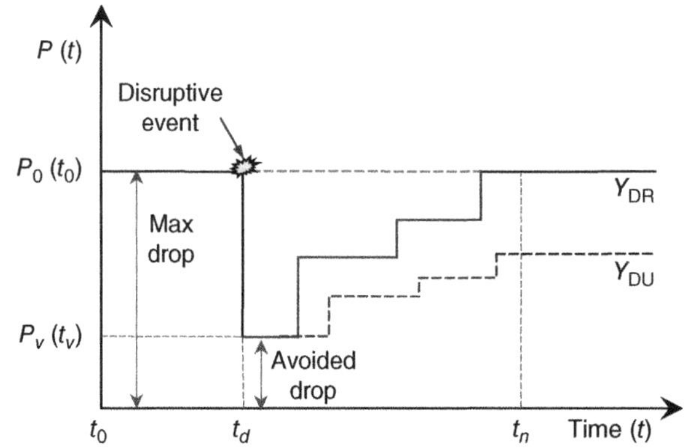

	Adaptation measures (solid line)			Natural recovery (dashed line)	
Step	Time in previous state	Performance level reached	Step	Time in previous state	Performance level reached
1	2	70	1	3	55
2	4	80	2	4	65
3	5	100	3	4	70

5.3 Model the settlement of a building foundation due to ground subsidence with a Gaussian degradation curve described by:

$$S(t) = 50e^{-\frac{(t-30)^2}{300}} - 2.489$$

where $S(t)$ represents the settlement (in millimeters) at time t (in years) since the construction of the building. If $P(t) = 100–S(t)$, calculate the performance loss (as indicated in the curve below) of the building in 30 years. Now, consider the potential increase in the ground subsidence due to increased groundwater extraction, leading to an additional settlement rate of 1 millimeter per year and calculate the performance loss. Plot the deterioration of building performance with time for both scenarios.

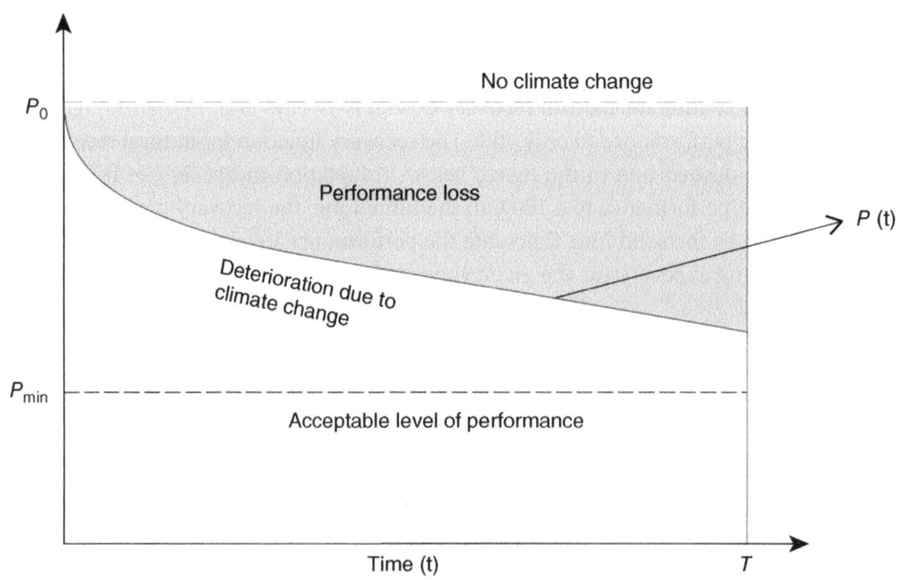

$$\text{Design resilience, } R_d = \frac{\int_{t=0}^{t=T} P(t)dt - P_{min} \times T}{(P_0 - P_{min}) \times T}$$

Now consider that the acceptable level of settlement (S_{max}) is only 80 mm (i.e., $P_{min} = 20$), calculate the resilience of the building to subsidence using the following equation for both the scenarios described above.

5.4 If the baseline performance of a system is given by $BP(t) = 80$ and the disruption induced deterioration and the following recovery can be represented by a single equation given by $AP(t) = (t-15)^2 + P_v$. If the deterioration due to the disruption started at t_d and the minimum performance level is $P_v = 50$, then determine t_d, and t_n in days and calculate the impacted area. Now considering similar disruption profiles are expected to occur for 15 years with an average frequency of 3 events per year, calculate the resilience of the system. Also, plot the resilience curve similar to the figure as shown below:

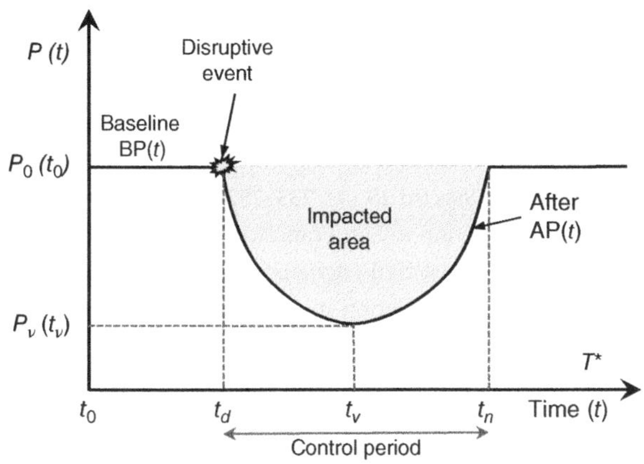

Resilience can be calculated using the following equation:

$$\psi = \frac{\int_{t_0}^{T^*} \mathrm{BP}(t)\mathrm{d}t - \sum_{i=1}^{N(T^*)} \mathrm{IA}_i(t_i)}{\int_{t_0}^{T^*} \mathrm{BP}(t)\mathrm{d}t}$$

References

Adger, W.N. (2000). Social and ecological resilience: are they related? *Progress in Human Geography* 24 (3): 347–364. https://doi.org/10.1191/030913200701540465.

Anderies, J.M., Mathias, J.D., and Janssen, M.A. (2019). Knowledge infrastructure and safe operating spaces in social–ecological systems. *Proceedings of the National Academy of Sciences* 116 (12): 5277–5284.

ASCE (2018a). Policy statement 418 – the role of the civil engineer in sustainable development. American Society of Civil Engineers. https://www.asce.org/issues-and-advocacy/public-policy/policy-statement-418---the-role-of-the-civil-engineer-in-sustainable-development/ (accessed 20 September 2018).

ASCE (2018b). Working definitions. American Society of Civil Engineers. http://ciasce.asce.org/working-definitions (accessed 20 September 2018).

ASCE (2021). Minimum design loads and associated criteria for buildings and other structures (ASCE/SEI 7-22). American Society of Civil Engineers. https://www.asce.org/publications-and-news/asce-7 (accessed 16 September 2024).

ASCE (2023). Pathways to resilient communities. American Society of Civil Engineers. https://www.asce.org/-/media/asce-images-and-files/advocacy/documents/pathways-to-resilient-communities-asce-toolkit.pdf (accessed 16 September 2024).

Bocchini, P. and Frangopol, D.M. (2011). Resilience-driven disaster management of civil infrastructure. In: *Computational Methods in Structural Dynamics and Earthquake Engineering* (ed. M. Papadrakakis, M. Fragiadakis, and V. Plevris), 1–11. Keynote Paper, Institute of Structural Analysis and Antiseismic Research, School of Civil Engineering, National Technical University of Athens (NTUA).

Bocchini, P., Frangopol, D.M., Ummenhofer, T., and Zinke, T. (2014). Resilience and sustainability of civil infrastructure: toward a unified approach. *ASCE Journal of Infrastructure Systems* 20 (2): 04014004.

Brundtland, G.H. (1987). *Our Common Future: Brundtland-Report.* Oxford: Oxford University Press.

Bruneau, M., Chang, S.E., Eguchi, R.T. et al. (2003). A framework to quantitatively assess and enhance the seismic resilience of communities. *Earthquake Spectra* 19 (4): 733–752.

Cardno, C. A. (2021). Floating fire station nears completion in San Francisco. American Society of Civil Engineers. https://www.asce.org/publications-and-news/civil-engineering-source/civil-engineering-magazine/article/2021/09/floating-fire-station-nears-completion-in-san-francisco (accessed 9 October 2024).

CIG (2018). Climate change. Climate Impacts Group. https://cig.uw.edu/learn/climate-change/ (accessed 20 September 2018).

Cimellaro, G.P., Reinhorn, A.M., and Bruneau, M. (2010a). Framework for analytical quantification of disaster resilience. *Engineering Structures* 32 (11): 3639–3649.

Cimellaro, G.P., Renschler, C., Reinhorn, A.M., and Arendt, L. (2016). PEOPLES: a framework for evaluating resilience. *Journal of Structural Engineering* 142 (10): 04016063.

City of San Francisco (2015). Guidance for incorporating sea level rise into capital planning in San Francisco: assessing vulnerability and risk to support adaptation. http://onesanfrancisco.org/sites/default/files/inline-files/Guidance-for-Incorporating-Sea-Level-Rise-into-Capital-Planning1.pdf (accessed 20 September 2018).

Combaz, E. (2014). *Disaster Resilience: Topic Guide.* Birmingham: Governance and Social Development Research Centre, University of Birmingham.

Deslatte, A., Koebele, E.A., Bartels, L. et al. (2023). Institutions, voids, and dependencies: tracing the designs and robustness of urban water systems. *International Review of Public Policy* 5: 180–222.

Henry, D. and Ramirez-Marquez, J.E. (2012). Generic metrics and quantitative approaches for system resilience as a function of time. *Reliability Engineering and System Safety* 99: 114–122.

Holling, C.S. (1973). Resilience and stability of ecological systems. *Annual Review of Ecology and Systematics* 4: 1–23.

IPCC (2015). Climate change 2014: Synthesis report. Intergovernmental Panel on Climate Change. http://ar5-syr.ipcc.ch/ipcc/ipcc/resources/pdf/IPCC_SynthesisReport.pdf (accessed April 2016).

IPCC (2019). *IPCC Special Report on the Ocean and Cryosphere in a Changing Climate* (ed. H.-O. Pörtner et al.), 755. Cambridge and New York, NY: Cambridge University Press https://doi.org/10.1017/9781009157964.

IPCC (2021). *Climate Change 2021: The Physical Science Basis. Contribution of Working Group I to the Sixth Assessment Report of the Intergovernmental Panel on Climate Change*, 2391. Cambridge and New York, NY: Cambridge University Press https://doi.org/10.1017/9781009157896.

IPCC (2023). Climate change 2023: synthesis report. Intergovernmental Panel on Climate Change. https://www.ipcc.ch/report/ar6/syr/ (accessed September 2024).

ISO (2018). ISO 37120:2018: sustainable cities and communities – indicators for city services and quality of life. International Organization for Standardization. https://www.iso.org/standard/68498.html (accessed 8 October 2024).

MacCarthy, J., Richter, J., Tyukavina, S. et al. (2024). *The latest data confirms: Forest fires are getting worse.* World Resources Institute https://www.wri.org/insights/global-trends-forest-fires (accessed 20 March 2025).

National Weather Service (2024). Temperature. https://www.weather.gov/source/zhu/ZHU_Training_Page/definitions/dry_wet_bulb_definition/dry_wet_bulb.html (retrieved 1 October 2024).

NOAA (2017). National centers for environmental information, state of the climate: global climate report for annual 2016. National Oceanic and Atmospheric Administration. https://www.ncdc.noaa.gov/sotc/global/201613 (retrieved 20 September 2018).

NOAA (2018a). Sea level rise viewer. National Oceanic and Atmospheric Administration. Office for Coastal Management. https://coast.noaa.gov/digitalcoast/tools/slr.html (accessed 20 September 2018).

NOAA (2018b). Surging seas. National Oceanic and Atmospheric Administration. Office for Coastal Management. https://coast.noaa.gov/digitalcoast/tools/surging-seas.html (accessed 20 September 2018).

NOAA (2024). Trends in atmospheric carbon dioxide (CO_2). National Oceanic and Atmospheric Administration. Global Monitoring Laboratory. https://gml.noaa.gov/ccgg/trends/ (accessed 2 April 2024).

OCCRI (2017). Third Oregon climate assessment report. Oregon Climate Change Research Institute. http://www.occri.net/publications-and-reports/third-oregon-climateassessment-report-2017/ (accessed September 2018).

O'Rourke, T.D. (2007). Critical infrastructure, interdependencies, and resilience. *The Bridge* 37: 22–29.

PPD (2011). PPD-8: national preparedness. Presidential Policy Directive. https://www.dhs.gov/presidential-policy-directive-8-national-preparedness (accessed 20 September 2018).

PPD (2013). PPD-21: critical infrastructure security and resilience. Presidential Policy Directive. https://obamawhitehouse.archives.gov/the-press-office/2013/02/12/presidential-policy-directive-critical-infrastructure-security-and-resil (accessed 20 September 2018).

Reddy, K.R., Janga, J.K., Verma, G., and Nagaraja, B. (2024). Integration and quantification of resilience and sustainability in engineering projects. *Journal of the Indian Institute of Science* https://doi.org/10.1007/s41745-024-00440-w.

Rinaldi, S.M., Peerenboom, J.P., and Kelly, T.K. (2001). Identifying, understanding, and analyzing critical infrastructure interdependencies. *IEEE Control Systems Magazine* 21 (6): 11–25.

Rose, A. and Liao, S.-Y. (2005). Modeling regional economic resilience to disasters: a computable general equilibrium analysis of water service disruptions. *Journal of Regional Science* 45 (1): 75–112.

Ruiz, S. (2020). Climate change will alter forests as we know them. Global Forest Watch. https://www.globalforestwatch.org/blog/forest-insights/climate-change-will-alter-forests-as-we-know-them/ (accessed 14 December 2024)

State of Washington (2017). Adaptation strategies for resilient cleanup remedies: a guide for cleanup project managers to increase the resilience of toxic cleanup sites to the impacts from climate change. https://fortress.wa.gov/ecy/publications/SummaryPages/1709052.html (accessed 20 September 2018).

The Rockefeller Foundation (2015). City resilience framework. https://www.rockefellerfoundation.org/wp-content/uploads/City-Resilience-Framework-2015.pdf (accessed 8 October 2024).

Timmerman, P. (1981). Vulnerability. Resilience and the collapse of society: a review of models and possible climatic applications. In: *Environmental Monograph*, 1. Canada: Institute for Environmental Studies, University of Toronto.

UNDRR (2015). Sendai framework for disaster risk reduction 2015–2030. United Nations Office for Disaster Risk Reduction. https://www.undrr.org/publication/sendai-framework-disaster-risk-reduction-2015-2030 (accessed 8 October 2024).

UNFCCC (2023). Technical dialogue of the first global stocktake: synthesis report by the co-facilitators on the technical dialogue. In *UN Climate Change Conference - United Arab Emirates*. United Nations Framework Convention on Climate Change. https://unfccc.int/documents/631600 (accessed 16 September 2024).

USACE (2014). Sea-level change curve calculator using the flood risk reduction standard for sandy rebuilding projects. United States Army Corps of Engineers. https://www.usace.army.mil/corpsclimate/Climate_Preparedness_and_Resilience/App_Flood_Risk_Reduct_Sandy_Rebuild/SL_change_curve_calc/ (accessed 20 September 2018).

USEPA (2016). Climate resilience evaluation and awareness tool (CREAT). United States Environmental Protection Agency. https://www.epa.gov/crwu/climate-resilience-evaluation-and-awareness-tool (accessed 20 September 2018).

USEPA (2017). National stormwater calculator with climate assessment tool. United States Environmental Protection Agency. https://toolkit.climate.gov/tool/national-stormwater-calculator%E2%80%94climate-assessment-tool (accessed 20 September 2018).

Section II

Sustainability and Resiliency Metrics and Assessment Tools

6

Sustainability and Resiliency Indicators, Metrics, and Assessment Tools

Chapter Goal and Objectives
The chapter aims to provide a comprehensive overview of the key attributes of sustainability and resiliency indicators, highlighting the tools and metrics used for their assessment and the importance of integrating these for holistic evaluation in environmental and engineering practices. The specific objectives are to (i) define and differentiate sustainability and resiliency indicators; (ii) explore various sustainability metrics and their applications in assessment; (iii) introduce sustainability assessment tools and their role in decision-making; (iv) discuss resiliency indicators and their significance in system design and evaluation; (v) examine resiliency metrics for quantifying system robustness and adaptability; (vi) present resiliency assessment tools for measuring system performance under stress; and (vii) explore integrated tools that combine sustainability and resiliency assessments for a more comprehensive analysis.

6.1 Introduction

In the first section of this book, we discussed the influence of human activities on the environment. We described their impacts on ecosystems, biodiversity and species survival, soil contamination, water and air quality, climate change, resource depletion, and others. These impacts on the environment, in turn, affect human health and their well-being, as well as the economy. Furthermore, current human activities and their impacts on the environment could compromise the development of future societies and individuals. These aspects signifies the importance of considering sustainability and resiliency in engineering projects.

Sustainability analysis is performed to identify and decide the most sustainable alternative among several, and often competitive, alternatives or solutions. As discussed in previous chapters, the triple bottom line (TBL) of sustainability encompasses environmental, social, and economic metrics with the goal of maximizing the overall positive contribution from these three dimensions. Similarly, the importance of resiliency and considering various dimensions of resiliency have been discussed in Chapter 5. However, to measure the degree of sustainability and resiliency in engineering projects, it is essential to define and apply quantitative sustainability and resiliency metrics. Section II of this book presents common indicators, metrics, and tools that are used to quantify and assess sustainability and resiliency of engineering projects.

Sustainable and Resilient Engineering: Drivers, Metrics, Tools, and Applications, Second Edition. Krishna R. Reddy, Claudio Cameselle, and Jeffrey A. Adams.

Companion website: www.wiley.com/go/reddy/sustainableresilientengineering2e

6.2 Attributes of Sustainability and Resiliency Indicators

Sustainability and resiliency indicators are measurable aspects of environmental, economic, or social dimensions and how they are affected under gradual or sudden shocks or stressors associated with potential engineering alternatives for a project (Bell and Morse 2012). Since they are measurable, these indicators can be estimated beforehand or monitored on a real-time or periodic basis. These are used to determine how a project or a project alternative and its sustainability and resilience characteristics may positively (or negatively) contribute to human health and the environment. A sustainability or resilience indicator should have the following attributes defined by the "SMART" acronym:

- *Specific* – The indicator should target a specific area of interest for the analysis. It identifies the "what," "when," or "how" associated with the project. As an example, an indicator for a project alternative may be the minimization of emissions of particulate matter less than 10 μm in diameter (PM_{10}). PM_{10} emissions can have a deleterious effect on the health of people with respiratory ailments, such as asthma or chronic lung disease, and can also lead to other long-term adverse health effects. A specific indicator may be the use of emission controls on heavy equipment at a project site.
- *Measurable* – The indicator should be capable of being counted, compiled, analyzed, or tested so that a data set can be collected and assessed to determine the degree of success. In our example of PM_{10} emissions, filters or other measurement devices can be deployed at or near a project to measure PM_{10} emissions. Baseline or ambient conditions may also be established to determine the degree to which the project may contribute to the measured indicator.
- *Actionable/achievable* – The indicator should have a clear performance target that is easily understandable and may be realistically achieved at a project. For instance, if heavy diesel-powered equipment is to be used at a project, it may be unrealistic to expect zero PM_{10} emissions; however, another performance standard, such as a 50% reduction compared to previous projects where emissions controls were not used, may be appropriate and achievable.
- *Relevant* – The indicator should be selected such that it has a meaningful contribution to the overall goal or strategy associated with the project. Many indicators can be selected for a given project; however, they should be critically assessed for their overall meaningful contribution to the sustainability or resiliency for a project. In the PM_{10} example, it is relatively easy to demonstrate that reduced PM_{10} emissions have a direct benefit to the project's environmental conditions as well as meaningful contributions to the economy and society by protecting public health, quality of life, and associated economic benefits.
- *Timely* – The indicator should be achieved within an appropriate timeframe and/or subjected to the time constraints of the project. For this example, the 50% PM_{10} reduction may be assigned to the life of the project or over a specific subset of time, such as a period when equipment operations will be the greatest and PM_{10} reductions are most necessary.

The key indicators, as discussed above, may be objective or subjective.

6.3 Sustainability Indicators

Sustainability can be viewed at global, continental, national, city, company (corporation), or specific project scale. Indicators are often defined and tracked to help assess progress toward sustainability or sustainable development. However, these indicators can vary widely depending on the scale at which sustainability is being considered. For example, the United Nations Department of Economic and Social Affairs (UNDESA) proposed 134 indicators to reflect sustainability at a global level (UNDESA 2007). Table 6.1 shows some of these UN measurable objective indicators for sustainable development.

Table 6.1 Indicators considered for UN Sustainable Development Goals.

	SDG	Indicators
1	No Poverty	Unemployment rate Poverty index Population below poverty line
2	Zero Hunger	Prevalence if undernourishment Prevalence of obesity in adult population Human trophic level Cereal yield Yield gap closure (% of potential yield) Exports of hazardous pesticides
3	Good Health and Well-Being	Maternal mortality rate Neonatal mortality rate Mortality rate, under age 5 Death rate due to cardiovascular disease, cancer, diabetes, or chronic respiratory disease in adults (age 30–70) Death rate attributable to household air pollution and ambient air pollution Traffic deaths Life expectancy at birth Adolescent fertility rate Universal health coverage Daily smokers Gap in health status by income
4	Quality Education	Primary enrollment Secondary completion rate Literacy rate Variation in science performance explained by socioeconomic status
5	Gender Equality	Ratio of female to male mean years on education received Ratio of female to male labor participation rate Seats held by women in national parliament Gender wage gap
6	Clean Water and Sanitation	Population using least basic drinking water services Population using least basic sanitation services Freshwater withdrawal Anthropogenic wastewater that receives treatment Population using safely managed drinking water Population using safely managed sanitation services
7	Affordable and Clean Energy	Population with access to electricity Population with access to clean fuel and technology for cooking CO_2 emissions from fuel combustion per total electricity output Renewable energy share in total final energy consumption

(*Continued*)

Table 6.1 (Continued)

	SDG	Indicators
8	Decent Work and Economic Growth	GDP growth Victims of modern slavery Adults with bank accounts Unemployment rate Fatal work-related accidents Guaranteed fundamental labor rights Youth not in employment, education, or training
9	Industry, Innovation, and Infrastructure	Rural population with access to all-season roads Population using internet Quality of trade and transport-related infrastructure Higher education universities ranking Expenditure on research and development Female share of graduates from STEM
10	Reduced Inequalities	Elderly poverty rate Gini coefficient: Deviation from perfect distribution of income among the population Palma ratio: Ratio between the share all income received by the 10% People with highest disposable income and that received by the 40% People with lowest disposable income
11	Sustainable Cities and Communities	Population urban population living in slums Access to improved water source Satisfaction with public transport Population with rent overburden Access of points of interests with 15 min walking distance
12	Responsible Consumption and Production	Municipal solid waste generated Electronic waste SO_2 emissions N_2 emissions Exports of plastic waste Nonrecyclable MSW
13	Climate Action	CO_2 emissions from fossil fuel combustion and cement production CO_2 emissions associated with imports and exports
14	Life below Water	Mean area protected in marine sites important for biodiversity Ocean health index Fishing: Overexploitation Marine biodiversity threats
15	Life on Land	Mean area protected in terrestrial sites important for biodiversity Red list index of species survival Permanent deforestation Terrestrial and freshwater biodiversity threats
16	Peace, Justice, and Strong Institutions	Homicides Unsentenced detainees Corruption perception index Children involved in child labor Press freedom index Access and affordability to justice Timeliness of administrative proceedings Persons held in prison
17	Partnerships for the Goals	Government spending on health and education Government revenue excluding grants Tax justice

Source: Adapted from Sachs et al. (2023).

Following the UN initiative, the European Union (EU) has proposed over 100 sustainability indicators (EC 2009). Many countries around the world have also developed national-level sustainability indicators. For example, the UK developed a national sustainable development strategy and proposed 68 indicators (DEFRA 2005). Many local authorities have also undertaken initiatives to address local sustainability and proposed sustainable development indicators; for instance, the city of London (UK) proposed a set of 61 indicators to complement the UK Sustainable Development Strategy (DEFRA 2005). Even businesses and corporations are under growing pressure from stakeholders to align with the principles of sustainable development, and they have developed eco-efficiency or sustainability indicators and tools to assess them. Global Reporting Initiative (GRI) is a framework for reporting corporate sustainability performance initiated by the US-based Coalition of Environmentally Responsible Economies (CERES) in partnership with the United Nations Environmental Program (UNEP), which proposed 79 indicators for sustainability assessment (GRI 2010). Numerous companies have used GRI to assess environmental, economic, and social dimensions of projects, products, and operations. For example, BASF, one of the largest chemical companies in the world, has developed a tool called "Socio-Eco-Efficiency tool-SEEBalance®" that encompasses life cycle assessment (LCA), economic life cycle costing, and social indicators (BASF 2019). In considering engineering projects with respect to sustainability, key indicators are essential to the evaluation of a project, whether they are considered objective or subjective indicators. All aspects of an engineering project may be considered on a life-cycle basis (considering raw material acquisition, material manufacturing, construction, etc.) or on an individual, discrete basis, whether this constitutes the design phase, the construction phase, the operational phase, or the final disposal phase. Additionally, any combination of these project phases may be considered when assessing sustainability.

Further, when considering the sustainable aspects of an engineering project, it is essential to consider indicators representative of all dimensions that constitute the TBL. Environmental indicators may include the following:

- Greenhouse gases and other air emissions
- Contributions to climate change
- Use of freshwater resources
- Impacts on soil
- Utilization of raw natural resources
- Impacts on surface water or groundwater
- Use of recycled/repurposed materials
- Overall waste generation
- Diversion of waste materials from or to landfill facilities.

Economic sustainability indicators that may be considered for the engineering projects may include the following:

- Direct and indirect job creation within the community
- Direct and indirect investment within the community
- Facilitation of government grants for the project and community as a whole
- Long-term tax and revenue generation within the enhanced community
- Degree of highest and best use achieved by the engineered project.

When compared to environmental and economic dimensions, social sustainability indicators have not been incorporated as extensively, nor have they been as developed or refined. In general, social sustainability is focused on the impacts of engineering activity on society, including the dimensions related to quality of life, diversity, cultural awareness, social cohesion, and harmony. Some key indicators of social sustainability include the following:

- Enhancement of community aesthetics
- Enhancement of quality-of-life features (e.g. improved transportation opportunities or recreational facilities)
- Public participation in decision-making
- Educational and job training opportunities
- Interaction between community groups
- Emotional ownership of the community in an engineering project
- Improved physical and mental health and well-being of members of the community
- Enhanced social opportunities for members of the community
- Strengthening or enhancement of existing community institutions (e.g. recreational organizations, charitable foundations, and houses of worship).

Table 6.2 shows a list of indicators to be applicable for the sustainability assessment in manufacturing companies selected by the Advanced Manufacturing Laboratory at Texas Tech University. The list of indicators includes the three dimensions: environment, economy, and social, considering the recommendations from GRI. The dimensions are decomposed into six aspects – Energy and Material Usage (EM), Emissions to Natural Environment (NE), Economic Performance (EP), Products (P), Workers (W), and Community Development and Social Justice (CS).

The sustainability of healthcare buildings can be analyzed with the 57 indicators outlined in Table 6.3, organized in four areas: environmental, sociocultural and functional, economy, and technical and site. These indicators were selected based on the HBSAtool-PT (de Fátima Castro et al. 2017), a sustainability assessment method developed for this type of buildings. The sustainability assessment can be performed assigning different weights or importance to the indicators through an opinion survey to all building stakeholders.

Table 6.4 contains a list of indicators for sustainability assessment of industrial organizations with the TBL approach. The compilation of indicators in Table 6.4 includes a total of 753 indicators derived from the 24 studies from 1998 to 2018. The indicators were reduced on average by 92.3%, with no essential information loss. This result confirms the homogeneity of the indicators used in the 24 studies, highlighting the uniformity and coherence in the selection for different studies in various countries, industries, and authors.

Table 6.5 shows a list of indicators for sustainable development that relate to fiscal sustainability. The selection of these indicators in specific cases depends on the available data or local problems or issues. There is no indicator that shows the global evolution of the entire capital stock. The most appropriate indicators are those that can help in the prediction of future trends, and hence the specific selection in each case is based on the available information and the applicability of the prediction.

Table 6.6 shows a series of indices that have been used to evaluate the sustainability of nations or states and that can be useful tools in designing policies that improve environmental quality, the

Table 6.2 Sustainability indicators in manufacturing.

Dimension: ENVIRONMENTAL	**Aspect 1: Energy and Material Usage (EM)**	Dimension: ECONOMIC & SOCIAL	EP5: Organization's openness to stakeholder review and participation
	EM1: Material usage (total and per unit of product)		**Aspect 4: Products (P)**
	EM2: Percent of materials used that are recycled input materials		P1: Percent of products designed for disassembly, reuse, or recycling
	EM3: Energy consumption (total and per unit of product)		P2: Percent of products with an environmental label
	EM4: Percent of renewable energy		P3: Percent of products with take-back policies in place
	EM5: Energy saved due to conservation and efficiency improvements		P4: Customer satisfaction
	EM6: Total water consumption		P5: Adherence to general guidelines for warning and safety labels
	EM7: Percent of recycled/reused water		**Aspect 5: Workers(W)**
	Aspect 2: Emissions to Natural Environment (NE)		W1: Lost workdays due to illness or injury
	NE1: Total greenhouse gas emissions by weight NE2: NOx, SOx, and other air emissions by weight		W2: Average hours of employee training per year W3: Employee job satisfaction rate
	NE3: Total water discharge volume		W4: Employee turnover rate
	NE4: Total solid waste weight		W5: Gender ratio
	NE5: Weight of hazardous waste		**Aspect 6: Community Development and Social justice (CS)**
	Aspect 3: Economic Performance (EP)		CS1: Community spending and charitable contributions
	EP1: Percent of suppliers without EHS violations		CS2: Number of community-company partnerships
	EP2: Investments in environmental protection		CS3: Percent of products consumed locally
	EP3: Investments in local suppliers		CS4: Ratio of company wage compared to local minimum wage
	EP4: Costs associated with EHS compliance		CS5: Percent of investment in human rights clauses

Source: Fan et al. (2010)/with permission of Institute of Electrical and Electronics Engineers.

Table 6.3 Sustainability indicators in healthcare buildings.

Areas	Categories	Indicators
A1 Environmental	C1 Environmental life cycle impact assessment	I1 Assessment of the buildings' life cycle impacts
	C2 Energy	I2 Primary energy consumption
		I3 Onsite energy production
		I4 Energy performance
	C3 Soil use and biodiversity	I5 Layout optimization
		I6 Soil sealing
		I7 Reuse of previously built or contaminated areas
		I8 Ecological protection of the site
		I9 Rehabilitation of the surrounding
		I10 Use of native plants
		I11 Heat island effect
	C4 Materials and solid waste	I12 Construction waste
		I13 Reused and recycled products and materials
		I14 Waste separation and storage
		I15 Responsible sourcing of materials
	C5 Water	I16 Potable water consumption
		I17 Recycling and reuse of effluents
		I18 Treatment of contaminated effluents
		I19 Water efficient equipment
A2 Sociocultural and functional	C6 Occupants' health and comfort	I20 Natural ventilation
		I21 Toxicity of finishing materials
		I22 Thermal comfort
		I23 Visual comfort
		I24 Acoustic comfort
		I25 Indoor air quality
	C7 Controllability by the user	I26 Ventilation and temperature control
		I27 Natural light adjustment
	C8 Landscaping	I28 Visual link with the surrounding landscape
	C9 Passive design	I29 Layout and orientation
		I30 Passive systems
	C10 Mobility plan	I31 Accessibilities
	C11 Space flexibility and adaptability	I32 Availability and accessibility of social areas

Table 6.3 (Continued)

Areas	Categories	Indicators
		I33 Space optimization
		I34 Space flexibility
		I35 Space adaptability
A3 Economy	C12 Life cycle costs	I36 Initial cost
		I37 Operational costs
	C13 Local economy	I38 Hiring local goods and services
	C14 Corruption avoidance plan	I39 Sustainable purchase policies
A4 Technical	C15 Environmental management systems	I40 Commissioning
		I41 Environmental management plan
		I42 Infection control
		I43 Reducing noise pollution
	C16 Technical systems	I44 Efficiency of lighting and air conditioning systems
	C17 Security	I45 Occupant safety
		I46 Responsible construction practices
	C18 Durability	I47 Materials of high strength and durability
		I48 Proper selection of furniture
	C19 Awareness and education for sustainability	I49 Education of occupants
		I50 Education of service providers
		I51 Satisfaction surveys
	C20 Skills in sustainability	I52 Integration in the team of a qualified sustainability expert
A5 Site	C21 Local community	I53 Local community development
	C22 Cultural value	I54 Heritage framework
	C23 Conveniences	I55 Accessibility to public transport
		I56 Low-impact mobility
		I57 Local amenities

Source: Costa et al. (2023)/MDPI/CC BY 4.0.

economy, or the development of society toward more sustainable models. However, as important as having a sustainability index, it is necessary to determine the usefulness of the indices. Table 6.6 shows the use and "success" of 24 indices based on the presence of these indices in the press worldwide, indicating the year of inception or design of the index and the number of years that such index has been used. The impact or success of each index can be an interesting result for the development of newer, more accurate, and useful indices.

Table 6.4 Sustainability indicators for industrial organizations.

Subjects – environmental indicators	Subjects – social indicators	Subjects – economic indicators
Electric energy	**Employees**	**Cost/expense**
Electric energy consumption (38)1	Number of employees (12)	Nonconformity cost (2)
Use of renewable energies (9)	Turnover index (10)	Expense with wages (10)
Water	Training of the employees (in hours) (15)	Expense with taxes (6)
Water consumption (31)	Level of the formal instruction (6)	Environmental expense (8)
Reuse and recycling (3)	Discrimination (7)	Operational expense (8)
Water acidification (6)	Wages and benefits (8)	**Profit**
Waste	Health and security (21)	Liquid profit (13)
Volume of solid waste (28)	Career and stability (6)	Financial indicators (5)
Volume of hazardous waste (9)	Participation in management decisions (4)	Added value (13)
Gases emission	Complaining (4)	Productivity (4)
Volume of atmospheric gases (42)	Deaths (4)	**Investments**
Product	Job creations (7)	Capital for investment (10)
Volume of recycling (6)	**Work**	Benefits for employees (5)
Volume of reuse (5)	Child labor (7)	Investment in R&D activities (8)
Durability level (11)	Satisfaction level (5)	Return on the assets (2)
Resources/Materials	Intensity level (5)	**Suppliers**
Consumption of natural resources (29)	Noise level (6)	Local suppliers (5)
Consumption of recycling materials (9)	Accidents/injuries (15)	Gross revenue
Hazardous materials (3)	Number of diseases (10)	Gross revenue value (14)
Effluents	Management quality (5)	**Shareholders/director's board**
Volume of liquid effluents (26)	**Clients/consumers**	Participation of the shareholders (8)
Labels and certificates	Satisfaction level (7)	Meetings of board directors (5)
Environmental labels and certificates (6)	Number of complaining consumers (5)	**Institutional**
Logistics	**Community and stakeholders**	Number of organizational units (7)
Transportation and logistics (6)	Engagement of the community (21)	Contributions or donations (2)
Reverse logistics (3)	Local partnerships (10)	Competitiveness/market (4)
Environmental spending/ investments	Investments to benefit community (8)	Expense with clients (5)
Environmental spending (14)	**Ethics**	
Environmental fines (3)	Ethical behavior (18)	
Environmental management system (9)		
Impacts/environment degradation		
Impacts/environment degradation (25)		
Soil		
Quality and use of soil (5)		
Total frequency (326)	Total frequency (224)	Total frequency (147)
Total of indicators (348)	Total of indicators (244)	Total of indicators (161)
Percentage of the compilation (93.7%)	Percentage of the compilation (91.8%)	Percentage of the compilation (91.3%)

Source: Feil et al. (2019)/MDPI/CC BY 4.0.

Table 6.5 General sustainability indicators that underline fiscal sustainability.

Concept	Indicator	Example of use
Economic capacity	Under-employment	Jacksonville, FL, US
	Adults with BA/BSc degree or higher	Jacksonville, FL US
	Poverty rate	Austin, TX, US
Economic sustainability	% of green jobs in the economy 3+ years	Portland, OR, US
	% of total new businesses surviving	Austin, TX, US
	% of total jobs in the public sector	Austin, TX, US
	Child poverty rate	Seattle, WA, US
Tax capacity	Housing affordability – repayments as % of income	Dublin, Ireland
	% of households spending over 30% of income on housing	Jacksonville, FL, US
	Taxes as % of average income	Austin, TX, US
Natural capital/ Green infrastructure	Level of reforestation	Bogota, Colombia
	Percentage area of tree canopy	Portland, OR, US
	Ecological status: water bodies	Dublin, Ireland
	Percentage of land surface impervious to water	Austin, TX, US Seattle, WA, US
Public infrastructure	% of roads in good condition	Bogota, Colombia
	Improved traffic flow	Tel Aviv, Israel
	Energy efficiency of public buildings	Buenos Aires, Argentina Dublin, Ireland Tel Aviv, Israel
Human capital	Public school dropout rate	Ilheus, Brazil
	High school graduation rate	Jacksonville, FL, US
	% adults at healthy weight	Portland, OR, US
	% 8th graders at healthy weight	Seattle, WA, US
	Babies with low birth weight	Ilheus, Brazil
Social capital	% reporting trust in city leaders or government	Jacksonville, FL, US Austin, TX, US
	Crime rate	Seattle, WA, US
	Racial disparities in juvenile courts	Austin, TX, US

Source: Greenwood (2018)/with permission of Springer Nature.

Table 6.6 Example list of sustainability indices for nations or states.

Index	Acronym	Starting year	Number of years reported
Big Mac Index	BMI	1988	25
Bribe Payers Index	BPI	1999	14
Carbon Footprint	CF	2001	12
Climate Change Performance Index	CCPI	2006	7
Commitment to Development Index	CDI	2003	10
Corruption Perception Index	CPI	1996	17
Debt to GDP ratio	None	1986	28
Democracy Score	DS	2004	9
Ecological Footprint	EF	1993	20
Environmental Performance Index	EPI	1996	17
Environmental Sustainability Index	ESI	2000	13
Failed States Index	FSI	2005	8
Gender Empowerment Index	GEi	1995	18
Gender Development Index	GDI	1995	18
Genuine Progress Indicator	GPI	1994	19
Global Competitiveness Index	GCI	1996	17
Global Hunger Index	GHI	2007	6
Global Peace Index	GIPI	2007	6
Happiness Index	HI	1987	26
Happy Planet Index		2006	7
Human Development Index	HDI	1990	23
Human Poverty Index	HPI	1997	16
Living Planet Index	LPI	1998	15
Mothers Index	MI	2001	12
Press Freedom Index	PFI	2002	11

Source: Adapted from Morse (2016, 2018).

Example 6.1

The USEPA has undertaken an initiative to redevelop an old industrial site into a residential complex. The following indicators were identified by the stakeholders to evaluate the sustainability of the project:

- Affordability of housing options
- Job creation and support for local businesses
- Energy efficiency
- Surface water quality deterioration
- Risk of contamination.

Identify which indicator belongs to which pillar of sustainability (environmental or economic or social).

Answers:

Indicator	Pillar of sustainability
Affordability of housing options	Social and economic
Job creation and support for local businesses	Economic and social
Energy efficiency	Environmental and economic
Surface water quality deterioration	Environmental and social
Risk of contamination	Environmental and social

6.4 Sustainability Metrics

The indicators presented above provide key variables that may be assessed when evaluating the degree of sustainability for a specific engineering project or a product. The indicators presented earlier may not be easily measurable. However, numerical values or quantitative characteristics may be integrated with these indicators so that they may be objectively and accurately assessed. As a result, metrics can now be connected to the indicators. Sustainability metrics are numerical values that may be used to assess specific indicators related to sustainability, and they are vitally important for performing objective analysis with respect to engineering project sustainability (Sikdar 2003). These metrics should be incorporated into a range of sustainability measurement tools, which are discussed in greater detail in subsequent chapters in this book.

The metrics that may be used to assess sustainability of engineering projects are, in many cases, straightforward. As a result, their ability to be accurately measured in many cases is well established. This is especially the case for economic and environmental sustainability. However, social sustainability indicators and metrics have not been as extensively defined or developed. Many of the social metrics can be evaluated only qualitatively or semi-quantitatively, which makes the determination of social impacts difficult.

Before some common metrics are presented, it is important to note that there is no standard established framework regarding an appropriate set of parameters to be used for sustainability evaluation of engineering projects, nor is there a consensus on what constitutes sustainable engineering. Additionally, there is a wide range of opinions regarding the degree to which individual metrics contribute to or affect sustainability. This is further reflected in the relatively wide range of scope inherent in several sustainability assessment tools that are presented in later chapters of this book. Further, there is no commonly accepted set of metrics used by engineering practitioners to evaluate whether an engineering project is sustainable.

Physical metrics may be used to assess the physical inputs and outputs of an engineering project alternative, including those focused or tailored for positive or negative contributions with respect to sustainability. Such examples include the following:

- Energy consumption (kWh or BTU)
- Renewable energy consumption (kWh or BTU or as a percentage of total energy consumption)
- Freshwater or recycled/reclaimed water consumption (gallons or liters)
- Air emissions (tons or kg)
- Greenhouse gas emissions (tons or kg)
- Carbon emission offset (tons or kg or a percentage of greenhouse gas emissions)
- Solid waste generation (tons or kg)
- Use of recycled solid materials (tons or kg).

Several of these indicators, including energy (nonrenewable or renewable), water (fresh or reclaimed), or air emissions, may be combined on a per unit basis (e.g. per treated unit mass/volume of soil or water). Of course, the indicator further may be coupled with time or a monetary unit to determine these metrics based on a unit time or a unit cost. Additionally, other actions may be quantitatively assessed, including credits/offsets of ecological restoration, increased real estate value, and preservation or restoration of natural resources.

Since the potential list of sustainability metrics for engineering projects is enormous, and because there is a lack of consensus or standard regarding key indicators and related metrics, there has been a growing dialogue between several sustainability-focused organizations and regulatory agencies regarding potential efforts for standardization. One example is the USEPA's Tool for the Reduction and Assessment of Chemical and Other Environmental Impacts (TRACI). This environment impact assessment method, utilized by several LCA software programs, includes the following nine environmental impact categories (USEPA 2016a; Bare 2011):

- *Global climate change impact category* – Reported as carbon dioxide (CO_2) equivalents
- *Acidification impact category* – Reported as sulfur dioxide (SO_2) equivalents
- *Eutrophication impact category* – Reported as nitrogen (N) equivalents
- *Ozone depletion impact category* – Reported as trichlorofluoromethane (CFC-11) equivalents
- *Photochemical smog formation impact category* – Reported as ozone (O_3) equivalents
- *Human health particulate matter (PM) impact category* – Reported as fine particulate matter ($PM_{2.5}$) equivalents
- *Human health cancer impact category* – Reported as comparative toxicity unit cancer (CTUcancer) equivalents
- *Human health noncancer impact category* – reported as comparative toxicity unit noncancer (CTUnoncancer) equivalents
- *Eco-toxicity impact category* – Reported as comparative toxicity unit eco-toxicity (CTUeco) equivalents.

Other impact categories, such as those associated with renewable energy and nonrenewable energy use may also be incorporated into an assessment when permissible by the LCA tool that is being used for an analysis.

Example 6.2

There is an integrated renewable energy project coming up consisting of solar, hydro, wind, and geothermal power plants in a developing country whose power needs are rising exponentially each year. In the context of this project, identify one most important indicator each in the environmental, economic, and social pillars of sustainability, justify why it can be the most important, and discuss ways in which impacts on these indicators can be optimized.

Answers:

Environmental Pillar: Greenhouse Gas Emissions

Justification: One of the key environmental indicators for a renewable energy project is the reduction of greenhouse gas (GHG) emissions, particularly CO_2. This is crucial because energy production from fossil fuel based energy usage is one of the largest global sources of GHG emissions, which are directly linked to climate change. Measuring the reduction in GHG emissions can indicate the project's contribution to mitigating global warming.

Optimization: To optimize this indicator, the project should:

- Maximize the efficiency of each renewable energy source to ensure the highest possible energy output per unit of GHG emitted (or, ideally, none at all).

- Use the latest technology with the smallest possible carbon footprint during manufacturing, deployment, and operation phases.
- Establish clear benchmarks and monitoring systems to track emission reductions and make data-driven improvements.

Economic Pillar: Levelized Cost of Energy (LCOE)
Justification: The LCOE is a critical economic indicator, representing the cost of building and operating a power generation plant over an assumed financial life and duty cycle. It is a comprehensive metric that helps compare different methods of electricity generation on a consistent basis. For a developing country, achieving a low LCOE is crucial to ensure that the shift to renewable energy is financially sustainable and does not place an undue burden on the economy.

Optimization: To improve the LCOE, the project can:

- Utilize a mix of funding sources, including low-interest loans and grants, to minimize capital costs.
- Optimize the scale of renewable energy production to benefit from economies of scale.
- Engage in long-term power purchase agreements to ensure predictable and stable revenue streams.

Social Pillar: Employment and Job Creation
Justification: Employment and job creation is a vital social indicator for the renewable energy project. The construction, operation, and maintenance of renewable energy facilities can provide significant employment opportunities in a developing country. It can also facilitate technology transfer and skill development, leading to sustainable community development and poverty alleviation.

Optimization: To enhance this social indicator, the project should:

- Prioritize hiring and training local workers, thereby investing in the community and increasing the project's acceptability.
- Collaborate with educational institutions to create training programs tailored to the skills needed for the renewable energy sector.
- Implement inclusive policies that enable diverse segments of the population to benefit from job opportunities.

These indicators align with the SMART criteria: they are Specific to the project's goals, Measurable through data collection and analysis, Achievable with the current technology and resources, Relevant to the broader goals of sustainability, and are Timely, with impacts that can be assessed and optimized over the project's lifecycle. By focusing on these key indicators and actively working to optimize them, the renewable energy project can significantly contribute to the sustainable development of the host country.

6.5 Sustainability Assessment Tools

Several general and specific assessment tools have been developed to account for a wide range of metrics in engineering projects (Ness et al. 2007). Although many tools focus primarily on environmental related metrics, tools are being developed or revised to incorporate economic and social metrics as well. These assessment tools take on many forms and range from simple to complex. Some have been developed to focus with a local context, while others have been developed to address a more global context. These tools also range in precision ranging from generic to specific frameworks or metrics; some measure qualitative metrics, while others are focused on quantitative measurement. Many tools exist in the public domain, while others have been proprietarily developed for in-house corporate work and for its use by consultants.

In the subsequent chapters of this Section II, the following general tools used for sustainability assessment are presented:

- Although it is not an assessment tool for sustainability, the ***material flow and budget analysis*** can be performed to determine the reserves of resources, consumption, and flows among extraction, consumption, and use of the materials. In general, the material flows are evaluated for a country or a region, and therefore, the import/export flows have to be considered. The analysis of the materials in use and the life expectancy of the products that contains these materials is an indicator of necessity of a material to a specific society and the amount of materials for recycling at the end of the useful life for respective products.
- The emission of greenhouse gases has a direct impact on the global warming and climate change. Thus, the ***GHG inventory*** and the **carbon footprint** are indicators of the impact of each activity or organization toward global warming and climate change. The reduction of the carbon footprint can be used as an indicator of environmental sustainability when deciding among different project alternatives or modifications of processes.
- One of the most powerful tools to assess the environmental impacts of a product or a process is the **LCA**. This tool can be used to perform a comprehensive analysis of all the environmental impacts of a product or a process across all the life cycle stages, from the extraction of raw materials, transformation, manufacturing, use, and disposal, including the transportation of materials or products between and within the life cycle stages. LCA can provide a detailed analysis of all the potential impacts and is an efficient tool to determine the critical stages or operations that result in major environmental impacts. Unfortunately, this analysis is complex and requires a significant amount of data, so it is important to define the boundaries of our system and the scope of our study to accurately identify the data required for the analysis.
- ***Streamlined life-cycle assessment (SLCA)*** is another sustainability assessment tool that was developed to overcome the difficulties and limitations of LCA. The ***SLCA*** is based on the identification of the most significant environmental impacts across the life cycle stages of a product. A matrix consisting of potential environmental impacts as the column headings and the different life cycle stages of the project/product as the row headings is created. The matrix is populated with values that range from 0 (for very high negative impact) to 4 (for no negative impact) relating the magnitude of impact of a particular life cycle stage to a specific environmental impact category. The sum of scores under an impact category in the matrix will give the global score for one alternative. Comparing these global scores for different alternatives quantifies the relative sustainability of the two products/alternatives. The graphical representation of the matrix either as a heat map or as a target plot will be useful to determine the most critical impacts on the life cycle stage of a product/project.
- ***Economic input–output (EIO) analysis*** is a powerful tool developed by Nobel Laureate Dr. Wassily Leontief. Many nations create input–output models of their economy that are useful in predicting the evolution of their economy and the definition of their national budgets. The EIO models can be applied to evaluate the environmental impacts and thereby perform an environmental sustainability analysis by defining the environment as an additional sector and quantifying the economic fluxes associated with the environmental impacts of various sectors. Alternatively, the EIO models can be defined in terms of material fluxes instead of economic fluxes. This method of analysis using material fluxes is called physical input–output (PIO) analysis. PIO models are more detailed than the aggregated economic models, and hence more difficult to construct, but the benefits of greater detail facilitate a more comprehensive analysis of environmental impacts.

- ***Quantitative environmental health risk assessment*** is a tool developed by USEPA to determine the risks from the exposure of chemicals to potential receptors (human health and ecosystem). The risk assessment method will allow for the evaluation of the risks with lower degree of subjectivity and lower uncertainty than the individual perception. Risks to humans and the environment are real and should be carefully evaluated, taking human perception into account.

In addition to the above general assessment tools, many specific tools have been reported for specific engineering applications. For example, Leadership in Energy and Environmental Design (LEED) is a sustainability rating tool developed for assessing the sustainable practices adopted in green buildings and further certifies those buildings to acknowledge the innovative and efficient use of resources while causing minimal harm to the human health and the environment (USGBC 2018). Envision™, like LEED, is another sustainability rating tool developed for assessing and acknowledging the sustainable initiatives used in constructing any civil infrastructure (ISI 2018). Likewise, Spreadsheets for Environmental Footprint Analysis (SEFA) (USEPA 2016b) and SiteWise™ (SURF 2018) are quantitative sustainability assessment tools developed for encouraging the incorporation of the principles of green and sustainable remediation in the remediation of contaminated sites. Additionally, the Social Sustainability Evaluation Matrix (SSEM) developed by Reddy et al. (2014) provides a structured framework for assessing the social dimensions of sustainable remediation. It categorizes social sustainability into four key dimensions: socio-individual, socio-institutional, socio-economic, and socio-environmental. Each dimension includes specific indicators, which are rated on a scale from −2 (unacceptable) to +2 (ideal), allowing for a nuanced evaluation of social outcomes. Although originally designed for remediation projects, the SSEM has since been adapted for broader engineering applications by modifying the indicators to align with project-specific contexts, demonstrating its flexibility and relevance across various domains. These frameworks are presented in greater detail throughout this book.

6.6 Resiliency Indicators

In the past decade, resilience assessment has gained prominence as a vital part of disaster management. It is widely recognized as a process used to evaluate the functionality and performance of complex systems exposed to disasters. "Resilience" itself is an abstract concept, and resilience assessment involves determining its value or level – complicating it further. Indicators, being more tangible than the abstract concept of resilience, can effectively highlight positive or negative changes in resilience. As a result, indicator-based resilience assessments allow stakeholders to practically analyze complex systems and make informed decisions. Identifying the appropriate indicators is crucial before conducting a resilience assessment. The most appropriate manner to understand resilience is by considering various dimensions of it; for example, some commonly perceived dimensions for resilience are environmental/ecological resilience, social resilience/community resilience, economic resilience, infrastructure resilience, and institutional resilience. Table 6.7 presents selected indicators for resilience evaluation in large-scale systems following this categorization.

The indicators presented in Table 6.7 are often applicable to larger scale assessments such as assessing the resilience of a city or a community or an ecosystem to unexpected shocks or stressors. A major portion of research on indicator-based resilience assessments is focused on evaluating community resilience or ecosystem resilience to natural disasters. For example, Sharifi (2016) provided a comprehensive list of resilience indicators used in various community resilience assessment tools. However, in engineering, we are often concerned about the resilience of a particular engineered

Table 6.7 Select resilience indicators commonly used to quantify resilience.

Dimension	Sub-Dim.	Indicators	Impact
Social	Population and demographics	% of elderly population (ages 65 and above)	–
		% children in the population	–
		People per sq. km of land area	–
		% Single parent household	–
		% population whose income is below minimum wage	–
		Ratio of nonminority to minority population	+
		% of population with sensory, physical, and mental disabilities	–
	Health	% population with health insurance	+
		% uninsured adults	–
		Psychosocial support facilities per 10 000 persons	+
		Hospital beds per 1000 people	+
		Distance to nearest health facility	–
		Number of social welfare home beds per 10 000 persons (including homeless shelters and nursing homes)	+
	Community	% of red cross volunteering organizations	+
		Budget of volunteer organizations	+
		No. of registered nonprofit organizations	+
		% Population with access to phone	+
		% Population with access to internet/TV/radio	+
		% Households with at least one vehicle	+
		Roads network density	+
		Distance to nearest transport network	–
		Net in and out migration	–
	Social protection	Public spending on social assistance as % of GDP	+
		No. of social advocacy organizations per 10 000 population	+
		No. of courses in disaster risk reduction (DRR) as part of the educational curriculum per schools and colleges	+
		National Historic Registry sites area per population	–
	Economic capacity	Median household income	+
		% Owner-occupied housing units	+
		% Employees not in farming, fishing, forestry, extractive industry, or tourism	+
Economic	Employment	% Labor force employed	+
		% Labor force employed by federal government	+
		Physicians/nurses/midwives per 10 000 persons	+
		% Female labor force participation	+
	Income	Average annual income per capita	+
		Disposable income per capita	+
		% Population with multiple source of income	+
		% Population living below the poverty line	–

Table 6.7 (Continued)

Dimension	Sub-Dim.	Indicators	Impact
		Gini coefficient	–
		Absolute difference between male and female median income	–
	Business activity	Ratio of large to small businesses	+
		Area of commercial establishment per 10 000 persons	+
		% Commercial establishments outside of high hazard zones to total commercial establishment	+
		% Population living in urban areas	–
		Total financial revenue from business	+
		Ratio of public finance income and expenditure	+
		Price of land per square meter	–
		TRESS index (to represent economic diversity)	–
		Energy use for residential, commercial, and industrial land use	–
		Water use by households in the area	–
		Food security rate	+
		Proportion of cultivated land area	+
		Crop insurance policies per square mile	+
		% Housing units covered by disaster insurance	+
Environmental	Air	No. of days per year with pollution concentration above the local norm or standard	–
		Health cost of air pollution ($/year)	–
		% Acid rain per total precipitation	–
		NO_x/SO_x emissions	–
		GHG emission per capita per year	–
	Soil	Soil erosion ton/year/person	–
		% Land that is not impervious	+
	Land	Normalized difference vegetation index (NDVI)	+
		Ratio of land that turned to urban areas to total land area	–
		% Forest cover	+
		% Land area that is developed open space	+
		% Land area that is arable (cultivated) land	+
		Ratio of built and unbuilt spaces	–
		% Land area under protected status	+
		% Area of beach	+
		% Population with access to a green urban area less than 300 m away.	+
		% Area of urban green space	+
		Degree of slope	–
		Areas of potential debris flow torrents and landslides	–

(*Continued*)

Table 6.7 (Continued)

Dimension	Sub-Dim.	Indicators	Impact
		% Land area that does not contain erodible soils	+
	Water	No. of rainy days	+
		No. of days with more than normal precipitation	−
		% Dam capacity	+
		Number of river miles whose water is useable	+
		Total water consumption	−
		% Ground water withdrawals (agriculture, public supply, industry)	−
	Natural disaster susceptibility	Frequency of natural disasters	−
	Waste	Amount of solid waste by source	−
		Recycling ratio of wastes	+
Institutional	Public institutions	% Population covered by a mitigation plan	+
		% Population living in the hazardous area	−
		% Employed persons working in public administration, defense, social security or municipal activities	+
		Net per capita spending for mitigation projects	+
		% Governmental financial resources to carry out risk reduction activities	+
		% Building infrastructure not in high hazard zones	+
		% Local schools, hospitals, and health facilities that remained operational during emergencies in past events	+
		% Population access to electricity/gas/safe water/radio/internet	+
		No. of fire/police/ambulance stations and shelters per 1000 people	+
		Existence of structural measures that physically prevent extreme events	+
		Average number of early warning centers per each independent zone	+
		% Land area that consists of windbreaks and environmental plantings	+
		% Vulnerable shoreline protected by dykes/embankments	+
		The GDP growth rate of environment administration (%)	+
		Fixed assets investment density (amount of investment per area)	+
		Proximity of county seat to nearest county seat within a Metropolitan Statistical Area	−
		Proximity of county seat to state capital	−
		Existence of government plans (i.e. disaster, climate adaptation)	+
	Recovery response	Microfinancing, cash aid, soft loans, loan guarantees available to affected households after disasters to restart livelihoods	+

Table 6.7 (Continued)

Dimension	Sub-Dim.	Indicators	Impact
		Crop insurance policies per square mile	+
		% Population under age 65 with health insurance	+
		% Housing units covered by disaster insurance	+
		Presidential disaster declarations divided by number of loss-causing hazard events	+
		Local government plan to support the restoration, protection, and sustainable management of ecosystems services	+
		Coastal setback, shoreline restoration	+
		Local government access to resources and expertise to assist victims of psychosocial impacts of disasters	+
Infrastructure	General	% Infrastructure area per unit area (transportation, park, power, water services [ha.])	+
		Megawatt hours per energy consumer	−
		Ratio of water production to demand	+
		Ratio of gas production to demand	+
		% Commercial establishments outside of high hazard zones ÷ total commercial establishment	+
	Housing	% Housing units that are not mobile homes	+
		% Average housing age	−
		Area of informal settlements (ha.)	−
		% Vacant units that are for rent	+
		Hotels/motels per 10 000 persons	+
		Distance to nearest public shelter	−
	Social infra.	Hospital beds per 10 000 persons	+
		% Area of community services (recreational facilities, parks, historic sites, libraries, museums)	+
		Average number of internet, television, radio, telephone, and telecommunications broadcasters per household	+
	Emergency response	No. of fire/police/ambulance stations and shelters per 1000 people	+
		Mean distance in meters from denser settlement areas to the nearest/hospital/ fire/police station	−
		No of firefighters/police officers/medical staff per 1000 population	+
	Transportation	Rail miles per square mile area	+
		Principle arterial miles per square mile	+
		Major road egress points per 10 000 persons	+
		Roads network density	+
		Mean distance in meters from denser settlement areas to the nearest airport (only airports operated by Avinor or the military to be included)	−
		No. of vehicle per 1000 persons	+

Source: Adapted from Assarkhaniki et al. (2020).

system (bridge, a building, or energy microgrids, among others) rather than the whole community, and hence the scale of assessments and the boundaries are also much smaller. Although the approach and indicators adopted for resilience evaluation, as presented in Table 6.7, might not be applicable in their entirety in evaluating engineering resilience, the concepts are still relevant to identify corresponding indicators that are appropriate for smaller scales. To reflect on this, Reddy et al. (2024) categorized the resilience of engineered systems, specifically infrastructure systems, into four main categories: The technical robustness of the system to unexpected events such as climate disasters and the corresponding cascading environmental, social, and economic consequences, also termed environmental, social, and economic resilience. This type of categorization will aid practicing engineers to consider not only the technical soundness of the system, which is the core principle of engineering, but also consider the impacts of achieving or not achieving such robustness at the cost of the environment, society, and the economy. An example set of indicators for performing such resilience assessment of levee systems is presented in the previous chapter (refer to Table 5.2). A similar approach to identifying the indicators considering various aspects of resilience should be adopted for various engineered systems to make a system truly "resilient and sustainable."

6.7 Resiliency Metrics

Similar to sustainability metrics, resilience metrics are variables that can either be measured (if quantitative) or tracked (if qualitative) to reflect key indicators. While there is a growing body of research defining metrics for assessing community resilience – whether for cities, counties, or large-scale systems – quantifying engineering resilience is comparatively less developed. As discussed in the previous chapter, which introduced the technical and cascading TBL dimensions, resilience metrics depend on the specific indicators identified for a given project.

In engineered systems, technical resilience indicators generally fall into two categories: performance-related and time-related properties. Performance-related properties, such as robustness, adaptability, redundancy, flexibility, and survivability, are determined by the system's structure. Time-related properties, like reparability, recoverability, rapidity, and resourcefulness, are shaped by the availability of maintenance resources. These indicators are often visualized using functionality curves, which illustrate how a system responds to shocks or stressors.

Figure 6.1 shows examples of different functionality curves, representing system responses based on adaptation plans and various types of shocks or stressors. Table 6.8 presents the corresponding resilience indicators that can be derived from these curves, tailored to both short-term extreme weather events (e.g. hurricanes, flash floods, landslides) and long-term climate change (e.g. temperature changes, sea level rise). The specific metrics for these indicators are typically project dependent and can be obtained through detailed engineering analysis (e.g. numerical modeling of structures under stress) or rated based on the practitioner's experience.

The sustainability metrics discussed in Section 6.4 can also be applied to quantify cascading indicators across environmental, social, and economic dimensions. For instance, the overflow of a small levee during a flood can impact the surrounding ecosystem, cause environmental damage, and lead to significant financial challenges for the local community. These impacts have well-established metrics, which can be used to quantify the relevant indicators, as outlined in Section 6.3. Additionally, if adaptive measures are implemented, these strategies will incur environmental and economic costs. Such costs, when compared to the costs of a standard design, can be assessed using conventional sustainability metrics. As discussed earlier for social sustainability, the metrics for cascading

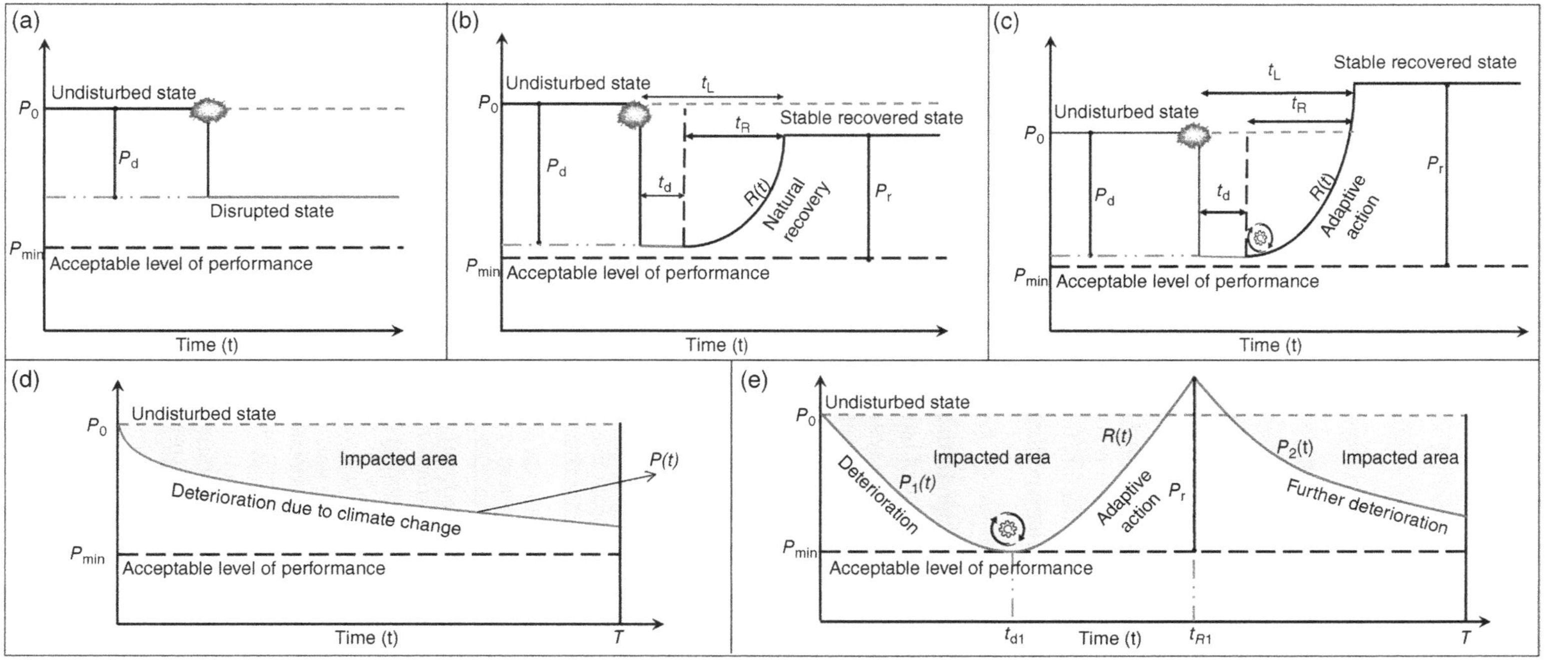

Figure 6.1 Functionality curves for engineering resilience quantification for different types of performance deterioration and recovery profiles: (a) Sudden disruption resulting in acceptable levels of performance degradation requiring no further adaptation; (b) sudden disruption resulting in significant performance degradation and partial recovery without any adaptive action; (c) sudden disruption resulting in significant performance degradation and recovery through adaptive actions resulting in recovered performance to be better than undisturbed performance; (d) long-term deterioration due to climate change under acceptable levels during the lifespan; and (e) long-term change resulting in significant deterioration within the lifespan, recovery through adaptive action, and continued deterioration from the stable recovered state thereafter. *Source:* Reddy et al. (2024)/with permission of Springer Nature.

social impacts have not been as extensively defined or developed. Many of the social metrics can be evaluated only qualitatively or semi-quantitatively, which makes the determination of such social impacts difficult. The indicator metrics commonly used for the evaluation of community resilience can help to an extent in quantifying cascading social impacts; however, research in this regard is still evolving.

6.8 Resiliency Assessment Tools

Tools for assessing the resilience of large-scale systems like communities and cities are numerous, as seen in Table 6.8; however, the same is not true for smaller-scale systems like engineered infrastructure. Although there are no generalized tools developed for resilience assessments of engineered systems, some application-specific tools do exist. However, most of these tools mainly focus on calculating the technical resilience metrics, while tools generally used for sustainability metrics can be applicable to specific indicators in measuring the cascading environmental, social, and economic metrics. Some application-specific tools for engineering resilience assessments are described below:

- The Resilience-based Earthquake Design Initiative (REDi™) rating system is a framework developed to assess and enhance the resilience of buildings during and after seismic events. Although originally focused on seismic resilience, the system has since been expanded to include ratings for wind and flood resilience as well. To achieve a REDi™ rating (platinum, gold, or silver), a building must meet the required criteria for that tier in three key categories: operational resilience, building resilience, and ambient resilience. Additionally, a Resilience Assessment is conducted to ensure that the specific resilience objectives for each rating – measured in terms of downtime and financial loss – are met. The framework also includes provisions for evaluating multiple hazards when applicable.

Table 6.8 Technical resilience indicators applicable to short-term shocks and long-term stressors.

Resiliency category	Indicators	Sudden extremes	Long-term deterioration
Resiliency of the design to the disruption encountered	Performance deterioration due to disruption (P_d)	✓	✓
	The time taken for the deterioration (t_0)	✗	✓
Resilience due to the proposed adaptive strategy or the natural recovery ability of the system	The time the system stays in the deteriorated state before any adaptation (t_d)	✓	✗
	The improvement in the performance of the system postrecovery (P_R)	✓	✓
	Time taken for the system to recover its performance to reach the desired levels (t_R)	✓	✓
	The functional time loss (t_L)	✓	✗

Source: Reddy et al. (2024)/with permission of Springer Nature.

- The Climate Resilience Evaluation and Awareness Tool (CREAT) is a software-based framework developed by the U.S. Environmental Protection Agency (EPA) to help water utilities assess and improve their resilience to climate change impacts. Originally designed to focus on water systems, CREAT assists utilities in evaluating vulnerabilities to various climate-related hazards, including extreme weather events, sea-level rise, and droughts. To complete a CREAT assessment, utilities must evaluate potential climate threats, identify critical assets, and explore adaptation strategies. The tool provides risk reduction options and estimates the costs and benefits of these measures. CREAT also includes a risk assessment process that quantifies the likelihood and consequences of different climate scenarios. By using CREAT, utilities can develop targeted adaptation plans that reduce their vulnerability and enhance their long-term operational resilience to climate impacts.
- The Resilience and Disaster Recovery (RDR) Tool Suite, developed by the U.S. Department of Transportation (USDOT), assists transportation agencies in making informed decisions about infrastructure investments by evaluating the return on investment (ROI) under various hazard conditions such as floods, storms, and earthquakes. It employs established concepts like Robust Decision-Making to address uncertainty in future scenarios. The RDR Tool Suite provides capabilities for ranking projects based on economic ROI using Benefit-Cost Analysis (BCA), Break-even Analysis, or BCA under Uncertainty. It accounts for factors such as reduced repair costs, faster recovery times, and improved network connectivity. Agencies can use default values or customize benefit and cost calculations to reflect specific data. Additionally, the tool includes features for assessing asset vulnerability, estimating capacity losses under different hazard conditions, and evaluating project performance across a range of potential hazards.
- The Resilient Node Cluster Analysis Tool (ReNCAT), developed by Sandia National Laboratories, USA, helps communities and planners identify optimal locations for microgrids to enhance resilience during grid outages. ReNCAT supports decision-making by balancing the need to provide critical services while minimizing investment costs. The tool evaluates trade-offs between maintaining critical infrastructure operations, such as hospitals and emergency services, and the cost of establishing microgrids. By analyzing power distribution networks, ReNCAT identifies sections that can be isolated into microgrids, ensuring that these areas remain powered even if the rest of the grid fails. The tool quantifies the "social burden" – the human effort required to access essential services during outages – and incorporates this into its recommendations, helping to prioritize investments that minimize both financial costs and social impacts. ReNCAT has been used in various projects, including in New Orleans, Puerto Rico, and San Antonio, to improve resilience by suggesting microgrid portfolios tailored to the unique needs of each community. It has been particularly effective in areas vulnerable to large-scale disruptions, where energy resilience is critical. ReNCAT is one such tool which not only considers the resilience of the grid structures but also takes into consideration the cascading economic and social impacts during final decision-making, as explained in the earlier sections.

The tools provided above are some examples of available tools to perform resilience assessments to achieve resilience in particular application areas. Additionally, researchers are continuously developing frameworks and tools for resilience quantification, and research in this area is ever-evolving. Apart from the technical resilience of engineered systems, to measure some of the cascading TBL metrics, tools mentioned in Section 5.6 can be used. For example, to quantify the additional environmental impacts of resilient design compared to a standard design can be quantified using tools like LCA and life cycle costing, respectively. If the metrics are not available for certain indicators, expert ratings or surveys can be used as a means of measuring such indicators.

6.9 Integrated Sustainability and Resilience Assessment Tools

Many professional engineering organizations, such as ASCE (2023a, b) and ITRC (2021), have established guidelines for incorporating sustainability and resilience into engineering project design and implementation. However, to fully embed these concepts from the outset, it is essential to quantify both sustainability and resiliency. While past frameworks, as described in earlier sections, were developed to measure these concepts, they primarily treated sustainability and resilience independently. Few efforts have been made to integrate the two holistically. The concept of viewing sustainability and resilience as distinct yet interdependent in engineering systems has not received adequate attention, and limited efforts were put into integrating these aspects and developing an integrated resiliency and sustainability assessment framework.

Professional organizations have developed several tools to assess sustainability and resilience in engineering. For example, the US Green Building Council created the Leadership in Energy and Environmental Design (LEED) rating system for buildings, while the Institute for Sustainable Infrastructure introduced the Envision™ checklist, widely used by civil engineers to evaluate project sustainability and resiliency. International tools include the UK's Building Research Establishment Environmental Assessment Method (BREEAM), Japan's Comprehensive Assessment System for Built Environment Efficiency (CASBEE), Germany's Deutsche Gesellschaft für Nachhaltiges Bauen (DGNB) rating system, and Green Globes™, developed by the Green Building Initiative. Sector-specific tools also exist, such as INVEST™, a rating system for transportation infrastructure developed by the US Department of Transportation and the Federal Highway Administration.

Despite their widespread use, the lack of standardization and transparency in many rating systems can lead to inconsistent results based on the assessor. To address these issues, researchers are developing integrated quantification frameworks that combine sustainability and resilience, enabling stakeholders to make more informed decisions (Reddy et al. 2024). Chapter 13 will provide an overview of the framework developed by Reddy et al. (2024), called the TQUALICSR (Tiered Quantification of Life Cycle Sustainability and Resilience) framework in this regard.

6.10 Summary

To define or quantify various dimensions of sustainability, and resiliency, indicators that can be expressed by metrics are needed. These indicators should be carefully selected and should be those that are meaningful, reliable, and quantifiable. We have addressed several key environmental metrics, including consumption of natural resources, climate change and greenhouse gas emissions, air pollution, water pollution, land use/degradation, energy consumption, and waste generation, among others. We have also addressed several key economic metrics and social metrics related to the anticipated returns and positive and negative side effects associated with economic activity. The key question is how to find a means to measure and properly account for the numerous metrics in a reasonable manner such that we can critically assess economic activities and/or find means to reduce impacts to identify more sustainable and resilient alternatives. Several general sustainability and resiliency assessment tools are identified, and many tools for very specific applications are being developed by engineering and practicing professionals. Further, resilience is a relatively abstract concept with multiple dimensions, and quantification metrics vary depending on the scale of the assessment. Large-scale systems like cities or communities have well-defined tools for resilience quantification, while the same is not true of smaller-scale systems like engineered infrastructure. We introduced some commonly used terms for technical resilience such as functionality curve-related metrics which can be used for the quantification of the same. Further, the concept of cascading impact indicators was also introduced to understand the impacts of resilience/nonresilience on the TBL. Such assessments helps in holistically integrating resilience concepts into engineered systems. Further integration of resilience and sustainability was rather less explored by researchers and engineers, and several application-specific rating-based tools in this regard are available, which were briefly discussed.

6.11 Questions

6.1 What is the most general definition of sustainability?

6.2 Describe the aspects of a project or product that may be considered on a life-cycle basis.

6.3 What are the most desirable characteristics of a sustainability or resiliency indicator? Explain briefly.

6.4 Mention five indicators for environmental, economic, and social sustainability.

6.5 What are the impact categories assessed in TRACI?

6.6 Select a project or a product and discuss actions that could be taken to optimize the effects of three environmental key indicators.

6.7 Select a project or a product and discuss actions that could be taken to optimize the effects of three economic key indicators.

6.8 Select a project or a product and discuss actions that could be taken to optimize the effects of three social key indicators.

6.9 Describe a scenario where trying to optimize a physical sustainability metric (i.e. recycling water use instead of fresh water) would NOT have a desirable effect on a project or a product.

6.10 Describe factors or activities you believe would be important to include in a greenhouse gas inventory assessment.

6.11 When might a Streamlined LCA be appropriate in place of a traditional LCA?

6.12 Discuss factors you believe would be important to include in an environmental health risk assessment.

6.13 Name some of the sustainability assessment tools.

6.12 Problems

6.1 Consider a sea wall being constructed to safeguard coastal communities from inundation. Considering climate change, what can be the potential technical, environmental, social, and economic resilience indicators for such a structure. List all possible indicators, and describe briefly in 3–4 sentences why you have chosen these indicators. Present your answer in the form of a table.

6.2 In the context of a street redevelopment project aimed at achieving sustainability, consider the integration of features such as LED lamps, bicycle tracks, green infrastructure, sustainable materials in the pavement, bike lanes, and pedestrian lanes with permeable pavements. Given these features, identify and justify all the resilience indicators that could effectively assess the resilience of this project. Identify indicators from technical, environmental, economic, and social resilience aspects. List all possible indicators and describe briefly in 3–4 sentences why you have chosen these indicators. Present your answer in the form of a table.

6.3 List potential resilience indicators that can be used to quantify technical, environmental, economic, and social resilience of a flyover project aimed at alleviating traffic burden in a given city. List possible indicators and discuss briefly in 3–4 lines why each indicator is important. Provide your answer in the form of a table.

6.4 Given below are the potential resilience indicators that are identified for a waste management strategy using multi-pronged approach including recycling, composting, incineration, and landfills.

- Resistance to climate change
- Adaptive capacity
- Impacts on wildlife and surrounding ecosystems
- Impacts on the local GDP
- Hunger index
- Degree of trust in the government
- Impacts on public health

Identify which indicator belongs to which aspect of resiliency (technical or environmental or economic or social). Discuss why certain indicators may or may not be relevant for assessing resilience in the context of this project.

6.5 A sustainable urban development initiative aims to develop a sustainable urban neighborhood, integrating residential, commercial, and recreational spaces. The project team has identified the following indicators to evaluate the sustainability of the project:

- Percentage of green space and biodiversity preservation
- Access to affordable housing options for diverse income groups
- Job creation and support for local businesses
- Energy efficiency
- Walkability and access to public transportation
- Waste generation and management

Identify which indicator belongs to which pillar of sustainability (environmental or economic or social). Discuss why certain indicators may or may not be relevant for assessing sustainability in the context of the project.

6.6 Consider a port construction project to improve the commodity imports and exports from a small country, thereby helping the economy and providing job opportunities to the local population. Identify one indicator each in the environmental, economic, and social pillars of sustainability and justify why this indicator can be the most important indicator for this project. Discuss ways in which one can optimize the selected indicators in the context of this project.

6.7 The Three Gorges Dam in China is the largest dam in the world with huge material inputs and related environmental impacts. However, the construction of this dam helped in providing a continuous source of water supply for both potable and nonpotable uses. In the context of this project, what can be the potential sustainability indicators in the economic and social aspects (at least two each), discuss whether this project has a positive/negative impact on the selected indicator, and justify your answer. In the context of TBL sustainability, discuss how this project can be an example for having different impacts (positive/negative) on each pillar of sustainability.

References

ASCE (2023a). *Standard Practice for Sustainable Infrastructure*. Reston, VA: American Society of Civil Engineers (ASCE) https://doi.org/10.1061/9780784415733.

ASCE (2023b). *Integrating Resilience and Sustainability into Civil Engineering Projects* (ed. C. Field and C. Zawislak). Reston, VA: American Society of Civil Engineers (ASCE) https://doi.org/10.1061/9780784484814.

Assarkhaniki, Z., Rajabifard, A., and Sabri, S. (2020). The conceptualisation of resilience dimensions and comprehensive quantification of the associated indicators: a systematic approach. *International Journal of Disaster Risk Reduction* 51: 101840.

Bare, J. (2011). TRACI 2.0: the tool for the reduction and assessment of chemical and other environmental impacts 2.0. *Clean Technologies and Environmental Policy* 13 (5): 687–696.

BASF (2019). SEEbalance®. Measuring sustainable development on a product level. https://www.basf.com/global/en/who-we-are/sustainability/management-and-instruments/quantifying-sustainability/seebalance.html (accessed 22 January 2019).

Bell, S. and Morse, S. (2012). *Sustainability Indicators: Measuring the Immeasurable?* Routledge.

Costa, M., de Fátima Castro, M., and Mateus, R. (2023). Methodology to identify and prioritise the sustainability aspects to be considered in the design of Brazilian healthcare buildings. *Applied Sciences* 13 (6): 3807.

de Fátima Castro, M., Mateus, R., and Bragança, L. (2017). Development of a healthcare building sustainability assessment method–Proposed structure and system of weights for the Portuguese context. *Journal of Cleaner Production* 148: 555–570.

DEFRA (2005). *Securing the Future- the UK Government Sustainable Development Strategy*. London: Department of Environment, Food and Rural Affairs (DEFRA) https://www.gov.uk/government/publications/securing-the-future-delivering-uk-sustainable-development-strategy (accessed 22 January 2019).

EC (2009). 2009 monitoring report of the EU sustainable development strategy. European Communities (EC). https://ec.europa.eu/eurostat/web/products-statistical-books/-/KS-78-09-865 (accessed 22 January 2019).

Fan, C., Carrell, J. D., & Zhang, H. C. (2010). An investigation of indicators for measuring sustainable manufacturing. In: *Proceedings of the 2010 IEEE International Symposium on Sustainable Systems and Technology* (pp. 1–5). IEEE.

Feil, A.A., Schreiber, D., Haetinger, C. et al. (2019). Sustainability indicators for industrial organizations: systematic review of literature. *Sustainability* 11 (3): 854.

GRI (2010). GRI Standards. Global Reporting Initiative (GRI). https://www.globalreporting.org/standards/ (accessed 22 January 2019).

Greenwood, D.T. (2018). Moving beyond traditional indicators of fiscal sustainability: examples from locally chosen indicators. In: *The Palgrave Handbook of Indicators in Global Governance* (ed. D. Malito, G. Umbach, and N. Bhuta). Cham: Palgrave Macmillan https://doi.org/10.1007/978-3-319-62707-6_14.

ISI (2018). Envision: Driving Success in Sustainable Infrastructure Projects. Institute for Sustainable Infrastructure (ISI). https://sustainableinfrastructure.org/envision (accessed 25 July 2018).

ITRC (2021). Sustainable Resilient Remediation SRR-1. Interstate Technology & Regulatory Council (ITRC), SRR team, Washington, D.C., USA. www.itrcweb.org. (accessed 18 October 2024).

Morse, S. (2018). Analysing the Use of Sustainability Indicators. In: Malito, D., Umbach, G., Bhuta, N. (eds) *The Palgrave Handbook of Indicators in Global Governance*. Palgrave Macmillan, Cham. https://doi.org/https://doi.org/10.1007/978-3-319-62707-6_19

Morse, S. (2016). Measuring the success of sustainable development indices in terms of reporting by the global press. *Social Indicators Research* 125: 359–375. https://doi.org/10.1007/s11205-014-0847-8.

Ness, B., Urbel-Piirsalu, E., Anderberg, S., and Olsson, L. (2007). Categorising tools for sustainability assessment. *Ecological Economics* 60 (3): 498–508.

Reddy, K.R., Sadasivam, B.Y., and Adams, J.A. (2014). Social sustainability evaluation matrix spiepr132 (SSEM) to quantify social aspects of sustainable remediation. In: ICSI 2014, pp. 831–841. https://doi.org/10.1061/9780784478745.078.

Reddy, K.R., Janga, J.K., Verma, G., and Nagaraja, B. (2024). Integration and quantification of resilience and sustainability in engineering projects. *Journal of the Indian Institute of Science* https://doi.org/10.1007/s41745-024-00440-w.

Sachs, J.D., Lafortune, G., Fuller, G., and Drumm, E. (2023). Implementing the SDG Stimulus. Sustainable Development Report 2023. Paris: SDSN, Dublin: Dublin University Press, 2023. 10.25546/102924.

Sharifi, A. (2016). A critical review of selected tools for assessing community resilience. *Ecological Indicators* 69: 629–647.

Sikdar, S.K. (2003). Sustainable development and sustainability metrics. *AIChE Journal* 49 (8): 1928–1932.

SURF (2018). SiteWise version 3.1. Sustainable Remediation Forum. http://www.sustainableremediation.org/library/guidance-tools-and-other-resources/sitewise-version-31 (accessed 25 July 2018).

UNDESA (2007). Indicators for Sustainable Development: Guidelines and Methodologies, 3rd Edition. United Nations Department of Economic and Social Affairs (UNDESA). United Nations, New York. ISBN 978-92-1-104577-2. https://sustainabledevelopment.un.org/index.php?page=view&type=400&nr=107&menu=1515 (accessed 22 January 2019).

USEPA (2016a). Tool for Reduction and Assessment of Chemicals and Other Environmental Impacts (TRACI). US Environmental Protection Agency. https://www.epa.gov/chemical-research/tool-reduction-and-assessment-chemicals-and-other-environmental-impacts-traci (accessed 25 July 2018).

USEPA (2016b). *Methodology & Spreadsheets for Environmental Footprint Analysis (SEFA)*. US Environmental Protection Agency. https://clu-in.org/greenremediation/methodology/# SEFA (accessed 25 July 2018).

USGBC (2018). *Leadership in Energy and Environmental Design*. U.S. Green Building Council. https://new.usgbc.org/leed (accessed 25 July 2018).

7

Material Flow Analysis and Material Budget

Chapter Goal and Objectives
The goal of this chapter is to provide a comprehensive framework for understanding the flow and budgeting of natural resources within a given system. The main objectives of the chapter are: (i) to analyze how human activities dominate the flows of materials and how the present consumption rate of materials will lead to the depletion of natural resources, affecting the ability of future generations to meet their own needs; (ii) to analyze the budget of natural resources by examining the extraction, consumption, and reserves of these resources in relation to sustainable development goals; (iii) to develop a methodology for constructing a material budget, offering a structured approach to quantify and balance resource inputs and outputs across various stages of production and consumption; (iv) to introduce material flow analysis (MFA) as a systematic tool for tracking material flows within specific systems (e.g. industrial sectors, regions), identifying inefficiencies, resource bottlenecks, and opportunities for improvement, with particular emphasis on its application in waste management, including understanding waste generation, types of waste produced, and opportunities for recycling and waste reduction; and (v) to explore how national material accounts (NMA) are used by countries to monitor resource use, assess environmental impacts, and guide policy-making for resource efficiency at the national level.

7.1 Introduction

The concept of sustainability is challenged by society's rapidly increasing use of nearly every resource on the planet. As an example, the increasing consumption of materials in the United States is presented in Figure 7.1. It emphasizes the trends of increased nonfuel resource use, with construction materials, such as aggregate, rock, and sand, constituting the majority of materials used. The consumption scenario in the United States is also prevalent in many other developed countries.

Figure 7.2 shows the ratio of the sum of anthropogenic flows to natural flows for the elements in the periodic table. Anthropogenic flows include mining and fossil fuel combustion, while natural flows include erosion and air-blown dust. The flow of elements on the left in the figure is dominated by anthropogenic use, whereas natural processes dominate the element flows on the right. The four metals that constitute over 98% of the mass of all metal utilized by society are highlighted in Figure 7.2. Copper is anthropogenic flow-dominated, aluminum is natural flow-dominated, and

Sustainable and Resilient Engineering: Drivers, Metrics, Tools, and Applications, Second Edition. Krishna R. Reddy, Claudio Cameselle, and Jeffrey A. Adams.

Companion website: www.wiley.com/go/reddy/sustainableresilientengineering2e

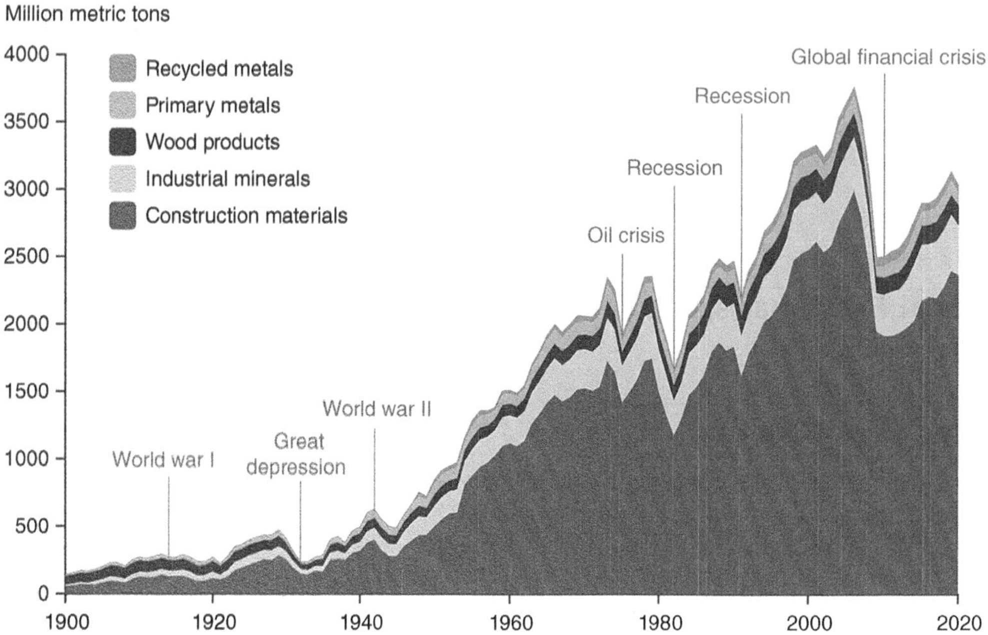

Figure 7.1 U.S. nonfuel material consumption, 1900–2020. *Source:* Center for Sustainable Systems (2024)/U.S. Material Use Factsheet/University of Michigan.

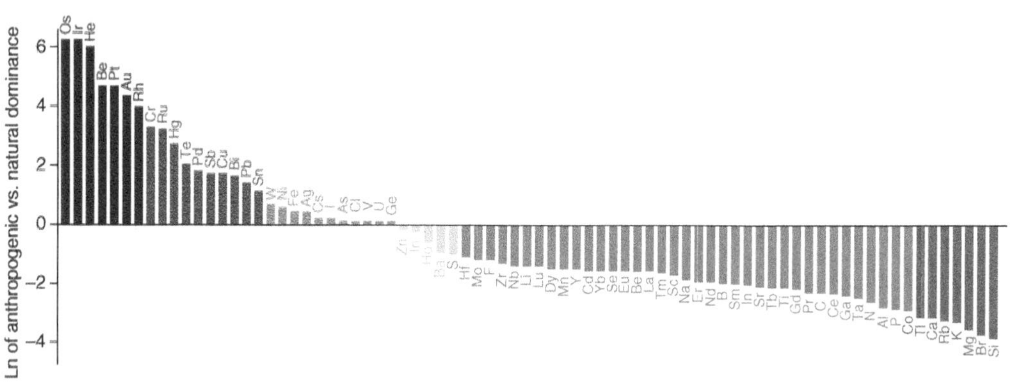

Figure 7.2 Ratio of anthropogenic/natural flows for elements. *Source:* Adapted from Klee and Graedel (2004).

iron and zinc are roughly evenly distributed among anthropogenic and natural flows. Figure 7.2 shows the fact that humans dominate the global cycles of about one-third of the elements in the periodic table.

7.2 Budget of Natural Resources

Before analyzing the flows and consumption of natural resources in material flow analysis (MFA), we need to define certain fundamental variables and parameters (Graedel and Allenby 2010). A budget is defined as the accounting of the receipts, disbursements, and reserves of the valued resource. The resource is stored in a container called a reservoir. The actual amount of the resource is called the pool size. The rate at which the resource enters and exits the reservoir is called flux. The inflow and outflow streams are called sources and sinks.

As an example, we can model the components of a budget as a water trough. If the total source and sink flows are not equal, the pool size (i.e. the quantity of the material of interest in the reservoir) will change over time as indicated in Eq. (7.1). For any reservoir, the change in contents can be readily calculated once all flows are known. Conversely, measuring the change in contents can aid in determining an unknown flow.

$$\Delta_r = \int_{t_1}^{t_2} \left(\sum_i F_i - \sum_o F_o \right) \mathrm{d}t \tag{7.1}$$

A cycle is defined as a system of connected reservoirs that transfer and conserve a specific resource. In most material flow analyses, there are several reservoirs, each with several linkages to other reservoirs. A complete analysis quantifies all flows and the contents of each reservoir. Figure 7.3 shows a generic cycle for the human-driven flows of a metal with several reservoirs (ore-environment, processing, fabrication, etc.) and the sources and sinks that interconnect the reservoirs. The actual cycles for metals are much more complex than what the simple example in Figure 7.3 suggests. Multiple reservoirs can be identified in production, fabrication, manufacturing, use, and waste facilities. The flows to other regions or countries must be considered since the analysis is done for a specific country or region (Chen and Graedel 2012). The inflows and outflows to the same reservoir are not necessarily the same, so the nickel in the reservoir increases or decreases, which may be accounted for with Eq. (7.1). The average residence time of a metal (or any other resource) in a reservoir can be determined with Eq. (7.2), where Ψ indicates the fraction of the constituent having a residence time between τ and $\tau + \mathrm{d}\tau$.

$$\tau_r = \int_{\tau}^{\tau + \Delta\tau} \tau \Psi(\tau)\,\mathrm{d}\tau \tag{7.2}$$

As an example, steel, which is widely used for a vast array of products, exhibits a significant range in residence time on a product-by-product basis – from a few years for kitchen gadgets to more than half a century for steel beams in buildings.

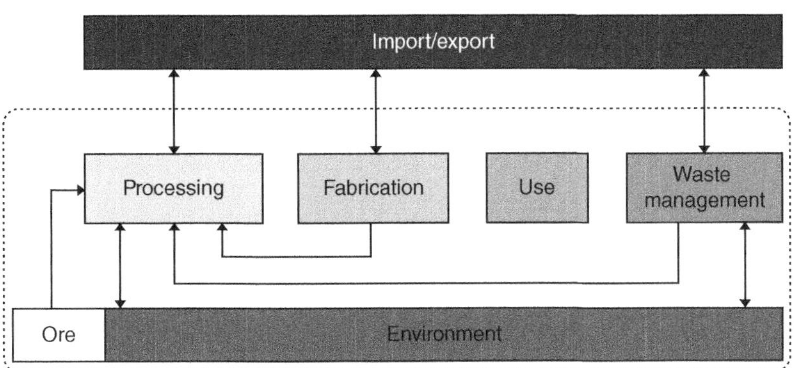

Figure 7.3 Generic cycle for the human-driven flows of a metal.

7.3 Constructing a Budget

The construction of a budget requires several tasks. The first step is the identification of the substance or substances to be evaluated, followed by reservoirs with the corresponding sinks and sources. Then, the limits or boundaries of the system may be defined. The second step requires the identification of the reservoir contents and the fluxes. This can be done using available data or specific measurements wherever possible, and estimations of resource content and fluxes can be used if no other source of objective data is available. This has to be done for all sinks and sources, or at least for the most relevant contributions. Small fluxes can be disregarded if their contribution is minimal. The analysis of material flows can be tracked at different organizational levels: for a single facility, groups of facilities, at a regional or national level, or globally. In principle, an MFA can be conducted at any spatial or organizational level where a reservoir and the associated sources and sinks can be identified and measured.

7.4 Material Flow Analysis

Figure 7.4a depicts the entire chromium cycle for Japan in 2000, including multiple reservoir analyses. Japan did not mine chromium, but imported large amounts in various forms, and fabricated, exported, used, and recycled it extensively (Graedel 2024). The fluxes of chromium among the different reservoirs can be seen in the Figure 7.4a, where the fluxes are expressed in Gg (gigagrams, or thousand metric tons) of chromium per year. An example of a subset budget is presented in Figure 7.4b, which illustrates chromium inputs and outputs at the waste management stage of the life cycle in Japan in 2000. Of the ~150 units discarded, about two-thirds were recycled within Japan, along with a small amount of scrap that was imported. A little more than one-third was lost to landfilling. Similar material cycles can be prepared for different materials at the country or project level.

Many materials are mined or produced in a small number of countries and are then transported and sold in many other countries. This is very common for many materials traded relatively freely around the world. Thus, a multilevel cycle can be constructed by characterizing the country-level cycles and then aggregating them to regional and global levels. If MFAs for the same reservoir in several different industrial ecosystems are constructed, flows can be directly compared, as in a study of copper use in the mid-1990s (Figure 7.5), where countries are represented as ecosystems. A similar comparison for copper use in the mid-1990s at the regional level may be performed (see Figure 7.6). North America, Europe, and Asia show a similar level of copper usage, whereas other regions (including Antarctica) use much less. When the analysis is performed in terms of the amount of copper per capita, the results are very different. Antarctica now has a clear lead, reflecting the large quantity of resources needed to maintain a high-technology lifestyle in an extremely cold environment.

Example 7.1

Consider the following process flow figure given below during cement production. If 1000 tons of raw material is input every hour. The following data is given:

- Halogen gas concentration in the raw material – 25 mg/kg of the material
- Fraction of halogens emitted into the atmosphere – 2%
- Fraction of halogens emitted as exhaust from second stage furnace – 90%

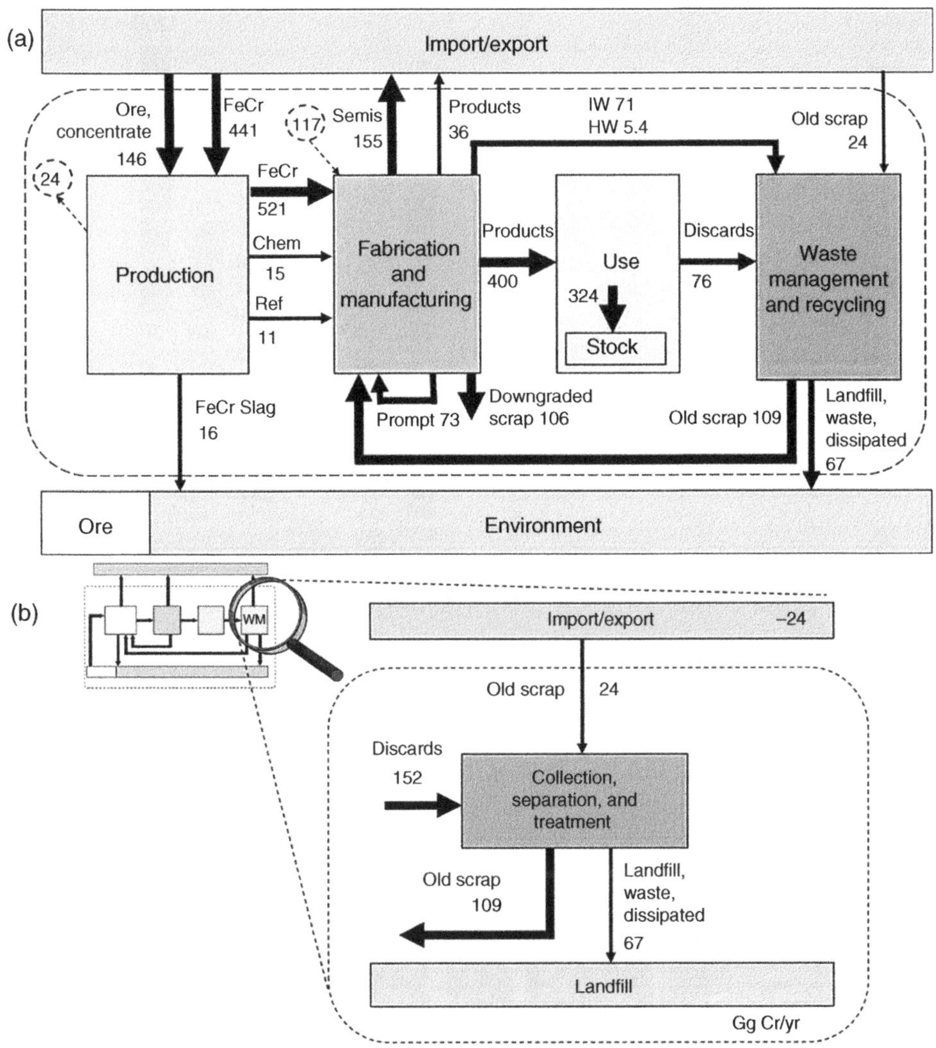

Figure 7.4 Japan chromium cycle in 2000. FeCr: ferrochromium; IW: industrial waste; HW: hazardous waste; Ref: refractories. (Gg Cr/yr). *Source:* Adapted from STAF Project, Yale University.

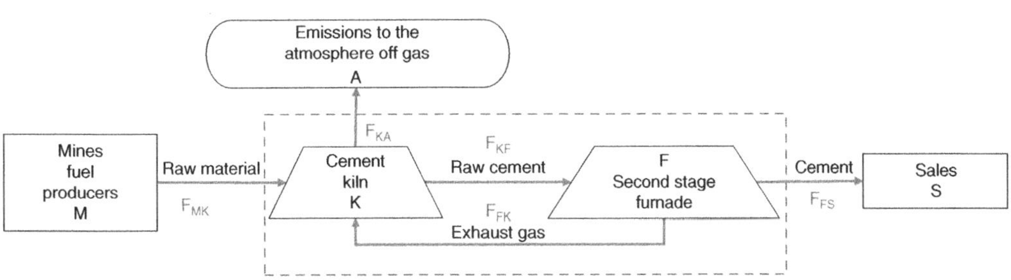

Considering the system is in equilibrium, calculate the concentration of halogen gas in the final product. Calculate the input and output fluxes of halogen in the cement kiln and the second-stage furnace.

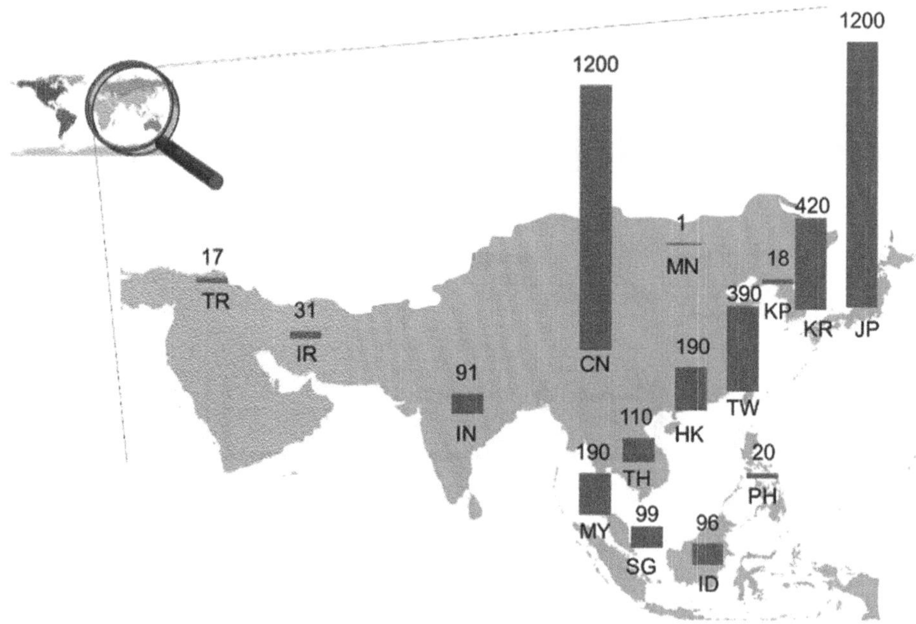

Figure 7.5 Country-level copper usage (Gg Cu/year).

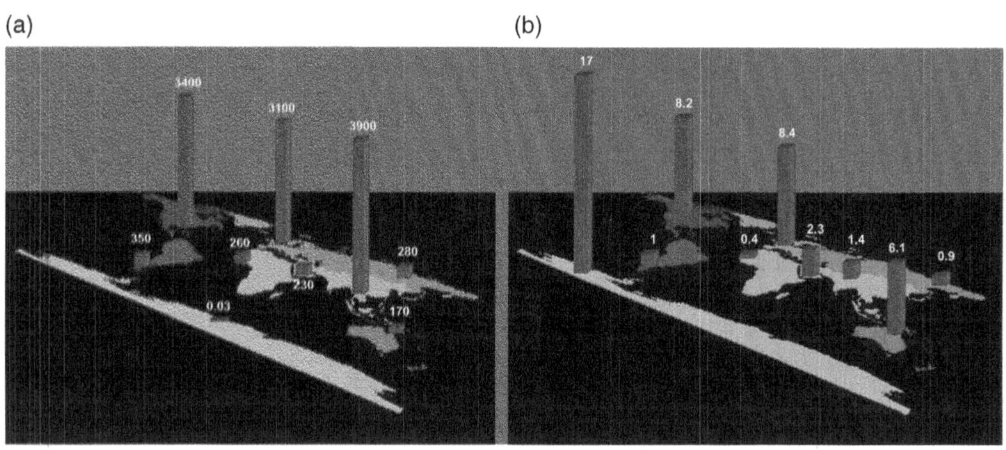

Copper entering use (Gg Cu/year) 1994

North America	3400
South America	350
Europe	3100
Africa	260
Antartica	0.03
Middle East	230
Asia	3900
Russia	280
Oceania	170

Copper entering use (kg Cu/capita year) 1994

North America	8.2
South America	1
Europe	8.4
Africa	0.4
Antartica	17
Middle East	2.3
Asia	1.4
Russia	0.9
Oceania	6.1

Figure 7.6 Copper use on a regional basis: (a) total use, (b) use per capita. *Source:* Reck et al. (2008). Reproduced with permission of American Chemical Society.

Solution:

The first step to calculate input/output fluxes in the system is the calculation of F_{MK}, the input flux of halogen gas from input M to kiln K.

$$F_{MK} = 1000\frac{\text{tons raw material}}{\text{hour}} \times \frac{1000\ \text{kg}}{1\ \text{ton}} \times 25\frac{\text{mg halogen gas}}{1\ \text{kg raw material}}$$
$$= 2.5 \times 10^7\ \text{mg halogen gas/hour}$$

Next, we can compute input/output fluxes at kiln K with the following equivalency:

$$F_{MK} + F_{FK} = F_{KF} + F_{KA}$$

Whose inputs are dependent on input flux, exhaust gas from furnace F, and generates outputs to the atmosphere and to the furnace. Knowing that 2% of halogens are emitted into the atmosphere, we have the following expressions for F_{KF} and F_{KA}:

$$F_{KF} = 0.98(F_{MK} + F_{FK})$$
$$F_{KA} = 0.02(F_{MK} + F_{FK})$$

Next, we compute input/output fluxes for second stage furnace F:

$$F_{KF} = F_{KF} + F_{FS}$$

And similarly, we can express F_{FK} and F_{FS} in terms of fractions of the input F_{KF}:

$$F_{FK} = 0.90(F_{KF}) = 0.90(0.98(F_{MK} + F_{FK}))$$
$$F_{FS} = 0.10(F_{KF}) = 0.02(0.98(F_{MK} + F_{FK}))$$

We can now obtain the value of F_{FK} in terms of F_{MK}

$$F_{FK} = 0.90(0.98(F_{MK} + F_{FK}))$$
$$F_{FK} = 0.882 \times F_{MK} + 0.882 \times F_{FK}$$
$$0.118 \times F_{FK} = 0.882 \times F_{MK}$$
$$F_{FK} = \frac{0.882}{0.118} \times F_{MK} = 7.46 \times F_{MK}$$

Input/output values

Calculation of final product F_{FS}

$$F_{FS} = 0.10(0.98(8.46 \times F_{MK})) = 2.07 \times 10^7\ \text{mg halogen gas/hour}$$

Calculation of input/output fluxes at cement kiln K

$$\text{Input } F_{MK} = 2.5 \times 10^7\ \text{mg halogen gas/hour}$$
$$\text{Input } F_{FK} = 7.46 \times F_{MK} = 18.65 \times 10^7\ \text{mg halogen gas/hour}$$
$$\text{Output } F_{KA} = 0.02(8.46 \times F_{MK}) = 0.42 \times 10^7\ \text{mg halogen gas/hour}$$
$$\text{Output } F_{KF} = 0.98(8.46 \times F_{MK}) = 20.73 \times 10^7\ \text{mg halogen gas/hour}$$

Input/output fluxes at second stage furnace F

$$\text{Input } F_{KF} = 20.73 \times 10^7\ \text{mg halogen gas/hour}$$
$$\text{Output } F_{FK} = 18.65 \times 10^7\ \text{mg halogen gas/hour}$$
$$\text{Output } F_{FS} = 2.07 \times 10^7\ \text{mg halogen gas/hour}$$

7.5 Material Flow Analysis: Wastes

An interesting application of MFA is the detailed analysis of waste management. As an example, Figure 7.7 depicts, on a global basis, the estimates of waste generation flow in seven categories during the mid-1990s. Industrial waste represents about 75% of the total, followed by municipal solid waste (12%) and construction and demolition debris (8%). A more detailed analysis permits an evaluation of the amount of metals in waste flows. Copper and zinc are valuable metals that can be found in wastes. In most wastes, the concentrations are very low, but copper is quite prevalent in electronics, and the use of both copper and zinc is relatively high in vehicles (Figure 7.8). This suggests that the flows of these two metals in some wastes are high enough to have a relatively high contribution to their respective MFAs.

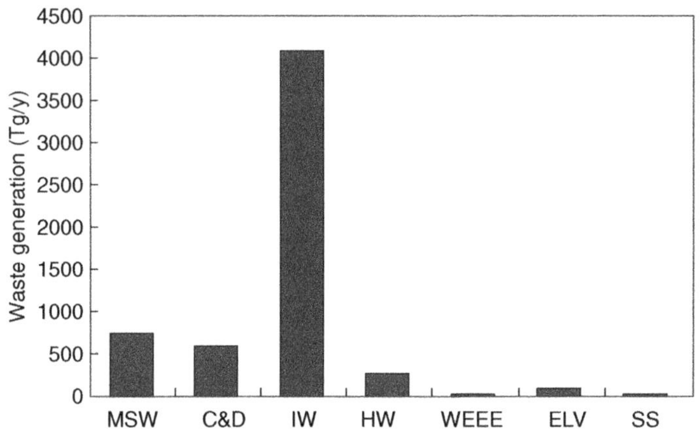

Figure 7.7 World waste generation flows in the 1990s. MSW: municipal solid waste, C&D: construction and demolition debris, IW: industrial waste, HW: hazardous waste, WEEE: waste of electrical and electronic equipment, ELV: end-of-life vehicles, SS: sewage sludge.

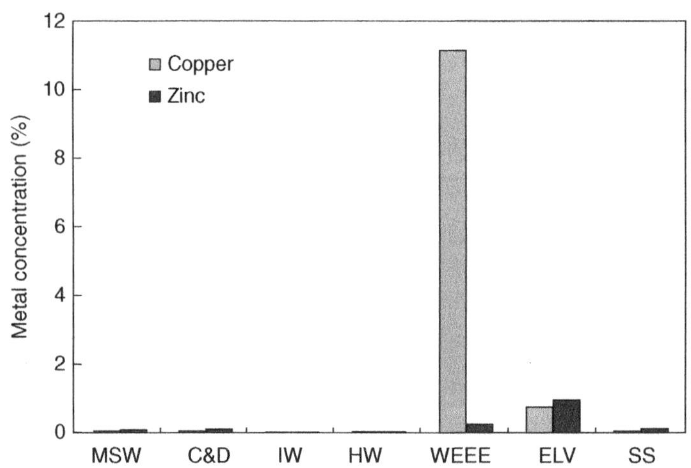

Figure 7.8 Copper and zinc concentration in wastes. MSW: municipal solid waste, C&D: construction and demolition debris, IW: industrial waste, HW: hazardous waste, WEEE: waste of electrical and electronic equipment, ELV: end-of-life vehicles, SS: sewage sludge.

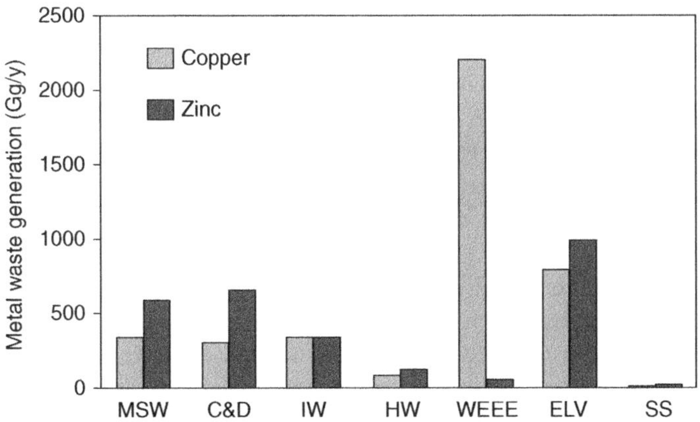

Figure 7.9 Global flows of copper and zinc in discarded materials (wastes). MSW: municipal solid waste, C&D: construction and demolition debris, IW: industrial waste, HW: hazardous waste, WEEE: waste of electrical and electronic equipment, ELV: end-of-life vehicles, SS: sewage sludge.

In calculating the MFAs, the product of total flow by metal concentration gives the total metal flow in several waste streams. Copper is mainly concentrated in electronics and vehicles, whereas zinc is more evenly distributed among several waste fractions (Figure 7.9). Thus, the results from the MFAs for copper and zinc suggest which materials may be more effectively recycled or repurposed compared to existing practices.

Material recycling is the primary alternative to extracting and processing new raw materials. The MFA for copper and zinc identifies the respective amounts of these metals in the different reservoirs, such as products in use and in wastes. However, it is important to note that these metals are incorporated into products with specific lifetimes; therefore, the metals will not be available for recycling until the useful life of these products has elapsed. Recycling streams incorporate several sequential processes, including development, use, discard, and reuse, with products having different lifetimes overlapping in the recycling stream.

An important question is to determine how much material is available in products in use that can serve as reusable resources in the future when the products are discarded, as well as how much of these respective materials can feasibly be extracted from the discarded products; and a second question is to determine when the materials will become available for resource extraction and reuse. In other words, we need to determine the resources in use, their age, their probable lifetime until entering a waste stream, and where they can be reprocessed and reused. Eq. (7.3) accounts for resources or stocks in use that correspond with the amount of resources in the reservoir "use" in Figure 7.3.

$$S(t) = \sum_{i=1}^{n} N_i(t) \cdot \frac{M_i(t)}{N_i(t)} \tag{7.3}$$

where

S is the in-use stock,

N is the number of units of a particular product (automobiles, etc.) that use the resource to provide services of some sort, and

M_i/N_i is the material intensity per unit of product, all summed over the i types of products addressed by the analyst.

As an example, the calculation of the amount of copper in use requires the amount or concentration of copper in respective products as well as the number of those products in use. The first step is the identification of the products with the major amounts of copper. A typical car contains 21 kg of copper, and a typical house contains 200 kg of copper. The main product categories include buildings, equipment, transportation, and infrastructure. Figure 7.10 depicts copper stocks per capita in products in use at three different levels: city, state, and country. Differences among the three levels exist because some reservoirs (off-shore oil platforms, military and naval equipment, etc.) are not presented at each level.

One of the most commonly used metals is iron. In a two-century study of iron stocks, Müller et al. (2006) demonstrate that about half of the mineable iron that was below ground in the United States in 1800 has been extracted, and most of that material is still in use. The total (above and below ground) stock of iron has increased over the last half-century because of extensive imports of iron in products (Figure 7.11). Iron stocks in use in the United States have continued to grow throughout

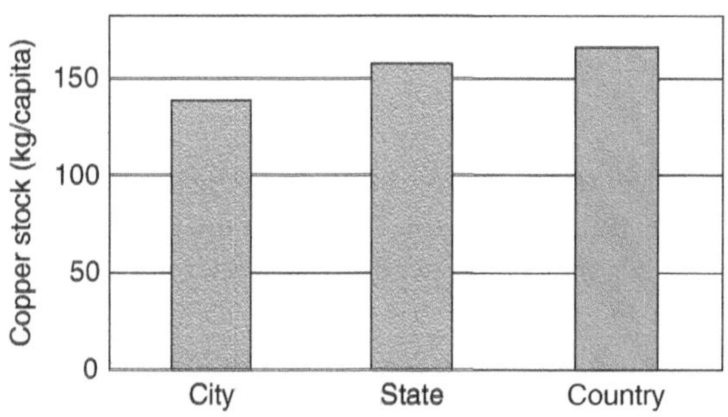

Figure 7.10 Copper in-use stock results at different spatial levels. City: New Haven, State: Connecticut, Country: USA. *Source:* Drakonakis et al. (2007)/with permission of Elsevier.

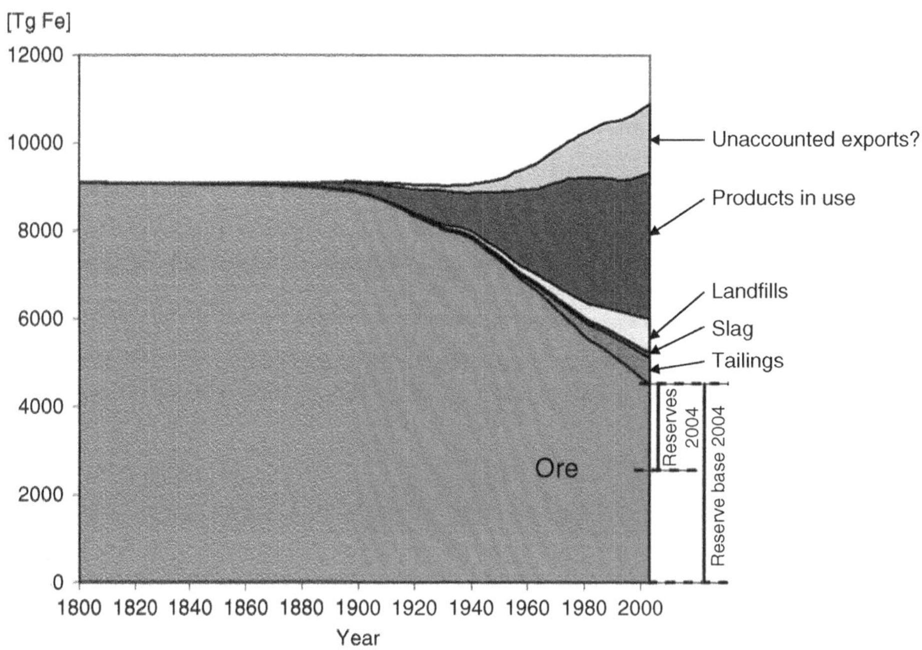

Figure 7.11 Distribution of iron stocks in the US. *Source:* Müller et al. (2006). Reproduced with permission of National Academy of Sciences, USA.

the twentieth century; the largest fraction is contained in building and infrastructure construction. On a per capita basis, the stock reached saturation around 1980, suggesting that by that point the United States had built about as many buildings, bridges, and other infrastructure products as were needed on a per-person basis, and new construction was performed at a steady rate with population growth or was merely replacing old infrastructure.

Example 7.2

How much ammonium will a person ingest by consuming 500 g of fish from a lake contaminated with approximately 150 tons of ammonium from a fertilizer factory spill? Given that the lake's volume is around $10^8\ m^3$, with a fish loading of 300 g per 1000 m^3, and the ammonium accumulation factor in fish is 3.2 L/kg, what is the approximate amount of ammonium ingested?

Solution:

Given Data:

Ammonium in lake = 150 kg
Lake volume = $10^8\ m^3 = 10^{11}$ L
Fish loading = 300 g per 1000 m^3 = 0.3 kg per 10^6 L
Ammonium accumulation factor = 3.2 L/kg
Fish consumed = 500 g = 0.5 kg

Step 1. Determination of ammonium concentration in water
Cwater = total ammonium in lake/lake volume
Cwater = 150 kg/10^{11} L = 1.5×10^{-6} kg/L = 1.5 mg/L

Step 2. Determination of ammonium in fish
Cfish = accumulation factor × Cwater
Cfish = 3.2 L/kg × 1.5 mg/L = 4.8 mg/kg

Step 3. Determination of ammonium ingested
Ammonium ingested = Cfish × fish consumed
Ammonium ingested = 4.8 mg/kg × 0.5 kg = 2.4 mg

Final answer:
The person will ingest approximately 2.4 mg of ammonium

7.6 National Material Account

The basic model for the national material account (NMA) was established by Matthews et al. (2000) and is presented in Figure 7.12. The NMAs attempt to count all flows crossing jurisdictional borders and those crossing the boundary between the ecosphere and the anthroposphere (except water and agricultural tillage). Minor flows are excluded even if they have significant environmental impacts. Material inputs from the environment are defined as the intentional extraction or movement of natural materials by human or human-controlled means. Material outputs to the environment are defined as material flows released from the economic system into the environment. Table 7.1 listed the categories of material inputs and outputs.

Hidden flows are associated with materials associated with extraction (mining) or processing that are discarded and hence not accounted for. For example, if a firm in Japan buys 1 kg of refined zinc, that zinc is mined as 50 kg of ore plus rock (2% zinc "ore grade"). It was then milled to remove the rock and leave 18 kg of zinc ore, smelted to produce 2 kg of impure zinc, and then refined to the

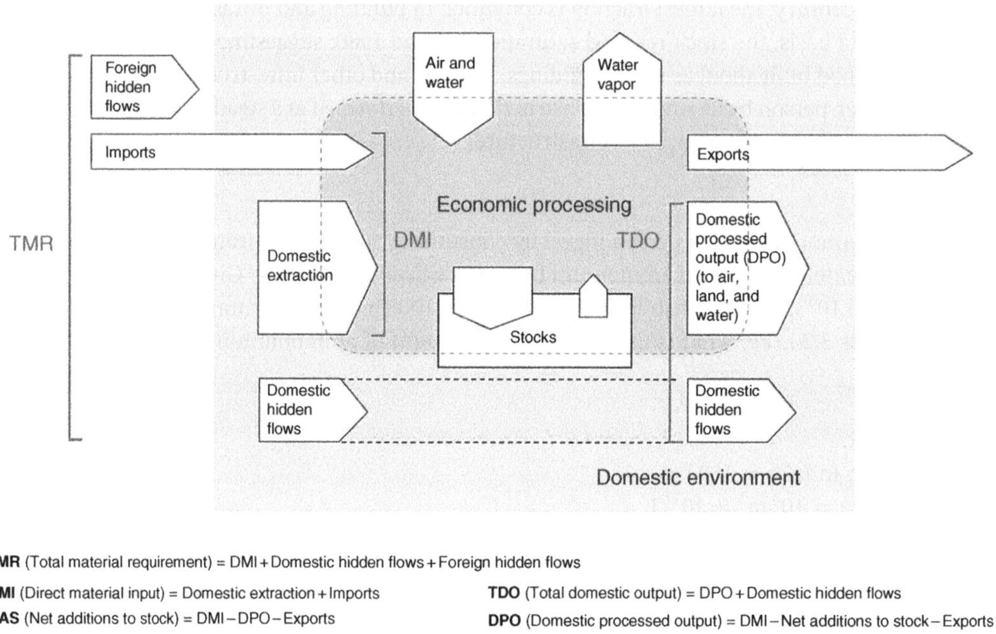

Figure 7.12 Basic model for a national material account. *Source:* Matthews et al. 2000/with permission of World Resources Institute. TMR: total material requirements, DMI: domestic material inputs, FHF: foreign hidden flows, DHF: domestic hidden flows, DE: domestic extraction, DPO: domestic processed output, TDO: total domestic output.

Table 7.1 Input and output material categories.

Material input categories	Material output categories
Domestic extraction (amount used)	Domestic processed output released to nature (e.g. carbon dioxide, herbicides)
Unused domestic extraction (domestic "hidden flows")	Disposal of unused domestic extraction (domestic "hidden flows")
Imports	Exports
Indirect flows associated with imports (foreign "hidden flows")	Indirect flows associated with exports

purchased finished product (1 kg of zinc). Of the original 50 kg that was extracted, 49 kg constitutes a hidden flow invisible to the purchaser of the refined zinc, but which nonetheless had a physical effect (requiring handling as mine tailings or refining waste), mostly in the country where the mining and refining took place. The NMA is a mass balance of the flows in Figure 7.12. If we define:

- DE = domestic extraction
- Direct material input: DMI = DE + imports
- Direct material consumption: DMC = DMI − exports
- DPO = domestic processed output
- TDO = total domestic output

DMI and DMC measure material use. Domestic hidden flows (DHF) account for unused domestic extraction. Foreign hidden flows (FHF) represent the upstream primary resource requirements of the imports. Therefore, the total materials requirements (TMR) of an economy are determined by Eq. (7.4), and the total material consumption (TMC) is determined by Eq. (7.5).

$$\text{DMI} + \text{DHF} + \text{FHF} = \text{TMR} \tag{7.4}$$

$$\text{TMR} - \text{exports} - \text{export HF} = \text{TMC} \tag{7.5}$$

Figure 7.13 shows an example of NMA for Austria in 1996. In Figure 7.13, in addition to the NMA flows explained in Figure 7.12, additional flows of water and air are bonded or mixed into that amount, accounting for nearly 40% of the additional mass. In the Austrian NMA, DMI amounts to about 22 metric tons per capita per year, and another 13 tons of air and water enter the production process. About half of the direct material input, 11 tons, is added to the existing material stocks. On the output side, there are three components: exports to other countries, additional outputs of air and water that result from economic processing, and direct processed output (DPO), consisting of wastes, emissions, and dissipatedly used materials. This amounts to 11 tons per capita per year. Since stocks are growing, all outputs taken together amount to only about two-thirds of all input. DPO makes up about half of this output. This result is similar to other studied countries – it is smaller than DMI – in the Austrian case, 13 tons per capita per year.

NMA is also a useful source of information for assessing the sustainability of the practices and activities in a country. It is often noted that large flows tend to have modest environmental impacts per unit weight, while small flows sometimes have significant impacts. This is often referred to as the "elephants and scorpions" analogy.

Equation (7.6) defines how NMAs may be used to estimate the environmental impact associated with the use and consumption of materials. The total impact consists of the product of the weight of the material used and its potential impact. This approach has been used occasionally, especially in the Netherlands, to evaluate progress in environmental improvement and to identify potential areas of improvement to reduce the impact on the environment.

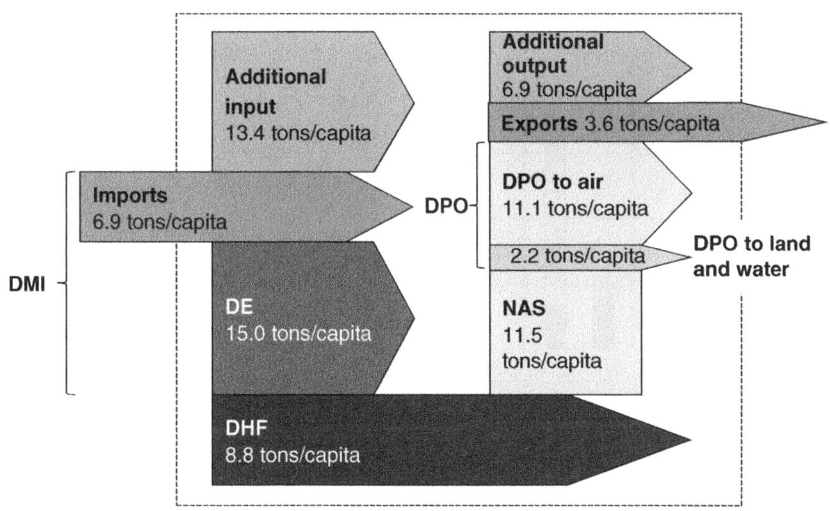

Figure 7.13 NMA for Austria in 1996. *Source:* Fischer-Kowalski (2000). Reproduced with permission of John Wiley & Sons.

$$T_{m,e} = W_m{}^{*}E_{m,e} \tag{7.6}$$

where

$T_{m,e}$ = total impact of material m of type e;
W_m = annual weight of material m used;
$E_{m,e}$ = material m impact of type e.

NMA flows and results can be compared for different regions or countries, and this may help in understanding the intrinsic characteristics of each country/society (Figures 7.14–7.17). Figure 7.14

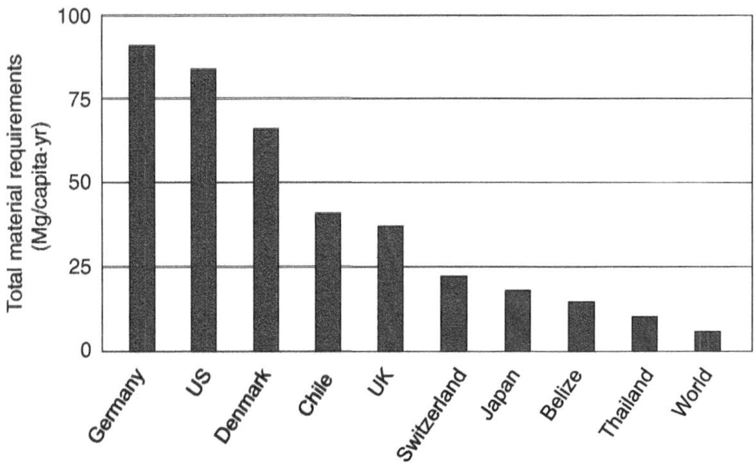

Figure 7.14 Comparison of total material requirements (TMR) across countries.

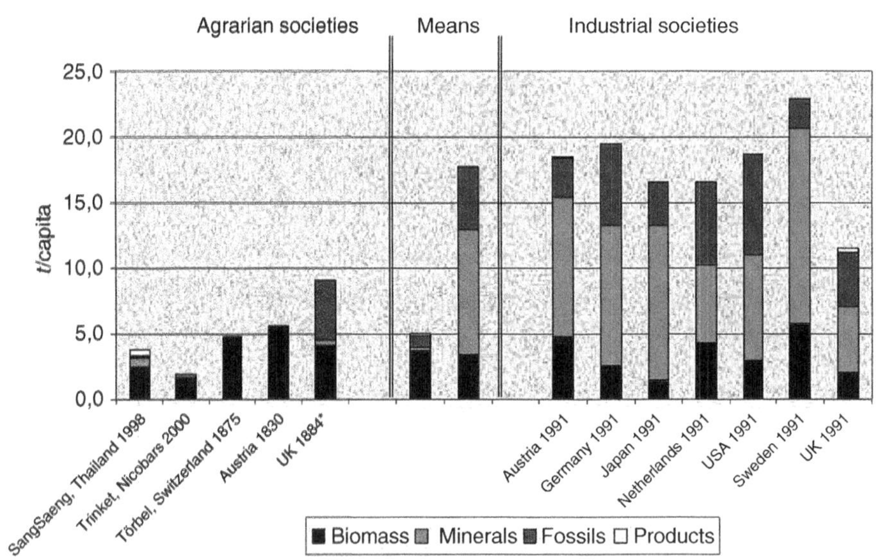

Figure 7.15 Metabolic profiles by socioecological regimes (DMC/capita). *Source:* Fischer-Kowalski (2000). Reproduced with permission of John Wiley & Sons.

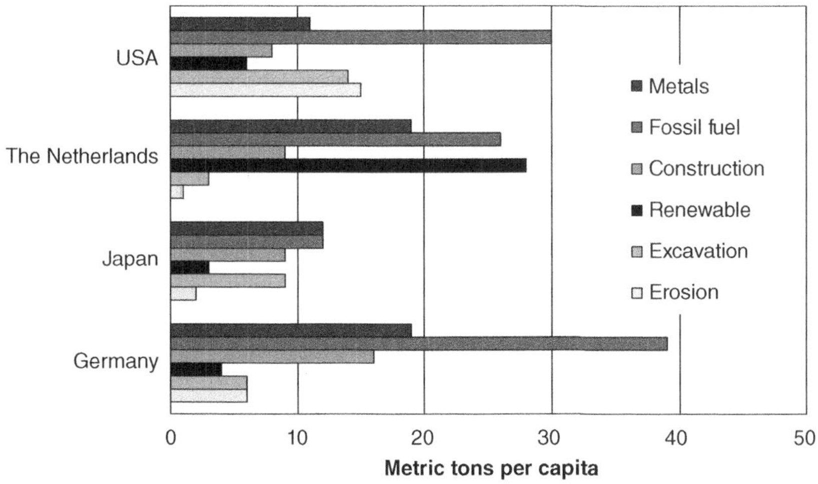

Figure 7.16 Sector contributions to TMR. *Source:* Matthews et al. (2000)/with permission of World Resources Institute.

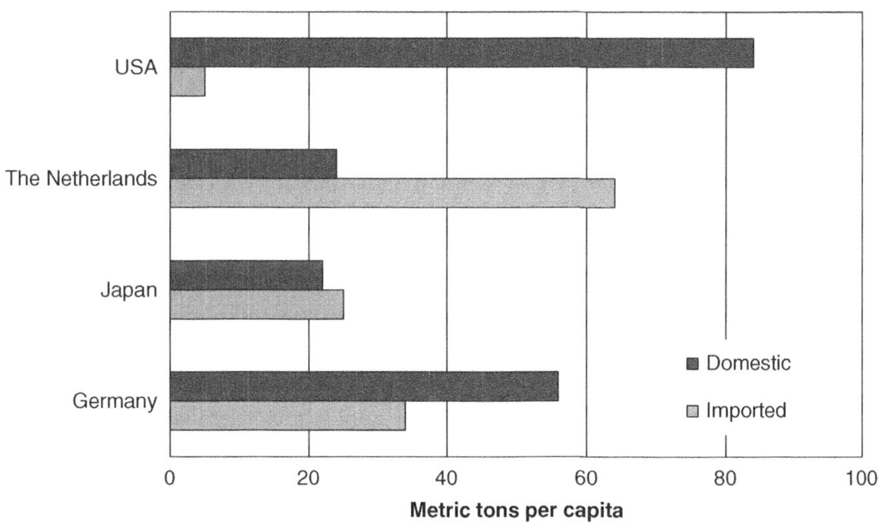

Figure 7.17 Domestic and foreign components of TMR. *Source:* Drakonakis et al. (2007)/with permission of Elsevier.

shows total mass requirements per capita in various countries. As depicted in the figure, there are enormous variations among countries; one explanation can be found in the respective standard of living of each country or society. Figure 7.15 demonstrates that industrial societies typically exhibit direct material consumption rates at least threefold greater than less developed, agrarian-focused societies. Most of the difference is accounted for by mineral usage and, in some countries, fossil fuel usage. Figure 7.16 shows the use of different types of materials in various countries. For example, Germany was the largest per capita user of fossil fuels in 1994, and the Netherlands led by far in renewable applications. Figure 7.17 compares the fraction of total material requirements that come from imports rather than domestic supplies. The differences among countries are striking.

The United States has a very small import fraction, whereas the Netherlands shows the opposite. Japan shows a balanced point between imports and exports. Finally, it is important to note that most NMAs do not include the flow of energy, water, greenhouse gases, or pollutants. The addition of these flows is important to assess the environmental impacts and sustainability of the related regions or countries. While making a comparison, it is important to make sure that the assessments include equivalent types of flows.

Material accounts of smaller geographical units are obviously embedded in and are part of larger units. For example, the city of London is embedded in the UK, and for the purpose of this analysis, the UK is assumed to be included in the European Union. As it can be anticipated for a city, virtually all of London's materials come from outside its boundaries. Both the UK and the European Union import about 20% of DMI. The most interesting thing about this comparison is that the average Londoner has a DMI of about half that of the average EU person. Does this suggest that cities really are more efficient from a materials standpoint? The likely answer to that question is no. The material analysis in a country includes industrial facilities and other major activities that account for large materials consumption and are often not accounted for in major cities because they are located out of the areas with the greatest population density.

7.7 Summary

Few anthropogenic element cycles have been adequately characterized; however, humans appear to dominate most of the element cycles that have been studied. Approximately one-third of the elements in the periodic table seem to be dominated by anthropogenic processes instead of natural processes. The analysis of the flows of materials is useful in understanding the extraction, transformation, uses, and trade of the most commonly used materials: metals and construction materials. Material cycle analysis reveals the potential for the recovery and reuse of materials, but different materials require different strategies. Characterizing anthropogenic material cycles provides a wealth of useful information relating to resource availability, environmental impacts, and policy options.

Industrial economies are becoming more efficient in their use of materials, but consumption and waste generation continue to increase. Material cycles reveal quite different behavior at different spatial levels. Imports associated with resource-intensive extraction and processing are increasingly shifted to less-developed countries. The extraction and use of fossil energy resources dominate output flows in all industrial countries. Finally, over time, material flows of all kinds increase as populations become more affluent.

7.8 Questions

7.1 Discuss how metals may be considered "anthro-dominated" or "natural-dominated"

7.2 What are examples of nonmetallic compounds that may be considered "anthro-dominated" or "natural-dominated"?

7.3 Discuss how a reservoir behaves if sources and sinks are in balance. Discuss if one is dominant compared to the other.

7.4 List reservoirs that may be considered for the manufacture of a product.

7.5 Consider crude oil extraction and refining. List reservoirs associated with these activities and discuss relative residence times within these reservoirs.

7.6 Describe the different tasks involved in the construction of a material budget.

7.7 What is material flow analysis? Explain briefly.

7.8 Define the terms budget, reservoir, pool size, flux in the context of material flow analysis.

7.9 Discuss how net consumers or net producers of a product would have a different material flow analysis.

7.10 How is the total material requirements and total materials consumption of an economy determined?

7.11 Describe how a material flow analysis may be used to devise effective or more optimal waste reclamation strategies.

7.12 Explain the application of material flow analysis in the context of waste management.

7.13 What is national material account? Explain briefly.

7.14 Describe how a national material account may be used to assess sustainability practices.

7.15 Discuss how subset material accounts (i.e. cities) may be used to assess national material accounts.

7.9 Problems

7.1 At a household level, identify different paths of nitrogen into and out of the household, draw a material flow diagram showing fluxes due to different processes and the possible stock reservoirs. Try to quantify input and output flows for all the processes you depict considering an average household.

7.2 In Figure 7.4 of the textbook carefully analyze the figure and describe different processes and the input and output fluxes from each process. Verify whether the flow diagram is in equilibrium or not. Additionally, solve for the following:

- If a new FeCr ore is discovered in Japan, increasing the input ore concentrate from 146 Gg/yr to 200 Gg/yr. Redraw the figure, and determine the increase in stock quantities, and the increase in waste output to the environment.
- If the existing FeCr ore is exhausting, quickly reducing the ore concentrate from 146 Gg/yr to 100 Gg/yr in 2001. Redraw the figure, and determine the reduction in stock quantities, and the decrease in waste output to the environment.

7.3 Convert the material flow diagram (steel cycle) given below for China into per capita and plot a similar material flow diagram for the United States. Compare with current consumption patterns in the United States and comment on the differences (Find references as required and cite the references).

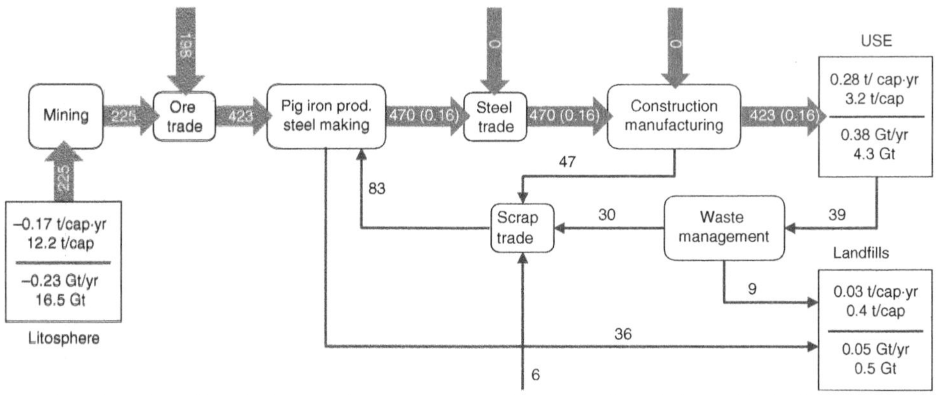

The steel cycle in China, 2010. Population 1.36 billion people. Absolute stocks and flows with recycling content, per capita stock by sector. Flows smaller than 2 Mt/yr and some markets have been omitted for better visibility. Color of flow additionally represents recycling content: gray 0%, green 100%. Units: flows in Mt/yr, content of recycled iron in brackets, stocks and stock changes as displayed.

7.4 For a project of your choice, identify the major materials that are used and draw material flow diagram from raw material extraction to waste disposal/emissions for one of the materials identified. Choose an appropriate boundary and use quantitative numbers for the input/output flows and stock from each process.

7.5 A manufacturing plant produces plastic bottles using injection-molding machines. The plant operates 24 hours a day, 7 days a week. Each injection-molding machine produces 1000 bottles per hour, with each bottle weighing 50 g. The raw material used for manufacturing the bottles is polyethylene (PE), with a density of 0.92 g/cm^3. The manufacturing process generates waste in the form of plastic scraps, with 5% of the raw material ending up as scrap. The scraps are collected and recycled back into the production process.

- Draw a material flow diagram illustrating the flow of raw materials, production, and waste generation in the plastic bottle manufacturing process. Show your boundaries.
- Calculate the total daily production of plastic bottles in kilograms.
- Determine the total daily amount of plastic scraps generated.
- Calculate the percentage of raw material that is recycled back into the production process.

7.6 An oil refinery processes 100 000 barrels of crude oil per day. The following data is given regarding the refinery plant → Initial stock: Gasoline – 10 000 barrels, Diesel – 5000 barrels, Jet fuel – 2000 barrels, Residual fuel oil – 3000 barrels.

- 80% of the crude oil is converted into gasoline and diesel.
- 15% of the crude oil is converted into jet fuel.
- 5% of the crude oil becomes residual fuel oil.
- Additionally, during the refining process, there is a 2% loss of material due to inefficiencies.
- The refinery has a storage capacity for each product: gasoline, diesel, jet fuel, and residual fuel oil.

Draw a material flow diagram showing the boundary, input and outputs of the refinery, and calculate:

- The daily production of each product: gasoline, diesel, jet fuel, and residual fuel oil.
- The total daily loss due to inefficiencies.
- Verify if the sum of all product outputs and the loss equals the total input of crude oil.
- Determine the change in stock for each product, considering the starting and ending inventory.

References

Center for Sustainable Systems (2024). U.S. material use factsheet. Pub. No. CSS05-18, University of Michigan.

Chen, W.Q. and Graedel, T.E. (2012). Anthropogenic cycles of the elements: a critical review. *Environmental Science & Technology* 46 (16): 8574–8586.

Drakonakis, K., Rostkowski, K., Rauch, J. et al. (2007). Metal capital sustaining a North American city: iron and copper in New Haven, CT. *Resources, Conservation and Recycling* 49 (4): 406–420.

Fischer-Kowalski, M. (2000). The end of the pipe – material outflows from industrial economies. Presentation at the Year 2000 Gordon Conference on Industrial Ecology: Engineering Global systems, New London, USA (11–16 June 2000).

Graedel, T. (2024). Stocks and Flows Project (STAF). Center for Industrial Ecology. Yale School of the Environment. Yale University. https://cie.research.yale.edu/research/stocks-and-flows-project-staf (accessed 7 June 2024).

Graedel, T.E. and Allenby, B.R. (2010). *Industrial Ecology and Sustainable Engineering*. Upper Saddle River, NJ: Prentice Hall.

Klee, R.J. and Graedel, T.E. (2004). Elemental cycles: a status report on human or natural dominance. *Annual Review of Environment and Resources* 29: 69–107.

Matos, G.R. (2022). Materials flow in the United States—A global context, 1900–2020: U.S. Geological Survey Data Report 1164, 23 p.

Matthews, E., Amann, C., Bringezu, S. et al. (2000). *The Weight of Nations: Material Outflows from Industrial Economies*. Washington, DC: World Resources Institute.

Müller, D.B., Wang, T., Duval, B., and Graedel, T.E. (2006). Exploring the engine of anthropogenic iron cycles. *Proceedings of the National Academy of Sciences* 103 (44): 16111–16116.

Reck, B.K., Müller, D.B., Rostkowski, K., and Graedel, T.E. (2008). Anthropogenic nickel cycle: insights into use, trade, and recycling. *Environmental Science & Technology* 42 (9): 3394–3400.

8

Carbon Footprint Analysis

Chapter Goal and Objectives
The chapter aims to provide a detailed overview of the carbon footprint concept and the tools, methodologies, and strategies for quantifying and reducing greenhouse gas (GHG) emissions. The chapter's objectives are to (i) define carbon footprint and its relevance to global warming potential (GWP); (ii) explain the process of measuring carbon footprints, including scope definition, emissions measurement, baseline establishment, target setting, offsetting, and verification; (iii) review standards for calculating carbon footprints within global and national frameworks; (iv) examine GHG inventory developments in the United States, focusing on key policies, including the USEPA's Greenhouse Gas Reporting Program; (v) highlight tools for GHG inventory and provide practical guidance; (vi) present a carbon footprint case study; and (vii) explore programs aimed at mitigating GHG emissions, including policies, technologies, and strategies.

8.1 Introduction

Most human activities require the consumption of energy and materials, resulting in the generation and emission of gases into the atmosphere that contribute to the greenhouse effect. Greenhouse gases (GHGs) naturally occur in the Earth's atmosphere and make life possible on Earth by trapping the Sun's energy and heating the surface of the planet to about 15 °C (58 °F). In atmosphere, the key natural GHGs are water vapor (H_2O) and CO_2. In their absence, the Earth's surface temperature would average approximately −18 °C (0 °F). However, changes in human activities in the last 150 years, such as rapid industrialization, massive use of fossil fuels, and deforestation, have resulted in increased levels of GHGs entering the Earth's atmosphere.

Several important GHGs of anthropogenic origin are listed in Table 8.1, including CO_2 and CH_4. These gases are responsible for the increasing greenhouse effect in the atmosphere and the acceleration of global warming and climate change. Considering the importance of climate change and its impact on the future generations, the inventory of GHGs has been proposed as a metric to measure the sustainability of the human activities.

Sustainable and Resilient Engineering: Drivers, Metrics, Tools, and Applications, Second Edition. Krishna R. Reddy, Claudio Cameselle, and Jeffrey A. Adams.

Companion website: www.wiley.com/go/reddy/sustainableresilientengineering2e

8.2 Global Warming Potential and Carbon Footprint

The emissions of anthropogenic GHGs are a result of various activities (IPCC 2007, 2014)). These GHG sources include agriculture, fossil fuel combustion, and industrial processes. Human activities are responsible for almost all of the increase in greenhouse gases in the atmosphere over the past 150 years (USEPA 2025). This includes emissions from burning fossil fuels for electricity, heat, and transportation, as well as deforestation and other land-use changes. Since 1750, humans have increased the abundance of carbon dioxide in the atmosphere by nearly 50% (Herring 2020). The emissions of CO_2 are mostly related to the combustion of fossil fuels and deforestation, which reduces the amount of CO_2 trapped in biomass. CH_4 emissions have been associated with livestock enteric fermentation (i.e. generated by cows), manure management, paddy rice farming, land use, and wetlands changes, landfill emissions, and losses in petroleum and natural gas extraction and transmission. Nitrous oxide (N_2O) has been used for its anesthetic and analgesic effects in the medical industry, as an oxidizer in rocketry, and in automobile racing to increase the power output of engines. Hydrofluorocarbons (HFCs), perfluorocarbons (PFCs), and sulfur hexafluoride (SF6) are manufactured gases commonly used in different industrial applications and products. SF6 is used as a gaseous dielectric medium for high voltage (35 kV or above) in electrical equipment applications. PFCs have been used in refrigeration units as replacements for CFCs (chlorofluorocarbons), which were banned to protect the stratospheric ozone layer. HFCs, PFCs, and SF6 have been identified in the Kyoto Protocol as gases of concern from the standpoint of the global warming.

Global warming potential (GWP) is a relative measure of how much heat a GHG traps in the atmosphere. GWP compares the amount of heat trapped by a certain mass of the gas, normalized by the same mass of carbon dioxide. Each gas has a different GWP, which is a measure of its "strength" in contributing to global warming as compared to carbon dioxide. For example, methane has a GWP of 29.8, which means it is 29.8 times more potent as GHG than carbon dioxide in a time horizon of 100 years. Thus, 1 g of methane contributes equivalently to the global warming as 29.8 g of CO_2.

Total GHG emissions are calculated and reported as carbon dioxide equivalents (eCO_2 or CO_2e) by accounting for the GWP of each gas as reported in Table 8.1.

Table 8.1 Greenhouse gases and global warming potential.

Greenhouse gas	Chemical formula	Lifetime (yr)	Global warming potential (time horizon)	
			20 yr	**100 yr**
Carbon dioxide	CO_2	Variable	1	1
Methane (fossil origin)	CH_4	12.4	82.5	29.8
Methane (non-fossil origin)			80.8	27.2
Nitrous oxide	N_2O	121	273	273
Hydrofluorocarbons (HFC)[a]	$C_xH_yF_z$	1.4–270	437–12 000	124–14 800
Perfluorocarbons (PFC)[a]	C_xF_z	1000–50 000	5210–8630	7390–12 200
Sulfur hexafluoride (SF6)	SF_6	3200	16 300	23 500

[a] See Table 2.1 for specific greenhouse gases.
Source: Adapted from IPCC (2023).

Table 8.2 Changes in global warming potentials in different assessments.

		GWP values for 100-yr time horizon				
Industrial designation or common name	**Chemical formula**	**Second assessment report (SAR)**	**Fourth assessment report (AR4)**	**Fifth assessment report (AR5)**		**Sixth assessment report (AR6)**
Carbon dioxide	CO_2	1	1	1	1	1
Methane (fossil origin)	CH_4	21	25	28	34	29.8
Methane (non-fossil origin)						27.2
Nitrous oxide	N_2O	310	298	265	298	273

Source: IPCC (1995, 2007, 2014, 2023).

The GHG inventory is the accounting of GHG emissions associated with the operations of an entity, a process, or a product, which alternatively is referred to as a carbon footprint. When calculating the total GHG emissions or the carbon footprint, it is essential to be aware of current emissions associated with the specific entity, process, or product for which the sustainability is being analyzed. Additionally, it is important to use the present values of the GWP that have been updated for major GHG gases (methane and N_2O) presented in the recent editions of the IPCC reports (Table 8.2). The carbon footprint calculation is very useful in helping to predict the future emissions and identifying major emission sources. The data may then be used to plan, design, implement, and monitor mitigation strategies to reduce respective carbon footprints.

8.3 Measuring Carbon Footprint

The determination of the carbon footprint (Scrucca et al. 2021) involves several steps as described below.

8.3.1 Define the Scope of Your Inventory

The first step is to establish the objective and purpose of measuring the carbon footprint. This facilitates the definition of the limits of the system to be evaluated. Once the limits of the system are defined, all the sources of GHGs must be listed and accounted. The limits of the system must also include the unit and the time period to be used for accounting of the GHG emissions. The unit could be a specific product, such as the manufacturing of a pair of shoes or a car, or it could be the generation of electricity in a power plant; in this case, the carbon footprint could be based on a specific unit of energy (kWh). The unit can also be a manufacturing process, such as the manufacturing of canned food, or an entity, such as an apartment building, residential subdivision, or a university, for a specific period of time.

8.3.2 Measure Emissions and Establish a Baseline

Once the limits of the system are defined, the next step is to identify the sources of GHGs and account for the emissions from each source. In this process, three groups of sources can be identified; these sources are usually referred to as Scope 1 through 3:

- *Scope 1* – Direct GHG emissions from the organization, originating from equipment, machinery, or processes controlled by the entity (e.g. boilers, furnaces, company-owned vehicles) at the location where the activity occurs.
- *Scope 2* – Indirect GHG emissions from the organization, associated with the generation of energy that is purchased for use in our system, e.g. emissions for purchased electricity, steam, heat, or cooling.
- *Scope 3* – Other indirect GHG emissions from the organization, associated with the value chain, resulting from its activities but coming from sources that are not owned or controlled by the entity. This "Scope 3" is further divided into "Upstream" (purchased goods and services, employee travel through fossil fuel-consuming means, etc.) versus "Downstream" (transportation and distribution, processing and use of products, end-of-life treatment, etc.).

To calculate the carbon footprint, the GHG emissions must be accounted for all the sources considered. This always must include Scopes 1 and 2, whereas Scope 3 is optional. The GHG emissions are determined using appropriate emission factors, for example, the amount of CO_2, CH_4, and/or N_2O for the combustion of fossil fuels (Table 8.3). Vehicular GHG emissions can be accounted for using the emission factors in Table 8.4, although the emissions may also be determined for the type of vehicle and number of passengers. The emission factor for electricity use must be determined locally since the total gas emissions largely depend on the type of energy used in its production (power plant, nuclear, hydroelectric, photovoltaic, etc.). Finally, all GHG emissions must be accounted in terms of CO_2e using the GWP in Table 8.2 to obtain the total carbon footprint expressed in CO_2e.

8.3.3 Develop Targets and Strategies to Reduce Emissions

The determination of the carbon footprint not only allows the calculation of the total GHG emission, but it also assists in identifying where these emissions are produced by analyzing the data of each GHG source. An action plan may be designed using this data to manage the emissions. This plan is often known as a Climate Change Action Plan. Carbon footprint reduction targets can either be absolute (e.g. a 10% reduction relative to the chosen baseline year) or intensity-based (e.g. a percentage reduction per unit produced or per full-time employee (FTE) of a services organization).

8.3.4 Offset Unavoidable Emissions

Some sources of GHG are difficult to reduce; these are usually associated with the production of energy. To compensate for these emissions, organizations can then purchase carbon credits (or "offsets") to compensate for any unavoidable emissions. Offsets are derived from projects that are specifically implemented to avoid emissions. The carbon footprint reduction due to those projects can be used to compensate for unavoidable emissions.

Table 8.3 GHG emission factor for stationary combustion.

Fuel type	Heating value (mmBtu per short ton)	CO_2 factor (kg CO_2 per mmBtu)	CH_4 factor (g CH_4 per mmBtu)	N_2O factor (g N_2O per mmBtu)	CO_2 factor (kg CO_2 per short ton)	CH_4 factor (g CH_4 per short ton)	N_2O factor (g N_2Oper short ton)
Coal and coke							
Anthracite coal	25.09	103.69	11	1.6	2602	276	40
Bituminous coal	24.93	93.28	11	1.6	2325	274	40
Sub-bituminous coal	17.25	97.17	11	1.6	1676	190	28
Lignite coal	14.21	97.72	11	1.6	1389	156	23
Fossil fuel-derived fuels (solid)							
Municipal solid waste	9.95	90.70	32	4.2	902	318	42
Plastics	38.00	75.00	32	4.2	2850	1216	160
Tires	28.00	85.97	32	4.2	2407	896	118
Biomass fuels (solid)							
Agricultural byproducts	8.25	118.17	32	4.2	975	264	35
Solid byproducts	10.39	105.51	32	4.2	1096	332	44
Wood and wood residuals	17.48	93.80	7.2	3.6	1640	126	63
	(mmBtu per scf)	(kg CO_2 per mmBtu)	(g CH_4 per mmBtu)	(g N_2O per mmBtu)	(kg CO_2 per scf)	(g CH_4 per scf)	(g N_2O per scf)
Natural gas							
Natural gas	0.001026	53.06	1.0	0.10	0.054 44	0.001 03	0.000 10
Biomass fuels (gaseous)							
Landfill gas	0.000485	52.07	3.2	0.63	0.025254	0.001552	0.000306
Other biomass gases	0.000655	52.07	3.2	0.63	0.034106	0.002096	0.000413

(Continued)

Table 8.3 (Continued)

	(mmBtu per gallon)	(kg CO_2 per mmBtu)	(g CH_4 per mmBtu)	(g N_2O per mmBtu)	(kg CO_2 per gallon)	(g CH_4 per gallon)	(g N_2O per gallon)
Petroleum products							
Aviation gasoline	0.120	69.25	3.0	0.60	8.31	0.36	0.07
Butane	0.103	64.77	3.0	0.60	6.67	0.31	0.06
Ethane	0.068	59.60	3.0	0.60	4.05	0.20	0.04
Motor gasoline	0.125	70.22	3.0	0.60	8.78	0.38	0.08
Naphtha (<401 °F)	0.125	68.02	3.0	0.60	8.50	0.38	0.08
Residual fuel oil no. 5	0.140	72.93	3.0	0.60	10.21	0.42	0.08
Special naphtha	0.125	72.34	3.0	0.60	9.04	0.38	0.08
Still gas	0.143	66.72	3.0	0.60	9.54	0.43	0.09
Biomass fuels (liquid)							
Biodiesel (100%)	0.128	73.84	1.1	0.11	9.45	0.14	0.01
Ethanol (100%)	0.084	68.44	1.1	0.11	5.75	0.09	0.01
Rendered animal fat	0.125	71.06	1.1	0.11	8.88	0.14	0.01
Vegetable oil	0.120	81.55	1.1	0.11	9.79	0.13	0.01

Source: USEPA (2015)/United States Environmental Protection Agency Public domain.

Table 8.4 GHG emission factors for mobile combustion.

Fuel type	CO_2 (kg/gal)
Aviation gasoline	8.31
Biodiesel (100%)	9.45
Diesel fuel	10.21
Ethanol (100%)	5.75
Kerosene-type jet fuel	9.75
Liquefied natural gas (LNG)	4.46
Liquefied petroleum gases (LPG)	5.68
Motor gasoline	8.78
Residual fuel oil	11.27

Source: USEPA (2015)/United States Environmental Protection Agency Public domain.

8.3.5 Independent Verification

The final step in the determination of carbon footprint reduction efforts is the verification of the process. The verification or certification is typically completed by an external organization. The external certification increases the credibility of the process, enhances stakeholder trust, and increases senior management confidence. This external verification benefits the public corporate image of the company or the organization and may be used in the marketing of the company's products or services to customers.

8.4 Standards for Calculating the Carbon Footprint

Various standard methods have been developed for carbon footprint analysis. However, it is important to use recognized standards and approaches when calculating carbon footprint so that the results can be compared with other entities, process, or products. If different standards are used, the results may vary and an accurate, direct comparison may not be possible.

One of the first developed standards is the GHG Protocol (ghgprotocol.org), a corporate accounting and reporting standard issued by the World Business Council for Sustainable Development and the World Resources Institute. This was the most widely used international accounting tool for government and businesses to quantify and manage GHG emissions. The first edition of this standard method was published in 2001. Later, in 2006, the International Organization for Standardization (ISO) developed ISO 14064-I: Specification with Guidance at the Organization Level for Quantification and Reporting of GHG Emissions and Removals. Currently, ISO 14064 is widely accepted in the determination of the carbon footprint (ISO 2018, 2019a,b).

ISO 14064 is organized into three parts. Part 1 provides guidance for organizational-level specifications regarding the quantification and reporting of GHG emissions and reductions. Part 2 includes specifications for guidance at the project level for quantification, monitoring, and reporting of GHG reductions or removal enhancements. Finally, Part 3 includes specifications and guidance for the validation and verification of assertions reflecting quantification, monitoring, and reporting activities.

The International Auditing and Assurance Standards Board (IAASB) is an independent standard-setting body that has developed an alternative standard for the evaluation of GHG emissions and carbon footprint estimation. The standard, called the *International Standard on Assurance Engagements*, was designed for engagements other than audits or reviews of historical financial information. Other methods and standards for GHG inventory analysis and carbon footprint analysis include the *Attest Engagements on GHG Information* developed by the American Institute of CPAs, the EU ETS (Emissions Trading System) in Europe, and the DEFRA guidelines in the United Kingdom.

The existence of several estimation standards is the result of multiple organizations or countries having moved to develop or adopt a standard to normalize GHG emissions and carbon footprint estimation in the absence of a widely recognized standard. Currently, ISO 14064 is considered the most appropriate, favored standard, although other standards are still preferred in specific countries or sectors. The ongoing use of several standards can create difficulty when side-by-side analyses or comparisons are desired; however, multiple standards/frameworks may be used for GHG and carbon footprint analyses to provide a check and confirm the results of the primary analysis.

8.5 GHG Inventory: Developments in the United States

The United States issued Executive Order (EO) 13514, "Federal Leadership in Environmental, Energy, and Economic Performance," which was signed by President Obama on 5 October 2009. EO 13514 specifically outlined how the United States Federal government was planning to approach the problem of GHG emissions. The GHG inventory in EO 13514 was organized in three scopes. Scope 1 includes GHG emissions from sources owned or controlled by a Federal agency. Scope 2 includes GHG emissions resulting from the generation of electricity, heat, or steam purchased by a Federal agency. Scope 3 includes GHG emissions from sources not owned or directly controlled by a Federal agency but related to agency's activities, such as vendor supply chains, delivery services, and employee travel and commuting.

The first step in the application of EO 13514 was the development of goals for GHG emissions reduction by the agency. Each agency was directed to prepare a comprehensive GHG inventory and, based on the results, was asked to define percentage reduction targets for Scope 1, 2, and 3 emissions with respect to 2008 emissions. EPA developed the tool Development of Agency Reduction Targets (DARTs) for agencies to use for the calculation of Scope 1 and 2 GHG emissions. The Federal Energy Management Program (FEMP) developed the Annual GHG and Sustainability Data Report, a tool for agencies to use for developing their comprehensive GHG inventories.

8.6 USEPA: Greenhouse Gas Reporting Program

The USEPA issued the Mandatory Reporting of GHGs Rule (74 FR 56260) in response to the FY2008 Consolidated Appropriations Act (H.R. 2764; Public Law 110-161). This rule entitled the large sources and generators of GHGs in the United States to report GHG data and other relevant information to USEPA. The purpose of the rule is to have an accurate inventory and timely data of GHG emissions to undertake future policy decisions. The rule is referred to as 40 CFR Part 98: Greenhouse Gas Reporting Program (GHGRP). This rule establishes mandatory annual reporting of GHG emissions for those suppliers or facilities that meet at least one of the following conditions: suppliers of certain products that would result in GHG emissions if released, combusted, or oxidized; facilities that inject CO_2 underground for sequestration or otherwise; and facilities that emit greater than 25 000 metric tons per year of GHGs.

The emissions for the year 2010 were reported to USEPA via the electronic GHG reporting tool electronic greenhouse gas reporting tool (e-GGRT) in September 2011. USEPA made the GHGRP data for the first year of the reporting period to be publicly available through their data publication tool in January 2012 and has been continuously uploading the annual emissions data ever since. This data is available at ghgdata.epa.gov/ghgp.

8.7 Tools for GHG Inventory

There are several tools and software programs available for the quantitative determination or estimates of GHGs/carbon footprint. Life cycle assessment (LCA) includes the inventory of GHGs in its procedure, but there are other specific software tools developed for the calculation of carbon footprint as well as online-based calculators and simple carbon footprint spreadsheets as listed below.

- *Climate leadership in parks tool (CLIP)* – A tool that can be used to estimate GHG emission from sources in the parks such as vehicle combustion and wastewater treatment. It aids in promoting ways and creating awareness among the users to reduce the emissions from parks.
- *Comparable emissions database (CEDB)* – A tool that calculates potential emissions due to moving of passengers and freight from a wide range of vehicles across all transportation modes.
- *Employee commute emissions calculator* – This tool helps users measure emissions from commuting of employees via different transportation modes.
- *EPA's e-GGRT* – A tool used for reporting of the GHG emissions under the GHGRP.
- *Durable goods calculator (DGC)* – This tool calculates emissions and benefits from energy saving from different disposal methods for disposal of 14 typical durable goods.
- *Waste reduction model (WARM)* – A tool to estimate the GHG emissions impacts of the waste management practices (e.g. landfilling, recycling, incineration, and composting).
- *Events emissions calculator* – A tool that can be used to estimate the GHG emissions from events based on the length, the commute of the people attending the event, modes of transportation, etc.
- *Fleet emissions calculator* – A tool used to measure the emissions from the fleet using the information on the type of vehicle and the distance traveled.
- *GHG protocol calculation tools and guidance* – A tool to estimate indirect emissions from the service sector (e.g. banks and hospitals) and office-based organizations.
- *GHG inventory guidance, calculator, and inventory management plan for small- and medium-sized organizations and service sector businesses* – A tool used to estimate emission from low-emitting businesses (e.g. service-industry company with 1000 branch offices, a company with one small manufacturing plant, and an owner of a single office building).
- *MOtor Vehicle Emission Simulator (MOVES)* – A tool to estimate emissions from on-road and nonroad mobile sources.
- *National recycling coalition's (NRC) environmental benefits calculator* – A tool that helps in determining the GHG and energy benefits from current waste disposal practices.
- *Northeast recycling council's environmental benefits calculator* – A tool used to calculate the environmental benefits based on waste management practices.
- *Recycled Content (ReCon) tool* – A tool used to estimate life cycle energy and GHG impacts of purchasing or manufacturing specified materials. The tool also calculates GHGs and energy benefits based on recycled content in those materials.
- *Shipping emissions calculator* – A tool used to measure emissions from the shipment of items based on the distance of shipping, mode of shipping, and weight of items.
- *Office footprint calculator™* – A tool used to calculate emissions from an individual's office-related activity (e.g. energy use and waste stream) and the resulting CO_2 emissions from it.
- *Optimization model for reducing emissions of GHGs from automobiles (OMEGA)* – A tool used to estimate the cost of technology for automobile manufacturers to achieve variable fleet-wide levels of vehicle GHG emissions.
- *Paper use emissions calculator* – A tool used to estimate emissions from paper use.
- *Personal GHG emissions calculator* – A tool used to estimate a person's or a family's GHG emissions from their daily activities.
- *GREET model: The GHGs, regulated emissions, and energy use in transportation model* – A life cycle modeling tool developed by Argonne National Laboratory to evaluate fuel combustion of various vehicles on a fuel-cycle/vehicle-cycle basis. The tool can also be used to evaluate total fuel consumption, energy use, GHG emissions, and emission of criteria pollutants.

8.8 Carbon Footprint Case Study

The University of Vigo has been measuring its carbon footprint since 2008 to assess the consumption of resources and the impact of the organization on the environment (UVigo 2024). The objective is not only to achieve a more sustainable organization but also to raise awareness within the university community and in society, promoting more sustainable practices in the everyday activities and ensuring that the University of Vigo is in a continuous process of environmental improvement.

The carbon footprint is calculated using the standard ISO 14064. It classifies GHG emissions into three scopes:

- *Scope 1* – Direct GHG emissions from the organization, e.g. heating systems, company-owned vehicles, etc. at the location where the activity occurs.
- *Scope 2* – Indirect GHG emissions from the organization, associated with electricity, steam, heat... from "non-renewable" sources (e.g. electricity consumption).
- *Scope 3* – Other indirect GHG emissions from the organization, but coming from sources that are not owned or controlled by the entity, e.g. water consumption, waste management and treatment, faculty, staff and student commuting, etc.

The carbon footprint is measured in tons of CO_2e, allowing for a comparison of emissions of various GHGs based on their GWP over a specified period, typically 100 years (Table 8.1).

The first step in determining the carbon footprint is to identify the sources of GHG and evaluate the emissions over a period of the year in a study. The sources considered at the UVigo were:

- Diesel fuel/heating oil consumption (heating system)
- Natural gas consumption (heating system)
- Biomass consumption (new heating system)
- Campus buildings (emissions derived from the construction of the University buildings and facilities).
- Electric power consumption
- Mobility of staff, faculty and students (Scope 3)
- Water consumption
- Paper consumption
- Waste generation.

The resulting emissions for each GHG can be calculated with Eq. (8.1):

$$A \times F_g = E_g \tag{8.1}$$

where

E_g = resulting emissions of GHGs
F_g = emission factor of GHGs
A = quantification of an activity in units.

As an example, the value of A could be the amount of natural gas consumed in a year (expressed in MWh), the cubic meters of diesel fuel/heating oil consumed, the consumption of electricity (kWh), or the distance traveled by personnel (faculty and staff) and students.

To calculate the actual GHG emission for CO_2, CH_4, and N_2O for the natural gas consumption, Eqs. (8.2–8.4) can be used (GHG Protocol 2024):

$$CO_2 \text{ emissions (kg/year)} = \text{annual natural gas consumption (MWh)} \times \text{emission factor } (182 \text{ kg } CO_2/\text{MWh}) \tag{8.2}$$

$$CH_4 \text{ emissions (kg/year)} = \text{annual natural gas consumption } \left(\text{m}^3\right) \times \text{emission factor } (0.018\text{kg } CH_4/\text{MWh}) \tag{8.3}$$

$$N_2O \text{ emissions (kg/year)} = \text{annual natural gas consumption } \left(\text{m}^3\right) \times \text{emission factor } (0.00036 \text{ kg } N_2O/\text{MWh}) \tag{8.4}$$

where

Annual natural gas consumption is the amount of natural gas combusted during the year from company records or directly measured by a natural gas flow meter (expressed in MWh).

The total emissions of each gas (carbon dioxide, methane, N_2O, SF6, HFCs, and PFCs), multiplied by respective GWPs (Table 8.1), are expressed in CO_2e. The total sum of these emissions represents the total carbon footprint of UVigo determined for a one-year period. The total GHG emissions, or carbon footprint, can be calculated using Eq. (8.5):

$$\text{Total GHG emissions } (CO_2e) = \sum_g E_g \cdot \text{GWP}_g \tag{8.5}$$

In calculating the carbon footprint, the UVigo decided to consider only CO_2 emissions, disregarding the possible emissions of CH_4 and N_2O associated with fuel combustion because they were negligible in the final value of the carbon footprint. Other GHG gases were not emitted or considered in the determination of the carbon footprint. The total amount of CO_2 emitted was determined, evaluating the sources of GHG multiplying by the emission factor. Table 8.5 lists the emission factors for the GHG sources identified in the University for 2023 for Scopes 1 and 2. The different nature and management of wastes resulted in the selection of different emission factors for each type of waste generated at the UVigo. The mobility of personnel (faculty and staff) and students was considered for evaluating Scope 3. A survey among all the members of the university community allows for the determination of the distance travelled and the vehicle used (bus, train, car, motorcycle). From these data and the emissions of each type of vehicle, the CO_2 emissions associated with the mobility were determined. Finally, the forested area in the campus acts as a sink for CO_2. The 83.1 ha of forest in the Campus were estimated: for 2023, a CO_2 absorption biocapacity of 359.8 tons of CO_2e/year, which corresponds to a total of 100.75 gha (global hectares in the ecological footprint) (Herva et al. 2011). The sink of CO_2 in the campus forested area was considered in the final reporting of the carbon footprint.

Figure 8.1 illustrates UVigo's carbon footprint for Scopes 1, 2, and 3 between 2008 and 2023. For a period of 11 years, from 2008 to 2018, the university's carbon footprint was around 23 000 tons per year, equivalent to 1 ton CO_2e per person. To reduce this footprint, UVigo implemented an ambitious emissions reduction program, making significant changes to lighting and heating systems, and installing renewable energy sources. Within just 5 years (2019–2023), this initiative led to a 26% decrease in the carbon footprint.

As Figure 8.1 shows, Scopes 1 and 2, directly impacted by the emissions reduction program, experienced a notable decline in the last 5 years, from 2019 to 2023. Scope 3, primarily influenced by staff and student transportation habits, demonstrated a smaller reduction, attributed to increased use of public transport and low/zero-emission vehicles. Key initiatives to reduce the carbon footprint

Table 8.5 Emission factors in calculating the CO_2e emissions from the sources identified at the University of Vigo in 2023.

GHG source	Emission factor	
University buildings	1500	g CO_2e/m^2 yr
Electric power consumption	273	g CO_2e/kWh
Natural gas consumption	182	g CO_2e/kWh
Diesel fuel consumption	2721	g CO_2e/l
Biomass consumption	1474	g CO_2e/kg
Water consumption	84	g CO_2e/m^3
Paper consumption	1840	g CO_2e/kg
Municipal waste		kg CO_2e/kg
Organic	0.59	
Packaging	0.196	
Paper and cardboard	0.0564	
Glass	0.21	
Batteries	0.05	
Hazardous waste and WEEE		kg CO_2e/kg
Halogenated solvents	0.0983	
Non-halogenated solvents	0.0983	
Aqueous solutions	0.116	
Acidic solutions	0.116	
Alkaline solutions	0.116	
Chromic mixture	0.116	
Contaminated solids	0.164	
Special	2.03	
Biohazardous	2.03	
Cytotoxic	2.03	
Oils	0.126	
Batteries	0.138	
Empty containers	0.078	
Photographic (developer+fixer)	0.116	
Paints and varnishes	0.164	
Pressurized containers	0.078	
WEEE	0.0375	

included generating electricity through 1000 kW of solar panels, with an additional 600 kW planned. Building lighting was improved by replacing old fixtures with LEDs. Heating systems in several buildings were upgraded from the traditional diesel/heating oil system to less carbon-intensive and polluting fuels (e.g. natural gas) or renewable energy sources, including biomass or geothermal systems. These measures resulted in a 13.7% decrease in external electricity

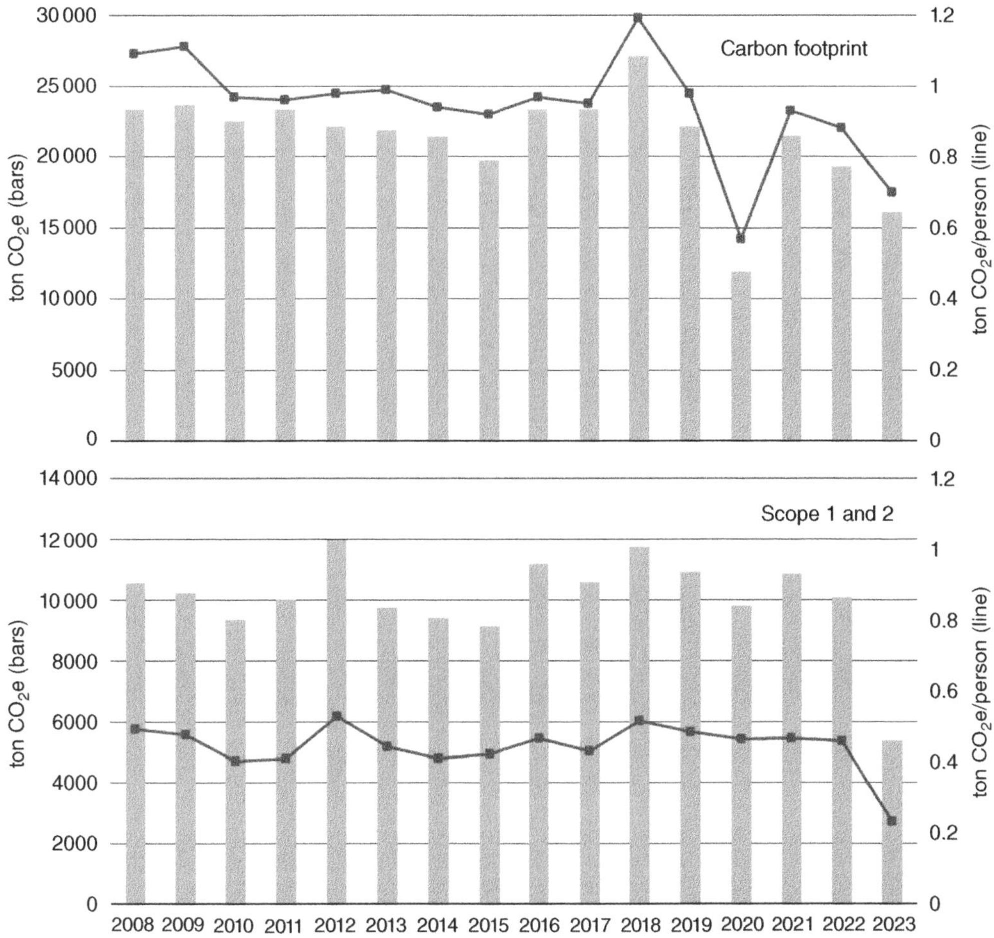

Figure 8.1 Carbon footprint for the UVigo from 2008 to 2023, as per standard ISO 14064. *Source:* Adapted from UVigo (2024).

consumption and a 70% reduction in diesel consumption. Consequently, the carbon footprint per person dropped from 1.19 tons CO_2e/person in 2018 to 0.88 tons CO_2e/person in 2023 (Figure 8.2 and Table 8.6).

The carbon footprint can be converted into ecological footprint determining the land area required to absorb the aforementioned GHG emissions. This metric calculates the amount of standard land measured in global hectares. The ecological footprint is defined as the ecologically productive land area or aquatic ecosystem (crops, pastures, forests, or aquatic ecosystems, also including the volume of air) necessary to maintain and satisfy the consumption of resources and energy used, as well as to absorb and assimilate the waste derived from a specific activity (Wackernagel and Rees 1998, 2016). The methodology to transform the carbon footprint into the ecological footprint uses a conversion factor of 0.28 gha/ton CO_2e as reported in the Global Ecological Footprint Factor: GFN Canadian 2005 tables (WRI & WBCSD 2004). Figure 8.3 shows the ecological footprint of the UVigo from 2008 to 2023.

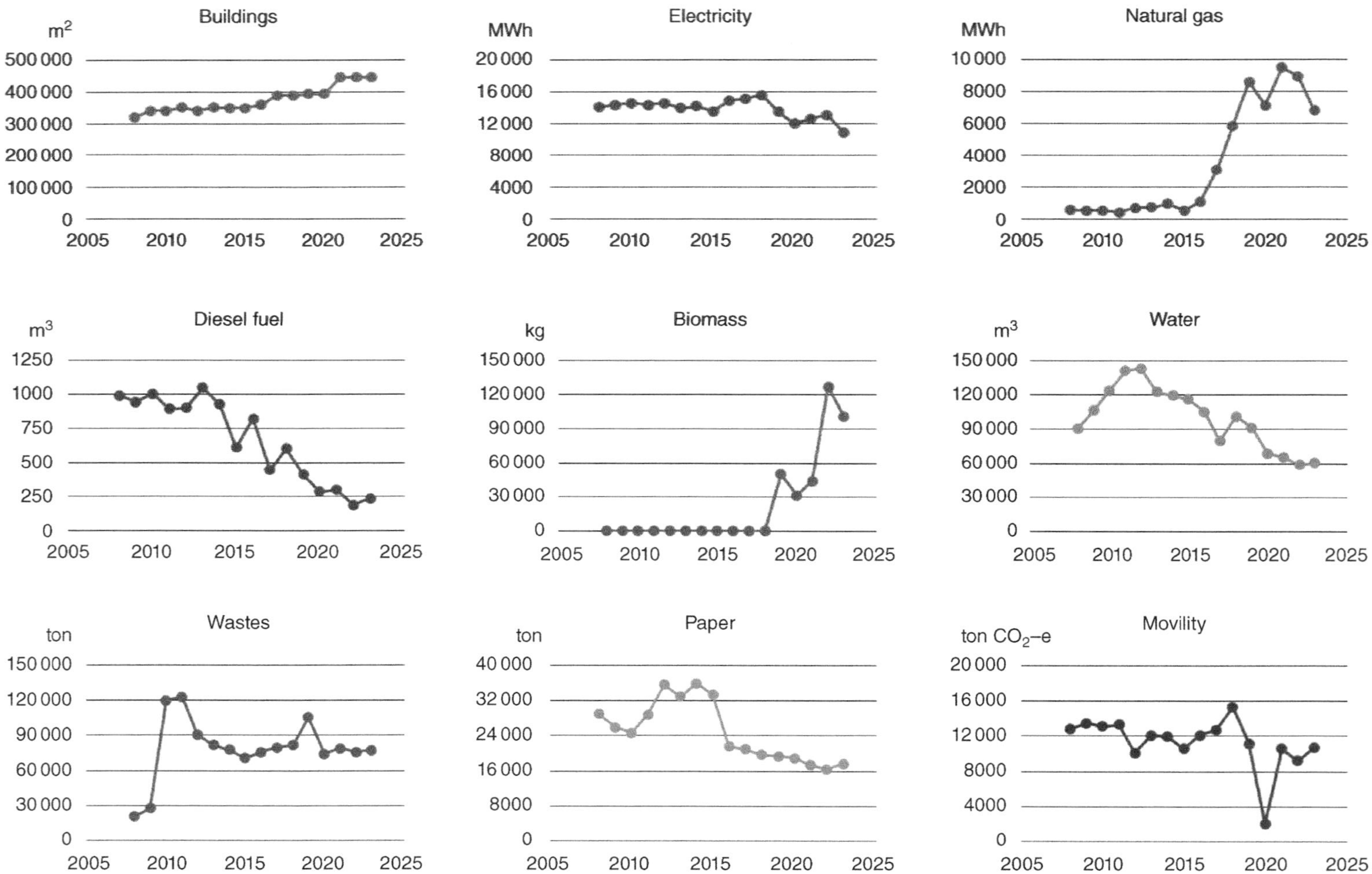

Figure 8.2 Sources of GHG included in the determination of the carbon footprint in the UVigo from 2018 to 2023. *Source:* Adapted from UVigo (2024).

Table 8.6 Generation of renewable energy at the University of Vigo.

Year	Biomass (kWh)	Photovoltaic (kWh)	Thermosolar (kWh)	Geothermal (kWh)
2019	200 000	10 000		800 000
2020	175 000	675 000		825 000
2021	562 000	925 000		875 000
2022	603 000	1 029 000	18 000	976 000
2023	575 000	1 258 000	18 000	948 000

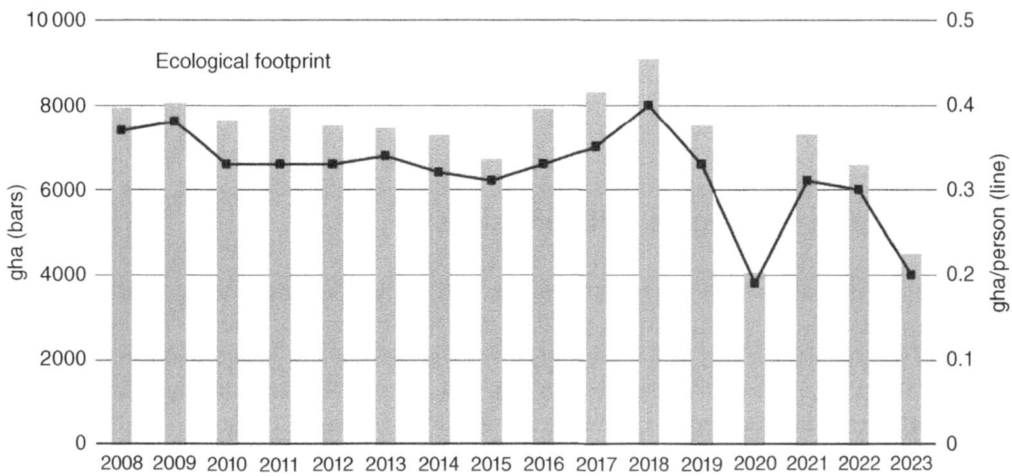

Figure 8.3 Ecological footprint for the UVigo from 2008 to 2023. *Source:* Adapted from UVigo (2024).

Example 8.1

Germany's GHG emissions for the year 1990 are as follows:

- Methane (CH_4): 5,682 Gg
- Nitrous oxide (N_2O): 226 Gg
- Carbon dioxide (CO_2): 1,014,155 Gg
- Hydrofluorocarbons (HFC-134a): 2,500 Gg

1) Calculate the total CO_2 equivalent emissions for Germany for the year in question.
2) Determine the percentage contribution of each gas to the total CO_2 equivalent emissions.

Solution:

The global warming potentials of different GHGs in CO_2e are as follows:
Methane: GWP = 28
Nitrous oxide: GWP = 265
Hydroflurocarbons: GWP = 1300

The CO_2 equivalent emissions can then be calculated as:
Methane = 5682 Gg × 28 = 159 096 Gg
Nitrous oxide = 226 Gg × 265 = 59 890 Gg
Hydroflurocarbons = 2 500 Gg × 1 300 = 3 250 000 Gg

Total CO_2e emissions
= 159 096 Gg + 59 890 Gg + 3 250 000 Gg + 1 041 155 Gg
= 4 510 141 Gg

Contribution of each gas to the overall GHG emission are calculated as follows:

GHG	Percentage contribution
Methane	3.5%
Nitrous oxide	1.3%
Hydrofluorocarbons	72.1%
Carbon dioxide	23.1%

8.9 Programs to Mitigate GHG Emissions

Organizations have developed several programs to mitigate GHG emissions and their effects on global warming and climate change (Dixit et al. 2016). The production of energy is one of the main sources of GHG emissions. Any action in energy saving and efficiency is going to reduce GHG emissions, but the main objective to achieve a significant reduction in GHG emissions is to change from a fossil fuel energy-based system to renewable energy with minimum GHG generation (Bölük and Mert 2014). Other potential actions for reducing GHG emissions include better practices in livestock management and agriculture activities. Cities are also a major source of GHG emissions mainly due to transportation and heating systems and electricity consumption, hence many cities try to implement action plans to mitigate the GHG generation locally (Gurney et al. 2021). The following are some selected programs aimed at mitigating GHG emissions:

1) AgSTAR Program
2) California Climate Action Registry
3) Climate Change Technology Program (CCTP)
4) Climate Friendly Parks
5) Energy Star
6) Global Climate Change Monitoring
7) Intergovernmental Panel on Climate Change
8) NCAnet
9) Pew Center on Global Climate Change
10) SF6 Emission Reduction Partnership for Electric Power Systems
11) US Global Change Research Program (USGCRP)
12) World Resource Institute (WRI).

Key negative impacts of CO_2 emissions to the atmosphere include global warming, climate change, and sea level rise. Climate models predict that the mitigation of the GHG emissions from industrial, power, and transportation sectors alone is not enough to combat climate change.

Figure 8.4 Carbon dioxide removal processes and technologies. *Source:* USDOE (2024)/US Department of Energy/Public domain.

Additional actions should consider the following aspects: (i) development of clean technologies to create decarbonized energy and industrial systems for the future; (ii) removal of significant amounts of CO_2 from the atmosphere to counterbalance the emissions from sectors difficult to decarbonize, such as agriculture and shipping, with the objective to achieve a zero net balance of carbon in the atmosphere; and (iii) removal of legacy CO_2 emissions from the atmosphere and oceans to stop or even reverse the negative impacts of climate change. It is estimated that gigatons of CO_2 must be removed from atmosphere and oceans per year until 2050 to induce meaningful reductions in CO_2 in the atmosphere and the resulting effects on climate change. To put this in perspective, 1 gigaton of CO_2 is about 40% of EU, 20% of USA, 35% of India, or 8% of China emissions, respectively, in 2023

Carbon dioxide removal includes a group of processes and technologies that can remove CO_2 from the oceans and atmosphere. The CO_2 is then stored and sequestered in geological or ocean reservoirs, but it also can be stored in bio-based materials or value-added products, such as low-carbon concrete. The available processes or technologies for carbon sequestration and storage are as follows (Figure 8.4):

1) Direct air capture with storage: removal of carbon dioxide directly from the atmosphere and storing it underground.
2) Soil carbon sequestration: the process of capturing and storing atmospheric carbon dioxide in the soil by growing photosynthetic plants and accumulation of the decomposed biomass in soil.
3) Biomass carbon removal and storage: capture of carbon dioxide from the atmosphere as biomass (plants), converting it into a usable product, and then storing the carbon in a durable form.
4) Enhanced mineralization: acceleration of the natural process of carbon dioxide absorption by minerals.
5) Ocean-based carbon dioxide removal: a group of processes and technologies to enhance and accelerate the ocean's natural ability to absorb and store carbon dioxide from the atmosphere.
6) Afforestation/reforestation: a nature-based process to store carbon from the atmosphere in newly planted trees.

Example 8.2

A technology firm has committed to becoming carbon neutral by 2040. The company's baseline carbon footprint in the year 2020 was 100 000 tons of CO_2e, encompassing Scope 1, 2, and 3 emissions. The firm aims to set a Science-Based Target (SBT) for 2030 that aligns with the 1.5°C pathway, which requires a 4.2% annual reduction in emissions.

- Determine the total percentage reduction required by 2030 to align with the 1.5 °C pathway.
- Calculate the total emissions target for 2030 based on the required percentage reduction.

- Develop an annual reduction strategy that outlines the percentage reduction needed each year to meet the 2030 target, considering an exponential decay model. Plot and show the decay with time.

Solution:

Total Percentage Reduction Required by 2030

Given a 4.2% annual reduction over 10 years (2020 to 2030), we first calculate the cumulative reduction over this period. Using the cumulative reduction percentage, we calculate the target emissions for 2030.

The total percentage reduction required by 2030 to align with the 1.5 °C pathway is:

$$\%\, reduction = 100 - \left(1 - \frac{r}{100}\right)^{10} \times 100 = 100 - \left(1 - \frac{4.2}{100}\right)^{10} \times 100 = 34.89\%$$

The company has to reduce it's emissions by 34.89% by 2030.

Calculate the Total Emissions Target for 2030

Emissions target for 2030 = 100 000 × (1−0.3489) = 65 111 tons CO_2e.

The annual reduction in absolute emissions is visualized below:

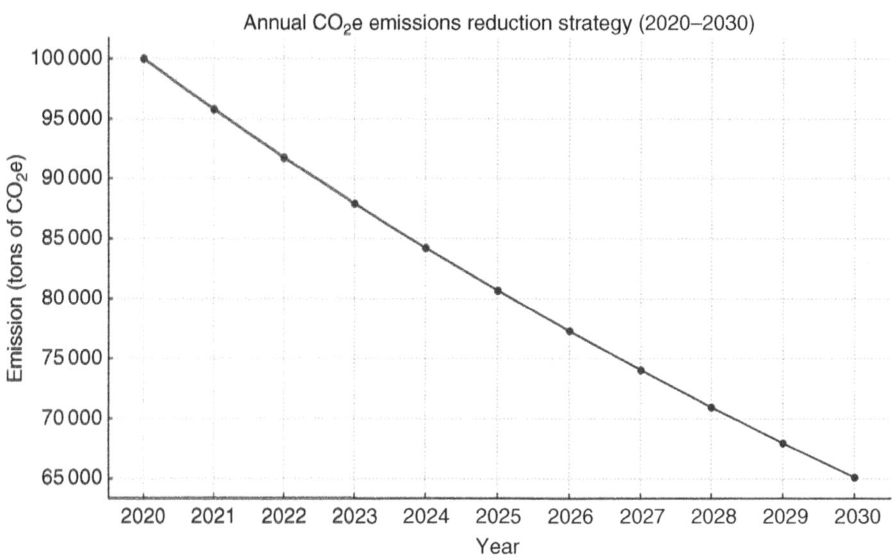

8.10 Climate Action Plans

Company or organization actions to achieve carbon neutrality involve a comprehensive approach that are commonly presented in climate action plans. These plans commonly address five key areas: GHG emissions reduction, zero waste, water efficiency, biodiversity enhancement, and transformative scholarship. Carbon neutrality, or net-zero GHG emissions, may be achieved by both reducing and offsetting emissions generated from operations, which include energy use, transportation, and travel (Table 8.7).

Table 8.7 Activities and strategies to achieve net-zero emissions.

Energy efficiency	Prioritize energy-saving measures. Adopt energy-efficient systems.
Renewable energy	Install onsite solar panels. Offsite power purchase agreements. Use heat pumps or geothermal energy sources.
Fleet electrification	Promote sustainable commuting options. Transition to low-emission or electric vehicles.
Zero waste	Reduce waste at the source through procurement and use decisions. Implement recycling and reuse programs. Use composting for organic waste.
Net-zero water	Reduce water consumption. Maximize the use of non-potable water. Stormwater management solutions. Upgrade water fixtures.
Biodiversity	Developing native ecosystems. Improve soil quality. Use permeable surfaces. Create a resilient landscape (mitigate heat islands).

One of the primary strategies to reach net-zero emissions is improving **energy efficiency** across an organization. Common approaches include upgrading existing lighting to LED systems and installing automated controls to ensure lights turn off when spaces are unoccupied. In addition, upgrades to more efficient HVAC systems, such as heat pumps or ground-source heat exchange systems, and switching to cleaner fuels (natural gas could be an alternative to other more contaminating fuels) can significantly reduce emissions. Adoption of sustainable building practices and standards, such as improved insulation and energy-efficient windows, further supports energy efficiency in both new constructions and retrofits. Finally, implementing modern, energy-efficient technologies helps reduce overall energy consumption and emissions.

Renewable energy sources play a crucial role in reducing an organization's reliance on fossil fuels. Onsite solar installations provide a direct source of renewable electricity. Additionally, entering into power purchase agreements (PPAs) allows an organization to source renewable energy from external providers that can substantially lower organizational emissions.

Reducing **transportation-related emissions** may include the encouragement or provision of transit-focused commuting options and fleet electrification. Encouraging the use of public transportation by offering incentives, promoting carpooling, biking, and walking, and gradually transitioning the organization vehicle fleet to low-emission or electric vehicles all contribute to lower transportation emissions. By supporting these methods, organizations can significantly reduce Scope 3 emissions associated with transportation to and from organization facilities.

Other actions for carbon neutrality include

Establishing a **zero-waste** organization involves reducing materials use, reducing waste generation, and implementing comprehensive waste management practices. Programs focused on minimizing waste at the source, including procurement policies that favor materials and supplies that minimize packaging, increasing recycling rates, and composting organic waste help the organization move toward zero waste. Reuse initiatives, like surplus programs and donation systems, further support waste reduction by providing alternative options for disposal.

Water conservation and effective stormwater management are also critical components of carbon neutrality. Reducing water consumption through high-efficiency fixtures, prioritizing non-potable water for secondary uses, and collecting stormwater for applications like irrigation and toilet flushing all reduce the demand for domestic water sources. Developing stormwater systems that capture, store, and repurpose rainwater also mitigates runoff volume and rate, helping to alleviate the burden on surrounding water infrastructure and runoff-based pollution and erosion.

Enhancing **biodiversity** in organization facilities not only supports local ecosystems but also improves environmental resilience. Developing green spaces that feature native plants, improve soil quality, and increase tree canopy coverage all contribute to healthy, biodiverse facilities. Replacing impermeable surfaces, such as asphalt, with permeable materials allows for groundwater recharge and reduces runoff. Additionally, a thoughtful landscape design can help mitigate the urban heat island effect, creating a more comfortable and sustainable environment for the community.

Through these coordinated strategies, any organization can make significant strides toward carbon neutrality, fostering a culture that values sustainability and contributes to global efforts to reduce GHG emissions.

8.11 Carbon Markets: Credits/Offsets

The Kyoto Protocol (1997) and the Paris Agreement (2015) are international accords aimed at reducing GHG emissions to limit global warming and mitigate the negative impacts of climate change. These agreements have increased pressure on governments, companies, and organizations to shrink their carbon footprint. In this context, carbon credits and carbon offsets emerged as practical tools to help achieve reductions in GHG emissions (Carbon Credits 2024).

Carbon credits are permits allowing a company to emit a specific amount of GHGs, typically one credit equating to one ton of CO_2 equivalent (CO_2e). Through carbon credits, governments set emission limits for companies and organizations, creating a financial incentive to reduce emissions over time. If a company emits less than its assigned limit, it can trade and/or sell excess credits in carbon markets, effectively monetizing its carbon emissions reduction efforts.

Carbon offsets, on the other hand, represent actions taken to reduce GHG emissions in one place to compensate for emissions generated elsewhere. Often used voluntarily, carbon offsets allow companies to compensate for their emissions by investing in projects like reforestation, renewable energy, or direct CO_2 capture. The goal of carbon offsets is to fund projects that reduce or remove an amount of GHGs equivalent to the company's emissions.

While both carbon credits and carbon offsets aim to reduce total GHG emissions, they operate differently. Carbon credits are government-issued allowances that companies can trade if they emit less than their assigned limit. Carbon offsets involve compensating for emissions by financing reduction projects, which can also be traded in carbon markets to offset emissions from other companies that have exceeded their targets.

8.12 Summary

A GHG inventory is an accounting of GHG emissions associated with the operations of the entity, process, or product. A carbon footprint is the total GHGs for an entity, process, or product expressed in CO_2 equivalents. Regulations and policies as well as calculation tools are evolving, and it is essential to calculate GHG emissions/carbon footprints to estimate current emissions, predict future emissions, and identify major emission sources. Carbon footprint analysis also facilitates planning, design, implementation, and monitoring of mitigation strategies. Numerous GHG inventory calculators are available in the public domain. The semiquantitative and quantitative outputs of these calculators may be used to identify mitigation initiatives for GHG emission reductions. Several of these initiatives have been adopted into comprehensive frameworks that have been promulgated by a wide variety of local, state, and Federal agencies.

8.13 Questions

8.1 What do you mean by carbon footprint?

8.2 What are the steps involved in the determination of carbon footprint? Explain briefly.

8.3 What are the three groups of sources of greenhouse gas emissions? Explain briefly.

8.4 What is the USEPA's greenhouse gas reporting program? What does it entail?

8.5 List any five tools used for determining the carbon footprint for different applications. Explain any two of the tools in detail.

8.6 What is a climate change action plan?

8.7 Name some of the programs aimed at mitigating the greenhouse gas emissions. Explain any two programs in detail.

8.14 Problems

8.1 A manufacturing company is assessing its carbon footprint to identify areas for sustainability improvements. The company's operations for a particular year resulted in the following emissions and energy consumptions:

- Electricity consumption: 5 000 000 kWh, with a grid emission factor of 0.7 kg CO_2e/kWh
- Natural gas usage: 500 000 cubic meters
- Company vehicles: Consumed 200 000 liters of gasoline
- Air travel for business: 1 000 000 km flown, with an average emission factor of 0.15 kg CO_2e/km per passenger
- Waste sent to landfill: 1500 tons, with an emission factor of 0.5 kg CO_2e/kg of waste.

Using the given data, answer the following:

1) Find the GHG emission factor for gasoline and natural gas and report the values with references.
2) Calculate the total CO_2e emissions for the company for the year in question.
3) Determine the percentage contribution of each source to the total CO_2e emissions. Draw a pie chart showing the contribution.
4) Identify and categorize these emissions into scope-wise emissions (Scope 1/2/3).
5) Calculate the scope-wise emissions of the company in the given year.

8.2 A multinational corporation is conducting a comprehensive carbon footprint analysis across its operations, supply chain, and product lifecycle. The following data have been collected for the fiscal year:

- Natural gas consumption – 1 000 000 MMBtu
- Company fleet fuel (diesel) – 500 000 gal
- Electricity purchased – 750 000 MWh
- Heating (purchased steam) – 200 000 MMBtu
- Business travel – 10 000 000 miles (air travel, average short-haul)
- Employee commuting – 5 000 000 miles (average car travel)
- Waste generated – 25 000 tons (non-recycled)
- Purchased goods and services – $500 000 000 (spend-based method, average emission factor of 0.1 kg CO_2e per dollar spent)

Emission Factors:

- Natural gas – 53.06 kg CO_2e/MMBtu
- Diesel – 10.21 kg CO_2e/gal
- Electricity – 0.4 kg CO_2e/kWh
- Steam – 50 kg CO_2e/MMBtu
- Air travel (short-haul) – 0.15 kg CO_2e/mile
- Car travel – 0.4 kg CO_2e/mile
- Waste – 1.2 kg CO_2e/kg

Using the given data, calculate the following:

1) Define Scope 1/2/3 emissions and categorize the emissions given under each scope.
2) Calculate the total CO_2e emissions for each scope and the overall carbon footprint for the fiscal year.
3) Determine the percentage contribution of each category within each scope to its total emissions; draw a pie chart showing category-wise contributions.

References

Bölük, G. and Mert, M. (2014). Fossil & renewable energy consumption, GHGs (greenhouse gases) and economic growth: evidence from a panel of EU (European Union) countries. *Energy* 74: 439–446.

Carbon Credits (2024). *The Ultimate Guide to Understanding Carbon Credits.* https://carboncredits.com/the-ultimate-guide-to-understanding-carbon-credits/ (accessed 12 November 2024).

Dixit, M.K., Culp, C.H., Fernandez-Solis, J.L., and Lavy, S. (2016). Reducing carbon footprint of facilities using a facility management approach. *Facilities* 34 (3/4): 247–259.

GHG Protocol (2024). *Calculation Tools and Guidance.* Greenhouse gas protocol. https://ghgprotocol.org/calculation-tools-and-guidance (accessed 8 August 2024).

Gurney, K.R., Liang, J., Roest, G. et al. (2021). Under-reporting of greenhouse gas emissions in US cities. *Nature Communications* 12 (1): 553.

Herva, M., Álvarez, A., and Roca, E. (2011). Sustainable and safe design of footwear integrating ecological footprint and risk criteria. *Journal of Hazardous Materials* 192 (3): 1876–1881.

IPCC (1995). IPCC second assessment. In: *Climate Change 1995.* A Report of the Intergovernmental Panel on Climate Change. IPCC.

IPCC (2007). Climate change 2007: synthesis report. In: *Contribution of Working Groups I, II and III to the Fourth Assessment Report of the Intergovernmental Panel on Climate Change* (eds. Core Writing Team, R.K. Pachauri, and A. Reisinger), 104. Geneva, Switzerland: IPCC.

IPCC (2014). Climate change 2014: synthesis report. In: *Contribution of Working Groups I, II and III to the Fifth Assessment Report of the Intergovernmental Panel on Climate Change*, 87 pp. Geneva, Switzerland: IPCC. https://archive.ipcc.ch/report/ar5/syr/.

IPCC (2023). Climate change 2023: synthesis report. In: *Contribution of Working Groups I, II and III to the Sixth Assessment Report of the Intergovernmental Panel on Climate Change* (eds. Core Writing Team, H. Lee, and J. Romero), 184 pp. Geneva, Switzerland: IPCC. https://doi.org/10.59327/IPCC/AR6-9789291691647.

ISO (2018). Greenhouse gases. Part 1: Specification with guidance at the organization level for quantification and reporting of greenhouse gas emissions and removals. Norm ISO 14064-1:2018. https://www.iso.org/standard/66453.html (accessed 29 October 2025).

ISO (2019a). Greenhouse gases. Part 2: Specification with guidance at the project level for quantification, monitoring and reporting of greenhouse gas emission reductions or removal enhancements. Norm ISO 14064-2:2019. https://www.iso.org/standard/66454.html (accessed 29 October 2025).

ISO (2019b). Greenhouse gases. Part 3: Specification with guidance for the verification and validation of greenhouse gas statements. Norm ISO 14064-3:2019. https://www.iso.org/standard/66455.html (accessed 29 October 2025).

Scrucca, F., Barberio, G., Fantin, V. et al. (2021). Carbon footprint: concept, methodology and calculation. In: *Carbon Footprint Case Studies. Environmental Footprints and Eco-design of Products and Processes* (ed. S.S. Muthu). Singapore: Springer. https://doi.org/10.1007/978-981-15-9577-6_1.

USDOE (2024). *Carbon Negative Shot. Office of Fossil Energy and Carbon Management.* US Office of Energy. https://www.energy.gov/fecm/carbon-negative-shot (accessed 8 August 2024).

USEPA (2015). *Emission Factors for Greenhouse Gas Inventories.* EPA Center for Corporate Climate Leadership. https://www.epa.gov/sites/production/files/2015-12/documents/emission-factors_nov_2015.pdf (accessed 8 March 2025).

UVigo (2024). Carbon Footprint and Ecological Footprint. Office of the Environment. University of Vigo. https://oma.webs.uvigo.es/Sostibilidade/CambioClimatico/PegadaEcoloxica.php# (accessed 6 August 2024).

USEPA (2025). *Global Greenhouse Gas Overview. United States Environmental Protection Agency.* Available at: https://www.epa.gov/ghgemissions/global-greenhouse-gas-overview.

Herring, D. (2020). Are humans causing or contributing to global warming? Climate.gov, Science and Information for Climate-Smart Nation, National Oceanic and Atmospheric Administration (NOAA). Available at: https://www.climate.gov/news-features/climate-qa/are-humans-causing-or-contributing-global-warming (accessed 3 July 2025)

WRI and WBCSD (2004). *The Greenhouse Gas Protocol, A Corporate Accounting and Reporting Standard,* Revisede. Washington, DC. USA: World Resources Institute and World Business Council for Sustainable Development ISBN 1-56973-568-9.

Wackernagel, M. and Rees, W. (1998). *Our ecological footprint: reducing human impact on the earth.* B.C., Canada: New society publishers.

Kyoto Protocol (1997). UNFCCC. Available online: http://unfccc.int/kyoto_protocol/items/2830.phpunfccc.int/resource/docs/convkp/kpeng.pdf (accessed 20 March 2025)

Paris Agreement (2015). UNFCCC. Available online: https://unfccc.int/sites/default/files/english_paris_agreement.pdf (accessed 20 March 2025)

Wackernagel, M. and Rees, W. (2016). *Our ecological footprint.* Macat International Limited.

9

Life Cycle Assessment

Chapter Goal and Objectives
The objective of this chapter is to provide a comprehensive overview of Life Cycle Assessment (LCA), detailing its methodology, tools, and applications while emphasizing its relevance in evaluating environmental impacts throughout the life cycle of products or systems. The specific objectives include the following: (i) define LCA and explain its importance in environmental analysis; (ii) outline the historical development and procedural framework of LCA; (iii) describe the four key phases of the LCA methodology: goal and scope definition, life cycle inventory (LCI), life cycle impact assessment (LCIA), and results interpretation; and (iv) present commonly used LCA tools and their practical applications.

9.1 Introduction

Life cycle assessment, or life cycle analysis (LCA), is a methodology developed to measure the impact of products, processes, or services on the environment, allowing for a measurement of the environmental sustainability of such products, processes, or services. LCA is now increasingly being used to assess engineering designs and/or processes and to facilitate the selection of more environmentally sustainable options. Before we delve into the specifics of LCA, let us discuss an example of sustainability measurement as it relates to alternative products. Consider a coffee cup – specifically, three product alternatives – a polyfoam cup, a coated paper cup, and a ceramic mug. Which type of cup do you think is more sustainable? At first glance, most people would argue that the ceramic cup is the most sustainable choice. Intuitively, one may think that the ceramic cup can be reused, while the polyfoam cup and the coated paper cup are most often disposed after one use. Further, many would argue that the coated paper cup is more sustainable than the polyfoam cup – it is generated from renewable sources (trees and paper), whereas the polyfoam cup is generated from nonrenewable petroleum products.

However, the purpose of a comprehensive sustainability assessment such as LCA is to look past initial intuition. Let us take a deeper look at several parameters associated with the life cycle of these three products. When comparing the processing water effluent, the manufacturing of a typical polyfoam cup generates approximately half of the effluent as compared to a ceramic cup; however, the fabrication of a polyfoam cup results in twice as many air emissions as those for a ceramic cup.

Sustainable and Resilient Engineering: Drivers, Metrics, Tools, and Applications, Second Edition. Krishna R. Reddy, Claudio Cameselle, and Jeffrey A. Adams.

Companion website: www.wiley.com/go/reddy/sustainableresilientengineering2e

A ceramic cup requires significantly more energy input for manufacturing than paper or polyfoam cups. Further, polyfoam cups are relatively easier to recycle as compared to coated paper cups, and it is nearly impossible to recycle a ceramic cup. Of course, a ceramic cup may be used hundreds, if not thousands of times and may be repurposed, as opposed to the typical single-use polyfoam and paper cups. Reuse does require washing, which in turn consumes energy (hot water generation) and effluent (wash water, detergents, etc.).

As you can see, once these pre-use, "operation," and post-use factors are incorporated, the answer as to the relative degree of sustainability considering these three options becomes unclear. This is the motivating factor behind conducting an LCA, with the goal of performing a structured and comprehensive environmental assessment of a product, process, or activity during its life cycle. A life cycle refers to the major activities in the course of the product's lifespan from its manufacture, use, and maintenance to its final disposal, including the raw material acquisition required for manufacturing of the product (USEPA 2006).

This chapter provides (i) various definitions of LCA, (ii) a brief history of LCA, and (iii) a systematic standardized framework for performing LCA. In addition, challenges encountered in different phases of LCA are also highlighted. Several software packages are available to perform LCA, but SimaPro software developed by PRé Consultants in the Netherlands is the most commonly used software to perform LCA. This chapter presents the main features of SimaPro, followed by a simple example highlighting the LCA methodology and software application.

9.2 Life Cycle Assessment

9.2.1 Definition and Objective

There are several definitions for LCA. According to the Society of Environmental Toxicology and Chemistry (SETAC), LCA is defined as follows:

> *"a process to evaluate the environmental burdens associated with a product, process, or activity by identifying and quantifying energy and materials used and wastes released to the environment; to assess the impact of those energy and materials used and releases on the environment; and to evaluate and implement opportunities to effect environmental improvements. The assessment includes the entire life cycle of the product, process, or activity, encompassing, extracting, and processing raw materials; manufacturing, transportation, and distribution; use, re-use, maintenance, recycling, and final disposal" (Consoli* 1993*).*

Specifically, the performance and completion of LCAs attempt to satisfy several specific goals with respect to the environment:

- The assessments serve to ultimately reduce or minimize the generation and magnitude of pollution as well as the conservation of nonrenewable resources.
- LCAs may be used to differentiate the impacts between two comparable products, as well as to assess design options for the same product. In doing so, there is a focus on attempting to maximize recycling of materials and the reduction of waste materials, which in turn can help preserve and conserve ecological systems.
- Further, by identifying processes or alternative products, LCAs can help drive the development or utilization of cleaner technologies.
- Finally, when unavoidable side effects are identified and quantified, measures can be implemented to abate or prevent environmental pollution.

LCA studies focus on the potential impacts that occur throughout a particular activity's life cycle – beginning with the acquisition of raw materials that form the activity's/product's inputs all the way through final disposal. This approach, referred to as "cradle to grave," considers all the environmental impacts that occur, including the use of natural resources, human health, and ecological consequences. Specifically, the side effects of production, use, and disposal, including related depletion or exhaustion of input resources, emissions, effluents, and wastes that could have a negative impact on the environment has to be properly considered. It is also useful to identify where specifically in a product or in an activity's life cycle that actions may be applied to target reductions in the environmental impact. Figure 9.1 presents a schematic that depicts the interaction of inputs and outputs with respect to these human activities.

9.2.2 Procedure

LCA determines the environmental impacts of products, processes, or services, through production, usage, and disposal (Hauschild et al. 2018). Several key steps with respect to a product are included in an LCA analysis. Figure 9.2 depicts the flow of these key LCA activities as well as the interaction between them and the inputs and outputs to be considered in the LCA process. These include the following:

- *Raw materials extraction and harvesting* – This includes emissions and waste generated during the raw material extraction and harvesting processes.
- *Materials processing* – Refinement of the materials into some stable, usable form.
- *Manufacturing of the product* – This includes energy and material inputs, as well as waste generated by the manufacturing process, and by the operation of equipment to generate the product.
- *Use of the product* – An appropriate usage timeline/service life, and the corresponding energy inputs and waste generation should be considered.
- *Disposal and/or recovery* – This includes impacts related to disposal, incineration, as well as other potential environmental impacts related to these tasks.
- *Re-manufacturing/recycling* – If a product is to be reformulated or repurposed, this includes all energy inputs and effluent/emissions outputs to get the product into a usable form by recycling.

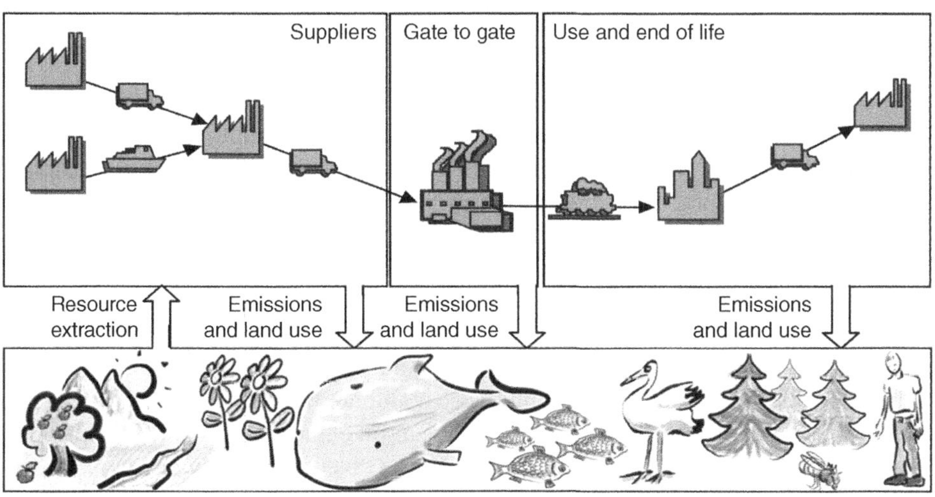

Figure 9.1 Environmental impacts of a product throughout a product life cycle. *Source:* With permission from Raul Carlsson, Swedish Research Institutes (RISE), Sweden 2018.

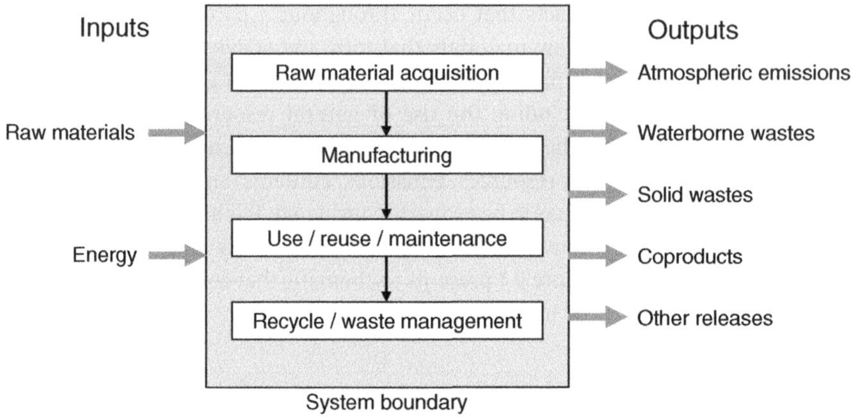

Figure 9.2 Inputs and outputs and stages in the life cycle of a product. *Source:* USEPA (2006)/Public domain.

- *Transportation* – All transportation activities, along with energy inputs and emissions between the above activities should also be accounted for.

As shown, inputs such as materials, energy, water, and air are considered in an LCA. The activities that utilize these inputs include materials acquisition; formulation, processing, and manufacturing; product distribution; product use; recycling of products, components, and materials; and waste management. Outputs that are considered include principal products, co-products, water effluents, airborne emissions, solid waste, and other environmental interactions.

Assessing the environmental burdens of a product, process, or service can be daunting. The conventional approach to completing an LCA is through the use of a process-based LCA method. In a process-based LCA, the inputs and outputs into a specific product or an activity are itemized. While performing a comprehensive LCA, it is important to perform several key steps, including a compilation of relevant inputs and outputs, evaluating potential environmental impacts, and interpreting the results in relation to the objectives of the particular study.

9.2.3 History

The origin of LCA can be dated in the early 1960s (Table 9.1), when concerns about the rapid depletion of fossil fuels and other natural raw materials arose. Industries began to account for energy requirements of processes and products, with the ultimate objective of reducing energy consumption wherever possible without loss of product quality. The LCA idea rapidly developed, and in 1963, the first LCA report about the energy requirements in chemical intermediates production was published. Then, in 1969, the Coca-Cola Company conducted an LCA study to determine the impact of the different beverage containers with the goal of identifying and adopting the most beneficial form. The Coca-Cola LCA procedure was later refined by the USEPA to create the first attempt of LCA standardization: the so-called "Resource and Environmental Profile Analysis."

During the 1980s, numerous attempts were undertaken around the world to refine and expand the LCA methodology to any field or activity. In 1992, as a result of the United Nations Conference on Environment and Development (the "Rio Summit"), the ISO 1400 standards evolved to establish a standard protocol for sustainable development. After some modifications (Table 9.2 and Figure 9.3), the present versions of the ISO standards for LCA were issued in 2006.

Table 9.1 History of life cycle assessment.

Early 1960s	Concerns about the rapid depletion of fossil fuels, which sparked interest in finding ways to account energy use and to project future resource supplies and use.
1963	Harold Smith published one of the first LCA report on energy requirements for the production of chemical intermediates at World Energy Conference.
Late 1960s	The studies in *The Limits to Growth* and *A Blueprint for Survival* initiated the trend on predicting finite resources in the world.
1969	The Coca-Cola Company conducted an LCA study on different beverage containers.
1970–1975	USEPA refined The Coca-Cola LCA methodology and created the *Resource and Environmental Profile Analysis.*
Late 1970s–1980s	Environmental concerns shifted to issues of hazardous waste management and solid waste. The life cycle logic was incorporated into emerging method of risk assessment and analyzing the environmental problems.
Late 1980s	A broad base of consultants and researchers across the globe has been further refining and expanding the LCA methodology.
1991	11 State Attorneys General in the USA denounced the use of LCA results to promote products until uniform methodology and consensus were reached on how environmental comparison can be advertised non-deceptively.
1992	ISO 1400 family grew out of ISO's commitment to support the objective of sustainable development discussed at the United Nations Conference on Environment and Development, in Rio de Janeiro.
1993	ISO launched the new technical committee, ISO/TC207, Environmental Management.
1997–2010	ISO developed the LCA standards.

Table 9.2 International Organization for Standardization's (ISO's) LCA and other related standards.

ISO 14040:2006[a)]	Environmental management – life cycle assessment – principles and framework.
ISO 14044:2006[a)]	Environmental management – life cycle assessment – requirements and guidelines.
ISO 14045:2012	Environmental management – eco-efficiency assessment of product systems – principles, requirements and guidelines.
ISO/TR 14047:2012	Environmental management – life cycle assessment – illustrative examples on how to apply ISO 14044 to impact assessment situations.
ISO/TS 14048:2002	Environmental management – life cycle assessment – data documentation format.
ISO/TR 14049:2012	Environmental management – life cycle assessment – illustrative examples on how to apply ISO 14044 to goal and scope definition and inventory analysis.
ISO 14046:2014	Environmental management – water footprint – principles, requirements and guidelines.
ISO/TS 14071:2014	Environmental management – life cycle assessment – critical review processes and reviewer competencies: additional requirements and guidelines to ISO 14044:2006.
ISO/TS 14072:2014	Environmental management – life cycle assessment – requirements and guidelines for organizational life cycle assessment.
ISO/TR 14073:2016	Environmental management – water footprint – illustrative examples on how to apply ISO 14046.

(*Continued*)

Table 9.2 (Continued)

ISO/TS 14074:2022	Environmental management – life cycle assessment – principles, requirements and guidelines for normalization, weighting and interpretation.
ISO/FDIS 14075	Environmental management – principles and framework for social life cycle assessment.
ISO/CD TS 14076	Eco-technoeconomic analyses: principles, requirements and guidelines

[a] Replaced ISO Standards 14 040, 14 041, and 14 042.

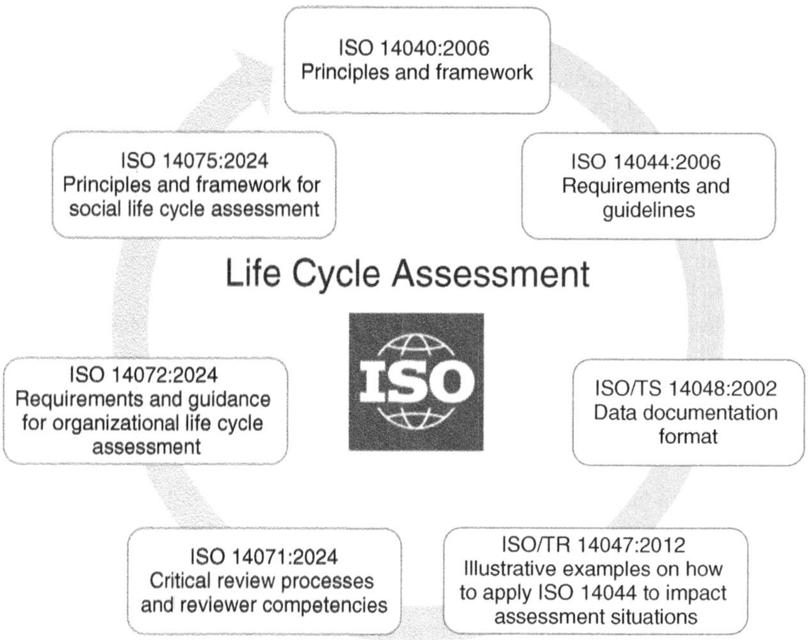

Figure 9.3 ISO standards for life cycle assessment.

9.3 LCA Methodology

The United Nations Environment Program (UNEP) and the ISO, a worldwide federation of national standards bodies, have standardized the LCA framework and guidance within the following standards:

- *ISO 14040:2006, Environmental Management – Life Cycle Assessment – Principles and Framework*: Provides a clear overview of the practice, applications, and limitations of LCA.
- *ISO 14044:2006, Environmental Management – Life Cycle Assessment – Requirements and Guidelines*: Provides guidance on Life Cycle Inventory (LCI) analysis, impact assessment, the interpretation phases of LCA, and data collection and quality issues.

These standard procedures replaced an earlier series of standard procedures on LCA, and the two standards were reviewed and reaffirmed in 2010.

As defined by the ISO standards, the LCA methodology framework consists of the following phases:

- *Goal and scope definition* – Identifying the product, process, or service to be assessed, a functional basis for comparison of alternative products or services being assessed, the required level of detail, and the boundaries, which are being adopted.
- *Inventory analysis* – Identifying the inputs, usually energy and materials, involved in the different stages of the life cycle as well as the outputs, namely the products, liquid, solid, and gaseous emissions/wastes, and land uses required to support the activity. The LCI is typically depicted in a flow chart.
- *Impact assessment* – Identifying qualitatively and, where possible, quantitatively, the effects of resource use, and the emissions generated. These are usually grouped into impact categories that can then be weighted based on significance or importance.
- *Interpretation* – Representing the result in the most informative manner with descriptions of the opportunities to reduce the impacts of the product, process, or service.

Figure 9.4 illustrates an LCA methodology framework that depicts the interaction between these key actions. More details on these various phases are provided in the following sections.

9.3.1 Goal and Scope Definition

The first step of an LCA is to define the scope and goal(s) of the study. LCA studies are commonly used to improve the design of a product or the components of an activity to be more protective of human health and the environment. The studies may also be used to select among a range of products or activities that are meant to achieve a need but where the choice incorporates environmentally protective features (Curran 2016). More specifically, some common goals include the following:

- *Assessment of environmental impacts of current products and activities* – These studies assess the overall burden to environmental health and the environment resulting from all inputs (e.g.

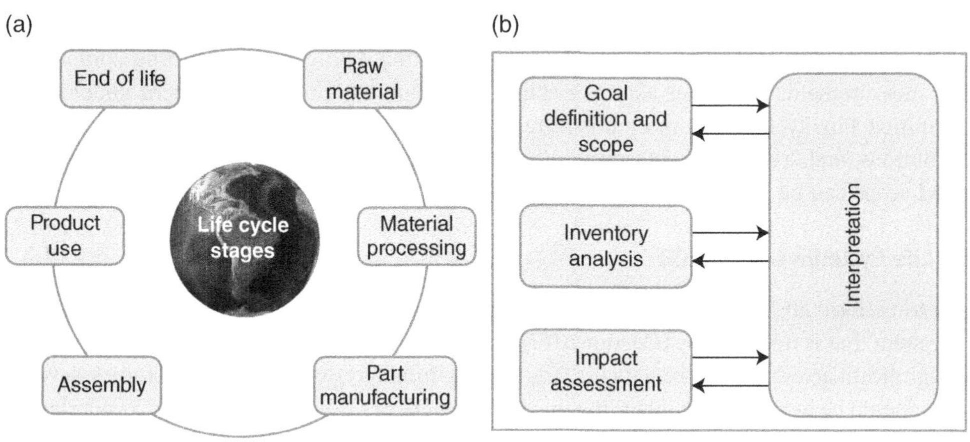

Figure 9.4 Life cycle assessment framework. (a) life cycle stages, (b) methodology.

materials and energy) and outputs (e.g. emissions and waste generation) associated with the manufacturing, use, and disposal of a product, including those associated with materials, transport, and packaging of the product. These studies also may be applied to a specific activity or a process. For the case of activities, baseline assessments may also be undertaken to determine the net effect of changes to the activity in question. In both cases, the information from the LCA can be used to guide the development of completely new products or activities.
- *Assessment of environmental impacts when selecting between several product and activity options* – These studies assess the relative burdens between a range of product or activity options including the "do-nothing" alternative. The benefits and drawbacks of these options are quantitatively assessed such that the overall environmental burdens compared between options may also be combined with social and economic considerations to make an appropriate decision.
- *Support of public policy initiatives and legislation* – Public policies and legislative actions often require the consensus of a range of stakeholders. However, it is often difficult to reach a consensus on meaningful pieces of legislation, as the interests of the stakeholders may be varied and often at odds with each other. LCA studies can often provide conclusive data analysis that may be desired or even necessary to convince groups of the merits of such initiatives. The studies can also be useful to guide the decisions and enforcement of such initiatives.
- *Data gap analysis* – Perhaps ironically or paradoxically, LCA studies can identify data gaps inherent in such studies; identifying areas where little knowledge may exist, or areas where such additional knowledge can prove to be informative with respect to the product, the activity, or the LCA study itself.

When determining the scope and the goal of the LCA, it is also important to assess who the target audience for the LCA may be. If the LCA is meant to support confidential or proprietary development decisions within an organization, proper precautions to be protective of sensitive information such as trade secrets should be taken. For public entity audiences, similar precautions should be considered with respect to sensitive aspects of the study, although it is likely the expectation that such assessments may become public. Further, for suppliers or producers that occupy a specific location or range on an activity or product value chain, it is important to assess upstream or downstream perspectives.

Additionally, it is important to set a range of study expectations when developing the LCA scope and goal. For instance, it should determine if all potential uses for a specific product/activity are to be considered or if a specific range or single use should be considered. Expectations regarding data accuracy, the expected statistical significance, and potential effects of data variance or inaccuracy should be determined at this stage. Procedures for handling data gaps, establishing compatible units of measurement, and other aspects regarding data quality and measurement should also be determined. Finally, feedback mechanisms regarding data quality assurance/control, the impact of assumptions, and "real-time" determinations if the assessed data satisfies the originally stated goals and scope can be implemented at this time.

9.3.2 Life Cycle Inventory (LCI)

In order to perform an LCA, it is critical to properly account for the inputs and outputs associated with a system that is under study (Curran 2012). This task is known as the LCI. The LCI incorporates a significant array of potential data into the study. Ultimately, to facilitate LCA studies, several key tasks should be incorporated, including the development of a flow diagram that incorporates all the key processes that are under study, the development of a data collection plan, the collection of data, and the evaluation of the results of the study, as well as the provision to report the results.

When assessing a system using an LCA, a *system boundary* must be defined first. The boundary serves to illustrate a separation of the system from its surroundings. The boundary encompasses the system, which is defined as a collection of materially and energetically connected operations that perform a defined function. The greater the environment lies to the outside of the system boundary; it acts as the source of all inputs into the system and acts as a sink for all outputs of the system. A schematic depiction of a system, its boundary, and the greater environment is depicted in Figure 9.5.

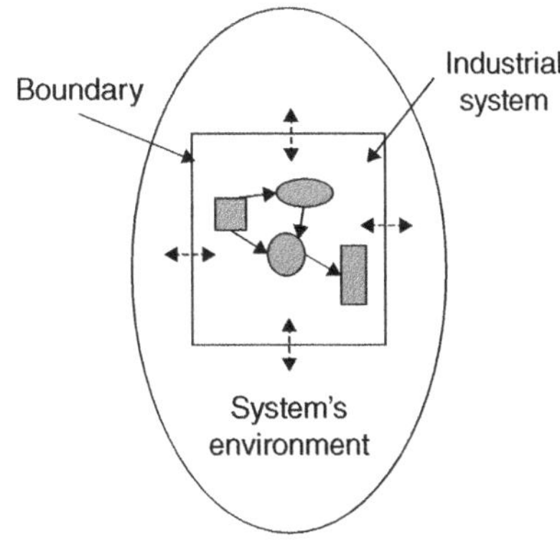

Figure 9.5 Industrial system, its boundaries and the inputs–outputs with the environment.

Inputs (i.e. raw materials, energy) and outputs (i.e. atmospheric emissions, waterborne wastes, solid wastes, coproducts, and other releases) must be properly identified during the inventory analysis. When considering the inputs and outputs, it is important to specifically determine the following:

- Which materials are used by the system?
- What are the quantities of the materials that are consumed?
- How much energy is consumed at various stages?
- What feedstock is used to generate energy?
- What wastes or emissions are produced?
- Are any of the wastes or emissions toxic?

Outputs occur from each stage of the life cycle; this is depicted in Figure 9.6. As shown in the figure, one of the processes or stages of the life cycle is selected; in this case, it is raw material acquisition. As indicated, the extraction of the raw materials may result in primary releases to air, water, or soil. Solids, in turn, may be recycled, incinerated, or landfilled. If materials are incinerated, ash (solid) and emissions (air) will be generated. The air emissions may result in air pollution or

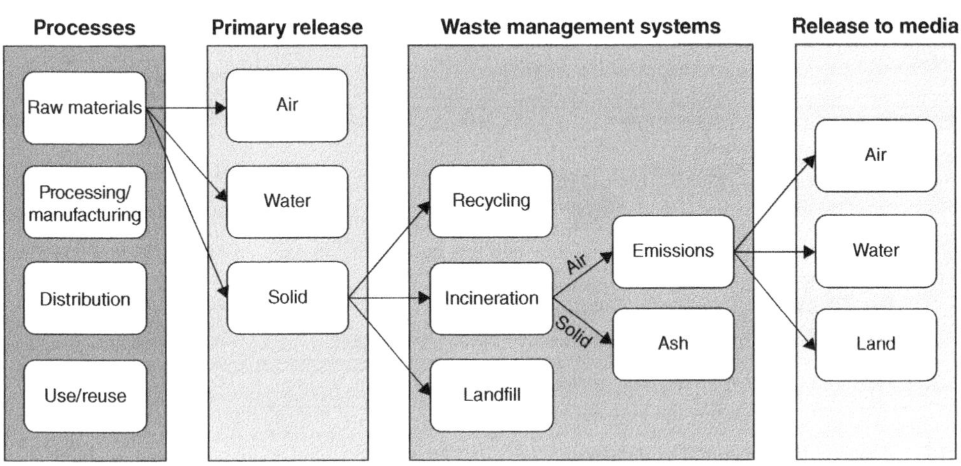

Figure 9.6 Outputs from raw materials.

subsequent deposition from airborne pollutants may be deposited into soil or surface water, leading to the impacts on these media.

An LCI incorporates inputs and outputs to track *material flows within systems*. For example, Manufacturer A produces intermediate materials from raw materials, such as plastic pellets from petroleum feedstock. Manufacturer B produces components from intermediate materials, such as cases for laptops using materials derived from Manufacturer A. Manufacturer C produces components or assembles final products using intermediate materials produced by Manufacturer B, such as laptop computers.

Another concept to be considered in LCI analyses is the *feedback loop*; that is, the reversion of material inputs and associated outputs to generate other inputs. For instance, coal and iron are used to generate steel, which may then be fabricated into a shovel. This shovel, in turn, is used to extract more iron ore and coal to make more shovels. These "reverse" flows should also be accounted for in the LCI.

In addition to inputs and outputs, the *specific stages for the life cycle* are also defined. Some of these include raw materials acquisition and processing, manufacturing, use, maintenance, and reuse, recycling, and waste management. Each of these specific activities requires specific inputs and results in outputs that may lead to environmental impacts. Figure 9.7 depicts an entire life cycle that incorporates successive steps of the raw materials producer, the manufacturer, and the consumer, including final product disposal.

When performing an LCI, it is important to determine an equivalent *unit of function*. For instance, a 12-oz aluminum can, a 20-oz plastic bottle, and a 2-l plastic bottle are all single containers that are used to package soft drinks; however, it is clear to see that the raw materials and resulting waste generation from the usage of these containers are different. Therefore, a functional unit needs to be defined. This allows for all data to be normalized to a baseline unit. Once selected, other similar products may be related to the functional unit, taking into consideration size, time, and intensity scales associated with the inputs and outputs as well as their relative effect on the environment.

The *treatment of data collection* is as important as including inputs, outputs, and system boundaries. It is very important to determine the goals for data quality, which can then be used to determine cost/benefit for accurate data collection, as well as data quality indicators, which serve as a benchmark for determining if the collected data is useful and of a sufficient quality for the study.

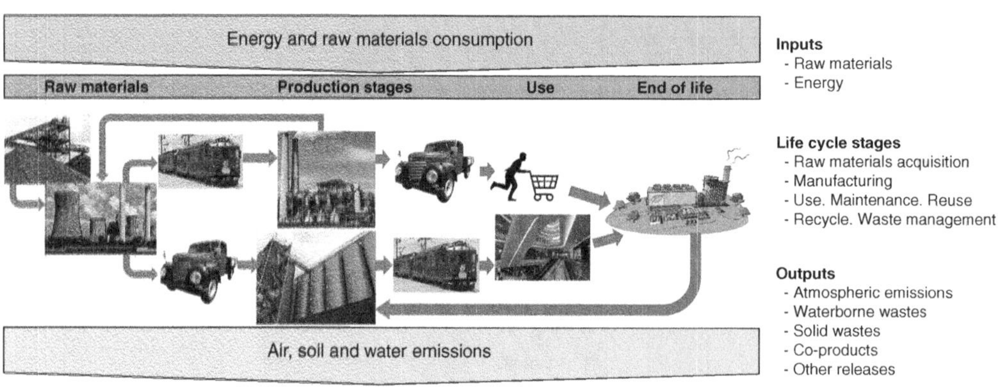

Figure 9.7 Stages of the entire life of a product.

The types and sources of data should be determined as well, including published research data, government statistics, laboratory results, "real-time" equipment sources, and a wide range of publicly available references. The data may be directly sampled, modeled, measured, based on regional or spatial assumptions or estimates, or general assumptions, especially where more refined data measurement or modeling may be difficult, time-consuming, or cost-prohibitive. Further, consideration should be given to a sensitivity analysis such that determinations can be made regarding the overall consequences to the study when data is erroneous, missing, or misrepresented. Finally, these decisions can also be used to determine when minor, secondary, or "background" data may be neglected or eliminated because it is of minor or no consequence to the quality of the overall study.

Finally, the results of the LCI must be analyzed, interpreted, and reported in a *format* that will be useful for the audience or stakeholders to understand its significance. The results may be used to determine direct scores between assessment alternatives or to identify trends, either due to the whole product or activity or associated with a specific input or an aspect associated with the product or activity. The resulting data may also be used to determine if spatial/geographical or temporal trends may exist with respect to the findings of the study. The context of the result with respect to the overall effects on human health or the environment may also be assessed. Clear representation of the data is also critical to demonstrate the findings of the study, whether the data and findings are presented in graphical or text form. At this time, informed decisions regarding product/activity design or selections among several alternatives may be undertaken.

Perhaps intuitively, the results of the LCI may also answer questions regarding the model. If there is a sense that the data and the results gleaned from the study appear to be counterintuitive, misleading, or even inaccurate, the model assumptions, including model boundary, inputs, outputs, and processes may be reconsidered to assess if the model and the underlying assumptions, data, and data collection are accurate. If omissions or errors regarding data or the model itself are identified, changes can be made to these assumptions, and a reassessment may be undertaken.

LCI analyses can be beneficial and powerful; however, there are *several unresolved problems* that should be recognized. First, the definition of a functional unit may be difficult where a product has multiple functions. Next, the system boundary must be selected carefully; if too broad, the LCI process may be unnecessarily complex or lack sensitivity; if too narrow, important inputs and outputs may be erroneously neglected. Additionally, modelers need to use discretion for when to stop adding selection criteria to the inventory analysis, or put differently, what may be neglected from the model and overall effects that need to be considered. Finally, discretion and judgment are essential in choosing how to best allocate environmental burdens.

9.3.3 Life Cycle Impact Assessment (LCIA)

The LCI stage is very important when a comprehensive LCA is performed. However, it is equally important to determine the context of the measured parameters, namely what their relative effects are to human health and the overall environment. These effects are measured and accounted in an LCIA. The objective of an LCIA is to evaluate the impacts on human health and the environment resulting from emissions, resource extraction, and other interventions from human activities and technical systems. To emphasize, an LCIA should address ecological effects and human health effects as well as resource depletion. Figure 9.8 depicts the elements of an LCIA.

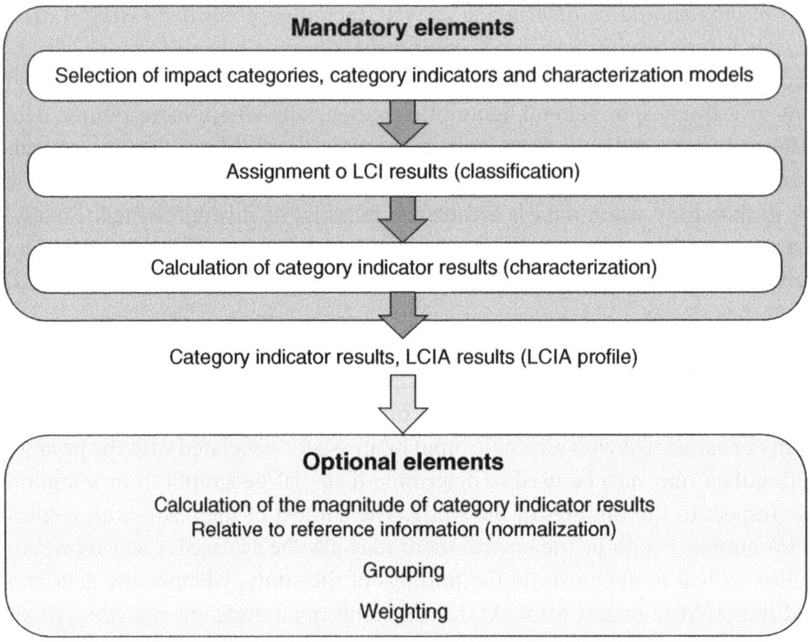

Figure 9.8 Elements of life cycle impact analysis (ISO 14040:2006). *Source:* Hemdi et al. (2009).

An LCIA evaluation aims to assess subjective descriptions of a process to a human activity in objective physical terms. The assessment is meant to determine and assess the relative differences in potential ecological and human health effects for each option associated with the activity. Some of the environmental impacts that are considered include the use of renewable or non-renewable resources, the generation of toxic substances, the release of greenhouse gas emissions or potentially ozone-depleting gases, the generation of liquid and solid wastes, the release of excess nutrients to the environment, and the detriment of land resources through mineral extraction or deforestation.

An LCIA evaluation differs from other evaluation techniques. For example, a *risk assessment* is used to assess rules and regulations about handling chemicals. An *environmental impact assessment* is used to assess the application of rules and regulations pertaining to improvements and enhancements of the built environment. A *cost–benefit analysis* is used to evaluate the meaningfulness of measures (with respect to financial considerations) to decrease environmental impacts. As discussed above, numerous environmental effects may be incorporated into an LCIA. These may be further categorized as local, regional, or global effects:

- On a local basis, some cause-and-effect examples include the generation, combustion, and disposal of chemicals and related chronic and acute health effects; industrial and household processes and the release of related chemicals to the environment; and chemicals in an industrial working environment and their effects with respect to human health and ecosystem functions.
- On a regional basis, some examples include the release of by-product gaseous emissions and the formation of photochemical or particulate smog; the release of chemical effluents and its effects on large bodies or water, including acidification or eutrophication, and the release of chemical compounds to the subsurface with the resulting fouling of surface and groundwater resources.

- On a global level, some relationships include the over exploitation of mineral wealth and the reduction of the potential utilization by future generations; the release of greenhouse gas emissions and their resulting effect on climate change; and the release of ozone-depleting chemicals and their potential effects on the protective ozone layer, resulting in greater surface levels of ultraviolet radiation.

One example of the cause and effect of these relationships is the effects of CFCs/Halons. When these emissions occur, chemical reactions release chlorine and bromine ions. When the *midpoint effect* is considered, these ions act to destroy ozone layer; and a midpoint analysis assesses the resulting depletion of the ozone layer. Taken further, an *endpoint analysis* may be performed with a broader perspective of the environmental impacts. In such analyses, the resulting effects of skin cancer incidents, crop damage, immune system suppression, damage to materials, damage to marine life, and human optical maladies, such as cataracts, are measured.

The US environmental protection agency (USEPA) Science Advisory Committee has attempted to *classify environmental effects* based on several criteria. First, greater weight is placed on larger-scale impacts as compared to smaller-scale impacts. The severity of a hazard is ranked by the relative danger posed; for instance, more toxic substances are of a greater concern than less toxic substances. The potential for exposure is also considered – well-sequestered substances pose less of a threat than easily mobilized substances. Finally, a greater weight or emphasis is applied to the "penalty for being wrong" – incidents that result in greater remediation times are of greater concern than those requiring less remediation times.

Applying actual effects to these categories, the EPA has classified the problems into high, medium, or low risk, as shown by the following examples:

- Relatively high-risk problems include habitat alteration and destruction, species extinction and loss of biological diversity, stratospheric ozone depletion, and global climate change.
- Relatively medium-risk issues include herbicides and pesticides in the environment, toxics, nutrients, biochemical oxygen demand and turbidity in surface waters, acid deposition, and airborne toxics.
- Relatively low-risk issues include oil spills, groundwater pollution, the presence of radionuclides in the environment, acid runoff to surface water, and thermal pollution.

The LCIA has four steps, as depicted in Figure 9.8: classification, characterization, normalization, and valuation (weighing).

Step 1 – Classification: Resource usage and emissions/releases are assigned to respective environmental impacts. It is common for multiple intertwined relationships to exist (e.g. sulfur dioxide emissions resulting in damage to surface water bodies and the built environment due to acid rain), and it is also common for many intertwined cause-and-effect ("stressors" and "impacts") chains.

Step 2 – Characterization: This is accomplished by the quantitative determination of the impacts resulting from the stress indicated by the LCI values by using the characterization equation, which is defined as follows:

$$S_j = \Sigma C_{i,j} \cdot E_i \tag{9.1}$$

where S_j = category stress indicator for category j; $C_{i,j}$ = "characterization factor" for species i and category j (i.e. what level of environmental stress of category j is caused by a unit mass of species i);

and E_i = mass flow identified for species i in the inventory assessment. When performing the characterization, it is critical to apply appropriate weighting to properly capture the impacts associated with inventory data. To accomplish this, it is important that the impact be defined and that methods exist to quantify the contribution of the impact. As presented in Chapter 8, greenhouse gas contributions may be expressed in terms of a global warming potential (GWP). As a base unit, carbon dioxide (and its associated impact is defined as 1; or alternatively, GWP is expressed as CO_2 equivalents. Methane, for instance, has a 28-fold greater effect as a greenhouse gas as compared to carbon dioxide on a molecular basis. Therefore, methane has a GWP of 28.

Step 3 – Normalization: During the normalization process, the impact values resulting from the characterization stage are related to a common set of reference values. During the normalization process, the normalization equation is used, which is defined as follows:

$$N_j = S_j/R_j \tag{9.2}$$

where N_j = normalized indicator; S_j = category stress indicator for category j; and R_j = reference value. Figure 9.9 depicts a graphical representation of the normalization process for two optional product designs for a variety of stress indicators. As shown, the normalized indicators associated with each category stress indicator may be summed to determine the cumulative effect of each option, and to determine the relative effects of different category stress indicators associated with each option.

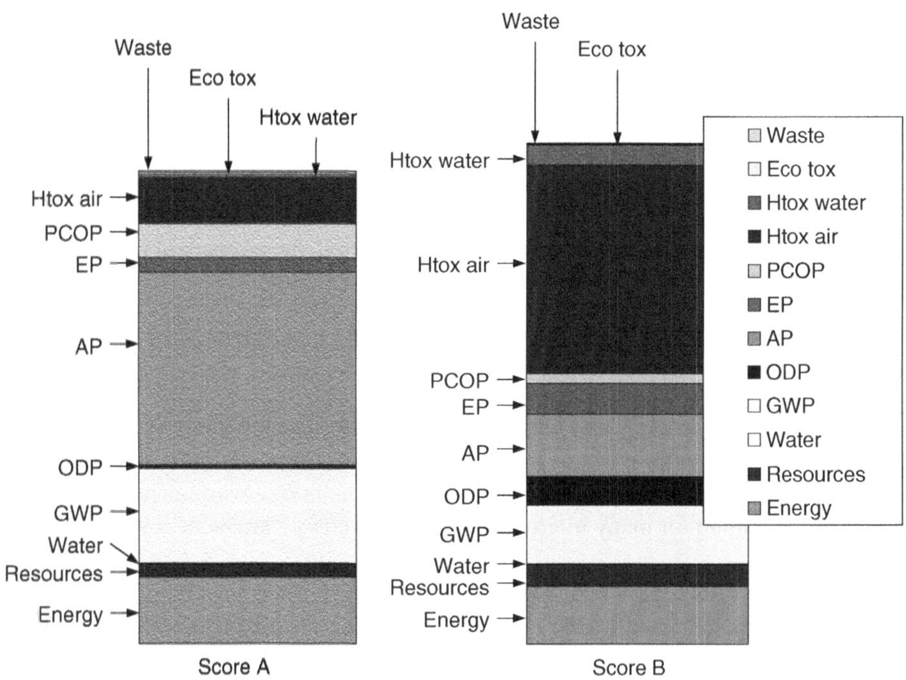

Figure 9.9 Valuation with normalization of two optional product designs.

Step 4 – Valuation process: During the valuation process, weighting factors are assigned to different impact categories based on their perceived relative importance. The valuation equation is applied in this step and is defined as follows:

$$W_j = \Omega_j / N_j \tag{9.3}$$

where W_j = weighted impact indicator; Ω_j = weighting factor; and N_j = normalized indicator. Once the weighted impact indicators have been computed, the overall LCA score may be calculated as follows:

$$I = \Sigma\, W_j \tag{9.4}$$

where I = LCA score; and W_j = weighted impact indicator.

Figure 9.10 summarizes the flow of the LCIA process. As shown in the figure, following the completion of an inventory assessment, associated resource usage and emissions/releases are assigned to respective environmental impacts during the classification stage, and the quantitative impacts resulting from the stress indicated by the LCI values are determined during the characterization stage. The assessed impacts are then normalized, where a valuation assessment is performed. Weighting is applied on a subjective basis, the resulting weighted factors are summed and an LCA score is determined.

As with LCI studies, the results of the LCIA may be assessed to determine if the results appear to be counterintuitive, misleading or erroneous. If so, the parameters may be reconsidered, including several of the somewhat subjective factors used for the quantitative assessments explained above. If

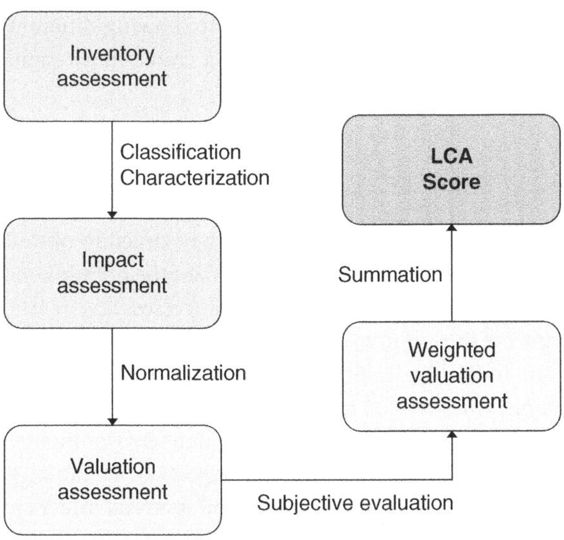

Figure 9.10 The LCIA sequence.

the data appears to be accurate, logical, and appropriate, accurate conclusions may be drawn, and recommendations may be derived and presented. However, it is important to acknowledge the assumptions, subjectivity, and potential limitations of the data due to these factors.

Several issues must be considered with respect to the use of LCIAs:

- LCIAs are subject to the heuristics inherent in predefined software methodologies. There could be differences in the results using two different software programs.
- There are effects associated with spatial variation. For example, impacts of manufacturing 1 kg of copper could be different in the United States, China, and Finland.
- There is the issue of capturing local environmental uniqueness, as well as the dynamics of the overall environment. For example, local air pollution emission could affect the nearby receptors non-uniformly depending on the wind direction. Such differences are not accounted; instead, average values are used.
- The weighting process can be quite subjective, resulting in appreciable variability between assessments. Individuals differ in their risk perception of various hazards, thereby deriving an acceptable set of weightings will be difficult.

9.3.4 Interpretation

Following the computational steps associated with the LCIA process, the results are reported and interpreted in a systematic way so that the need and opportunities to reduce the negative impacts of the product or process or service on the environment may be identified and mitigated. Sensitivity and uncertainty analysis results could help make better interpretation of the results leading to better decision-making.

Example 9.1
For a typical residential building, draw a comprehensive life cycle flowchart showing different materials used and different life cycle stages involved. Show boundaries for a cradle to gate and cradle to grave assessment.

Solution:
Boundaries:

1) Cradle to Gate Assessment focuses on the environmental footprint from the extraction of raw materials through to the completion of construction. This analysis emphasizes the energy consumption, resource utilization, and emissions inherent in the production processes. It is instrumental in identifying areas where improvements in resource efficiency can significantly reduce the overall environmental impact during the initial stages of a building's life cycle.
2) Cradle to Grave Assessment, on the other hand, broadens the scope to encapsulate the building's end-of-life phase, including demolition, waste management, and the subsequent environmental implications of disposal, recycling, or reuse. This assessment offers a holistic view of a building's ecological footprint, highlighting the significance of sustainable design and material selection throughout its lifespan.

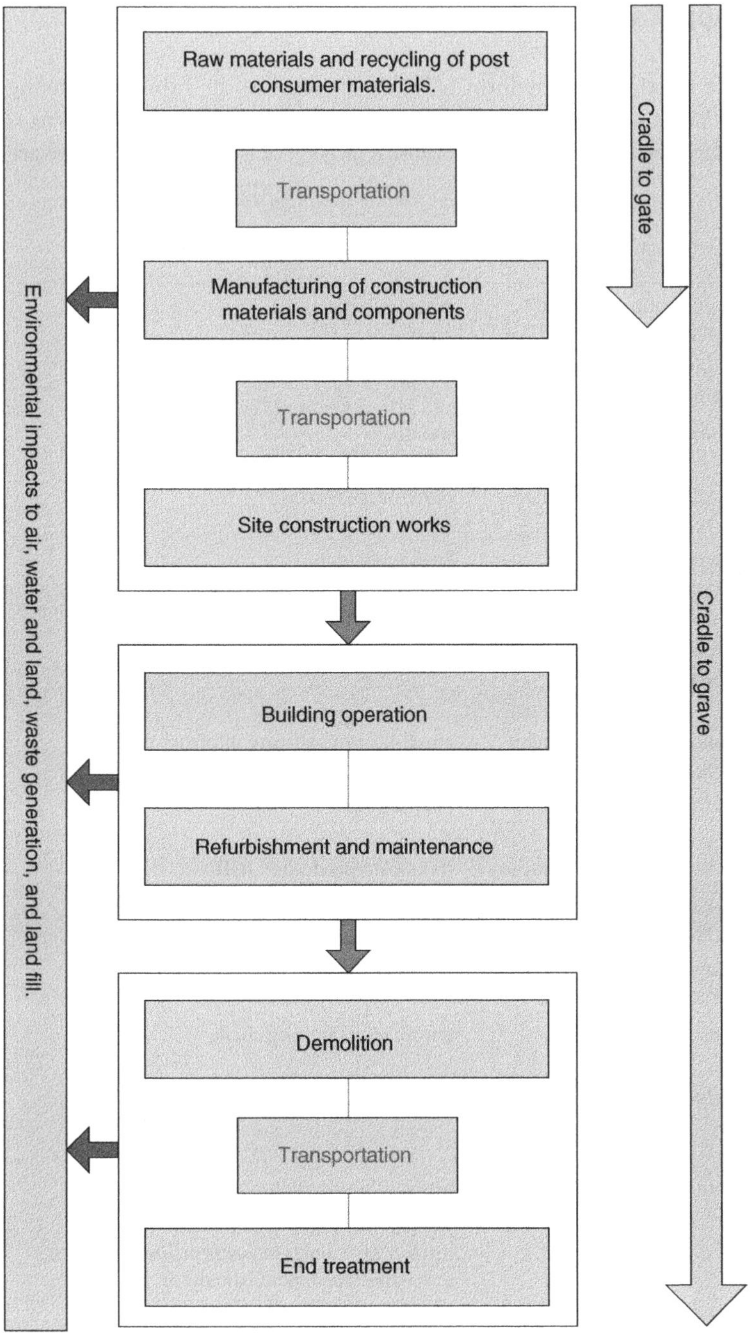
Raw materials and recycling of post consumer materials.
Transportation
Manufacturing of construction materials and components
Transportation
Site construction works
Building operation
Refurbishment and maintenance
Demolition
Transportation
End treatment
Environmental impacts to air, water and land, waste generation, and land fill.
Cradle to gate
Cradle to grave

9.4 LCA Tools and Applications

Several software packages are developed to perform LCA, as summarized in Table 9.3. Among these, SimaPro developed by PRé Consultants in the Netherlands is commonly used. SimaPro is software specifically designed for the LCA of an organization, a project, or a product. This software

Table 9.3 Software packages and solutions for LCA.

BEES	Building for environmental and economic sustainability NIST, USA.	nist.gov/bees/
Eco-Bat	Eco balance assessment tool HEIG-VD, Switzerland.	eco-bat.ch
EcoScan	Environmental impact and cost of products TNO, The Netherlands.	tno.nl
EIME	Environmental improvement made easy bureau-veritas	bureauveritas.com
EPD tools suit	Enthought python distribution enthought scientific computing solutions	enthought.com
GaBi	GaBi product sustainability software thinkstep global headquarters, Germany.	gabi-software.com
JEMAI-LCA Pro	Japan environmental management association for industry	jemai-lca-pro.software.informer.com
openLCA	Open source life cycle assessment software	openlca.org
LTE/OGIP	Energie und Stoffflüsse Switzerland.	rb-i.ch/de/produkte/OGIP
SimaPro	Life cycle assessment software PRé consultants, The Netherlands.	Simapro.com
TEAM™	Life cycle assessment software ecobilan, France.	ecobilan.pwc.fr
The boustead mode	Boustead consulting UK.	boustead-consulting.co.uk
US life cycle inventory database	National renewable energy laboratory.	nrel.gov/lci/
ECOPRO	Ecoperformance profile Signum, Switzerland.	https://sinum.com/en/services/ecopro-efficiency
OEKO-TEX	OEKO-TEX® impact calculator Switzerland.	https://www.oeko-tex.com/en/our-standards/oeko-tex-step/impact-calculator
ECO-iT	ECO-it – sustainable design software PRè consultants, Netherlands.	https://www.environmental-expert.com/software/version-eco-it-sustainable-design-software-182649
GEMIS	Global emission model for integrated systems Oko Institute, Germany.	https://www.oeko.de
Umberto LCA+	Umberto ifo Hamburg, Germany.	https://www.ifu.com/de/lca-software

makes sustainability decisions measurable; thus, the LCA results can be used to implement changes at various levels, so the organization or product evolves to more sustainable conditions.

The SimaPro software requires the inventory of relevant energy and material inputs to the process and the outputs to the environment. The software is designed to assess the potential impacts associated with identified inputs and releases. Several methodologies and databases can be used for the assessment: TRACI, BEES, or ReCiPe are commonly used.

TRACI, "Tool for Reduction and Assessment of Chemicals and Other Environmental Impacts," is an environmental impact assessment tool developed by the USEPA. This tool allows the evaluation of factors for LCIA, industrial ecology, and sustainability metrics. Those factors quantify the potential impacts of the project inputs and outputs in specific categories, including ozone depletion, climate change, acidification, eutrophication, smog formation, human health impacts, and ecotoxicity. The factors are normalized against the TRACI database, so the importance of impact in each category can be compared. Other notable tools that can be used in this regard include BEES (the Building for Environmental and Economic Sustainability) from NIST and ReCiPe, established by the Dutch National Institute for Public Health and the Environment (RIVM).

The last step in the LCA is the interpretation of the program results by the user, thus allowing for informed decisions for the project toward more sustainable conditions. SimaPro incorporates several features that help in the interpretation of the LCA results for a specific project and the comparison among several alternate solutions for the project.

The data for input and output flows in the project can be processed in SimaPro, using formulas designed by the user. The data and plots can be exported to a spreadsheet, so the user can adapt the results into presentations with specific formats or purposes. However, the more useful features are related to the presentation/comparison of results and the sensitivity analysis of data. The so-called pedigree matrix quantifies the uncertainty of the amount of a specific input or output, allowing for easier detection of errors or sensitive points. The Parameters test tab permits the user to vary the values of different parameters, which is especially useful in the performance sensitivity analysis.

SimaPro presents results in several standard graphs. As an example, the "Network" is a graphical representation of a model that permits the identification of the main sources of impacts in a graphical way using thin or thick arrows to represent the flow of materials and impacts through all the stages of the project.

LCAs can be applied to make decisions regarding the environmental sustainability of alternate processes or designs as well as optimize the components of a process or design for improved sustainability. Several detailed examples of the applications of LCA in engineering projects are provided in Chapter 21.

Example 9.2

Emissions due to the biogas energy produced using two types of feedstock are given below:

Chemical	Feedstock 1 (wt. %)	Feedstock 2 (wt. %)
CO_2	80	65
CH_4	10	30
CFC	0.05	0
N_2O	4	5
NO	5.95	0

Characterization factors for different environmental impacts are provided below:

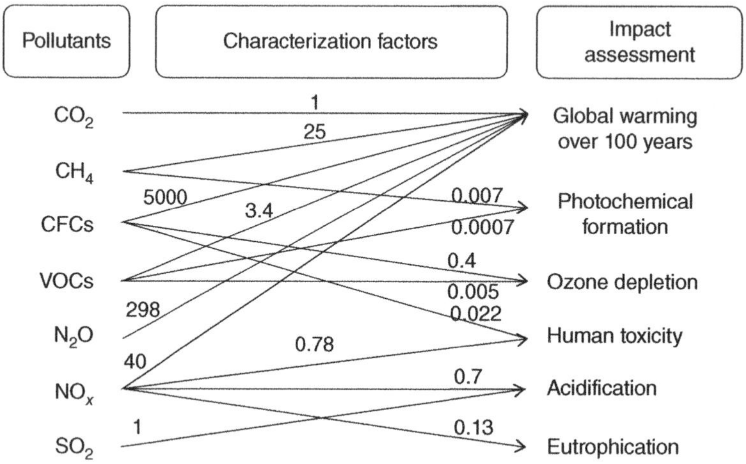

Compare the impacts of both types of feedstock in generating 1 MWh of electricity. Burning feedstock 1 generates about 0.8 kWh/kg, while feedstock 2 generates 1.5 kWh/kg. Burning 1 kg of feedstock 1 generates 500 g of emissions (including all the gases mentioned above), while feedstock 2 generates 800 g of emissions (including all the gases mentioned above).

Solution:

Feedstock 1
Quantity that needs to be burnt to generate 1 MWh of electricity = 1250 kg
Total gas emissions from generating 1 MWh of electricity = 625 kg

Gases	% Emission	Total emission (kg)
CO_2	80	500
CH_4	10	62.5
CFC	0.05	0.3125
N_2O	4	25
NO	5.95	37.1875

Impact assessment	CO_2	CH_4	CFC	N_2O	NO	Total impact
Global warming over 100 years	500	1562.5	1562.5	7450	1487.5	12 562.5
Photochemical smog formation	-	0.44	-	-	-	0.44
Ozone depletion	-	-	0.125	-	-	0.125
Human toxicity	-	-	0.0069	-	29.0	29.0069
Acidification	-	-	-	-	26.0	26.0
Eutrophication	-	-	-	-	4.8	4.8

Feedstock 2
Quantity that needs to be burnt to generate 1 MWh of electricity = 667 kg
Total gas emissions from generating 1 MWh of electricity = 533.6 kg

Gases	% Emission	Total emission (kg)
CO_2	65	346.84
CH_4	30	160.08
CFC	0	0
N_2O	5	26.68
NO	0	0

Impact assessment	CO_2	CH_4	CFC	N_2O	NO	Total impact
Global warming over 100 years	346.84	4002	0	7950.64	0	12 299.48
Photochemical smog formation	-	1.12	-	-	-	1.12
Ozone depletion	-	-	0	-	-	
Human toxicity	-	-	0	-	0	-
Acidification	-	-	-	-	0	-
Eutrophication	-	-	-	-	0	-

A bar chart comparing the relative impact of generating electricity using each feedstock type on different environmental indicators is presented below:

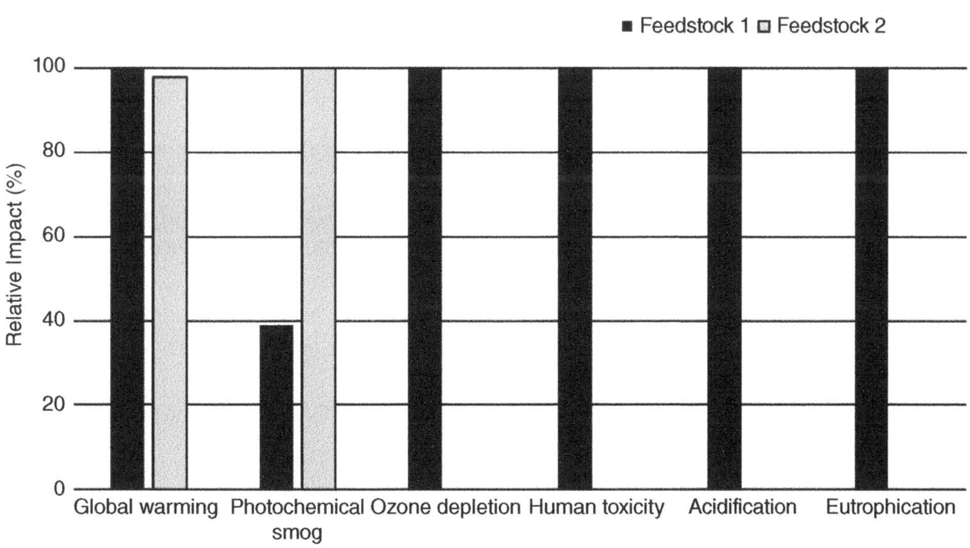

9.5 Summary

LCA is a useful means to measure the impact of products, processes or services in the environment, allowing for a measurement of the sustainability of such products, processes or services. LCA studies are commonly used to improve the design of a product or the components of an activity to be more protective of human health and the environment.

The first step of an LCA is to define the goal(s) and scope of the study. As demonstrated in this chapter, selecting the system and system boundary for an LCA is of prime importance. However, there is an inclination to draw a broad boundary with the goal of developing a comprehensive model. This can lead to problematic side effects with respect to the lack of data, complexity, or expense associated with performing such studies. As a result, the models can become unworkable, hence, system boundaries should be carefully selected to avoid such issues. The next step is to prepare an LCI of all materials and processes involved in each life cycle stage. LCI results could be used to assess the environmental impacts, but LCIA is a systematic process to evaluate the impacts on human health and the environment resulting from emissions, resource extraction, and other interventions from human activities and technical systems. Finally, LCIA results are interpreted to address ecological effects and human health effects as well as resource depletion.

The mere use or attempt for use of LCA can be beneficial to raise awareness with respect to measures that may be used to mitigate environmental impacts associated with activities or products. Therefore, regardless of the problems that may arise for LCA complexity, it is still beneficial to pursue more efficient and useful LCA for projects.

9.6 Questions

9.1 What is the purpose of an LCA?

9.2 What are the major phases involved in LCA methodology established by ISO?

9.3 Why is LCA important in assessing the sustainability of a product, a process, or an organization? Explain your answer with an example.

9.4 Name some of the major life cycle stages involved in a typical life cycle of a product.

9.5 Select a common household product and two alternatives. Using a qualitative framework, discuss the degree of sustainability for manufacture, operation, maintenance, and disposition of the product and the alternatives.

9.6 What is a functional unit? Why is it an essential part of life cycle analysis?

9.7 What do you mean by a system boundary in the context of LCA? What is the significance of the system boundary in LCA?

9.8 What are the effects of a "small" system boundary? What are the effects of a "large" system boundary?

9.9 How can a data gap effectively be used to assess an LCA and the product/project that is being assessed?

9.10 Discuss how a feedback loop is handled in an LCI.

9.11 Describe the importance of assessment format in preparing an LCI.

9.12 What are the four key steps of an LCIA? Discuss in detail the importance of each step.

9.13 Name three life cycle impact assessment methods mentioned in this chapter.

9.14 Provide detailed examples of local, regional, and global effects that may be identified during an LCIA.

9.7 Problems

9.1 Following emissions were expected due to the generation of each MWh of electricity generated in the state of Illinois.

Gas	kg/MWh
CO_2	536.5
CH_4	0.0103
NO_x	1.2784
CO	0.262
SO_2	2.31
PM	0.063

The TRACI characterization factors for the gases mentioned above are as given below:

Emissions	CO_2	CH_4	NO_x	CO	SO_2	PM
Global warming air (kg CO_2e / kg substance)	1	25	–	–	–	–
Acidification air (kg SO_2 eq / kg substance)	0	–	0.7	–	1	–
HH particulate air ($PM_{2.5}$ eq / kg substance)	–	–	0.0072	0.00035	0.061	1
Eutrophication air (kg N eq / kg substance)	–	–	0.04429	–	–	–
Eutrophication water (kg N eq / kg substance)	–	–	0.291	–	–	–
Ozone depletion air (kg CFC-11 eq / kg substance)	–	–	–	–	–	–
Smog air (kg O_3 eq / kg substance)	–	0.014	24.79	0.056	–	–

The Science and Engineering Laboratories (SEL) building at the UIC has an annual energy consumption of around 800 MWh. Calculate the annual impact on the above-listed indicators for this building. Consider that, as part of its climate action plan, UIC plans to replace 50% of the energy consumed by SEL with geothermal energy. Calculate the revised emissions and the corresponding impacts on each category. Find any source for data regarding emissions per MWh of geothermal energy generated and use that to calculate the impacts due to geothermal energy generation.

9.2 Average CO_2 emissions from a car are 114.1 g CO_2e/km, average particulate matter emitted is approximately 16.589 mg/km, and average NO_x emissions are 2.57 g/km. Assume the car travels for 1 million km in its lifetime.

- What are mid-point and end-point methods for LCIA? State clearly the definitions and differences.
- Calculate both midpoint and endpoint impacts on all relevant indicators due to the given car in its lifetime. Tabulate and present your results.

Note: The characterization factors for various emissions for midpoint impacts and conversion factors to convert midpoint impacts to endpoint impacts according to the ReCiPe method can be found at https://www.rivm.nl/en/documenten/recipe2016cfsv1120180117.

9.3 Following data regarding GHG emissions during different life cycle stages is given:

Emission source	**Coefficient**	**Value**	**Unit**
Production emissions vehicle body	$e_{prod,body}$	4.56	kg CO_2e/kg
Production emissions battery pack	$e_{prod,bad}$	83.50	kg CO_2e/kWh
Gasoline – WtW	e_{wtw}	2.83	kg CO_2e/l
Diesel – WtW	e_{wtw}	3.18	kg CO_2e/l
LPG – WtW	e_{wtw}	2.00	kg CO_2e/l
CNG – WtW	e_{wtw}	2.68	kg CO_2e/kg
CBG – WtW	e_{wtw}	0.749	kg CO_2e/kg
H_2 – WtW	e_{wtw}	9.13	kg CO_2e/kg
Conventional electricity – WtW	e_{wtw}	0.401	kg CO_2e/kWh
Renewable electricity – WtW	e_{wtw}	0.036	kg CO_2e/kWh
Recycling emissions vehicle body	$e_{recyc,body}$	−2.93	kg CO_2e/kg
Recycling emissions battery pack	$e_{recyc,bad}$	−48.4	kg CO_2e/kWh
Total lifetime traveled distance	$s_{lifetime}$	230 000	km
Energy density	ξ_{bat}	0.200	kWh/kg

Following data regarding modeling parameters for different car types is given

Energy source	**Vehicle model**	**Peak motor power**	**Curb weight**	**Fuel consumption**	**Electricity consumption**	**Battery gross capacity**
Gasoline	Volkswagen Passat 2.0 TSI	206 kW	1653 kg	7.2 l/100 km	0 kWh/100 km	0 kWh
Diesel	Mercedes-Benz C 300 d	180 kW	1635 kg	4.9 l/100 km	0 kWh/100 km	0 kWh
CNG	Audi A4 40 g-tron	125 kW	1665 kg	3.9 kg/100 km	0 kWh/100 km	0 kWh
CBG	Audi A4 40 g-tron	125 kW	1665 kg	3.9 kg/100 km	0 kWh/100 km	0 kWh
H_2	Toyota Mirai	135 kW	1900 kg	0.79 kg/100 km	0 kWh/100 km	2 kWh

Energy source	Vehicle model	Peak motor power	Curb weight	Fuel consumption	Electricity consumption	Battery gross capacity
Gasoline PHEV Conv.	Volkswagen Passat GTE	160 kW	1730 kg	1.4 l/100 km	11.5 kWh/ 100 km	13 kWh
Gasoline PHEV Renew.	Volkswagen Passat GTE	160 kW	1730 kg	1.4 l/100 km	11.5 kWh/ 100 km	13 kWh
Diesel PHEV Conv.	Mercedes-Benz C 300 de	225 kW	2060 kg	1.4 l/100 km	18.7 kWh/ 100 km	13.5 kWh
Diesel PHEV Renew.	Mercedes-Benz C 300 de	225 kW	2060 kg	1.4 l/100 km	18.7 kWh/ 100 km	13.5 kWh
BEV Conv.	Tesla Model 3 Standard Plus	239 kW	1684 kg	0 l/100 km	14.3 kWh/ 100 km	58 kWh
BEV Renew.	Tesla Model 3 Standard Plus	239 kW	1684 kg	0 l/100 km	14.3 kWh/ 100 km	58 kWh

Calculate the life cycle cumulative emissions of each car model at different life cycle stages. Show clearly the calculations for at least three model types and plot the results of all the models as a bar chart. The bar chart should look like the figure below:

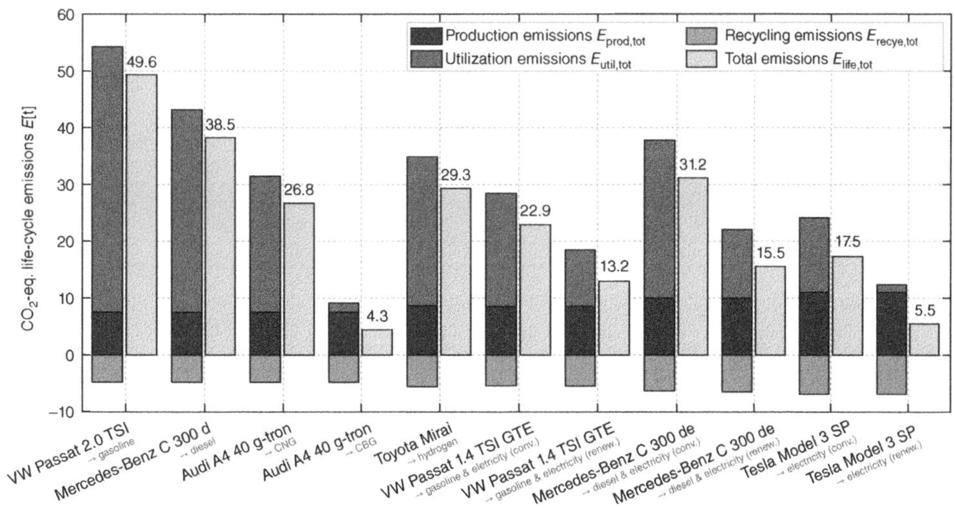

Note: You can refer to the publication below, and do your calculations. Your answers should match with the answers given in the manuscript: Buberger et al. (2022).

References

Buberger, J., Kersten, A., Kuder, M. et al. (2022). Total CO_2-equivalent life-cycle emissions from commercially available passenger cars. *Renewable and Sustainable Energy Reviews* 159: 112158.

Consoli, F. (1993). Guidelines for life-cycle assessment: a "code of practice. In: *Proceedings of the 5th Technical Workshop on LCA Sponsored by SETAC at Sesimbra*, Portugal (31 March – 3 April 1993). Pensacola, FL: Society of Environmental Toxicology and Chemistry (SETAC).

Curran, M.A. (ed.) (2012). *Life Cycle Assessment Handbook: A Guide for Environmentally Sustainable Products*. Wiley.

Curran, M.A. (ed.) (2016). *Goal and Scope Definition in Life Cycle Assessment*. Springer.

Hauschild, M., Rosenbaum, R.K., and Olsen, S.I. (2018). *Life Cycle Assessment – Theory and Practice*. Cham: Springer International Publishing.

Hemdi, A.R., Saman, M.Z.M., and Sharif, S. (2009). An indicator for measuring sustainable product design: a review and future research. In: *Conference on Scientific & Social Research CSSR08'09* (14–15 March 2009). Paper number: 8695303.

USEPA (2006). *Life Cycle Assessment: Principles and Practice*. U.S. Environmental Protection Agency. Report No: *EPA/600/R-06/060*.

10

Streamlined Life Cycle Assessment

Chapter Goal and Objectives
The goal of this chapter is to present and evaluate streamlined life cycle assessment (SLCA) as an efficient alternative to traditional life cycle assessment (LCA) while highlighting its methodology, potential expansions, and real-world applications. Specific objectives are (i) to explain the differences and similarities between SLCA and conventional LCA, emphasizing SLCA's retained core elements and streamlined procedures; (ii) to detail the steps and processes involved in performing an SLCA; (iii) to explore the potential for expanding SLCA to address additional issues, such as health and safety impacts; (iv) to provide examples of SLCA applications in various industries to demonstrate its practicality and effectiveness; and (v) to discuss other potential applications and scenarios where SLCA could be utilized for broader sustainability assessments.

10.1 Introduction

The usefulness of a full-scale life cycle assessment (LCA) to determine the degree of sustainability of an organization, project, or product was discussed in Chapter 9. However, such a full-scale study can be significantly time-consuming and expensive. In fact, an LCA typically utilizes significantly expensive software and databases, and the output of the assessment may be questioned since the LCA results often show some variation depending on the software/database used. Moreover, an LCA requires the itemization of inputs (material and energy resources) and outputs (emissions and wastes to the environment) for a given step in producing a product. Even for a very simple product, this process-based LCA method can quickly spiral into an overwhelming number of inputs and outputs to be considered.

As seen in Chapter 9, even for a simple case of a disposable paper cup, the LCA might list the paper and glue for the materials, the electricity or natural gas for operating the machinery to form the cup for the inputs, scrap paper material, waste glue, and low-quality cups that become waste for the outputs. For a more complex product such as an automobile, with over 20 000 individual parts, or a process such as electricity generation, the complexity is much higher. When considering the broad life cycle perspective, this same task must be done across the entire life cycle of the materials for the manufacturing and use of the product.

Sustainable and Resilient Engineering: Drivers, Metrics, Tools, and Applications, Second Edition. Krishna R. Reddy, Claudio Cameselle, and Jeffrey A. Adams.

Companion website: www.wiley.com/go/reddy/sustainableresilientengineering2e

Again, in the case of the drinking paper cup, one needs to identify the inputs, such as pulp, water, and dyes to make the paper, the trees and machinery to make the pulp, and the forestry practices to grow and harvest the trees. Similarly, one needs to include inputs and outputs for packaging the cup for shipment to the store, the trip to the store to purchase the cups, and the result of throwing the cup in the trash, eventually leading to landfilling or incineration.

Overall, the LCA method can be so complex and expensive, and it may become unworkable or cost-prohibitive. Furthermore, it can be quite inefficient and even contentious among stakeholders. Even though LCAs can be significantly beneficial for both their output and their ability to increase awareness of environmental impacts, they are not commonly used as a routine assessment tool.

How can the benefits of the LCA process be realized without the drawbacks related to cost, time, and complexity? The idea is to develop a method that retains the quintessential aspects of the LCA methodology while eliminating some of the complexities associated with it to dramatically improve efficiency. This alternative method is called streamlined life cycle assessment (SLCA) (Weitz et al. 1999; Graedel and Allenby 2010). The goal of SLCA is to retain the breadth, approach, and awareness-raising aspects of a full-scale LCA (as detailed in Chapter 9), but simplifying the process such that the overall efficiency of the process is greatly improved. This chapter provides SLCA methodology and simple applications and highlights other applications.

10.2 Streamlined LCA (SLCA)

To conduct a product-related SLCA, it is important to identify the key life cycle stages in SLCA. It should be noted here that "product" could also be "engineered system." As depicted in Figure 10.1, the SLCA includes the following stages: premanufacture, product manufacture, product packaging, product use, and product recycling/disposal. Additionally, it is important to identify the key environmental concerns to be included in the SLCA, as depicted in Figure 10.2. These concerns include the following: materials choice, energy use, solid residues, liquid residues, and gaseous residues. Both life cycle stages and environmental concerns listed here are similar to those considered in the LCA method discussed in Chapter 9.

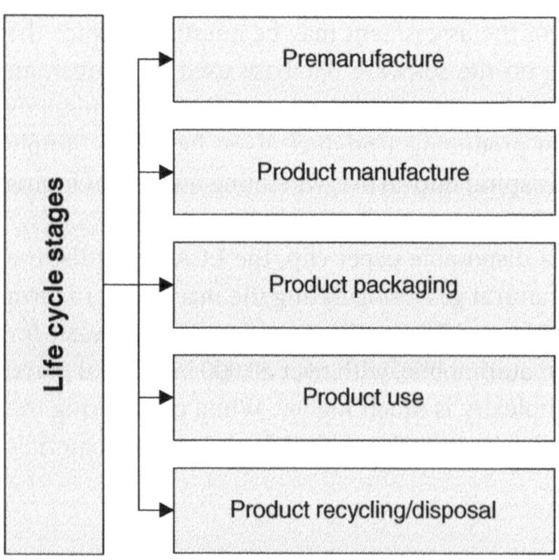

Figure 10.1 Life cycle stages for product LCA.

Figure 10.2 Environmental concerns for product life cycle assessment.

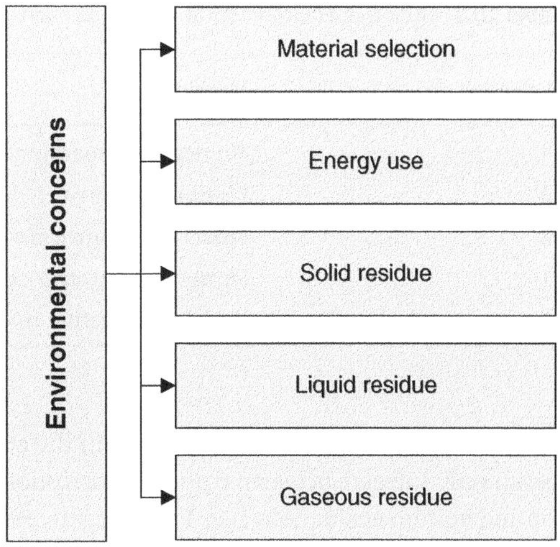

Once the key life cycle stages and environmental concerns have been identified, a matrix may be established to account for the resulting environmental concerns with each respective life cycle stage. This matrix, called the environmentally responsible product (ERP) matrix, is shown in Figure 10.3. The columns are populated by the respective environmental concerns (depicted along the horizontal axis), and the life cycle product stages are shown on the vertical axis (each with a corresponding row). Each resulting square in the matrix represents the specific environmental concern associated with a specific stage of the life cycle. For instance, the middle square or cell of the matrix (3, 3) corresponds to the solid residue that results from product packaging.

The matrix is populated with numerical values based on the relative environmental impact associated with an activity of the product's life cycle. The goal is to efficiently determine an entry for each element in the matrix as an "informed estimate" of the overall life cycle impact. One possible

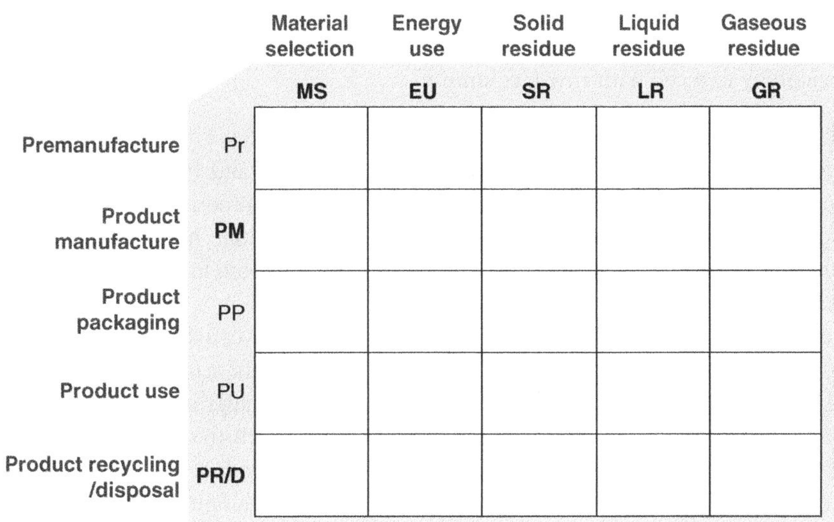

Figure 10.3 Matrix for accounting the environmental impacts of each life cycle stage.

Table 10.1 Relative environmental impact of each activity of a product life cycle.

Value	Meaning
4	No negative environmental impact
3	Minimal negative environmental impact (i.e. less than the expected average)
2	Moderate negative environmental impact (i.e. about the expected average)
1	Substantial negative environmental impact (i.e. above the expected average)
0	Very high negative environmental impact

format, which seems to be very effective in the characterization of the environmental impacts, is to assign only integers between 0 and 4; descriptions of these values are presented in Table 10.1. The top and bottom evaluations (i.e. 4 – no negative environmental impact; 0 – very high negative environmental impact) tend to be obvious in their assignment. If those numbers are not assigned for a specific impact and stage of life cycle, the choices are restricted to 3, 2, and 1 – better, equal to, or poorer than expected.

For example, while considering material choice for packaging (row 3; column 1), the cell would receive a value of 4 if no packaging were used for the specific product. A score of 0 would be assigned if the package used three or more materials that were not recyclable. Scores of 3, 2, or 1 would be assigned by the relative performance of a number of factors, including the use of recycling packaging materials, minimization of material diversity, existence of recycling infrastructure, and the relative involvement of the packaging engineer with the product design team to identify other efficiencies.

Once each of the matrix cells has been populated, the overall ERP rating can be computed by summing all of the matrix cell values as shown in Eq. (10.1):

$$R_{\text{erp}} = \Sigma M_{i,j} \tag{10.1}$$

where

R_{erp} = product rating;
$M_{i,j}$ = value corresponding to a cell with row i, column j.

When all the cells within the matrix are populated, the total score of this 5×5 matrix is the result of cumulatively added values of all cells, resulting in the value between 0 and 100 for the matrix, with a similar meaning as shown in Table 10.1. If the total score is 100, the product shows minimal environmental impact. For the total score of 0, the product shows a very high impact. Any intermediate value defines the degree of sustainability of the product and the room for improvement to get a better product with less impact.

Figure 10.4 depicts the graphical representation of the SLCA matrices for two different products, using a color-coded system instead of the 0 to 4 range of numeric values for the scoring. Figure 10.4a shows a completed matrix for an environmentally undesirable product, considering the low-scoring values in most of the matrix cells. The overall score is 45 out of 100. Figure 10.4b shows a matrix for a relatively more ERP, earning a score of 73 out of 100. Note that even a product with a high score retains areas where improvement is possible; perhaps because the original design was forced to use a toxic material or an alternative end-of-life approach was too costly. Additionally, Figure 10.5 shows an alternative graphical representation – this “rose” or “target” plot shows the product

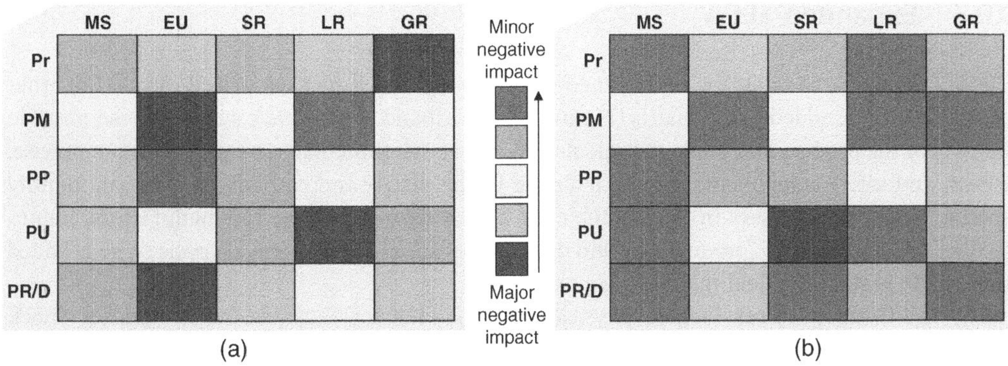

Figure 10.4 Graphical representation of the SLCA matrix using coded colors instead of integer values, for (a) an irresponsible product, and (b) a more sustainable product.

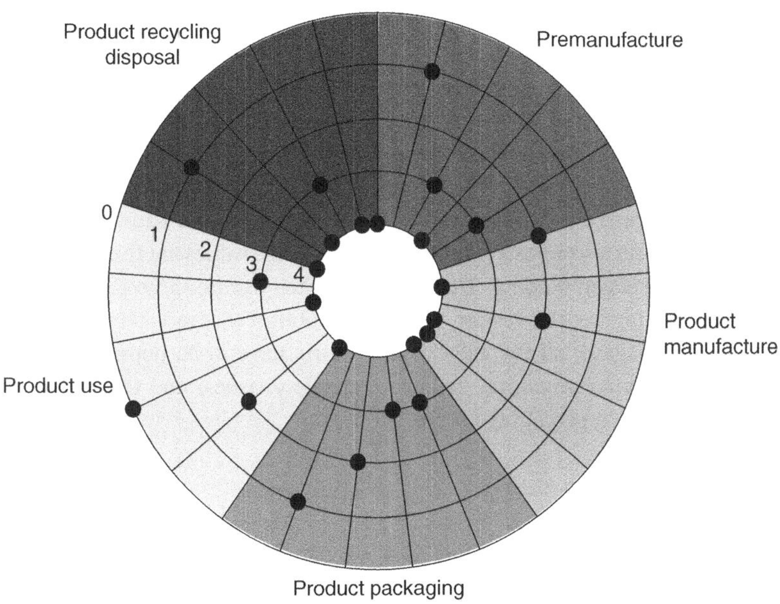

Figure 10.5 Target plot for the graphical representation of the SLCA matrix scoring. It is the same product as Figure 10.4b.

environmental concerns and life cycle categories in radial form. The target plot is a circle divided into five shaded sectors that correspond with the five stages in the product life cycle: premanufacture, manufacture, packaging, product use, and recycling. The scoring for each environmental impact and each stage of the product life is represented as a black dot over five concentric circles. These circles represent the scoring from 0 (the outer circle) to 4 (the inner circle). Points are closer to the center or "target" for life cycle categories with less relative environmental impact. A target plot is a graphical alternate means for displaying the matrix results, with high scores near the center and low scores near the edge (as with a rifle-range target). In this plot, it is easier to identify groupings and extreme values.

10.3 Expanded SLCA

The SLCA matrix considers the environmental impacts of the product across its life cycle, but other categories can be added to the matrix (Figure 10.3) to expand the life cycle stages and the possible impacts of the product and make the calculation of its environmental sustainability more precise. Health and safety components may be included in the matrix and calculations. We can identify specific health/safety issues in the five life cycle stages described before (premanufacture, manufacture, product delivery, product use, and disposal/recycling), and a new life cycle stage is added now to the matrix: the field maintenance or service.

The health/safety issues include physical hazards, chemical hazards, noise hazards, shock hazards, and ergonomic hazards (Figure 10.6). The health and safety issues can be organized into a matrix (Figure 10.7) where scores are again assigned to cells using a scale between 0 and 4. This is a 6 × 5 matrix, the maximum scoring is 120, and the final scoring is normalized by dividing it by 120 and expressing as percentage to obtain a final score ranging from 0 to 100. Figure 10.8 shows an example of the completed matrix for a transportation product. The reason for the fractional scores is that the four major subsystems in the product were evaluated separately, and the results were averaged. In this case, the values in each cell are summed on a row and column basis. The total score is 94.8 out of 120, which corresponds to a final score of 79% after normalization. It is clear that chemical hazard is the most problematic of all of the health and safety issues (as seen by the column scores), and product manufacture is the most problematic of the life cycle stages (as seen by the row scores).

Next, the health and safety matrix can be integrated in the product SLCA matrix. Thus, the values assigned to each cell are summed up on a row basis (Figure 10.9), and these values are then normalized to result in an average value between 0 and 4. The values are then inserted into the product SLCA matrix (originally shown in Figure 10.3) as an additional column (Safety). The resulting matrix is depicted in Figure 10.10. Again, all the cells in the matrix have to be populated with the corresponding value for all the environmental, health, and safety issues, and they may be summed and normalized to determine a percentage score.

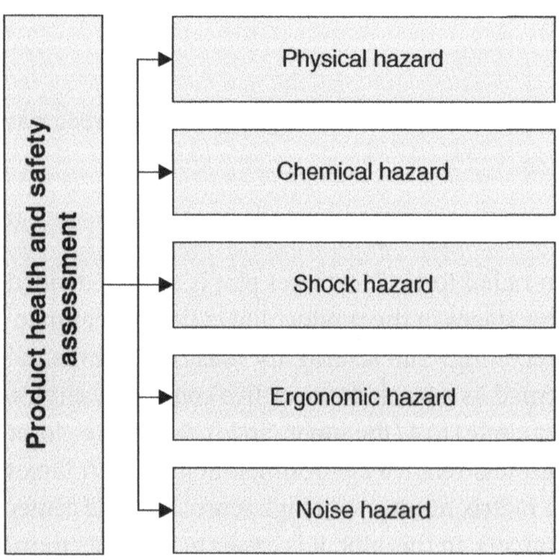

Figure 10.6 Product health and safety assessment.

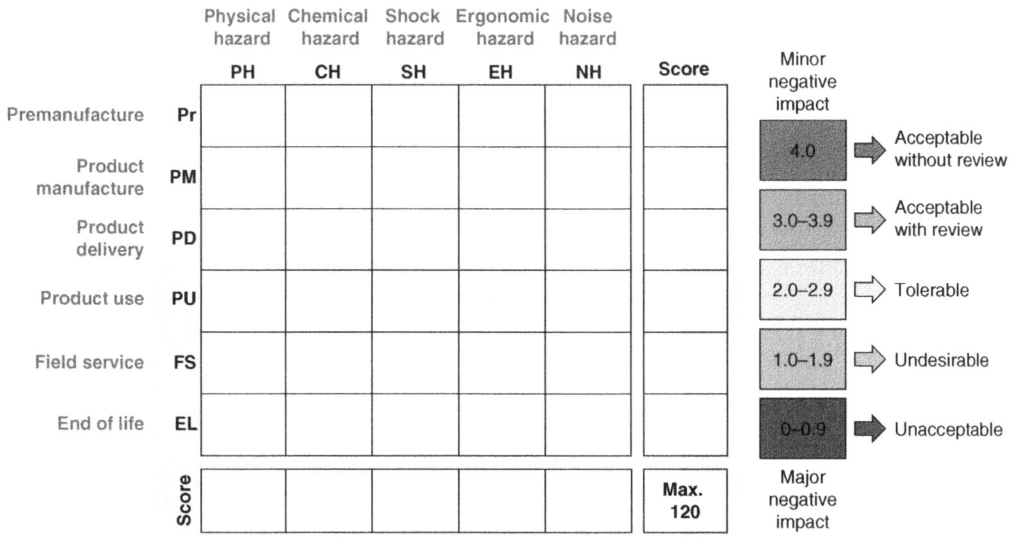

Figure 10.7 SLCA health and safety matrix.

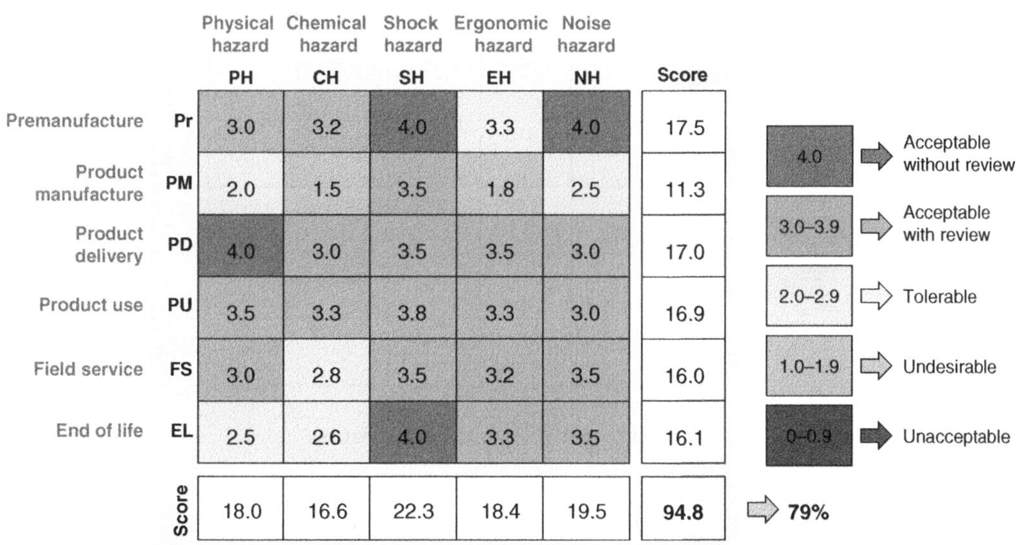

Figure 10.8 Health and safety matrix for a transportation product.

An example of an expanded SLCA matrix for a specific product is shown in Figure 10.11. The total score after normalization is 54%, indicating clear opportunities for improvement in the sustainability of this product. As shown in Figure 10.11, the premanufacture and end-of-life stages need the most improvement, and material selection and safety are of more concern than other environmental, health, and safety issues.

Safety matrix

		Physical hazard PH	Chemical hazard CH	Shock hazard SH	Ergonomic hazard EH	Noise hazard NH	Average
Premanufacture	Pr	2	3	1	4	1	2.2
Product manufacture	PM	2	2	3	2	3	2.4
Product delivery	PD	3	3	2	2	4	2.8
Product use	PU	3	4	2	3	1	2.6
Field service	FS	1	1	4	0	2	1.6
End of life	EL	2	0	1	3	3	1.8

Product matrix

		Material selection MS	Energy use EU	Solid residue SR	Liquid residue LR	Gaseous residue GR	Safety	Score
Premanufacture	Pr						2.2	
Product manufacture	PM						2.4	
Product delivery	PD						2.8	
Product use	PU						2.6	
Field service	FS						1.6	
End of life	EL						1.8	
	Score						13.4	

Figure 10.9 Integration of the safety matrix into the product SLCA matrix.

10.4 Simple Example of SLCA

An example of the application of SLCA is presented in Table 10.2 and Figures 10.12 through 10.14. This example depicts the SLCA of two typical automobiles manufactured in the 1950s and the 2000s (Graedel and Allenby 2010). Table 10.2 depicts several characteristics assigned to both automobile models with regard to the materials, fuel efficiency, use of catalytic devices for exhaust gases, and the use of ozone-depleting gases for air conditioning. The values in Table 10.2 suggest important changes in the manufacturing of automobiles from the 1950s to the 2000s toward more sustainable products, but the SLCA can give a better picture of the real changes. Figure 10.13 shows the relative scoring for the two models in the SCLA matrix, whereas Figure 10.14 shows the scoring depicted on

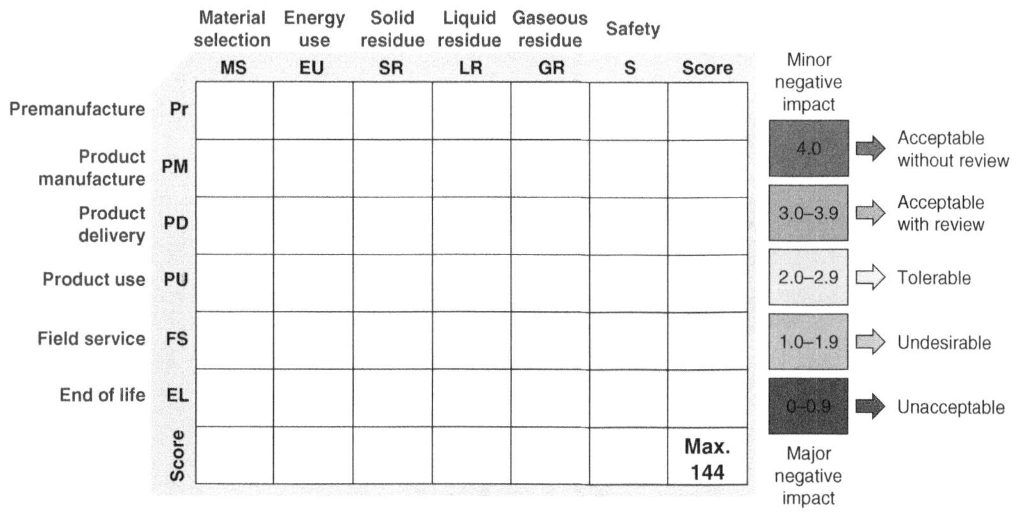

Figure 10.10 Expanded product matrix (6 × 6). The total score is now the result of summing all the matrix element values and dividing by 144 to get the percentage score.

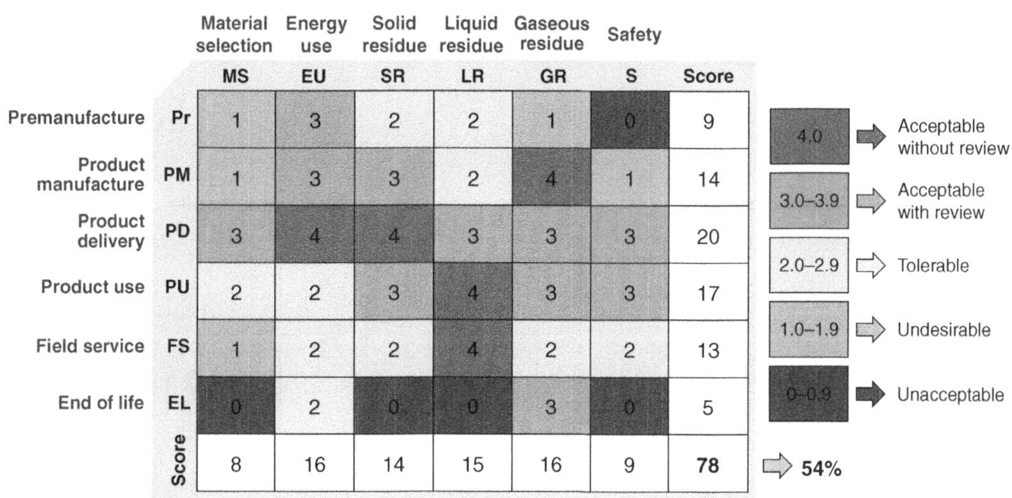

Figure 10.11 Completed example of the expanded SLCA matrix for a specific product.

Table 10.2 Materials and fuel efficiency for automobile models (Graedel and Allenby 2010).

	1950s	2000s
Materials		
Plastic	0	101
Aluminum	0	68
Steels	1290	793
Fluids	96	81
Other	515	391
Total (kg)	1901	1434
Fuel efficiency (mi/gal)	15	27
Exhaust catalyst	No	Yes
Air conditioning	CCF-12	HFC-134a

Figure 10.12 Typical 1950s and 2000s automobile.

1950s Automobile

	MS	EU	SR	LR	GR
Pr	2	2	3	3	2
PM	0	1	2	2	1
PP	3	2	3	4	2
PU	1	0	1	1	0
PR/D	3	2	2	3	1

Score: 46

2000s Automobile

	MS	EU	SR	LR	GR
Pr	3	3	3	3	3
PM	3	2	3	3	3
PP	3	3	3	4	3
PU	1	2	2	3	2
PR/D	3	2	3	3	2

Score: 68

Figure 10.13 SLCA matrix for two automobiles (scoring as per Table 10.1).

a target plot. As shown in these figures, there is a significant difference in the relative environmental impacts associated with these two car models. The 1950s automobile earned a global score of 46. As an example, some individual scoring in the matrix is notable. Matrix element 2,1 is 0 because of the use of toxics in manufacture. Element 1,3 is 3 resulting from the generation of waste rock in metal mining. Element 5,3 is 2 because the steel in the automobile is highly recyclable, even if the design for recycling was not performed. Element 3,4 is 4 because product delivery involves no liquid waste. Element 4,4 is 0 because no exhaust emissions equipment was present.

The matrix for the 2000s automobile shows a total score of 68 – much better than the 1950s car but still offering much opportunity for improvement. The individual impacts of the two cars can be seen

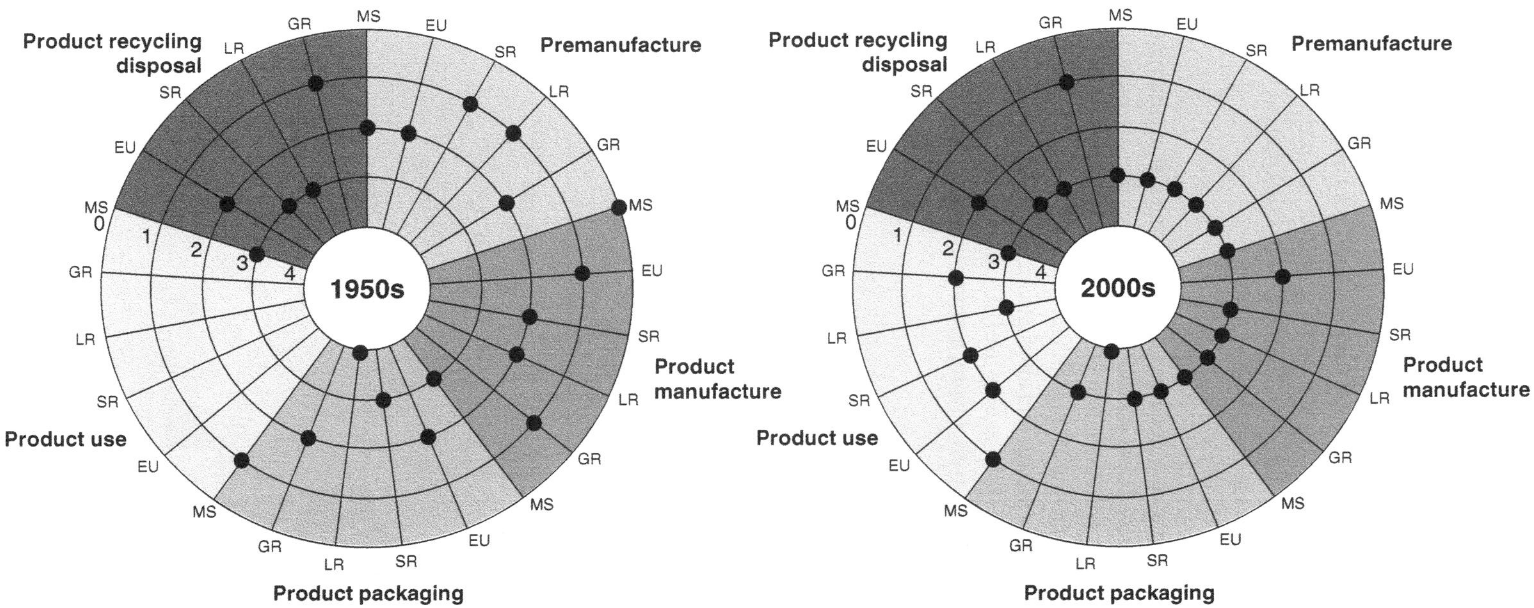

Figure 10.14 Target plot for the SLCA for two automobiles. *Source:* Adapted from Graedel et al. (1995).

in Figure 10.14. The target plot for 1950 shows that one could improve the score by addressing almost any stage of the life cycle. The 2000s target plot clearly indicates that the use of life cycle stage and, to a lesser extent, the recycling life cycle stage are where improvements may be realized.

Example 10.1

The City of Chicago has been making several efforts to improve pedestrian infrastructure. In this exercise, the Navy Pier Flyover, an elevated path in the City's Lakefront Trail, is analyzed for its sustainability versus the previous situation with no bridge for pedestrians. The elevated path can successfully eliminate conflicting points for traffic and pedestrians in the intersection of Grand Avenue and Illinois Street adjacent to Navy Pier.

Answer:

In this exercise, two SLCA matrices for the no-bridge intersection alternative and the existing design alternative are presented. These are the life cycle stages that are relevant including material production/extraction (MPE), transportation of materials (TM), construction of facility (CF), facility use (FU), maintenance of facility (MF), and end of life (EL). These were selected based on relevance to the Navy Pier Flyover project and alternatives. The life cycle stages are assessed based on the following: material selection (MS), energy use (EU), solid residue (SR), liquid residue (LR), gaseous residue (GR), and safety (S), with 4 being a minor impact/ a better alternative and 0 being a major impact/worse alternative.

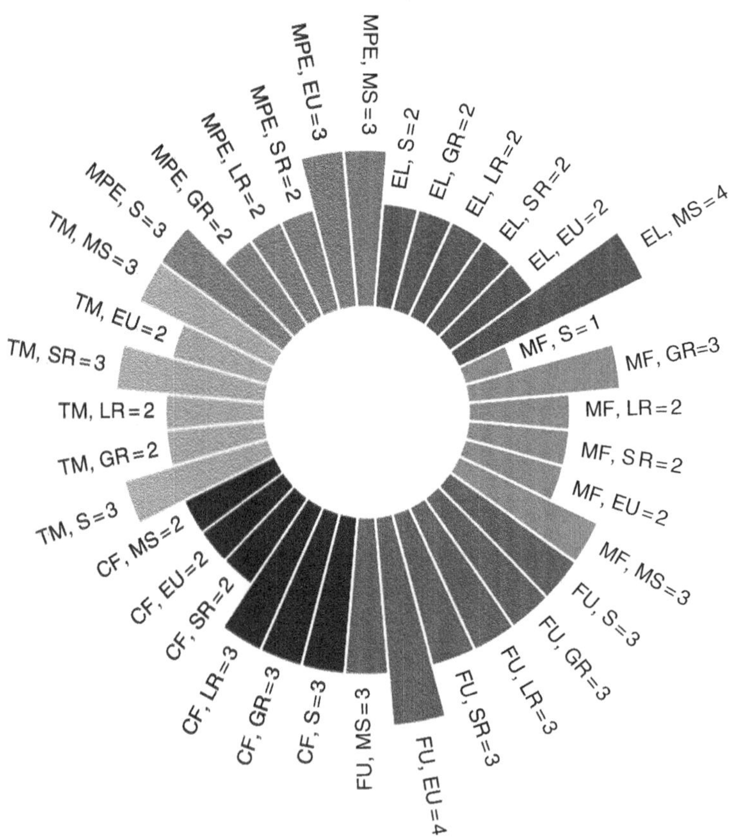

Figure E.10.1 Target plot for alternative one.

Alternative 1: No-bridge intersection improvements

	MS	EU	SR	LR	GR	S	Score
MPE	3	3	2	2	2	3	15
TM	3	2	3	2	2	3	15
CF	2	2	2	3	3	3	15
FU	3	4	3	3	3	3	19
MF	3	2	2	2	3	1	13
EL	4	2	2	2	2	2	14
Score	18	15	14	14	15	15	91

The overall score alternative 1 received was a 91, following was the score breakdown: MPE: 15, TM: 15, CF: 15, FU: 19, MF: 13, and EL: 14. The criteria have the overall score of MS: 18, EU: 15, SR: 14, LR: 14, GR: 15, and S : 15. Some of the lower ratings for the design included MF/S, considering that the maintenance of this facility could introduce the potential of harm to the users depending on downtime and ability of cyclists/pedestrians to maneuver around traffic/cars. One of the higher ratings was the FU/EU ratings; since this is a typical no-bridge intersection, the energy being used for FU is relatively low considering this does not need, for example, multiple LEDs to light up the bridge like other alternatives.

Alternative 2: Existing Design of Navy Pier Flyover

	MS	EU	SR	LR	GR	S	Score
MPE	2	1	2	2	2	3	12
TM	2	3	3	3	3	3	17
CF	2	2	2	2	1	3	12
FU	3	2	2	2	3	2	14
MF	3	3	2	2	3	3	16
EL	2	2	2	2	2	2	12
Score	14	13	13	13	14	16	83

The overall score alternative 2 received was 83, following was the score breakdown: MPE: 12, TM: 17, CF: 12, FU: 14, MF: 16, and EL: 12. The criteria have the overall score of MS: 14, EU: 13, SR: 13, LR: 13, GR: 14, and S: 16. This received a lower scoring than the first alternative.

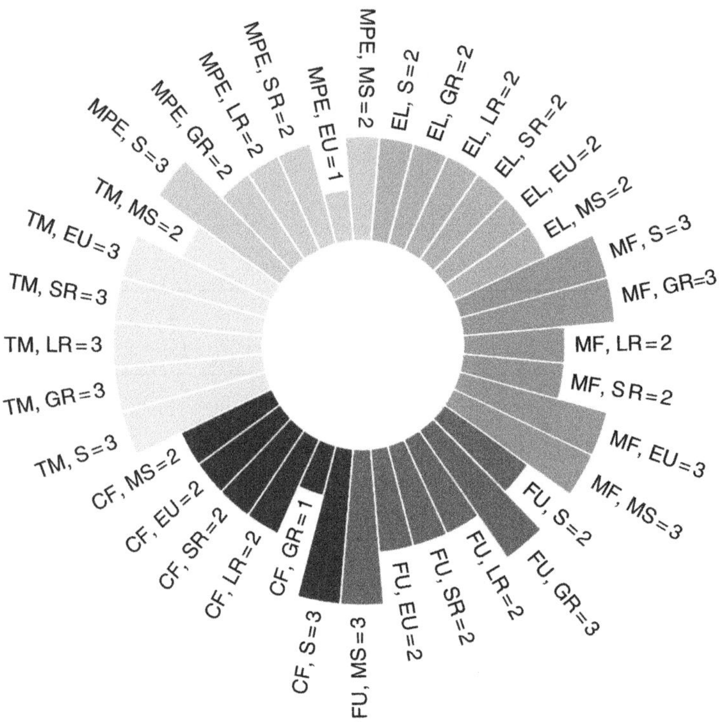

Figure E.10.2 Target plot for alternative two.

10.5 Applications of SLCA

SLCAs are not only useful for products, but they may also be very useful to assess processes and facilities. As an example, consider several common life stages associated with a generic process – resource provisioning, process implementation, primary process operation, complementary process operation, and process termination (Figure 10.15). Once again, similar environmental

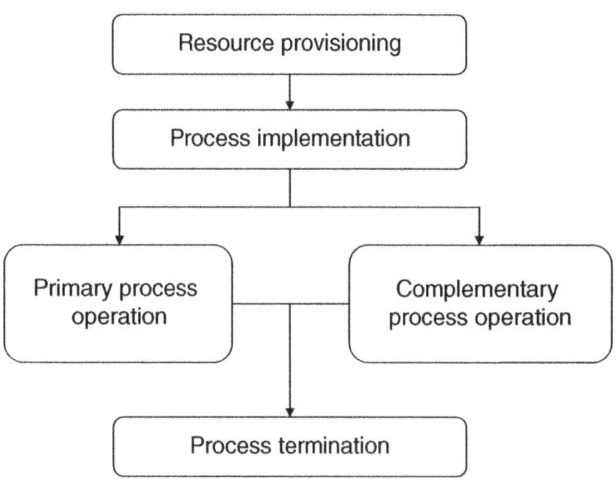

Figure 10.15 Life stages of an industrial process.

concerns may be applied as in our product example – material selection, energy use, solid residues, liquid residues, and gaseous residues. Figure 10.16 shows a matrix constructed with these respective life cycle stages and environmental concerns.

Table 10.3 and Figures 10.17 and 10.18 show an example of an SLCA comparison – once again for 1950s and 2000s automobiles – but this time focused on the manufacturing processes. The

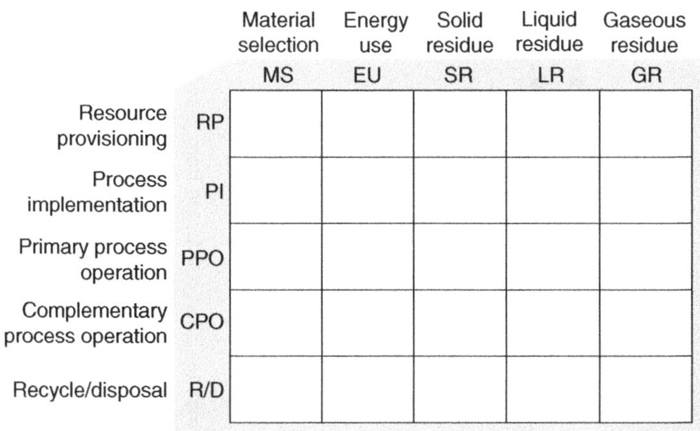

Figure 10.16 Environmentally responsible process matrix.

Table 10.3 Characteristics of the generic automobile manufacturing process.

	1950s	**2000s**
Material source	Virgin	Some recycled
Energy source	Very large	Large
Metal trimming	Common	Common
Metal cleaning	CFCs	Detergents
Metal plating	Ubiquitous	Common
Consumables recycling	No	Yes
VOC emissions	Large	Modest

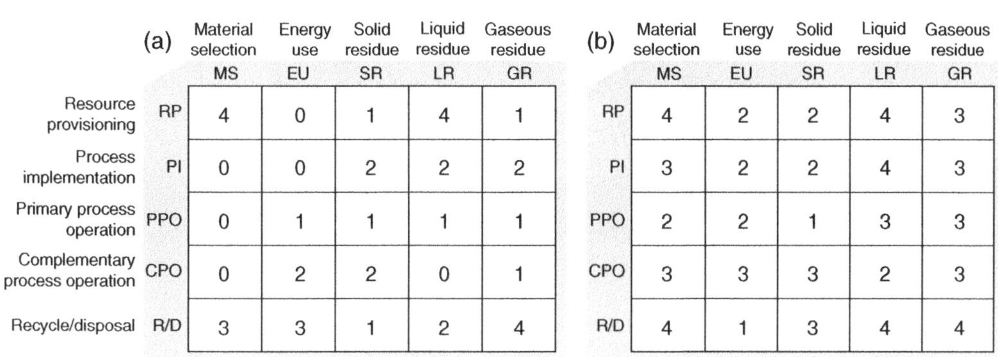

Figure 10.17 SLCA matrix for the manufacturing process of (a) 1950s and (b) 2000s generic automobiles.

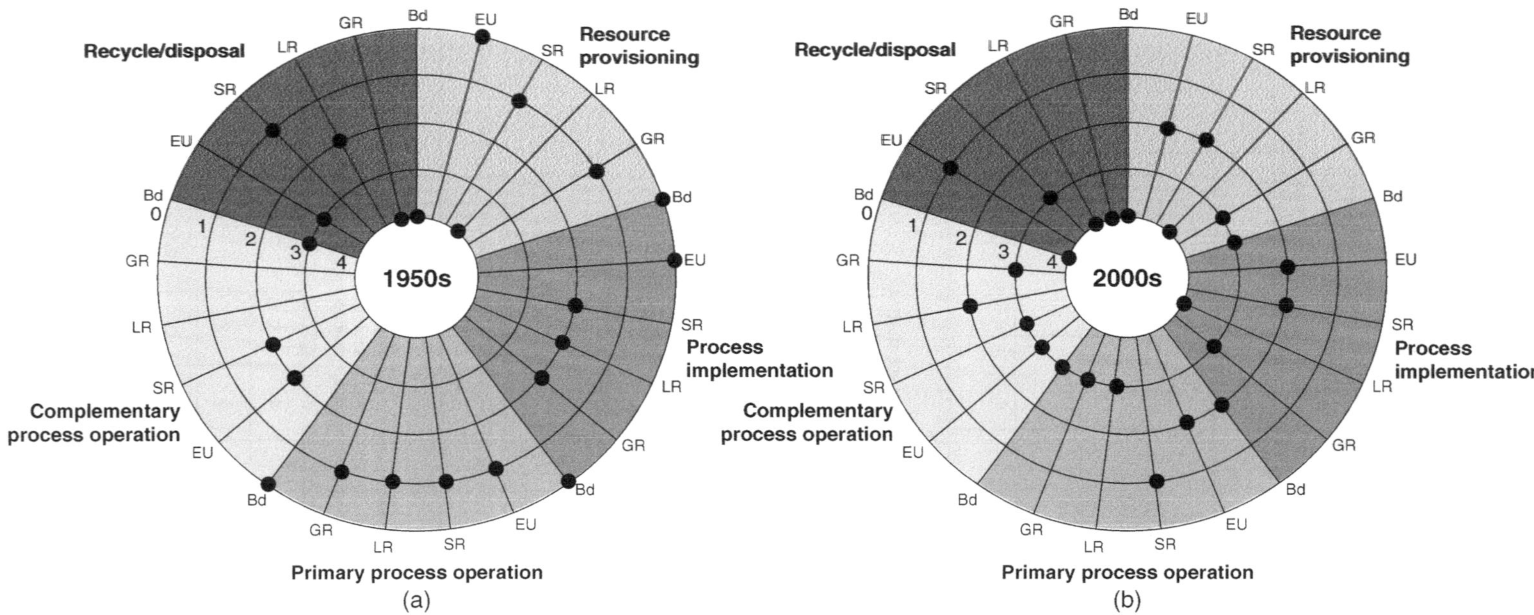

Figure 10.18 Target plot for the manufacturing process of (a) 1950s and (b) 2000s generic automobiles. *Source:* Adapted from Graedel and Allenby (2010).

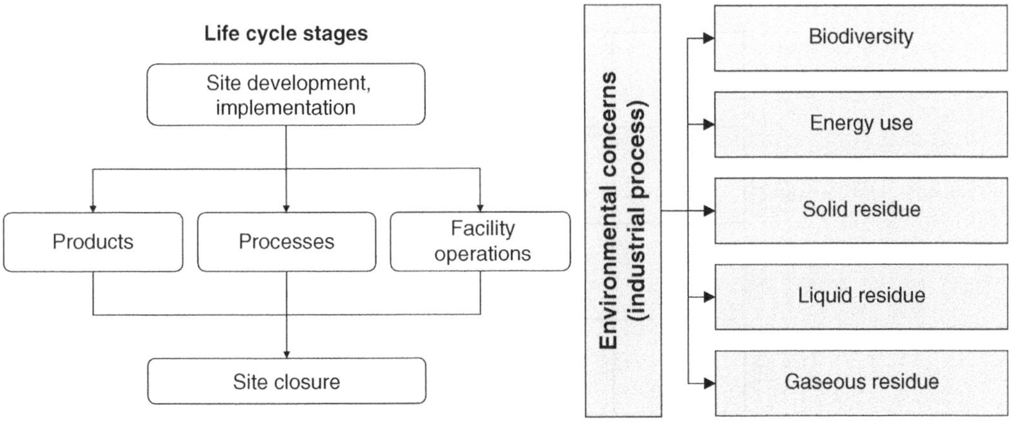

Figure 10.19 Life cycle stages and environmental concerns for an industrial facility.

improvement in the manufacturing process of automobiles is evident in Figure 10.18. The scoring points are spread all over the target plot for the manufacturing process of the 1950s, whereas the points tend to concentrate in the center for the 2000s manufacturing process, indicating a higher score and lower environmental impacts.

SLCAs can be applied to a variety of facets or positions on product/service value chains. As another example, consider the life cycle stages of an industrial facility (Figure 10.19). Some important stages include site development/implementation, facility operations, products, processes, and site closure. Again, these stages may be combined with related environmental concerns – biodiversity, energy use, solid residues, liquid residues, and gaseous residues for SLCA. A matrix for the SLCA of the facility is constructed, similar to those for a product or a process SLCA (Allenby and Graedel 1999). The matrix is completed with the individual scores for each life cycle and environmental concern to obtain the absolute assessment or a comparative assessment as shown in previous examples. In the case of SLCA of industrial facilities, the information from the SLCA for the products and processes of such facility can be integrated to obtain the assessment of the facility, as shown in Figure 10.20.

SLCA can also be easily adapted to assess social aspects of products or designs as well. Chapters 20–22 provide example applications of SLCA in different ways for alternate engineering designs.

Example 10.2

This exercise explores the aspects of unsustainable farming methods, food insecurity, and food shortages within the Back of The Yards neighborhood (Chicago, USA). To address these challenges, two projects are considered: A hydroponics garden and community gardens. These projects try to find a practical solution for the immediate needs of food and food security in the neighborhood. Using the SLCA, assess the sustainability of the two alternatives: hydroponic and community gardens.

Answer:

Project 1: Hydroponic Gardens

The hydroponic gardening method typically has lower emissions due to reduced water usage and fertilizer runoff. However, they still contribute to emissions through energy consumption for lighting, heating, water circulation and transportation of materials. The total environmental impact score is 40.

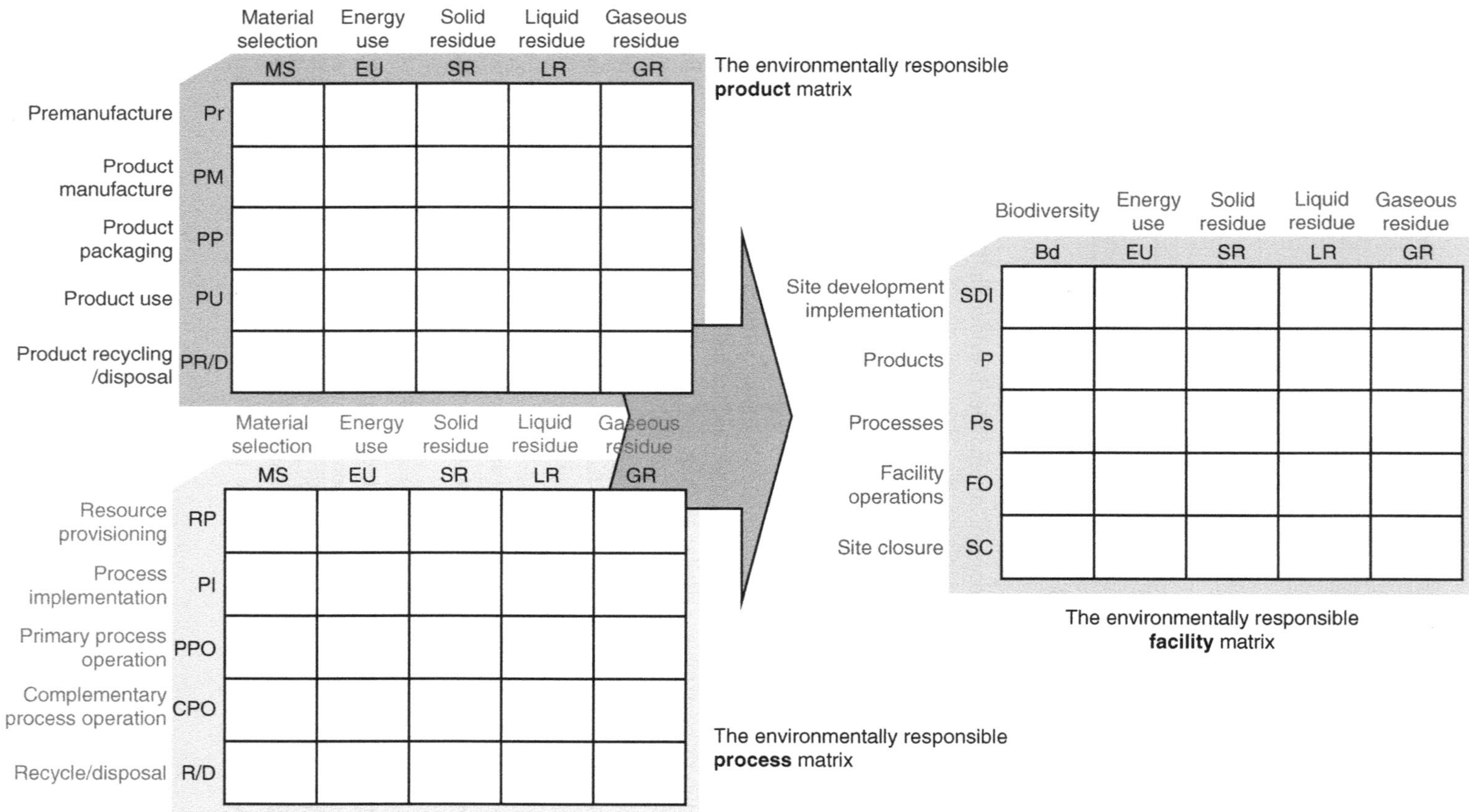

Figure 10.20 Integration of SLCA matrix for products and processes in the facility SLCA matrix.

Project 2: Community Gardens

Community Gardens have lower emissions compared to large-scale agriculture however, their emissions contributions are derived from gas powered equipment (lawn mower/tiller), transportation of supplies and produce, and energy use for irrigation systems. The total environmental impact score is 54.

Based on this analysis, community gardens would be the most sustainable option between the two alternative projects or scenarios.

Environmental matrix: Hydroponic gardens

Environmental impacts	Plant growing materials	Structural materials	Plant growing energy	Transportation (materials)	Score
Water usage	3	3	3	3	12
Energy usage	3	3	0	0	6
Emissions	2	3	0	0	5
Wastewater	2	3	3	4	12
Materials	1	1	2	1	5
Score	11	13	8	8	40

Environmental matrix: Community gardens

Environmental impacts	Plant growing materials	Structural materials	Plant growing energy	Transportation (materials)	Score
Water usage	4	4	3	4	15
Energy usage	1	4	4	0	9
Emissions	1	4	4	0	9
Wastewater	2	4	3	3	12
Materials	2	4	3	0	9
Score	10	20	17	7	54

Target Plots

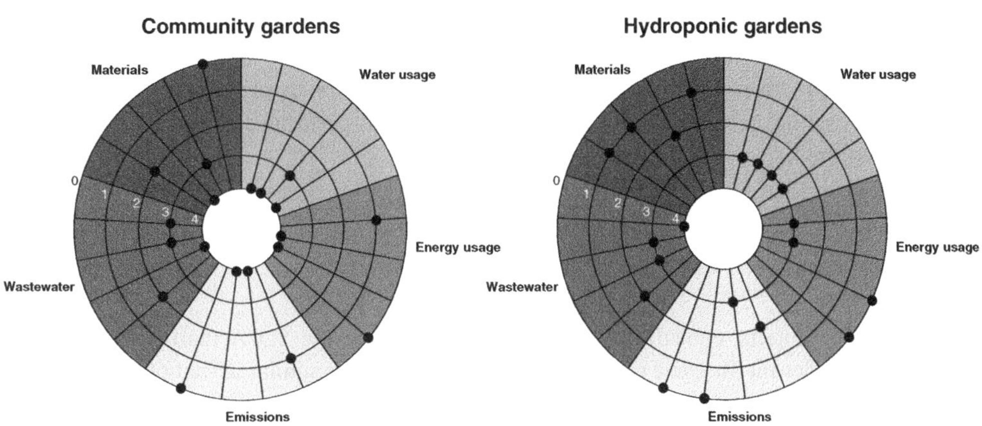

10.6 Summary

SLCA can be done efficiently and can reveal most or all the concerns identified by traditional LCA. Matrix plots or target plots facilitate the identification of opportunities for improvement. The concept may be extended in a straightforward manner to address the health and safety considerations of processes and facilities.

As demonstrated, SLCAs are relatively straightforward and efficient, and they can be used to assess or reveal most or all of the concerns raised during the course of a more comprehensive "traditional" LCA analysis. Graphical representations of these matrices, whether in the form of heat-mapped matrices or as target plots, can identify opportunities for improvement with respect to processes, activities, or products. The concepts discussed with respect to SLCAs can be extended to a range of processes, facilities, or other aspects of the product/service value chain. However, the results of SLCA can vary based on an individual's subjective judgment of the potential and importance of environmental concerns at each life cycle stage.

10.7 Questions

10.1 What is SLCA?

10.2 Discuss factors that may lead you to choose an SLCA over a traditional LCA.

10.3 What are the key life cycle stages in SLCA?

10.4 Assuming a product category such as environmental impacts of manufacturing and a scale from 0 to 4. List for each score from 0 to 4 characteristics that would be assigned to each numerical value.

10.5 Select a common household product. Using the example shown in Figure 10.3, prepare an SLCA matrix for the product. Calculate the corresponding score for the product.

10.6 Prepare a rose diagram of the SLCA matrix results.

10.7 Discuss specific health and safety metrics that may be included in an SLCA matrix.

10.8 Consider an SLCA that identified a clear advantage of product A over product B. Assuming product A is more expensive to produce, prepare a memo that describes the SLCA and why product A should be selected. Discuss how the SLCA was performed and why aspects other than cost should be considered.

10.8 Problems

For the following projects, consider two options and complete the SLCA matrix as shown in Figure 10.13 with relevant life cycle stages, and plot the target plots similar to that of Figure 10.14.

10.1 Federal Administration incentivizes the development of new cement materials and formulations with the objective to reduce the emissions of CO_2 associated with cement manufacturing and use. In this exercise, the SLCA methodology is used to assess the environmental impacts of two alternatives for cementitious materials: OPC-NS (ordinary Portland cement with natural sand) and PLC-F-CS (Portland limestone cement with fly ash and combined sand).

10.2 Sewage water can be reused after the appropriate treatment in a wastewater treatment plant, but only if the final effluent meets the quality characteristics for secondary uses. These possible uses include irrigation, bathroom sinks, and others. The final effluent in the wastewater treatment plant (WWTP) after the secondary treatment (biological treatment) requires a disinfection step to meet the quality for secondary uses. In this exercise, two alternative disinfection processes are considered: chlorination/dichlorination and UV treatment. Using the SLCA, assess the environmental sustainability of the two alternatives for water disinfection.

10.3 The emissions of greenhouse gases have led to global warming and climate change that increase the frequency and intensity of storms. In such intense storms, the sewer system can be saturated, and the excess of stormwater may affect the population, infrastructure, and environment. This exercise considers the development of green infrastructure for the control of stormwater. Two alternative scenarios are considered: bioswales and permeable pavement. Using the SLCA method, assess the sustainability of green infrastructure alternatives.

10.4 Miami–Dade County (MDC), a pivotal hub of economic activity and international trade, faces escalating threats from rising sea levels and intensified storm surges. In response to these looming threats, MDC has initiated various mitigation strategies to assess coastal storm risks and propose viable solutions. Two alternative options are considered as protection from sea level rise and storm surges: Dirt Levee and Concrete Storm Wall. Using the SLCA method, assess the sustainability of both alternative options.

10.5 Contaminated soils pose a significant risk to ecosystems and human health. The remediation technologies should be effective for the removal/elimination of the contaminants while minimizing the impact in the site. This exercise examines the remediation of a contaminated site using two alternative technologies: electrokinetic remediation and emulsified zero valent iron method for the removal of contaminated soil and groundwater at an oil refinery in Lincoln, Nebraska. Assess the sustainability of the two alternative treatments with the SLCA method.

10.6 The Illinois Tollway Authority recently completed reconstructing a 14-mile stretch of Interstate 90 from Schaumburg to Elgin, Illinois, USA. Several innovative sustainable practices were implemented in this project, including the use of jointed concrete pavement with a recycled aggregate subgrade. In this exercise, two options for the materials used in the construction of the road are considered: the use of virgin aggregate as subgrade or the use of recycled aggregates. Using the SLCA method, determine the sustainability of the two alternatives.

References

Allenby, B.R. and Graedel, T.E. (1999). Defining the environmentally responsible facility. In: *Measures of Environmental Performance and Ecosystem Condition* (ed. P.C. Schulze), 42–62. Washington, DC: National Academies Press.

Graedel, T.E. and Allenby, B.R. (2010). *Industrial Ecology and Sustainable Engineering*. Upper Saddle River, NJ: Prentice Hall.

Graedel, T.E., Allenby, B.R., and Comrie, V.R. (1995). Matrix approaches to abridged life cycle assessment. *Environmental Science & Technology* 29 (3): 134–139. https://doi.org/10.1021/es00003a751.

Weitz, K., Sharma, A., Vigon, B. et al. (1999). Streamlined life-cycle assessment: a final report from the SETAC North America streamlined LCA workgroup. Society of Environmental Toxicology and Chemistry (SETAC) and SETAC Foundation for Environmental Education.

11

Economic Input–Output Life Cycle Assessment

Chapter Goal and Objectives
The goal of this chapter is to explore the economic input–output life cycle assessment (EIO-LCA) model, its results, and challenges, while comparing it with conventional life cycle assessment (LCA) and physical input–output (PIO) analysis, and demonstrating its applications through example projects. The main objectives of the chapter are to (i) explain the economic input–output (EIO) model and its integration into life cycle assessment (EIO-LCA); (ii) analyze the results of EIO-LCA, addressing the interpretation challenges and uncertainties in its outcomes; (iii) present example projects that apply the EIO-LCA model; (iv) compare and contrast the EIO-LCA method with conventional life cycle assessment (LCA); and (v) discuss the differences between economic input–output (EIO) analysis and physical input–output (PIO) analysis.

11.1 Introduction

As presented in Chapter 9, a full-scale life cycle assessment (LCA) is the best option for analyzing a project or a product because it can give a complete picture of the environmental impacts associated with a product, an activity, or an organization. However, the full-scale LCA can get very complex and prohibitive. Even for the environmental assessment of a very simple product, the full LCA may incorporate an overwhelming number of inputs that make the process extremely complicated. The streamlined life cycle assessment (SLCA) concept was presented in Chapter 10 as an efficient alternative to full-scale LCAs, which can be a simplified approach toward achieving the same general goals of LCA.

An alternative to LCA and SLCA, an economic input–output (EIO) model with LCA, often known as EIO-LCA, can be used. An EIO model is used to represent the monetary transactions between industry sectors in a mathematical form. It is a means to assess what outputs (goods or services) of an industry are used as inputs or consumed by other industries. This EIO-LCA model was designed to assess the environmental impacts of a product or a project while overcoming the complexity of the full-scale LCA as well as the issues associated with boundary definition and circularity.

Sustainable and Resilient Engineering: Drivers, Metrics, Tools, and Applications, Second Edition. Krishna R. Reddy, Claudio Cameselle, and Jeffrey A. Adams.

Companion website: www.wiley.com/go/reddy/sustainableresilientengineering2e

11.2 EIO Model

EIO models are useful to analyze and predict the behavior of the economy in a country (Rose 1995). Many nations create EIO models of their economy at various degrees of specificity or frequency. For example, the U.S. EIO models are created every five years and represent the transactions among and between approximately 400 industry sectors. National budgets are based on National Financial Accounts (NFAs), which are obtained from EIO models. Accuracy in NFAs is critical; even small errors can result in budget estimates that are erroneous by wide margins. These errors, if they occur, can have notable consequences because these budgets are used to allocate funding and other similar expenditures.

When determining NFAs and gross domestic products (GDPs), it is critical to understand the inputs and outputs related to various sectors of the economy. A method to do this, the input–output analysis, was invented by Russian-American economist and Nobel laureate Dr. Wassily Leontief. Born in Germany in 1905, Dr. Leontief developed this method to analyze the intricacies of NFAs, and this information can be used to make predictions about the economic effects of a change in actions in different economic sectors (Leontief 1986). Specifically, his research focused on how an economic change in a given sector could affect other sectors. The importance of this cannot be understated, as the heart of the EIO model is the "input–output transaction table." An assumption in this model is that each sector produces a homogeneous output from a product composition and price standpoint, as explained below.

Figure 11.1 shows an example of a transaction table for two sectors (Graedel and Allenby 2010). As mentioned above, for this description, it is assumed that each sector produces a homogeneous

From \ To		Producers: Sector 1	Producers: Sector 2	Final demand	Total output
Producers	Sector 1	z_{11}	z_{12}	y_1	x_1
Producers	Sector 2	z_{21}	z_{22}	y_2	x_2
Import		i_1	i_2		

Imports are added to the columns (inputs into a sector)

From \ To		Producers: Sector 1	Producers: Sector 2	Export	Final demand	Total output
Producers	Sector 1	z_{11}	z_{12}	e_1	y_1	x_1
Producers	Sector 2	z_{21}	z_{22}	e_2	y_2	x_2

Exports are added to the rows (output of a sector)

Figure 11.1 Input–output transaction table for a two-sector economy showing the imports and exports for each sector.

product (i.e. a homogeneous composition and price). Sectors appear in both rows and columns. In Figure 11.1, z_{ij} defines intra- and inter-industry flows; y_i indicates flows entering use, or final demand; and x_i indicates total production, or output, of a sector. Columns represent the input, and rows represent the output of a sector, respectively. Imports are entered to the columns (represented as inputs) of a sector, while exports are entered to rows as outputs of a sector.

In constructing a basic EIO model for a two-sector economy, the total production for Sector 1 is presented in Eq. (11.1):

$$x_i = a_{ij}x_i + y_i \tag{11.1}$$

where:

a_{ij} = flow from sector i to sector j,
x_i = production of sector i,
y_i = final demand of sector i.

The same equation presented in matrix notation (Eq. 11.2) can be used to define the total production of the two sectors:

$$\mathbf{x} = \mathbf{A}\mathbf{x} + \mathbf{y} \tag{11.2}$$

where

$\mathbf{A}$ is the 2×2 matrix populated by a_{ij} values,
$\mathbf{x}$ is the vector populated by x_i values,
$\mathbf{y}$ is the vector populated by y_i values.

Constant technical coefficients z_{ij} are equal to $a_{ij} \times x_j$. Mass balance markets x_i are equal to the addition across rows of z_{ij} and y_i: $x_1 = z_{11} + z_{12} + y_1$; $x_2 = z_{21} + z_{22} + y_2$. This is equivalently represented as $\mathbf{x} = \mathbf{A}\mathbf{x} + \mathbf{y}$ (Eq. (11.2)), where $\mathbf{x}$ = the production vector; $\mathbf{y}$ = the final demand vector, and $\mathbf{A}$ = the technical coefficient matrix. Now, to solve, $\mathbf{x} = \mathbf{A}\mathbf{x} + \mathbf{y}$, first, $\mathbf{A}$ is subtracted from the identity matrix $\mathbf{I}$, resulting in $(\mathbf{I} - \mathbf{A})\mathbf{x} = \mathbf{y}$. Pre-multiplying with $(\mathbf{I} - \mathbf{A})^{-1}$ results in $\mathbf{x} = (\mathbf{I} - \mathbf{A})^{-1} \times \mathbf{y}$, where $(\mathbf{I} - \mathbf{A})^{-1}$ is the Leontief inverse.

EIO models are typically presented in a large-scale matrix form. Each economy or industrial sector is represented in the matrix as a row and a column. The intersection of the row and the column for a specific sector is the output from this sector that is used as an input in the same sector. This format provides two helpful characteristics. First, it can be determined if the output of a given industry sector is used as an input for the same sector (mathematically, this is represented as a non-zero value along the diagonal of a matrix). Second, using linear algebra computations, the total effects of changes (direct and indirect) can be calculated for a specific economy.

EIO models have several important uses and applications when studying an economy. First, they can be used to study the changes in supply, demand, and other relevant changes in an economy. Additionally, they can identify when economic shifts have occurred or when outputs from specific industry sectors have increased or decreased. This is quite a common scenario over a given time period, such as the instance when output increases from service-based sectors within advanced economies, coupled with a noticeable shift from low-tech manufacturing outputs. Further, EIO models can be used to aid decision-makers in determining cause-and-effect relationships within an economy, especially when connections or linkages have not been previously understood or observed between seemingly unrelated sectors. Pronounced or drastic changes in economies can also be mapped in such a manner.

EIO modeling does have some limitations. First, as an economic-based model, it is not particularly useful in tracking physical transactions, such as materials usage or quality of recycled materials. The solutions provided by the models can be limited in nature – an example of an EIO model outcome would be to determine the impact of producing 1 kg of metal instead of $1 worth of the same metal. Further, there are questions on how materials and energy flows can be incorporated into related models.

11.3 EIO-LCA

The limitations of EIO models may be overcome through the coupling with LCAs. For instance, an additional row and column can be added to incorporate the output and input of the "environment" sector. In this way, the utilization of natural resources as well as the effects of emissions and waste generation can be added to the EIO model, as well as the influences on and resulting from interactions with other sectors. Such an approach also eliminates two major difficulties of conventional LCA studies – boundary definition and circularity effects. First, the boundaries of an LCA study are, by definition, very broad and inclusive because all inputs and outputs among all industry sectors are included. Second, the circularity effects are considered in the assessment because self-consumption in the sectors is included in the matrix.

Formally speaking, the combination of an EIO and a LCA (EIO-LCA) is a mathematical method to estimate the amount of energy and materials consumed and the emissions generated from all economic activities (Matthews and Small 2000; Nakamura and Kondo 2009). The method uses industry transactions, namely, purchases of materials among industries and the direct environmental emissions from each industry, to estimate the total amount of emissions throughout the life cycle of a product, service, or sector. This mathematical procedure can determine the effect of changing the output of a single sector in the whole system just using economic and environmental data. The method can be applied to any economy or economic system if the information on the transactions between sectors is available. Each application of the method defines an EIO-LCA model.

With sufficient computing power readily available such that the requisite large-scale matrix manipulations could be performed in a short span of time, researchers at the Green Design Institute of Carnegie Mellon University developed a computing application of Leontief's method in the mid-1990s. The researchers subsequently developed an EIO-LCA computational tool capable of quickly evaluating a commodity, service, or representative supply chain. The tool provides information about the impacts of products, materials, services, or industries with respect to resource consumption and emissions throughout a supply chain. As an example, this tool permits to evaluate the effects of producing an automobile, including not only the impacts in the final assembly facility but also those associated with natural resources extraction (e.g. mining and materials extraction), fabrication of parts, etc., that constitute an entire supply chain.

EIO-LCA models may be applied to state or national economies at various levels of detail. Each national EIO model is developed using a base of publicly available resource use and emissions data. Since the mid-1990s, the method has been applied to develop economic models of the United States (on a federal basis and individual state basis), Canada, and several member countries of the European Union. Table 11.1 provides a listing of several EIO-LCA models.

Table 11.1 Examples of EIO-LCA models eiolca.net.

United States models	2002 Benchmark producer price
	2002 Benchmark purchaser price
	1997 Industry Benchmark producer price
	1997 Industry Benchmark purchaser price
	1992 Industry Benchmark
	1992 Industry mini
	1997 Pennsylvania (PA) state
	1997West Virginia (WV) state
	1997 PA + WV combined state
International models	2002 Canada industry account
	1995 Germany industry account
	2002 Spain industry account
Customized models	Build-a-product option based on the US 1997 model
	Build-a-sector option based on the US 1997 model
	MATLAB models (may require license)

11.4 EIO-LCA Model Results

11.4.1 Interpretation of Results

When interpreting the results of EIO-LCA models, it is recommended to focus first on interpreting the economic results and then focus on a single economic metric of choice. The environmental information about resource consumption and emission metrics must be interpreted in a similar way. Additionally, the environmental metrics can be combined and interpreted in conjunction with the economic results to obtain more detailed information about the direct and indirect impact of each product or sector; therefore, it is important to understand both sets of results.

When the interpretations are made, several assumptions made in these models should be kept in mind. First, these models are linear. The model results represent the impacts through the production phase of the sector, but the use phase and the end-of-life phases are not directly included in the results. However, additional analyses with the EIO-LCA method could be performed to account for these life cycle stages. Further, many assumptions are incorporated in the creation of the impact vectors, i.e. the values for material consumption, emissions, and environmental effects. The input–output (IO) models used in the development of EIO-LCA models are defined for the economy of a single nation. However, imports and exports represent a major contribution to any economy's transactions. In the IO models, the imports are implicitly assumed to have the same production characteristics as similar products made in the country of study.

11.4.2 Uncertainty

When discussing the use and applicability of these EIO-LCA models, it is important to discuss sources of uncertainty. First, the models are typically constructed using available data, which in the best case is one or two years old, and that data is going to be used to develop a model for the following year. The economic data may vary widely in relatively short periods of time. In a similar fashion, environmental data can vary significantly over time due to operational changes in industrial processes and process efficiency, the regulatory framework with respect to emissions or waste generation, or production levels. Further, there is uncertainty not just with respect to changes in data over time but also with respect to the original or baseline data collected for the model. The data used in EIO-LCA models is typically obtained from surveys and forms elaborated by governments for statistical purposes and submitted to the industries. The uncertainty in sampling, the degree of response from participants, biases inherent in these responses, missing or incomplete data, estimations or guessing incorporated into the specific responses, etc., can serve as a source of uncertainty in the modeling. Further, the baseline data may also be incomplete or have an unknown factor affecting some degree of the data. Often, the unknown factor or factors may later become known and incorporated, which can skew results due to the deviation with respect to the baseline data. A good example is toxicological or toxic release data.

Not only may there be errors, unknowns, or omissions in the baseline data, but it can often be difficult to aggregate and/or characterize the data. Often, important collected data may be difficult to categorize considering the EIO sectors of the IO matrix. For example, electricity consumption in commercial buildings may be aggregated based on industry sector (e.g. accounting, medical, engineering, and legal) or can be organized by the type of activity, such as office buildings, retail, etc. Therefore, the aggregation and allocation of data have been done considering the most appropriate industry sector and the objective of the study. Further, the aggregation of sectors may introduce uncertainty. One of the most interesting characteristics of the EIO-LCA method is its capacity to represent or estimate the impacts of a change in the demand of a specific industrial sector. An industry sector includes a collection of various industry types, and this is defined in the selected model. Thus, the IO model and the aggregation of industrial activities in sectors may lead to uncertainty due to how well a specific industry is modeled based on the aggregation into sectors.

11.4.3 Other Issues and Considerations

Several other issues are of importance when an EIO-LCA model is being constructed. For instance, each EIO-LCA model uses the currency of the country of origin, and the monetary values are based on when the model was constructed. Rates of inflation within a country as well as currency exchange rates between different countries over time may have a significant effect on the accuracy or applicability of the model. Another example is the correct use of producer versus purchaser prices. For many goods, producer prices can be significantly less than purchaser prices. As an example, producer prices associated with leather goods are typically 35% of the final purchaser prices in the United States. Further, EIO-LCA models are incomplete in analyzing environmental effects. There is a very limited number of usable metrics to be included in EIO-LCA models. Further, there is no data in a format to be incorporated in the models for environmental impacts such as habitat destruction, non-hazardous waste generation, and/or nontoxic water pollution. Finally, the EIO-LCA method estimates the resource consumption and emissions associated with the life cycle of an industry sector. Therefore, the EIO model and results are representative of the inventory stage of the LCA, but the actual environmental or human health impacts associated with the emissions and consumption are not included.

11.5 Projects Using the EIO-LCA Model

The input–output transactions table represents the monetary flow between sectors. These transaction tables are constructed from "make" and "use" data, which represents the sales and purchases between two particular sectors. Transactions include the purchases between two sectors and the total demand that is the sum of the purchases of one sector with all the others. Further, the sum of value-added (not inter-industry purchases) and final demand is GDP. From the input–output transactions table, a technical requirements matrix is created by dividing each column by the total sector input matrix. Entries represent direct inter-industry purchases per dollar of output. An example of a transactions table is shown in Table 11.2.

Table 11.3 presents an example of a two-sector EIO model. As shown in the table, reading across row 1, Sector 1 sells \$150 of output to Sector 1 itself, \$500 of output to Sector 2, and \$350 of output to consumers. In the first column, Sector 1 purchases \$150 of output from Sector 1 itself, \$200 of output from Sector 2, and adds \$650 of value to produce its output. Similar readings may be made with respect to Sector 2. The transaction flows are presented in the Transactions Matrix in Table 11.4.

A requirements matrix is presented in Table 11.5. The matrix entries are calculated as $A_{ij} = X_{ij}/X_j$, where X equals output, i equals intra-sector or inter-sector output for Sector i, and j equals total

Table 11.2 Example of transactions table.

Output from sectors	Input to sectors 1	2	3	n	Intermediate output O	Final demand F	Total output X
1	x_{11}	x_{12}	x_{13}	x_{1n}	O_1	F_1	X_1
2	x_{21}	x_{22}	x_{23}	x_{2n}	O_2	F_2	X_2
3	x_{31}	x_{32}	x_{33}	x_{3n}	O_3	F_3	X_3
n	x_{n1}	x_{n2}	x_{n3}	x_{nn}	O_n	F_n	X_n
Intermediate input I	I_1	I_2	I_3	I_n			
Value added V	V_1	V_2	V_3	V_n		GDP	
Total input X	X_1	X_2	X_3	X_n			

Table 11.3 EIO example for two sectors.

	1	2	Final demand
1	150	500	350
2	200	100	1700
Value added	650	1400	2050

Table 11.4 Transactions matrix for a two-sector EIO example.

	Sector 1	Sector 2	Final demand	Total output
Sector 1	150	500	350	1000
Sector 2	200	100	1700	2000
Value added	650	1400	2050	
Total input	1000	2000		

Table 11.5 Requirements matrix for a two-sector EIO example.

	Sector 1	Sector 2
Sector 1	150/1000 = 0.15	500/2000 = 0.25
Sector 2	200/1000 = 0.2	100/2000 = 0.05

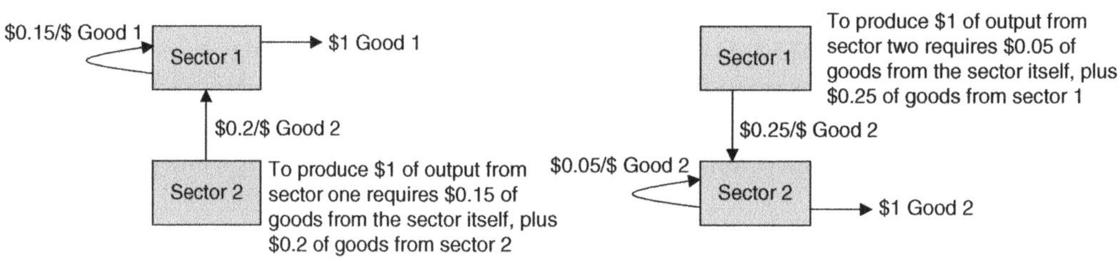

Figure 11.2 Production of two goods (Good 1 and Good 2) in a two-sector EIO model.

sector output for Sector j. As shown in the flowchart in Figure 11.2, for the production of Good 1 in our two-sector model, to produce \$1 of output from Sector 1 requires \$0.15 of goods from Sector 1 and \$0.20 of goods from Sector 2. For the production of Good 2, as shown in Figure 11.2, to produce \$1, \$0.05 of goods are required from Sector 2 and \$0.25 is required from Sector 1. The Leontief Inverse matrix is then calculated using $[\mathbf{I}-\mathbf{A}]^{-1}$.

Final demand is next added. For example, to determine the effects of an additional \$100 of demand from Sector 1, using the eq. $\mathbf{X} = [\mathbf{I}-\mathbf{A}]^{-1}\,\mathbf{y}$, the resulting total outputs of Sector 1 and Sector 2 are \$125.40 and \$26.40, respectively, or \$151.80 total. The direct intermediate inputs of \$15 for Sector 1 and \$20 for Sector 2 for an output of \$100, or \$135 total.

Further, environmental effects may be added such as the sector-level environmental impact matrices R may be constructed, which represents the effect per monetary output for a particular sector. As an example, assume 100 g of hazardous waste generation per dollar in Sector 1 and 5 g of hazardous waste generation per dollar in Sector 2. Introducing these values into our two-sector model, the waste generation is accounted for as outflows from each sector, as shown in our revised flowcharts in Figure 11.3.

Finally, based on the matrix calculations and calculating waste production B as the product of matrices B and X, 12 540 g of hazardous waste is generated by Sector 1, and 132 g of hazardous waste is generated by Sector 2, for a total hazardous waste generation of 12 672 g.

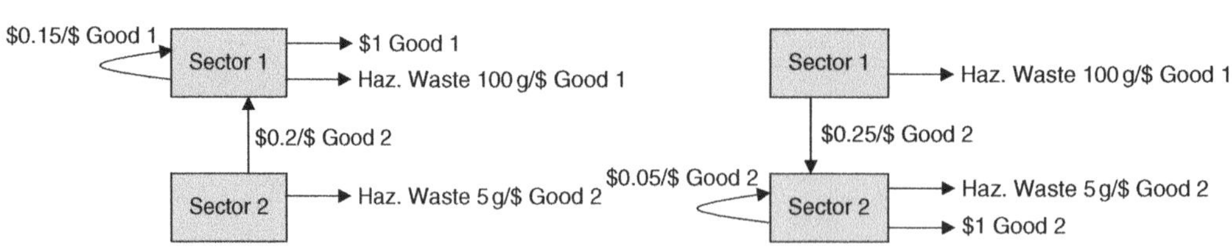

Figure 11.3 Production of Good 1 and Good 2 in a two-sector EIO model, including the costs of the generated waste.

11.6 Conventional LCA versus EIO-LCA

The following illustrates characteristics, strengths, and drawbacks of both conventional LCA and EIO-LCA:

Conventional LCA

- Provides detailed and specific analysis of processes
- Provides product comparisons
- Allows for process improvements or weak point analyses
- Suitable for future product development assessments
- Subjective establishment of the system boundary
- Tend to be time-sensitive and costly
- Difficult design of new processes
- Uses proprietary data
- Modeling cannot be replicated if confidential data is used
- There is uncertainty in the data.

EIO-LCA

- Provides economy-wide, comprehensive assessment in which all direct and indirect environmental effects may be included
- Allows for sensitivity analyses and scenario planning
- Use publicly available data, allowing for independently reproducible results
- Suitable for future product development assessments
- Incorporates information on every commodity in the economy
- Some product assessments contain aggregate data
- There is a difficulty in linking dollar values to physical units
- Economic and environmental data may reflect past practices
- Process assessments can be difficult
- Imports are treated as US products
- Can be difficult to apply to an open economy with substantial noncomparable imports
- Non-US data availability can be a problem
- There is uncertainty in the data.

With respect to boundary definition, there is always a question of how to set the model boundaries for an LCA. For instance, "conventional" LCAs include all processes, but when time and financial limitations may constrain the extent and detail of the model, it is important to draw a boundary such that the most important processes are captured. For an EIO-LCA, the boundary, by definition, is the entire economy along with the interrelationships between industrial sectors. Additionally, in an EIO-LCA, the products described by a sector represent an "idealized" or "average" product – not a real specific product. Further, when implementing an EIO-LCA model on a large scale, a matrix (incorporating hundreds of rows and columns representing the sectors of the US economy) may be incorporated. These can be augmented with sector-level environmental impact coefficient matrices that represent the amount of impact per dollar of output. Assuming this matrix **R** incorporates these impact values on its diagonal, the environmental impact **E** may be calculated as follows:

$$\mathbf{E} = \mathbf{R}\,[\mathbf{I} - \mathbf{A}]^{-1}\,\mathbf{y} \tag{11.3}$$

where:

I = identity matrix;
y = vector representing final desired demand;
A = requirements matrix.

As a brief example, consider the switch from paper to electronic versions (e.g. a CD) for a set of conference proceedings. In calculating costs and assuming a \$3 unit cost of a CD (also assuming 500 copies, or \$500), this compares to the production of 200 pages for 300 people, or 60 000 pages at 5 cents per page, or \$3000. Table 11.6 shows the associated environmental costs that may be factored, and Table 11.7 shows a listing of several physical environmental effects.

In further demonstrating how complex the input–output vectors can be, Figure 11.4 represents the supply chain for the manufacture of an automobile as well as a component engine. The vectors are presented in terms of actual cost (in dollars) as well as a relative percentage of the total cost. As shown, there may be a significant number of entries.

Table 11.6 Economic effects of using paper or CDs for conference proceedings.

Paper		CDs	
Sector	**\$**	**Sector**	**\$**
Pulp mills	3120	Magnetic media	1500
Logging	500	Misc. plastics	250
Industrial chemicals	400	Wholesale trade	170
Wholesale trade	260	Plastics	110
Sawmills	230	Industrial chemicals	80
Forestry products	150	Tracking services	70
Crude petroleum	140	Electric components	60
Tracking services	130	Paper mills	50
Electric utilities	130	Electric utilities	50

Table 11.7 Environmental effects of using paper or CDs for conference proceedings.

Emission	Paper (lb)	CDs (lb)
SO_2	35	2
Particulates	10	0.1
GWP	3000	600
RCRA	175	60
TRI chemicals	20	2

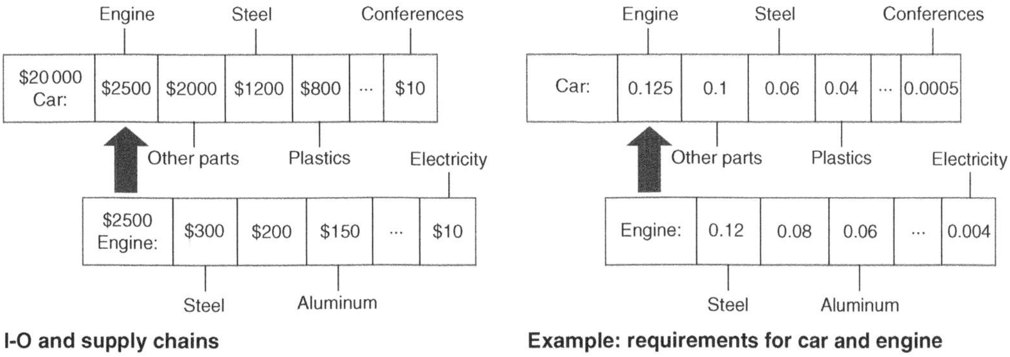

Figure 11.4 Input–output in supply chains for automobile and engines.

11.7 EIO versus Physical Input–Output (PIO) Analysis

In addition to EIO analyses, similar analyses may be performed using physical units as opposed to monetary inputs. Physical input–output (PIO) analyses are often quite detailed and require a significant amount of effort as compared to the typically highly aggregated nature of EIO analyses.

Figure 11.5a shows a typical PIO transaction table and the basic PIO model for a two-sector economy. When creating the model, two assumptions are commonly made. First, it is assumed that the

From \ To		Producers: Sector 1	Producers: Sector 2	Final demand	Total output
Producers	Sector 1	z_{11}	z_{12}	y_1	x_1
Producers	Sector 2	z_{21}	z_{22}	y_2	x_2

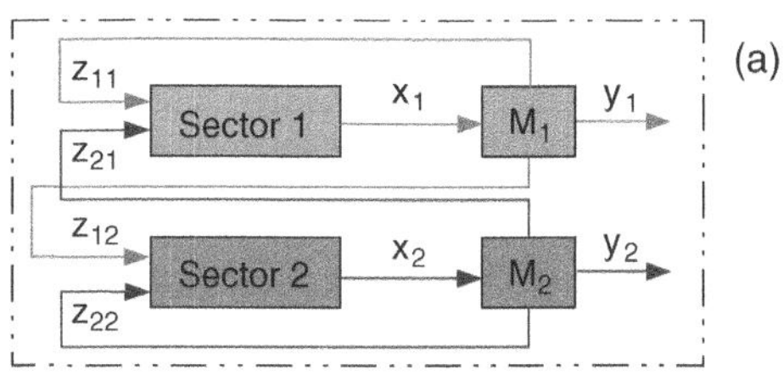

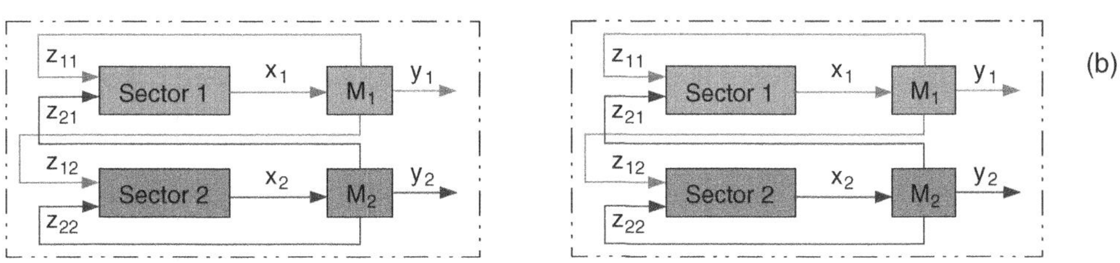

Assumption 1:
Conservation of mass holds across markets

$$x_1 = z_{11} + z_{12} + y_1$$
$$x_2 = z_{21} + z_{22} + y_2$$

Assumption 2:
Technical coefficients a_{ij} are constant

$$a_{11} = z_{11}/x_1 \quad a_{12} = z_{12}/x_2 \quad a_{21} = z_{21}/x_1 \quad a_{22} = z_{22}/x_2$$
$$\Leftrightarrow$$
$$z_{11} = a_{11} \cdot x_1 \quad z_{12} = a_{12} \cdot x_2 \quad z_{21} = a_{21} \cdot x_1 \quad z_{22} = a_{22} \cdot x_2$$

Figure 11.5 Basic PIO model for a two-sector economy. (a) Physical input–output transaction table. (b) Assumption 1 and 2.

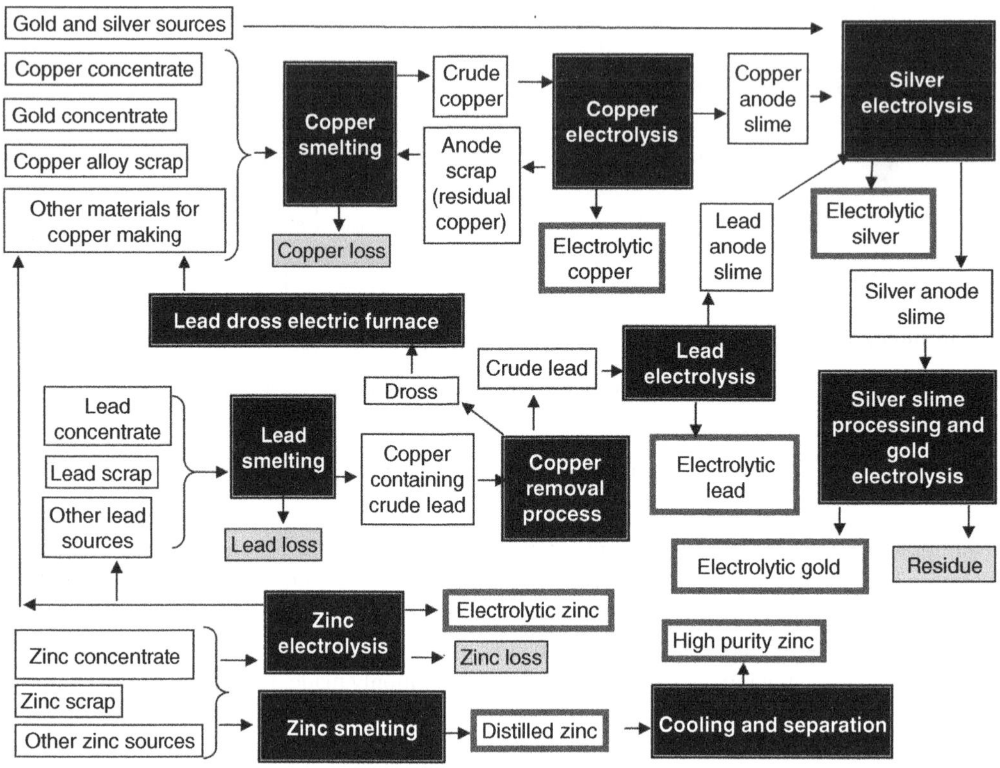

Figure 11.6 Flow of metals among production process. *Source:* Nakamura et al. (2008). Reproduced with permission of American Chemical Society.

conservation of mass holds across all markets. Second, it is assumed that technical coefficients a_{ij} are constant (Figure 11.5b).

As an example, suppose a decision is being considered that would replace leaded solder (60% tin and 40% lead) with nonlead solder (95.5% tin, 3.9% silver, and 0.6% copper). To illustrate this use and its effect, Figure 11.6 depicts a flow of metals among production processes. In considering the example further, Figure 11.5a also depicts a PIO for Japan's integrated metals production. As shown, columns represent the inputs of a metal industry sector, while the rows depict the outputs of a metal industry sector. Following the computations of the PIO, the results can be summarized in Figure 11.7. Figure 11.7 depicts the implications of non-lead solder – first, the percent change in needed supplies of the constituent metals and second, the demand for silver when considering both types of solder. Figure 11.8 further depicts the final destination of silver for different industry uses considering both types of solder.

In building PIO models and analyses, there are several challenges that should be considered with respect to data. First, it is often difficult to match economic value to physical flows – even if prices were known, it is difficult to accurately decouple economic data because it is often aggregated into a different format than physical units. As a result, physical data needs to be used, but these are often not directly available for use in modeling. Additionally, there are many aspects of the economy that cannot be measured directly in physical data terms. Services, for example, cannot accurately be assigned physical units. As a result, a mixed, or a hybrid, system can be used where physical units are important for accurate modeling.

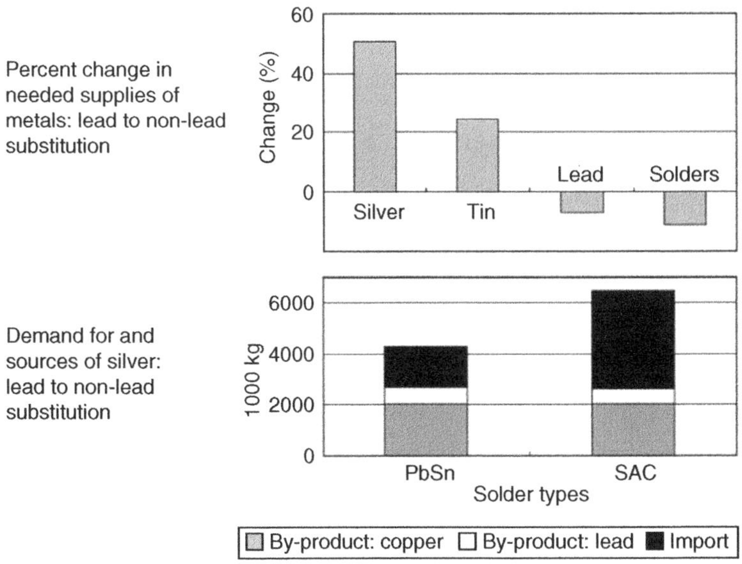

Figure 11.7 Implications of non-lead solder (SAC). *Source:* Nakamura et al. (2008). Reproduced with permission of American Chemical Society.

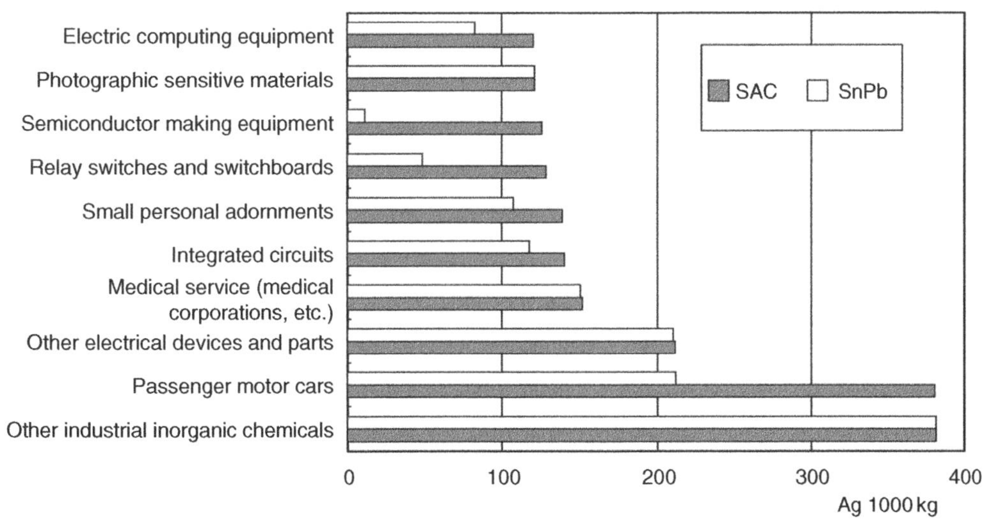

Figure 11.8 Final destination of silver under alternative solder approaches (SAC: non-lead solder). *Source:* Nakamura et al. (2008). Reproduced with permission of American Chemical Society.

11.8 Summary

National material accounts describe the inputs and outputs of entire societies, and they also indicate both indirect and direct flows. EIOs further model these activities by measuring and modeling domestic monetary exchanges among industrial sectors. EIO-LCA is a mathematical method that analyzes the effect of changing the output of a single sector using existing economic and environmental data. The use of EIO-LCA models does in fact offer several advantages over conventional LCA modeling. Finally, PIOs describe the domestic physical material exchanges that occur among industrial sectors.

11.9 Questions

11.1 What is an economic input–output model? Why was this developed?

11.2 Who developed the method of input–output analysis?

11.3 What are the major advantages of EIO-LCA over conventional LCA?

11.4 List some limitations associated with EIO modeling.

11.5 What are some of the limitations associated with an EIO-LCA model?

11.6 Discuss in detail some sources of uncertainty that may occur during the use of EIO-LCA models.

11.7 What are the major assumptions commonly made in PIO models?

11.10 Problems

11.1 The input–output coefficient matrix of steel, automobile, and rail industries is as given below:

$$\mathbf{I} = \begin{bmatrix} 0.33 & 0.66 & 0.08 \\ 0.5 & 0.25 & 0.25 \\ 0.12 & 0.33 & 0.33 \end{bmatrix}$$

The first row corresponds to the requirements of the steel industry, while second and third rows correspond to the requirements of the automobile and rail industries, respectively. Answer the following questions:

- Considering that there is no external demand, what is the proportion of outputs required from the three industries to operate self-sustainably (closed-loop)?
- Assuming that there is an external demand of 30, 50, and 100 M units of steel, auto, and rail components, respectively. Calculate the required output from the three industries (open-loop).

11.2 Consider a simple economy with two sectors: sector A and sector B. \$1 worth of sector A products need \$0.20 of sector A products and \$0.50 worth of sector B products. \$1 worth of sector B products need \$0.70 worth sector A products and \$0.10 sector B products. If the exports from sectors A and B are \$30 000 and \$10 000, respectively.

- Express this problem in an input–output model.
- How much worth products should be produced from sectors A and B to meet the total demand?

11.3 There are three interdependent sectors: steel, automobile, and semiconductors. The input–output table below shows the inter-sectoral dependencies:

From \ To	Steel	Automobile	Semiconductor
Steel	0.3	0.4	0.1
Automobile	0.5	0.2	0.2
Semiconductor	0.2	0.1	0.3

The exports from each sector are as follows: steel: \$40 000, automobile: \$20 000, and semiconductors: \$10 000. Determine the total amount of production required from the steel, automobile, and semiconductor sectors to meet both inter-sectoral and export demands.

References

Graedel, T.E. and Allenby, B.R. (2010). *Industrial Ecology and Sustainable Engineering*. Upper Saddle River, NJ: Prentice Hall.

Leontief, W.W. (1986). *Input-Output Economics*. Oxford University Press.

Matthews, H.S. and Small, M.J. (2000). Extending the boundaries of life-cycle assessment through environmental economic input-output models. *Journal of Industrial Ecology* 4 (3): 7–10.

Nakamura, S. and Kondo, Y. (2009). *Waste Input-Output Analysis: Concepts and Application to Industrial ecology*, vol. 26. Springer Science and Business Media.

Nakamura, S., Murakami, S., Nakajima, K., and Nagasaka, T. (2008). Hybrid input-output approach to metal production and its application to the introduction of lead-free solders. *Environmental Science and Technology* 42 (10): 3843–3848. https://doi.org/10.1021/es702647b.

Rose, A. (1995). Input-output economics and computable general equilibrium models. *Structural Change and Economic Dynamics* 6 (3): 295–304.

12

Environmental Health Risk Assessment

Chapter Goal and Objectives
The goal of this chapter is to provide a comprehensive overview of environmental health risk assessment, focusing on the methodologies used for hazard identification, dose–response, exposure assessment, and risk characterization, while also highlighting the potential need for ecological risk assessments in certain contexts. The specific objectives are (i) to explain the emergence of the risk era and its relevance to environmental health; (ii) to explain the key steps in environmental health risk assessment, including hazard identification, dose–response assessment, exposure assessment, and risk characterization; (iii) to explore risk management strategies in the context of environmental health; and (iv) to discuss the potential necessity of ecological risk assessments in specific scenarios.

12.1 Introduction

As presented in Chapter 9, when performing a life cycle impact assessment, the final objective is to evaluate ecological and human health effects as well as resource depletion resulting from emissions, resource extraction, and other material fluxes across the life cycle of a product. The results of life cycle impact assessment (LCIA) for different product alternatives are useful for the relative comparison of potential damage on human health and the environment of different products. However, this method does not provide a quantitative indicator of the absolute risk or actual damage that may have been caused by a product. Moreover, LCIA involves many stressors, locations, and impact categories, which makes it impossible to perform an accurate risk assessment. In this context, it is necessary to develop a method or a tool to perform a quantitative environmental risk assessment so that the risks can be accurately determined.

The objective of this chapter is to analyze how risks can be evaluated as a necessary step for developing methods of risk management. It is essential to develop methods or tools to determine human health risks resulting from exposure to the environment and workplace to be able to minimize the negative mid-term or long-term effects on health. Considering the interdisciplinary field of this topic, synergic working groups should be encouraged where scientists, engineers, policymakers, and regulatory agencies work together in developing the methods and rules to minimize the risk to human health and prevent future damages at the environmental, economic, and societal levels.

Sustainable and Resilient Engineering: Drivers, Metrics, Tools, and Applications, Second Edition. Krishna R. Reddy, Claudio Cameselle, and Jeffrey A. Adams.

Companion website: www.wiley.com/go/reddy/sustainableresilientengineering2e

12.2 Emergence of the Risk Era

As increasing amounts of chemicals exhibiting harmful or toxic characteristics were utilized by industry and in consumer products, a greater need arose to evaluate the risks of human exposure to these compounds. The US Congress issued a directive to study the institutional means for risk assessment. In 1981, the Committee of Institutional Means for Assessment of Risks to Public Health was formed. The conclusions of its work were reported in the eponymous *RED BOOK: Risk Assessment in the Federal Government: Managing the Process*, published in 1983 (National Research Council 1983). Experts from the administration, industry, and academia formed a committee, which reviewed available literature on risk assessment and analyzed the regulations and operations of the federal agencies in relation to the use of harmful substances. The conclusions and recommendations in the *Red Book* have been a valuable contribution for risk assessment and environmental health.

During the 1970s, public interest increased with respect to the effects of technology on the environment and public health. By that time, evidence had been accumulated regarding the carcinogenic effects of some commonly used substances and materials, such as asbestos. The concerns of the public regarding health and environmental risk and safety motivated politicians and regulatory agencies to act in developing objectives and scientific-based methods to estimate the risks to human health posed by these products and activities. The study conducted by the Committee of Institutional Means for Assessment of Risks to Public Health was focused primarily, although not exclusively, on the increased risk of cancer due to exposure to chemicals in the environment. One key finding of the study was the necessity of separating risk assessment methods from risk management. Further, the study identified the necessity of performing risk analyses based on uniform guidelines for all regulatory agencies, and a uniform framework of risk assessment should include the following four steps: hazard identification, dose–response assessment, exposure assessment, and risk characterization. The commission also suggested the necessity of a single organization to be designated to perform risk assessment for all regulatory agencies.

12.3 Risk Assessment and Management

As discussed above, risk assessment should be differentiated from risk management. Risk assessment can be defined as the process to identify risk and facilitate risk reduction for individuals by implementing control systems and regulations for those hazards identified in the environment. The management of risk is the process of balancing available policy alternatives to select the most appropriate regulatory action. This is done by considering the data from the risk assessment and environmental, economic, and social concerns to reach a decision that permits the minimization of effects resulting from exposure to hazardous materials or activities. Figure 12.1 establishes the differentiation between risk assessment and management and includes the four steps of risk assessment, which can be defined as follows:

- Hazard identification is the identification of those chemicals with a causal link to particular diseases or health effects.
- Dose–response assessment is the relation between exposure to specific concentrations of hazards and the probability of developing a specific disease or health problem.
- Exposure assessment is the determination of the degree of exposure to hazards before and after the application of risk management actions.

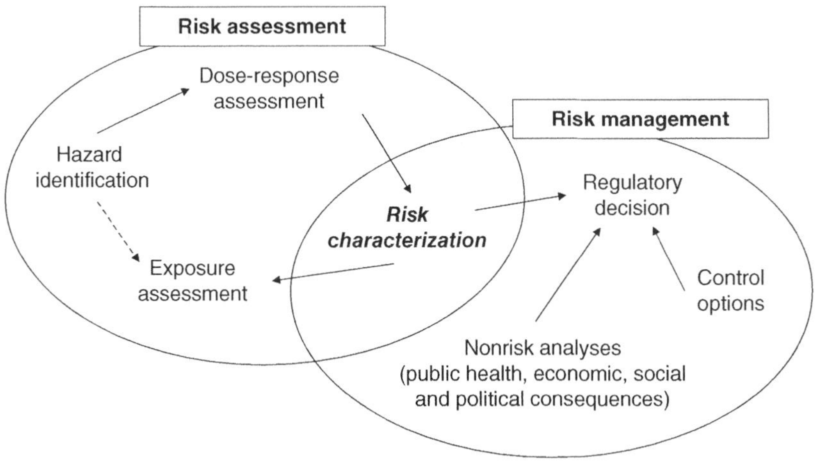

Figure 12.1 Risk assessment and management.

- Risk characterization is the definition of the nature and intensity of the risks to the human health and the environment, including the degree of uncertainty of the effects after exposure to particular hazards.

12.3.1 Hazard Identification

A hazard is anything that can produce conditions that generate damage in health or safety (USEPA 2023). In general, we identify hazards as chemicals that can produce an accident (e.g. gasoline may explode or burn at a gas station) or affect our health through exposure (e.g. organic solvents inhaled in a workplace and smoke from a cigarette). Other elements or situations can be identified as hazards if they can threaten our safety or health, for instance, the storage of harmful chemicals close to a community.

The first step in risk assessment is the identification of those substances or chemicals that exhibit any specific toxic effect; in other words, the first step involves identifying if a causal link exists to particular health effects at environmentally relevant concentrations. The problem with the identification of the toxic capacity of chemicals is the huge number of chemicals that are available. More than 50 000 chemicals exist, and it is reported that each year approximately 1000 new chemicals are synthesized. As a result, the identification of the risks for humans and the environment associated with exposure to these chemicals is daunting due to the limitation of available data to assess the possible effect of chemicals in different scenarios. Additionally, the presence of multiple compounds may have an additive or multiplicative relationship with respect to risk. Further, some chemicals at very low concentrations may exert a significant risk over a mid- or long-term period (Neale et al. 2015).

Substances that are suspected to show specific toxic effects for living organisms are listed as chemicals of potential concern (COPCs). With COPCs, precautions are encouraged when managing these substances until further studies definitively conclude if a particular substance causes health problems or is linked with any specific disease. It is also important to identify if exposure to a specific substance results in critical health effects, such as the development of cancer or serious noncancer maladies. For these compounds, additional study is needed to establish if scientific evidence exists of a specific critical health effect or effects.

The identification of a hazard is not a problem in itself. The important thing is the probability of a dangerous incident, an accident, or an exposure that may occur due to the hazards. This defines risk as the probability of an incident of exposure associated with a hazard that may occur.

Once a chemical has been identified as being capable of producing harmful or toxic effects, the next step is the determination of the concentration in the environment and the possible means of exposure that can produce those harmful effects. Exposure to chemicals may take place by contact with skin, eyes, ingestion, or inhalation. The concentration of chemicals or contaminants in air, water, soil, vegetables, food, etc., is indicative of the level of risk for humans and ecosystems.

12.3.2 Dose–Response Assessment

The health response to specific doses of chemicals helps to define the toxicity of the studied compound. Thus, it is possible to determine the relationship between the magnitude of the administered, applied, or absorbed dose and the probability of occurrence and magnitude of health effect(s). Figure 12.2 shows the possible route through which a chemical may enter the body and the possible organs where related toxic effects may appear, as well as the ways the chemical can be excreted from the body.

There are several methods of determining how an organism responds to a specific dose of a contaminant/chemical. The dose of the contaminant is the amount of contaminant received by the individual, and it is commonly expressed as milligrams (mg) of contaminant per kilogram (kg) of body mass. The most common and objective method of knowing the safe dosage is through laboratory testing of animals, such as rodents or other small mammals. The exposition of rodents to specific doses of a chemical or increasing doses may determine the risk of exposure to that chemical. The results of the toxicity test may be reported as follows:

- *Effective concentration (EC)* – The received concentration by the subject necessary to induce a specific reaction or effect, e.g. leg paralysis.

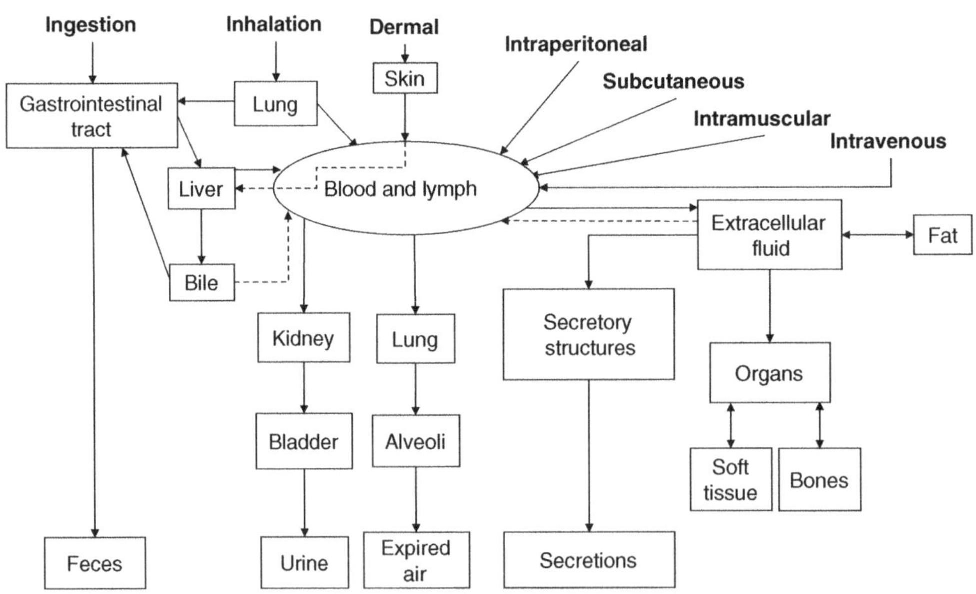

Figure 12.2 Exposition to hazards and effects on body.

- *Lethal concentration (LC)* – The necessary concentration to produce the death of the subject in a specific period of time. Usually, it is reported as LC_{50}, which is the dose to produce the death of half of the individuals in the test.
- *Lowest observable effect concentration (LOEC)* – The lowest dose resulting in a detectable specific health effect in the subject.
- *No observed adverse effects level (NOAEL)* – The maximum dose with no measurable negative effect in the subject.
- *Maximum allowable toxicant concentration (MATC)* – The maximum dose that does not provoke any significant negative impact in the subject.

Then, the laboratory results can be extrapolated to humans and the ecosystem to determine the maximum limits of exposure to prevent negative impacts on human or environmental health. Alternative methods include medical examinations or checkups for people commonly exposed to a chemical or a specific environment (e.g. factory workers who may be exposed to organic solvents in the air). Thus, the analysis of the impact of diseases on a specific group of people who may be exposed to a chemical or a specific environment may help determine if a specific compound or group of compounds may induce the development of specific diseases.

Figure 12.3 shows a typical dose–response curve for a specific substance used to identify the risk of carcinogenic effects compared with the frequency of tumors when there is exposure to a noncarcinogen. The comparison of the two plots may be used to determine if the tested substance is carcinogenic or not carcinogenic, as well as to identify the dose where carcinogenic effects begin to appear. For carcinogens (Figure 12.3), slope factor (SF) or cancer potency factor (CPF) is determined. For noncarcinogens, the reference dose (RfD) is determined. The RfD, as defined by USEPA (1993), is an estimate, with uncertainty spanning approximately an order of magnitude, of daily oral exposure to the human population (including sensitive subgroups) that is likely to be without an appreciable risk of deleterious effects during a lifetime. RfD is commonly reported in units of milligrams per kilogram per day (mg/kg d). In the determination of the RfD, the NOAEL is considered, and the uncertainty factor (UF) is incorporated, which accounts for the different effects of toxicity between humans and animals used in laboratory toxicity tests and variation among several human populations. USEPA uses the RfD as the limit of exposure to chemicals to assure little or no adverse health effects during a lifetime duration.

12.3.3 Exposure Assessment

When a substance is determined to have a negative health effect, and there is a causality between exposure to a substance and the development of specific disease(s), it is important to assess exposure on a qualitative or quantitative basis. Exposure to a harmful chemical may occur through the medium or media where the chemical is present: air, water, or soil. It is necessary to determine the concentration in the medium to determine the risk. The receptor can be any organism that contacts the medium. A contaminant in the air can be inhaled by humans or by animals; if it is present in the water, the aquatic organisms would be directly affected, but any other human or animal may be exposed to the chemical via ingestion, dermal absorption, or inhalation of vapors emanating from impacted water. Soil impacts may directly affect plants, but humans and animals may also be exposed when these plants are introduced into the food chain.

In the exposure assessment, the concentration of the chemical is critical to determine the risk of exposure, but it has been demonstrated that chronic, continuous exposure to even low concentrations may have negative effects that could be worse than short-term, acute exposure to high concentrations. There is a cumulative exposure effect that needs to be considered when evaluating risk. Furthermore, some classes of contaminants tend to be accumulated within organisms

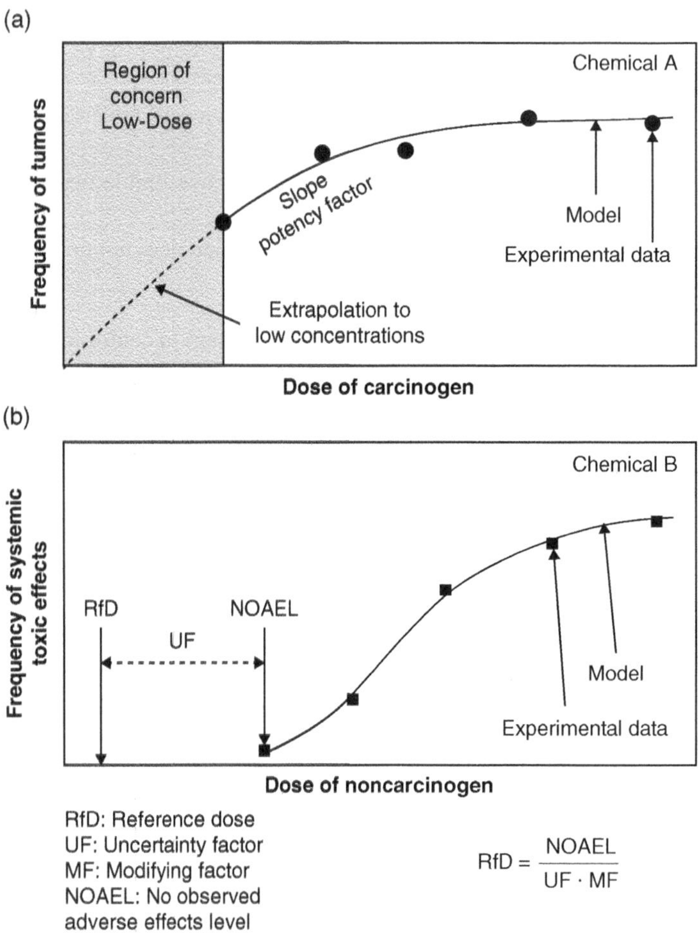

Figure 12.3 Determination of reference dose for (a) Carcinogens, and (b) non-carcinogens. *Source:* USEPA (1993)/US Environmental Protection Agency/Public domain.

(bioaccumulation). For instance, some metals or organics tend to accumulate within fatty tissues, and the toxic effects appear when the concentration within the organism reaches a specific level, even if the concentration in the environment is much lower. This is defined as the accumulation factor, which is the ratio of the contaminant in an organism to the concentration in the ambient, steady-state environment, where the organism can intake the contaminant via the exposure pathways described above.

The quantification of exposure must include the concentrations in the medium, the possible intake, and the time period of exposure. As an example, exposure to air pollutants associated with automobile exhaust is going to be more significant in policemen or bridge toll collectors who may be exposed for several hours on a repetitive basis, as compared to other people who are infrequently exposed to exhaust and for short periods.

Equation (12.1) is the general formula for estimating chemical exposures by inhalation.

$$I = \frac{C \cdot \mathrm{CR} \cdot \mathrm{ET} \cdot \mathrm{EF} \cdot \mathrm{ED}}{\mathrm{BW} \cdot \mathrm{AT}} \tag{12.1}$$

where

I is the chemical intake (mg/kg d)
C is the chemical concentration (mg/m^3)
CR is the contact rate (m^3/h)
ET is the exposure time (h/day)
EF is the exposure frequency (day/year)
ED is the exposure duration (years)
BW is the body weight (kg)
AT is the average time (days)

12.3.4 Risk Characterization

Risk characterization is a complex process where all the data from the risk assessment is considered. Thus, the data from hazard identification and dose–response assessment, as well as the exposure assessment may help in the characterization of risk, keeping in mind that there will be in general a degree of uncertainty. The risk characterization should include the list of identified hazards, their effects, modes of exposure, transportation, and methods of prevention. This information will be used to develop the risk assessment.

The risk characterization can be defined as the likelihood of adverse effects on human health due to exposure to a hazard and defines the maximum exposure level to avoid such negative effects. Thus, the cancer risk equation (Eq. 12.2) permits the determination of the likelihood of developing a tumor due to exposure to a substance with carcinogenic potential. The cancer risk calculated in Eq. (12.2) is defined as the probability of an individual developing cancer:

$$\text{Cancer risk (CR)} = I \times \text{CPF} \tag{12.2}$$

where

I = chronic daily intake averaged over 70 years (mg/kg yr)
CPF or SF = cancer slope (or potency) factor obtained from the toxicity assessment (mg/kg/day) as shown in Figure 12.3a.

The likelihood of noncancerous health effects is presented in Eq. (12.3). The hazard quotient (HQ) is defined as the ratio of the potential exposure to the substance and the level at which no adverse effects are expected.

$$\text{HQ} = I/\text{RfD} \tag{12.3}$$

where

I = exposure level (or chemical intake) obtained from exposure assessment
RfD = reference dose obtained from the toxicity assessment for the same exposure pathway as I.

When the HQ is less than or equal to one ($\text{HQ} \leq 1$) adverse noncancer effects are not likely to occur from the exposure pathway and to that particular chemical assuming the exposure conditions factored into Eq. (12.3). An HQ greater than 1 indicates that a risk has exceeded the assumed threshold for a noncancerous health effect to occur given how much an exposure concentration exceeds the RfD. An HQ greater than 1 means that exposure to a specific chemical is higher than RfD and that increasing values of HQ means increasing the potential for adverse effects.

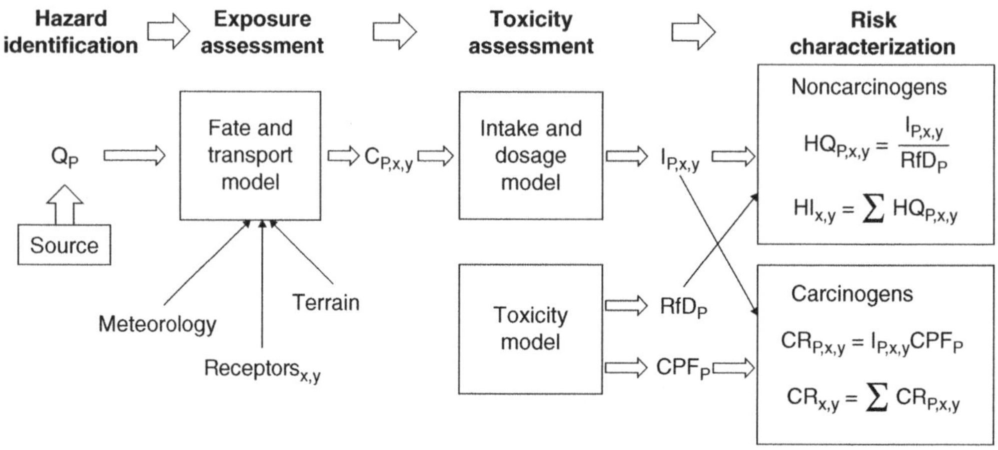

Figure 12.4 Risk assessment sequence. *Source:* Willis and Keller (2007). Reproduced with permission of Elsevier.

If an individual is exposed to several substances with health risk potential for the same organ, the HQ for each chemical can be summed to evaluate the total noncarcinogenic effects. The sum of the HQ for each chemical is called the hazard index (HI). Similarly, exposures below 1.0 for HI likely will not result in adverse health effects over a lifetime of exposure and would commonly be considered acceptable. It should be noted that the HI is only an approximation of the aggregate effect of several chemicals' exposure on the body because different substances might have their negative effects by different (i.e. nonadditive) mechanisms.

While the concept of risk assessment can readily be appreciated, in practice it is extremely challenging. Figure 12.4 is from a study relating land use policy to exposure to air toxics. A detailed computer model, diagrammed here, was required for the analysis. The variables on the diagram are as follows: Q = volatilization rate per unit area; C = contaminant concentration, I = chemical intake; HQ = hazard quotient; HI = hazard index; RfD = reference dose; CR = carcinogenic risk; SF = cancer slope (or potency) factor. Each of these variables must be quantified to carry out the computation.

12.4 Ecological Risk Assessment

The ecological risk assessment can be defined as the process to determine how likely it is that the environment would be impacted due to exposure to one or more stressors, such as contaminant chemicals, climate change, diseases, changes in land use, or other factors (USEPA 2017). The process requires several steps and phases of gathering information, analyzing, and proposing conclusions as depicted in Figure 12.5.

The first step is planning, which can be considered as an initial step preceding the actual ecological risk assessment. The main objective and activity of planning is to define the scope of the study and to gather basic information as described below:

- Define the individuals, species, communities, population, or ecosystems affected that are going to be the target of the risk assessment;

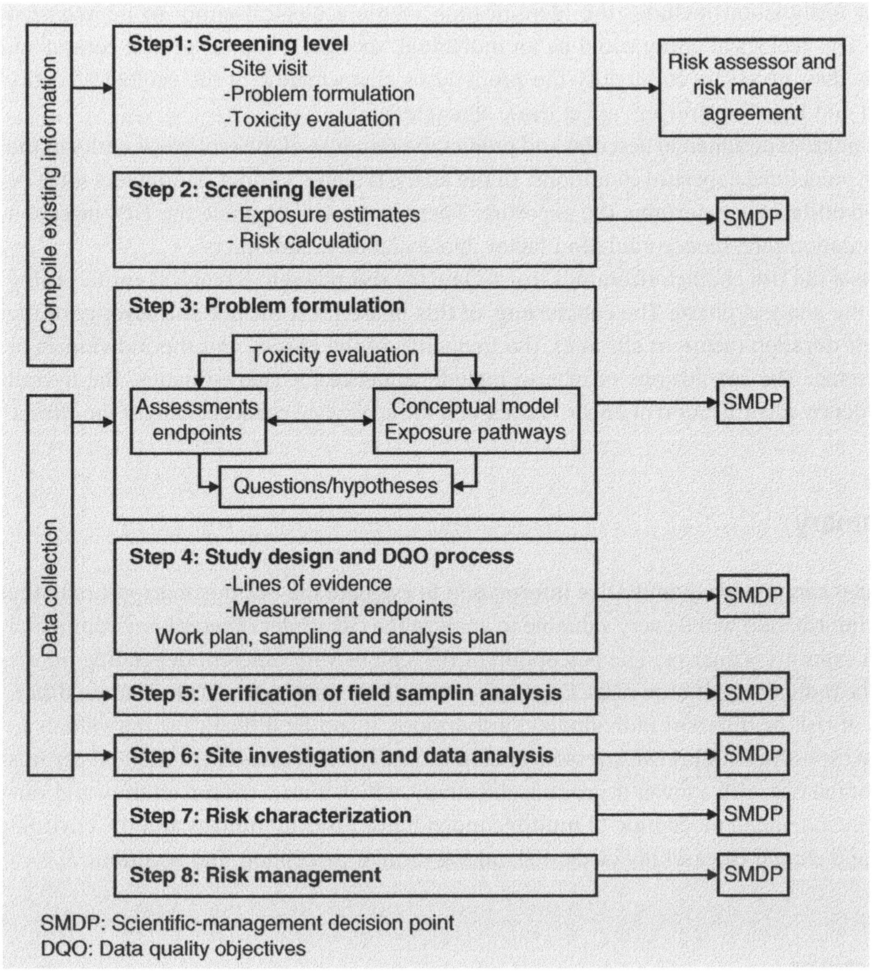

Figure 12.5 Ecological risk assessment process. *Source:* USEPA (2016)/US Environmental Protection Agency Public domain.

- Define the stressors to be analyzed: contaminant chemicals, microbiological (diseases) or biological (invasive species) elements, radiation, physical factors (changes in habitats, flood, etc.), and others;
- Define the source of the stressors: point sources (discharge of an industrial effluent) or dispersed sources (acid rain);
- Determine the exposure pathways (air, water, soil, waste, and food) and routes (inhalation, skin contact, ingestion, etc.);
- Determine the mechanisms of accumulation/assimilation/metabolization of the contaminants in the living organisms;
- Define the ecological impacts of the stressors: diseases and health disorders, tumors, mortality, reproduction effects, etc.;
- Define the duration of the impact: chronic, periodic or intermittent, immediate, and acute.

The ecological risk assessment, once the planning and scope are defined, is comprised of three sequential steps: problem formulation, analysis, and risk characterization.

The problem formulation includes the identification of the ecological entity to be analyzed and protected. The ecological entity could be an individual, species, group of species, ecosystem, or habitat. It is then necessary to identify the property or characteristic of the ecological entity to be protected and how the stressors affect those characteristics.

The analysis phase is designed to describe and predict the response of the ecological entity to the stressors under predefined exposure conditions. In the analysis phase, various parameters must be calculated to quantitatively determine the exposure. These parameters include the HQ, ingestion rate, bioaccumulation rate, bioaccumulation factor, bioavailability, and others.

The objective of the risk characterization is to estimate the risk posed to ecological entities using the data from the analysis phase. The conclusions of this phase must include the severity of the effects and their duration (acute or chronic), the frequency of the effects, and the individuals or population affected. The conclusions must also include evidence for risk estimates, the overall degree of confidence, identification of uncertainties, and the ecological consequences of the effects.

12.5 Summary

A risk assessment can provide quantitative information to evaluate the risk due to exposure to the stressors. This information will be very valuable to analyze the risk under different environmental conditions and exposure scenarios. The perception of risk can be very different depending on the familiarity of the individual with the risks. Exposure to a well-known situation can result in different valuations of risk by different individuals. Furthermore, it can be difficult for individuals to assess risk from exposure to unknown stressors or conditions. The risk assessment method permits an evaluation of the risk with a lower degree of subjectivity, which lowers the uncertainty and variation among the individual perception of multiple individuals. Risks to humans and the environment are real and should be carefully evaluated, taking human perception into account.

12.6 Questions

12.1 What is risk assessment and what are the four steps involved in risk assessment?

12.2 What is hazard identification and what does it encompass?

12.3 Discuss the terms of Eq. (12.1), and how you may select "baseline" representative values for modeling.

12.4 Discuss the differences between acute and chronic health or ecological effects.

12.5 Explain briefly how is a dose–response assessment performed.

12.6 What are the different metrics used for reporting the toxicity of a chemical?

12.7 What are the different exposure routes for contaminant exposure to human and ecological entities?

12.8 What are some compounds or exposure scenarios where bioaccumulation effects are prominent?

12.9 Define the terms: cancer risk, HQ, and RfD.

12.10 When is the hazard quotient approach used?

12.11 Discuss how the presence of multiple chemicals could affect a cancer risk or HQ calculation.

12.12 What are the steps involved in ecological risk assessment? Explain briefly.

12.7 Problems

12.1 A 90-day study on non-carcinogenic chemical exposure to rats revealed the following results:

Dose (mg/kg day)	Observation	Effect level
0	Control	–
2	No significant difference between control and treated animals	No observed effects
10	1.5% decrease in the gain of body weight (not considered to be biologically significant)	No observed adverse effects
100	15% decrease in the gain of body weight	Lowest dose of observed adverse effects

Answer the following questions:

1) Define the reference dose concept and explain various factors used to calculate the reference dose.
2) Using the given experimental data and USEPA guidelines to determine the uncertainty factor, calculate the reference dose of the non-carcinogenic chemical to human exposure, and justify the usage of any uncertainty or modifying factors.

12.2 Outdoor benzene concentration in an industrial area is about 10 μg/m^3. Volume of air intake by an average human is around 6 l/min. For an individual who was living in this area for five years and was in contact with the outdoor atmosphere for an average of eight hours per day on weekdays and five hours per day on weekends, calculate the total chemical intake in milligrams. If the person weighs around 80 kg, calculate the chronic daily intake (mg/kg day) and the cancer risk. (Inhalation slope factor for benzene: 0.1 (mg/kg day)$^{-1}$).

12.3 If the chemical mentioned in Problem 12.2 is a non-carcinogen instead of benzene and if the concentration of the chemical in the atmosphere is 2 μg/m^3, calculate the HQ using the reference dose obtained in Problem 12.1.

References

National research Council (1983). *Risk Assessment in the Federal Government: Managing the Process.* National Academies Press.

Neale, P.A., Ait-Aissa, S., Brack, W. et al. (2015). Linking in vitro effects and detected organic micropollutants in surface water using mixture-toxicity modeling. *Environmental Science & Technology* 49 (24): 14614–14624.

USEPA (1993). *Reference Dose (RfD): Description and Use in Health Risk Assessments.* US Environmental Protection Agency. https://www.epa.gov/iris/reference-dose-rfd-descriptionand-use-health-risk-assessments#1.1 (accessed May 2017).

USEPA (2016). *Superfund Ecological Risk Assessment 8-Step Process.* US Environmental Protection Agency. https://archive.epa.gov/reg5sfun/ecology/web/html/8stepera.html (accessed November 2024).

USEPA (2017). *Risk Assessment.* US Environmental Protection Agency. https://www.epa.gov/risk (accessed May 2017).

USEPA (2023). *Human Health Risk Assessment.* US Environmental Protection Agency. https://www.epa.gov/risk/human-health-risk-assessment (accessed November 2024).

Willis, M.R. and Keller, A.A. (2007). A framework for assessing the impact of land use policy on community exposure to air toxics. *Journal of Environmental Management* 83 (2): 213–227.

13

Emerging Sustainability and Resiliency Assessment Tools

Chapter Goal and Objectives

The goal of this chapter is to explore emerging environmental, economic, and social assessment tools and examine how their integration fosters a comprehensive sustainability and resilience assessment framework. The major objectives are (i) to identify and discuss emerging environmental assessment tools; (ii) to examine emerging economic assessment tools such as life cycle costing, cost–benefit analysis, and ecosystem services valuation; (iii) to review social assessment tools like environmental justice tools and social vulnerability assessment tools; (iv) to explore tools that integrate environmental, economic, and social aspects for sustainability assessment; and (v) to present emerging tools that integrate sustainability and resilience.

13.1 Introduction

As discussed in Chapters 6 through 12, several initiatives and approaches have been developed toward the assessment and quantification of the degree of sustainability associated with a product/project. The qualitative tools that we have discussed earlier do not thoroughly assess environmental, economic, and social impacts. However, several actions taken in the form of best management practices could aid in incorporating some degree of sustainable practice. Semiquantitative tools rule out the ambiguity and uncertainty in sustainability assessment by incorporating ratings or scores to identify relative impacts among different alternatives. However, they may be of limited utility due to subjectivity of the person who is performing the assessment. Quantitative tools like life cycle assessment (LCA) are perhaps the most standardized and suitable way of assessing environmental sustainability. However, they do have limitations such as the large amount and detailed data required for inventory analysis and reliability of the data, among others. In lieu of this, many new and innovative sustainability assessment tools with specific applicability are being developed. This chapter identifies and summarizes some of these emerging tools that try to address the issues and challenges that are being faced with the existing tools for sustainability assessment and highlights the major improvements in these tools.

Sustainable and Resilient Engineering: Drivers, Metrics, Tools, and Applications, Second Edition. Krishna R. Reddy, Claudio Cameselle, and Jeffrey A. Adams.

Companion website: www.wiley.com/go/reddy/sustainableresilientengineering2e

13.2 Environmental Assessment Tools/Indicators

The importance of environmental protection and awareness toward resource depletion has gained significant importance over the last two decades. Further, the grand challenges of global climate change and population growth have made sustainable development and practices imperative in building and civil infrastructure. The same principles are generally being applied to engineering and technology in general to realize sustainable development. This, in fact, has spurred a great interest in the development of tools to assess the environmental impacts of a product/ project and the associated sustainability. Over the years, several qualitative tools, such as simple yes/no tools or spreadsheets, semiquantitative tools, such as relative rating/scoring tools, and quantitative tools, such as LCA, have been developed for environmental impact assessment. Several new environmental assessment tools (qualitative, semiquantitative, and quantitative) have been developed specifically for the impact assessment of building and urban development. Some of the qualitative and semiquantitative tools are BREEAM (Building Research Establishment Environmental Assessment Method) Communities, CASBEE-UD (Comprehensive Assessment System for Built Environment Efficiency - Urban Development), SBTool^{PT}-UP (Sustainability Assessment Tool for Urban Planning), and Pearl Community Rating System (PCRS). Leadership in Energy and Environmental Design (LEED) is specifically developed to assess green buildings, while Envision is developed to assess the sustainability of any other civil infrastructure. A detailed explanation regarding the LEED and its application to green building design is presented in Chapter 16. Likewise, the Envision rating system is described in Chapter 17.

The BREEAM tool was launched in 1990 in the United Kingdom by the Building Research Establishment (BRE). It was developed for the assessment of green buildings at first and was later extended for the assessment of urban community, hence, the name BREEAM Community. Likewise, CASBEE-UD launched in 2007 by Japan Sustainable Building Consortium and the Japanese Green Building Council is a tool for the assessment of the environmental performance of buildings. SBTool^{PT}-UP was established by the International Initiative for a Sustainable Built Environment in 2005. SBTool^{PT} specifically assesses the building sector in Portugal. Later, the scope of the tool was also expanded to include the development of sustainable urban planning and design. PCRS, developed by the Abu Dhabi Urban Planning Council in 2010, is the first sustainability assessment tool in the Middle East. PCRS also included "culture" as a fourth dimension of sustainability apart from environmental, economic, and social dimensions. PCRS has been developed based on the differences and similarities between BREEAM and LEED and by also accounting for the spatial and cultural dimensions of UAE (Ameen et al. 2015). Similar rating tools for specific applications such as roadways exist. Examples include GreenLITES (Green Leadership In Transportation Environmental Sustainability) developed by the New York State Department of Transportation; Greenroads developed by the Washington State Department of Transportation; and INVEST (Infrastructure Voluntary Evaluation Sustainability Tool) developed by the Federal Highway Administration (FHWA) of the USA.

Carbon footprint is one of the major environmental indicators for the assessment of environmental sustainability. Since the building sector is highly material and energy intensive, it induces a much larger carbon footprint than many other sectors. However, the embodied greenhouse gas (GHG) emissions in buildings and other features of the built environment are often unaccounted. This is particularly important when trying to select a low-carbon footprint material for building construction. Realizing this difference, Chen and Ng (2015) established a comprehensive model that accounts for a building's embodied GHG emissions, which was integrated into one of the most widely accepted building environmental assessment schemes, LEED. Furthermore, several carbon footprint calculators have been developed or are being continuously developed for specific applications (E.g., Asphalt pavement embodied carbon tool (asPECT), developed by TRL

(Transportation Research Laboratory) Limited; Carbon calculator for foundation structures and ground improvement methods, developed by EFFC (European Federation of Foundation Contractors), and DFI (Deep Foundations Institute); Embodied Carbon in Construction Calculator (EC3), developed jointly by several industry partners including Skanska, and C Change Labs - later being managed and continuously developed by a non-profit organization called Building Transparency).

In addition to the carbon footprint tools, specific LCA tools are also being developed for different applications (e.g. a sustainability calculator for constructions containing geosynthetics, developed by International Geosynthetics Society (IGS); Pavement Life-cycle Assessment Tool (PaLATE), developed by the Consortium on Green Design and Manufacturing at the University of California, Berkeley; SiteWise™ developed jointly by the US Navy, US Army, and USACE for sustainable remediation of contaminated sites; Spreadsheet for Environmental Footprint Analysis (SEFA) developed by USEPA for sustainable remediation of contaminated sites).

The existing sustainability assessment methods focus more on environmental aspects with limited or less attention toward economic and social aspects of sustainability. A multicriteria decision analysis (MCDA) method that incorporates a wide range of issues in the analysis is essential. Over the years, many researchers have focused on developing computer-based sustainability assessment tools that involve a life-cycle approach in assessing environmental impacts. In a similar manner, Kumanayake and Luo (2017) proposed a conceptual framework with an associated model for developing an automated sustainability assessment tool. The tool assists in building material-related decision-making in early design stages within a computerized environment based on embodied carbon emissions and life-cycle cost with an LCA approach. Numerous studies have shown that embodied energy can account for about 40% to 60% of the total energy use of low-energy buildings (Kumanayake and Luo 2017). The type of materials used in the construction of a structure has a significant impact on its embodied energy content and carbon emissions due to the large quantity of materials used. This concept can also be extended to environmental impact assessment in other sectors/domains.

The framework developed by Kumanayake and Luo (2017) includes an input module, assessment module, and output module, as shown in Figure 13.1. The goal of the proposed framework is to perform sustainability assessment of a building. The criterion for environmental performance is taken as the embodied carbon. The life-cycle stages include raw material extraction, material manufacturing, transportation, construction, maintenance, and demolition. All of these stages contribute to embodied carbon emission of buildings in many ways. The life-cycle cost is selected as the criterion for economic performance, and from the various cost components associated with the building life cycle, initial investment cost, maintenance cost, and deconstruction cost, which are directly relevant to material-related decisions, are considered. The input module is comprised of building modeling software (Autodesk), which is further integrated with a building material database developed using structured query language (SQL). This provides the necessary inventory required for the construction of the entire building as conceptualized in Autodesk.

The assessment module, as shown in Figure 13.1, identifies the scope and impact assessment method to be used for LCA. The quantitative data regarding the environmental performance is obtained for the different life-cycle stages, including material extraction, manufacturing, and transportation. Likewise, the direct costs incurred in each of the life-cycle stages are estimated (specific to the country) to evaluate the economic performance. Finally, in the output module, the environmental and economic performances are quantified, and the integrated performance score of the building (regarding both environmental and economic aspects) is computed. Suitable weights for each criterion were assigned, and the results are aggregated into a single index to finally compute the "Building Sustainability Assessment Index." It should be noted that the proposed framework does not include the social aspects of sustainability. However, the entire framework provides a

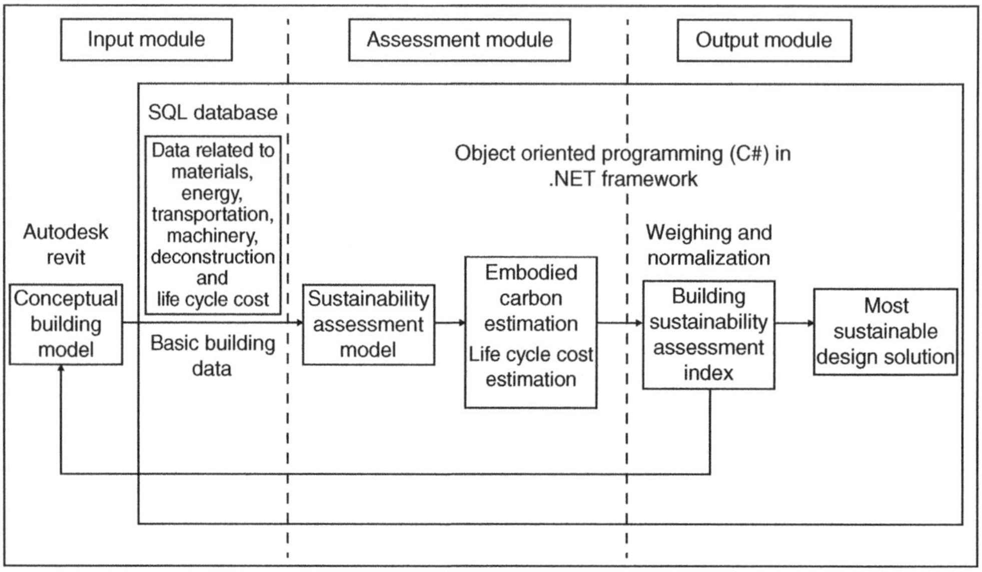

Figure 13.1 Schematic of the framework for integrated sustainability assessment model developed by Kumanayake and Luo (2017) for assessing building sustainability. *Source:* Kumanayake and Luo (2017) with permission of Elsevier.

good and a reasonably accurate assessment of the environmental and economic sustainability of the building.

13.3 Economic Assessment Tools

Development of innovative and practical sustainability assessment tools has been increasing in recent years. In addition, the use of a life-cycle approach in the environmental, economic, and social sustainability evaluation is given due importance. However, the appropriate integration and coherence in a combined sustainability assessment involving LCA, life-cycle costing (LCC), and cost–benefit analysis (CBA) is lacking. The interdependency of environmental and economic concerns and their related costs and benefits needs to be accurately considered to achieve an accurate and meaningful life-cycle sustainability assessment. In a similar manner, Hoogmartens et al. (2014) developed a framework to integrate LCA, LCC, and CBA, which is an element of LCC. A schematic of the framework developed by Hoogmartens et al. (2014) illustrating the interactions between different individual assessment tools is shown in Figure 13.2. This figure shows that within widely recognized tools, such as LCA, LCC, and CBA, many other methodologies may be distinguished that focus on different aspects of sustainability namely, environmental LCA (eLCA), social LCA (sLCA), financial LCC (fLCC), environmental LCC (eLCC), full environmental LCC (feLCC), societal LCC (sLCC), financial CBA (fCBA), environmental CBA (eCBA), and social CBA (sCBA) (Hoogmartens et al. 2014).

13.3.1 Life-Cycle Costing

Life-cycle cost is similar to life-cycle analysis except that it involves evaluating the impacts of different processes and materials in the life-cycle stages in terms of monetary value. The costs within the life cycle of a product or a project would include the direct costs such as the cost of materials,

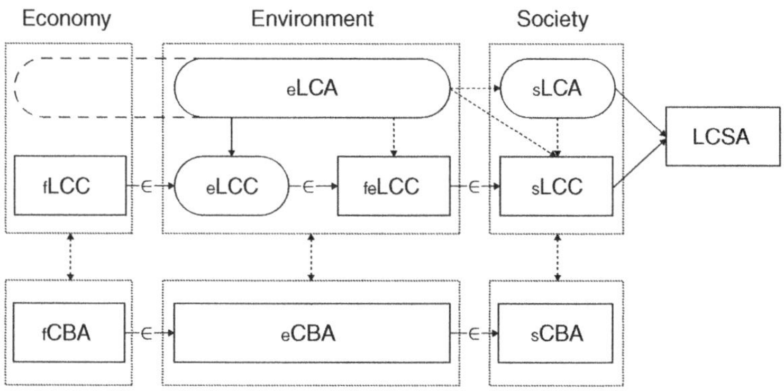

Figure 13.2 Schematic of the integrated framework for life-cycle sustainability assessment. *Source:* Hoogmartens et al. (2014). Reproduced with permission of Elsevier.

labor, equipment, and transportation. Indirect costs, which are often linked with environmental and societal aspects, can also be evaluated by assessing the monetary impacts on the submethodologies that were mentioned earlier.

Specifically, to deal with environmental, social, and financial concerns, four LCC types are used. The definitions and a brief explanation about each of these LCC types, as presented in Hoogmartens et al. (2014), are given below.

Conventional LCC assessments that focus on private investments from a firm or a consumer are categorized as fLCCs. Since economic lifetime is a major factor, the fLCC often does not account for the complete life cycle. An example of fLCC includes investment costs, research and development (R&D) costs, and sales revenues (all of which are negative costs).

eLCCs build upon data of fLCC and extends it to life-cycle costs borne by other firms. The focus of an eLCC is on the real internalized cash flows with no conversion to monetary measures from environmental emissions (e.g. implementation of CO_2 taxes, waste disposal costs, and global warming adaptation costs).

feLCCs are an extension of the eLCC with environmental costs that can be identified by methods such as eLCA. It is important to note that the conversion of environmental impacts into monetary values is not straightforward and must be performed carefully.

Lastly, in the sLCC, the assessment is performed on the costs incurred by anybody in the society, be it today or in the future. In addition, the social impacts associated with all the life cycle stages of a product are considered. Public health impacts must be quantified and converted into monetary values, which is often a challenge in carrying out this type of assessment.

13.3.2 Cost–Benefit Analysis

In simple terms, CBA is a method to evaluate the net monetary value at a discount rate associated with a product or project. If the net value is positive, the product/project achieves profitability. The CBA can be influenced by financial, environmental, and social concerns, and accordingly, there are three types of CBA: namely, fCBA, eCBA, and sCBA, respectively. There can always be a combination of the three CBA types mentioned for an overall CBA analysis.

The fCBA is a tool for estimating the profitability of an individual firm based on direct costs and cash flows in a project. This CBA, however, does not reflect the environmental and social aspects in the CBA that are essential for a more comprehensive sustainability assessment. The eCBA is the tool for analyzing the indirect costs and profitability caused by environmental impacts. This is obtained by expressing the damage caused by environmental impacts in terms of their equivalent monetary value. Some of the examples of costs pertaining to eCBA include pollution costs (e.g. GHG emissions), damage to ecosystems, groundwater pollution, and surface water pollution. In addition, the eCBA integrates financial aspects as well. The sCBA assesses a product/project by considering the societal benefits and well-being. However, the measure of the welfare of the society is in terms of the monetary value to be able to compare this aspect across different product/project alternatives. Some of the examples of social benefits include the development of commercial and recreational centers for communities, reduction in noise pollution, reduction in congestion, increase in employment opportunities, and increased safety in the community. Similar to eCBA, the sCBA integrates the information of fCBA and eCBA since it comprises both financial and environmental aspects.

13.4 Ecosystem Services Valuation Tools

Ecosystem services are the benefits that ecosystems provide to people (Millennium Ecosystem Assessment 2005). In recent years, there has been an increasing importance given to the integration of ecology, economics, and geography to support the decision-making on ecosystem services. However, ecosystem service assessments need to be quantifiable, replicable, credible, flexible, and affordable (Bagstad et al. 2013). A variety of decision support tools (DSTs) have been developed in support of a more systematic assessment of ecosystem services. Numerous approaches for integrating ecosystem services into both public- and private-sector decision-making processes are being developed. Some of the tools developed are generalizable, while many others are site-specific. Further, these tools may differ in the way they handle economic valuation, spatial and temporal representation of services, and incorporation of existing biophysical models (Bagstad et al. 2013). The tools used for the assessment of ecosystem services constitute the evaluation of resource conservation, land-use planning, and hydrologic and ecological modeling that are, in fact, important aspects of sustainable development. A study by Bagstad et al. (2013) presented an extensive critical review of 17 ecosystem services assessment tools that assess, quantify, model, value, and/or map ecosystem services by applying these tools to a common site. The 17 tools reviewed are listed as follows:

- Ecosystem Services Review (ESR)
- Integrated Valuation of Ecosystem Services and Tradeoffs (InVEST)
- Artificial Intelligence for Ecosystem Services (ARIES)
- LUCI
- Multiscale Integrated Models of Ecosystem Services (MIMES)
- EcoServ
- Co$ting Nature
- Social Values for Ecosystem Services (SolVES)
- Envision
- Ecosystem Portfolio Model (EPM)
- InFOREST
- EcoAIM
- ESValue

- EcoMetrix
- Natural Assets Information System (NAIS)
- Ecosystem Valuation Toolkit
- Benefit Transfer and Use Estimating Model Toolkit.

The tools were assessed against eight evaluation criteria that are important tool-characteristics desirable of an analytical ecosystem services tool. The eight criteria are listed as follows: (i) quantification and uncertainty, (ii) time requirements, (iii) capacity for independent application, (iv) level of development and documentation, (v) scalability, (vi) generalizability, (vii) nonmonetary and cultural perspectives, and (viii) affordability, insights, and integration with existing environmental assessment. In general, the authors concluded that the tools evaluated differed greatly in their performance against the eight evaluation criteria used.

13.5 Environmental Justice Tools

Environmental justice as defined by USEPA is "the fair treatment and meaningful involvement of all people regardless of race, color, national origin, or income with respect to the development, implementation, and enforcement of environmental laws, regulations, and policies" (USEPA 2024a). Environmental justice forms one of the prime aspects of social sustainability, as it deals with ensuring equal distribution of facilities and benefits to all people in the nearby community or the society in general. There have been some legal tools developed as a part of the formulation of the idea of environmental justice by the USEPA (Ewall 2012). However, several other tools, such as the Environmental Justice Strategic Enforcement Assessment Tool (EJSEAT), were developed for the US Environmental Protection Agency (USEPA) Office of Enforcement and Compliance Assurance. This tool helps identify areas with potentially high and adverse environmental and public health problems. The EJSEAT uses a simple method to identify such areas by using 18 federally recognized and managed databases (EJSEAT 2018).

The Environmental Justice Screening and Mapping Tool (EJSCREEN) is another peer-reviewed tool, with a web-based geographic information system, developed by EPA for nationally consistent environmental justice (EJ) screening and mapping. This tool builds upon the National Environmental Justice Advisory Committee report on EJ screening and prior EPA experiences. EJSCREEN is a "pre-decisional screening tool" that highlights the places with potential EJ concerns. To facilitate comparing the potential environmental burden of a specific area with state or US averages, the current version of EJScreen v2.3 provides 13 Environmental Justice indexes and 13 supplemental indexes (USEPA 2024b). The EJ indexes and supplemental indexes are the combination of environmental indicators with the demographic index and a supplemental demographic index of a given area, respectively. Census block groups of population size in the range of 600–3000 people, which serve as a standard unit of analysis in EJScreen, were used to develop these indexes. The entire United States is currently divided into 242 336 block groups. The EJScreen dataset comprises 7 socioeconomic demographic indicators and 13 environmental indicators. In addition, supplemental indexes, derived from the EJScreen methodology, utilize five-factor supplemental demographic indexes, in contrast to the two-factor demographic indexes. The most recent EJScreen version also includes health- and climate-related indicators. The equations used for calculating the EJ indexes and supplemental indexes, illustrating the relationship between demographic factors and environmental conditions, are presented below (USEPA 2024b):

$$\text{Demographic Index} = (\%\ \text{People of Color}_z + \%\ \text{Low Income}_z)/2$$

$$\text{Supplemental Demographic Index} = (\%\ \text{Low Income}_z + \%\ \text{Unemployed}_z + \%\ \text{Limited English Speaking}_z + \%\ \text{Less than HS Education}_z + \%\ \text{Low Life Expectancy}_z)/5$$

$$\text{EJ Index} = (\text{Environmental Indicator Percentile for Block Group}) \times (\text{Demographic Index for Block Group})$$

$$\text{Supplemental Index} = (\text{Environmental Indicator Percentile for Block Group}) \times (\text{Supplemental Demographic Index for Block Group})$$

Subscript z indicates the z-scores for the socioeconomic indicators, which is a statistical measure showing the distance of an indicator from the state or national mean. Z-scores can be calculated as the ratio of the difference between the value for the block group and the state or national average to the standard deviation of a given indicator at a state or national level. The 13 EJ Index and supplemental index names are as follows (USEPA 2024b):

- Particulate matter (PM2.5)
- Ozone
- Nitrogen dioxide
- Diesel particulate matter (DPM)
- Toxic releases into the air
- Traffic proximity
- Lead paint
- Superfund proximity
- Risk management plan facility proximity
- Hazardous waste proximity
- Underground storage tanks
- Wastewater discharge
- Drinking water noncompliance.

EJSCREEN aids users in pinpointing areas with challenges, such as minority and/or low-income populations (vulnerable or disadvantaged communities) and potential environmental quality concerns. However, several environmental issues, like drinking water quality and indoor air quality, are not yet covered, with adequate quality, coverage, or resolution, in comprehensive national databases. Additionally, EJScreen data for small areas carry significant uncertainty, and even for indicators that assess risks or hazards directly, the data has significant uncertainty.

13.6 Integrated Sustainability Assessment Tools

The development of sustainability assessment models has evolved over time from less accurate qualitative tools to more reliable quantitative tools. However, there is still a great scope for integrating the three aspects (environmental, economic, and social) of sustainability for a complete and adequate assessment of sustainability. An integrated sustainability assessment model or a tool is the one that integrates life-cycle environmental impacts, life-cycle cost (including the indirect costs and benefits), and life-cycle social impacts. A few frameworks exist that integrate the three aspects of sustainability. Fuzzy Evaluation for Life Cycle Integrated Sustainability Assessment (FELICITA), a novel model developed by Kouloumpis and Azapagic (2018), integrates LCA, LCC, and sLCA into a fuzzy inference framework. The use of fuzzy logic in sustainability assessments is most desirable due to the imprecise and uncertain data that is dealt with in the assessment. A good collection of articles on the use of fuzzy logic for LCA to determine the environmental impacts with uncertain data is presented in Kouloumpis and Azapagic (2018).

Fuzzy logic is a method used to describe the degree of truth when the values of variables cannot be determined exactly, which is mostly the case in sustainability assessment. The nature of the variables is often descriptive rather than numerical. The input data for the FELICITA model are the values for the sustainability indicators estimated through LCA, LCC, and sLCA. After the application of the fuzzy logic and the subsequent inference of the data, the individual sustainability indicators are aggregated into composite LCA, LCC, and sLCA indexes. This process is repeated, using the composite indicators as the inputs to obtain an overall life-cycle sustainability indicator for each alternative considered and finally determine the most sustainable option. A schematic representation of the FELICITA model is shown in Figure 13.3. A case study describing the functionality and applicability of the FELICITA model is presented in Kouloumpis and Azapagic (2018).

The FELICITA model aids in resolving the issues with uncertain data with regard to sustainability indicators by using fuzzy inference techniques. The model can address a differing number of indicators within the three sustainability aspects without affecting the overall results. The model

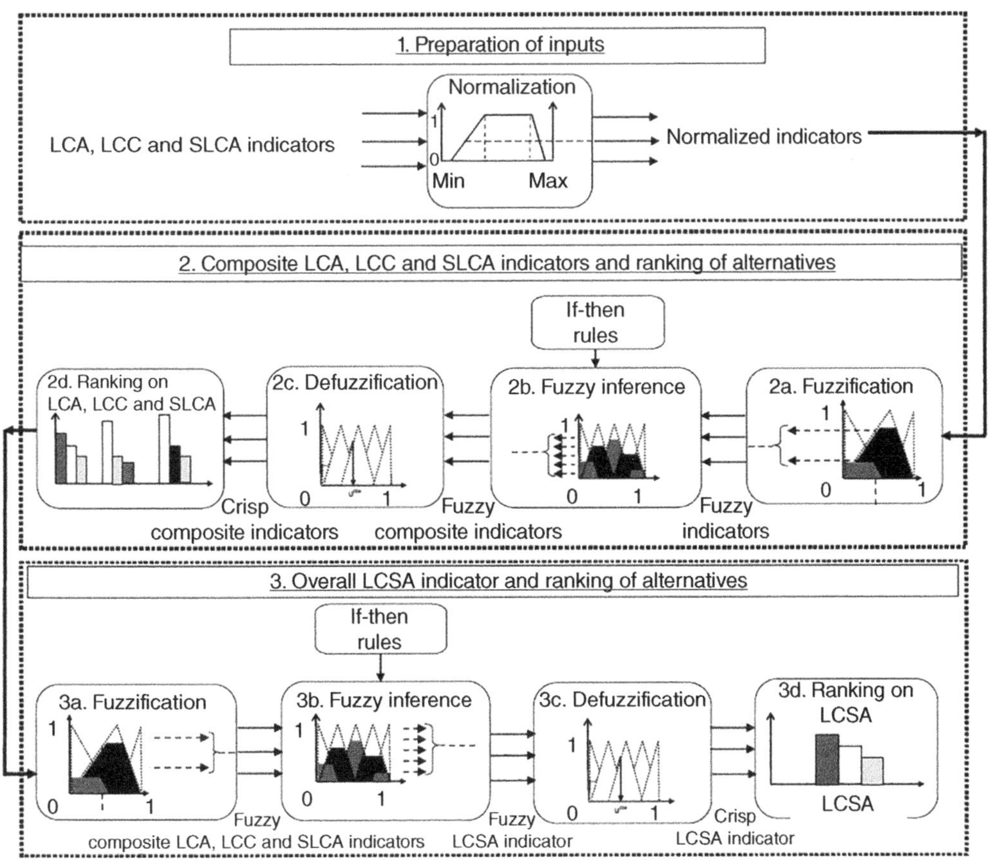

Figure 13.3 Schematic of the FELICITA model. *Source:* Kouloumpis and Azapagic (2018). Reproduced with permission of Elsevier.

can also avoid the need for elicitation of preferences for different indicators and aspects of sustainability, reducing the subjectivity in the decision-making process. The model is capable of accommodating different stakeholder preferences through the choice of different rule bases (neutral, optimistic, or pessimistic). Its further advantage is that it can be easily replicated and adjusted to reflect various changes in the assessment process. The FELICITA model can serve as a useful tool for integrated life cycle sustainability assessments. It can be used by companies for integrated sustainability assessment of the products and services as well as by policymakers to examine the consequences of policies on the overall sustainability.

The Quantitative Assessment of Life Cycle Sustainability (QUALICS) is another integrated sustainability assessment framework/tool that quantifies the triple bottom line aspects (environmental, economic, and social) with a life-cycle perspective. The QUALICS framework is based on combining two multicriteria evaluation methods namely the Integrated Value Model for Sustainable Assessment (MIVES) and analytic hierarchy process (AHP). MCDA is a vital component of sustainability assessment tools as it allows for assessing the uncertainty associated with the data used for sustainability assessment and also identifies the relevance and/or importance of each criterion used in sustainability evaluation. Specifically, the combination of two multicriteria methods MIVES and AHP has proved to be an effective tool to support decision-making in the selection of sustainable options. MIVES is a methodology that combines the concepts of MCDA and value engineering. It uses a generic value function that standardizes each indicator and the AHP method to determine the weights for the different criteria and indicators, which essentially make up the requirements of the sustainability assessment. A detailed description of the QUALICS framework and its application to sustainability assessment of remedial options for contaminated site remediation is presented in Trentin et al. (2019).

13.7 Integrated Sustainability and Resilience Assessment Tools

Some efforts to integrate resilience and sustainability in the past include studies by Roostaie et al. (2019), who proposed an integrated resilience and sustainability assessment framework for buildings, relying heavily on existing rating tools like LEED and Envision™ as the means of quantification (Roostaie et al. 2019; Roostaie and Nawari 2022). Grafakos et al. (2017) attempted an indicator-level integration of sustainability and resiliency for assessing low-carbon technologies. However, this kind of integration makes it difficult to distinguish between these terms. To address this, Reddy et al. (2024) have developed an integrated quantification framework for resilience and sustainability called the TQUALICSR (Tiered Quantification of Life Cycle Sustainability and Resiliency). A brief overview of this framework is provided here.

The TQUALICSR framework categorizes variables representing sustainability/resilience into three hierarchical levels: requirement, criteria, and indicators, similar to that of the QUALICS framework mentioned before. The TQUALICSR framework allows for considering various quantification methods in the form of tiers as follows: Tier-1 (qualitative, e.g. inclusion of a best management practice (Y/N)), Tier-2 (semiquantitative, e.g. surveys/ratings), and Tier-3 (quantitative, e.g. LCA, LCC, etc.). This tiered approach allows for flexibility in selection and quantification of indicators. The integrated quantification methodology, as seen in Figure 13.4, starts with a resilience assessment, which includes quantification of technical resilience as well as the cascading environmental, social, and economic indicators, as described in Chapters 5 and 6, using different tiered metrics depending on tool availability. The resilience quantification should be performed for various disasters of importance depending on the location. This assessment should provide the stakeholders an idea on the resilience of different alternatives for a proposed project. All the alternatives should be subjected

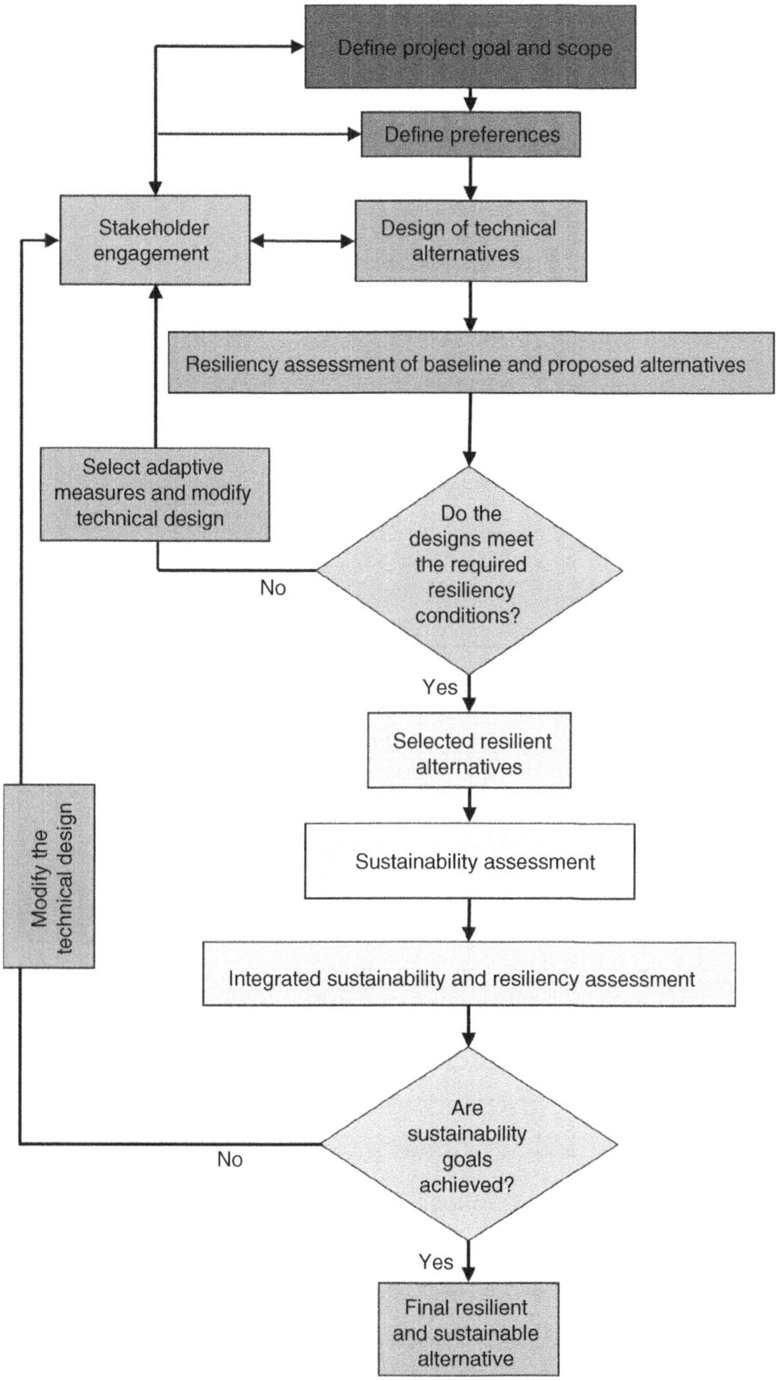

Figure 13.4 Methodology flowchart of the TQUALICSR framework. *Source:* Reddy et al. (2024)/Springer Nature.

to the resilience assessment to ensure that they fulfill all stakeholder preferences for resilience. This exercise results in a set of resilient alternatives for a given project.

Following the resilience assessment, the resilient alternatives should then be subjected to a comprehensive life cycle sustainability assessment. The sustainability assessment includes identification

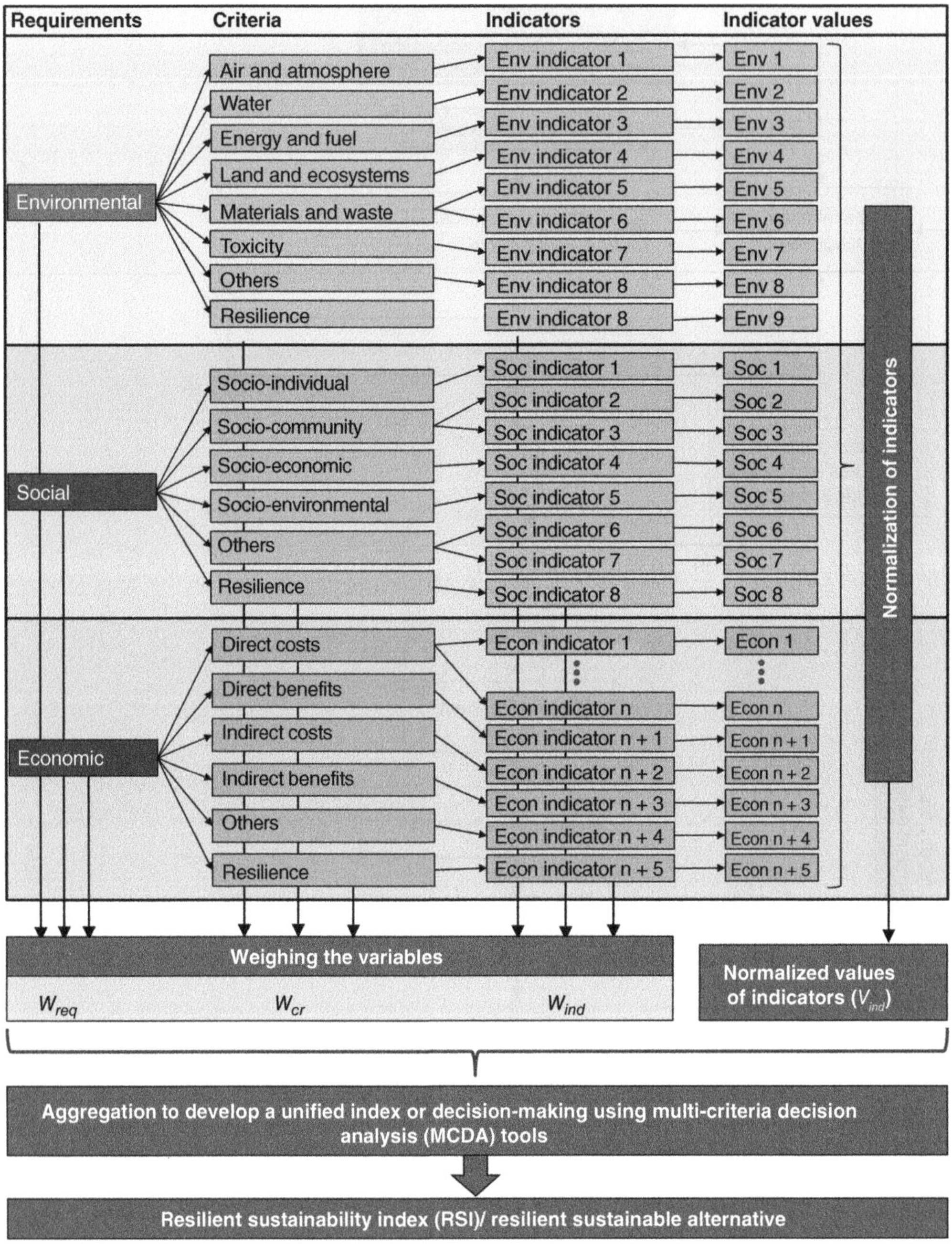

Figure 13.5 Integration methodology for the TQUALICSR framework. *Source:* Reddy et al. (2024)/Springer Nature.

of project-specific indicators and corresponding metrics. The cascading impact indicators quantified during the resilience assessment should be included as one of the criteria in order to holistically consider both sustainability and resiliency. Following the quantification of indicators, the values obtained should then be normalized and integrated, as depicted in Figure 13.5. A detailed description of this framework and case studies on the application of this framework can be found in Reddy et al. (2024).

13.8 Summary

Numerous tools have been developed to estimate/quantify the sustainability of a product or project. Some of the tools are specific in their applicability, while others can be extended to evaluate sustainability in several other domains. Several new qualitative and quantitative tools have been developed in recent years. Most of the recent efforts are focused on incorporating the LCA approach into an integrated sustainability assessment framework. Although environmental sustainability assessment is well established, there is still a major need for adequate and appropriate inclusion of economic and social aspects. Several qualitative tools developed for the environmental assessment of buildings have been developed. Further, many tools are also being developed for systematic assessment of ecosystem services, which are an essential part of sustainable development. Some of the tools that have been developed rigorously account for economic sustainability assessment by including CBA with LCC. This way the tools account for broader economic impacts that are not immediately apparent. In order to realize social sustainability, it is essential to involve all the stakeholders in the decision-making process and make all the voices heard. In this regard, several federally recognized tools such as EJSEAT and EJSCREEN have been developed to uphold environmental justice. Limited tools or frameworks are available for the integrated quantification of resilience and sustainability. While research in this area is still developing, frameworks like TQUALICSR offer promising approaches for addressing this gap.

13.9 Questions

13.1 Name some of the tools developed for the environmental sustainability assessment of building and urban planning and development. Explain these tools briefly.

13.2 Name any 10 ecosystem services assessment tools.

13.3 List and discuss the importance of criteria used to evaluate ecosystem services assessment tools.

13.4 What is life-cycle cost analysis? How is it different from cost–benefit analysis?

13.5 What are the different components/aspects of life-cycle cost analysis? Explain briefly.

13.6 What are the different components/aspects of cost–benefit analysis? Explain briefly.

13.7 What is meant by environmental justice? How is it related to sustainability assessment or the sustainable development of a community?

13.8 Name some of the tools used by USEPA to evaluate environmental justice.

13.9 List the indicator categories into which the EJSAT data is divided. Explain these categories briefly.

13.10 List the 11 environmental justice indexes used by EJSCREEN tool developed by USEPA.

13.10 Problems

13.1 A key and ongoing priority for automobile makers is to enhance fuel efficiency. One effective method is to reduce vehicle weight. To achieve this, manufacturers have increasingly turned to high-strength, lighter-weight materials like aluminum to replace traditional steel. However, aluminum is more costly than steel – approximately \$3000 per ton compared to \$750 per ton for steel. Aluminum-intensive vehicles (AIVs) are projected to reduce weight by replacing half of the steel with stronger aluminum in a ratio of 1 ton of steel to 0.8 ton of aluminum, which is expected to improve fuel efficiency by around 20%.

Assume:

- Current vehicles achieve 25 miles per gallon of gasoline, and gasoline is priced at \$3.50 per gallon.
- Conventional vehicles cost \$20 000 to produce, including \$1000 for steel and \$250 for aluminum.
- AIVs are identical to current vehicles in all respects except for the substitution of lighter aluminum for steel.
- All vehicles are assumed to travel 100 000 miles over their lifetime.
- Tons are measured as short tons (2000 pounds each).

Answer the following:

- Between conventional vehicles and AIVs, which option is more cost-effective over the life cycle given the above data? Create a visual aid to illustrate life cycle cost differences between steel-based vehicles and AIVs.
- What changes in the prices of steel, aluminum, and gasoline would need to occur to shift the cost advantage to the other vehicle type?

13.2 The socioeconomic and few environmental indicators for a location in the United States are given in table below.

Indicator	Value	State percentile	National percentile
Particulate matter 2.5 ($\mu g/m^3$)	9.39	73	81
Ozone (ppb)	73.7	89	91
Nitrogen dioxide (NO_2) (ppb)	16	86	97
Diesel particulate matter ($\mu g/m^3$)	0.475	96	95
Toxic releases to air (toxicity-weighted concentration)	7700	77	90
Indicator	**Value**	**State average**	**National average**
People of color	86%	39%	40%
Low income	45%	28%	30%
Unemployment rate	13%	6%	6%
Limited English speaking households	8%	4%	5%
Less than high school education	16%	10%	11%
Under age 5	6%	6%	5%
Over age 64	15%	17%	18%

Assuming a state-level standard deviation of 21% and national level standard deviation of 23% for all the socioeconomic indicators. Calculate the demographic index and supplemental demographic index for the location. Further, calculate the EJ and Supplemental EJ indexes for all the indicators.

13.3 A state health agency wants to identify communities for targeted intervention programs in your city to reduce exposure to diesel particulate matter (DPM), a known pollutant contributing to respiratory health issues. They have requested an analysis of a 5-mile radius around an industrial zone located at coordinates (34.05223, −118.24368). Use EJScreen to evaluate the DPM exposure percentile and Hazardous waste proximity index (HWP) for a 5-mile radius around the specified coordinates.

- What is the percentile rank for DPM exposure in this area?
- What is the percentile rank for the hazardous waste proximity index (HWP)?
- Comment on various environmental and socioeconomic indicators of the location

13.4 The indicator values for different environmental, social, and economic indicators obtained during a sustainability assessment performed on four different alternatives are presented in the table below:

Requirement	Criteria	Indicator	Unit	Alternative 1	Alternative 2	Alternative 3	Alternative 4
Environmental	Air	Ozone depletion	kg CFC-11 eq	0.3	0.3	0.2	0.1
		Global warming	kg CO_2e	2 882 073.7	2 623 403.5	3 705 080.9	3 373 158.2
		Smog	kg O_3 eq	243 265.9	217 265.1	294 341.0	274 466.6
	Water	Acidification	kg SO_2 eq	22 051.2	19 008.1	17 598.9	14 947.6
		Eutrophication	kg N eq	3501.9	1835.3	5670.0	3153.7
	Human health	Carcinogenics	CTUh	0.1	0.1	0.1	0.1
		Noncarcinogenics	CTUh	0.4	0.4	0.4	0.4
		Respiratory effects	kg PM2.5 eq	2160.7	1917.7	1525.4	1389.5
	Land	Ecotoxicity	CTUe	7625.2	5 135 998.3	374 831.2	7 817 289.0
		Fossil fuel depletion	MJ surplus	6 960 664.6	5 941 067.0	2 472 307.9	2 517 126.9
Social	SI	Health and safety	Rating	3.0	10.0	5.0	13.0
	SC	Public services	Rating	5.0	12.0	5.0	12.0
	SE	Product value	Rating	3.0	6.0	2.0	6.0
	SEn	Aesthetics	Rating	−12.0	−8.0	−1.0	0.0
Economic	DC	Material	USD	2.9	1.7	3.8	2.2
	IC	Environmental cost	USD	126 811.2	115 429.8	163 023.6	148 419.0

Note – Abbreviations: SI – socio-individual, SC – socio-community, SE – socioeconomic, SEn – socio-environmental, DC – direct costs, IC – indirect costs.

Calculate the integrated sustainability indexes for all the four alternatives, considering equal weights. The sustainability index can be calculated using the equations given below:

$$V_{cr} = \sum_{1}^{n} V_{ind} \times W_{ind} \tag{13.1}$$

$$V_{req} = \sum_{1}^{m} V_{cr} \times W_{cr} \tag{13.2}$$

$$V_{final} = \sum_{1}^{k} V_{req} \times W_{req} \tag{13.3}$$

where V_{ind}, V_{cr}, and V_{req} are the values of each indicator, criteria, and requirement, respectively, while W_{ind}, W_{cr}, and W_{req} are their corresponding weightages. V_{final} will give the final integrated sustainability index based on the entire analysis. n, m, and k are the number of variables in the indicator, criteria, and requirement categories, respectively.

Note: It is important to normalize indicator values on a standardized scale (say 0–1), in order for an appropriate comparison.

13.5 In Problem 13.3, adjust the weights of the different requirements by prioritizing one requirement over the others – for example, giving greater weight to environment than to social and economic or vice versa. Comment on the observed differences resulting from these adjustments. Briefly discuss the importance of appropriately weighing different variables during a sustainability assessment.

References

Ameen, R.F.M., Mourshed, M., and Li, H. (2015). A critical review of environmental assessment tools for sustainable urban design. *Environmental Impact Assessment Review* 55: 110–125.

Bagstad, K.J., Semmens, D.J., Waage, S., and Winthrop, R. (2013). A comparative assessment of decision-support tools for ecosystem services quantification and valuation. *Ecosystem Services* 5: 27–39.

Chen, Y. and Ng, S.T. (2015). Integrate an embodied GHG emissions assessment model into building environmental assessment tools. *Procedia Engineering* 118: 318–325.

EJSEAT (2018). *Environmental Justice Strategic Enforcement Assessment Tool (EJSEAT)*. http://www.nam.org/Issues/Energy-and-Environment/Affordable-Energy/Domestic-Energy/Environmental-Justice-Strategic-Enforcement-Assessment-Tool/ (accessed 23 October 2018).

Ewall, M. (2012). Legal tools for environmental equity vs. environmental justice. *Sustainable Development Law and Policy* 13 (1): 4–13.

Grafakos, S., Enseñado, E.M., and Flamos, A. (2017). Developing an integrated sustainability and resilience framework of indicators for the assessment of low-carbon energy technologies at the local level. *International Journal of Sustainable Energy* 36 (10): 945–971.

Hoogmartens, R., Van Passel, S., Van Acker, K., and Dubois, M. (2014). Bridging the gap between LCA, LCC and CBA as sustainability assessment tools. *Environmental Impact Assessment Review* 48: 27–33.

Kouloumpis, V. and Azapagic, A. (2018). Integrated life cycle sustainability assessment using fuzzy inference: a novel FELICITA model. *Sustainable Production and Consumption* 15: 25–34.

Kumanayake, R. and Luo, H. (2017). Development of an automated tool for buildings' sustainability assessment in early design stage. *Procedia Engineering* 196: 903–910.

Millennium Ecosystem Assessment (2005). *Guide to the Millennium Assessment Reports.* https://www.millenniumassessment.org/en/Index-2.html (accessed 20 September 2018).

Reddy, K.R., Janga, J.K., Verma, G., and Nagaraja, B. (2024). Integration and quantification of resilience and sustainability in engineering projects. *Journal of the Indian Institute of Science* 104 (2): 435–488. https://doi.org/10.1007/s41745-024-00440-w.

Roostaie, S. and Nawari, N. (2022). The DEMATEL approach for integrating resilience indicators into building sustainability assessment frameworks. *Building and Environment* 207: 108113.

Roostaie, S., Nawari, N., and Kibert, C.J. (2019). Sustainability and resilience: a review of definitions, relationships, and their integration into a combined building assessment framework. *Building and Environment* 154: 132–144.

Trentin, A.D.S., Reddy, K.R., Kumar, G. et al. (2019). Quantitative assessment of life cycle sustainability (QUALICS): framework and its application to assess electrokinetic remediation. *Chemosphere* 230: 92–106.

USEPA (2024a). *Environmental Justice.* US Environmental Protection Agency. https://www.epa.gov/environmentaljustice (accessed 15 October 2024).

USEPA (2024b) *EJ and Supplemental Indexes in EJScreen.* US Environmental Protection Agency. https://www.epa.gov/ejscreen/ej-and-supplemental-indexes-ejscreen (accessed 15 October 2024).

Section III

Sustainable and Resilient Engineering Practices

14

Emerging Technologies for Sustainable and Resilient Engineering

Chapter Goal and Objectives
This chapter explores key innovations that enhance sustainability and resilience in engineering practices. The objectives are to present: (i) emerging smart technologies like AI (artificial intelligence), big data, sensors, IoT (internet of things), 3D printing, and digital twinning, which improve efficiency, monitoring, and decision-making; (ii) the use of engineered materials, such as recycled waste and nanomaterials, for promoting a circular economy; (iii) resource conservation in the energy sector through alternative energy sources and innovative storage technologies; and (iv) nature-based solutions, including bio-based materials, biomimicry, green infrastructure, and synthetic biology as essential strategies for integrating sustainability and resiliency into engineering systems.

14.1 Introduction

In this chapter, we explore how cutting-edge innovations can reshape the landscape of engineering practices, fostering environmental stewardship and driving sustainable and resilient solutions. Emerging smart technologies, including artificial intelligence (AI), are being developed to address crucial problems such as resource usage optimization, improvements in energy efficiency, and enhancement of predictive maintenance. With fossil fuels nearing depletion, alternative sources of energy and material generation are essential for conserving natural resources and promoting sustainability. Recycling and repurposing waste materials through engineering can reduce waste and conserve resources. Additionally, nature-based solutions and engineering with nature through technologies like synthetic biology (SynBio) and biomimicry can offer sustainable ways to address contemporary challenges without harming the environment. Although some of these technologies are still in their early stages and may have limitations such as data privacy issues, job security concerns, and environmental impacts, they hold the potential to significantly contribute to sustainable and resilient engineering. When properly implemented, each of these technologies can pave the way toward sustainable and resilient engineering practices, improving our planet, its inhabitants, and the economy. While previous chapters introduced the concepts of sustainable and resilient engineering and tools to quantify these aspects, this chapter provides an overview of emerging technologies that can potentially enhance sustainability and resilience in engineering practices and support the achievement of sustainable development goals (SDGs).

Sustainable and Resilient Engineering: Drivers, Metrics, Tools, and Applications, Second Edition. Krishna R. Reddy, Claudio Cameselle, and Jeffrey A. Adams.

Companion website: www.wiley.com/go/reddy/sustainableresilientengineering2e

14.2 Emerging Smart Technologies

Smart technologies refer to devices or systems that use digital technology, connectivity, and often AI to operate interactively and automatically. This section focuses on how emerging technologies in this domain can help contribute to achieving sustainability and resiliency in engineering practices.

14.2.1 Big Data and Artificial Intelligence

AI enables machines to "learn from experience, adjust to new inputs, and perform tasks that typically require human intelligence or more" (Russell and Norvig 2020). As seen in Figure 14.1, although the term AI has been used for several decades, the development of these technologies has been especially rapid in the past decade. Machine learning and deep learning technologies have been a primary focus of research and development, which are capable of understanding and processing a wide variety and quantity of data. In today's world, with multiple intertwined challenges and crises of increasing complexity and uncertainty, the rapid development of AI as an emerging technology offers unprecedented potential to address these issues.

AI technologies offer two main benefits. First, AI automates important but repetitive and time-consuming tasks, allowing humans to focus on higher-value work. Second, AI uncovers insights trapped in massive amounts of unstructured data, such as videos, photos, written reports, business documents, social media posts, and emails, which previously required human management and analysis.

At the heart of sustainable and resilient engineering lies the need for informed decision-making, a domain where Big Data excels. Big Data refers to the enormous volumes of data generated from myriad sources, including sensors, devices, and social media, as well as the ability to manage these

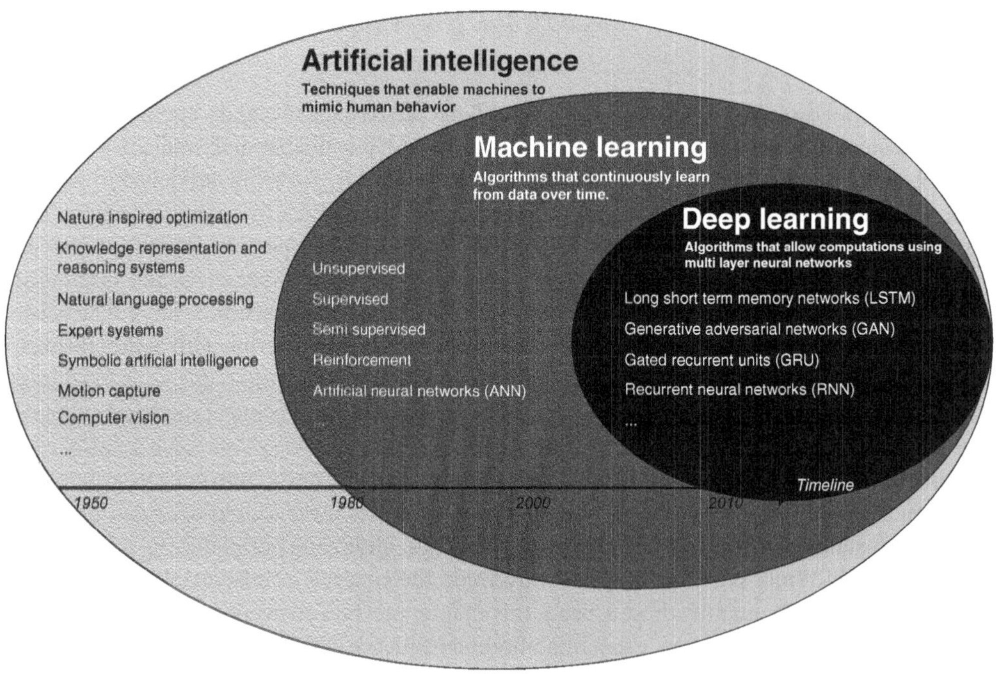

Figure 14.1 The evolution of artificial intelligence. *Source:* Janga et al. (2023) with permission of Elsevier.

data sets, to analyze them, and to draw meaningful conclusions from them (Sagiroglu and Sinanc (2013) and Wu et al. (2013)). This data, characterized by its volume, velocity, and variety, offers unprecedented insights into complex systems.

AI serves as a potent tool in interpreting and leveraging Big Data. With capabilities ranging from machine learning to predictive analytics, AI transforms raw data into actionable insights. In sustainable engineering, for instance, AI algorithms can predict energy consumption patterns, leading to more efficient usage. These algorithms can be used to design or optimize time-of-use tariffs by season or time-of-use, many of which are already in place around the world, and are likely to continue their emergence as a globally accepted standard as energy costs rise, carbon regulations are introduced, and the use and demand for renewable energy continues to grow. Interconnecting power generation sites will allow operators to manage and orchestrate the energy system to prioritize or engage specific power sources or actions, which in turn will reduce operational costs. AI-driven applications may be used to direct the power systems to switch to batteries when electricity prices are high or when grid power usage reaches certain grid current limits. Further, AI-driven predictive maintenance in industrial settings enhances equipment longevity and reduces waste. Aside from this example, many more applications of Big Data and AI can contribute to SDGs, as presented in Figure 14.2 (UN 2024). Further, leveraging AI and big data for predictive risk analysis, infrastructure failure assessment, real-time situational awareness, post-event recovery monitoring, and rapid impact assessment enhances disaster-resilience in engineering systems.

Figure 14.2 Ways in which big data and analytics can contribute to SDGs. *Source:* UN (2024), https://www.un.org/en/global-issues/big-data-for-sustainable-development / United Nations.

Despite numerous benefits, integrating AI and Big Data comes with its own set of challenges. One significant hurdle is data privacy and security. As vast amounts of data are collected and processed, ensuring the privacy and security of this data is paramount. Another challenge is the digital divide. Ensuring equitable access to the benefits of these technologies is crucial for inclusive sustainable development. Moreover, the ethical use of AI, avoiding biases in algorithms and ensuring transparency in decision-making processes, is essential for maintaining public trust. Looking forward, the role of Big Data and AI in sustainable and resilient engineering will only grow in importance.

However, it is imperative that the growing use of AI is accompanied by robust ethical standards and equitable access to technology. Continuous innovation, coupled with responsible governance, will ensure that the integration of Big Data and AI not only advances sustainable and resilient engineering but also contributes to a sustainable future for all. Additionally, a growing concern is focused on the enormous amounts of power used by data centers required to build and operate AI systems, especially with the development of generative AI technologies, which require enormous amounts of data to train the models, questioning the sustainability of using such systems. To address this, some researchers suggest integrating data centers with sustainable energy sources and using energy-saving technologies, such as low-power servers and auxiliary energy-saving devices, to increase the energy efficiency of data centers and reduce their environmental and economic footprint (Rong et al. 2016).

14.2.2 Sensors and Remote-Sensing Technologies (Satellites, UAVs, GIS)

Remote sensing refers to the process of gathering information about an object or phenomenon without direct physical contact. Unlike in situ or on-site observations, remote sensing relies on various technologies such as satellites, drones, or sensors to collect data from a distance. Data obtained through the process of collecting information about the Earth's surface, waters, and atmosphere using ground-based, airborne, and satellite remote sensing platforms is called as Earth observation (EO) data. This approach allows us to study and monitor Earth's surface, atmosphere, and oceans, providing valuable insights for scientific research, environmental monitoring, and resource management. These technologies offer unprecedented capabilities in environmental monitoring, resource management, and infrastructure development, contributing significantly to sustainable and resilient engineering practices across various sectors. Figure 14.3 presents a list of different types of EO data, and the sustainable development goals whose progress can be measured by the use of such data.

Satellites equipped with advanced sensors provide a comprehensive view of the Earth's surface, facilitating the monitoring of environmental parameters on a global scale. For instance, satellite-based sensors can detect changes in land cover, vegetation health, and water quality. The data obtained from these sensors informs land use planning, biodiversity conservation efforts, and water resource management strategies. Studies such as that by Jiang et al. (2019) demonstrate the effectiveness of satellite remote sensing in monitoring and assessing the impacts of land use changes on ecosystem services, highlighting its relevance in sustainable land management practices.

Unmanned aerial vehicles (UAVs) offer a more localized and flexible approach to remote sensing, particularly in areas where satellite imagery may lack the desired resolution. UAVs equipped with multispectral cameras and light detection and ranging (LiDAR) sensors can capture high-resolution imagery and generate detailed elevation models of terrain. These capabilities are invaluable in assessing slope stability, erosion risks, and habitat fragmentation, which are essential considerations in sustainable and resilient infrastructure development and land use planning (Möller et al. 2018).

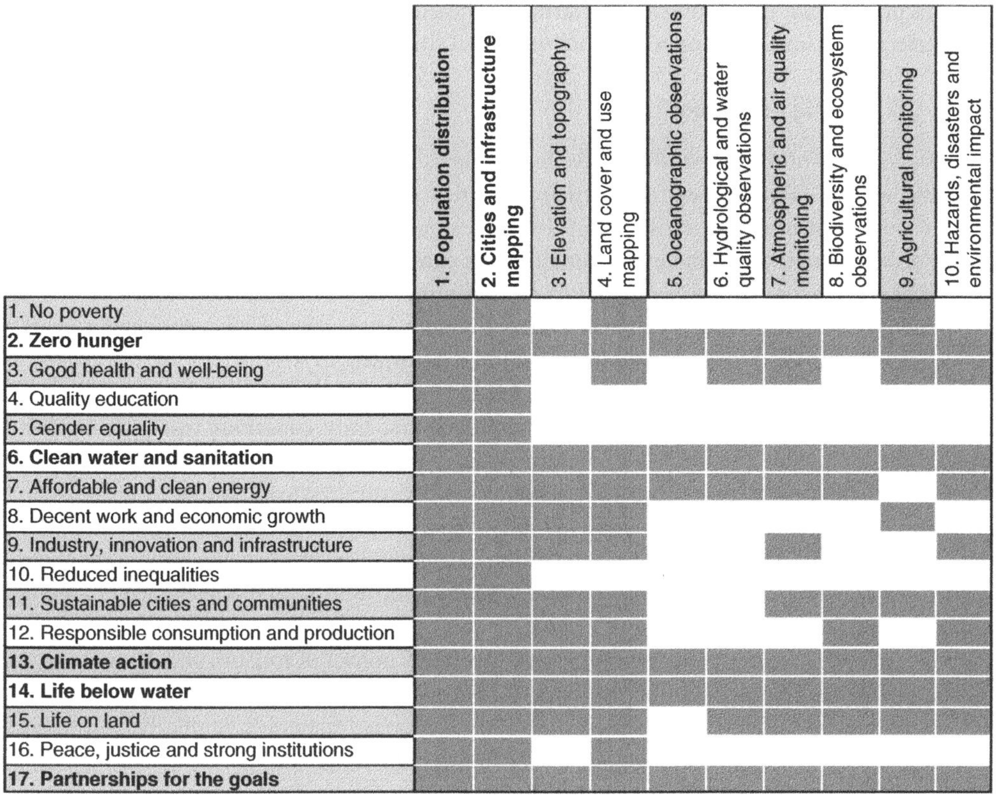

Figure 14.3 SDGs that are measurable by Earth observation data. *Source:* Ferreira et al. (2020)/Springer Nature.

Geographic information systems (GIS) serve as powerful tools for storing, integrating, and analyzing spatial data collected from various sources, including satellites, UAVs, and ground-based sensors. GIS enables the creation of detailed maps and spatial models, facilitating informed decision-making processes in sustainable and resilient engineering projects. For example, GIS-based analysis of urban heat islands and air pollution hotspots helps identify areas in need of green infrastructure interventions to improve air quality and enhance urban resilience (Han et al. 2020).

Sensors play a vital role in monitoring resource consumption, energy efficiency, and waste generation in buildings, industrial facilities, and transportation systems. For instance, smart meters equipped with sensors provide real-time data on energy and water usage, enabling users to identify inefficiencies and implement measures to reduce consumption (Zhang et al. 2019). Remote-sensing technologies contribute to optimizing agricultural practices, a critical aspect of sustainable and resilient food production. Satellite imagery and UAV-based farm equipment enable farmers to monitor crop health, detect pests and diseases, and optimize irrigation scheduling.

Moreover, sensors and remote-sensing technologies play a crucial role in developing resilient infrastructure systems capable of withstanding environmental hazards and climate change impacts. For example, remote-sensing data is used to assess flood risks, landslide susceptibility, and coastal erosion vulnerabilities, informing the design and planning of infrastructure projects (Zhang et al. 2021b). Additionally, sensors embedded in infrastructure assets, such as bridges and dams, provide real-time monitoring of structural health and performance, enabling early detection of defects and timely maintenance interventions (Ghandehari et al. 2019).

14.2.3 3D Printing and Robotics

Robotics and 3D printing are transformative technologies with immense potential to revolutionize sustainable and resilient engineering practices. Robotics applications may enhance efficiency and reduce resource consumption across various industries by automating tasks and streamlining processes. For instance, in manufacturing, robotic arms can optimize material usage and minimize waste generation by precisely executing tasks with minimal variance and within minimal margin of error. This capability proves invaluable in optimizing energy usage and enhancing overall resource efficiency, while maintaining the technical robustness.

The advent of 3D printing has revolutionized manufacturing processes by offering flexibility, customization, and material efficiency. This approach not only conserves resources but also enables the production of complex geometries that were previously unattainable. Moreover, 3D printing facilitates decentralized production. This decentralization minimizes logistics and transportation-related carbon emissions, reducing the environmental footprint associated with mass production and distribution. Further, the use of robots/3D printers to generate components of structures may help avoid man-made errors, thereby contributing to the resilience of an engineering system.

Several real-world examples highlight the transformative impact of robotics and 3D printing on sustainable engineering practices. In the construction sector, companies are utilizing robotic 3D printers to fabricate building components onsite using recycled materials. Similarly, researchers are leveraging 3D printing technology in the healthcare industry to create customized medical devices and prosthetics with minimal material waste.

The synergy between robotics and 3D printing opens new possibilities for sustainable and resilient engineering innovation. By leveraging both technologies' strengths, manufacturers can achieve higher production efficiencies while reducing human labor and material waste. Further, the construction industry can immensely benefit both in terms of efficiency and effectiveness by leveraging these emerging technologies.

14.2.4 Digital Twinning

Digital twinning involves creating precise digital replicas of physical entities, utilizing advancements in computational sciences and sensor-based, real-time data. Initially applied in manufacturing, digital twins can simulate various living and nonliving systems. These virtual models allow for the dynamic representation and monitoring of complex systems, enhancing the ability to predict behavior and optimize performance across different conditions (Tzachor et al. 2022). The scope of digital twins has expanded to include diverse applications such as public health, planetary science, urban systems, and climate forecasting.

In the context of sustainable engineering, digital twins hold considerable promises for achieving SDGs, as shown in Figure 14.4. They can enhance resource efficiency, support the safe innovation of green technologies, and facilitate comprehensive monitoring. For instance, digital twins can optimize water supply systems by integrating smart technologies like leak detection sensors and predictive maintenance tools, which improve performance and reduce waste.

A case study in this regard is the digital twin for the 30 000-person town of Herrenberg in Germany (Dembski et al. 2020). The digital twin incorporates a detailed 3D model of the built environment, urban mobility simulations, wind flow simulations, and sensor data to optimize urban

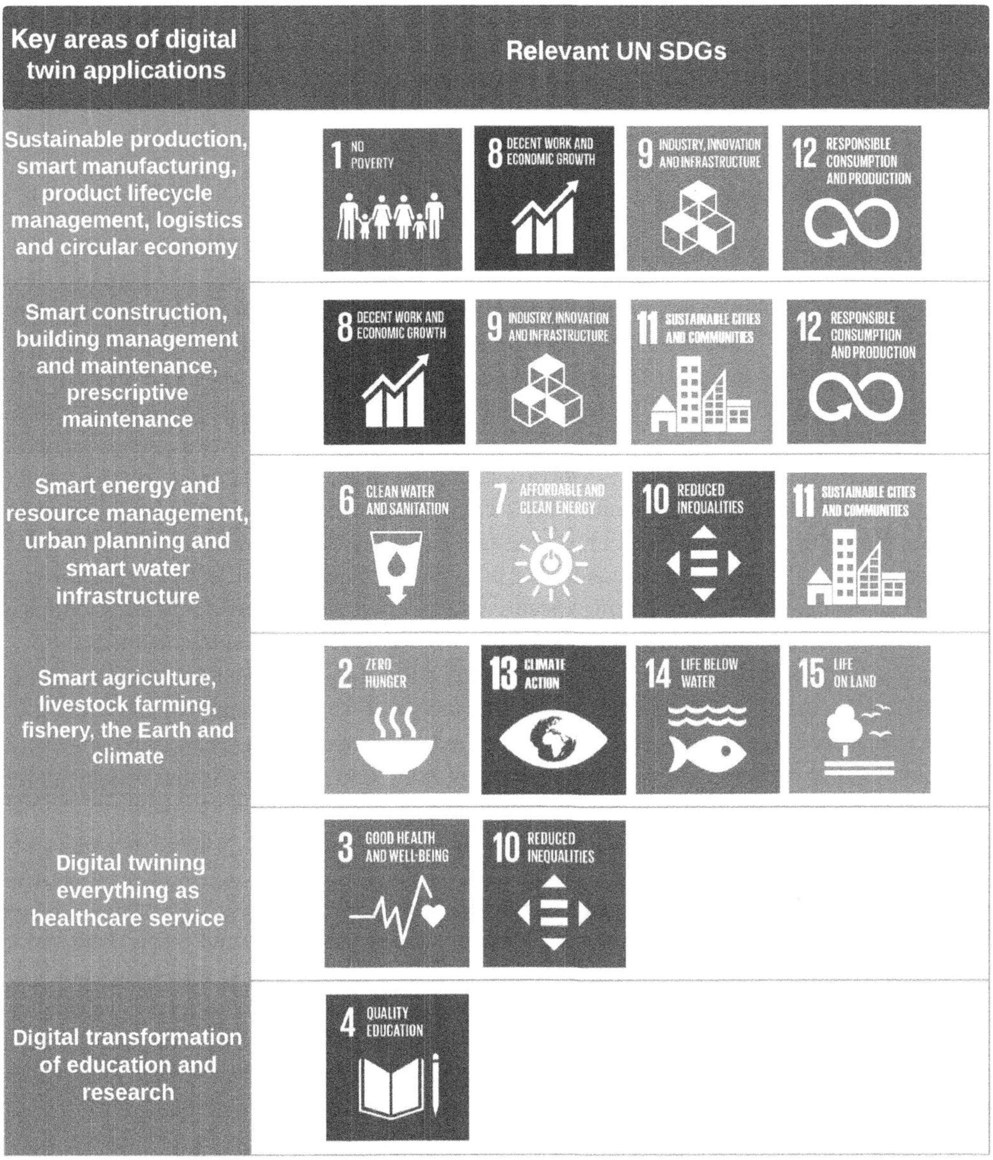

Figure 14.4 The potential of digital twins to contribute to the 17 SDGs. *Source:* Hassani et al. (2022)/ MDPI/Public domain.

planning. It was used to engage citizens in participatory processes, allowing them to visualize the impact of urban planning decisions in real time. This approach enabled a more democratic, transparent, and sustainable urban development by facilitating collaboration between stakeholders and improving decision-making processes. Furthermore, digital twins for infrastructure systems with real-time monitoring can simulate potential hazards to these systems, allowing for an evaluation of their resilience to such hazards. Hence, the use of digital twinning technology can enable sustainable and resilient engineering practices and has the potential to significantly contribute to the achievement of SDGs set by the UN.

However, several challenges remain, including the need for sufficient data availability, costs associated with remote sensing, addressing socio-technical complexities, and ensuring responsible and equitable deployment. Overcoming these barriers is essential to fully leverage digital twins for sustainable and resilient development.

14.2.5 Internet of Things

IoT is a network of physical objects – devices, vehicles, buildings, and other items – embedded with electronics, software, sensors, and connectivity, enabling these objects to collect and exchange data (Gubbi et al. 2013). IoT devices gather, exchange, and utilize data to autonomously make intelligent decisions. IoT enables sustainable and resilient engineering by enabling smarter resource management and reduced environmental footprints. By integrating sensors and connected devices, IoT facilitates real-time data collection and analysis, optimizing processes across various sectors to promote sustainability, some of which are described below.

Transportation – In the transportation sector, IoT can support the development of autonomous vehicles and smart traffic management systems. These technologies can help reduce emissions by optimizing routes, managing traffic flow, and improving fuel efficiency. For instance, smart traffic lights can adjust in real time based on traffic conditions, reducing idle time for vehicles and, consequently, emissions. IoT-enabled vehicle monitoring systems can track performance metrics, leading to timely maintenance and better fuel management, further contributing to reduced carbon footprints.

Energy – IoT can play a pivotal role in the energy sector through the implementation of smart grids. Smart grids utilize real-time data to enhance energy distribution, improve efficiency, and integrate renewable energy sources. IoT devices, such as smart meters and sensors, provide detailed insights into energy consumption patterns, enabling utilities to manage supply and demand more effectively. This not only helps in reducing energy wastage but also promotes the use of sustainable energy sources. Moreover, IoT can facilitate demand response programs where consumers are incentivized to reduce or shift their energy usage during peak times, balancing the load on the grid and reducing the need for fossil fuel-based power plants.

Agriculture – In agriculture, IoT technologies like smart irrigation systems and supply chain monitoring tools can help minimize resource wastage and enhance productivity. Smart irrigation systems use sensors to monitor soil moisture levels and weather conditions, providing water only when necessary. This precision agriculture approach conserves water and ensures optimal plant growth. IoT-enabled supply chain monitoring helps track the journey of produce from farm to table, reducing food waste by ensuring timely transportation and storage conditions. Additionally, IoT can aid in pest management by providing real-time data on pest activity, allowing for targeted interventions and reducing the need for chemical pesticides.

Water management – IoT solutions in water management can contribute to significant reductions in water wastage and pollution. Smart water meters and sensors can detect leaks and monitor water quality in real time, ensuring efficient water usage and preventing contamination. IoT-enabled systems can also optimize the operation of water treatment plants by monitoring various parameters and automating processes, leading to more efficient and sustainable water management practices.

While IoT offers numerous benefits for sustainable and resilient engineering, there are challenges that need to be addressed. Data security and privacy concerns are paramount, as the proliferation of connected devices increases the potential for cyber threats. Ensuring interoperability among various IoT devices and platforms is also crucial for seamless integration and effective operation. Additionally, the environmental impact of producing and disposing of IoT devices themselves must be considered, advocating for sustainable design and recycling practices.

14.3 Engineered Materials for Circular Economy

Traditionally, product life cycles have been configured in a linear path, with key milestones that have included resource extraction, product creation, and disposal. The circular economy disrupts this flow, prioritizing the extended use of materials and products. Sharing, leasing, reuse, repair, refurbishment, and recycling become the key milestones, maximizing the value extracted from existing resources. This minimizes waste generation, reduces carbon emissions, and fosters long-term economic growth.

At its core, the circular economy is about resource efficiency. The goal is to design and utilize materials in a way that minimizes waste and conserves finite natural resources. Recycling plays a crucial role, transforming used materials into valuable resources for new products. Furthermore, advancements in nanotechnology offer exciting possibilities. By engineering waste materials at the nanoscale, we can create novel nanomaterials with unique properties, further extending the life cycle of resources within the circular economy. A brief discussion on recycled waste materials and nanomaterials for engineering purposes is presented in this section.

14.3.1 Recycled Waste Materials

Industrial and other waste materials typically disposed of in landfills are increasingly the focus of research for potential beneficial applications in several engineering fields. The reuse of such materials not only reduces waste generation but also conserves raw or virgin materials that would otherwise be used for these applications. Examples of the reuse of waste materials are especially extensive in Civil and Environmental Engineering fields. For example, industrial waste materials like steel slag, fly ash, biochar, construction and demolition waste, glass, and plastic are being researched and/or have been deployed in several engineering applications such as pavement stabilization, innovative concrete mixes, soil and groundwater remediation, and water treatment, among others.

Additionally, recycled materials are finding applications in the mechanical, electronics and electrical, and manufacturing industries. In mechanical engineering, reclaimed metals and polymers are being used to produce components and assemblies, reducing the need for virgin materials. The electronics and electrical industries are incorporating recycled metals and plastics into new products, thereby decreasing electronic waste (E-waste). Further, research on the innovative use of various waste materials, the development of relevant policies, implementation of extended producer responsibility (EPR) laws and the introduction of incentives for reusing waste materials are needed to truly achieve the goal of a circular economy.

14.3.2 Nanomaterials

Nanotechnology, a rapidly evolving field, has revolutionized various aspects of science and engineering, offering innovative solutions to pressing global challenges. One of its most promising applications lies in developing nanomaterials, which possess unique properties and functionalities due to their nanoscale dimensions.

Nanomaterial-based catalysts facilitate efficient pollution remediation processes by enhancing the degradation of pollutants in air, water, and soil. Nanoparticles of titanium dioxide (TiO_2) demonstrate photocatalytic activity, breaking down organic pollutants and purifying contaminated water sources (Umar and Aziz 2013). The use of nanoscale zero-valent iron and other nanomaterials for contaminant soil and groundwater remediation has received increased attention in the past decade (Zhao et al. 2016; Cameselle and Reddy 2019). Moreover, nanomaterials enable the development of lightweight, high-strength materials for energy-efficient transportation and infrastructure. Carbon nanotubes (CNTs) and graphene, with their mechanical properties and conductivity, hold promise for manufacturing lightweight yet durable materials. These advancements contribute to reducing energy consumption and minimizing the environmental footprint of infrastructure systems. Further, nanomaterials enhance the efficiency of solar cells and improve the performance of energy storage devices such as batteries and supercapacitors (Zhao et al. 2011). Nanostructured electrodes in batteries enable faster-charging rates and longer lifecycles, supporting the integration of intermittent renewable energy sources into the grid.

The emergence of nanomaterial-based technologies has spurred investment in research and development, driving economic growth. Moreover, nanomaterials facilitate the transition to a circular economy by enabling the recycling and reuse of materials. Functionalized nanoparticles enable efficient separation and purification processes, promoting the recovery of valuable resources from waste streams. Furthermore, nanocomposites and nanocoating prolong the lifespan of products, reducing the need for frequent replacements and conserving resources over time.

14.4 Resource Conservation in Energy Sector

Resource conservation is a fundamental concept in sustainable engineering, emphasizing the responsible management and preservation of natural resources. It involves designing products and processes with minimal resource inputs, implementing energy-efficient technologies, and promoting sustainable practices. Integrating resource conservation and the circular economy into sustainable engineering practices is essential for achieving holistic sustainability outcomes. This section discusses how resource conservation can be achieved by choosing alternative energy sources and innovative energy storage technologies.

14.4.1 Alternative Energy Sources

Deposits of fossil fuels, such as coal, oil, and natural gas, are being exploited and depleted at an accelerating rate due to increasing energy demands from rapid urbanization and industrialization (IEA 2022). Additionally, energy generation from fossil fuels generates GHG emissions, exacerbating climate change. In order to address these issues, it is necessary to adopt alternative energy sources that can conserve fossil fuels and reduce GHG emissions related to energy production.

Numerous alternative energy sources are being developed and deployed to replace fossil fuel-based energy generation (Ang et al. 2022). Although the adoption of some renewable energy sources like solar, wind, and hydropower is increasing across the globe, fossil fuels still constitute the majority of energy generated around the world (USEIA 2023). Nevertheless, these nonfossil fuel energy alternatives each exhibit certain advantages and disadvantages, as shown in Table 14.1. Because of these inherent costs and benefits, it is necessary to perform appropriate assessments to quantify the sustainability of adopting an alternative energy source specific to the location under consideration. Additionally, it is necessary to ensure that potentially viable alternative sources provide reliable sources of energy. Further, several technologies, including hydrogen and tidal energy, may exhibit potential for large-scale utilization but are still in the nascent stages of research and development. Further research on such technologies is essential to provide society with reliable alternatives to fossil fuel-based energy.

Table 14.1 Pros and cons of various alternative energy sources to fossil fuels.

Energy source	Pros	Cons
Nuclear	• No greenhouse gas emissions during generation of electricity • Only a small amount of fuel is required	• Nonrenewable • Generates hazardous nuclear waste • Poses constant threat to the safety of the surrounding communities • High upfront costs
Hydro	• Low carbon technology • Generates large amounts of energy • No pollution	• High upfront costs • Effected by drought conditions • Displaces numerous communities • Ecological damage due to habitat loss
Solar	• No carbon emissions • Abundant • No pollution • Can be installed in individual buildings	• High upfront costs • End-of-life waste disposal • Intermittent generation (only day time) • Dependent on weather conditions
Wind	• No carbon emissions • Cheap • No pollution	• Viable only at windy locations which can be remote, leading to high transmission costs • Can be dangerous to birds and other ecosystems • End-of-life waste disposal
Geothermal	• Low carbon emissions • Low maintenance costs • No pollution • Minimal ecological impact • Can be installed in individual buildings	• High upfront costs • Only viable at locations with suitable underground temperatures
Biofuels (ethanol, biodiesel, and biogas)	• Abundant • Can be used in vehicles	• Unsustainable as crops used are mostly food crops and can lead to food shortage • Pollutes air
Biomass (wood)	• Produced from waste materials (forest residues, sawdust, etc.)	• Leads to carbon emissions • Pollutes air • Can be unsustainable if forests are cut down extensively
Biogas from composts, and landfills	• Cheap • Reduces waste • Avoids carbon emissions from organic waste	• Leads to carbon emissions, but better than disposal of waste • Supply is limited, cannot be a large-scale alternative
Hydrogen	• Abundant • Can be used in vehicles	• Still in initial stages • Needs energy to generate hydrogen • Lower generation efficiency
Wave/tidal	• Low carbon technology • Can serve as coastal protection against high tides • Predictable	• High upfront costs • Disturbs ocean ecology • Still in initial stages • Intermittent source • Only possible close to the shore potentially increasing transmission costs

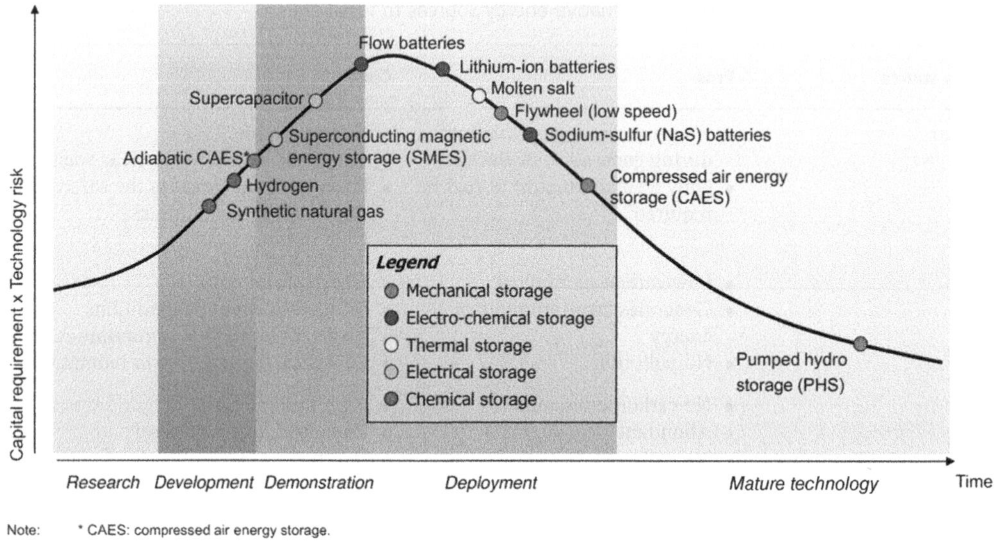

Figure 14.5 Maturity of various energy storage technologies. *Source:* Moore and Shabani (2016)/MDPI/Public domain.

14.4.2 Innovative Energy Storage Technologies

Several renewable energy sources like solar, wind, hydro, and tidal are dependent on physical and temporal factors, such as weather and climate conditions, seasons, and time of the day leading to intermittent generation of energy. Hence, it is necessary to store the excess energy generated during periods of high generation and low demand to ensure that grid stability utilizes generated energy while minimizing waste. Numerous energy storage technologies are available, varying in terms of storage duration and energy type (Zhang et al. 2021a). As shown in Figure 14.5, these technologies are at different stages of maturity, with pumped hydro storage, where water is pumped to a higher elevation using excess energy and stored as potential energy, being the most mature. Figure 14.6

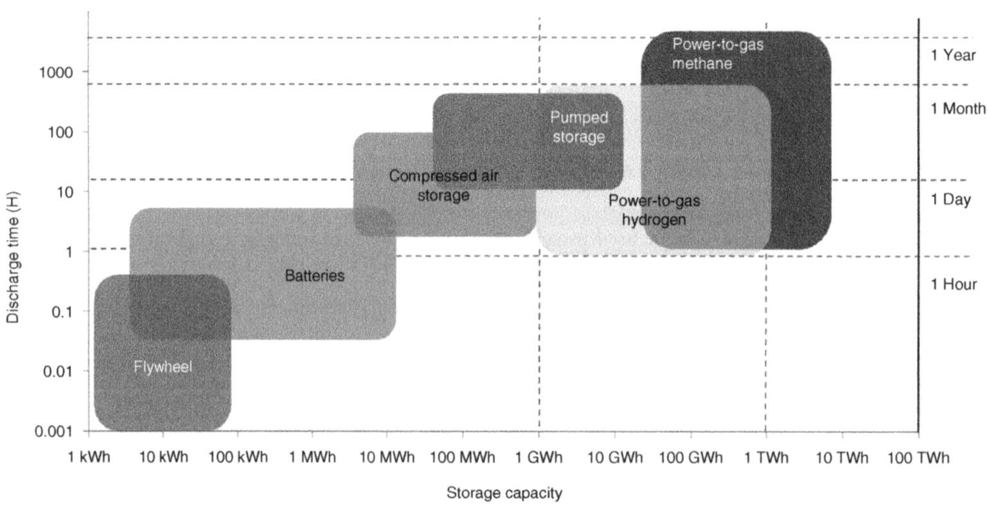

Figure 14.6 Storage capacity and discharge times of various technologies. *Source:* Moore and Shabani (2016)/MDPI/Public domain.

presents the capacity and duration of some of these technologies. Depending on the discharge time and storage capacities, these emerging technologies can significantly reduce energy waste, thereby contributing to the sustainability of the energy sector.

14.5 Nature-Based Solutions

14.5.1 Bio-Based Materials

Bio-based materials are substances derived from renewable biological resources such as plants, animals, and microorganisms. Due to their biological source, they inherently possess high carbon content, are biodegradable, and have a low ecological footprint. They often offer comparable or superior performance to conventional materials, making them viable alternatives for numerous engineering applications. Bio-based materials can be utilized across disciplines and media to minimize environmental impact, promote social responsibility, and optimize economic longevity. Some general categories of biomaterials include bioplastics, biocomposites, bio-based lubricants and chemicals, bioadhesives, and bio-based construction materials. These materials are typically sourced from plants, animals, and microorganisms and can also be synthesized through laboratory processes to convert renewable biomass into functional materials.

In 2018, 600 million tons of construction debris was generated in the United States (USEPA 2020), about double the amount of generated municipal solid waste. The use of bio-based materials can reduce the waste generated. Materials such as wood and plant-based alternative concrete are sustainable options for the construction industry. Some architectural bio-based materials include timber, cork, mycelium, and hempcrete. While entire projects may not yet be feasible with these materials, partial replacement is possible. Challenges in utilizing bio-based materials in construction include material longevity, strength, upscale functional performance, and long-term durability. However, with continuous research and application, it is possible to custom apply such materials wherever feasible.

In 2022, the United States consumed 20 million barrels of petroleum daily. Using biofuels can help reduce dependency on fossil fuels and decrease greenhouse gas emissions. Bio-based materials support the circular economy by allowing for smoother transitions back into the cycle. Plastics, derived from petroleum-based chemicals, negatively impact our health, aquatic systems, and wildlife. Plastics pose significant environmental concerns throughout their lifecycle, from production to degradation. By switching to bioplastics, CO_2 emissions, the prevalence of microplastics in nature, and the quantity of landfill-disposed waste can be reduced.

Although the current use of bioplastics benefits the environment, only 3% of plastics are bio-based, with the remaining 97% being petroleum-based (Mori 2023). One of the most common bioplastics is polylactic acid (PLA), used in packaging materials, disposable tableware, food containers, and 3D printing filaments. Derived from corn starch and sugarcane, PLA is formed through fermentation, purification, and polymerization processes. PLA production uses 65% less energy than the production of conventional plastics, generates 68% fewer greenhouse gases, and contains no toxins. By using compostable and renewable bioplastics like PLA, industries can reduce reliance on conventional plastics like PET (polyethylene terephthalate), thus contributing positively to the environment.

The use of bio-based materials presents numerous opportunities to enhance sustainability in engineering practices. These materials offer a renewable and biodegradable alternative to conventional resources, helping reduce carbon footprints and environmental impact. They can also drive innovation in material science, leading to the development of new products and applications.

Additionally, bio-based materials support the circular economy by facilitating the recycling and reuse of resources. As an example, biochar, produced through the thermal decomposition of organic matter in the absence of oxygen, is used in various sectors. It is applied for contaminant removal as a nutrient and fertilizer source in agriculture and as an amendment in cement for construction practices (Woolf et al. 2010; Reddy and Chetri 2021).

However, challenges remain in the widespread adoption of bio-based materials. Upscaling production to meet industrial needs without compromising sustainability is a significant hurdle. Economic factors, such as the cost of raw materials and production processes, also play a crucial role. Moreover, there is a need for comprehensive policies and incentives to promote research, development, and adoption of bio-based materials. Overcoming these challenges requires a concerted effort from researchers, policymakers, and industry stakeholders. With continuous advancements and collaborative efforts, bio-based materials can play a pivotal role in achieving sustainable engineering practices and contributing to a greener future.

14.5.2 Biomimicry

Biomimicry, or the practice of learning from and mimicking nature's forms, processes, and ecosystems to create sustainable and resilient designs, is a transformative approach to engineering and architecture. This approach holds immense potential for creating more efficient, sustainable, resilient, and aesthetically pleasing designs by leveraging strategies honed by natural evolution. Some examples of biomimicry are presented below:

- *Architectural Innovations:*
  - The *Eastgate Centre in Harare, Zimbabwe,* uses a passive cooling system inspired by termite mounds, which naturally regulate temperature. This design significantly reduces energy consumption for heating and cooling.
  - *The Gherkin in London* mimics the structure of a Venus flower basket sponge. Its aerodynamic shape and diagrid pattern enhance stability and ventilation, demonstrating biomimicry's role in creating energy-efficient, and safe buildings.
  - *The Eden Project in Cornwall, UK,* features geodesic domes inspired by natural structures like dragonfly wings and soap bubbles, providing an energy-efficient and visually striking environment for plant conservation.
- *Materials:*
  - Nature-inspired materials such as *mycelium-based composites* offer sustainable alternatives to traditional building materials like concrete and plastic. Mycelium, the root structure of fungi, is biodegradable, strong, and can be grown into specific shapes, making it ideal for sustainable construction (Alaneme et al. 2023).
  - Bio-based materials like bamboo and hemp are used for their strength, flexibility, and rapid growth, making them ideal for sustainable and resilient building practices.
- *Resource Efficiency:*
  - Sunflower-inspired solar panels mimic the heliotropic movements of sunflowers, optimizing light absorption and energy generation.
  - Water collection systems inspired by the Namib Desert beetle enhance resource efficiency by mimicking the beetle's ability to harvest moisture from the air (Yu et al. 2021).
- *Biophilic Design:*
  - Integrating natural elements such as green roofs, living walls, and natural lighting into architectural designs can improve occupant well-being and productivity. This connection

to nature, or biophilic design, is increasingly recognized for its psychological and physiological benefits.

While biomimicry offers numerous benefits, challenges remain in its widespread adoption. These include the need for interdisciplinary collaboration, overcoming technical limitations, and the initial costs of developing and implementing biomimetic designs. However, as research advances and technology improves, the opportunities for biomimicry to contribute to sustainable and resilient development will grow significantly.

14.5.3 Green Infrastructure

Green infrastructure refers to a network of natural and seminatural systems that deliver a wide range of ecosystem services, integrating nature with urban environments to enhance sustainability. This approach includes elements such as green roofs, permeable pavements, rain gardens, and urban forests, which collectively manage stormwater, reduce urban heat islands, improve air and water quality, and promote biodiversity. Incorporating green infrastructure can create more sustainable and resilient urban areas that efficiently manage resources, reduce environmental impact, and improve the quality of life for urban residents. Further details about green infrastructure are provided in Chapter 17.

14.5.4 Synthetic Biology

Synthetic Biology (SynBio) involves designing and constructing new biological parts, devices, and systems. It combines biology and engineering, allowing for the creation of organisms with new abilities not found in nature. Techniques include DNA synthesis, gene editing, and the creation of synthetic cells. SynBio can significantly contribute to sustainable and resilient engineering by enabling the production of biofuels, reducing waste, and creating biodegradable materials. Engineered microbes can convert agricultural waste into useful products, synthesize bio-based chemicals, and replace harmful manufacturing processes with eco-friendly alternatives (French 2019). These innovations can decrease GHG emissions, reduce pollution, and promote the use of renewable resources.

However, the use of SynBio raises several ethical and safety concerns. The introduction of genetically modified organisms into the environment can pose significant ecological risks, including unintended consequences on natural ecosystems. There is also trepidation of potential misuse of SynBio technologies for bioterrorism, which underscores the need for strict regulatory frameworks and oversight. Additionally, the long-term impacts of these technologies are not yet fully understood, necessitating continuous monitoring and robust governance to ensure that they contribute positively to sustainable development.

14.6 Summary

Emerging technologies are transforming engineering practices to foster environmental stewardship and sustainable solutions. Smart technologies such as AI, big data, sensors, remote sensing, 3D printing, robotics, digital twinning, and the IoT play crucial roles in optimizing resource use, improving energy efficiency, and enhancing robust predictive maintenance. Recycling and repurposing waste materials support circular economy, while nanomaterials offer innovative solutions for various applications, including environmental remediation and creation of lightweight & high-strength materials for energy efficiency. Alternative energy sources and energy storage technologies reduce reliance on fossil fuels and mitigate the associated greenhouse gas emissions, while also providing energy resilience to communities. Additionally, nature-based solutions like bio-based materials, biomimicry, and SynBio offer significant potential for resource conservation, efficient design practices, and energy efficiency. These emerging technologies hold significant potential to advance sustainable and resilient engineering practices despite certain limitations and challenges.

14.7 Questions

14.1 What are the key benefits of AI in sustainable engineering practices?

14.2 How does big data contribute to informed decision-making in sustainable engineering?

14.3 What challenges do you face when integrating AI and big data into engineering?

14.4 What are the implications of the digital divide in the context of smart technologies for sustainability?

14.5 How do ethical considerations impact the implementation of AI in sustainable engineering?

14.6 How do satellites and UAVs (drones) contribute to remote sensing in sustainable practices?

14.7 What roles do GIS play in your sustainable engineering projects?

14.8 How can sensors help monitor and improve energy efficiency in buildings and industrial facilities?

14.9 What are the applications of 3D printing in sustainable and resilient engineering, and how does it reduce resource consumption?

14.10 How do digital twins enhance resource efficiency and support green technologies?

14.11 What challenges might you encounter when deploying digital twins in sustainable engineering?

14.12 How can you apply digital twinning to optimize water supply systems?

14.13 How can the IoT facilitate smarter resource management in various sectors?

14.14 What are the advantages and disadvantages of using recycled waste materials in engineering applications?

14.15 How do nanomaterials enhance pollution remediation and energy-efficient infrastructure?

14.16 What are some innovative uses of nanotechnology in achieving sustainability goals?

14.17 What are the advantages and disadvantages of various alternative energy sources compared to fossil fuels?

14.18 What are some examples of nature-based solutions for sustainable and resilient engineering?

14.19 How does biomimicry contribute to developing sustainable and resilient engineering solutions?

14.20 How does the circular economy model promote sustainability in engineering practices?

14.8 Problems

14.1 A smart building equipped with IoT sensors optimizes energy consumption by monitoring and controlling heating, ventilation, and air conditioning systems. The building has a total energy consumption of 500 000 kWh per year. After implementing IoT technology for energy management, the building reduces its energy consumption by 15%.

- Calculate the annual energy savings in kWh.
- If the cost of electricity is $0.12 per kWh, determine the annual cost savings from the energy reduction.
- Discuss how these savings could contribute to the building's sustainability goals.

14.2 A solar installation generates 300 000 kWh annually. The battery system has a storage capacity of 150 kWh and can discharge 90% of its capacity to meet peak demand.

- If the community consumes an average of 500 kWh daily, how many days could the battery power the community?
- Calculate the cost savings from using stored energy instead of grid energy at $0.15 per kWh for a month (30 days).
- Discuss the sustainability benefits of using this energy storage system.

14.3 A large metropolitan city implements green roofs (20 km^2) and urban forests (30 km^2). Green roofs absorb 25 tons of CO_2/km^2 annually, and urban forests absorb 40 tons of CO_2/km^2 annually.

- Calculate the total CO_2 absorbed by both green roofs and urban forests.
- If permeable pavements cover 50 km^2 and reduce storm water runoff by 50%, determine the runoff reduction from an estimated 500 000 m^3 annual runoff.
- If the economic value of CO_2 absorption is $50 per ton, calculate the total economic benefit from the CO_2 absorbed.
- Discuss the implications of these nature-based solutions for urban sustainability.

References

Alaneme, K.K., Anaele, J.U., Oke, T.M. et al. (2023). Mycelium based composites: a review of their bio-fabrication procedures, material properties and potential for green building and construction applications. *Alexandria Engineering Journal* 83: 234–250.

Ang, T.Z., Salem, M., Kamarol, M. et al. (2022). A comprehensive study of renewable energy sources: classifications, challenges and suggestions. *Energy Strategy Reviews* 43: 100939.

Cameselle, C. and Reddy, K.R. (2019). Elemental iron and other nanotechnologies for soil remediation. In: *Engineering Tools for Environmental Risk Management–4: Risk Reduction Technologies and Case Studies* (ed. K. Gruiz, T. Meggyes, and E. Fenyvesi), 465–495. CRC Press.

Dembski, F., Wössner, U., Letzgus, M. et al. (2020). Urban digital twins for smart cities and citizens: the case study of Herrenberg, Germany. *Sustainability* 12 (6): 2307.

Ferreira, B., Iten, M., and Silva, R.G. (2020). Monitoring sustainable development by means of earth observation data and machine learning: a review. *Environmental Sciences Europe* 32 (1): 1–17.

French, K.E. (2019). Harnessing synthetic biology for sustainable development. *Nature Sustainability* 2 (4): 250–252.

Ghandehari, M., Mosavi, A., Jafarzadeh, H., and Fatahi, M. (2019). Remote sensing and sensor technologies for structural health monitoring of civil infrastructures: a review. *Remote Sensing* 11 (10): 1193.

Gubbi, J., Buyya, R., Marusic, S., and Palaniswami, M. (2013). Internet of Things (IoT): a vision, architectural elements, and future directions. *Future Generation Computer Systems* 29 (7): 1645–1660.

Han, Z., Li, M., Li, X. et al. (2020). Review of remote sensing applications for urban heat islands and their mitigation strategies. *Remote Sensing* 12 (3): 442.

Hassani, H., Huang, X., and MacFeely, S. (2022). Enabling digital twins to support the UN SDGs. *Big Data and Cognitive Computing* 6 (4): 115.

International Energy Agency (IEA) (2022). World Energy Outlook 2022. IEA. https://www.iea.org/reports/world-energy-outlook-2022 (accessed 2 September 2024).

Janga, J.K., Reddy, K.R., and Raviteja, K.V.N.S. (2023). Integrating artificial intelligence, machine learning, and deep learning approaches into remediation of contaminated sites: a review. *Chemosphere* 345: 140476.

Jiang, B., He, Y., Bai, Y., and Xu, M. (2019). Application of satellite remote sensing in monitoring and assessing impacts of land use changes on ecosystem services: a review. *Remote Sensing* 11 (5): 528.

Möller, M., Hochleithner, S., Rutzinger, M. et al. (2018). Unmanned aerial vehicle (UAV) photogrammetry for mapping erosion and deposition landforms on alpine glaciers: potential and challenges. *Remote Sensing* 10 (3): 408.

Moore, J. and Shabani, B. (2016). A critical study of stationary energy storage policies in Australia in an international context: the role of hydrogen and battery technologies. *Energies* 9 (9): 674.

Mori, R. (2023). Replacing all petroleum-based chemical products with natural biomass-based chemical products: a tutorial review. *RSC Sustainability* 1 (2): 179–212.

Reddy, K.R. and Chetri, J. K. (2021). Use of biochar for sustainable environmental remediation. In: *Advances in Construction Materials and Sustainable Environment: Select Proceedings of ICCME 2020*, 1–10. Singapore: Springer Singapore.

Rong, H., Zhang, H., Xiao, S. et al. (2016). Optimizing energy consumption for data centers. *Renewable and Sustainable Energy Reviews* 58: 674–691.

Russell, S.J. and Norvig, P. (2020). *Artificial Intelligence: A Modern Approach*, 4e. Pearson.

Sagiroglu, S. and Sinanc, D. (2013). Big data: a review. In: *2013 International Conference on Collaboration Technologies and Systems (CTS)*, 42–47. IEEE.

Tzachor, A., Sabri, S., Richards, C.E. et al. (2022). Potential and limitations of digital twins to achieve the sustainable development goals. *Nature Sustainability* 5 (10): 822–829.

Umar, M. and Aziz, H.A. (2013). Photocatalytic degradation of organic pollutants in water. *Organic Pollutants-Monitoring, Risk and Treatment* 8: 196–197.

United Nations (UN) (2024). Big Data for Sustainable Development. https://www.un.org/en/global-issues/big-data-for-sustainable-development (accessed 10 October 2024).

United States Energy Information Administration (USEIA) (2023). International Energy Outlook 2023. https://www.eia.gov/outlooks/ieo/ (accessed 11 March 2025).

United States Environmental Protection Agency (USEPA) (2020). Advancing Sustainable Materials Management: Facts and Figures Report. https://www.epa.gov/facts-and-figures-about-materials-waste-and-recycling/advancing-sustainable-materials-management (accessed 11 March 2025).

Woolf, D., Amonette, J.E., Street-Perrott, F.A. et al. (2010). Sustainable biochar to mitigate global climate change. *Nature Communications* 1 (1): 56.

Wu, X., Zhu, X., Wu, G.Q., and Ding, W. (2013). Data mining with big data. *IEEE Transactions on Knowledge and Data Engineering* 26 (1): 97–107.

Yu, Z., Zhang, H., Huang, J. et al. (2021). Namib desert beetle inspired special patterned fabric with programmable and gradient wettability for efficient fog harvesting. *Journal of Materials Science & Technology* 61: 85–92.

Zhang, H., Song, J., Niu, L. et al. (2019). Review on sensors for building energy efficiency monitoring and control. *Energy Procedia* 158: 3972–3977.

Zhang, Z., Ding, T., Zhou, Q. et al. (2021a). A review of technologies and applications on versatile energy storage systems. *Renewable and Sustainable Energy Reviews* 148: 111263.

Zhang, Q., Wang, Y., Cheng, Q. et al. (2021b). Remote sensing technologies for landslide detection and monitoring: a review. *Remote Sensing* 13 (3): 501.

Zhao, X., Sánchez, B.M., Dobson, P.J., and Grant, P.S. (2011). The role of nanomaterials in redox-based supercapacitors for next generation energy storage devices. *Nanoscale* 3 (3): 839–855.

Zhao, X., Liu, W., Cai, Z. et al. (2016). An overview of preparation and applications of stabilized zero-valent iron nanoparticles for soil and groundwater remediation. *Water Research* 100: 245–266.

15

Sustainable and Resilient Energy Engineering

Chapter Goal and Objectives

The goal of this chapter is to identify and evaluate sustainable and resilient energy alternatives, focusing on the impacts of various energy generation methods and exploring strategies for clean energy adoption. The main objectives are (i) to assess the environmental impacts of energy generation, including air emissions, solid waste generation, water resource use, and land resource use; (ii) to examine the benefits and challenges of nuclear energy, particularly regarding waste management and its carbon-neutral profile; (iii) to explore various clean energy technologies, including solar, wind, water, geothermal, and biomass energy; (iv) to review the economic considerations of alternative energy sources, with emphasis on the Levelized Cost of Energy and RPSs; and (v) to provide a comparative analysis of conventional and renewable energy sources in terms of sustainability and resiliency.

15.1 Introduction

A diverse range of energy sources is critical for any healthy nation and economy. Over the past couple of centuries, much of the developed world's energy has been derived from fossil fuels – coal, oil, and natural gas. While these energy sources have proven to be useful, as described throughout this text, numerous detrimental side effects are produced by their utilization as energy sources. Such detrimental effects are related to the emission of contaminants and greenhouse gases. Moreover, many fossil fuel reserves are concentrated in a few countries, leading to significant financial and environmental impacts related to materials transport to manage supply and demand. As a result, the economies of many developed and developing countries are directly affected by supplies and prices of fossil fuels in the open market. Because of the economic and environmental stresses associated with fossil fuel use, an increased focus has been placed on renewable energy sources.

Figures 15.1 and 15.2 and Table 15.1 depict several key statistics pertaining to energy sources and consumption in the United States. Figure 15.1 shows the percentages of respective energy sources in the United States in 2022 (USEPA 2024). As shown, an overwhelming majority of energy sources are constituted by fossil fuels – about 83%. An additional 8.5% of energy is derived from nuclear fuels. Despite the significant attention that sustainable energy sources have received over the past decade, they still only constitute 8.5% of energy sources.

Sustainable and Resilient Engineering: Drivers, Metrics, Tools, and Applications, Second Edition. Krishna R. Reddy, Claudio Cameselle, and Jeffrey A. Adams.

Companion website: www.wiley.com/go/reddy/sustainableresilientengineering2e

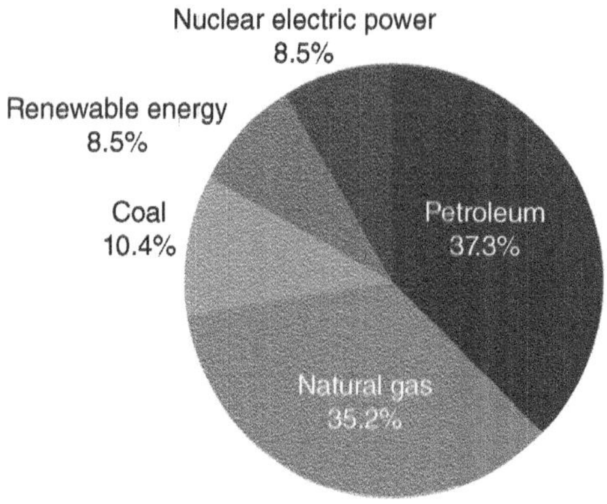

Figure 15.1 Energy sources in the United States in 2022.

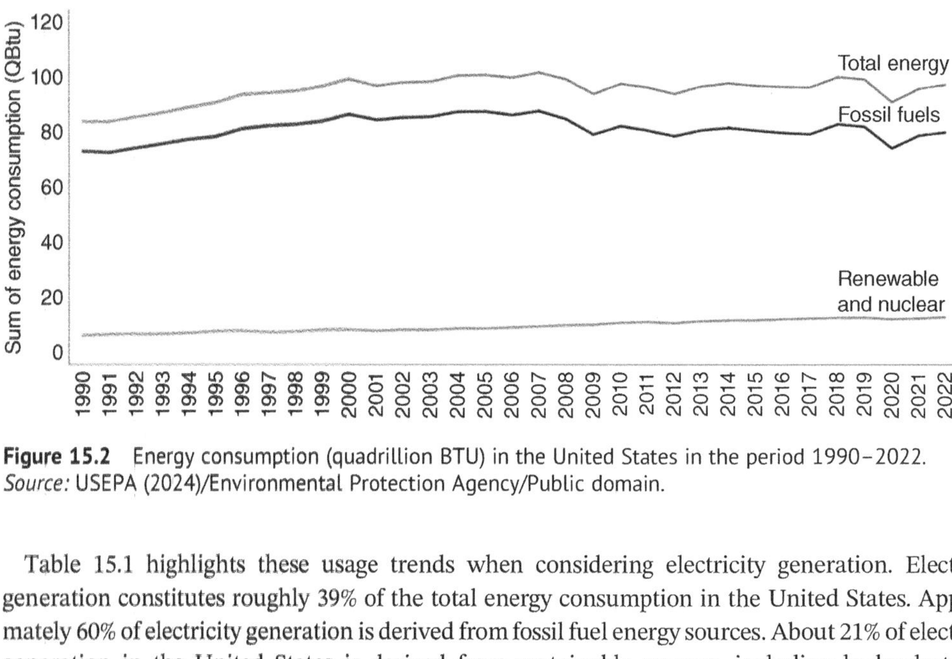

Figure 15.2 Energy consumption (quadrillion BTU) in the United States in the period 1990–2022. *Source:* USEPA (2024)/Environmental Protection Agency/Public domain.

Table 15.1 highlights these usage trends when considering electricity generation. Electricity generation constitutes roughly 39% of the total energy consumption in the United States. Approximately 60% of electricity generation is derived from fossil fuel energy sources. About 21% of electricity generation in the United States is derived from sustainable sources, including hydroelectricity, biomass, wind, geothermal, and solar (EIA 2024). Figure 15.2 depicts the resulting energy consumption with respect to energy source categories. As shown, fossil fuels have steadily remained as the primary energy source as compared to renewable sources and even nuclear energy sources.

This chapter examines environmental impacts of energy generation using fossil fuels, explores nuclear energy, discusses strategies for clean energy, and finally presents various renewable energy sources for sustainability.

15.2 Environmental Impacts of Energy Generation

Energy is essential for civilization, especially for urban civilizations that consume significant amounts of energy in homes, commercial centers, heating systems, and transportation, as well as in the industrial production of goods. However, the generation, transportation, and utilization

Table 15.1 US utility-scale electricity generation by source in 2023.

Energy source	Billion kWh	Share of total
Total – all sources	4178	
Fossil fuels (total)	2505	60.0%
Natural gas	1802	43.1%
Coal	675	16.2%
Petroleum (total)	16	0.4%
Petroleum liquids	12	0.3%
Petroleum coke	5	0.1%
Other gases	11	0.3%
Nuclear	775	18.6%
Renewables (total)	894	21.4%
Wind	425	10.2%
Hydropower	240	5.7%
Solar (total)	165	3.9%
Photovoltaic	162	3.9%
Solar thermal	3	0.1%
Biomass (total)	47	1.1%
Wood	31	0.8%
Landfill gas	8	0.2%
Municipal solid waste (biogenic)	6	0.1%
Other biomass waste	2	0.1%
Geothermal	16	0.4%
Pumped storage hydropower	−6	−0.1%
Other sources	10	0.2%

Source: EIA, https://www.eia.gov/tools/faqs/faq.php?id=427&t=3.

of energy sources result in several key environmental impacts. Many of these are direct effects and are quite intuitive, while others are more indirect and not commonly considered. Several key effects – air emissions, water resource use, water discharges, solid waste generation, and land resource use – are discussed in the subsequent sections.

15.2.1 Air Emissions

The generation and emission of airborne pollutants are a primary side effect of energy-related activities. About 83 of anthropogenic greenhouse gas emissions result from energy-related activities; most of these are in the form of carbon dioxide resulting from the consumption of fossil fuels (USEPA 2024). More than one-half of these emissions result from large-scale, stationary sources, while approximately one-third are generated from transportation-related sources. In further consideration of these sources, electrical power plants result in the largest percentage of these emissions. Fossil fuel-fired power plants are responsible for the generation of 67% of the United States' sulfur dioxide emissions, 23% of nitrogen oxide emissions, and 40% of anthropogenic carbon dioxide emissions (USEPA 2024). These emissions can lead to several detrimental effects on the environment, including smog, acid rain, and other aesthetic and public-health-threatening consequences. Additionally, the emission of greenhouse gases can contribute to climate change. To further demonstrate the emission of greenhouse gases, Figure 15.3 presents a summary of these emissions by energy sector. As shown, fossil fuel combustion constitutes the highest percentage of these emissions. Figures 15.4 through 15.7 further describe the sectors that utilize fossil fuels and result in the generation of emissions. As depicted in these figures, quite intuitively, electrical generation and transportation activities consume the most fossil fuels and result in the highest percentages of emissions.

Coal has long been considered the "dirtiest" of the fossil fuels, or equivalently, results in the greatest generation of deleterious emissions. When coal is burned, carbon dioxide, sulfur dioxide,

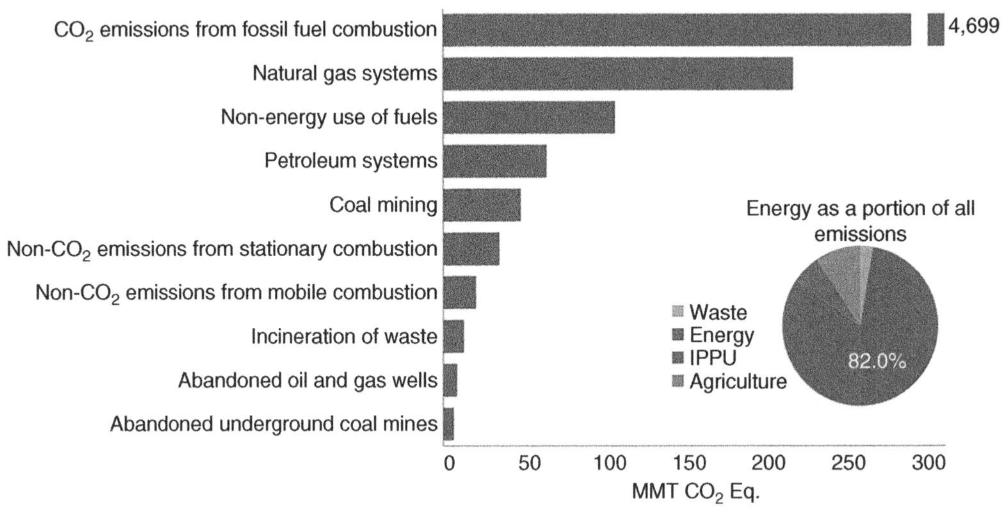

Figure 15.3 Emission of greenhouse gases expressed as CO_2 equivalents by energy production sector in the United States in 2022.

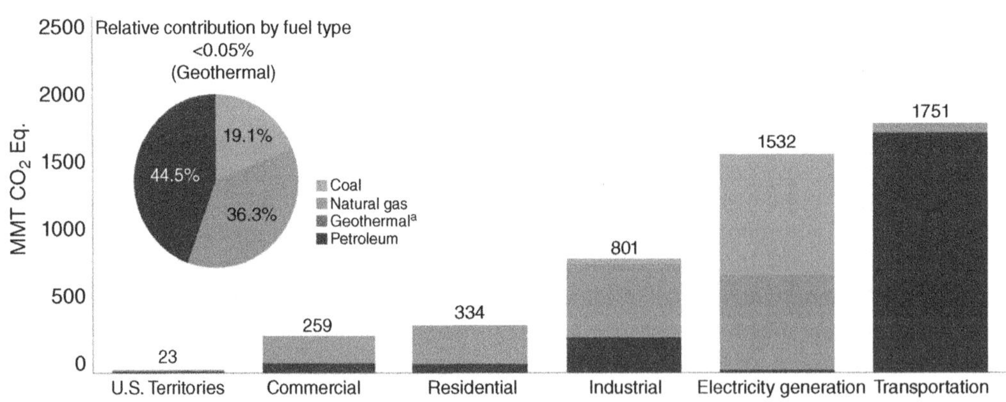

Figure 15.4 Emission of greenhouse gases (as CO_2 equivalent) in 2022 from fossil fuel combustion by sector and fuel type.

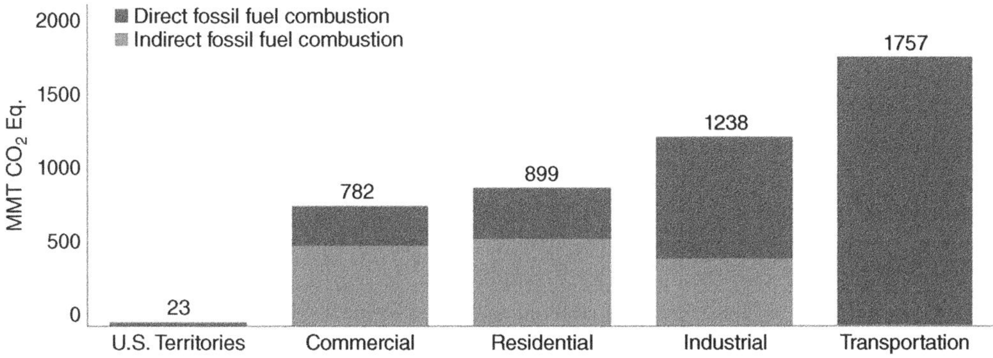

Figure 15.5 End-use sector emissions of greenhouse gases (as CO_2 equivalent) in 2022 from fossil fuel combustion.

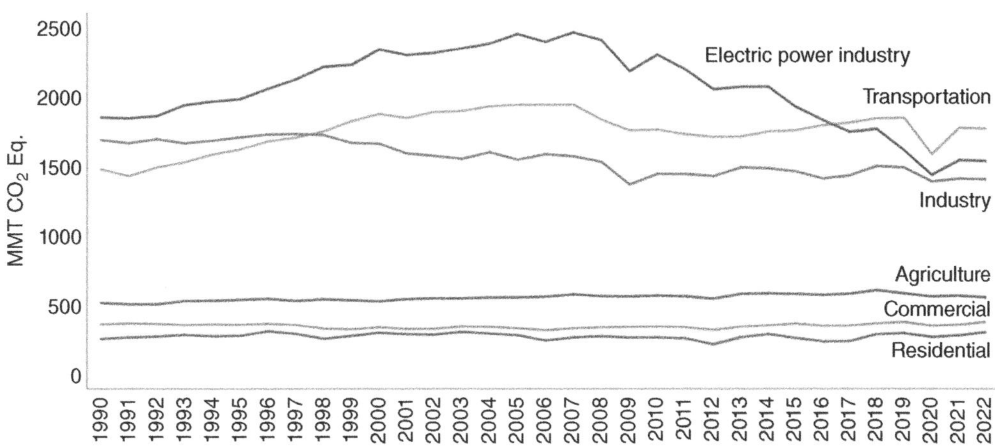

Figure 15.6 Emission of greenhouse gases (as CO_2 equivalent) allocated to economic sectors.

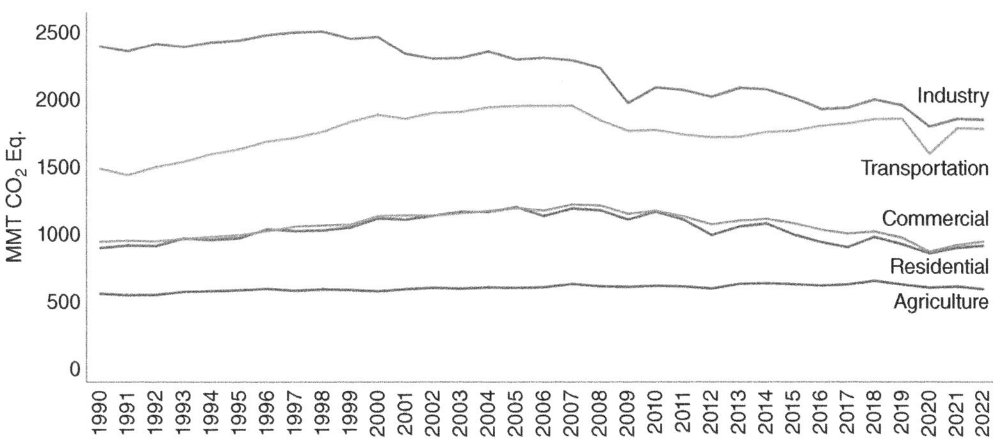

Figure 15.7 Emission of greenhouse gases (as CO_2 equivalent) from electricity distributed to economic sectors.

nitrogen oxides, and mercury compounds are released as emissions into the atmosphere (Miller et al. 2004; Pacyna et al. 2006). Because of these deleterious compounds, coal-fired boilers are required to have control devices in place to reduce the amount of emissions that are released into the atmosphere. For a megawatt-hour (MWh) of electrical generation, 2249 pounds of carbon dioxide, 13 pounds of sulfur dioxide, and 6 pounds of nitrogen oxides are generated (USEPA 2024). The mining, processing, and transportation of coal from a mine to a power plant generates additional emissions. Further, because of its common presence with coal, methane is often vented into the atmosphere during the mining and processing of coal. As previously discussed, methane is a more potent greenhouse gas when compared to carbon dioxide.

Though less commonly used, oil is considered a "cleaner" alternative for use in electricity generation. The combustion of oil results in the emission of nitrogen oxides, sulfur dioxide, carbon dioxide, methane, and mercury compounds, although the amounts of sulfur dioxide and mercury compounds can vary widely depending on the relative content of sulfur and mercury present in the oil being consumed. Emissions are far more favorable for oil as compared to coal because a MWh of electrical generation by oil generates 1672 pounds of carbon dioxide, 12 pounds of sulfur dioxide, and 4 pounds of nitrogen oxides (EIA 2016). However, oil wells and oil collection equipment also result in the emission of methane. Further, oil extraction, transportation, and refining are highly energy-intensive processes that result in the consumption of vast quantities of natural gas or fuels, resulting in additional emissions.

Because of these concerns and regulations related to emissions, natural gas, considered to be the "cleanest" of all the fossil fuels, is being increasingly used for electricity generation. Nevertheless, natural gas consumption does result in the generation of nitrogen oxides and carbon dioxide. Further, when not completely burned or when leaking from pipelines or other similar facilities, methane can be emitted directly into the atmosphere. Again, emissions are far more favorable as compared to coal and oil; for a MWh of electrical generation, 1135 pounds of carbon dioxide, 0.1 pounds of sulfur dioxide, and 1.7 pounds of nitrogen oxides are generated (EIA 2016). As shown, natural gas combustion results in half of the emission of carbon dioxide and significantly less sulfur dioxide and nitrogen oxides when compared to coal combustion. Although the extraction and transportation of natural gas results in emissions, these are also lower as compared to oil and coal (Entrekin et al. 2011; Brandt et al. 2014).

15.2.2 Solid Waste Generation

The processing and consumption of fossil fuels, particularly coal and oil, result in the generation of solid waste materials in addition to gaseous emissions. During the mining of coal, significant quantities of soil and rock overburden are generated. These materials, referred to as tailings, require active management and disposal. The burning of coal also results in ash. This waste, typically consisting of metal oxides and alkalis, can constitute up to 10% of the combusted coal. Further, solid wastes are generated during processing as well as during emissions control through the use of "scrubbers" when pollutants are removed from stack gas. Much of this waste needs to be disposed of in landfills or in abandoned mines, although some of this is being recycled and diverted into useful products, including cement and building materials (Deonarine et al. 2013; Yao et al. 2015). Oil refining also creates solid waste and is commonly in the form of wastewater sludge and other solids that can contain high concentrations of heavy metals and other toxic compounds (Hu et al. 2013). Further, solid residues can be generated during the combustion process during electricity generation.

15.2.3 Water Resource Use

Fuel processing and electricity generation require the use of significant quantities of water in various stages of the process: extraction, transportation, and energy production. Water is used primarily as a heat transfer substrate, and it is commonly converted from liquid form into steam to drive turbines and electrical generator equipment and is also used as a coolant. However, such use can result in detrimental effects on the resource itself or the greater environment (Miller et al. 2004).

When coal is processed at the mine, significant quantities of water are used to remove impurities. Stormwater can also pick up impurities when it comes in contact with uncovered coal stockpiles. When discharged, the impurity-laden wash water can harm the environment, including land and water bodies, due to the presence of heavy metals. Commonly, this process is closely monitored and requires permitting and/or regulatory oversight. Water is also used for steam generation and cooling purposes at the power plant (Spang et al. 2014). Extraction of the required water (e.g. river, lake, or groundwater) can affect aquatic ecosystems. Additionally, once consumed in the power generation or cooling cycle, pollutants can accumulate in the water. When released into surface waters, these pollutants can also have a negative effect on ecosystem health. Further, thermal pollution (e.g. the release of warm or hot water into a surface water body) can have a detrimental effect on ecosystem health.

Electricity generation-related oil and natural gas combustion results in similar detrimental effects that occur with coal combustion, namely pollutant loading/release and thermal pollution during the release of wastewater. Additionally, oil drilling/extraction utilizes significant quantities of water. As with coal, this water use and release can result in the contamination of surface water bodies or groundwater, typically through the release of hydrocarbons. Further, transportation of oil is a relatively risky process; water-borne transportation is often required, and spills can have catastrophic effects on wildlife and the aquatic environment.

15.2.4 Land Resource Use

The extraction, generation, and disposal of wastes associated with electricity also require the use of applicable land resources. First, the energy sources themselves must be extracted from the Earth, either in the form of mines or extraction wells. In addition to consuming large quantities of land, these extraction activities are extremely invasive and result in unavoidable alteration or destruction of natural habitat areas, including deforestation, soil erosion, and impacts to surface water and groundwater. These materials also need space for processing (e.g. refineries, processing plants) to yield usable materials. Electricity generation plants also occupy large tracts of land, both for the structures needed for generation, control, and transmission and for the storage of fuel materials (Giampietro et al. 2014). Finally, the wastes generated need to be properly stored or disposed of, commonly in landfills, abandoned mines, or specialized storage facilities.

15.3 Nuclear Energy

Before we discuss the advantages and drawbacks of several renewable energy sources, it is important to discuss nuclear energy. Nuclear energy has been a popular energy source worldwide; however, because of real or perceived dangerous side effects, it has been arguably the most controversial energy source on the planet.

The theory behind nuclear energy is relatively straightforward – when atoms are split, significant energy is liberated, and this energy may be captured to generate electricity. This process is known as nuclear fission. In a nuclear power plant, nuclear fission occurs within a nuclear reactor. Because of its relative ease of use in the fission process, uranium is commonly used for nuclear fuel. The uranium atom is relatively large and relatively easy to split, securing it as a popular choice for fission. Because uranium is a mined resource (derived from mineral ores), it is not considered renewable; however, a relatively minute amount of uranium is needed to derive vast amounts of energy as compared to other sources.

Figure 15.8 provides a more detailed schematic of how energy is derived from a nuclear source. In a nuclear reactor, fuel rods (consisting of uranium pellets) are placed into a water vessel. A reaction is induced in which the uranium atoms are split, releasing energy in the form of heat. The prodigious amount of heat that is liberated generates steam, which in turn is directed to a turbine. The turbine powers an electrical generator, creating electricity. The steam cools and is transformed back into liquid water. It is circulated back into the system, where the cycle repeats. In many cases, excess heat is released into the atmosphere via cooling towers.

Nuclear energy has several attractive benefits. As a result, nuclear reactors are used in many developed countries to generate electricity (NEI 2017). The United States has 104 nuclear reactors, accounting for approximately one-fifth of the electrical generation in the United States. The use of nuclear fuel does not result in the burning of fuels; therefore, zero greenhouse gas emissions are generated (EIA 2016). Further, nuclear fuel is extremely efficient – one ton of uranium can produce 40 million kWh of electricity, which is equal to approximately 16 000 tons of coal or 80 000 barrels of oil. Not only are the emissions associated with fossil fuel use eliminated, but also the fuel used to transport a ton of uranium issignificantly less than the energy required to transport equivalent energy-producing quantities of coal or oil (Corner et al. 2011).

Nevertheless, nuclear energy also has several unique drawbacks. The mining and refining of uranium ore requires significant amounts of energy input, but the main drawback is related to the spent nuclear fuel; once the uranium has been used up in a nuclear reactor, highly radioactive

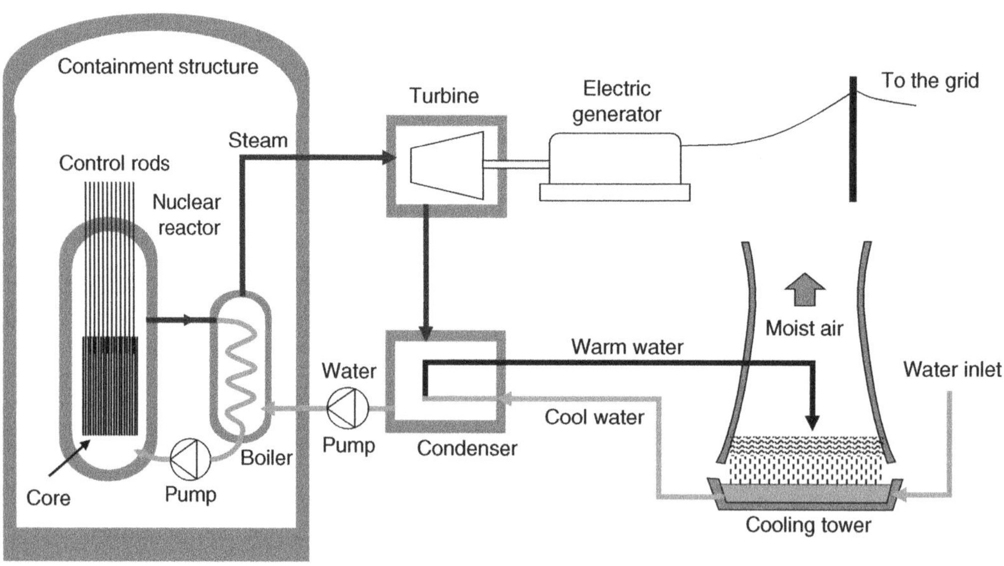

Figure 15.8 Sketch of a nuclear power plant.

waste remains. These wastes will remain dangerously radioactive for centuries, and the proper, safe disposal of these materials is still the subject of much contentious debate (Krauskopf 2013). Further, nuclear fuels can be refined into weapons-grade materials; therefore, the proliferation of nuclear energy for peaceful purposes can be used to develop weapons-grade material for nuclear weapon arsenals.

Nuclear energy electricity generation, like the use of fossil fuels, can result in thermal pollution when wastewater is discharged into the environment. Fortunately, steam and wastewater from the cooling process are not radioactive, as they do not contact the fuel rods. Nevertheless, stormwater runoff at mines/processing facilities can contain heavy metals and/or trace amounts of radioactive materials. In addition, major risks for the environment are associated with the handling of solid waste from the nuclear reactor. Every 18 to 24 months, nuclear power plants must shut down so that the spent uranium fuel rods can be moved from the reactor. Although no longer useful as a fuel material, this material is still highly radioactive. Currently, rods are stored at the facility where they were used, typically in steel and concrete-lined vaults. Approximately 2000 metric tons of these materials are produced annually in the United States. Additionally, much of the electrical generation equipment also becomes radioactively contaminated. All of these materials will remain dangerously radioactive for centuries. Although the packaging, transport, storage, and disposal of these wastes are strictly regulated by the US Department of Transportation and the US Nuclear Regulatory Commission, a permanent storage solution has not yet been determined (Ewing and Von Hippel 2009).

15.4 Strategies for Clean Energy

The consumption of fossil fuels for energy production, especially for transportation and electric energy generation, has significant negative side effects on the environment and public health. Thus, several strategies have been developed to counteract these negative effects. Some of these include the following:

- *Energy efficiency and reduced consumption* – Implementation of best management practices and technologies to reduce energy consumption.
- *Pollution controls* – Minimization of emissions through the adoption and implementation of new technologies and controls.
- *Carbon capture and sequestration* – Interim or long-term diversion, capture, and sequestration of carbon-laden emissions, either into water bodies or underground.
- *Renewable energy* – Increased use of energy from renewable sources coupled with a reduction of fossil fuel consumption.

While these strategies are broad-based, several specific initiatives can be implemented to pursue these goals. Some of these initiatives are as follows:

- *Energy efficient buildings* – Buildings account for 30% of all greenhouse gas emissions in the United States. These emissions can be reduced through the adoption of a range of technology-focused strategies, including the use of more efficient heating, air conditioning, and lighting systems, as well as the adoption of architectural principles and materials selection that can reduce the need for these systems. Figure 15.9 shows the use of electric energy in the American home; almost 50% is associated with heating and cooling systems and appliances (EIA 2017a).

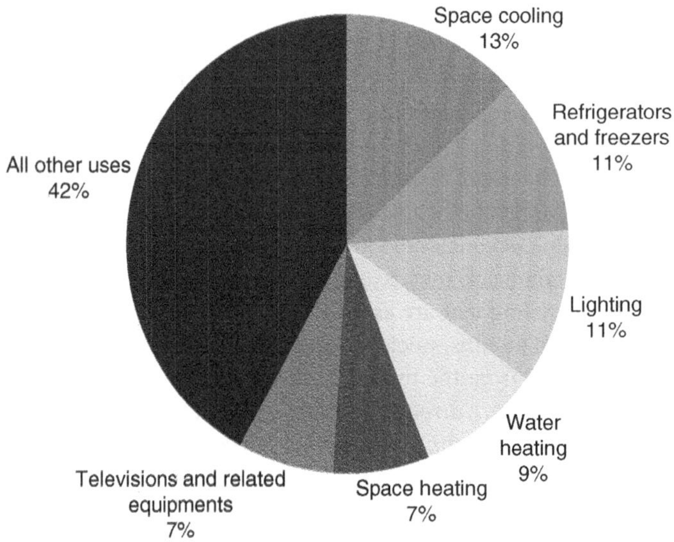

Figure 15.9 Residential major uses of electricity in the United States, 2015. *Source:* EIA (2017a).

- *Green vehicles* – Cars, trucks, buses, airplanes, trains, and other vehicles account for almost one-third of energy consumption in the United States, and they result in approximately one-third of greenhouse gas emissions. Vehicles are increasingly being developed that can use less energy. Cars are becoming more fuel efficient due to the incorporation of lighter materials and more efficient engines, resulting in less fuel consumption and related emissions per mile traveled. Alternative fuel vehicles, including hybrid vehicles or those that operate on electricity (Figure 15.10), standard battery cells, or hydrogen fuel cells, are eliminating the need for petroleum fuels. Flexible fuel vehicles run on gasoline but can accommodate up to 85% ethanol mixtures, which are considered a renewable fuel.

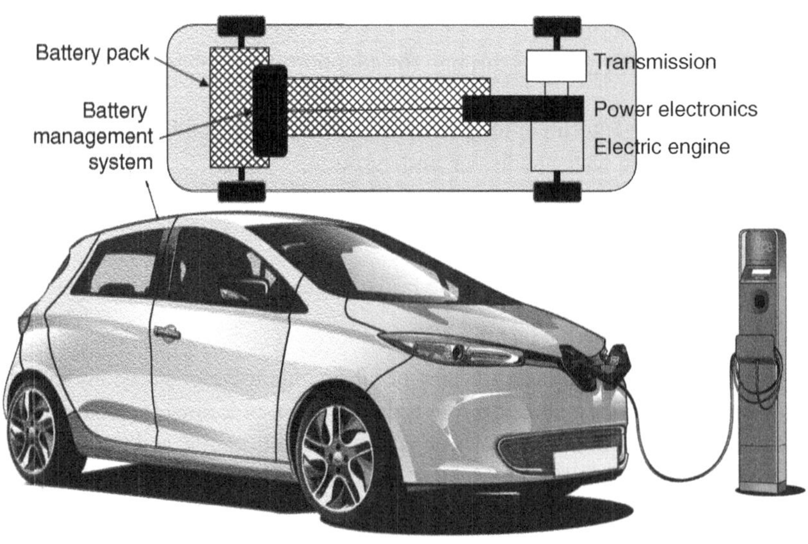

Figure 15.10 Electric car.

- *Carbon capture and underground storage* – While there is a growing desire to move toward renewable or zero-emission sources for electricity generation, it will be decades before these technologies will replace fossil fuels in these applications on a grand scale. In the meantime, there is an increased focus on the development of ways to capture carbon dioxide from power plants and factories and safely store it underground to prevent its release into the atmosphere. This can be accomplished through carbon capture and sequestration. The carbon dioxide-containing emissions can be diverted deep underground into rock formations, where they can be stored safely on a permanent basis. These storage reservoirs are typically placed more than a half-mile below the ground surface. Monitors are used to ensure that the captured carbon dioxide does not escape toward the surface. More details are provided in Chapter 20.

15.5 Renewable Energy

The adverse effects caused by the massive use of nonrenewable fossil fuels have resulted in an increasing interest in the adoption of energy derived from low or zero-emission renewable energy sources (Dincer 2000). The goal of the use of such energy sources is to reduce the overall carbon footprint associated with energy consumption. Common renewable energy sources include solar, wind, water, geothermal, biomass, or landfill gas/biogas. Energy derived from renewable sources can result in reduced emissions and air pollution, lower energy bills, enhanced economic development and technological advancement, improved reliability and security of the nation's energy system, diversification of energy sources, and less reliance on imported energy sources, which often consist of fossil fuels (Twidell and Weir 2015).

However, the consideration of renewable energy sources should be approached with the same caution when considering fossil fuel sources – no single technology exists that meets all needs or eliminates all problems. When considering such systems, it is important to assess the availability of such resources, consider the costs of those systems, examine direct and indirect costs, consider permitting requirements associated with facility development and siting, the concerns of project stakeholders, and available financing or economic incentives (Evans et al. 2009).

15.5.1 Solar Energy

Solar energy is energy derived from the light of the sun (EIA 2017b). It is a renewable and practically infinite source of energy. This energy can be harnessed in one of three ways:

Photovoltaic cells – These cells, typically contained in familiar solar panels, absorb sunlight and directly convert it into electricity.

Solar thermal technology – Sunlight is gathered and focused (typically through the use of mirrors) to generate steam, which in turn power turbines and electrical generators. Once the water cools, it can be recirculated and converted to steam again. Often, another material is used (such as molten salt) to transfer heat to water. The molten salt can retain heat for a long period of time, allowing these systems to operate during nighttime hours.

Passive solar heating – Architectural finishes, including properly sized/sited windows, awnings and window covers, can utilize solar heating in the wintertime and deflect sunlight in the summertime, lessening the requirement for heating and cooling derived from other sources.

Figure 15.11 Solar energy for electricity production, (a) photovoltaic, (b) solar thermal, and (c) photovoltaic plant.

Figure 15.11 shows the available technologies for electricity production from solar energy, and Figure 15.12 depicts areas suitable to the use of solar energy resources for electricity generation.

Solar reserves are located everywhere in the United States, although some areas receive more light than others, depending on the climate and latitude. In general, the greatest reserves are located in the southwestern portion of the United States. It has been reported that enough sunlight falls on a 100-mile by 100-mile area to meet the electrical needs of the entire country. Solar systems do not result in air emissions, and water discharges are negligible. Water may be used in closed-loop form within a solar thermal system. Additionally, negligible solid wastes are generated: mostly associated with equipment repair and de-commissioning as well as fabrication of components. With respect to land resources, solar systems require a large area. However, photovoltaic systems can be placed on existing structures. Solar thermal systems do require larger areas, depending on the array of mirrors required. Although the systems do not create waste, the installation of these systems can affect wildlife habitat due to their presence in previously undeveloped or wild land areas.

15.5.2 Wind Energy

Wind energy has been used for thousands of years. Uneven heating of the Earth plus the Earth's rotation results in the generation of wind due to the movement of warmer and cooler air masses. From antique windmills to modern wind turbines, harnessing wind power has been beneficial for civilization. Wind turbines capture the passing wind; the spinning blades in turn spin an electrical generator and convert it to electricity. Windmills and wind farms are best suited in areas of consistent winds, including hilltops/ridges, open plains, and shorelines. Large wind farms are commonly connected to the electricity grid, allowing for the transmission of the generated electricity. Figures 15.13 and 15.14 show areas of wind energy potential on a private, community, and utility-scale level, respectively.

Wind systems can be used in various parts of the country, although the best areas include the following states: North Dakota, Texas, Kansas, South Dakota, Montana, Wyoming, Nebraska, Oklahoma, Minnesota, Iowa, Colorado, New Mexico, California, Wisconsin, and Oregon. Wind turbines are best deployed in plains and mountain passes within these areas. For instance, winds are

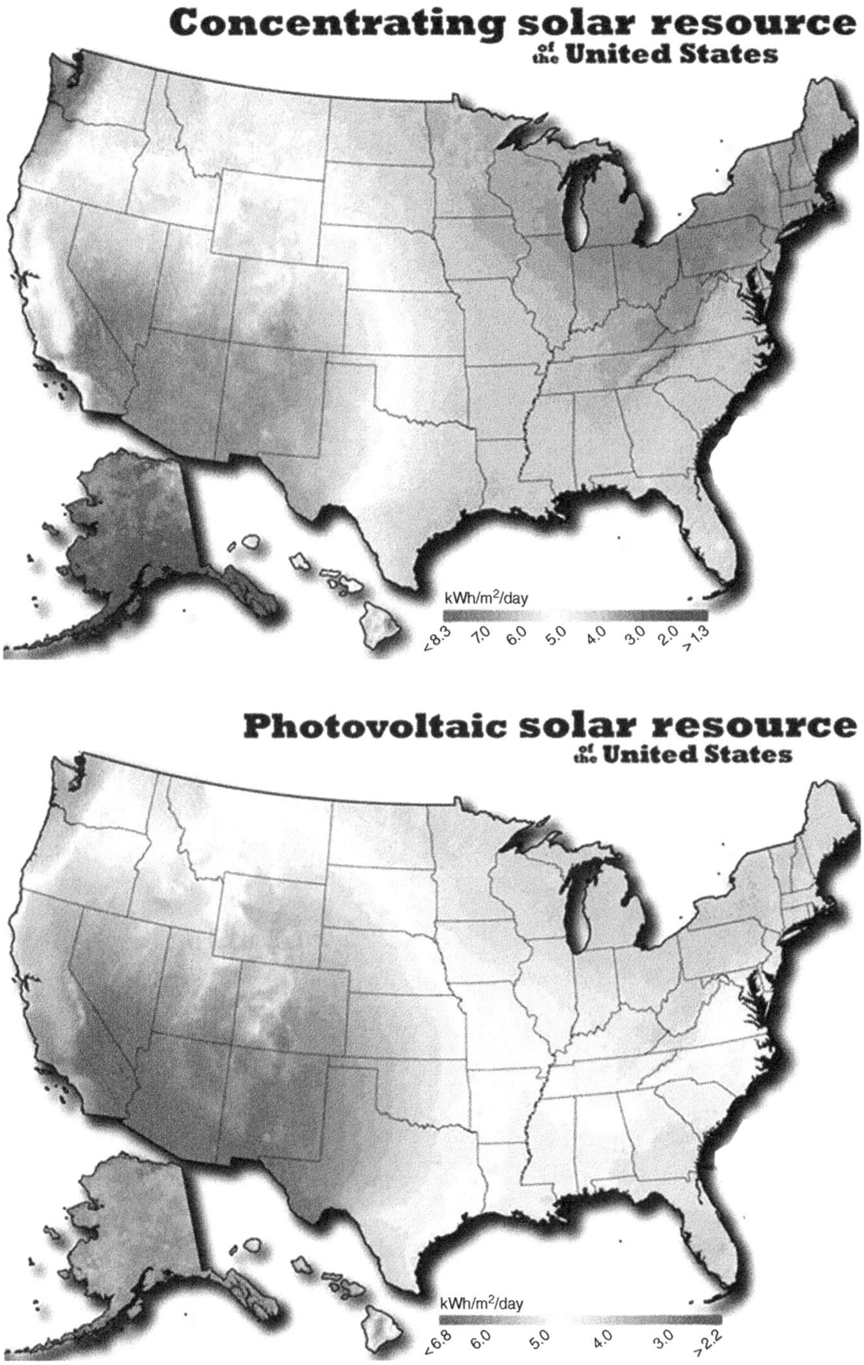

Figure 15.12 Suitable zones in the United States for the use of photovoltaic energy or concentrating solar energy for electricity production. *Source:* EIA, https://www.eia.gov/energyexplained.

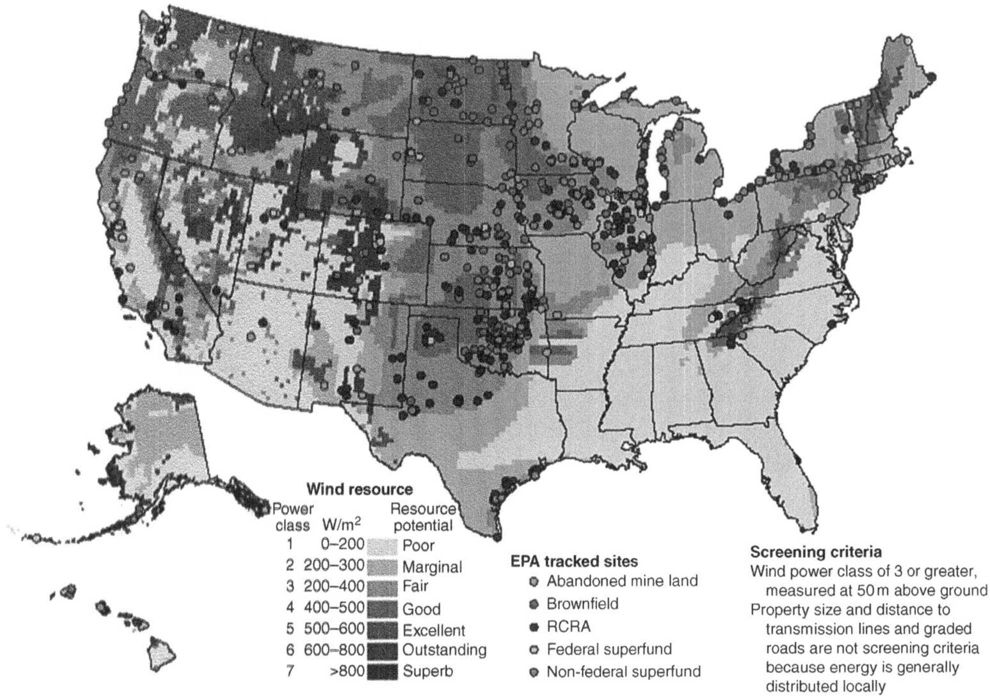

Figure 15.13 Potential areas for off-grid wind energy generation. *Source:* EIA, https://www.eia.gov/energyexplained.

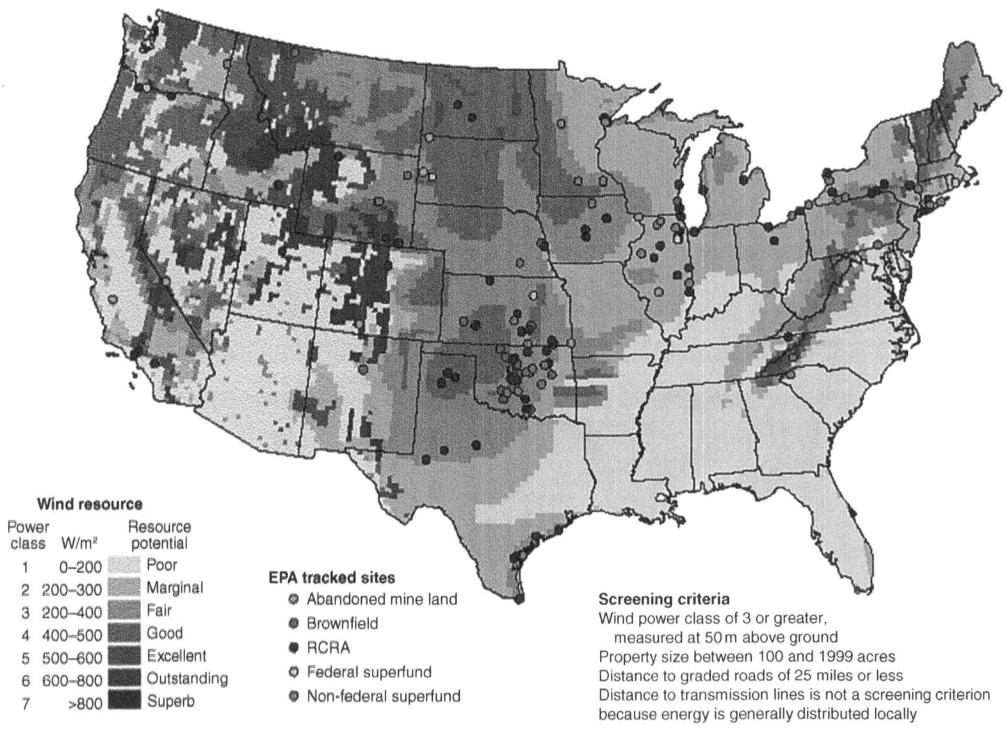

Figure 15.14 Potential areas for wind energy generation for community uses. *Source:* EIA, https://www.eia.gov/energyexplained.

consistent and strong enough in the Rocky Mountain and Great Plains states to meet between 10% and 25% of energy requirements in these states (Bradbury et al. 2014). Wind turbines require little to no water, other than for cleaning of blades. They do not generate air emissions, water discharges, or any appreciable solid wastes. Wind farms do require large areas of land or offshore space, and they may be considered aesthetically displeasing from a noise and visual standpoint. They may be incorporated into large, open spaces (e.g. cattle grazing areas) with little effect on land use. However, they have been known to increase bird and bat mortality, and if installed improperly, they can result in soil erosion.

15.5.3 Water Energy

Flowing water also contains energy that can be harnessed for useful purposes. As with the case of wind power, this technology has been harnessed for centuries, from riverside mills to large-scale hydroelectric dams. Water energy may be harnessed for electrical generation in several ways, but all rely on the flow of rushing water to spin an electrical generator. Hydroelectric dams capture energy from flowing river water (Figures 15.15 and 15.16). Dam operators can control the flow of this water by restricting the water flow rate, which will often result in the creation of lakes/reservoirs behind the dams. These water bodies also can be used as valuable resources for domestic water, recreation, and wildlife habitat (Kao et al. 2015). Wave power can be captured along an ocean coastline by special buoys or floating devices. Tidal power can also be used: ebb or flood tides flowing out or in toward a coastline can also be captured for electricity generation (Uihlein and Magagna 2016).

Water-derived power does not result in air emissions or solid waste generation; however, water-derived energy does come with several pitfalls. First, altered water flow can result in excessive vegetation, which, when decaying, can result in methane production. The use of dams can also greatly affect the flow and environment of rivers, leading to deleterious effects on migrating fish populations. Additionally, thermal vertical gradients can develop, affecting oxygen levels throughout the water column and adversely affecting wildlife and aquatic populations. Further, large-scale

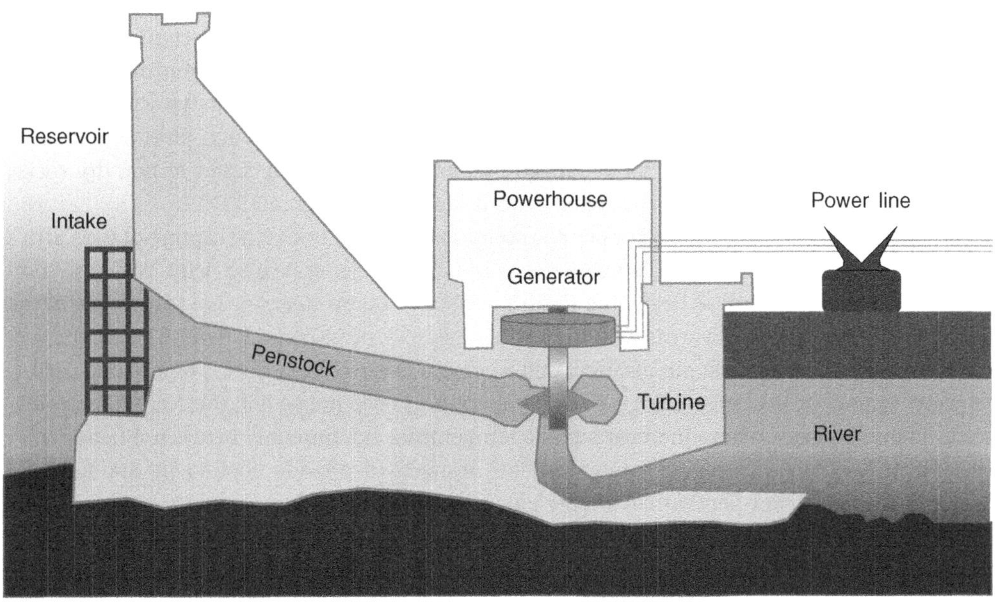

Figure 15.15 Schematic diagram of a hydraulic power plant.

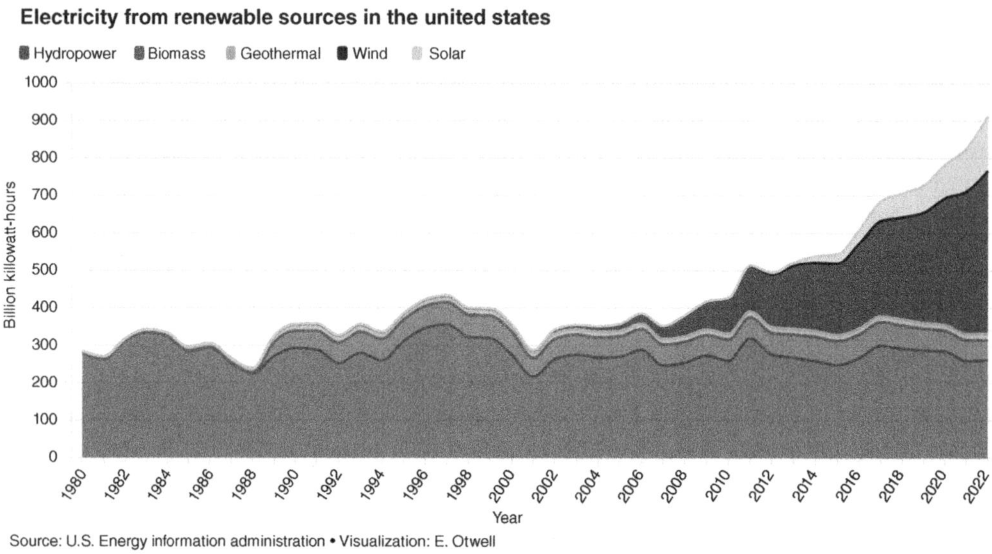

Figure 15.16 Production of hydropower and other renewable energies in the United States in the period of 1980–2022. *Source:* By Visualization: E. Otwell – How one device could help transform our power grid. By Alexandra Witze, August 24, 2023, in en:Science News. Article lists data source on chart as en:U.S. Energy Information Administration., Public Domain, https://commons.wikimedia.org/w/index.php?curid=136504207.

discharges from dams, either controlled during high flow periods or uncontrolled breaches or failures, can have adverse consequences on downstream areas. Water discharges do occur, and thermal pollution may result, but no contaminant loading occurs.

15.5.4 Geothermal Energy

Because temperatures generally increase with depth below the ground surface, in many cases, energy (geothermal energy) can be captured and used. Geothermal energy is continuously produced beneath the Earth's surface due to the extreme heat contained in molten rock (magma) within the Earth's core. In some cases, this heat is exposed to water, resulting in hot water or steam generation. This can be piped to the surface for electrical generation through turbine operation. Alternatively, geothermal energy can be tapped by piping and exposing water to hot, dry rocks, where the heated water can be re-circulated toward the surface (Figure 15.17).

Geothermal energy can be utilized under two scenarios. The first one is the capture of the Earth's heat in geothermal power plants to generate steam and, in turn, generate electricity. Wells are commonly drilled one to two miles below the Earth's surface to pump steam or hot water toward the surface. This is best applied in areas with hot springs, geysers, or volcanic activity. The second scenario is through the use of heat pumps. Heat that is present close to the surface is tapped and used to heat water or provide heat to buildings. These temperatures need not be hot; this technology is feasible in many locations where the near-surface temperature is commonly between 50 and 60 °F. The systems used for heating in winter can be reversed to provide cooling in the summer. Figures 15.18–15.20 show geothermal energy potential considering these scenarios.

Geothermal reserves exist everywhere although practically speaking, areas with known hot springs, geysers, and volcanic activity may be most easily and cost effectively utilized, including

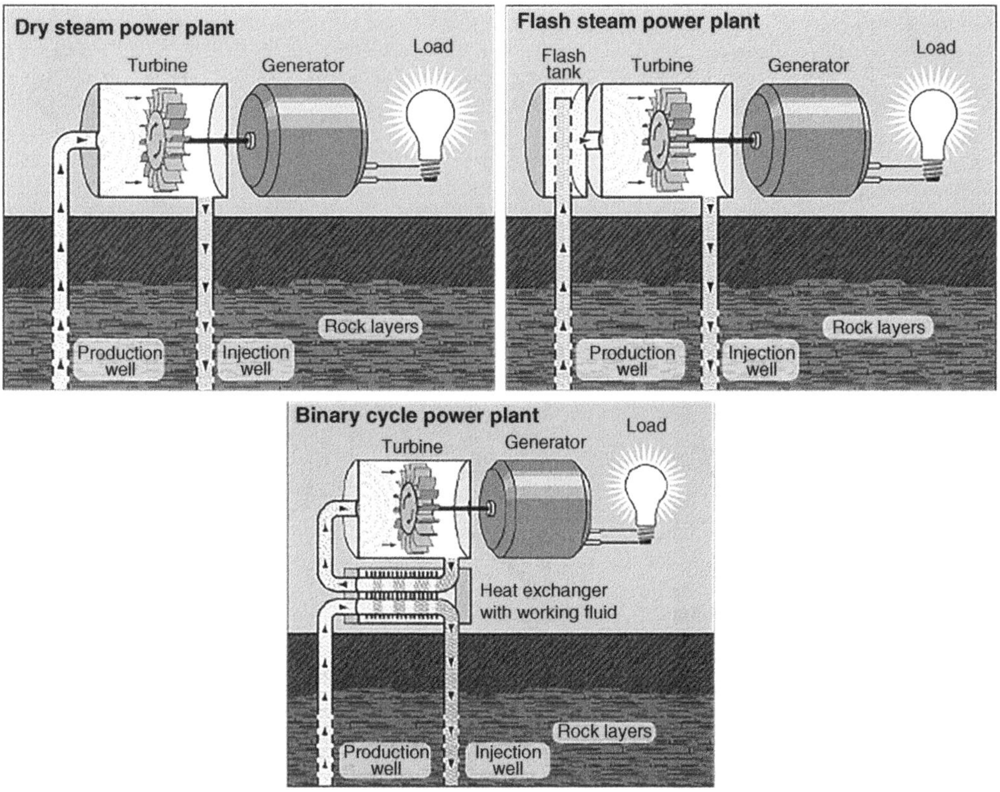

Figure 15.17 Geothermal energy power plants. *Source:* US EIA.

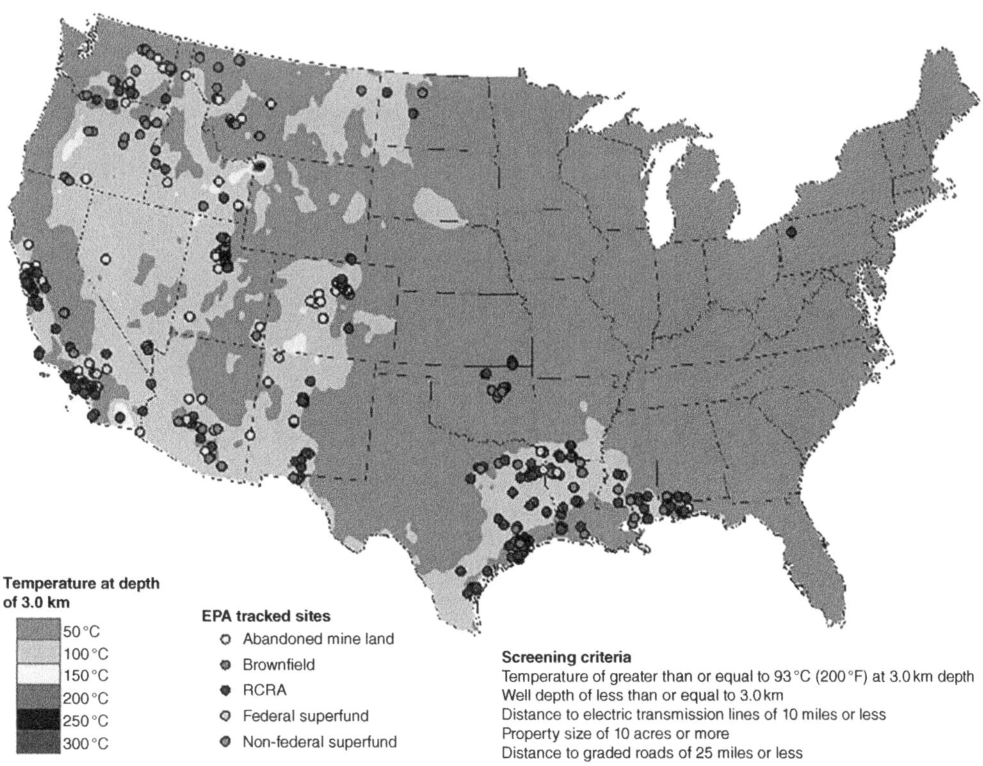

Figure 15.18 Potential sites for geothermal binary power plant facilities. *Source:* USEPA (2024)/ Environmental Protection Agency/Public domain.

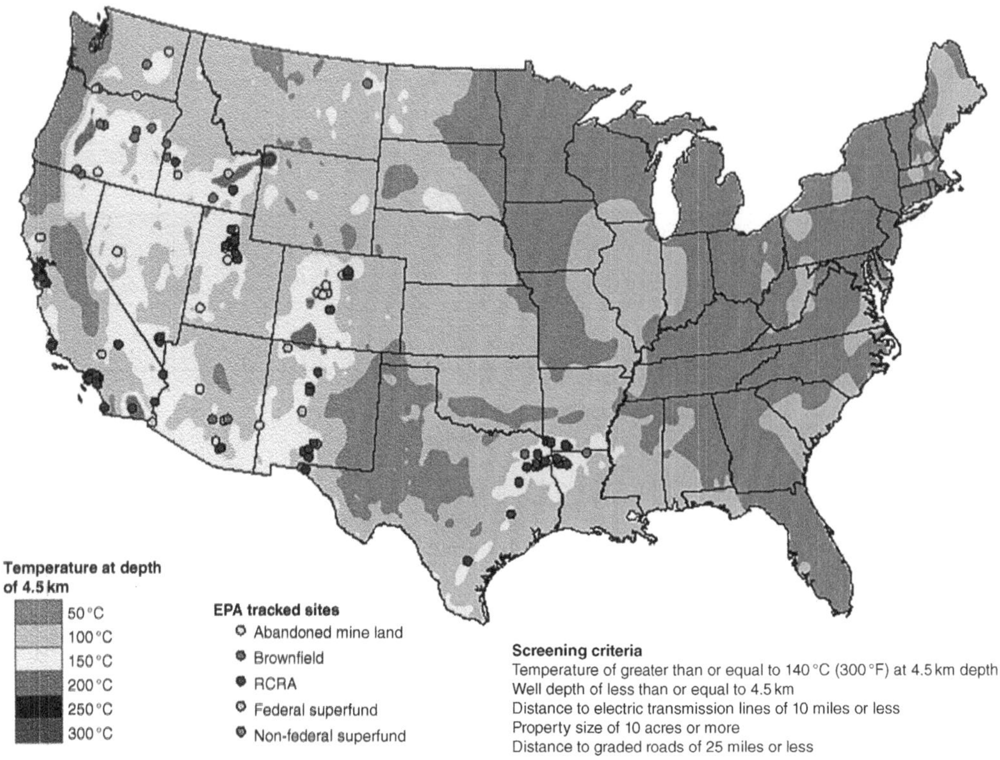

Figure 15.19 Potential sites for geothermal flash power plant facilities. *Source:* USEPA (2024)/Environmental Protection Agency/Public domain.

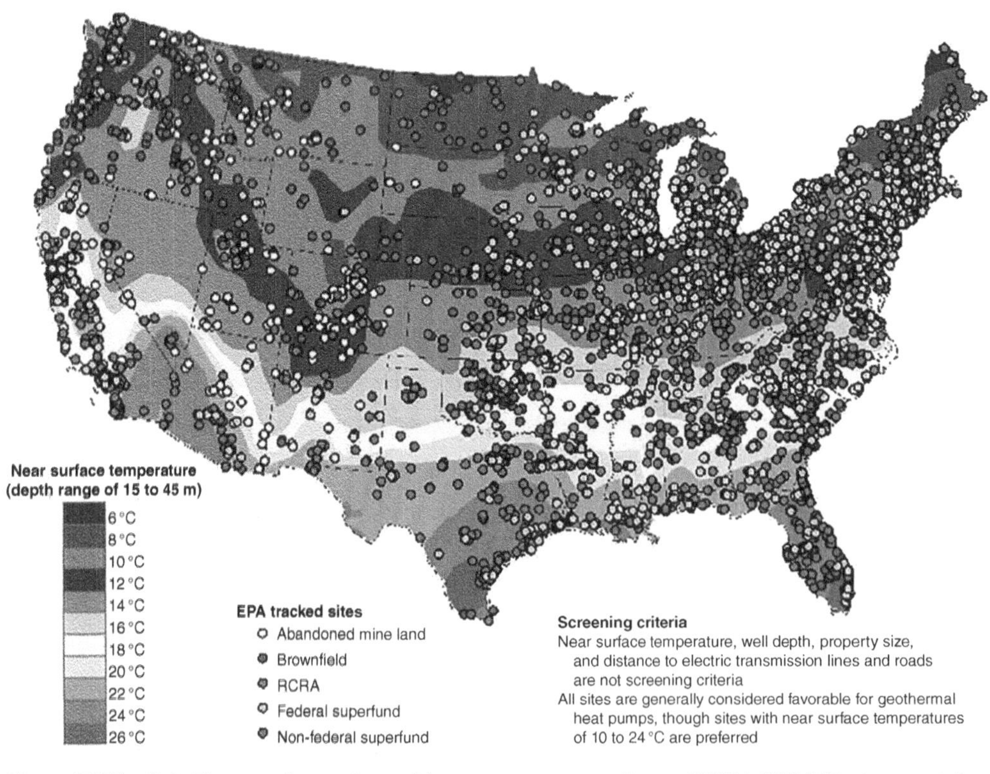

Figure 15.20 Suitable areas for geothermal heat pump systems. *Source:* USEPA (2024)/Environmental Protection Agency/Public domain.

portions of California, Oregon, Nevada, Arizona, Washington, Idaho, Montana, Alaska, Utah, and Hawaii. In 2003, a total of 2300 MW was generated from geothermal sources. It is believed that at least 20 000 MW of sources may be effectively used.

Inherently, geothermal energy uses water resources; however, the hot water and steam are often injected back in a quasi-closed loop process. If water is used to extract heat from dry rocks, water will need to be used from an outside source. Additionally, the water is commonly injected into the subsurface and is free of contaminants, though well drilling may result in localized contamination to the subsurface, depending on the method of drilling used. Geothermal energy does not result in any air emissions or solid waste generation. In general, there is no need for a significant use of land resources; although if extracted groundwater is not re-injected, land subsidence may occur. However, in large geothermal facilities, a significant impact can be observed in the subsoil and groundwater; when significant amounts of energy are dissipated or removed from the subsoil, especially in those sites with no geothermal activity, important changes in the temperature of groundwater can be measured. Florea et al. (2017) reported temperature changes in the subsoil and groundwater from 5 to 17 °C, depending on the size of the geothermal system and the flow of groundwater. These temperature changes may have an enormous impact on the geochemistry of groundwater with premature carbonate precipitation and pore soil clogging. Further research on this topic is necessary to evaluate the environmental impact of geothermal systems.

15.5.5 Biomass Energy

Biomass energy is derived from materials from plants and animals. Often these materials may be burned directly or transformed into other substances that can be burned to obtain energy. Wood or peat are common examples of materials that can be directly burned to obtain energy (IEA 2007). Biomass contains stored energy derived from sunlight converted during photosynthesis. While the burning of these materials releases carbon dioxide and cannot be considered a zero- or low-emission energy source, these plants absorb and sequester carbon dioxide, fixating it as the plant grows leaves, flowers, branches, and stems. Many different types of biomass may be used for energy, including wood chips, corn, and garbage materials. Some types can also be converted to liquid fuels (biofuels) for use in transportation activities, including plant and animal fat-based biodiesel and corn, sugar cane, and rice straw-derived ethanol. Although these materials may come from the primary crop yield (such as corn kernels), in many cases, biomass sources consist of plant or animal waste products, including peanut shells, sludge, manure, corn husks, and plant clippings. Because of the carbon sequestration, these may be considered carbon neutral. Additionally, because of the commonly rapidly growing cycles, these materials are considered renewable (McKendry 2002).

In many cases, biomass cannot be utilized exclusively as a fuel; it may need additional fuels to burn efficiently. This process, called co-firing, often utilizes coal as a co-fuel (Dai et al. 2008). The coal use allows for higher-temperature combustion, and the use of biomass greatly reduces net emissions (Koppejan and Van Loo 2012). Figures 15.21 and 15.22 show biomass potential considering the power and refining basis.

While the biomass, which consists of crop yield material or "fresh" waste, may be burned directly, methane, derived from decaying waste products, may also be burned to derive energy. Trash that ends up in landfills will produce methane as it decomposes. Methane can also be generated and captured from farm digesters – large tanks that generate methane from decaying manure. The methane is often purified or combined with other combustible gases and is burned to generate heat or electricity. Figure 15.23 depicts the pipe system for the collection and recovery of methane from landfills. Figure 15.24 shows the potential sites for landfill gas energy recovery as well as instances

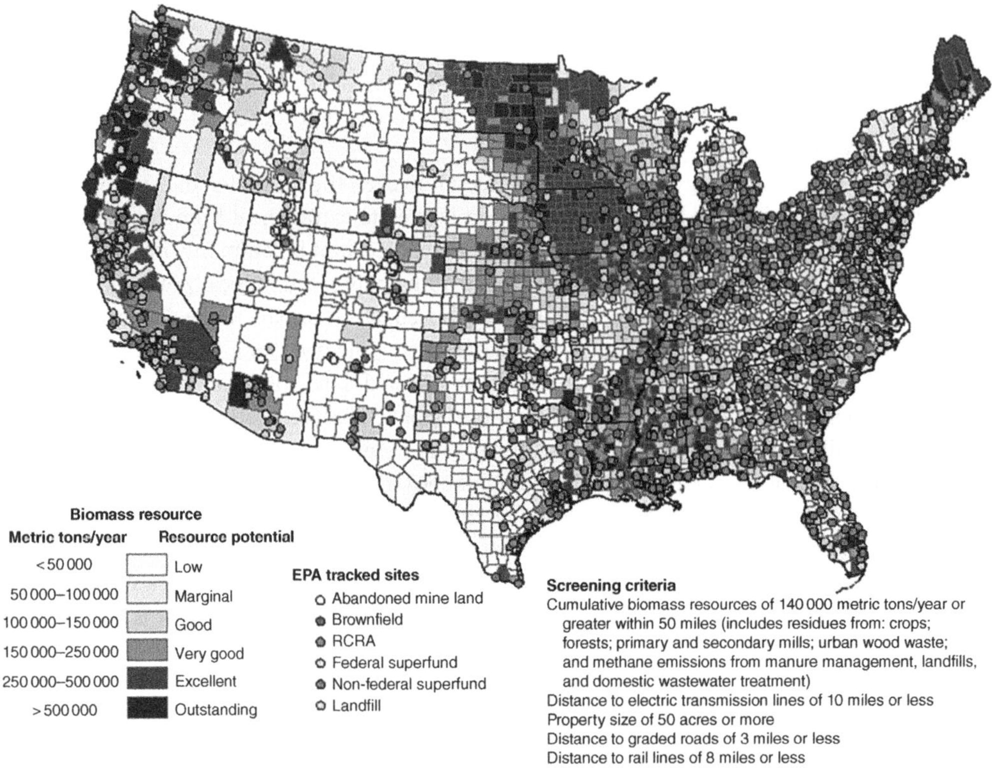

Figure 15.21 Potential areas for biopower facility siting. *Source:* USEPA (2024)/Environmental Protection Agency/Public domain.

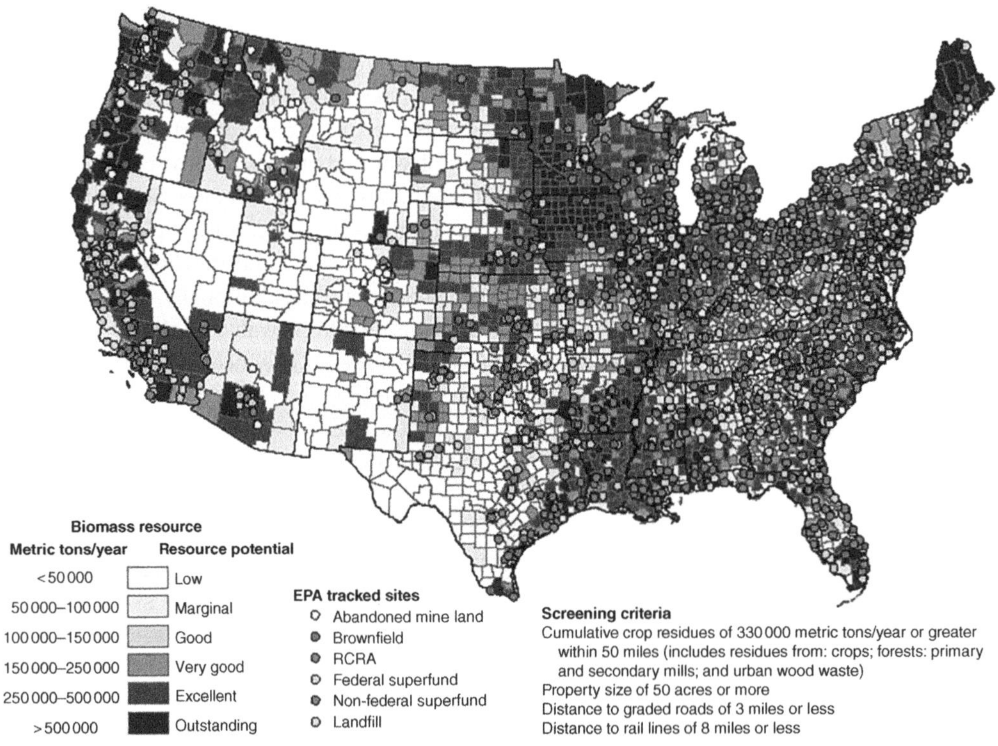

Figure 15.22 Potential areas for biopetroleum facility siting. *Source:* USEPA (2024)/Environmental Protection Agency/Public domain.

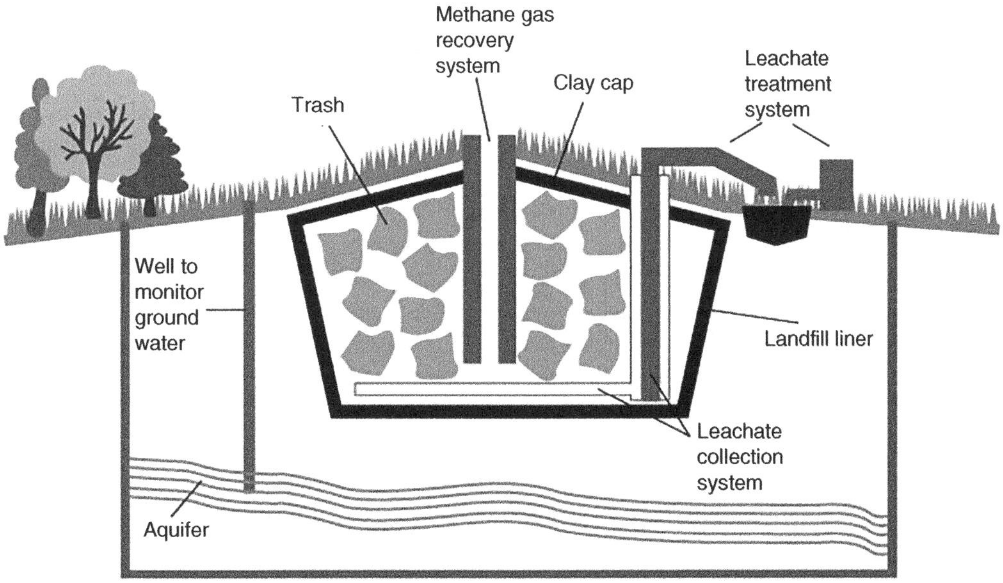

Figure 15.23 Collection of landfill gas (methane) for energy recovery.

Figure 15.24 Potential sites for landfill gas energy recovery. *Source:* USEPA (2024)/Environmental Protection Agency/Public domain.

of current activity. Municipal solid waste (MSW) may also be incinerated to generate heat or electricity; however, doing so results in the generation of carbon dioxide, sulfur dioxide, and nitrogen oxides comparable to emissions generated during the combustion of coal.

With respect to biomass, only 14 million dry tons of a currently estimated 590 million tons are available for utilization. Biomass sources are available throughout the United States. The burning of biomass results in the generation of air emissions, including nitrogen oxides and sulfur dioxide; however, the relative amounts are heavily dependent on the types of biomass burned. These emissions tend to be less than those associated with coal and therefore, as previously described, biomass can be attractive to co-fire with coal. Biomass burning has similar water re-use requirements as fossil fuel burning, namely, water needs to be extracted for steam generation and cooling. Cooling water also needs to be discharged into the surface water systems after its intended use. As with fossil fuels, water extraction and discharge can affect aquatic habitats by disrupting water levels and introducing thermal gradients. Discharged water may also result in pollutant loading of receiving waters. Solid waste (ash) is also generated during burning; however, the concentrations and mass of deleterious compounds (e.g. heavy metals) are much lower as compared to fossil fuels (IRENA 2012). From a land perspective, biomass requires productive agricultural areas for cultivation as well as areas associated with power generation. The cultivation of biomass-related products can incorporate sustainable procedures, including reduced use of agricultural chemicals and farming methods that reduce the dependence on water use and are focused on soil conservation.

Landfill gases, similar to biomass, result in the generation of air emissions, including nitrogen oxides and sulfur dioxide; however, the relative amounts are heavily dependent on the quality of the landfill gas burned. The emissions are often similar in constituency and magnitude to those associated with natural gas. Again, since landfill gas is the result of decaying organic matter, it is considered part of the natural carbon cycle, and the emission of carbon dioxide is considered neutral from a carbon standpoint. Water utilization as well as water and solid waste generation are similarly low in magnitude as compared to natural gas. Since landfill gases evolve from land use already devoted to waste-related activities, landfill gas combustion does not impact land resources.

15.6 Economic Considerations

In this chapter, we have primarily focused on the environmental effects and considerations for a range of energy generation alternatives. It is equally important to consider the economics of energy generation alternatives, which is necessary to select for appropriate technologies. To perform an economic analysis, both upfront capital investments as well as long-term operating and end-of-life disposition costs must be considered. A useful metric used in this analysis is the Levelized Cost of Energy (LCOE), which represents the net present value (NPV) of all capital and operating costs, including capital investment, operations and maintenance costs, fuel expenses, and decommissioning costs, divided by the discounted stream of energy output, in which both the costs and the energy production are annualized (Hansen 2019; Shen et al. 2020; Loth et al., 2022). It is expressed mathematically as (USDOE 2015):

$$LCOE = \frac{\sum_{t=1}^{n} \frac{I_t + M_t + F_t}{(1+r)^t}}{\sum_{t=1}^{n} \frac{E_t}{(1+r)^t}}$$

where

I_t = investment expenditures in year t
M_t = operation and maintenance expenditures in year t
F_t = fuel expenditures in year t
E_t = energy generation in year t
r = discount rate
n = project lifetime (years).

By factoring in full life-cycle costs and energy output, then converting to an annualized unit cost, LCOE allows for a direct, "face-to-face" economic comparison of energy generation technology alternatives, including renewable sources, nuclear, and fossil fuels. This approach is important, as it captures the combination of upfront and long-term costs, eliminating bias against technologies where capital costs constitute a greater proportion of expenses, but long-term operating costs may be lower due to negligible fuel costs, such as with solar power generation. Further, the LCOE approach allows for an analysis of cost trends when comparing alternatives. As an example, in 2010, both onshore wind and solar PV generation LCOEs were significantly higher than those associated with fossil fuel generation; however, due to significant declines in capital costs, by 2023, both solar PV and onshore wind LCOEs were notably lower than fossil fuel-generated LCOEs (IRENA, 2024).

While LCOE captures explicit economic costs, it is important to note that side effects of energy generation, such as the generation of greenhouse gases (GHGs) and other emissions, are often not captured and included in direct economic costs. These emissions; however, do impart costs (commonly referred to as externalities or external costs) to parties "external" to economic transactions related to power generation, i.e. the generator and the consumer (Owen, 2004), such as those related to air pollution and diminished public health or effects on crop yields. These external social and environmental costs can be substantial.

To address these externalities, agencies and governments have considered two general approaches to carbon pricing, often with the dual goals of raising revenue and to induce reductions in carbon emissions. The first is a carbon tax. With a carbon tax, a price or "tax" is set on a unit of carbon emissions, which then provides an economic incentive for emitters to reduce their emissions (Frank, 2014). With a cap-and-trade approach, a threshold level of emissions is established, allowing emitters to buy and sell emissions credits; essentially, a carbon tax sets a price and the resulting emissions are unknown, whereas a cap-and-trade framework sets a limit on emissions and relies on the market to establish the unit price of emissions (Frank, 2014). These approaches may also be combined into a hybrid approach, which may combine emissions limits with price floors or ceilings (Goulder and Schein, 2013).

Where carbon tax and cap-and-trade frameworks attempt to reduce carbon emissions and encourage the use of cleaner energy generation technologies through pricing, another approach attempts to diversify the mix of energy generation technologies by policy mandate. Renewable Portfolio Standards (RPSs) are regulatory policies that require electricity producers to supply a certain amount of their energy from renewable sources (Barbose et al., 2015). These standards aim to promote deploying renewable energy technologies, reduce GHG emissions, and stimulate economic growth in the clean energy sector. By creating a requirement for use, RPS provides an incentive for producers to construct and operate renewable-based generation facilities or purchase electricity generated from other facilities. Numerous US states and countries have adopted RPS frameworks. While RPS use has been noted to potentially increase consumer utility costs, potential benefits of

RPS applications may include reduced emissions, generating fuel and technology diversity, price stability, and economic development (Heeter et al., 2014).

15.7 Sustainability and Resiliency Considerations

Renewable energy sources offer notable and obvious advantages over traditional fossil fuel sources with respect to environmental and social health. In recent years, technological advancements have resulted in substantial decreases in equipment and generation costs, leading to several technologies now offering an economic advantage over fossil fuels. With lower costs and more favorable environmental and social attributes, renewable sources provide a clear sustainability advantage over conventional energy sources like coal and oil.

As conventional fossil fuel sources have ceded their historic economic cost advantages to renewable energy technologies, these effects are magnified by the fact that renewable energy sources do not rely on finite fuel resources located in intermittent locations across the planet. They offer the potential to generate energy anywhere the wind blows, water flows, or the sun shines. It is, however, important to note a couple of key factors that affect renewable energy sources and must be considered. First, several renewable technologies rely upon raw materials that face the same intermittent spatial and geographical concerns as fossil fuels. Deposits and reserves of materials, such as rare earth metals, may be located in areas of political instability or social/human rights concerns. Second, diurnal, seasonal, or long-term conditions may affect the generating capacity of some renewable technologies, such as solar, wind, or hydroelectric. Fortunately, these may be augmented with other generation capabilities, such as nuclear facilities, or may be coupled with battery storage arrays.

Over the long term, especially as cost advantages continue to emerge, renewable energy sources are in a strong position to meet global energy demands. Their scalability, flexibility, and independence from several limitations (see Table 15.2 for a comparative matrix) that affect fossil fuel applications allow renewables to provide a sustainable and resilient energy approach for decades to come.

15.8 Summary

While a diverse range of energy sources is necessary for a nation's economic health and quality of life, an overreliance on fossil fuels – coal, oil, and natural gas – has resulted in numerous detrimental side effects on the environment, including solid wastes, liquid wastes, and the generation of air emissions, including greenhouse gases. The overutilization of fossil fuels has taxed already depleting reserves, leaving the economies of developed and developing countries susceptible to market and pricing pressures.

Although there has been an increased focus on renewable energy sources and interest in nuclear power sources, an overwhelming majority of energy sources utilized in the United States are still constituted by fossil fuels. Nevertheless, sustainable energy sources – including hydroelectricity, biomass, wind, geothermal, and solar, can be utilized to greatly reduce wastes, emissions, and other stressors to the environment. Further, the appropriate use of these technologies based on the geographic/special availability of natural resources/phenomena from which they can be derived can lead to energy independence derived from a diverse mix of energy sources, providing a greater range of energy sources in the energy market. This can lead to improved macroeconomic health as well as

Table 15.2 Comparison of conventional and renewable energy sources for sustainability and resiliency.

Indicator	Conventional energy sources	Renewable energy sources
Resource availability	Finite fuel resources that will be depleted over time.	Virtually infinite resources that are sustainable if properly managed.
Environmental impact	High GHG and pollutant emissions; solid waste generation during fuel extraction, refining, and combustion; notable public health and environmental effects.	Minimal GHG and pollutant emissions and waste generation during operational phase.
Carbon footprint	Varies during construction phase; significant during operational phase.	Varies during construction phase; low to negligible during operational phase, especially for solar, wind, and hydroelectric power.
Economic viability	Historically cheaper than renewables; however, recently at parity or higher comparative LCOE costs; efforts to capture costs of carbon emissions (i.e. carbon taxes, cap-and-trade) eventual resource depletion expected to increase cost disadvantages.	Rapid cost reductions due to technological advancements have resulted in favorable LCOE values compared to traditional energy sources.
Energy security	Many fuel sources (e.g. petroleum and natural gas) are located in areas with political instability and/or social volatility; supplies are vulnerable to geopolitical risks and price volatility.	Generally promotes energy independence by utilizing local resources; however, some raw materials, such as rare earth metals, are located in areas with political and/or social volatility.
Infrastructure	Well-established transport, generation, and transmission due to legacy systems and applications but infrastructure is aging in many regions and requires upgrades and maintenance.	Requires new infrastructure, especially for generation and transmission; initial costs can be high but offset by elimination of infrastructure upgrade costs for legacy systems.
Social impacts	Pollution affects public health; resource extraction/transmission often results in land diminution and community displacement; notable legacy environmental justice issues.	Reduced public and environmental health risks; potential for economic development due to generation capacity and transmission development; however, resource extraction could result in community displacement.
Reliability	High reliability given proven technology and independence from weather/astronomical conditions (i.e. wind speed, diurnal/seasonal variations); however, social/political/economic volatility can disrupt supply chains.	Subject to diurnal/seasonal variations (e.g. solar and wind) but mitigated by battery storage and other time-independent operational strategies as well as portfolio management, such as combining nuclear with solar, wind, etc.
Technological advances	Legacy technologies provide potential for incremental efficiency and cost improvements; recent technological focus on carbon capture and storage.	Ongoing rapid advances in efficiency, storage, and integration technologies for a variety of generation alternatives.

localized economic stimulus through lower energy prices and job creation as these energy sources are developed and utilized. Finally, transition to renewable energy generation from traditional fossil fuel generation approaches offers several clear advantages with respect to public and environmental health, economic development, social impact, infrastructure, reliability, and energy independence/resilience.

15.9 Questions

15.1 What are the environmental impacts of energy generation?

15.2 Name some of the major sectors that utilize large amounts of energy.

15.3 Rank the order of the three major fossil fuel energy sources based on their carbon emissions.

15.4 Describe the contributions to air emissions from different fossil fuel energy sources.

15.5 What are the three major nonrenewable fossil fuel sources of energy?

15.6 Name a few sources of renewable energy.

15.7 What are the advantages and challenges of renewable energy sources?

15.8 Explain the principles behind the generation of nuclear energy and geothermal energy.

15.9 Discuss the benefits and drawbacks of nuclear energy use.

15.10 What are the few strategies developed to curb the negative impacts of energy generation using nonrenewable sources?

15.11 Compare and contrast strategies used to reduce energy demand.

15.12 Discuss some potential reuse options of solid waste generated from fossil fuel energy source use.

15.13 Discuss some of the environmental side effects that may arise from renewable energy source use.

15.14 Discuss potential sources and uses of biomass as an energy source.

15.15 Select three energy generation alternatives and qualitatively discuss features that would affect the inputs for levelized cost of energy (LCOE) analyses.

15.16 Compare and contrast the resiliency of alternate renewable energy sources, assuming a range of changing weather patterns resulting from a changing climate.

15.10 Problems

15.1 The average greenhouse gas (GHG) emissions per kWh of electricity generated for a given region are provided below: Geothermal: 6.0 g CO_2e, Hydropower: 1.0 g CO_2e, Nuclear: 3.7 g CO_2e, Solar: 18 g CO_2e, Wind: 7.0 g CO_2e, Coal: 850 g CO_2e, Natural gas: 500 g CO_2e, Oil: 900 g CO_2e. The energy mix of a hypothetical region is given in Table A. The current annual electricity demand of the region is 143 TWh growing at an average of 3% per annum. Considering the growth rate remains the same for the foreseeable future, calculate the total GHG emissions from the region due to electricity consumption over the next 10 years.

Table A Electricity mix of the region under consideration.

Source of energy	Current mix (% of total consumption)
Geothermal	0.5
Hydropower	5
Nuclear	27
Solar	10
Wind	1
Coal	10
Natural gas	46
Oil	0.5

15.2 In order to control GHG emissions and fossil fuel depletion, the region described in the above problem aims to limit the usage of fossil fuel-based electricity to under 10% within the next 50 years. The region plans to adopt a phased approach, as shown in Table B, to reach its target. Calculate the total GHG emissions reduced if this plan is effectively implemented.

Table B Target electricity mix to limit fossil fuel-based electricity to be adopted by the region.

Source of energy	Phase-1 target (% of total consumption by the end of next 5 yr)	Phase-2 target (% of total consumption by the end of next 20 yr)	Phase-3 target (% of total consumption by the end of next 50 yr)
Geothermal	1	3	5
Hydropower	5	5	5
Nuclear	26	29	30
Solar	15	25	35
Wind	5	10	15
Coal	10	5	1
Natural gas	38	23	9
Oil	0	0	0

15.3 If the transformation plans are implemented at a slower rate than expected and the region is able to finish phase-1 in 15 years and phase-2 in 50 years, calculate the reduction in GHG emissions.

References

Barbose, G., Bird, L., Heeter, J. et al. (2015). Costs and benefits of renewables portfolio standards in the United States. *Renewable and Sustainable Energy Reviews* 52: 523–533.

Bradbury, K., Pratson, L., and Patiño-Echeverri, D. (2014). Economic viability of energy storage systems based on price arbitrage potential in real-time US electricity markets. *Applied Energy* 114: 512–519.

Brandt, A.R., Heath, G.A., Kort, E.A. et al. (2014). Methane leaks from North American natural gas systems. *Science* 343 (6172): 733–735.

Corner, A., Venables, D., Spence, A. et al. (2011). Nuclear power, climate change and energy security: exploring British public attitudes. *Energy Policy* 39 (9): 4823–4833.

Dai, J., Sokhansanj, S., Grace, J.R. et al. (2008). Overview and some issues related to co-firing biomass and coal. *The Canadian Journal of Chemical Engineering* 86 (3): 367–386.

Deonarine, A., Bartov, G., Johnson, T.M. et al. (2013). Environmental impacts of the Tennessee Valley Authority Kingston coal ash spill. 2. Effect of coal ash on methylmercury in historically contaminated river sediments. *Environmental Science & Technology* 47 (4): 2100–2108.

Dincer, I. (2000). Renewable energy and sustainable development: a crucial review. *Renewable and Sustainable Energy Reviews* 4 (2): 157–175.

EIA (2016). *Carbon Dioxide Emissions Coefficients*. U.S. Energy Information Administration. https://www.eia.gov/environment/emissions/co2_vol_mass.cfm (accessed 12 April 2017).

EIA (2017a). *Electricity Explained – Use of Electricity*. US Energy Information and Administration. Independent statistics and analysis. https://www.eia.gov/energyexplained/ (accessed April 2017).

EIA (2017b). *Energy from the Sun*. US Energy Information and Administration. Independent Statistics and analysis. https://www.eia.gov/energyexplained (accessed 12 April 2017).

EIA (2024). *What is U.S. Electricity Generation by Energy Source?* US Energy Information Administration. https://www.eia.gov/tools/faqs/faq.php?id=427&t=3 (accessed 12 December 2024).

Entrekin, S., Evans-White, M., Johnson, B., and Hagenbuch, E. (2011). Rapid expansion of natural gas development poses a threat to surface waters. *Frontiers in Ecology and the Environment* 9 (9): 503–511.

Evans, A., Strezov, V., and Evans, T.J. (2009). Assessment of sustainability indicators for renewable energy technologies. *Renewable and Sustainable Energy Reviews* 13 (5): 1082–1088.

Ewing, R.C. and Von Hippel, F.N. (2009). Nuclear waste management in the united states – starting over. *Science* 325 (5937): 151–152. https://doi.org/10.1126/science.1174594.

Florea, L.J., Hart, D., Tinjum, J., and Choi, C. (2017). Potential impacts to groundwater from ground-coupled geothermal heat pumps in district scale. *Groundwater* 55 (1): 8–9.

Frank, C. (2014). Pricing carbon: a carbon tax or cap-and-trade? Brookings – Planet Policy. https://www.brookings.edu/articles/pricing-carbon-a-carbon-tax-or-cap-and-trade/ (accessed 11 December 2024).

Giampietro, N., Aspinall, R.J., Ramos-Martin, J., and Bukkens, S.G.F. (2014). *Resource Accounting for Sustainability Assessment. The Nexus Between Energy, Food, Water and Land Use*. New York, NY: Routledge.

Goulder, L.H. and Schein, A.R. (2013). Carbon taxes versus cap and trade: a critical review. *Climate Change Economics* 4 (3): 1350010.

Hansen, K. (2019). Decision-making based on energy costs: comparing levelized cost of energy and energy system costs. *Energy Strategy Reviews* 24: 68–82.

Heeter, J., Barbose, G., Bird, L. et al. (2014). *Survey of State-Level Cost and Benefit Estimates of Renewable Portfolio Standards*. National Renewable Energy Laboratory. NREL/TP-6A20-61042, LBNL-6589E, May.

Hu, G., Li, J., and Zeng, G. (2013). Recent development in the treatment of oily sludge from petroleum industry: a review. *Journal of Hazardous Materials* 261: 470–490.

IEA (2007). *Bioenergy Project Development and Biomass Supply*. Paris, France: International Energy Agency.

IRENA (2012). *Biomass for Power Generation. IRENA Working Paper. Renewable Energy Technologies: Cost Analysis Series*. Abu Dhabi: International Renewable Energy Agency.

IRENA (2024). *Renewable Power Generation Costs in 2023*. Abu Dhabi: International Renewable Energy Agency.

Kao, S.C., Sale, M.J., Ashfaq, M. et al. (2015). Projecting changes in annual hydropower generation using regional runoff data: an assessment of the United States federal hydropower plants. *Energy* 80: 239–250.

Koppejan, J. and Van Loo, S. (ed.) (2012). *The Handbook of Biomass Combustion and Co-Firing*. Routledge.

Krauskopf, K. (2013). *Radioactive Waste Disposal and Geology*, vol. 1. Springer Science & Business Media.

Loth, E., Qin, C., Simpson, J.G., and Dykes, K. (2022). Why we must move beyond LCOE for renewable energy design. *Advances in Applied Energy* 8: 100112.

McKendry, P. (2002). Energy production from biomass (part 1): overview of biomass. *Bioresource Technology* 83 (1): 37–46.

Miller, P.J., Atten, C.V., and Bradley, M.J. (2004). *North American Power Plant Air Emissions. Commission for Environmental Cooperation of North America*. Quebec, Canada: Montreal.

NEI (2017). *World statistics. Nuclear Energy Around the World*. Nuclear Energy Institute. https://www.nei.org/Knowledge-Center/Nuclear-Statistics/World-Statistics (accessed 12 April 2017).

Owen, A.D. (2004). Environmental externalities, market distortions and the economics of renewable energy technologies. *The Energy Journal* 25 (3): 127–156.

Pacyna, E.G., Pacyna, J.M., Steenhuisen, F., and Wilson, S. (2006). Global anthropogenic mercury emission inventory for 2000. *Atmospheric Environment* 40 (22): 4048–4063.

Shen, W., Chen, X., Qiu, J. et al. (2020). A comprehensive review of variable renewable energy levelized cost of electricity. *Renewable and Sustainable Energy Reviews* 133: 110301.

Spang, E.S., Moomaw, W.R., Gallagher, K.S. et al. (2014). The water consumption of energy production: an international comparison. *Environmental Research Letters* 9 (10): 105002.

Twidell, J. and Weir, T. (2015). *Renewable Energy Resources*. Routledge.

Uihlein, A. and Magagna, D. (2016). Wave and tidal current energy–a review of the current state of research beyond technology. *Renewable and Sustainable Energy Reviews* 58: 1070–1081.

USDOE (2015). *Levelized Cost of Energy (LCOE)*. United States Department of Energy. https://www.energy.gov/sites/prod/files/2015/08/f25/LCOE.pdf (accessed 12 December 2024).

USEPA (2024). *Inventory of U.S. Greenhouse Gas Emissions and Sinks: 1990–2022*. U.S. Environmental Protection Agency, EPA 430-R-24-004. https://www.epa.gov/ghgemissions/inventory-us-greenhouse-gas-emissions-andsinks-1990-2022 (accessed 12 December 2024).

Yao, Z.T., Ji, X.S., Sarker, P.K. et al. (2015). A comprehensive review on the applications of coal fly ash. *Earth-Science Reviews* 141: 105–121.

16

Sustainable and Resilient Materials, Waste Management, and Circular Economy

Chapter Goal and Objectives
This chapter aims to explore sustainable and resilient approaches in materials selection, eco-design, and waste management strategies, emphasizing circular economy principles, sustainability, and resiliency to address environmental, economic, and social impacts. The major goals are: (i) to discuss sustainable materials, with a focus on eco-design using green materials and frameworks like Environmental Product Declarations and Extended Producer Responsibility; (ii) to analyze the generation and types of waste, both hazardous and nonhazardous, while evaluating their environmental impacts; (iii) to present the common waste management hierarchy, including pollution prevention, waste minimization, reuse, recycling, energy recovery, and landfilling; (iv) to evaluate the integrated waste management approach, promoting the combination of strategies tailored to local conditions; (v) to explore sustainable waste management through the lens of the triple-bottom-line (environmental, economic, social) considerations; (vi) to highlight the role of the circular economy in driving sustainable waste management practices; and (vii) to address the need for resiliency in waste management systems to withstand climate change and other sudden shocks or stressors.

16.1 Introduction

The increasing demand for resources and rapid industrialization have created significant challenges in managing limited resource availability, the waste generated, and the corresponding environmental impacts. Finite resources, such as fossil fuels and minerals, are being extracted and consumed at an accelerating pace, resulting in reduced availability for future generations. These extracted resources are used to produce materials essential for various engineering applications, including construction, electronics, vehicles, and more.

With the growing rate of urbanization and industrialization, the demand for raw materials for engineering applications continues to rise, further depleting available reserves. Compounding this issue, materials used for these applications generate significant amounts of waste at the end of their intended life cycle. Traditional waste management strategies, which have often prioritized disposal over prevention, have worsened problems such as resource depletion, environmental pollution, and greenhouse gas emissions. Recognizing these challenges, sustainable materials and integrated

Sustainable and Resilient Engineering: Drivers, Metrics, Tools, and Applications, Second Edition. Krishna R. Reddy, Claudio Cameselle, and Jeffrey A. Adams.

Companion website: www.wiley.com/go/reddy/sustainableresilientengineering2e

waste management approaches have gained prominence. These approaches focus on minimizing waste at its source, promoting resource recovery, and reducing environmental footprints.

Beyond sustainability, material selection and waste management strategies must also incorporate resiliency to address vulnerabilities arising from changing climate conditions and sudden disruptions, such as natural disasters or supply chain interruptions. Integrating resiliency ensures that waste management systems can adapt to and recover from such challenges while maintaining functionality and environmental performance. This chapter presents the interconnected themes of sustainable materials, waste generation, integrated and sustainable waste management, and the circular economy while underscoring the importance of incorporating resiliency to meet the dual imperatives of sustainability and adaptability in the face of climate change and other extreme impacts.

16.2 Sustainable Materials

The growing emphasis on sustainable development and environmental stewardship has spurred the adoption of innovative frameworks and strategies for advancing material sustainability across all sectors. Key approaches such as eco-design, green materials, Environmental Product Declarations (EPDs), and Extended Producer Responsibility (EPR) integrate sustainability principles into product design and material selection to minimize resource depletion and pollution. Eco-design promotes environmentally conscious practices throughout the product lifecycle, while green materials offer renewable, recyclable, or biodegradable alternatives to conventional materials. EPDs provide standardized and transparent assessments of a product's environmental impact, enabling stakeholders to make informed, sustainable decisions. At the same time, EPR legislation shifts the responsibility for waste management to producers, encouraging the development of eco-friendly products and efficient waste-handling systems. Collectively, these approaches represent a transformative shift toward sustainable and resilient materials, addressing environmental challenges and driving innovation across industries.

16.2.1 Eco-design and Green Materials

Eco-design integrates environmental considerations into product development at every process stage, from manufacturing to disposal, prioritizing the use of sustainable, non-toxic, and environmentally friendly materials. The concept has existed for decades, and with the changing climate and its associated environmental impacts, the need for adopting eco-design across all sectors has never been higher. Measures such as minimizing energy consumption and enhancing energy efficiency are crucial to improve the environmental performance of systems. Various strategies have been established to address environmental impacts early in project development within engineering and design. For instance, the Design for the Environment (DfE) approach facilitates sustainable product innovation, while Life Cycle Assessment (LCA) provides a detailed overview of a project's overall environmental impact (Vallet et al. 2013). These strategies help identify the environmental impacts at different stages and inform decision-making, including selecting methods, materials, and processes to meet sustainability goals. As industries and policies prioritize sustainability, integrating eco-design principles will be crucial in achieving environmental goals.

Green materials, derived from natural sources, often have a minimal environmental footprint and are typically renewable, biodegradable, or recyclable, thus reducing waste and decreasing reliance on nonrenewable resources (Khan et al. 2021). Examples of green materials include bamboo, which is used for flooring, furniture, and wall coverings; corks, used as insulation from heat and sound; concrete made from fibers of hemp plants; recycled steel, which requires significantly lower energy for production than new steel; and recycled plastics.

These materials are essential in advancing eco-design across sectors such as construction, packaging, clothing, electronics, energy, and water treatment sectors. Incorporating green materials into eco-design can be achieved through strategies like LCA, which evaluates the environmental impact of alternative materials, emphasizing the benefits of promoting sustainable development. Despite the availability of many green and alternative materials, their adoption across industries remains limited. Lowering production costs could significantly boost their use. Advancing research on these materials, combined with policy reforms to support eco-design, is essential for reducing costs and driving wider adoption.

16.2.2 Environmental Product Declarations (EPDs)

EPDs quantify the environmental impacts of a product throughout its lifetime based on LCA. EPDs are developed according to ISO 14025, while the LCA are performed based on ISO 14040, and product category rules are obtained from ISO 14027. EPDs aim to communicate credible and transparent information regarding a product's environmental performance to manufacturers, distributors, and customers. Verified by an independent third party, declarations comply with pre-established standard operating rules specific to each product category to ensure comparability.

EPDs are categorized as Type III environmental communication instruments. Compared to Type I – environmental labeling (ISO 14024) and Type II – self-declared environmental claims (ISO 14021), which are more popularly used methods, EPDs are considered more reliable environmental communication instruments. The international EPD system was launched in 1999 and was the first EPD program to be developed worldwide. Type I and Type II environmental communication instruments lack transparency and reliability compared to the Type III EPDs. However, they are still not as prevalent as environmental labeling (Type I) (Minkov et al. 2015). This is because general program instructions and product category rules under ISO 14025 for the EPDs differ among schemes, as the Standard cannot provide detailed solutions for every case, given that different sectors need specific methodological solutions. Further, Type III declarations require large data sets and accurate information about the environmental impacts of products, which can burden small businesses and vendors.

Government incentives or campaigns to encourage using EPDs have not successfully attracted companies to pursue EPDs as their environmental communication instrument. Studies across various countries and various sectors conducted on companies of different sizes show that companies often learn about EPDs through technical consultancy or the market (Hunsager et al. 2014; Ibáñez-Forés et al. 2016), highlighting the importance of changes in policies to encourage the use of EPDs as a tool for communicating the environmental performance of their product.

16.2.3 Extended Producer Responsibility

EPR is a legislative tool that holds producers responsible for the environmental impact of their products throughout the product's lifecycle, extending to the post-consumer stage. By making manufacturers accountable for the costs of managing their products at the end of life, EPR provides a financial incentive to them to design environmentally friendly products, as the cost of end-of-life management would be lower. EPR was first formally introduced in Sweden by the Ministry of Environment in 1990 and then spread across various EU state members. The first EU regulation adopting the EPR was Packaging and Packaging Waste Directive (94/62/EC) in 1994. This directive aimed to reduce the environmental impact of packaging waste by implementing systems for the return, collection, and recycling of packaging materials. Later, the EPR became a fundamental part of the Waste Framework Directive (2008/98/EC). After formalization and implementation in EU, EPR has now become popular worldwide.

EPR operates on the principle that, as the manufacturers have the greatest control over product design and marketing, they must also bear the greatest responsibility to reduce emissions, pollution, toxicity and waste associated with the product. EPR requires manufacturers to internalize waste management costs in their product pricing and ensure the safe handling of their products, thus shifting waste management responsibilities from governments to producers (Cai and Choi 2019). Producers must either take back their used products and manage them through reuse and recycling or delegate this responsibility to a third party, such as Producer Responsibility Organizations (PROs), funded by producers to manage products post-use.

EPR is a sustainable environmental management principle that drives innovation in policy, product design, processes, supply chains, and technology. EPR legislation has been implemented globally in various sectors, such as e-waste, battery laws, and packaging. In the United States, there is no national EPR law, instead, certain state-level legislations cover sectors like-waste, and packaging waste. However, adopting a national-scale approach could harmonize these policies and minimize discrepancies (Hickle 2014).

The effectiveness of EPR has been demonstrated across multiple countries and sectors, where it has significantly improved waste management systems (Milanez and Bührs 2009; Shooshtarian et al. 2021; Niza et al. 2014) with stakeholders advocating for expansion and standardization of EPR programs. Despite its benefits, inconsistent regulations and financial barriers to smaller producers have been a much-discussed drawback of EPRs (Hanisch 2000). Addressing these issues through collaborative efforts and policy harmonization is essential for realizing EPR's full potential in driving sustainable material selection, production, and waste management.

16.3 Waste Generation and Problems

Waste is the inevitable by-product of all human activities. Virtually every activity generates some type of physical by-product. When the materials that result from any activity are no longer useful or have degraded, such that they no longer fulfil their original or otherwise useful purpose, they are classified as waste material. These wastes are commonly classified based on their respective generative sources. Some of these activities include those associated with residences, commercial and retail businesses, institutions, construction and demolition activities, municipal services, water/wastewater treatment, and incineration facilities. Several industrial processes generate waste materials, including industrial construction and demolition, fabrication, manufacturing, chemical refining and synthesis. Additionally, these waste materials may be classified depending on their inherent characteristics and physical properties (USEPA 2015). The storage, handling, transport, and disposal of these materials may be handled in a variety of ways, and new methods are under consideration for making these processes more sustainable.

Waste management is an area of increasing importance. Waste generation, both hazardous and nonhazardous wastes, continues to increase at an appreciable rate (USEPA 2017a). For instance, industries generate and dispose of over 7.6 billion tons of industrial solid wastes each year. Over 40 million tons of this material is considered hazardous. Additionally, both nuclear and medical wastes are increasing in quantity on an annual basis.

Proper waste management is critical to achieve sustainability. In this section, waste types and their impacts are discussed, following which, various waste management methods and approaches are discussed in the subsequent sections.

16.3.1 Types of Waste

The Resource Conservation and Recovery Act (RCRA) has statutorily defined solid wastes as any garbage or refuse, sludge from a wastewater treatment plant, residues from an air pollution control facility, and other discarded material, including solid, semisolid, liquid, or contained gaseous material resulting from commercial, industrial, mining, and/or agricultural operations and community activities. Solid wastes may be further categorized as nonhazardous or hazardous wastes (USEPA 2015). Regardless of this categorization, the generation of solid waste represents pollution and, in many cases, an unnecessary waste of resources. In particular, hazardous waste generation may lead to environmental degradation, degradation of natural resources, health problems, and premature death through chronic or acute exposures.

16.3.2 Nonhazardous Waste

Nonhazardous waste typically is composed of trash or garbage, generated from residential households, commercial offices, and other similar sources. These materials are generally classified as municipal solid waste (MSW). Industrial solid wastes such as mining wastes and coal combustion residuals are nonhazardous materials resulting from the manufacture and production of products by different manufacturing industries.

Regulations pertaining to the handling and disposal of nonhazardous waste are presented in 40 CFR Parts 239–259. These are commonly referred to as RCRA Subtitle D regulations (USEPA 2017b). These regulations prohibit the open dumping of solid waste, establish criteria for MSW landfills and other solid waste disposal facilities, and mandate the development of comprehensive plans for the management of MSW as well as nonhazardous industrial solid waste. Further, because these wastes are not considered hazardous, they present a potential for recycling and reuse of these materials as potentially valuable resources. The programs enacted by industries and government agencies are increasingly facilitating these actions.

16.3.3 Hazardous Waste

As opposed to nonhazardous solid waste, hazardous solid wastes consist of materials that are dangerous or potentially harmful to human health and the environment. Hazardous waste is also addressed by RCRA regulations; the legal framework in place for hazardous waste and its classification, identification, management, and disposal is described in 40 CFR Parts 260 through 279. This framework is also commonly referred to as RCRA Subtitle C regulations (USEPA 2017b). These regulations control hazardous waste from the time they are generated until their final disposal; a timeframe commonly referred to as "cradle to grave."

16.3.4 Effects and Impacts of Waste

Waste generation and management can have many adverse effects on public health and the environment. A wide range of chemicals is present within waste materials, many of which pose a threat to public health and the environment. Air pollution and greenhouse gas emissions could be a major problem. Leachate may be generated from these materials and, if not controlled, can enter the soil and groundwater, leading to contamination of these resources. When released into the environment, contaminants can alter the chemistry of receiving waters, leading to the destruction of aquatic life and aquatic ecosystems. Complex species on the food chain are also impacted, as species they prey upon are reduced or eliminated. Additionally, contaminants that enter the food chain through plants or microbiological organisms bioaccumulate and are passed on to other organisms following the food chain (Parfitt et al. 2010).

While RCRA was developed to handle both hazardous and nonhazardous wastes, many sites have been adversely impacted through previous land use or disposal practices (USEPA 2018). For instance, more than 36 000 environmentally impacted sites have been identified for potential consideration and addition to the Comprehensive Environmental Response, Compensation, and Liability Act (CERCLA) National Priority List (NPL). A total of over 1400 sites have been placed on this list, necessitating cleanup due to contamination that poses an acute, imminent threat to public health and the environment. The USEPA has also identified approximately 2500 additional contaminated sites that will require remediation. The US Department of Defense maintains approximately 19 000 sites, many of which have been extensively environmentally impacted. Further, approximately 400 000 underground storage tanks have been confirmed or are suspected to be leaking contents, affecting both soil and groundwater (USEPA 2018). Remediation of contaminated sites is a daunting task, and more about this will be discussed in Chapter 19.

16.4 Waste Management

As mentioned, wastes are being generated at an increasing rate. Population growth and urbanization, coupled with increased industrial, commercial, and institutional activity, contribute to the increased waste production, as well as rapid economic growth and industrialization throughout the developing world. These socioeconomic dynamics have led to increased consumption of raw materials, processed goods, and services. Although increased economic activity has, without question, improved the quality of life for billions of people, drastic environmental consequences have also resulted. It is necessary to properly manage a range of wastes in order to protect public health and the environment while also ensuring steady economic growth.

Figure 16.1 depicts a schematic for the hierarchy of waste management. As shown, the pyramid shape indicates less-favored options toward the bottom (e.g. disposal, energy recovery), with more preferred alternatives higher up the pyramid, including waste minimization and prevention.

At the top of the pyramid is the most desirable alternative, waste prevention. Because prevention reduces the challenges of waste management, it is a fundamental goal of all waste management strategies. Several technologies can be employed throughout the manufacturing, use, or postuse portions of product life cycles to eliminate waste, and in turn, reduce or prevent pollution. Some strategies include the use of manufacturing methods that incorporate less hazardous or harmful materials, the use of modern leakage detection systems for material storage, innovative chemical neutralization technologies to reduce reactivity, or water-saving technologies that reduce the need

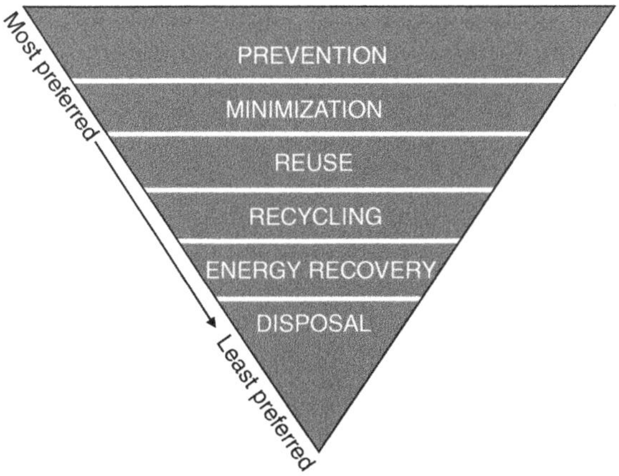

Figure 16.1 Hierarchy diagram on waste management.

for freshwater inputs. Pollution prevention can also be accomplished by the practice of green chemistry. All of these waste management options are discussed in detail in the following sections.

16.4.1 Pollution Prevention

One ideal approach for pollution prevention (P2) consists of actions that contribute to reducing the generation of hazardous substances or dissipation of natural resources (Bishop 2002). There are two goals associated with a P2 strategy: (i) every atom that enters a manufacturing process should leave the process as a saleable product (Trost 1991); and (ii) every erg of energy used in manufacturing should produce a desired material transformation. In pursuing these dual goals, it is important to assess current or future processes associated with waste generation. Some tasks that may be performed include constructing a process flow diagram, evaluating environmental impacts and issues, identifying pollution prevention opportunities, analyzing alternatives, and documenting results. In performing these assessments, P2 analyses are targeted at single facilities and processes, not products. Further, P2 is generally focused on the reduction of discarded or dissipated materials, energy, and water as opposed to finding suitable alternatives for their use. When processes have been assessed, P2 actions may result in process modification, technology modification, better housekeeping practices, substitution of inputs, and on-site and off-site reuse. Further, these actions may be applied to multiple scales. On a micro-scale level, chemicals and materials may be modified. On the mesoscale, the existing process may be modified. Finally, on a macro-scale, beneficial reuses of waste products may be identified.

A P2 strategy provides several benefits with respect to waste-generating processes and subsequent waste management. Significant financial savings can be achieved with respect to compliance costs, energy and materials purchases, waste treatment, and disposal costs. With less waste and emissions generation, work places are often made healthier. Because of less waste handling and disposal, legal liabilities are often reduced. Additionally, using an active process to reduce environmental impacts often leads to a positive association with a proactive commitment to protecting the environment.

An active P2 strategy differs in several ways from mandated environmental compliance. First, when used properly, P2 can be used as a proactive core organizing principle, leading to efficiencies throughout a process value chain, as opposed to often inefficient reactive or "retrofitting" measures to achieve compliance. The use of P2 encourages a systematic evaluation of a production process, which also can lead to other positive unintended side effects or discovery of other efficiencies. Further, because it results in a holistic view of a particular process, a proactive P2 strategy can often enhance profitability through reduced costs and efficiency gains (USEPA 2017c).

16.4.2 Green Chemistry

Green chemistry is a vital aspect of pollution prevention, and it is the practice of chemical science and manufacturing that is sustainable, safe, and nonpolluting, consuming minimum amounts of energy and material resources while producing virtually no waste (Sheldon et al. 2007). Figure 16.2 demonstrates a schematic of the process, including the use of renewable feedstocks (including those from sources such as biomass), proper controls, internal recycling, and the minimization of nondegradable wastes when they cannot be avoided.

With respect to feedstock, biomass sources are especially attractive to replace petroleum. Not only is petroleum a depleting resource, but also partially oxidized biomass materials can actually be more easily applied to processes when compared to partially oxidized petroleum materials. Ideally, feedstocks are renewable, pose no toxicity issues, and may be converted to a desired end-product in a few steps, resulting in 100% yield or 100% atom economy. The percentage of atom economy (Trost 1991) is defined as the total mass of the product divided by the total mass of the reactant (Figure 16.3).

When considering chemical processes, green chemistry focuses on improvements in several areas. First, it involves the elimination of toxic materials or processes. Second, consideration is given to the use of existing chemical synthesis processes by using safer, less polluting processes as well as identifying means to make reagents greener, safer, or less likely to result in adverse side effects that can lead to pollution (Bozell 2008). In pursuing these goals, green synthetic chemistry focuses on finding ways to make new chemicals and new ways to make existing chemicals (Figure 16.4). This involves either using existing feedstocks in processes that are more benign or using alternative feedstocks in similar processes.

Ultimately, green chemistry is driven by several key principles (Anastas and Warner 1998). It focuses on the prevention of waste as opposed to its treatment. Energy use, the use of toxic reagents and products, and the use of solvents/extractants are all minimized. Atom economy is practiced, and an emphasis is placed on the use of renewable raw materials. Finally, there is a greater emphasis on catalysis than stoichiometry. Several of these key principles are inherent to the goals of green engineering (Anastas and Zimmerman 2003; Tang et al. 2008).

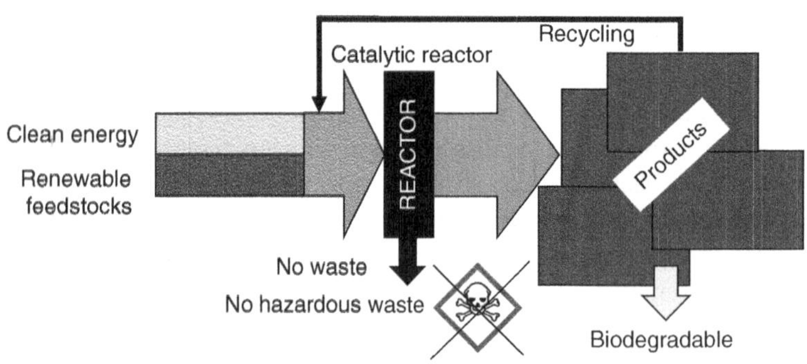

Figure 16.2 Production process based on green chemistry transformation of renewable feedstock.

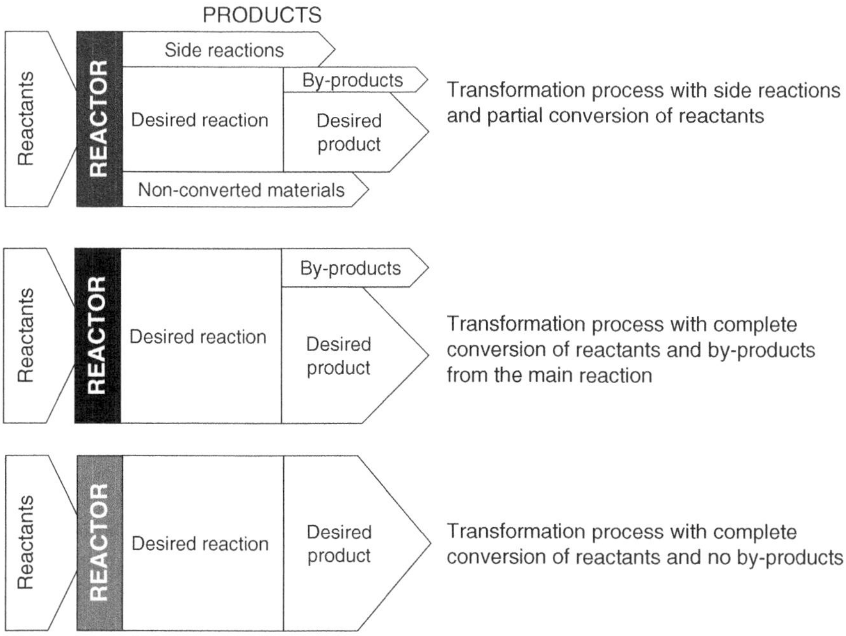

Figure 16.3 Objectives of the atom economy: transform every atom of reactant in saleable product increasing conversion and selectivity of the transformation process and decreasing side reactions and waste generation.

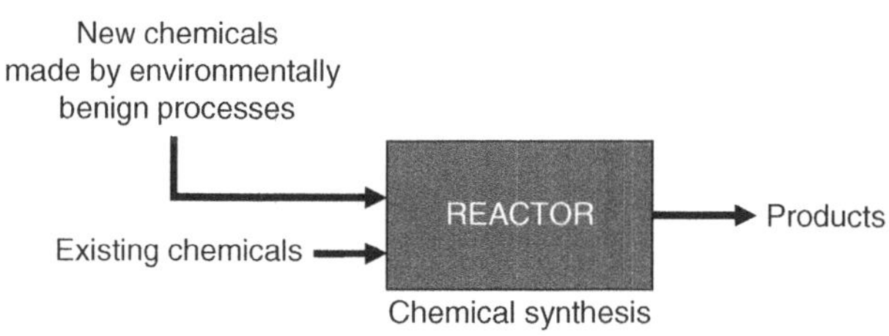

Figure 16.4 Substitution of existing chemicals by new chemicals made by environmentally benign processes.

16.4.3 Waste Minimization

While waste prevention or green chemistry techniques are viewed as superior alternatives with respect to waste management, waste minimization also offers an attractive alternative. Several strategies may be used to minimize and/or reduce waste generation (Clark 2012). First, manufacturing processes and products may be redesigned to use less material and energy. Moreover, increasing the efficiency of the transformation process results in less side products and less waste production, which, in turn, improves the economy of the global process. The modifications in the manufacturing process may reduce the utilization or production of less harmful chemicals, so waste management will incur relatively minor risks, and waste treatment may be less expensive with more options available for recycling of nonhazardous waste. The manufacturing processes may also be redesigned such that less waste and/or pollution is produced. The implementation of a quality system to minimize production errors and out-of-specification products will result in less waste generation. Products may also be designed such that they are easier to repair, reuse, remanufacture, compost, or recycle. Additionally, packaging systems may be redesigned to eliminate unnecessary materials. Finally, end-of-use strategies may be implemented to facilitate a greater focus on recycling or repurposing – either through incentive-based disposal options (i.e. fee per bag) or by establishing cradle-to-grave responsibility for manufacturers with respect to wastes generated during manufacturing (Fiksel and Fiksel 1996).

16.4.4 Reuse/Recycling

The use of recycled materials shows many advantages not only for the environment and the conservation of natural resources, but for the global economy of the production processes. When usable materials have already been created, it is also beneficial to focus on means to reduce the fabrication of identical materials when existing versions can be captured and reused. Reusing items decreases the use of material and energy resources as well as emissions and environmental degradation associated with manufacturing. Recycling captures the materials for refabrication, albeit with a reduction or elimination of the need for material inputs. With reuse, products are captured, cleaned, or processed, and re-entered into the market for an identical use. Figure 16.5 shows the energy consumption associated with different types of 350 ml containers, and Table 16.1 shows the energy savings in obtaining materials from wastes compared to raw materials (USEPA 2016).

Recycling may consist of primary, closed-loop recycling, or secondary recycling. Wastes may be classified either into preconsumer (internal to manufacturing processes) or postconsumer (captured after use in the marketplace) streams. With respect to household wastes, the materials may be sorted or not sorted prior to diversion to a material recovery facility (MRF). Further, materials recycling fees may be charged either on a set action basis ("pay as you throw") or on a unit basis ("fee-per-bag"). Some selected materials may be recycled as follows:

- *Biodegradable wastes* – Includes green wastes and other wastes that can easily breakdown in the natural environment – these materials may be composted on an individual or municipal basis.

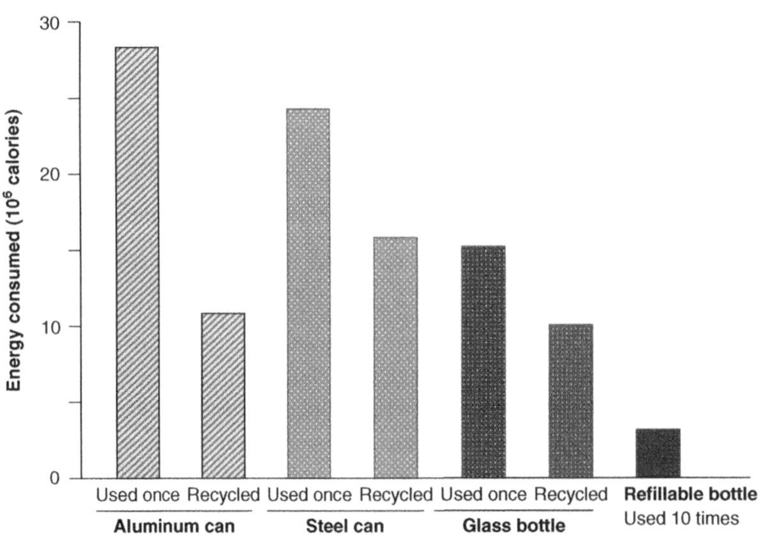

Figure 16.5 Energy consumption for the production of different types of 350 ml containers.

Table 16.1 Energy saving for recycled materials compared to the original raw material.

	Energy consumption (kcal) in the production of 1 kg		
Material	**Raw materials**	**Recycled materials**	**Energy saving (%)**
Glass	1200	800	35
Iron	10 300	5100	50
Paper	3700	1100	70
Polyethylene	4500	500	89
Aluminum	47 000	1400	97

- *Paper* – Recycling of paper is easy, effective, and substantially reduces energy and water inputs as well as pollution. Additionally, increased focus is being placed on the reduction of chlorine-based bleaching agents with hydrogen peroxide or oxygen products.
- *Metals* – The recycling of metals, such as aluminum or steel, reduces the need for mining and use of ore materials. Inputs and waste products may be greatly reduced.
- *Plastics* – Only approximately 4% of plastics are recycled. Further, recycling of plastics is relatively difficult, and it is often cheaper to fabricate new plastic products.

Nevertheless, recycling and reuse should be encouraged. Governments can facilitate these practices by providing subsidies or other financial incentives, as well as attach fees or taxes to the use of virgin materials. Additionally, public interest groups as well as private groups should continue to encourage these practices. Many industrial wastes are suitable to be used as construction materials, consuming large quantities of wastes for beneficial purpose and thus avoiding the need for new landfills (Sharma and Reddy 2004).

16.4.5 Energy Recovery

Waste materials may also be utilized for the generation of energy. Commonly, this is done with waste-to-energy incinerators. Approximately 600 of these incinerators exist worldwide, although most are located in Great Britain. Figure 16.6 illustrates a typical waste-to-energy incinerator as a part of comprehensive MSW management.

Waste-to-energy incineration offers several advantages (Astrup et al. 2015). First, the combustion greatly reduces the volume of waste materials, thereby reducing the need for landfill capacity. The pollution of water can be sufficiently minimized and hazardous waste residues are consolidated into a small volume by incineration. The resulting energy may be sold, reducing the overall cost associated with incineration. Additionally, some facilities are equipped to recover and sell metallic materials. Finally, modern facilities and controls greatly reduce resulting air emissions. However, incineration has some disadvantages. First, facilities are relatively expensive to construct and permit. They can often be expensive to operate as compared to local landfills with short transportation routes. These facilities often face significant public opposition, especially because older facilities do in fact contribute significant amounts of air emissions. Even new facilities are responsible for emitting carbon dioxide and other compounds (Lombardi et al. 2015). Further, there can be "competition" with recycling facilities, as incinerators divert combustible materials that could otherwise be recycled. Finally, to maintain operations, incinerators need input materials; therefore, they provide little to no incentive to reduce waste generation.

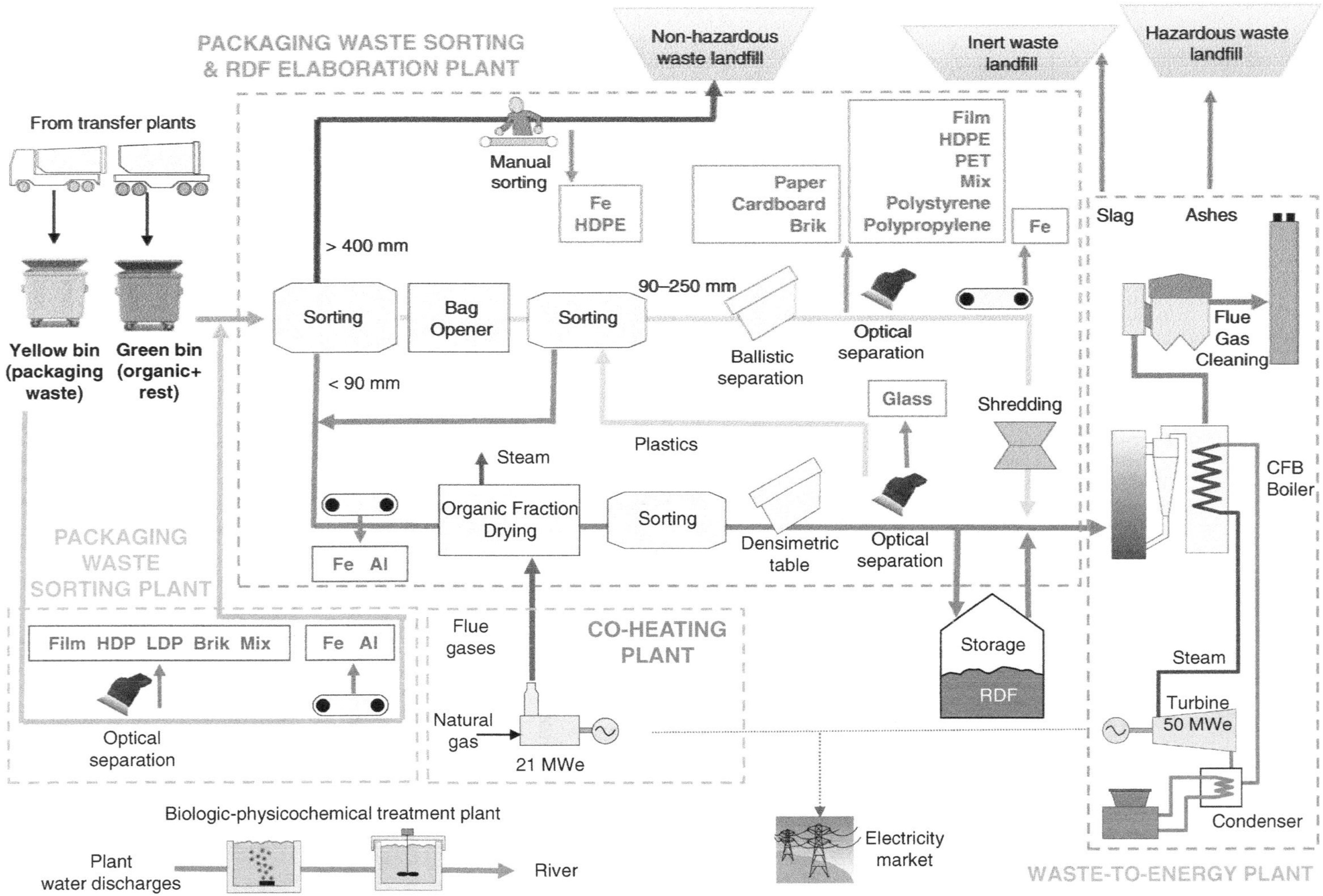

Figure 16.6 MSW management plant (SOGAMA, Galicia, Spain).

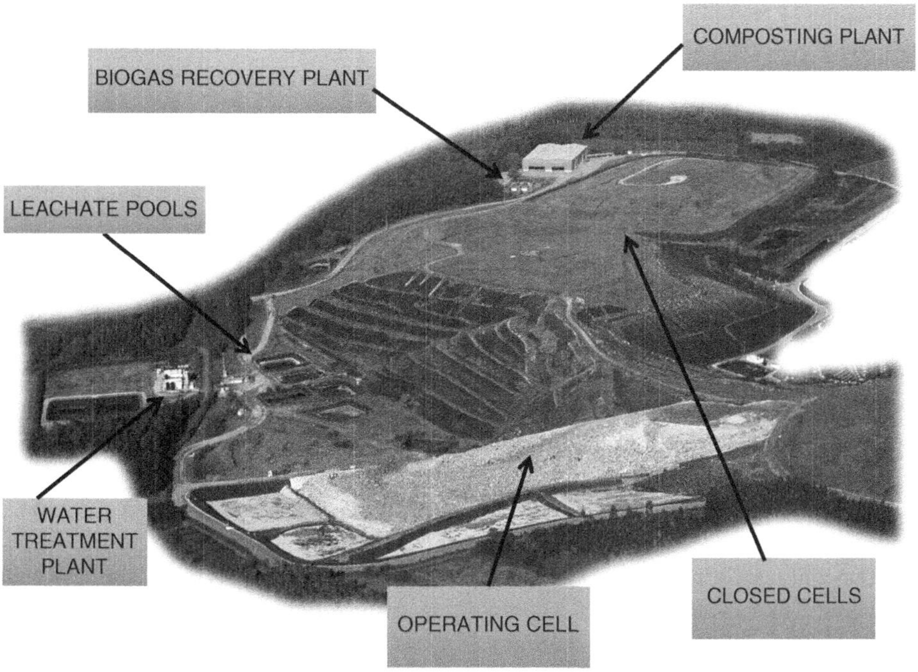

Figure 16.7 Landfill for nonhazardous waste. *Source:* SOGAMA, Galicia, Spain.

16.4.6 Landfilling

Solid waste of all types is quite commonly placed into landfill facilities (Zaman 2010; Alam and Alam 2014). While open dumps in the United States have essentially been eliminated, they still exist elsewhere. These facilities have been replaced in the United States with engineered landfills. Figure 16.7 shows a typical landfill facility, including protective and monitoring systems. Landfills are engineered structures meeting the regulatory requirements and consist of liner and cover systems, and leachate and gas collection and removal systems (Sharma and Reddy 2004). Engineered landfills eliminate the need to burn garbage materials and, when constructed properly, minimize odor and potential soil and groundwater pollution. Engineered landfills can be constructed quickly and can handle large quantities of waste, and modular design allows for efficient increases in capacity. They can be sited on lands that offer marginal benefits for other purposes, are relatively low cost to operate, and can be reclaimed for other purposes when landfill space is filled. However, landfills use does have several drawbacks. They result in the generation of noise, traffic, dust, and other nuisance conditions. Air emissions, both from equipment and from decaying waste, can result in contributions to greenhouse gases. Decomposition occurs slowly, limiting potential reuses on reclaimed land following filling. Finally, as with incineration, there is no incentive to reduce waste generation.

16.5 Integrated Waste Management

The waste management pyramid, or "hierarchy," presented in Figure 16.1 intuitively offers an optimal system to minimize the negative effects of waste generation at all levels; however, the system has its own limitations. First, it has no measurable scientific basis. Second, it does not address combinations of technologies. Finally, it does not address cost considerations. Because so many potential management techniques (each with several notable advantages) exist for dealing with waste materials, it is often appropriate to implement an integrated management approach (McDougall et al. 2008). In this approach, a combination of strategies is selected and used as appropriate. Figure 16.8 depicts how such a strategy may be applied for waste management for a combination of waste stream types. When considering hazardous wastes, it is also advantageous to convert materials into less hazardous forms or implement means to assure safe long-term storage.

When managing nonhazardous wastes, the approach techniques may be prioritized as shown in Figure 16.9. The first priority is the primary pollution/waste prevention. It encompasses making change in the industrial processes to eliminate use of harmful chemicals and use of less harmful materials. Reduce the use of material in packaging and products. To make products that are durable, recyclable, reusable, or are easier to repair.

The second priority is secondary pollution/waste prevention. It consists of reuse, repair, recycling/composting, and use of recycled products. The last priority is the waste management, which consists of treating waste to reduce toxicity. Incineration of waste to reduce the volume of the waste generated. The final option is the landfilling of waste.

When managing hazardous wastes, the approach techniques may also be prioritized as follows: The first priority is to produce less waste by changing industrial processes to eliminate hazardous waste production and recycle/reuse of hazardous waste. The second priority is to convert it to less hazardous material. This can entail natural decomposition, incineration, thermal treatment, chemical, physical, or biological treatment, and dilution in air and water. The last priority is to put waste in perpetual storage, such as underground injection wells, surface impoundments, landfilling of waste, and underground salt formations.

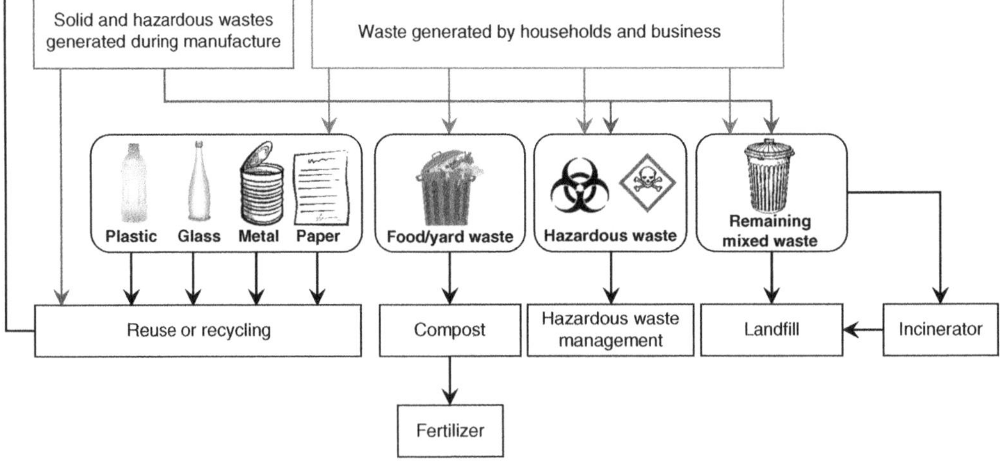

Figure 16.8 Integrated approach for minimization and recycling of wastes.

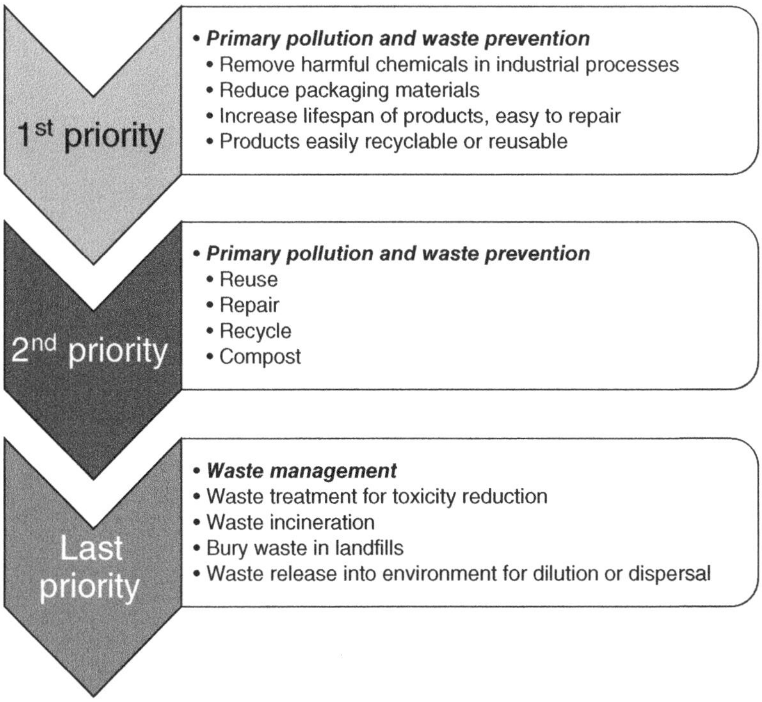

Figure 16.9 Priority operations for minimization and recycling of waste.

Hazardous waste materials can be processed and converted into less harmful materials. They may be detoxified using a variety of methods, including physical methods, chemical methods, bioremediation, phytoremediation, the use of nanomagnets, incineration, or plasma arc torch methods (Figure 16.10). Figure 16.11 illustrates the use of phytoremediation to transform hazardous waste materials into less hazardous materials. Phytoremediation is easy to establish, inexpensive, uses

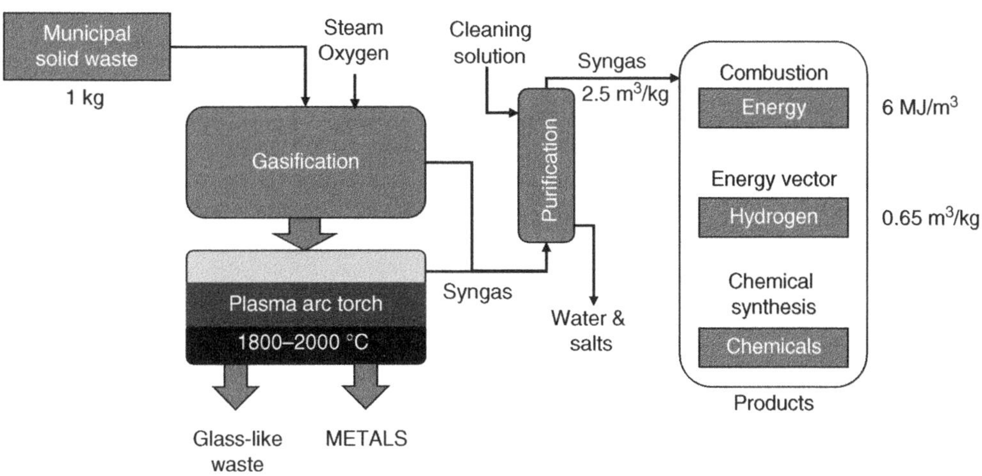

Figure 16.10 Waste gasification by plasma arc torch technology.

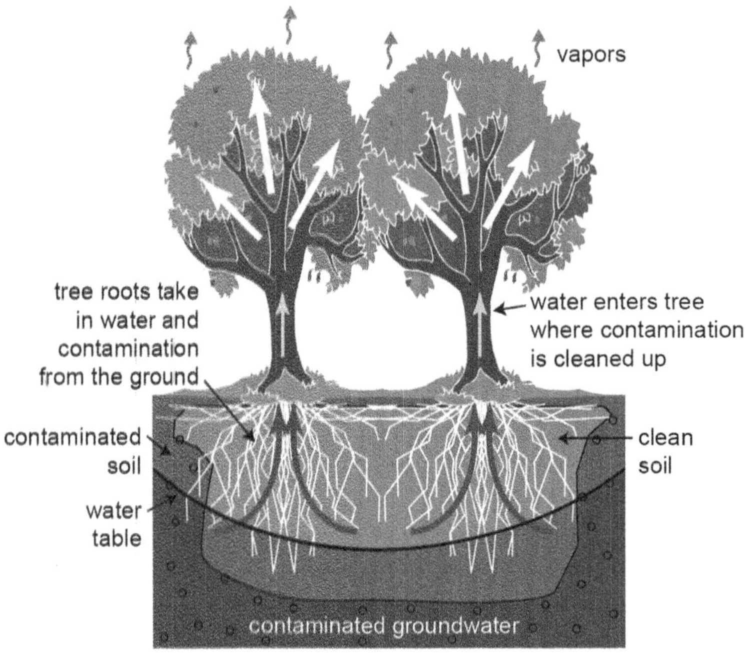

Figure 16.11 Phytoremediation technology for the cleanup of contaminated land. *Source:* USEPA (2012)/ United States Environmental Protection Agency/Public domain.

low energy inputs, can reduce materials dumped into landfills, and produces relatively little air pollution (Batty and Dolan 2013). On the other hand, it is slow, effective only to a depth that can be reached by roots, allows for emission of some toxic materials through plant respiration, and can render the plant materials themselves toxic.

Hazardous wastes can also be stored within the subsurface. Some methods include deep-well disposal, surface impoundments, and hazardous waste landfills (Blackman Jr 2016). Deep wells can be safe at appropriate sites. They are easy to implement, has a relatively low cost, and wastes can often be retrieved if necessary. However, they may lead to leaks or surface spills, failure of casings, and fractures within rock formations can lead to unanticipated migration. Surface impoundments are also low cost and easy to implement. The disposed waste can also be retrieved if necessary. These facilities can be constructed quickly and can be secured when implementing appropriate liner technology. However, groundwater can be impacted if liners fail, and volatile materials may escape and enter the atmosphere. Finally, landfills may be specially designed for hazardous waste, however, many protective measures should be incorporated to minimize the potential for releases into the environment (Sharma and Reddy 2004). Figure 16.12 depicts a typical hazardous waste landfill facility with clay and flexible membrane liners, leachate collector systems, and landfill covers to contain and control the waste.

16.6 Sustainable Waste Management

Lately, there is greater emphasis on sustainable waste management (Ai and Leigh 2017). Further, there is a growing interest in environmental justice awareness and advocacy for economically depressed areas where a disproportionate share of hazardous waste disposal facilities and other related facilities are located. As a result, sustainable waste management aims to be environmentally

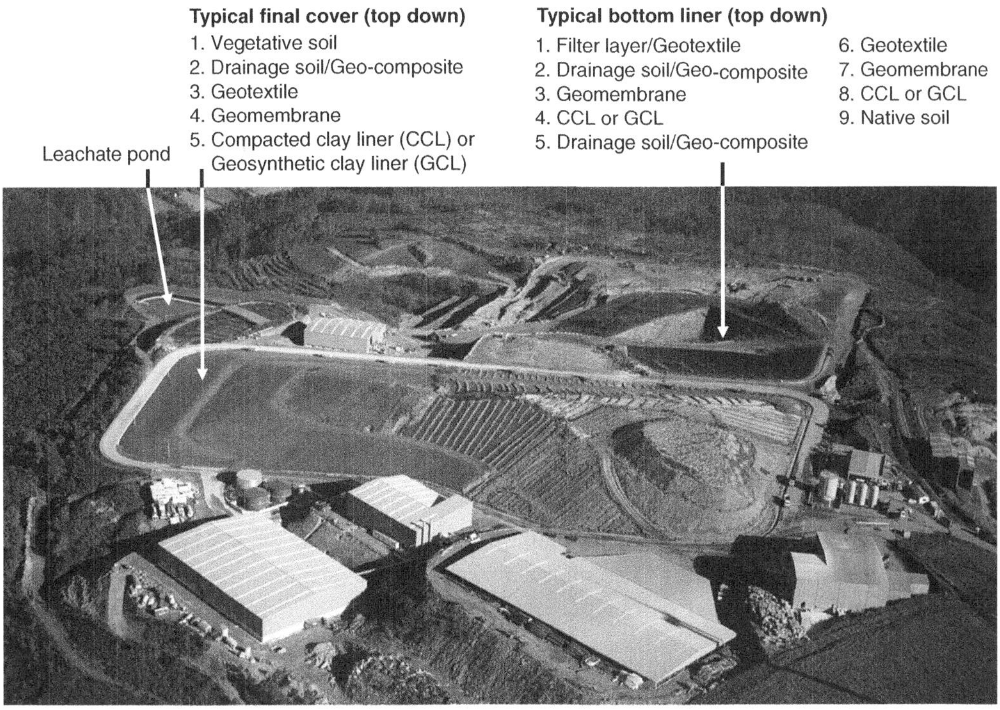

Figure 16.12 Hazardous waste landfill. *Source:* SOGAMA, Galicia, Spain.

effective, economically affordable, and a socially acceptable strategy for waste management. The integrated waste management strategies presented earlier provide an acceptable means to achieve sustainable waste management. Alternatively, the use of LCAs can also be coupled with integrated waste management strategies, thereby leading to sustainable waste management (Laurent et al. 2014a,b).

As described in earlier chapters, waste management typically occupies the final step in a product's life cycle. Therefore, an LCA boundary can be drawn around waste management steps. An LCA model for solid waste calculates total energy consumption, emissions to air or water, final solid waste, and the overall economic costs associated with these effects. Figure 16.13 presents a schematic of an LCA for integrated waste management. Inputs consist of waste, energy, other materials, and financial costs. The integrated waste management framework consists of activities such as biological treatment, thermal treatment, landfilling, recycling, collection, and sorting. Products include secondary materials, compost, and useful energy capture. Finally, outputs include air emissions, water emissions, and residual landfill material. The LCA captures key metrics, such as net energy consumption, air emissions, water emissions, residual landfill volume, recovered materials, and secondary materials (such as compost). The goal of the LCA is to devise more sustainable strategies, including more useful products, fewer emissions, less final inert waste, and less energy consumption. Additionally, economic sustainability is a goal as well. Further, using an existing waste management strategy as a baseline, the LCA can be used to make comparisons among potential strategic alternatives, allowing for the selection of the optimal waste management strategy based on the needs of the local environment, economy, and the society. Ultimately, the selection may be made based on a single criterion (e.g. available landfill space); multiple criteria where more than

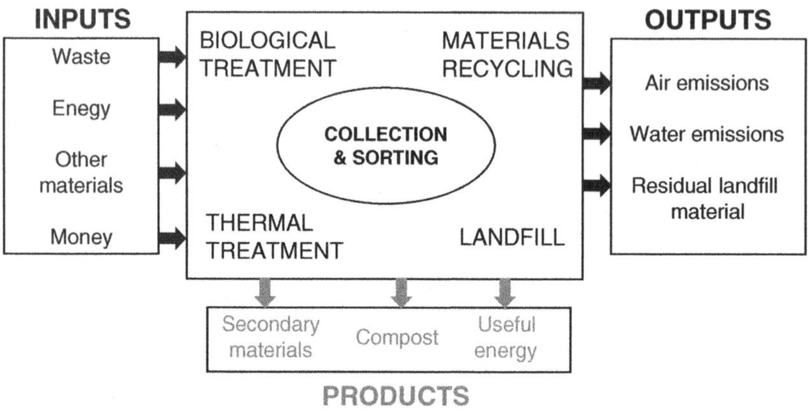

Figure 16.13 Life Cycle Assessment of an integrated waste management.

one issue are important (e.g. energy consumption and air emissions); "less is better," where one option is superior in all categories, or an impact analysis, where categories are combined to assess an overall effect, such as global warming.

16.7 Circular Economy

Construction and product consumption are often defined in a linear framework. Projects are manufactured or constructed, which includes refinement of raw materials, development of intermediate products, and manufacturing or erection of a useable product or project. The project has a design life during which it is operated and occasionally repaired or renovated to extend the useful life. At some point, it is removed from useful service, at which time it is disposed and reclaimed to some extent. This make/use/dispose pattern is referred to as the linear economy.

As an alternative, there is growing interest in the pursuit of a "circular economy." A circular economy is an industrial system that is restorative or regenerative by intention and design in that it replaces the linear "make/use/dispose" model with one that incorporates new raw materials on a limited or "as-needed" basis. The circular economy focuses on restoration, places greater emphasis on the use of renewable energy, minimizes or eliminates the use of toxic chemicals, and looks to minimize waste generation through optimized designs of materials, products, systems, and business models (Ellen MacArthur Foundation 2013).

The ultimate goal of the circular economy is to eliminate waste through better product or project design. Products and projects are designed and optimized for a cycle of disassembly and reuse, which also look to achieve savings in labor and energy inputs. Second, the circular economy approach looks to differentiate between consumable goods and durable goods/projects. Because of their rapid use cycle, consumable goods should be designed with the goal of safe and useful components and end-products so they may be returned safely to the environment. Durable goods/projects that may contain materials that are less friendly to the environment should be designed to maximize reuse alternatives. Finally, steps should be taken to incorporate renewable energy sources into the economic cycle in order to decrease resource dependence and increase economic resilience (Ellen MacArthur Foundation 2013).

16.8 Resiliency Considerations

16.8.1 Resilient Materials

Climate change and its extreme impacts are severely affecting the infrastructure worldwide. With the impacts expected to only increase in the following years, today's infrastructure must be resilient to such challenges. Materials used in engineering applications should withstand and adapt to extreme events such as climate impacts, severe weather, and natural disasters while maintaining their structural and functional integrity over prolonged periods. These materials must be durable, adaptable to diverse conditions, and have low environmental impacts.

Feedstock or raw material selection is critical in designing resilient materials for specific purposes. Prioritizing renewable and recyclable resources can further reduce the environmental impacts. However, these materials must still meet the necessary structural requirements, ensuring their expected functionality is not compromised. Moreover, using resilient materials must provide multifunctional advantages to avoid being susceptible to disruptions. The ability to reuse or recycle these materials will further reduce their environmental footprint. Designs inspired by natural systems can enhance the structural integrity of resilient materials (Chatterjee and Layton 2020). Additionally, innovations such as self-healing materials (Zhang et al. 2020), and composite materials that offer synergistic properties could be considered during the design process.

However, the challenge lies in balancing performance and cost and ensuring they are sustainable. Research, innovation, and investment in practices are necessary to address these challenges and make resilient materials durable and their use and production scalable and sustainable.

16.8.2 Resilient Waste Management

Waste management is becoming increasingly challenging with urbanization on the rise, leading to increased waste generation and burdening the existing waste management systems. These challenges are further exacerbated during extreme climate impacts or natural disasters, which damage the existing logistic systems and hinder the utilization of available facilities to their full capacity while simultaneously increasing the volume of waste through generation of disaster debris (Mamashli et al. 2021). Hence a resilient waste management strategy should consist of a two-pronged approach: Reducing and effectively managing the excess waste generated during a disaster and improving the resilience of waste management facilities to climate change impacts or other external stressors.

Disaster Debris Management: Appropriate legislation is crucial to address waste management during extreme events. In the United States, specific disaster waste management plans are the responsibility of local municipalities. Federal Emergency Management Agency (FEMA) incentivizes cost-sharing programs to encourage municipalities to prepare for debris management (Brown et al. 2011). The United States Environmental Protection Agency (USEPA) has also launched web-based tools and databases such as USEPA's Suite of Disaster Debris Management and Disposal (DDMD), which provide technical information on waste handling, disposal options, facilities, and regulations.

Waste management is often a low priority during a disaster event in most developing countries. Poor waste disposal planning has frequently resulted in accidents at disposal facilities, including structural failures and fires. The United Nations has developed several guides and tools designed explicitly for disaster waste management in developing countries, such as "Solid Waste Management in Emergencies," "Initial Clearing and Debris Removal" from the UN Joint Environmental Unit, and "Environmental Needs Assessment in Post Disaster Situations." However, these guidelines and tools are limited to technical information and fail to cover legislative considerations and organizational planning. While resilient and durable facilities are crucial for tackling the waste management crisis often seen during extreme events, it is essential to incentivize reducing waste generation. Tools and guidelines can also be leveraged to enhance waste management processes, including monitoring and collection, ensuring greater efficiency and adaptability. Changes to the existing waste management systems are essential to ensure the systems are adaptable and capable of handling large volumes of waste, particularly during extreme events.

Resilient Waste Management Facilities: Building resilient waste management facilities is a critical component of a sustainable and adaptive waste management strategy. These facilities must be designed to withstand and adapt to the impacts of extreme events, ensuring uninterrupted functionality while minimizing risks to the environment and public health. For example, landfills are engineered structures designed to minimize environmental pollution from waste disposal. However, if the design of various landfill components – such as the final soil cover system, landfill slopes, and gas extraction system – does not account for climate-related or other unexpected external stressors, failures can result in devastating impacts on the environment and public health. Certain regulations, such as the Resource Conservation and Recovery Act (RCRA) regulation no. 258.11, which discourages landfill sites from being located in 100-year floodplains, can help prevent landfill failures and mitigate waste management challenges during extreme events.

Resilience assessment of waste management systems, as discussed in Chapter 5, to extreme events, using engineering analyses or expert guidance, can help in ensuring the resilience of such waste management systems (e.g. Janga and Reddy 2024). By integrating such assessments into design and operational frameworks, waste management systems can be made more robust and better equipped to handle the growing challenges of climate change-related natural disasters.

16.8.3 Resilient Circular Economy

Despite the benefits, moving toward a circular economy faces significant scalability challenges and vulnerabilities associated with unforeseen environmental, economic, or social disruptions. Systems within the circular economy, such as supply chains and industrial systems, remain susceptible to upheaval from changing climate and extreme events. Future advancements in the promising concept of circular economy must incorporate resiliency by design to effectively address these disruptions.

A resilient circular economy must be adaptable in resource flows and production cycles to address disruptions in supply chains. Furthermore, the technical robustness of recycled or upcycled materials should not be compromised, compared to the use of virgin materials in the face of disruptions. Decentralized, local resource supply systems could help reduce reliance on global systems prone to disruptions. Technologies such as the Internet of Things (IoT), blockchain, and artificial intelligence (AI), as discussed in Chapter 14, can enable real-time monitoring and predictive analytics for resource flows, supply chain vulnerabilities, and system bottlenecks (Rejeb et al. 2022). Digital twins of circular economy systems could simulate potential disruptions and guide proactive measures to mitigate their impacts (Preut et al. 2021).

16.9 Summary

The push for sustainable materials has driven the adoption of eco-design, green materials, Environmental Product Declarations (EPDs), and Extended Producer Responsibility (EPR) to integrate sustainability into product lifecycles, reduce environmental impact, and promote innovation. These strategies collectively minimize resource depletion, enhance waste management, and support the transition to sustainable materials across industries.

Waste is generated by virtually every human activity. These wastes may be classified into a variety of hazardous and non-hazardous waste categories. Regardless of the classification, they may be managed using a variety of approaches. These approaches may be thought of as a "pyramid," with less-favored options toward the bottom (e.g. disposal, energy recovery), and more preferred alternatives higher up the pyramid, including minimization and prevention.

Since several potential management techniques (each with several notable advantages) exist for managing waste materials, it is often appropriate to implement an integrated management approach that utilizes a combination of different strategies for waste management.

The integrated waste management concept, combined with an LCA approach, can improve the degree of sustainability of waste management. The incorporation of LCA results in a more informed, scientific-based approach to strategic development than other arbitrary approaches. As with other LCA applications, more comprehensive data can lead to more accurate LCA modeling, but it will also result in more complicated modeling. Using these methods as well as others, integrated waste management strategies may be developed that can better meet the needs of the local population and the greater environment. Consideration of the pursuit of a circular economy should be of the highest priority to prevent waste generation and economic resilience.

The increasing impacts of climate change and other disruptions, such as geo-hazards, underscore the urgent need to adopt resilient and sustainable practices in materials, waste management, and circular economy systems. Resilient materials are engineered to endure extreme events and adapt to varying conditions, ensuring long-term performance and reliability. Similarly, resilient waste management systems are crucial for addressing the growing waste crisis, particularly during extreme events. These systems require adaptive strategies, robust infrastructure, and supportive legislation to remain effective under stress. The circular economy presents a transformative model to reduce resource extraction and waste generation by prioritizing material reuse. However, to be truly effective, it must integrate resilience to disruptions such as climate change, extreme weather events, geo-hazards, and supply chain vulnerabilities.

16.10 Questions

16.1 What are the major waste management options in waste management hierarchy?

16.2 What constitutes municipal solid waste?

16.3 What is the major focus of pollution prevention strategy for effective waste management?

16.4 Discuss how wastes are classified using the RCRA regulatory framework.

16.5 What are the characteristics that are used to determine if waste is hazardous under the RCRA framework?

16.6 Why was there a need to develop the CERCLA regulatory framework?

16.7 What are some of the site classifications covered by CERCLA?

16.8 Name three advantages of pursuing pollution prevention (P2) strategies.

16.9 Discuss the benefits of using renewable feedstocks from a green chemistry standpoint.

16.10 Describe in detail the advantages and drawbacks of recycling programs with respect to specific waste stream materials.

16.11 Discuss specific strategies to render hazardous waste less harmful or nonhazardous.

16.12 From an integrated waste management approach, what are some strategies for waste treatment/management? What are the benefits and drawbacks of these strategies?

16.13 What is sustainable and resilient waste management, and how can it be achieved?

16.11 Problems

16.1 A hypothetical city generates an average of 0.30 kilograms of municipal solid waste (MSW) per person per day. The current population of the city is 19.3 million people, growing at an average rate of 3% per year. The percentage of waste managed through different methods and the average composition of waste generated in the city are given below:

Table a Proportion of waste managed by different methods by the city

Management technique	Fraction of MSW
Recycled	25%
Composted	10%
Incinerated	20%
Landfilled	45%

Table b Average waste composition of the city

Waste component	Composition (%)
Kitchen waste	55
Paper (Non-Recyclable/Recyclable)	8/16
Plastics (Non-Recyclable/Recyclable)	3/8
Metals	1
Glass	4.5
Textile	2
Others	2.5

Answer the following questions:

a) Calculate the total amount of waste generated by the city in one year at the current rate.
b) Determine the total amount of waste that will be generated in the next year, considering the 3.0% increase in waste generation per person.
c) Identify the recyclables in the MSW in Table b, and what is their fraction of the total MSW generated.

16.2 For the data in Problem 16.1, answer the following questions:

a) What is the amount of energy saved if all the recyclables are recycled? (Use Table 16.1 for data on energy savings by recycling different materials)
b) If the landfilled waste has a methane potential of 30 tonnes per 1000 tonnes of waste and 1 m^3 of methane generates 10 kWh of electricity, calculate the amount of energy that can replace conventional energy sources, considering 80% of the methane generated in landfills can be effectively captured and converted to electricity.

16.3 The greenhouse gas (GHG) emissions corresponding to each management technique are given in the table below:

Management technique	Greenhouse gas emissions
Recycled	90 kg CO_2e per tonne of recycled waste
Composted	350 kg CO_2e per tonne of MSW composted
Incinerated	400 kg CO_2e per tonne of MSW incinerated
Landfilled	800 kg CO_2e per tonne of MSW landfilled

For the data in Problem 16.1, answer the following questions:

1) Calculate the GHG emissions per annum from each management method and the total GHG emissions per annum from MSW management in the city at current generation rates
2) The city aims to increase the recycling and composting rate of MSW. Assuming the city can increase the recycling rate by 12% in the next year and the composting rate by 7%, determine the percentage of GHG emissions reduced due to the diversion of these waste streams from landfilling.

References

Ai, N. and Leigh, N.G. (2017). *Planning for Sustainable Material and Waste Management*. PAS Report 587, American Planning Association, Chicago, IL.

Alam, P. and Alam, M. (2014). Bioreactor landfills: new trends in landfill design. *Applied Sciences and Engineering Research* 3 (1): 187–193.

Anastas, P.T. and Warner, J.C. (1998). *Green Chemistry: Theory and Practice*. New York, NY: Oxford University Press.

Anastas, P.T. and Zimmerman, J.B. (2003). The twelve principles of green engineering. *Environmental Science & Technology* 38: 94A–101A.

Astrup, T.F., Tonini, D., Turconi, R., and Boldrin, A. (2015). Life cycle assessment of thermal waste-to-energy technologies: review and recommendations. *Waste Management* 37: 104–115.

Batty, L.C. and Dolan, C. (2013). The potential use of phytoremediation for sites with mixed organic and inorganic contamination. *Critical Reviews in Environmental Science and Technology* 43 (3): 217–259.

Bishop, P.L. (2002). *Pollution Prevention: Fundamentals and Practice*. New York: McGraw-Hill.

Blackman, W.C. Jr. (2016). *Basic Hazardous Waste Management*. CRC Press.

Bozell, J.J. (2008). Feedstocks for the future–biorefinery production of chemicals from renewable carbon. *CLEAN–Soil, Air, Water* 36 (8): 641–647.

Brown, C., Milke, M., and Seville, E. (2011). Disaster waste management: a review article. *Waste Management* 31 (6): 1085–1098.

Cai, Y.J. and Choi, T.M. (2019). Extended producer responsibility: a systematic review and innovative proposals for improving sustainability. *IEEE Transactions on Engineering Management* 68 (1): 272–288.

Chatterjee, A. and Layton, A. (2020). Mimicking nature for resilient resource and infrastructure network design. *Reliability Engineering & System Safety* 204: 107142.

Clark, J.H. (2012). *Chemistry of Waste Minimization*. Springer Science & Business Media.

Ellen MacArthur Foundation (2013). *Towards the Circular Economy 1*. Ellen MacArthur Foundation.

Fiksel, J. and Fiksel, J.R. (1996). *Design for Environment: Creating Eco-Efficient Products and Processes*. McGraw-Hill Professional Publishing.

Hanisch, C. (2000). Is extended producer responsibility effective? *Environmental Science & Technology* 34 (7): 170A–175A.

Hickle, G.T. (2014). An examination of governance within extended producer responsibility policy regimes in North America. *Resources, Conservation and Recycling* 92: 55–65.

Hunsager, E.A., Bach, M., and Breuer, L. (2014). An institutional analysis of EPD programs and a global PCR registry. *The International Journal of Life Cycle Assessment* 19: 786–795.

Ibáñez-Forés, V., Pacheco-Blanco, B., Capuz-Rizo, S.F., and Bovea, M.D. (2016). Environmental Product Declarations: exploring their evolution and the factors affecting their demand in Europe. *Journal of Cleaner Production* 116: 157–169.

International Standard ISO 14025 (2006) Environmental labels and declarations — Type III environmental declarations — Principles and procedures.

International Standard ISO 14040 (2006) Environmental Management—Life Cycle Assessment—Principle and Framework.

International Standard ISO/TS 14027 (2017) Environmental labels and declarations — Development of product category rules.

Janga, J.K. and Reddy, K.R. (2024). Integrated resiliency and sustainability assessment of bioreactor landfills. *Indian Geotechnical Journal* 54 (3): 1117–1135.

Khan, N., Jhariya, M.K., Raj, A. et al. (2021). Eco-designing for sustainability. In: *Ecological Intensification of Natural Resources for Sustainable Agriculture* (ed. M.K. Jhariya, R.S. Meena, and A. Banerjee). Singapore: Springer.

Laurent, A., Bakas, I., Clavreul, J. et al. (2014a). Review of LCA studies of solid waste management systems–Part I: Lessons learned and perspectives. *Waste Management* 34 (3): 573–588.

Laurent, A., Clavreul, J., Bernstad, A. et al. (2014b). Review of LCA studies of solid waste management systems–Part II: Methodological guidance for a better practice. *Waste Management* 34 (3): 589–606.

Lombardi, L., Carnevale, E., and Corti, A. (2015). A review of technologies and performances of thermal treatment systems for energy recovery from waste. *Waste Management* 37: 26–44.

Mamashli, Z., Nayeri, S., Tavakkoli-Moghaddam, R. et al. (2021). Designing a sustainable–resilient disaster waste management system under hybrid uncertainty: a case study. *Engineering Applications of Artificial Intelligence* 106: 104459.

McDougall, F.R., White, P.R., Franke, M., and Hindle, P. (2008). *Integrated Solid Waste Management: A Life Cycle Inventory*. Wiley.

Milanez, B. and Bührs, T. (2009). Extended producer responsibility in Brazil: the case of tyre waste. *Journal of Cleaner Production* 17 (6): 608–615.

Minkov, N., Schneider, L., Lehmann, A., and Finkbeiner, M. (2015). Type III Environmental Declaration Programmes and harmonization of product category rules: status quo and practical challenges. *Journal of Cleaner Production* 94: 235–246.

Niza, S., Santos, E., Costa, I. et al. (2014). Extended producer responsibility policy in Portugal: a strategy towards improving waste management performance. *Journal of Cleaner Production* 64: 277–287.

Parfitt, J., Barthel, M., and Macnaughton, S. (2010). Food waste within food supply chains: quantification and potential for change to 2050. *Philosophical Transactions of the Royal Society B: Biological Sciences* 365 (1554): 3065–3081.

Preut, A., Kopka, J.P., and Clausen, U. (2021). Digital twins for the circular economy. *Sustainability* 13 (18): 10467.

Rejeb, A., Suhaiza, Z., Rejeb, K. et al. (2022). The Internet of Things and the circular economy: a systematic literature review and research agenda. *Journal of Cleaner Production* 350: 131439.

Sharma, H.D. and Reddy, K.R. (2004). *Geoenvironmental Engineering: Site Remediation, Waste Containment, and Emerging Waste Management Technologies*. Wiley.

Sheldon, R.A., Arends, I., and Hanefeld, U. (2007). *Green Chemistry and Catalysis*. Wiley.

Shooshtarian, S., Maqsood, T., Wong, P.S. et al. (2021). Extended producer responsibility in the Australian construction industry. *Sustainability* 13 (2): 620.

Tang, S.Y., Bourne, R.A., Smith, R.L., and Poliakoff, M. (2008). The 24 principles of green engineering and green chemistry: "improvements productively". *Green Chemistry* 10 (3): 268–269.

Trost, B.M. (1991). The atom economy – a search for synthetic efficiency. *Science* 254: 1471–1477.

USEPA (2012). Citizen's Guide to Phytoremediation. EPA 542-F-12-016, Office of Solid Waste & Emergency Response, Washington, DC.

USEPA (2015). *Waste Classification*. US Environmental Protection Agency. ISBN: 978-1-84095-601-6.

USEPA (2016). *Advancing Sustainable Materials Management: 2014 Fact Sheet*. US Environmental Protection Agency.

USEPA (2017a). *Advancing Sustainable Materials Management: Facts and Figures*. US Environmental Protection Agency. https://www.epa.gov/smm/advancing-sustainable-materialsmanagement-facts-and-figures (accessed 12 April 2017).

USEPA (2017b). *Resource Conservation and Recovery Act (RCRA) Laws and Regulations*. US Environmental Protection Agency. https://www.epa.gov/rcra (accessed 12 April 2017).

USEPA (2017c). *Pollution Prevention (P2)*. US Environmental Protection Agency. https://www.epa.gov/p2 (accessed 12 April 2017).

USEPA (2018). *Superfund: National Priorities List (NPL)*. U.S. Environmental Protection Agency. https://www.epa.gov/superfund/superfund-national-priorities-list-npl (accessed 25 July 2018).

Vallet, F., Eynard, B., Millet, D. et al. (2013). Using eco-design tools: an overview of experts' practices. *Design Studies* 34 (3): 345–377.

Zaman, A.U. (2010). Comparative study of municipal solid waste treatment technologies using life cycle assessment method. *International Journal of Environmental Science & Technology* 7 (2): 225–234.

Zhang, W., Zheng, Q., Ashour, A., and Han, B. (2020). Self-healing cement concrete composites for resilient infrastructures: a review. *Composites Part B: Engineering* 189: 107892.

17

Sustainable and Resilient Buildings

Chapter Goal and Objectives
The goal of this chapter is to explore sustainable and resilient building practices by examining green building history, concepts, and components, as well as carbon-neutral approaches, while considering environmental, social, and economic aspects and addressing resiliency to ensure safety against climate impacts. The main objectives are (i) to provide an overview of the history and evolution of green building practices; (ii) to discuss key green building concepts such as energy efficiency, renewable energy use, embodied carbon emissions and carbon neutrality, and sustainable materials; (iii) to analyze the green building rating systems, focusing on Leadership in Energy and Environmental Design(LEED) certification; (iv) to understand how buildings can be truly sustainable by considering broader environmental, social, and economic factors; and (v) to highlight the need for buildings to be resilient to climate impacts exacerbated by climate change while also being sustainable.

17.1 Introduction

Structures are an essential element for the existence and survival of human civilization. On a most basic level, they provide shelter for humans from the often harsh and dangerous effects of weather. With increased complexity, they provide a means of function for virtually every human activity and institution. The design and operation of structures can also have a profound influence, both positive and negative, on human health and the environment. As greater awareness of the environmental function and impacts of buildings evolves, "green building" approaches that utilize more resource-efficient design, implementation, and decommissioning of buildings and infrastructure have continued to grow in popularity.

The US Environmental Protection Agency (USEPA) defines green building and/or high-performance building as the practice of creating structures and using processes that are environmentally responsible and resource-efficient throughout a building's life cycle, from siting to design, construction, renovation, operation, maintenance, and demolition (USEPA 2013). These design principles are incorporated into the traditional design approach concerning economy, durability, utility, aesthetics, and comfort. Considering all life-cycle stages from conceptual design to eventual decommissioning or demolition, this design approach seeks to optimize resource inputs and/or the resulting

Sustainable and Resilient Engineering: Drivers, Metrics, Tools, and Applications, Second Edition. Krishna R. Reddy, Claudio Cameselle, and Jeffrey A. Adams.

Companion website: www.wiley.com/go/reddy/sustainableresilientengineering2e

environmental impacts from the use of the built environment. These inputs and impacts include energy, water, materials, natural resources, waste, air pollution, water pollution, indoor pollution, heat islands, stormwater runoff, noise, harm to human health, and environmental degradation/contamination (Retzlaff 2009).

Green buildings are designed to reduce the overall impact of the built environment on human health and the natural environment by utilizing efficient resource use, such as water, energy, and construction materials, protecting occupant health and enhancing productivity, and reducing waste, pollution, and other deleterious environmental effects. Some examples of these practices include the use of renewable, recycled, and repurposed materials in construction; incorporation of healthier indoor environments, such as through the use of low-emissions materials and natural lighting; and green landscaping principles, including use of native plant palettes, drought-tolerant plants, and the use of recycled/reclaimed water for irrigation. This chapter presents the history and importance of green buildings, green building concepts/components, green building rating systems (e.g. Leadership in Energy and Environmental Design [LEED]), and general principles of green buildings/infrastructure design and operation.

Green buildings aim to reduce environmental impacts throughout their life cycle, including construction, operation, and end-of-life waste management. However, achieving true sustainability requires addressing all three pillars of sustainability: environmental, social, and economic factors. Truly sustainable buildings are designed and constructed to minimize adverse effects on the environment, promote social well-being, and support economic viability. In addition, the increasing frequency and intensity of extreme events driven by climate change are challenging current building standards. This underscores the need for buildings to be resilient and capable of withstanding such challenges while maintaining minimal impact on the environment, society, and the economy. A truly sustainable and resilient building harmonizes these priorities, ensuring adaptability and long-term sustainability.

17.2 Green Building History

Several design approaches considered to be green have been used for millennia. For instance, architectural designs addressing local climate patterns and using local, renewable, and indigenous materials and building practices have been used throughout the history of human civilization. Yet, a greater contemporary focus on these principles arose out of increased environmental awareness spurred by incidents and conditions of the 1960s and 1970s. Further, economic shocks arising from the oil crises of the 1970s inspired significant research and activity to improve energy efficiency and to utilize renewable energy resources. Green building principles continued to evolve and became more formalized in the 1990s as architects, engineers, planners, and government agencies attempted to formalize procedures, guidelines, and codes. Several milestones that occurred during this period include the following:

- In 1989, the American Institute of Architects (AIA) formed the Committee on the Environment;
- In 1992, the USEPA funded the publishing of the Environmental Resource Guide, which was published by the AIA;
- In 1992, the USEPA and the United States Department of Energy launched the ENERGY STAR program;
- In 1992, the city of Austin, Texas, launched the first local green building program;
- In 1993, the United States Green Building Council (USGBC) was founded;
- In 1998, the USGBC launched the pilot "1.0" program of LEED.

17.3 Why Build Green?

Because buildings consume such a significant percentage of natural resources for construction and ongoing operation, green building design and operation principles can facilitate significant reductions in resource consumption and waste generation. For instance, in the United States, buildings account for 39% of total energy use, 12% of water consumption, 71% of total electricity consumption, and 39% of carbon dioxide emissions (Figure 17.1). Green building practices and technologies are estimated to reduce energy use by 30–50%, reduce carbon emissions by 35%, reduce water use by 40%, and reduce solid waste by 70% (Figure 17.2). Thoughtful incorporation of green building strategies can result in energy savings, reductions in materials use and waste generation, and operational cost savings while creating attractive and appealing occupational conditions.

Green building design and operation can provide several other environmental benefits. First, these principles can be used to enhance adjacent or nearby ecosystems and protect biodiversity. Indoor and outdoor air and water quality can be improved by appropriate material selection and installation of control systems. Waste streams can be reduced during construction and operation, which can lead to materials savings and resource use. From an economic standpoint, green building can reduce operating costs, create, expand, and shape markets for green products and

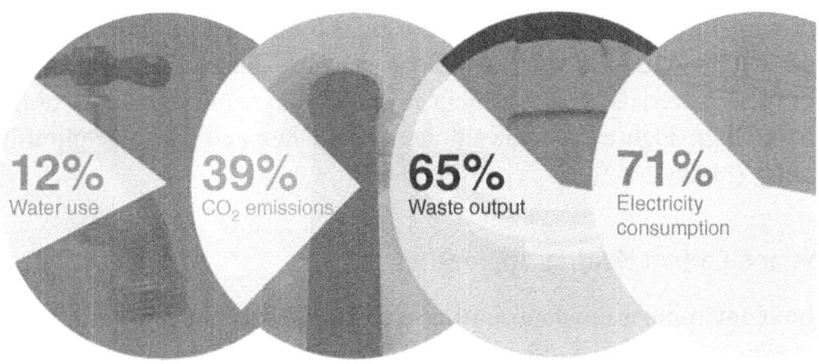

Figure 17.1 Impact of buildings in water and electricity consumption, waste generation, and carbon emissions.

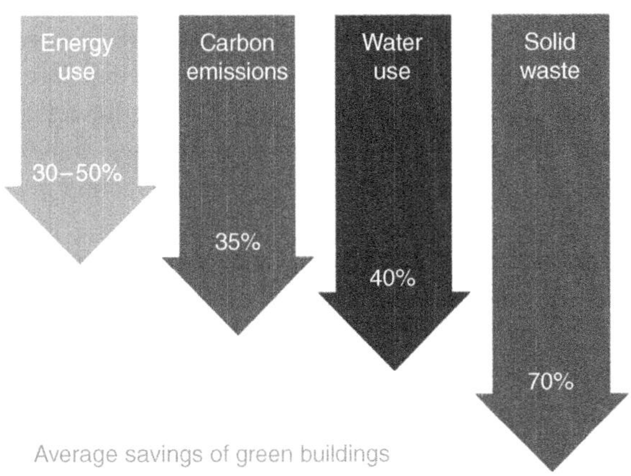

Figure 17.2 Average savings on energy, water, solid waste, and carbon emissions of green buildings.

services, improve occupant productivity, and optimize life-cycle economic performance. From a social standpoint, green building practices can enhance occupant comfort and health, heighten aesthetic qualities, minimize strain on local infrastructure, and improve overall quality of life (Willard 2012). Buildings may be called sustainable buildings when environmental, economic, and social impacts are considered.

17.4 Green Building Concepts

The technologies, methods, and design of green buildings incorporate a range of concepts. Some of these include energy efficiency, the incorporation of energy from renewable resources, water efficiency, environmentally protective and beneficial building materials and specifications, waste reduction, toxics reduction, indoor air quality, and smart growth and sustainable development principles. These strategies may be applied to a range of structures, including single-family and multifamily residential structures, retail and commercial buildings, schools, government facilities, industrial buildings, healthcare facilities, and government/institutional facilities.

Because of the materials-intensive nature of building construction, a key component of green building is the careful consideration of how construction and demolition materials are generated and managed (Kibert 2016). Some of the principal building materials warranting consideration include concrete, aggregate, cement, asphalt, steel, zinc, copper, and aluminum. Many of these materials are recyclable and may be reused in subsequent construction applications. Because of increased global demand for these materials, design principles that lead to reduced material usage can achieve significant cost savings. Figure 7.1 shows how material demand has precipitously increased over the past century (Matos 2012).

17.4.1 Embodied Carbon and Carbon Neutral Approaches

The significant role of the built environment in contributing to global greenhouse gas emissions is widely acknowledged. In particular, the impact of buildings can be categorized into two interconnected divisions: emissions from building operations, such as lighting and heating, and those arising from physical construction activities, including material production, transportation, assembly, maintenance, and waste management. Embodied carbon generally refers to the carbon emissions associated with the materials and construction processes throughout the entire life cycle of a building or infrastructure. This includes emissions generated during material extraction, processing, transportation to the site, and the construction process itself. In essence, embodied carbon represents the carbon footprint of a project prior to its operational phase. It also accounts for emissions from maintenance, eventual demolition, and the transportation and recycling of demolition waste. Unlike operational carbon, emissions resulting from electricity, heating, and lighting, among others, embodied carbon focuses solely on emissions tied to the materials and construction processes (Akbarnezhad and Xiao 2017).

The concept of embodied carbon has gained prominence over the last decade, particularly in the context of buildings. Embodied carbon in building materials accounts for approximately 11% of global carbon emissions, owing to the use of resource-intensive materials like concrete and steel, among others. To design buildings that are truly green or sustainable, it is necessary to develop innovative strategies that reduce embodied carbon emissions without compromising the structural safety and resilience of buildings in the face of adverse events. Recent advancements have shown promise in this regard. For instance, the use of less resource-intensive materials such as industrial byproducts (e.g. fly ash, blast furnace slag) in concrete production has demonstrated potential in reducing embodied carbon (de Brito and Kurda 2021). Additionally, the reuse of steel reinforcement bars and other construction and demolition waste can further mitigate embodied carbon.

Another increasingly popular approach to reduce carbon emissions from buildings is the pursuit of carbon neutrality. Achieving carbon neutrality involves balancing the carbon emissions generated throughout a system's life cycle so that the net emissions are effectively zero. In the context of buildings, this requires addressing both embodied carbon emissions and operational emissions, which include those arising from energy use, maintenance, and other activities during the building's lifespan (Lee et al. 2023).

Minimizing embodied carbon emissions is a critical step toward carbon neutrality. By reducing emissions during the production and use of materials, such as incorporating low-carbon alternatives like fly ash or biopolymers in concrete and reusing materials like steel and demolition waste, the overall carbon footprint of the building can be significantly reduced. However, achieving carbon neutrality also requires offsetting residual emissions through sustainable practices such as the adoption of renewable energy systems like solar or wind power and improving the energy efficiency of a building. At the end of a building's life cycle, recycling and reusing materials from demolition further minimize waste and offset embodied carbon. Achieving carbon neutrality by addressing embodied carbon and adopting sustainable practices is a transformative step toward creating green buildings.

17.5 Components of Green Building

Figures 17.3–17.6 depict green building concepts, including conceptual drawings of stormwater harvesting, solar power, hot water generation and circulation, and geothermal systems, respectively. Through increased efficiency and incorporation of renewable, zero-emission energy sources, these systems can lessen or eliminate the use of fossil fuels or fresh/potable water resources. Figure 17.7 depicts several green building concepts that can be incorporated into structures. These include the following:

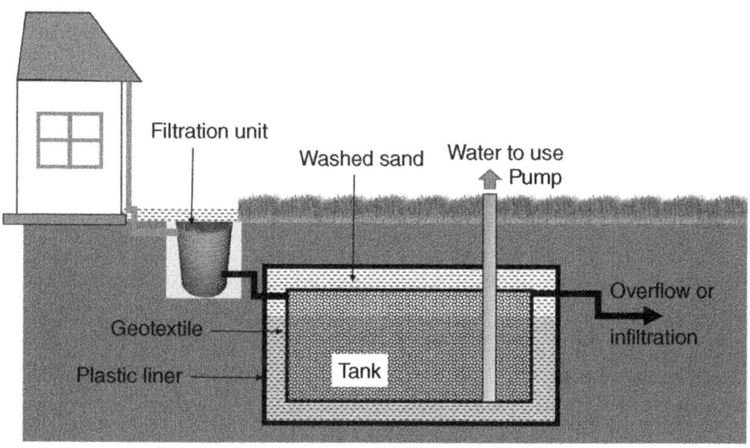

Figure 17.3 Features of green buildings: stormwater harvesting.

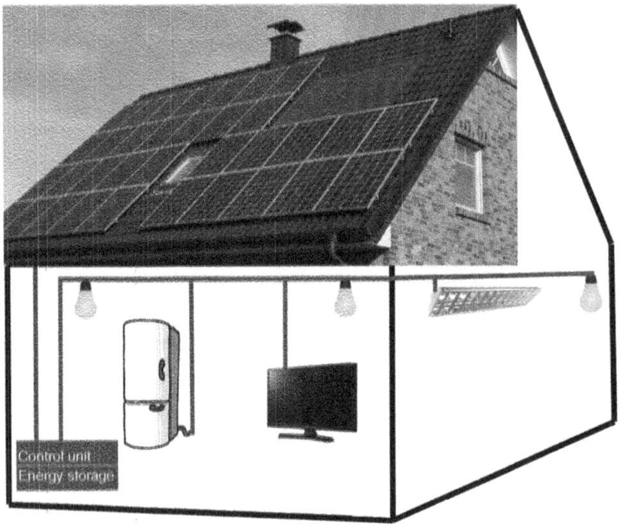

Figure 17.4 Features of green buildings: solar power.

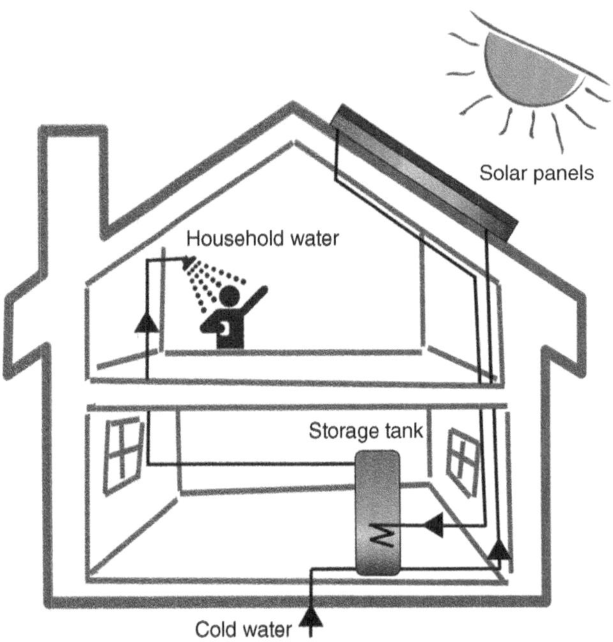

Figure 17.5 Features of green buildings: hot water generation with solar energy and circulation system.

- Home orientation to optimize solar exposure;
- High-efficiency low E (low emissivity) glass systems reduce losses of warm or cool air;
- Rain gardens to collect, control, and utilize stormwater;
- Native landscaping that reduces the need for irrigation or agricultural chemicals;
- Interior water conservation measures, such as low-flow plumbing fixtures and dual-flush or low-flow toilets;

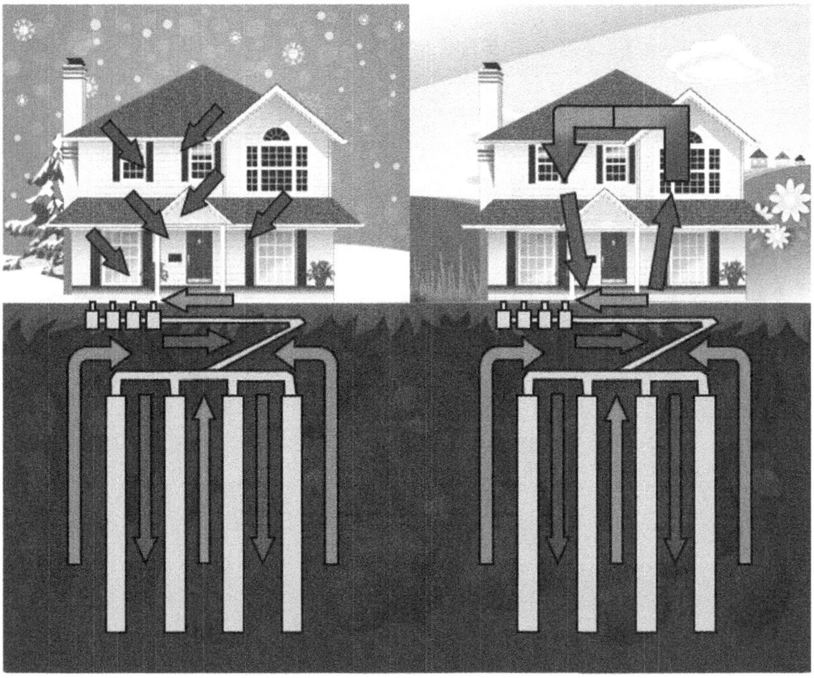

Figure 17.6 Features of green buildings: geothermal energy system.

- High-efficiency mechanical and heat ventilation and air-conditioning (HVAC) systems that reduce reliance on energy derived from nonrenewable sources;
- Energy-efficient appliances;
- Insulated floors and walls that reduce loss of warm air or cool air;
- Recycled building materials, such as wood for framing or decking, plastics and metals for finished surfaces, and concrete/asphalt for paving materials;
- Other considerations, such as low VOC (volatile organic compounds) paints, "whole-house" attic ventilation fans, green flooring materials, energy-efficient lighting systems, and smart control systems.

17.6 Green Building Rating – LEED

As the interest in green/sustainable buildings has evolved, objective assessment tools that may be used to determine the degree of green and sustainable design have been developed (Gowri 2004). Several rating systems have been developed, some of which are listed as follows:

- Green Star – Australia
- Protocollo Itaca – Italy
- Green Mark – Singapore
- Building Research Establishment Environmental Assessment Method (BREEAM) – United Kingdom
- Leadership in Energy and Environmental Design (LEED) – United States

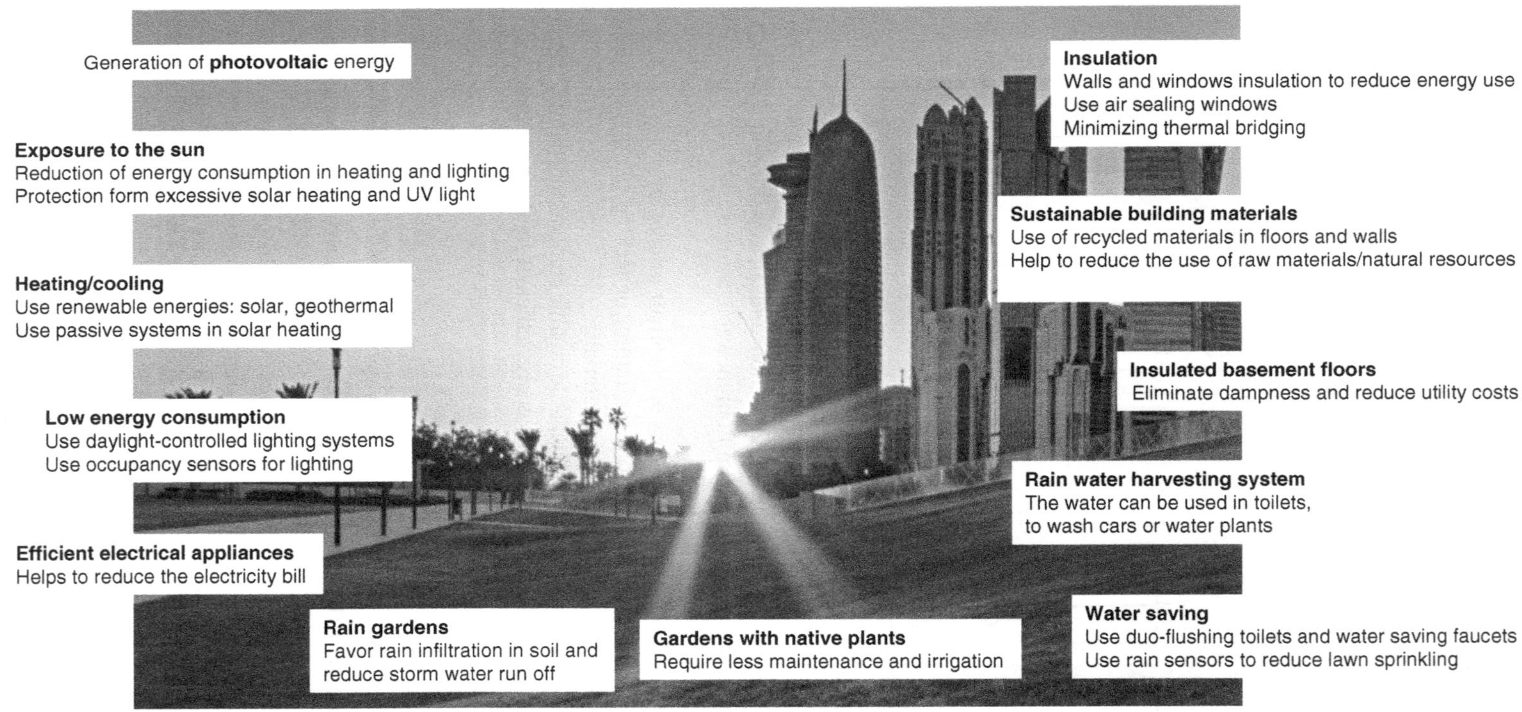

Figure 17.7 Features of green buildings.

In this Chapter we will specifically focus on LEED, which was developed by the USGBC as a consensus-based system to provide an independent, third-party verification that structures or communities were designed and constructed with the goal of achieving high performance in key areas of human and environmental health. Some of these areas include sustainable site development, water savings, energy efficiency, material selection, and indoor environmental quality (USGBC 2018). Several of these key parameters are described as follows.

Integrative process – This aspect of the rating system goes beyond any checklist and is imperative to any green building design. LEED encourages integration from the beginning of a project and early stages of building design while clarifying owner's aspirations, building performance goals, and project needs. An integrative process identifies the interactions among all the systems of the building and the site. The credit is given to a comprehensive approach toward integrating project team members of different systems and components to achieve mutual advantages, thereby leading to achieving high levels of building performance, human comfort, and environmental benefits. LEED promotes collaboration to enhance the efficiency and effectiveness of every system involved in the building design.

Locations and transportation – Buildings have an unavoidable effect and connection to their surrounding community. LEED encourages construction on previously developed or infill sites and away from environmentally sensitive areas. An emphasis is placed on building construction near existing infrastructure, community resources, and transit systems, as well as in locations that promote walking, physical activity, and time spent outdoors.

Sustainable sites – LEED discourages development on previously undeveloped land, seeks to minimize a building's impact on ecosystems and waterways, encourages regionally appropriate landscaping, including the use of native species, rewards smart transportation choices, and promotes control and reduction of stormwater runoff, erosion, light pollution, heat island effects, and construction-related pollution.

Water efficiency – Since a wide range of building systems use potable water, LEED encourages efficient and optimized use of water through conservation, use of reclaimed/recycled water, use of water-saving fixtures and appliances, and water-saving vegetation and irrigation systems that reduce water consumption.

Energy and atmosphere – LEED encourages energy-saving strategies, including energy-efficient design and construction, monitoring, energy-efficient appliances, enhanced refrigerant management systems, and lighting, renewable/zero emission energy resource use, and other measures, such as use of materials that reduces energy or thermal losses.

Materials and resources – LEED encourages the selection of sustainably grown, harvested, produced, and transported products and materials. LEED also promotes waste reduction/diversion, reuse, and recycling and places a specific emphasis on waste reduction at a product's source. LEED encourages the demonstration of reduced environmental impacts by reusing existing building resources or demonstrating a reduction in material use through life-cycle assessment. From a long-term impact perspective, LEED also promotes to reduce the release of persistent, bioaccumulative, and toxic chemicals associated with the life cycle of building materials.

Indoor environmental quality – The USEPA estimates that people spend approximately 90% of their day indoors. Therefore, indoor air quality has a major impact on overall health. LEED promotes strategies that promote healthy indoor environments, including improved indoor air quality, naturally ventilated spaces, increased use of natural light, reduced exposure to indoor-based air emissions (e.g. smoking, cooking), enhanced views, enhanced acoustics, and improved thermal comfort.

Innovation – LEED encourages innovative design by providing credit for the use of strategies and technologies that improve a building's performance well beyond a "baseline" performance if such strategies were not used. Rewards are also given to projects that utilize a LEED Accredited Professional on the project team to ensure a holistic, integrated approach to design and construction.

Regional priority – USGBC's regional councils, chapters, and affiliates have identified the most important local environmental concerns, and six LEED credits addressing these local priorities have been selected for each region of the country. Projects can earn credit by achieving one or more of these regional priorities.

LEED may be applied to all life-cycle aspects of a project or may be applied to a range of projects and project types, including new construction, existing building operations and maintenance, commercial interiors, core and shell, schools, retail, healthcare, homes, and neighborhood development (Figure 17.8). Points are tallied and based on the performance in the categories described above (also shown in Figure 17.9); the projects may earn one of the following four designations (Figure 17.10):

- *Certified* – 40–49 points
- *Silver* – 50–59 points
- *Gold* – 60–79 points
- *Platinum* – 80–110 points.

To pursue certification, a project is registered, and the progress is documented and applied for certification. Some examples of building types eligible for the certification include offices, retail and service establishments, libraries, schools, museums, religious institutions, hotels, and residential buildings of four or more habitable stories. LEED points are awarded on a 100-point scale, and

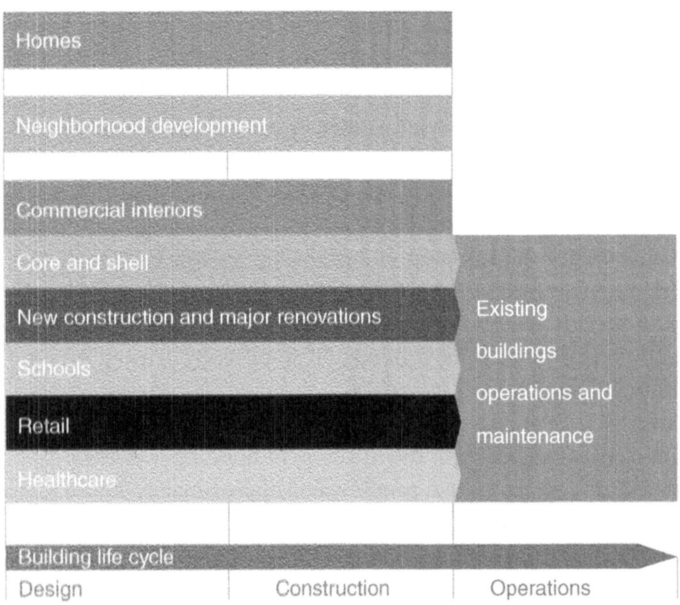

Figure 17.8 LEED system is applied to any type of building and address the complete lifetime of a building, from design and construction to use and decommissioned.

LEED v4 for BD+C: new construction and major renovation

Project checklist

Project name:

Date:

Y	?	N	Type	Item	Points
			Credit	Integrative process	1

Y	?	N	Type	Location and transportation	16
0	0	0			
			Credit	LEED for neighborhood development location	16
			Credit	Sensitive land protection	1
			Credit	High priority site	2
			Credit	Surrounding density and diverse uses	5
			Credit	Access to quality transit	5
			Credit	Bicycle facilities	1
			Credit	Reduced parking footprint	1
			Credit	Green vehicles	1

Y	?	N	Type	Sustainable sites	10
0	0	0			
Y			Prereq	Construction activity pollution prevention	Required
			Credit	Site assessment	1
			Credit	Site development - protect or restore habitat	2
			Credit	Open space	1
			Credit	Rainwater management	3
			Credit	Heat island reduction	2
			Credit	Light pollution reduction	1

Y	?	N	Type	Water efficiency	11
0	0	0			
Y			Prereq	Outdoor water use reduction	Required
Y			Prereq	Indoor water use reduction	Required
Y			Prereq	Building-level water metering	Required
			Credit	Outdoor water use reduction	2
			Credit	Indoor water use reduction	6
			Credit	Cooling tower water use	2
			Credit	Water metering	1

Y	?	N	Type	Energy and atmosphere	33
0	0	0			
Y			Prereq	Fundamental commissioning and verification	Required
Y			Prereq	Minimum energy performance	Required
Y			Prereq	Building-level energy metering	Required
Y			Prereq	Fundamental refrigerant management	Required
			Credit	Enhanced commissioning	6
			Credit	Optimize energy performance	18
			Credit	Advanced energy metering	1
			Credit	Demand response	2
			Credit	Renewable energy production	3
			Credit	Enhanced refrigerant management	1
			Credit	Green power and carbon offsets	2

Y	?	N	Type	Materials and resources	16
0	0	0			
Y			Prereq	Storage and collection of recyclables	Required
Y			Prereq	Construction and demolition waste management planning	Required
			Credit	Building life-cycle impact reduction	5
			Credit	Building product disclosure and optimization - environmental product Declarations	2
			Credit	Building product disclosure and optimization - sourcing of raw materials	2
			Credit	Building product disclosure and optimization - material ingredients	2
			Credit	Construction and demolition waste management	2

Y	?	N	Type	Indoor environmental quality	16
0	0	0			
Y			Prereq	Minimum indoor air quality performance	Required
Y			Prereq	Environmental tobacco smoke control	Required
			Credit	Enhanced indoor air quality strategies	2
			Credit	Low-emitting materials	3
			Credit	Construction indoor air quality management plan	1
			Credit	Indoor air quality assessment	2
			Credit	Thermal comfort	1
			Credit	Interior lighting	2
			Credit	Daylight	3
			Credit	Quality views	1
			Credit	Acoustic performance	1

Y	?	N	Type	Innovation	6
0	0	0			
			Credit	Innovation	5
			Credit	LEED accredited professional	1

Y	?	N	Type	Regional priority	4
0	0	0			
			Credit	Regional priority: specific credit	1
			Credit	Regional priority: specific credit	1
			Credit	Regional priority: specific credit	1
			Credit	Regional priority: specific credit	1

Y	?	N	Totals	Possible points:	110
0	0	0			

Certified: 40 to 49 points, **Silver:** 50 to 59 points, **Gold:** 60 to 79 points, **Platinum:** 80 to 110

Figure 17.9 LEED scoring system for new constructions.

Figure 17.10 LEED certification level based on the scoring system.

credits are weighted to reflect their potential environmental impacts. A total of 10 bonus points are available, including those associated with regional-specific parameters. A project must satisfy all prerequisites and earn a minimum number of points to be certified.

The Green Building Certification Institute (GBCI) administers LEED certification for all commercial and institutional projects registered under any LEED rating system. By participating in LEED, design professionals, operations professionals, and project proponents have the ability to objectively assess the degree of green and sustainable design and operation their project may achieve. While the environmental benefits that may be achieved are explicitly understood, LEED certification can also achieve enhanced financial performance. LEED-certified buildings are commonly designed to achieve lower operating costs through more efficient energy and water use, reduced waste generation and related costs, and reduced occupational health impacts. Additionally, LEED-certified structures are often eligible for tax incentives, enhanced or favorable zoning regulations, and other local and national incentives. Potential tenants often consider LEED certification an attractive benefit, which can lead to a premium for rental income. Finally, because of reduced operating costs and greater income potential, the underlying asset is more valuable in the marketplace for the project owner.

LEED credentials are sought by a range of technical and market professionals, including structural and landscape architects, interior designers, engineers, facility managers, real estate professionals, contractors, lenders, and government officials. A range of available LEED credentials include the following:

- LEED Green Associate
- LEED AP Building Design and Construction
- LEED AP Operations and Maintenance
- LEED AP Interior Design and Construction
- LEED AP Homes
- LEED AP Neighborhood Development
- LEED AP Without Specialty
- LEED Fellow.

State and local governments are increasingly adopting LEED for publicly owned and funded buildings, including governmental, educational, and other institutional uses. Various federal agencies, including the Department of Defense, Department of Agriculture, Department of Energy, and Department of State, have also developed LEED initiatives. Furthermore, many LEED projects are in design or operation around the world.

17.6.1 LEED Case Study: HBS Tata Hall Executive Education Center

Tata Hall at Harvard Business School (HBS), located on Harvard Way, Boston, Massachusetts, USA, is a seven-story, 153 700-square-foot building that provides living and learning spaces for HBS's Executive Education Program (Figure 17.11). The building includes 180 bedrooms with corresponding group lounges, two classrooms with a seating capacity of 99 each, as well as project and seminar rooms and administrative spaces. Sustainable construction and operational practices were effectively implemented, which contributed to earning a Platinum certification under LEED v3, with a score of 82 out of 110 points (Figure 17.12). A detailed scorecard is shown in Figure 17.13. The building design integrated the principles of sustainability, incorporating environmental strategies to minimize its negative impacts. The site was designed to minimize stormwater runoff and create a comfortable outdoor environment. Building covering was designed to minimize the requirements of energy to achieve internal comfort for occupants. The capture of daylight creates well-lit spaces for students and personnel with a solar control system during specific periods of daylight hours to minimize excessive solar heating and create an appropriate visual environment. The plumbing work was designed to save potable water, and the high-efficiency HVAC system provides a comfortable environment, controlling and minimizing energy consumption (Harvard 2024).

Descriptions of the building's various features that supported its Platinum certification are as follows:

- *Sustainable sites (22/26)* – The landscape design incorporates strategies to manage stormwater effectively, reduce runoff, and filter pollutants such as sediments and phosphorous, vital due to the site's proximity to the Charles River. The building features a river-facing lawn and other

Figure 17.11 View of the Tata Hall Executive Education Center building at HBS (Harvard 2024). *Source:* Sonia Smith/Shutterstock.

LEED® Facts

Harvard Business School
Tata Hall Executive Education Center

Location	Boston, MA
Rating System	LEED-NC v3
Certification	Platinum
Total Points	82/110
Sustainable Sites	22/26
Water Efficiency	6/10
Energy and Atmosphere	28/35
Materials and Resources	6/14
Indoor Environmental Quality	11/15
Innovation and Design	6/6
Regional Priority	3/4

Project metrics

48%	water savings compared to an Energy Policy Act of 1992 baseline
43%	reduction in energy costs compared to the baseline standard (ASHRAE 90.1-2007), estimated via energy modeling
5.2%	of energy use (by cost) is provided by an on-site renewable energy system (PV)
92%	of regularly occupied areas have access to views
90%	of individual spaces, including bedrooms, have individual lighting controls
90%	of individual spaces, including bedrooms, have individual thermal comfort controls

Figure 17.12 LEED facts and project metrics for Tata Hall at Harvard Business School.

intimate outdoor gathering spaces. Further, the project minimized hardscapes and maximized vegetation, contributing to the reduction of the urban heat island effect, enhanced by a high-albedo roof membrane and shaded areas. Native plants were used to reduce irrigation needs, further supporting water conservation goals.

- *Water efficiency (6/10)* – Water-saving fixtures, such as low-flow toilets and efficient irrigation, contribute to a 48% reduction in potable water use compared to the Energy Policy Act of 1992 baseline. This approach includes efficient plumbing and water-saving appliances, supporting the building's commitment to minimizing its environmental impact.
- *Energy and atmosphere (28/35)* – Tata Hall's HVAC system integrates high-efficiency fan coils, condensing boilers, and an enthalpy recovery system to optimize energy use. Occupancy sensors and CO_2 monitoring in heavily used spaces helps reduce the energy consumption for HVAC system. These elements collectively achieve a 43% energy cost reduction compared to baseline standards, partially supported by on-site solar photovoltaic systems contributing 5.2% of total energy. The double-skin façade reduces glare and regulates solar heat gain, optimizing both energy efficiency and occupant well-being.
- *Materials and resources (6/14)* – Emphasizing sustainability in material use, 31% of materials used in the construction of the Tata Hall were recycled, and 16% were regionally sourced. The use of Forest Stewardship Council (FSC)-certified wood further aligns with sustainable resource management and construction processes, such as use, reuse, and recycling of construction waste, achieving 88% waste diversion from landfills.
- *Indoor environmental quality (11/15)* – Ensuring a healthy indoor environment was central to the design, with features such as low-VOC materials, effective ventilation, and extensive natural lighting facilitated through the double façade structure of the building. Individual lighting and thermal controls for over 90% of occupied spaces also help in enhancing comfort and productivity.

LEED FOR NEW CONSTRUCTION & MAJOR RENOVATIONS (V2009)

ATTEMPTED: 75, DENIED: 0, PENDING: 0, AWARDED: 82 OF 110 POINTS

SUSTAINABLE SITES	22 OF 26
SSp1 Construction Activity Pollution Prevention	Y
SSc1 Site Selection	1 / 1
SSc2 Development Density and Community Connectivity	5 / 5
SSc3 Brownfield Redevelopment	0 / 1
SSc4.1 Alternative Transportation-Public Transportation Access	6 / 6
SSc4.2 Alternative Transportation-Bicycle Storage and Changing Rooms	1 / 1
SSc4.3 Alternative Transportation-Low-Emitting and Fuel-Efficient Vehicles	3 / 3
SSc4.4 Alternative Transportation-Parking Capacity	2 / 2
SSc5.1 Site Development-Protect or Restore Habitat	0 / 1
SSc5.2 Site Development-Maximize Open Space	1 / 1
SSc6.1 Stormwater Design-Quantity Control	1 / 1
SSc6.2 Stormwater Design-Quality Control	1 / 1
SSc7.1 Heat Island Effect, Non-Roof	0 / 1
SSc7.2 Heat Island Effect-Roof	1 / 1
SSc8 Light Pollution Reduction	0 / 1

WATER EFFICIENCY	6 OF 10
WEp1 Water Use Reduction-20% Reduction	Y
WEc1 Water Efficient Landscaping	2 / 4
WEc2 Innovative Wastewater Technologies	0 / 2
WEc3 Water Use Reduction	4 / 4

ENERGY AND ATMOSPHERE	28 OF 35
EAp1 Fundamental Commissioning of the Building Energy Systems	Y
EAp2 Minimum Energy Performance	1 / 0
EAp3 Fundamental Refrigerant Mgmt	1 / 0
EAc1 Optimize Energy Performance	16 / 19
EAc2 On-Site Renewable Energy	3 / 7
EAc3 Enhanced Commissioning	2 / 2
EAc4 Enhanced Refrigerant Mgmt	2 / 2
EAc5 Measurement and Verification	3 / 3
EAc6 Green Power	2 / 2

MATERIALS AND RESOURCES	6 OF 14
MRp1 Storage and Collection of Recyclables	Y
MRc1.1 Building Reuse-Maintain Existing Walls, Floors and Roof	0 / 3
MRc1.2 Building Reuse, Maintain 50% of Interior	0 / 1
MRc2 Construction Waste Mgmt	2 / 2
MRc3 Materials Reuse	0 / 2
MRc4 Recycled Content	2 / 2

MATERIALS AND RESOURCES	CONTINUED
MRc5 Regional Materials	1 / 2
MRc6 Rapidly Renewable Materials	0 / 1
MRc7 Certified Wood	1 / 1

INDOOR ENVIRONMENTAL QUALITY	11 OF 15
IEQp1 Minimum IAQ Performance	Y
IEQp2 Environmental Tobacco Smoke (ETS) Control	1 / 0
IEQc1 Outdoor Air Delivery Monitoring	1 / 1
IEQc2 Increased Ventilation	1 / 1
IEQc3.1 Construction IAQ Mgmt Plan-During Construction	1 / 1
IEQc3.2 Construction IAQ Mgmt Plan-Before Occupancy	1 / 1
IEQc4.1 Low-Emitting Materials-Adhesives and Sealants	1 / 1
IEQc4.2 Low-Emitting Materials-Paints and Coatings	0 / 1
IEQc4.3 Low-Emitting Materials-Flooring Systems	0 / 1
IEQc4.4 Low-Emitting Materials-Composite Wood and Agrifiber Products	0 / 1
IEQc5 Indoor Chemical and Pollutant Source Control	1 / 1
IEQc6.1 Controllability of Systems-Lighting	1 / 1
IEQc6.2 Controllability of Systems-Thermal Comfort	1 / 1
IEQc7.1 Thermal Comfort-Design	1 / 1
IEQc7.2 Thermal Comfort-Verification	1 / 1
IEQc8.1 Daylight and Views-Daylight	0 / 1
IEQc8.2 Daylight and Views-Views	1 / 1

INNOVATION IN DESIGN	6 OF 6
IDc1.1 Innovation in Design	1 / 1
IDc1.1 Innovation in Design	0 / 1
IDc1.2 Innovation in Design	1 / 1
IDc1.2 Innovation in Design	0 / 1
IDc1.3 Innovation in Design	1 / 1
IDc1.3 Innovation in Design	0 / 1
IDc1.4 Innovation in Design	0 / 1
IDc1.4 Innovation in Design	1 / 1
IDc1.5 Innovation in Design	0 / 1
IDc1.5 Innovation in Design	1 / 1
IDc2 LEED® Accredited Professional	1 / 1

REGIONAL PRIORITY CREDITS	3 OF 4
SSc3 Brownfield Redevelopment	0 / 1
SSc6.1 Stormwater Design-Quantity Control	1 / 1
SSc7.1 Heat Island Effect, Non-Roof	0 / 1
SSc7.2 Heat Island Effect-Roof	1 / 1
EAc2 On-Site Renewable Energy	1 / 1
MRc1.1 Building Reuse-Maintain Existing Walls, Floors and Roof	0 / 1

TOTAL	82 OF 110

Figure 17.13 LEED v3 scorecard for the Tata Hall at Harvard Business School.

- *Innovation in design (6/6)* – Innovation credits reflect Tata Hall's green roof, a pioneering feature on the HBS campus that provides insulation, manages stormwater, and mitigates heat gain through evapotranspiration, and effectively reduces the urban heat island effect.
- *Regional priority (3/4)* – Regional priority credits recognize the project's targeted approach to local environmental challenges. The Tata Hall project with its green roof and on-site solar power project earned three out of the four credits in this category.

17.7 Sustainable Buildings

While green buildings aim to minimize environmental impacts, achieving true sustainability requires incorporating broader social and economic considerations. The LEED rating system, introduced earlier in this chapter, primarily focuses on making buildings environmentally responsible. However, true sustainability goes beyond environmental objectives to include minimizing social and economic impacts on all the stakeholders involved. As seen in Figure 17.14, a sustainable building is broader in scope compared to a green building, i.e. the latter's focus is solely to minimize environmental impacts, while a sustainable building aims to minimize impacts across all three dimensions of sustainability throughout the life cycle of a building.

For instance, adding a green feature like rooftop solar panels might reduce environmental impact but could impose significant upfront costs on stakeholders, reducing viability. To address such challenges, engineers must adopt practices like community and stakeholder engagement and perform cost-benefit analyses, as discussed in earlier chapters. These practices can be integrated using frameworks such as QUALICS, outlined in Chapter 13, to quantitatively evaluate a building's sustainability across environmental, social, and economic dimensions.

The American Society of Civil Engineers (ASCE) has recently published a standard with comprehensive guidelines for creating environmentally responsible, socially equitable, and economically viable infrastructure (ASCE 2023). This standard emphasizes a life cycle approach to the infrastructure sector, encouraging leadership to assess and mitigate impacts across environmental, economic, and social domains. Further, details on the focus of the standard are provided in Chapter 18; meanwhile, the push toward sustainable practices in built environment is evident through the development of such comprehensive standards.

17.8 Resiliency Considerations

Extreme weather events, including storms, hurricanes, floods, and wildfires, are becoming more frequent and intense due to climate change. The IPCC and USEPA project the greenhouse gas emissions to continue rising for several decades, even under optimistic scenarios, as discussed in Chapter 2. These rising greenhouse gas emissions will further exacerbate global warming and climate-change impacts. This necessitates the design and construction of buildings to be resilient to changing climate.

As climate patterns become increasingly unpredictable, extreme events such as stronger winds and severe flooding pose increasing risks to the structural integrity of buildings. To address this, modern building design must account for the nonstationary nature of the climate, ensuring that structures are resilient to unexpected shocks and stressors. However, this resilience cannot come at the cost of sustainability or vice versa. Overly conservative designs that demand excessive

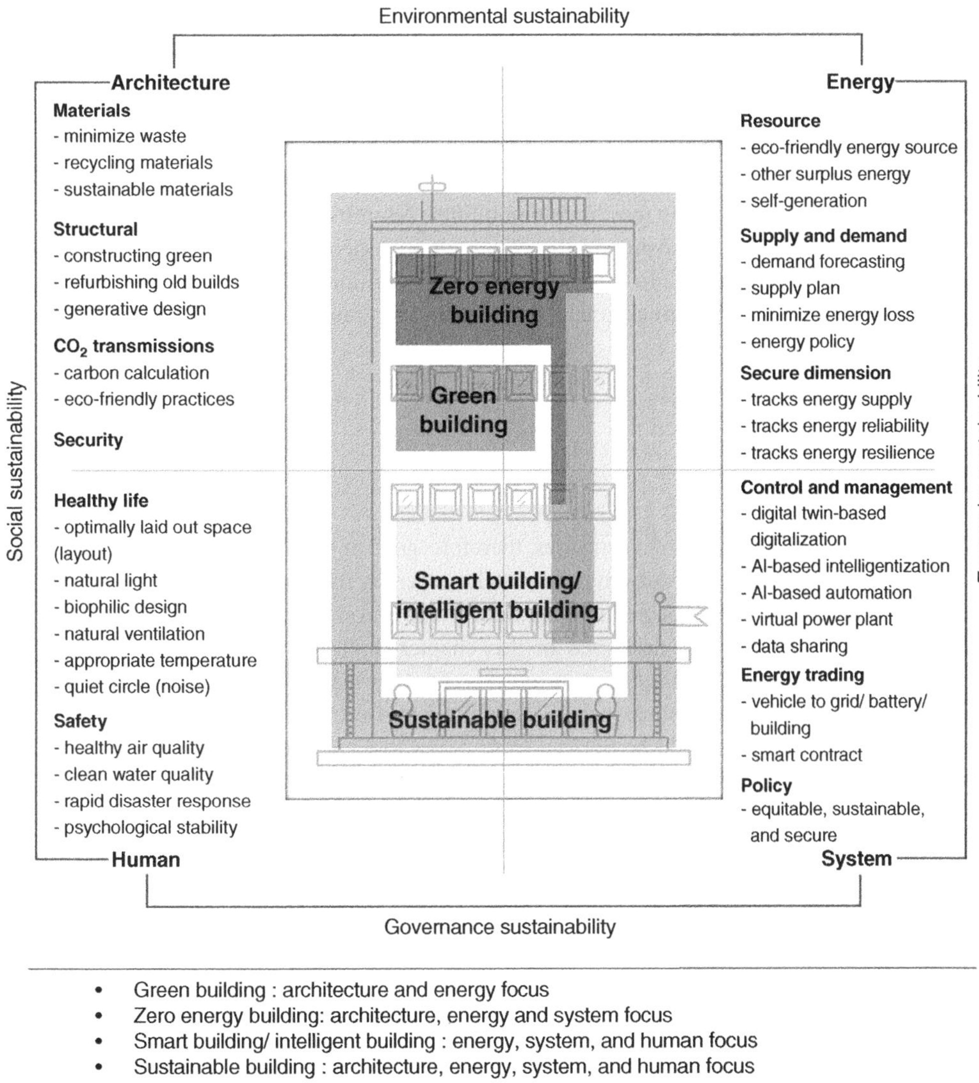

Figure 17.14 Various evolved building concepts. *Source:* Park et al. (2024)/MDPI/Public domain.

materials or higher construction costs may not only be economically unfeasible but could also exacerbate climate change through increased resource consumption and embodied carbon. To truly achieve the required balance, resilience measures must go beyond structural safety to include strategies for mitigating cascading impacts on communities, ecosystems, and economies. By proactively addressing both resilience and sustainability, buildings can be designed to adapt to the evolving climate while minimizing their contribution to future environmental challenges. Frameworks such as the Tiered Quantitative Assessment of Life Cycle Sustainability and Resilience (TQUALICSR), as discussed in Chapter 13, provide systematic approaches to balance these goals, enabling planners to incorporate resilience and sustainability from the earliest stages of design.

It is important to integrate strategies that enhance resilience, such as designing structures to endure stronger storms and the corresponding wind speeds, implementing site-specific stormwater management systems to mitigate runoff from heavy rainfall, and use of materials resilient to temperature changes and extreme flooding events, among others, into sustainable buildings (USGBC 2010). The incorporation of flexible, climate-responsive designs is another key aspect of climate resilience. For example, buildings in regions facing extreme heat can benefit from advanced natural ventilation systems, as described by Holmes and Hacker (2007), and high-albedo roofs. In areas with increasing wildfire risks, selecting fire-resistant materials and maintaining defensible space around properties can significantly reduce vulnerabilities. Adopting such strategies to mitigate climate impacts while not compromising on the sustainability aspects of a building is crucial. Ultimately, integrating resilience into the design and operation of buildings ensures their functionality and safety in a changing climate, safeguarding communities and promoting long-term sustainability.

17.9 Summary

Buildings can often last decades, even centuries; therefore, good or bad design will live on in use for many years. In recent years, greater environmental awareness has led to an increased focus on green and sustainable building design and operation. Green and sustainable design and operation, whether applied to buildings as described in this chapter or to infrastructure as described in Chapter 18, should follow several guiding principles. First, development sites and rights-of-way should be chosen to minimize disturbance to the natural environment. To the extent practicable, material and energy inputs should be derived from renewable resources as well as recycled and repurposed materials. Designs and operations should be pursued that select nonhazardous materials to a practicable degree, incorporate processes and systems that maximize mass, energy, and space efficiency, and minimize waste generation in all aspects of the project.

A truly sustainable building must go beyond environmental stewardship, addressing broader social and economic dimensions. Moreover, the increasing frequency and intensity of climate-related impacts demand that buildings be designed with resilience to withstand such challenges. In achieving this, it is essential to prioritize minimizing cascading environmental, social, and economic consequences, ensuring the development of buildings that are both sustainable and resilient.

Repetitive or universal, "one-size-fits-all" project designs should be avoided; instead, designs that incorporate the unique attributes of a given development site, including terrain, climate, and directional exposure should be strongly considered and incorporated into design. Designs should optimize maintenance and refurbishment while incorporating flexibility for future renovation or expansion. Finally, systems should be designed to enable and encourage recycling at the end of the design/operational life.

17.10 Questions

17.1 What is the definition of a green building as per USEPA?

17.2 What are the advantages of green buildings over conventional buildings?

17.3 Mention some of the important components of a green building.

17.4 Describe some key green building concepts and how they can be incorporated into buildings or regional development.

17.5 Name some of the rating systems developed for the certification of green buildings.

17.6 What are the key parameters involved in LEED certification of green buildings?

17.7 Describe in detail how green building principles can be incorporated into ongoing building operations.

17.8 Discuss the LEED credentials that are available to practicing professionals.

17.9 What are the four major designations earned by green buildings through LEED certification?

17.11 Problems

17.1 The LEED credits for new building design and construction in water efficiency (WE) are as follows:

- Outdoor water use reduction (WEc1)
- Indoor water use reduction (WEc2)
- Cooling tower water use (WEc3)
- Building level water metering (WEc4)

Check the criteria for awarding points in these credits on the LEED website. Now, considering the following two scenarios, calculate the total points in the WE credits for each scenario:

Scenario 1: The building (nonhealthcare) achieves a 51% reduction in outdoor water use compared to baseline by installing rain sensors and using native plants. Inside, it achieves a 22% reduction in water use compared to baseline with low-flow fixtures and water-efficient toilets. The cooling tower system is optimized for water treatment and reuse, achieving 15% reduced water consumption compared to a standard system. Additionally, a building-level water metering system tracks all water use.

Scenario 2: The building (nonhealthcare) achieves a 35% reduction in outdoor water use through a combination of drip irrigation and drought-resistant plants. Indoor water use is reduced by 19% with low-flow fixtures in restrooms and water-efficient faucets in kitchens. However, there is no cooling tower system or building-level water metering.

17.2 Two buildings, Building-A and Building-B, are designed with different strategies to achieve LEED certification.

Building-A focuses on passive solar design, energy-efficient windows, and lighting, achieving 15% energy savings over the baseline model. It has no renewable energy sources, but it includes a high-efficiency HVAC system with real-time energy meters. Building A also performs standard commissioning.

Building-B incorporates high-performance insulation, an automated energy management system, and a small solar installment, achieving 10% of its energy needs from on-site renewables. It achieves 18% energy savings through energy-efficient fixtures and undergoes enhanced commissioning but lacks building-level energy metering.

- Based on LEED energy and atmosphere credit criteria, which building is likely to score higher? Justify your answer using specific credit criteria. Make suitable assumptions as needed.
- Recommend one additional strategy for each building that could improve their scores in the EA (Energy and Atmosphere) category. Explain how each strategy aligns with LEED goals.

17.3 The Bullitt Center in Seattle, USA is recognized as one of the greenest commercial buildings, employing various strategies such as rainwater capture, rooftop solar panels, and waterless composting toilets. Review the building's sustainable design elements and research its likely LEED certification level. Identify at least three LEED credit categories where the Bullitt Center likely scores highest. Explain your choices with reference to specific design features of the building.

17.4 A high-rise office building in a hot, arid urban area faces water shortages and rising temperatures. The design team must focus on water efficiency (WE) and energy and atmosphere (EA) credits to make the building climate-resilient. The design currently achieves the following:

- The greenery development project surrounding the office building aims to restore 28% of the natural habitat in the disturbed area.
- Diverts 55% of the demolition waste by appropriate reuse for on-site filling and generates 9.5 lbs/ft^2 of waste for all new construction activities.

Calculate the expected points in relevant credits based on the above mentioned features.

References

Akbarnezhad, A. and Xiao, J. (2017). Estimation and minimization of embodied carbon of buildings: a review. *Buildings* 7 (1): 5.

ASCE (2023). *Standard Practice for Sustainable Infrastructure (73-23)*. Reston, VA: American Society of Civil Engineers (ASCE) https://doi.org/10.1061/9780784415733.

de Brito, J. and Kurda, R. (2021). The past and future of sustainable concrete: a critical review and new strategies on cement-based materials. *Journal of Cleaner Production* 281: 123558.

Gowri, K. (2004). Green building rating systems: an overview. *ASHRAE Journal* 46 (11): 56.

Harvard (2024). HBS Tata hall executive education center. Harvard Energy and Facilities. Harvard University. https://www.energyandfacilities.harvard.edu/green-building-resource/leed-case-studies/tata-hall-hbs (accessed 4 November 2024).

Holmes, M.J. and Hacker, J.N. (2007). Climate change, thermal comfort and energy: meeting the design challenges of the 21st century. *Energy and Buildings* 39 (7): 802–814.

Kibert, C.J. (2016). *Sustainable Construction: Green Building Design and Delivery*. Wiley.

Lee, G., Avelina, N., Rim, D. et al. (2023). Systematic review of carbon-neutral building technologies (CNBTs) by climate groups and building types. *Journal of Building Engineering* 78 (1): 107627.

Matos, G.R. (2012). *Use of Raw Materials in the United States from 1900 through 2010*. Reston, VA: U.S. Geological Survey.

Park, S., Cho, K., and Choi, M.I. (2024). A sustainability evaluation of buildings: a review on sustainability factors to move towards a greener city environment. *Buildings* 14 (2): 446.

Retzlaff, R.C. (2009). Green buildings and building assessment systems: a new area of interest for planners. *CPL Bibliography* 24 (1): 3–21.

USEPA (2013). Sustainable design and green building toolkit for local governments. US Environmental Protection Agency. Report No. EPA-904-B-10-001.

USGBC (2010). *Green buildings and climate resilience.* https://www.usgbc.org/resources/green-building-and-climate-resilience-understanding-impacts-and-preparing-changing-conditi (accessed 26 November 2024).

USGBC (2018). *LEED v4 rating system – green building rating system.* www.usgbc.org/leed (accessed 1 September 2018).

Willard, B. (2012). *The New Sustainability Advantage: Seven Business Case Benefits of a Triple Bottom Line.* New Society Publishers.

18

Sustainable and Resilient Civil Infrastructure

Chapter Goal and Objectives
The goal of this chapter is to explore and analyze sustainable and resilient civil infrastructure, emphasizing best practices for sustainability, innovative infrastructure solutions, and the necessity of resilience against climate impacts and other extreme shocks. The major objectives are (i) to define the principles of sustainable infrastructure and their application in civil engineering projects; (ii) to discuss the components of civil infrastructure and their sustainability requirements; (iii) to explain the Envision™ Sustainability Rating System for evaluating civil infrastructure; (iv) to identify various standard practices that can be useful in improving the sustainability of infrastructure; (v) to provide examples of built infrastructure projects such as airports and wastewater treatment plants to showcase sustainable approaches; (vi) to explore the various options in green infrastructure projects for stormwater management, including green roofs, permeable pavements, rainwater harvesting, and more; and (vii) to highlight the importance of ensuring infrastructure resilience to climate impacts and sudden shocks.

18.1 Introduction

Infrastructure consists of the basic physical and organizational structures needed for the operation of a society or an enterprise or the services and facilities necessary for an economy to function (Hayes 2005). Infrastructure can be further classified as "hard" or "soft." Soft infrastructure refers to institutions required for economic, public health, cultural, and social standards of developed nations, such as those that are associated with the financial systems, education systems, health-care systems, government systems, and law enforcement systems, as well as institutions that provide emergency services.

Hard infrastructure refers to the large physical networks necessary for the functioning of a nation. Transportation infrastructure consists of road and highway networks, bridges, tunnels, culverts, railroads, canals, seaports, and airports. Energy infrastructure includes the electrical transmission grid and related generation stations, natural gas pipelines, and petroleum pipelines. Water and environment infrastructure consists of the drinking water supply, wastewater treatment facilities, stormwater systems, irrigation systems, flood control systems, coastal structures and facilities, municipal solid waste collection and sorting facilities, landfills, incinerators, and

Sustainable and Resilient Engineering: Drivers, Metrics, Tools, and Applications, Second Edition. Krishna R. Reddy, Claudio Cameselle, and Jeffrey A. Adams.

Companion website: www.wiley.com/go/reddy/sustainableresilientengineering2e

recycling facilities. Telecommunications infrastructure includes telephone networks, television and radio transmission facilities, data transmission and Internet facilities, and communications satellites/relay systems.

While civil engineering traditionally has been focused on delivering technical solutions to meet society's infrastructure needs, there is a growing interest to seek sustainable infrastructure solutions given increased recognition of the finite nature of natural resources and a recognized growth in demand for these resources due to ever-growing population (Fenner et al. 2014). Further, sustainable solutions are sought - given the increased demands from the regulatory environment, a recognition to address aging infrastructure cost effectively, and the need to address these factors with limitations in government funding to support critical and necessary infrastructure programs (Frangopol and Liu 2007). Since infrastructure plays a critical role in providing a physical linkage between the social and economic aspects of a functioning society, and sustainability is focused on deriving benefits with respect to environmental, social, and economic conditions, there is a natural inspiration toward sustainable infrastructure approaches.

Chapter 17 dealt with the design and operational considerations for sustainable and resilient buildings. While this chapter focuses on civil infrastructure in general, it also addresses the principles and importance of sustainable infrastructure, including the use of the Envision™ tool for rating the sustainability of civil infrastructure projects. Examples of sustainable practices in civil infrastructure are provided to illustrate how sustainability can be integrated into these projects. Additionally, resilience strategies to mitigate the impacts of climate change on built infrastructure are emphasized, highlighting the importance of promoting both sustainable and resilient infrastructure.

18.2 Principles of Sustainable Infrastructure

In his book *Strategy for Sustainability: A Business Manifesto* (Werbach 2009), environmentalist and sustainability expert Adam Werbach defined a successful sustainability strategy as "different and much bigger than just green." Sustainable strategies should focus not only on effects and interactions with the natural environment but also on interactions with social, economic, and cultural dimensions. Because of the inherent large scale of major infrastructure projects, virtually any design or operational aspect of these projects will have a profound impact on these dimensions. Further, with increased anticipated impacts of climate change, infrastructure designs need to be adaptive to account for variances and extremes for key design variables that may be affected during the design life of infrastructure projects.

As outlined in the previous chapter, with respect to the design of sustainable infrastructure, regardless of the type, several guiding principles are useful and should be followed (Dasgupta and Tam 2005; Sarté 2010), including sites that protect the natural environment, materials selection that emphasizes renewable resources and recycled materials, waste minimization, and flexible, adaptive designs. As infrastructure is constructed or rehabilitated, sustainable designs should be sought that strive to be more durable and resilient and increase the probability of full operation under a wider range of extreme conditions. To facilitate this approach, new rating systems have emerged to assess the degree of sustainability for these projects. These rating systems aim to assess performance in extreme conditions as well as superior overall performance and expected durability within and beyond a conventional service life. An ideal sustainability rating system for infrastructure should encourage the conservation and restoration of resources and natural systems while also considering the impacts on socio-economic systems.

18.3 Civil Infrastructure

As stated in the introduction, infrastructure consists of facilities and networks that are necessary to allow society to function. These facilities may be highly visible or barely noticeable facilities in cities, towns, and open areas. Regardless of their degree of visibility or obscurity, objectively speaking, the condition of civil infrastructure in the United States desperately needs improvement. Civil infrastructure includes bridges, dams, roads, water supply/treatment, waste management, and other similar facilities. Figure 18.1 shows the civil infrastructure report cards issued by the American Society of Civil Engineers (ASCE) over the past few decades. In summary, ASCE concluded that the United States infrastructure earned an overall "C" grade in 2025, indicating a mediocre condition overall (Figure 18.2). Furthermore, it was estimated that approximately

Category	1988	1998	2001	2005	2009	2013	2017	2021	2025
Aviation	B–	C–	D	D+	D	D	D	D+	D+
Bridges		C–	C	C	C	C+	C+	C	C
Boardband									C+
Dams		D	D	D+	D	D	D	D	D+
Drinking water	B–	D	D	D–	D–	D	D	C–	C–
Energy			D+	D	D+	D+	D+	C–	D+
Hazardous waste	D	D–	D+	D	D	D	D+	D+	C
Inland waterways	B–		D+	D–	D–	D–	D	D+	C–
Levees					D–	D–	D	D	D+
Ports						C	C+	B–	B
Public parks				C–	C–	C–	D+	D+	C–
Rail				C–	C–	C+	B	B	B–
Roads	C+	D–	D+	D	D–	D	D	D	D+
Schools	D	F	D	D–	D	D	D+	D+	D+
Solid waste	C–	C–	C+	C+	C+	D–	C+	C+	C+
Stormwater								D	D
Transit	C–	C–	C–	D+	D	D	D–	D–	D
Wastewater	C	D+	D	D–	D–	D	D+	D+	D+
Overall GPA	**C**	**D**	**D+**	**D**	**D**	**D+**	**D+**	**C–**	**C**

The 2025 report card for America's infrastructure

C OVERALL GPA

D+	Aviation	B	Ports
C	Bridges	C–	Public parks
C+	Boardband	B–	Rail
D+	Dams	D+	Roads
C–	Drinking water	D+	Schools
D+	Energy	C+	Solid waste
C	Hazardous waste	D	Stormwater
C–	Inland waterways	D	Transit
D+	Levees	D+	Wastewater

Figure 18.1 Infrastructure report grading the state of main US infrastructure in 2025 and its evolution from 1988. *Source:* ASCE, https://infrastructurereportcard.org.

Comprehensive assessment of America's infrastructure

- Grades ranged from B (ports) to D (stormwater, transit) in 2025.
- No category received a D– for the first time since 1998.
- Eight sectors improved, many stuck at D– or D for years.
- Improvements driven by government & private investments.
- Energy and rail downgraded due to capacity, demand, and safety concerns.
- Broadband introduced as a new category, rated C+.
- Nine categories remain at D, indicating more work needed.

OVERALL GPA
C
2025

Key Findings

- Federal investments have positively influenced U.S. infrastructure.
- Low-rated categories saw gradual improvement.
- Nearly half of the 18 evaluated sectors received higher grades.
- Overall grade improved from C– to C.
- Ongoing investment is needed for long-term stability and efficiency.

Figure 18.2 Overall grade of the US infrastructure in 2025. *Source:* ASCE, https://infrastructurereportcard.org.

$9.1 trillion was needed for infrastructure investment over the next ten years as per current estimates. It is important to note that this investment is needed just to recover the performance of the systems to a good-repair conditions, without adjusting for inflation. Beyond this investment, new infrastructure and upgrades and maintenance of existing systems are needed to enhance quality of life (QL), promote economic competitiveness, and support job creation.

18.4 Envision™: Sustainability Rating of Civil Infrastructure

To meet the demand for an objective sustainability rating system for civil infrastructure, ASCE has been partnering with the American Council of Engineering Companies (ACEC), the American Public Works Association (APWA), and formed the Institute for Sustainable Infrastructure (ISI), which further collaborated with the Zofnass Program for Sustainable Infrastructure at Harvard University to develop Envision™. The Envision™ rating system, which assesses triple bottom line parameters, provides an objective framework for assessing the degree of sustainability of civil infrastructure projects, while describing the elements of sustainable projects. Envision™ also identifies means to optimize project designs, as well as to communicate the benefits of sustainable project aspects to designers, regulators, and project owners (ASCE 2018).

The purpose of Envision's web-based platform is to enhance the sustainability of the nation's infrastructure, excluding occupied buildings. Envision™ assesses civil infrastructure using two parallel criteria. First, the "pathway contribution" is assessed. In considering the needs of the community to be served by the civil infrastructure project under consideration, the assessment can select an appropriate solution that aims to "do the right thing" for the given community. Dialogue between the project team and proponents and the community-at-large is highly encouraged to identify key factors important to the community. Second, the "performance contribution" is assessed, in which the assessment is focused on how the project design and delivery can achieve the desired degree of performance. With this criterion, the emphasis is on "doing things right" by identifying and incorporating best management practices (BMPs) and encouraging higher degrees of sustainability with respect to the project.

Envision™ may be used for projects that exhibit a wide range of size and complexity. Instead of providing a uniform set of methods for projects, Envision™ incorporates the specific needs and circumstances of a community to optimize the performance of a project for that particular community. Additionally, as a voluntary assessment framework, Envision™ may either be used for internal self-assessment or for audit and/or independent verification purposes. By recognizing the progress and contributions to sustainable principles, the use of Envision™ is intended to pursue balanced approaches that maximize triple bottom line parameters, leading to conservation and restoration of natural resources and ecological systems as well as the strengthening of the social fabric of communities (Pedersen 2012; Behr 2014).

Envision™ provides a comparison of infrastructure alternatives across a range of life cycle categories. The results of an Envision™ assessment may be incorporated into a sensitivity analysis that assesses design variables of potential infrastructure alternatives with respect to the triple bottom line, allowing project stakeholders to optimize a project's outcome in these categories. Costs and benefits may be assessed over the life cycle, including environmental benefits and performance objectives, to achieve higher levels of sustainability. It is also intended to critically assess the entire life cycle of a project, including decommissioning activities.

Envision™ may be applied to numerous civil infrastructure categories, including roads, bridges, pipelines, railways, airports, power transmission systems, telecommunication towers, dams, levees, solid waste landfills, water supplies and conveyance systems, wastewater treatment plants, and public spaces. During the evaluation, Envision™ assesses the overall sustainability of a project not simply as a singular improvement but in terms of its overall contribution to the community that is served. Further, it is not intended to replace existing performance rating systems; rather, it is meant to provide an essential context for rating results. Envision™ incorporates four stages of assessment as follows:

- In Stage 1, a self-assessment checklist is applied, and a focus is placed on the education of the sustainability aspects of a particular infrastructure.
- In Stage 2, the third-party objective rating of the project is emphasized, allowing project proponents to submit a particular project for verification and public recognition. This stage includes guidance with respect to scoring. It is essential that sustainability achievements associated with a project are properly documented at this stage and that the independent third-party verification be performed by a qualified individual or entity.
- Additional stages (Stages 3 and 4) are focused on accounting for complex/multi-stage projects as well as optimization of a particular project during design, respectively.

The Stage 2 scorecard metrics are divided into five sections or categories: quality of life (QL), leadership (LD), resource allocation (RA), natural world (NW), and climate and resilience (CR), as shown in Table 18.1. A maximum of 64 credits can be earned that amounts to a maximum of 1000 points that may be achieved. Each credit is described in a two-page summary, as shown in Table 18.1, that includes the intent and purpose of the particular metric, the basis and documentation of the evaluation, the achieved score, and a summary of how a higher score may be achieved. More details can be found in the Envision™ Manual (ASCE 2018).

18.4.1 Envision™ Case Study: Berryessa Transit Center

The Berryessa Transit Center in northeastern San Jose, California, constructed and operated by the Santa Clara Valley Transportation Authority, is a key multi-modal transportation hub that enhances connectivity in the congested San Francisco Bay Area (Figure 18.3). The center addresses community transportation needs by connecting local transit systems and providing equitable access to services for low-income individuals, youth, elderly, and those with disabilities. The center includes a Bay Area Rapid Transit (BART) commuter rail station, a Valley Transportation Authority (VTA) bus hub, bicycle storage, pedestrian and cyclist trails, an EV-ready parking garage, and shuttle drop-off points. Public spaces feature local art, a contemplative garden, and a restored stream and wildlife habitat with native plantings to improve water quality in the natural ecosystem. Developed as part of the first phase of the VTA's BART Silicon Valley Program, this center will ultimately connect to Downtown San Jose and Santa Clara, providing frequent, reliable service to nearly two million county residents. By 2030, it is expected to serve up to 25 000 daily BART passengers and enable efficient travel to San Francisco in 60 minutes. These features bolstering the social and environmental aspects of sustainability have helped the project earn Envision® Platinum award, scoring nearly 50% of the total points. The overall Envision® scorecard of the project is presented in (Figure 18.4).

Table 18.1 Categories for the ENVISION™ rating system.

1	Quality of life	Wellbeing	QL1.1 Improve community quality of life
2			QL1.2 Enhance public health and safety
3			QL1.3 Improve construction safety
4			QL1.4 Minimize noise and vibration
5			QL1.5 Minimize light pollution
6			QL1.6 Minimize construction impacts
7		Mobility	QL2.1 Improve community mobility
8			QL2.2 Encourage sustainable transportation
9			QL2.3 Improve access and wayfinding
10		Community	QL3.1 Advance equity and social justice
11			QL3.2 Preserve historic and cultural resources
12			QL3.3 Enhance views and local character
13			QL3.4 Enhance public spaces and amenities
14	Leadership	Collaboration	LD1.1 Provide effective leadership and commitment
15			LD1.2 Foster collaboration and teamwork
16			LD1.3 Provide for stakeholder involvement
17			LD1.4 Pursue byproduct synergies
18		Planning	LD2.1 Establish a sustainability management plan
19			LD2.2 Plan for sustainable communities
20			LD2.3 Plan for long-term monitoring and maintenance
21			LD2.4 Plan for end-of-life
22		Economy	LD3.1 Stimulate economic prosperity and development
23			LD3.2 Develop local skills and capabilities
24			LD3.3 Conduct a life-cycle economic evaluation
25	Resource allocation	Materials	RA1.1 Support sustainable procurement practices
26			RA1.2 Use recycled materials
27			RA1.3 Reduce operational waste
28			RA1.4 Reduce construction waste
29			RA1.5 Balance earthwork on site
30		Energy	RA2.1 Reduce operational energy consumption
31			RA2.2 Reduce construction energy consumption
32			RA2.3 Use renewable energy
33			RA2.4 Commission and monitor energy systems
34		Water	RA3.1 Preserve water resources
35			RA3.2 Reduce operational water consumption
36			RA3.3 Reduce construction water consumption
37			RA3.4 Monitor water systems

Table 18.1 (Continued)

38	Natural world	Siting	NW1.1 Preserve sites of high ecological value
39			NW1.2 Provide wetland and surface water buffers
40			NW1.3 Preserve prime farmland
41			NW1.4 Preserve undeveloped land
42		Conservation	NW2.1 Reclaim brownfields
43			NW2.2 Manage stormwater
44			NW2.3 Reduce pesticide and fertilizer impacts
45			NW2.4 Protect surface and groundwater quality
46		Ecology	NW3.1 Enhance functional habitats
47			NW3.2 Enhance wetland and surface water functions
48			NW3.3 Maintain floodplain functions
49			NW3.4 Control invasive species
50			NW3.5 Protect soil health
51	Climate and resilience	Emissions	CR1.1 Reduce net embodied carbon
52			CR1.2 Reduce greenhouse gas emissions
53			CR1.3 Reduce air pollutant emissions
54		Resilience	CR2.1 Avoid unsuitable development
55			CR2.2 Assess climate change vulnerability
56			CR2.3 Evaluate risk and resilience
57			CR2.4 Establish resilience goals and strategies
58			CR2.5 Maximize resilience
59			CR2.6 Improve infrastructure integration

Source: Adapted from ASCE (2018).

Figure 18.3 Berryessa Transit Center in northeastern San Jose, California, USA.

	Indicator	Level	Points	Max.
Quality of Life	Improve Community Quality of Life	Restorative	26	26
	Enhance Public Health & Safety	Restorative	20	20
	Improve Construction Safety	None	0	14
	Minimize Noise & Vibration	Superior	6	12
	Minimize Light Pollution	Superior	6	12
	Minimize Construction Impacts	Conserving	8	8
	Improve Community Mobility	Restorative	14	14
	Encourage Sustainable Transportation	Restorative	16	16
	Improve Access & Wayfinding	Conserving	14	14
	Advance Equity & Social Justice	Improved	3	18
	Preserve Historic & Cultural Resources	Conserving	12	18
	Enhance Views & Local Character	Improved	1	14
	Enhance Public Space &Amenities	Conserving	11	14
	Innovate or Exceed Credit Requirements	Not Pursued	0	0
	Total		**137**	**200**
Leadership	Provide Effective Leadership & Commitment	Conserving	18	18
	Foster Collaboration & Teamwork	Superior	12	18
	Provide for Stakeholder Involvement	Superior	9	18
	Pursue Byproduct Synergies	Not Pursued	0	18
	Establish a Sustainability Management Plan	Superior	12	18
	Plan for Sustainable Communities	Restorative	16	16
	Plan for Long-Term Monitoring & Maintenance	Conserving	12	12
	Plan for End-of-Life	Not Pursued	0	14
	Stimulate Economic Prosperity & Development	Conserving	20	20
	Develop Local Skills & Capabilities	Not Pursued	0	16
	Conduct a Life-Cycle Economic Evaluation	Superior	10	14
	Innovate or Exceed Credit Requirements	Not Pursued	0	0
	Total		**109**	**182**
	Support Sustainable Procurement Practices	None	0	12
	Use Recycled Materials	None	0	16
	Reduce Operational Waste	Improved	4	14
	Reduce Construction Waste	Superior	10	16
	Balance Earthwork On Site	Enhanced	4	8
	Reduce Operational Energy Consumption	Not Pursued	0	26
	Reduce Construction Energy Consumption	None	0	12

	Indicator	Level	Points	Max.
Resource Allocation	Use Renewable Energy	None	0	24
	Commission & Monitor Energy Systems	Improved	3	14
	Preserve Water Resources	Restorative	12	12
	Reduce Operational Water Consumption	None	0	22
	Reduce Construction Water Consumption	Improved	1	8
	Monitor Water Systems	Conserving	12	12
	Innovate or Exceed Credit Requirements	Not Pursued	0	0
	Total		**46**	**196**
Natural World	Preserve Sites of High Ecological Value	NA	0	0
	Preserve Wetland & Surface Water Buffers	Restorative	20	20
	Preserve Prime Farmland	NA	0	0
	Preserve Undeveloped Land	Restorative	24	24
	Reclaim Brownfields	NA	0	0
	Manage Stormwater	None	0	24
	Reduce Pesticide & Fertilizer Impacts	Superior	5	12
	Protect Surface & Groundwater Quality	Restorative	20	20
	Enhance Functional Habitats	Enhanced	5	18
	Enhance Wetland & Surface Water Functions	Restorative	20	20
	Maintain Floodplain Functions	Restorative	14	14
	Control Invasive Species	Restorative	12	12
	Protect Soil Health	Restorative	8	8
	Innovate or exceed credit requirements	Not Pursued	0	0
	Total		**128**	**172**
Climate and Resilience	Reduce Net Embodied Carbon	Not Pursued	0	20
	Reduce Greenhouse Gas Emissions	None	0	26
	Reduce Air Pollutant Emissions	Improved	2	18
	Avoid Unsuitable Development	Enhanced	6	16
	Assess Climate Change Vulnerability	Conserving	20	20
	Evaluate Risk & Resilience	Not Pursued	0	26
	Establish Resilience Goals & Strategies	Not Pursued	0	20
	Maximize Resilience	Not Pursued	0	26
	Improve Infrastructure Integration	Restorative	18	18
	Innovate or exceed credit requirements	Not Pursued	0	0
	Total		**46**	**190**
	Overall Total	**PLATINUM**	**466**	**940**

Figure 18.4 Envision® score card for the Berryessa Transit Center in California, USA.

The features that helped in achieving the said score in each category are as follows:

- ***Quality of Life (137/200)*** – The Berryessa Transit Center achieved high scores in the QL category by addressing local transportation needs and enhancing community accessibility in the San Francisco Bay Area. By providing BART and VTA bus connections, as well as cyclist and pedestrian-friendly infrastructure, the transit center promotes sustainable commuting options. It achieved a *Restorative* level for "Community Mobility" and *Superior* for "Access and Wayfinding," improving transportation access for low-income, elderly, and disabled residents. The incorporation of public art and a contemplative garden honors local culture and fosters community pride, creating a space that enhances local character.
- ***Leadership (109/182)*** – Strong leadership was demonstrated through the project's emphasis on community collaboration and sustainable planning from the start. Engaging the Muwekma Ohlone Indian Tribe in designing the contemplative garden honored local culture and educated the public on indigenous history. By integrating sustainability principles during early planning, the project ensured a lasting positive impact on the community. Additionally, it created numerous job opportunities during construction and in long-term operations, boosting local economic resilience and fostering sustainable growth.
- ***Resource Allocation (46/196)*** – While the RA category scored lower, the transit center did incorporate sustainable practices in some areas. The center's effective handling of construction waste and water resources helped to a certain level, yet it fell short of a higher score due to the limited use of sustainable materials and renewable energy systems.
- ***Natural World (128/192)*** – Scoring highly in the NW category, the project undertook extensive ecological restoration. By removing 4.5 acres of concrete structures and replacing them with a 200-ft wide creek habitat restoration, which included realigning 920 ft of the creek, the transit center achieved a *Restorative* level for indicators such as "Preserve Wetland & Surface Water Buffers" and "Control Invasive Species." Replanting with native vegetation created new habitats and improved biodiversity, while stormwater management features, such as bioretention basins, reduce runoff and help manage water quality.
- ***Climate and Resilience (46/190)*** – Although the project did not score as much in this category, it did include certain carbon-reduction and climate resilience measures. The center achieved better ratings in certain indicators by incorporating electric vehicle (EV) charging stations, motion-sensor lighting, and solar panels, which help to mitigate its carbon footprint. Flood resilience was addressed through bioretention basins and drainage systems, enhancing the center's preparedness for extreme weather events.

18.5 Sustainable Infrastructure Practices

Envision™ provides a comprehensive framework for assessing the sustainability of infrastructure. However, it is often challenging to identify and adopt the best practices that enhance the sustainability of infrastructure, thereby allowing it to score higher in rating systems like Envision™. National and international agencies, such as ASCE, United States Army Corps of Engineers (USACE), and the United Nations Environmental Program (UNEP), have developed comprehensive guidelines to address the challenges of integrating sustainable practices into infrastructure development. These generic guidelines can be useful to the stakeholders to include sustainable practices in infrastructure development from the preliminary stages. The ASCE standard for sustainable infrastructure outlines some key principles in line with Envision™ rating categories that can help infrastructure projects become more sustainable and resilient (ASCE 2023). Listed below are the general practices recommended by the ASCE for the development of sustainable infrastructure:

Quality of Life

- Social equity and environmental justice should be the fundamental considerations during all the stages of infrastructure development through practices that expand mobility access and enhance safety and efficiency.
- Infrastructure development should prioritize the conservation of historic and cultural resources while also supporting the community's needs and goals.
- Address public health and safety concerns beyond regulatory requirements.

Leadership

- Share proven innovative solutions with the industry to promote sustainability practices.
- Ensure stakeholder engagement to address environmental, economic, and social issues related to the infrastructure project.
- Foster skill development and capacity building through job creation and training programs.
- Analyze all the alternatives and choose the most sustainable solution through comprehensive life cycle assessments.
- Manage long-term impacts through monitoring and using sustainability metrics and rating tools.

Resource Allocation

- Where possible, prioritize usage of materials recycled or recovered on-site or recycled by external sources; further usage of locally sourced materials through practices like effective balancing of earthwork should be prioritized.
- To the extent possible, reuse the existing infrastructure as long as it is safe, functional, and aligns with the project's sustainability goals.
- Reduce waste generation and avoid usage of hazardous materials in the construction and maintenance of the infrastructure project to the extent possible; in the case where hazardous materials have to be used or generated during the lifetime of the project, their use shall be monitored, and safe disposal should be ensured at the end of life.

Natural World

- The location of the project should be selected to minimize impacts on crucial ecological systems like conserved forests and wildlife habitats, among others.
- To the extent possible, the use of brownfield sites for infrastructure development should be encouraged.
- Avoid the introduction of invasive species during the development of infrastructure.
- Any means through which nearby waterbodies can be contaminated due to the project's activities should be mitigated.

Greenhouse gas (GHG) emissions

- At the least, a 15% reduction in GHG emissions compared to the baseline should be targeted through the reduction of embodied carbon emissions and operational emissions.
- Energy efficiency should be prioritized to achieve this through the adoption of sustainable practices like sustainable and renewable energy sources.

- To the extent possible, recycled or locally sourced materials should be used to reduce the embodied carbon emissions of the materials involved.

In addition to the above-mentioned practices to enhance the sustainability of infrastructure projects, ASCE also recommends practices to enhance climate resilience and minimize the economic impacts of the infrastructure projects. In similar lines, UNEP has outlined ten guiding principles to promote infrastructural sustainability, as seen in Figure 18.5, for policymakers involved in infrastructure projects (UNEP 2022). The USACE has also provided an elaborate guidebook delineating guidelines on sustainable practices in civil works related to infrastructure projects (USACE 2019). Such guiding principles are useful in driving sustainability in infrastructure projects and should be followed to ensure that projects meet their intended purpose while minimizing adverse impacts on the environment, society, and the economy.

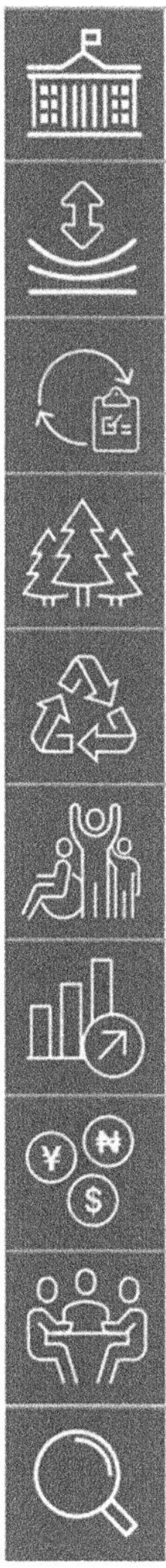

Figure 18.5 The ten guiding principles to sustainable infrastructure. International good practice principles for sustainable infrastructure/UNEP/Public domain.

18.6 Built-Infrastructure Projects: Examples

Civil infrastructure is multifaceted and can include bridges, dams, roads, water supply/treatment, stormwater management, waste management, and other similar facilities. To demonstrate how sustainability principles may be pursued across various types of civil infrastructure systems, few examples are provided herein.

18.6.1 Transportation Infrastructure Projects

Transportation infrastructure refers to the fundamental facilities and systems that support the movement of goods, animals, and humans from one place to another. It encompasses essential routes and networks such as roads, railways, airports, and waterways that facilitate economic activities, connect communities, and enhance the comfort and efficiency of daily life. These systems are vital for ensuring the proper functioning of economies and serving both national and global mobility needs. The transportation sector is a major contributor to global GHG emissions, and with growing transportation needs - investments in the construction of new infrastructure and maintenance of old infrastructure are expected to grow further. Sustainability in transportation infrastructure projects is therefore crucial for climate change mitigation. As an example of sustainable airports, an essential infrastructural component of the aviation industry, is provided in the following subsection.

18.6.1.1 Sustainable Airports

Sustainable practices in airport infrastructure should focus on several key areas, including sustainable energy management, use of sustainable materials in construction to reduce embodied carbon, achieving water efficiency, providing accessible gateways to ensure social equality, and efficient land use. Some of the sustainable practices in airport infrastructure, as identified by Greer et al. (2020), based on the current academic and industry literature are describe below.

- Sustainable construction practices should include using recycled and locally sourced materials for airfield pavements and terminal buildings and implementing comprehensive waste management strategies, including recycling and reducing construction waste.
- Energy management can be done by implementing energy monitoring systems, using energy-efficient HVAC and lighting, and exploring microgrids for on-site energy generation. Airports should utilize on-site renewable energy sources like solar photovoltaic (PV) systems plants to reduce greenhouse gas emissions. Further, use of energy-efficient building designs and low-emission ground vehicles should be encouraged.
- Comfort and health should be addressed by monitoring and improving indoor air quality in terminals and control towers and designing buildings to optimize natural lighting and ventilation, reducing the need for artificial heating and cooling.
- Water conservation can be achieved through water-efficient fixtures, reducing irrigation demand, and using alternative water sources like rainwater and greywater, while water quality should be managed by treating stormwater and de-icing fluid runoff to prevent contamination of local water bodies.
- Efficient land use and minimizing habitat disruption during construction and operation should be prioritized, along with promoting sustainable transportation options such as EVs and public transport integration.
- A holistic life-cycle assessment evaluating the triple bottom line sustainability impacts of airport infrastructure, engaging various stakeholders, is necessary.

18.6.2 Water and Wastewater Treatment Projects

A public drinking water system, or public water system (PWS), is a system that provides water for human consumption through at least 15 service connections or regularly serves (at least 60 days per year) at least 25 individuals (USEPA 2024). PWSs are responsible for providing drinking water to 90% of the population of the United States. Groundwater serves as a source for drinking water, either for public supply systems or for private well systems. Groundwater is recharged as precipitation infiltrates through permeable ground surfaces and migrates through soil and rock into aquifers. Many communities derive water sources from underground aquifers. Water supply wells are constructed by drilling through soil until the aquifers are reached. Under most conditions, pump systems are used to extract these groundwater sources to the surface for treatment, storage, and conveyance. While groundwater is generally consumed through in-house treatment units, drinking water supplied by PWSs has stringent treatment guidelines and requires extensive infrastructure for the treatment and transit of the treated water.

Indiscriminate disposal of wastewater generated through various human activities can lead to environmental contamination issues and public health concerns. Therefore, wastewater generated from residential, commercial, or industrial facilities needs extensive treatment before disposal into natural water bodies. Herein, we provide some general sustainable practices that can enhance the sustainability of various water and wastewater treatment infrastructure systems.

18.6.2.1 Sustainable Water Treatment

The treatment requirements for water and wastewater vary depending on the influent water quality and the desired effluent quality standards. Various alternative technologies are available to address these needs at different treatment levels. These include screening and sedimentation for removing suspended solids, aerobic and anaerobic digestion for microbial removal, and advanced methods like disinfection or membrane filtration to further remove contaminants of concern.

Recent advancements have introduced nanotechnologies to enhance various treatment stages. For instance, nano-adsorbents for chemical contaminants and nano-materials with anti-microbial properties can used for disinfection to avoid the formation of toxic byproducts, while thin-film nanocomposite membranes have demonstrated improved treatment efficiency and sustainability benefits (Qu et al. 2013). Additionally, 3D printing techniques are being applied to improve membrane design, minimizing deficiencies, reducing waste generation, and promoting reuse (Yusuf et al. 2020). Innovative disinfection methods, such as solar-based technologies, are gaining traction to avoid any residual chlorine in drinking water and minimize energy consumption (Jeon et al. 2022).

The choice of treatment technologies depends on specific use cases and treatment objectives, but significant sustainability gains can only be achieved through the efficient operation of treatment plants. Advanced AI-based optimization and reinforcement learning algorithms can facilitate better decision-making through real-time monitoring and predictions, improving treatment efficiency and reducing energy consumption (Padrón-Páez et al. 2020). Furthermore, the use of remote-sensing technologies to identify deficiencies in treatment systems is gaining traction in the industry.

Effectively managing waste generated during water and wastewater treatment, particularly sludge (or biosolids), offers additional sustainability benefits. A prominent practice in the water treatment industry is the land application of biosolids, where they are applied to soil to enhance its condition or serve as fertilizer for crops and vegetation. This practice not only reduces waste but also promotes resource recovery (USEPA 2022). However, recent focus on per- and polyfluoroalkyl substances (PFAS), also called forever chemicals, and their presence in trace amounts in biosolids has raised concerns over their land application. Nevertheless, advancements in technologies for treating PFAS offer potential solutions to address these challenges.

By integrating such approaches, the overall sustainability of water and wastewater treatment infrastructure can be enhanced through reduced resource consumption, improved energy efficiency, minimized waste generation, and a stronger focus on protecting public health.

18.6.2.2 Reuse of Treated Wastewater

The reuse of treated wastewater for non-potable applications, such as irrigation, industrial processes, and landscape maintenance, significantly enhances sustainability by reducing the demand for freshwater resources. This practice helps conserve drinking water supplies for essential uses, minimizes wastewater discharge into natural water bodies, and lowers the environmental impact of water and wastewater treatment infrastructure. By integrating advanced treatment technologies to meet safety standards, treated wastewater reuse not only supports water resource management but also contributes to a circular water economy, promoting resilience against water scarcity and climate change (Kandou et al. 2023; Reddy et al. 2023).

18.6.3 Stormwater Treatment/Green Infrastructure Projects

Stormwater runoff is generated when precipitation events result in surface flows over land or impervious surfaces, including paved streets, parking lots, and building rooftops. During overland flow, stormwater accumulates debris, chemicals, sediment, and other pollutants that could adversely affect the quality of receiving waters if the runoff is not treated prior to discharge. Large areas of connected impervious surfaces exist, and continued development and ongoing conversion toward impervious surfaces can dramatically increase the volume and rate of stormwater discharges, leading to increased pollutant delivery, flooding, and erosion as well as decreased groundwater recharge. As a result, the impacts of stormwater runoff within urbanized areas can be extensive (Lee and Bang 2000).

Many stormwater discharges are considered point sources and require coverage under an NPDES (National Pollutant Discharge Elimination System) permit. Understanding the impacts of urban stormwater is important to preserve and protect aquatic ecosystems and human health. The goal of sustainable stormwater infrastructure is to manage the small, frequent storms that account for most pollutant loading (Barbosa et al. 2012). A range of systems collectively called BMPs are available to control and treat these flows as well as to attempt to mimic pre-urbanization stormwater runoff to the extent practicable.

To assess the effectiveness of green infrastructure in protecting our water resources, researchers must link the performance of green infrastructure initiatives such as those listed above to water quality and ecological outcomes in receiving waters. Research indicates that green infrastructure and low impact development (LID) initiatives offer an effective approach to improving receiving water quality and flow. Further, a growing knowledge base is increasingly being shared by the regulatory and development communities with respect to the costs and benefits of sustainable stormwater control systems.

For example, sustainable stormwater infrastructure and its use to control flow volumes, rates, and pollutant loading provide many beneficial effects to the physical environment, including reduced pollutant loading and reduced alteration of flow rates, velocities, and volumes to receiving surface water bodies. Groundwater conditions are also improved; stormwater infrastructure can induce groundwater recharge, mimicking rates that would be expected in a natural setting prior to the addition of impervious surfaces or other additions to the built environment. Additionally, bioretention facilities can mitigate pollutant loading to the subsurface. The use of vegetation in planter boxes, rain gardens, wetlands, bioswales, green roofs, and tree canopies can facilitate improved air quality and sequester carbon dioxide from the atmosphere. These techniques can also moderate ambient air temperatures in the vicinity of the built environment, leading to a reduced reliance on HVAC output and a reduction in related energy consumption. Natural habitats may also be enhanced – open space can be preserved or enhanced, providing urban wildlife habitat, increasing biodiversity, and propagating native species in a controlled manner. Finally, sustainable stormwater infrastructure enhances urban QL through aesthetic improvements and dedication of open space to enhance physical activity and public health. Nevertheless, the pros and cons of any planned infrastructure under short-term or long-term conditions should be evaluated.

A brief description of these green stormwater infrastructure options is presented below.

18.6.3.1 Green Roofs

Green roofs incorporate actively growing vegetation into roof surfaces to reduce stormwater runoff volumes, regulate building temperatures, reduce urban heat island effects (Figures 18.6 and 18.7). Active research is underway with respect to the selection of plant species, propagation and establishment methods, plant succession, carbon sequestration potential, and water and nutrient requirements (Oberndorfer et al. 2007).

18.6.3.2 Permeable Pavements

Permeable pavements may be used as an alternative to impermeable concrete and asphalt (Figures 18.8 and 18.9). These materials allow a portion of stormwater runoff to infiltrate into the ground surface instead of contributing to overland flow and runoff. Typically, these materials incorporate specially designed air-entrained concrete or the use of discrete permeable paving blocks. Permeable pavements are among the most extensively studied and well-understood infrastructure practices. Assuming geotechnical and geologic conditions will not be compromised, the use of these materials addresses infiltration capacity and pollutant removal efficiency, adding to the

Figure 18.6 Green roof helps cool the building and minimize water run-off.

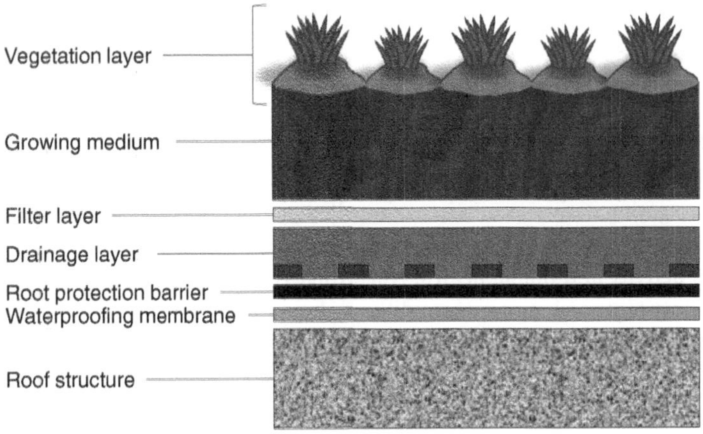

Figure 18.7 Green roof structure.

Figure 18.8 Permeable pavement. *Source:* USEPA.

long-term beneficial effects on surface water quality. Additionally, when systems are properly designed, they can be protective of groundwater quality (Scholz and Grabowiecki 2007).

18.6.3.3 Rainwater Harvesting

Recent droughts and potential impacts of climate change have generated interest in rainwater harvesting as an approach to water conservation as well as stormwater management (Figures 18.10 and 18.11). Research is being conducted on sizing methodologies, public perceptions, and the water quality of harvested rainwater. Ironically, in several generally arid western states, the legality of rainwater harvesting is questionable with respect to water rights. These longstanding and, in many cases, antiquated statutes are fortunately being revisited to address rainwater harvesting potential (Boers and Ben-Asher 1982).

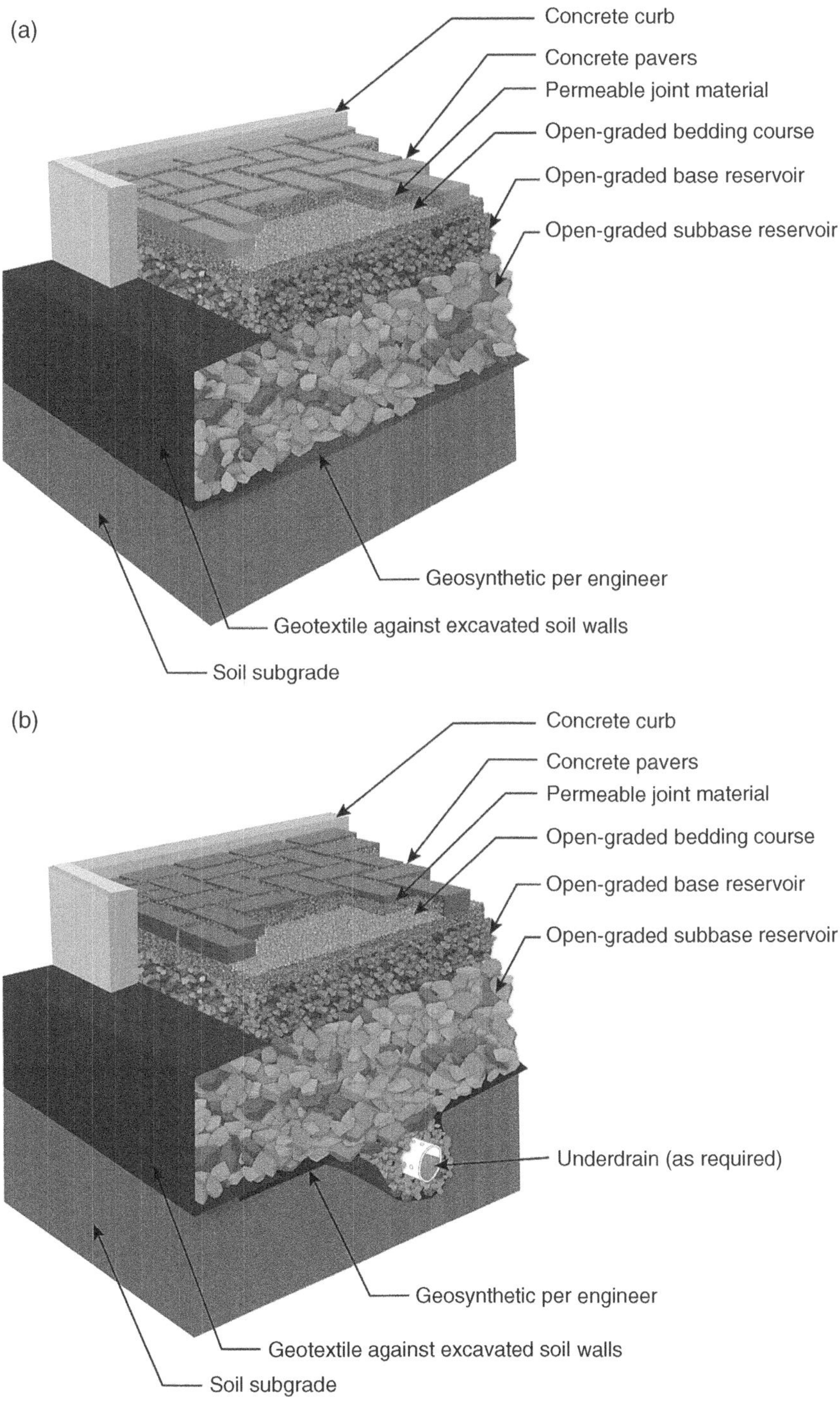

Figure 18.9 Typical structure for permeable pavement. (a) Full infiltration and (b) partial infiltration. *Source:* Courtesy of ICPI Smith (2017).

Figure 18.10 Rainwater harvesting. *Source:* USEPA.

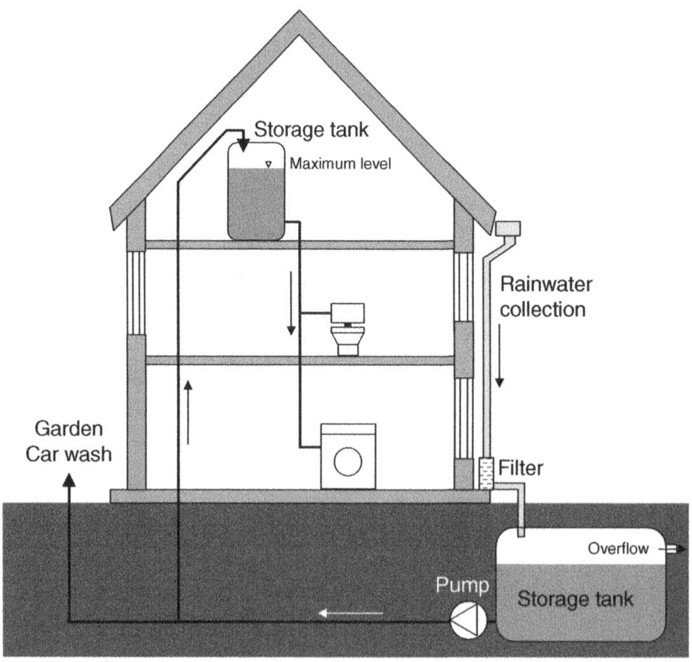

Figure 18.11 Functioning of a rainwater harvesting system.

18.6.3.4 Rain Gardens and Planter Boxes

Rain gardens and planter boxes are both forms of bioretention (Figure 18.12). The concept of bioretention is simple – stormwater runoff is captured and infiltration rates are controlled to provide a sufficient residence time within a treatment zone, allowing microorganisms within soil material to degrade a range of pollutants. Bioretention areas provide a source for groundwater recharge, pollutant removal, and runoff detention as well as an effective means for stormwater management where open space is limited. Research is underway to assess the factors that can be incorporated into bioretention design and operation (Davis et al. 2009).

18.6.3.5 Bioswales

Similar in many ways to rain gardens and planter boxes, bioswales are another form of bioretention. These swales are used to convey stormwater within a shallow, vegetation-covered channel. Bioswales can be effective in attenuating peak flows, reducing flow velocity, reducing pollutant loads, and enhancing biodiversity (Figures 18.13 and 18.14). Performance is sensitive to design parameters that are incorporated (Mazer et al. 2001).

18.6.3.6 Constructed Wetlands and Tree Canopies

Constructed wetlands are designed to facilitate stormwater collection and detention on a grander scale into a wetlands habitat. Constructed wetlands are complex systems with many design parameters; proper selection of species, topography, and other physical parameters are essential to create a successfully operating project (Figures 18.15 and 18.16). Previous studies have examined not only wetland performance and design but also wetland management and evaluation. Tree canopies in an urban environment can effectively mitigate air pollution, reduce runoff quantity, and reduce energy use while enhancing habitat and aesthetics (Figure 18.17). Additionally, research is being performed on the interaction between urban soils and urban trees (Wu et al. 2015).

Figure 18.12 Rainwater runoff collection in a rain garden.

Figure 18.13 Treatment of water runoff in a bioswale. *Source:* USEPA.

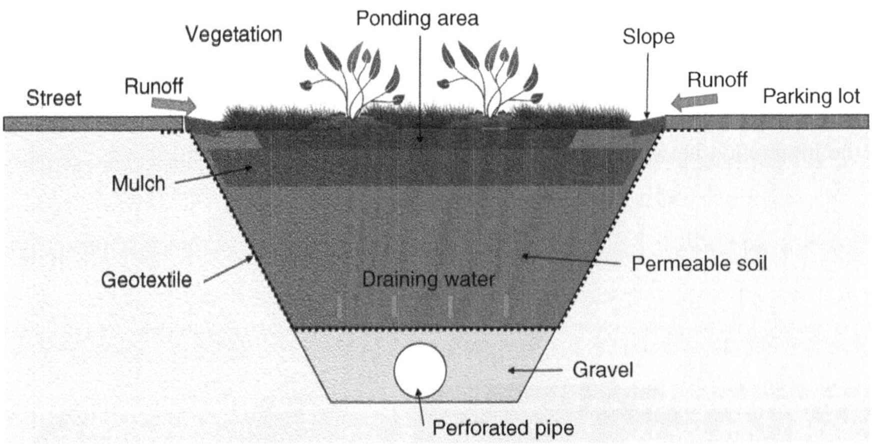

Figure 18.14 Bioswale structure.

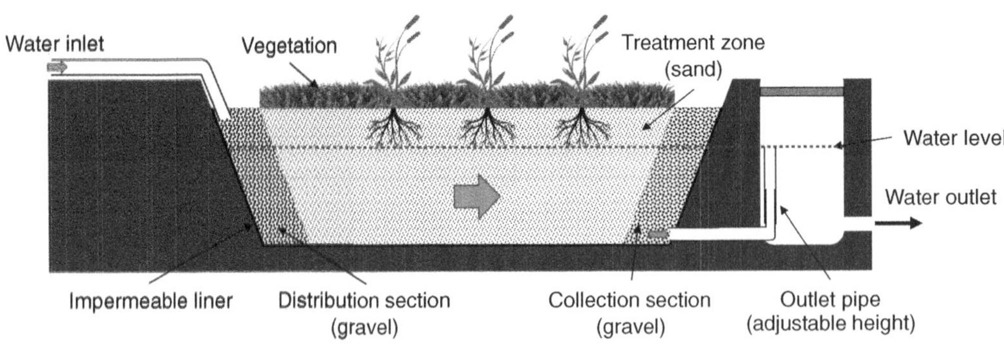

Figure 18.15 Structure of constructed wetlands for water treatment.

Figure 18.16 Wastewater treatment in a constructed wetland. *Source:* Ecocelta, Spain.

Figure 18.17 Tree canopies reduce storm water runoff and mitigate pollution.

18.7 Resiliency Considerations

Built infrastructure is susceptible to climate change impacts more than any other engineering system, and the cost of failure of infrastructure projects is extreme. It is important to ensure the resilience of civil infrastructure to the increasing intensity and frequency of climate-change impacts. According to the ASCE standard for sustainable infrastructure, a resiliency assessment should be conducted iteratively before and after site selection of the infrastructure project, considering any natural hazards and their increasing impacts due to climate change (ASCE 2023). Such an assessment should include the following steps: (i) identification of potential threats and hazards, (ii) infrastructure vulnerability assessment to the identified hazards, and (iii) risk assessment to mitigate the impact of hazards and meet stakeholders' objectives. The standard further provides guidelines to improve resilience as described below:

- Infrastructure sites should be selected to minimize any potential climate-related threats to the project, and the selected site should consider corresponding hazard loads throughout the life cycle of the project.
- Resilience goals should be established depending on current and projected climate change scenarios. Resilience strategies should be devised and implemented to meet these goals and mitigate the impacts of climate change projection or anticipated threats or hazards on the infrastructure project, environment, society, and the economy.

According to ASCE, the objective of a resilient infrastructure system should be to minimize exposure, impact, and risk related to any potential threats or hazards, which include climate-related disasters. While these guidelines are generic in nature and provide a way forward, practical frameworks are necessary to assess the resiliency at various stages throughout the life cycle of the infrastructure project. The resilience assessment methods discussed in Chapter 6 can be useful for this purpose. To integrate the resilience and sustainability aspects of the infrastructure projects – thereby minimizing the environmental, social, and economic implications of resilience – frameworks such as the TQUALICSR (Tiered Quantitative Assessment of Life Cycle Sustainability and Resilience) framework described in Chapter 13 can be useful.

18.8 Summary

Both "soft" and "hard" infrastructure is critically necessary for the function of society. While civil engineering has traditionally been focused on delivering technical solutions to meet society's needs and to protect the environment, only recently has increased awareness of nature and natural resources inspired a search for sustainable infrastructure. Several additional key factors have affected the design of civil infrastructure, including additional demands from the regulatory environment, a focus on cost-effective solutions to aging infrastructure, and the desire to accomplish this with limitations in government funding.

There is a well-established connection between sustainability concepts and environmentally focused engineering practices, and interest continues to grow with respect to the benefits of sustainability in other disciplines. As infrastructure provides a critical physical linkage between the social and economic aspects of a functioning society and sustainability is focused on deriving benefits with respect to environmental, social, and economic conditions, there is a natural connection to devise sustainable infrastructure. With growing concern about climate change impacts on infrastructure projects, resilience to such unexpected events should be an integral part of sustainable infrastructure.

18.9 Questions

18.1 What are the two major classifications of infrastructure and what do they comprise?

18.2 What is the principle behind sustainable infrastructure?

18.3 Discuss in detail the long-term benefits associated with sustainable infrastructure approaches.

18.4 Why has existing infrastructure received poor ratings during infrastructure assessments?

18.5 Who developed the Envision™ rating tool?

18.6 What is the purpose of the Envision™ tool?

18.7 Discuss the importance of the stages of Envision™ ratings.

18.8 What are the different methods to achieve sustainable water infrastructure?

18.9 What do you mean by green roofs and bioswales? Explain briefly.

18.10 What is the concept behind bioretention systems?

18.11 Discuss the importance of ongoing maintenance activities for tree canopies and constructed wetlands.

18.12 Discuss the technical challenges that may be encountered during the use of stormwater treatment systems (permeable pavers, bioswales, and bioretention) and discuss the strategies to mitigate these challenges.

18.10 Problems

18.1 A city is experiencing increased flooding due to extreme rainfall events and urban runoff. To mitigate these effects, the city plans to implement a green infrastructure strategy involving the construction of bioswales, green roofs, and permeable pavements.

Details regarding the catchment area, planned green infrastructure, and design storm event are as follows:

Catchment area characteristics:

Total catchment area: 100 hectares
Current impervious surface area: 60%
Soil type: clay loam (hydraulic conductivity: 10 mm/h)
Coefficient of runoff for impervious surfaces: 0.9
Coefficient of runoff for pervious surfaces (excluding green infrastructure): 0.5

Green infrastructure specifications:

Green roofs:

Area to be covered: 5 hectares

Depth of the substrate: 0.1 m

Field capacity of the substrate: 0.15 m^3/m^3

Permeable pavements:

Area to be replaced: 15 hectares

Porosity: 0.33

Infiltration rate: 500 mm/h

Bioswales:

Area to be covered: 10 hectares, depth: 0.5 m, side slope: 3:1 (H:V)

Vegetation type: native grasses (Manning's n: 0.30), longitudinal slope: 1.5%

Use the following equation to calculate flow velocity, $v = \frac{1.49}{n} R_h{}^{2/3} S^{1/2}$

Design storm event:

Intensity: 25 mm/h, duration: 2 hours, return period: 50 years

Calculate the following:

- Total runoff volume from the catchment for the design storm event before the implementation of green infrastructure.
- Expected reduction in runoff volume due to each type of green infrastructure.
- Total runoff volume from the catchment for the design storm event after the implementation of green infrastructure.
- Percentage reduction in peak runoff due to the implementation of green infrastructure.

References

ASCE (2018). *Envision*. American Society of Civil Engineers. https://www.asce.org/envision/ (accessed September 2018).

ASCE (2023). *Standard Practice for Sustainable Infrastructure (73-23)*. Reston, VA: American Society of Civil Engineers (ASCE). https://doi.org/10.1061/9780784415733.

Barbosa, A.E., Fernandes, J.N., and David, L.M. (2012). Key issues for sustainable urban stormwater management. *Water Research* 46 (20): 6787–6798.

Behr, C. (2014). A value-based rating system for envision. In: *ICSI 2014: Creating Infrastructure for a Sustainable World* (ed. J. Crittenden, C. Hendrickson, and B. Wallace), 744–754. Long Beach, CA: American Society of Civil Engineers.

Boers, T.M. and Ben-Asher, J. (1982). A review of rainwater harvesting. *Agricultural Water Management* 5 (2): 145–158. https://doi.org/10.1016/0378-3774(82)90003-8.

Dasgupta, S. and Tam, E.K. (2005). Indicators and framework for assessing sustainable infrastructure. *Canadian Journal of Civil Engineering* 32 (1): 30–44.

Davis, A.P., Hunt, W.F., Traver, R.G., and Clar, M. (2009). Bioretention technology: overview of current practice and future needs. *Journal of Environmental Engineering* 135 (3): 109–117. https://doi.org/10.1061/(ASCE)0733-9372(2009)135:3(109).

Fenner, R.A., Cruickshank, H.J., and Ainger, C. (2014). Sustainability in civil engineering education: why, what, when, where and how. *Proceedings of the Institution of Civil Engineers: Engineering Sustainability* 167 (5): 228–237. https://doi.org/10.1680/ensu.14.00002.

Frangopol, D.M. and Liu, M. (2007). Maintenance and management of civil infrastructure based on condition, safety, optimization, and life-cycle cost*. *Structure and Infrastructure Engineering* 3 (1): 29–41.

Greer, F., Rakas, J., and Horvath, A. (2020). Airports and environmental sustainability: a comprehensive review. *Environmental Research Letters* 15 (10): 103007.

Hayes, B. (2005). *Infrastructure: A Guide to the Industrial Landscape*. New York: W. W. Norton and Co. Ltd. http://www.gettextbooks.com/isbn/9780393349832/.

Jeon, I., Ryberg, E.C., Alvarez, P.J., and Kim, J.H. (2022). Technology assessment of solar disinfection for drinking water treatment. *Nature Sustainability* 5 (9): 801–808.

Kandou, V., Janga, J.K., Verma, G. et al. (2023). Environmental sustainability assessment of treated wastewater reuse: a case study. *Sustainability and Climate Change* 16 (5): 378–393.

Lee, J.H. and Bang, K.W. (2000). Characterization of urban stormwater runoff. *Water Research* 34 (6): 1773–1780.

Mazer, G., Booth, D., and Ewing, K. (2001). Limitations to vegetation establishment and growth in biofiltration swales. *Ecological Engineering* 17 (4): 429–443. https://doi.org/10.1016/S0925-8574(00)00173-7.

Oberndorfer, E., Lundholm, J., Bass, B. et al. (2007). Green roofs as urban ecosystems: ecological structures, functions, and services. *BioScience* 57 (10): 823–833. https://doi.org/10.1641/B571005.

Padrón-Páez, J.I., Almaraz, S.D.L., and Román-Martínez, A. (2020). Sustainable wastewater treatment plants design through multiobjective optimization. *Computers & Chemical Engineering* 140: 106850.

Pedersen, T.A. (2012). Institute for sustainable infrastructure envision™ rating system: applicability to water infrastructure projects. *Proceedings of the Water Environment Federation* 2012 (15): 1849–1856.

Qu, X., Brame, J., Li, Q., and Alvarez, P.J. (2013). Nanotechnology for a safe and sustainable water supply: enabling integrated water treatment and reuse. *Accounts of Chemical Research* 46 (3): 834–843.

Reddy, K.R., Kandou, V., Havrelock, R. et al. (2023). Reuse of treated wastewater: drivers, regulations, technologies, case studies, and Greater Chicago Area experiences. *Sustainability* 15 (9): 7495.

Sarté, S.B. (2010). *Sustainable Infrastructure: The Guide to Green Engineering and Design*. Wiley.

Scholz, M. and Grabowiecki, P. (2007). Review of permeable pavement systems. *Building and Environment* 42 (11): 3830–3836. https://doi.org/10.1016/j.buildenv.2006.11.016.

Smith, D.R. (2017). *Permeable Interlocking Concrete Pavements*, 5e. Interlocking Concrete Pavement Institute (ICPI).

UNEP (2022). International good practice principles for sustainable infrastructure. https://www.unep.org/resources/publication/international-good-practice-principles-sustainable-infrastructure (accessed 27 November 2024).

USACE (2019). Civil works sustainable infrastructure practices guidebook. https://www.publications.usace.army.mil/Portals/76/Users/182/86/2486/EP%201100-2-2.pdf (accessed 27 November 2024).

USEPA (2022). Basic information about biosolids. https://www.epa.gov/biosolids/basic-information-about-biosolids#uses (accessed 27 November 2024).

USEPA (2024). Information about Public Water Systems. https://www.epa.gov/dwreginfo/information-about-public-water-systems (accessed 4 December 2024).

Werbach, A. (2009). *Strategy for Sustainability: A Business Manifesto*. Harvard Business Press.

Wu, H., Zhang, J., Ngo, H.H. et al. (2015). A review on the sustainability of constructed wetlands for wastewater treatment: design and operation. *Bioresource Technology* 175: 594–601.

Yusuf, A., Sodiq, A., Giwa, A. et al. (2020). A review of emerging trends in membrane science and technology for sustainable water treatment. *Journal of Cleaner Production* 266: 121867.

19

Sustainable and Resilient Land Management

Chapter Goal and Objectives

Protecting our land is vital for the survival of humanity. This chapter highlights resilient and sustainable management practices for land by integrating innovative remediation approaches, comprehensive frameworks, and robust indicators and metrics that address soil and land pollution challenges while promoting long-term environmental and societal benefits. Specifically, this chapter presents (i) key concerns with land management and (ii) pollution problems and remediation approaches, emphasizing the need to incorporate resiliency and sustainability principles; (iii) sustainable remediation frameworks, focusing on their methodologies, metrics, and tools for assessing sustainability and resilience; (iv) case studies that demonstrate the application of sustainable and resilient land remediation strategies; and (v) challenges and opportunities in implementing sustainable remediation practices, including technological, regulatory, and socio-economic barriers.

19.1 Introduction

Land degradation is driven by natural processes and human activities, including deforestation, urbanization, overgrazing, unsustainable agricultural practices, and industrial development, which diminish land productivity and ecological balance. Among these, soil and groundwater pollution are particularly alarming, resulting from the infiltration of hazardous chemicals, heavy metals, and organic pollutants due to industrial releases/discharges, agricultural runoff, and improper waste disposal. These pollutants reduce soil fertility, disrupt ecosystems, and compromise groundwater quality, threatening public health and exacerbating water scarcity challenges. Addressing these interconnected issues is critical to achieving sustainable and resilient land management.

The United States Environmental Protection Agency (USEPA 2024) oversees the cleanup of more than 400 000 contaminated sites nationwide, including superfund program sites and those sites under the Resource Conservation and Recovery Act (RCRA). Around 23 000 sites require urgent action considering the concentration levels and hazardousness of the contaminants. The contaminants encountered include but are not limited to organic compounds, heavy metals, and radionuclides. A variety of sources can cause subsurface contamination, as depicted in Figure 19.1. Some activities that lead to contamination include land disposal of solid and liquid wastes, accidental

Sustainable and Resilient Engineering: Drivers, Metrics, Tools, and Applications, Second Edition. Krishna R. Reddy, Claudio Cameselle, and Jeffrey A. Adams.

Companion website: www.wiley.com/go/reddy/sustainableresilientengineering2e

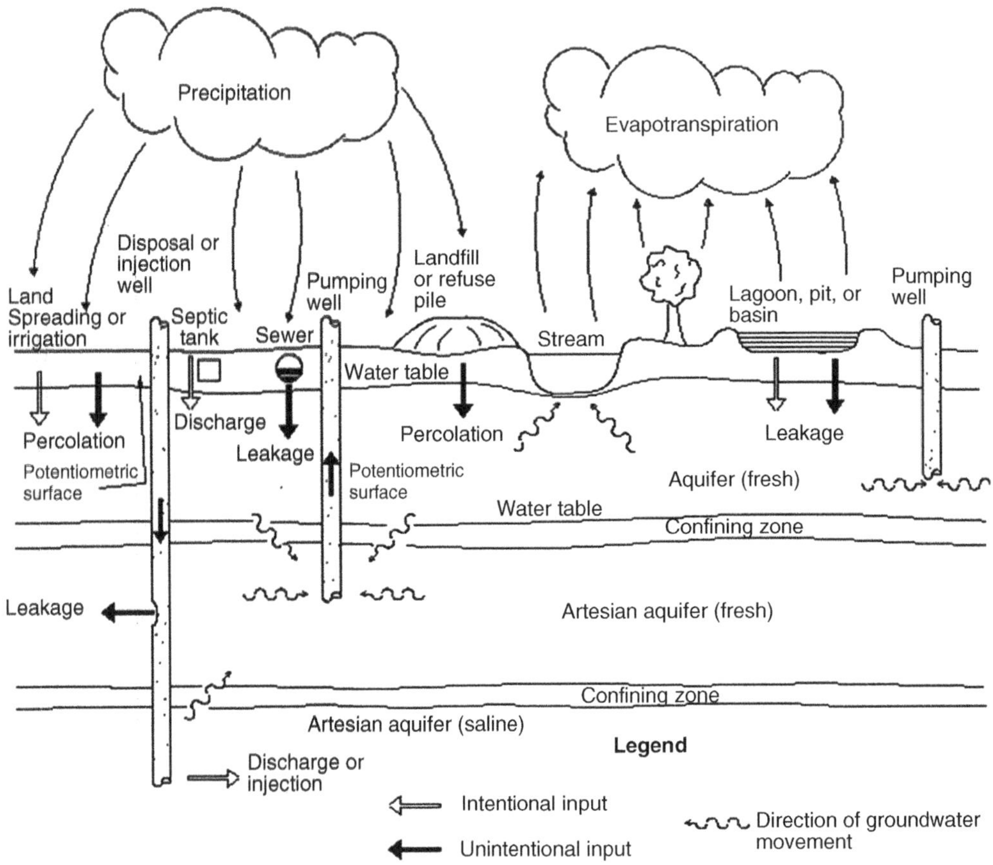

Figure 19.1 Sources of subsurface contamination.

spills, waste dumps, and landfills, leakage from underground storage tanks (USTs), and mining operations (leaching of the spoil material, milling wastes, etc.). It has been estimated that the cleanup cost of all those sites may be over $500 billion, depending on the contaminant nature and concentration and the cleanup targets. Besides, 7400 additional sites have been identified with Per- and Polyfluoroalkyl Substances (PFAS) contamination, which need more resources and intervention (EWG 2024).

Traditional, risk-based site remedial approaches have not always been sustainable because they often do not account for broader environmental impacts such as extraction and use of natural resources, waste generation, energy used, transportation of equipment and materials, and related greenhouse gas (GHG) emissions for on- and off-site operations. These approaches do not explicitly account for the net environmental benefit when all relevant environmental parameters are considered. To address this, principles of "green remediation" and "sustainable remediation" have emerged. "Sustainable remediation" is defined as a remedy or combination of remedies whose net benefit on human health and the environment is maximized through the judicious use of limited resources (Baker et al. 2009). On the other hand, "Green remediation" is defined as the practice of considering all environmental effects of remedy implementation and incorporating options that maximize the net environmental benefit of cleanup actions (USEPA 2008).

Green and sustainable remediation (GSR) has emerged as a remediation strategy (involving one or more remediation technologies) whose net benefit to human health and the environment is maximized through an optimized selection of remediation technologies and resource use that consider how the community, global society, and the environment would benefit, or be adversely affected by its implementation (ITRC 2011a,b).

This chapter presents an overview of sustainability decision frameworks, metrics, and assessment tools. This is followed by a discussion and analysis of case studies with respect to sustainability and the degree of success achieved with each of the respective studies. Finally, an outlook of the future evolution of this innovative approach to environmental remediation is presented. More detailed information can be found in Reddy and Adams (2015).

19.2 Land Management Problems

Land is crucial for livelihoods that depend on agriculture, forestry, rangelands, and aquatic environments such as rivers, lakes, and coastal areas. However, due to diverse political, social, and economic influences, the intense exploitation of natural resources to meet the demands of a rapidly expanding global population and economy has inadvertently led to the mismanagement and degradation of land and ecosystems. As presented in the latest report from the United Nations, over 40% of Earth's total land area is degraded or has diminished in natural productivity, primarily due to human activities. This degradation directly impacts half of the global population and threatens about half of the world's gross domestic product (GDP), amounting to approximately US$44 trillion (UNCCD 2022). The United Nations Convention to Combat Desertification (UNCCD) defines land degradation as the reduction or loss of the biological or economic productivity and complexity of rainfed cropland, irrigated cropland, or range, pasture, forest, and woodlands resulting from land uses or a process or combination of processes, including those arising from human activities and habitation patterns, such as soil erosion by wind or water, deterioration of the soil's physical, chemical, and biological properties, and the long-term loss of natural vegetation (UNCCD 1994). Land degradation is a complex process induced by both natural and man-made drivers. Natural drivers, specifically drought and salinity, are primary contributors. Man-made activities, including improper farming practices, overgrazing, deforestation, pollution, urbanization, and rapid population growth, also play significant roles.

Although global climate change and man-made activities may accelerate land degradation, the impact of land use changes by humans is significantly greater than that of natural causes. Land degradation is generally characterized by a decline in soil nutrients, a decrease in vegetation cover, changes in soil's physical, chemical, and biological properties, and pollution of water resources from soil contamination. Both natural and man-made drivers are briefly explained below:

Overgrazing is one of the major causes of land degradation globally, contributing to 35% of all the drivers (UNCCD 2017) and reducing the quality and extent of vegetation cover, which is crucial for land health. For instance, Hilker et al. (2014) found that overgrazing is a primary contributor to the degradation of the Mongolian steppe and is partially responsible for desertification. On the other hand, a few studies indicate no significant difference in land degradation between optimal communal grazing and overgrazing. Overgrazing near watering points is especially damaging, degrading the land around wells in villages. This leads to decreased soil organic matter, weakened physical properties, and lower resilience, particularly if vegetation recovery during normal rainfall periods is inadequate.

Deforestation is the purposeful clearing of forested land for nonforest uses such as agriculture, contributing to 30% of all causes of land degradation (UNCCD 2017). It leads to increased atmospheric carbon dioxide, as trees that absorb CO_2 for photosynthesis are cut down or burned, releasing stored carbon. With fewer trees available to absorb CO_2, unabsorbed GHGs accumulate in the atmosphere, accelerating global warming and climate change. Deforestation also threatens biodiversity, particularly in tropical forests that are home to numerous plant and animal species, leading to the risk of species extinction. Moreover, deforested land becomes vulnerable to erosion, ecosystem destruction, and desertification. Trees play a crucial role in binding soil particles and maintaining soil quality. Without vegetation, the soil loses its structural strength, aeration, water retention capacity, and biological activity significantly, resulting in land degradation. Thus, it is evident that beyond causing land degradation, deforestation has other significant environmental impacts.

Unsustainable agricultural activities significantly contribute to land degradation. Globally, a large portion of land is used for agriculture, and while traditional farming methods generally exert minimal pressure on these resources, modern farming practices often lead to substantial environmental issues. Practices such as heavy tilling, monocropping, and row cropping can negatively affect soil quality. More sustainable practices such as contour farming and even some no-till systems may also affect the quality of soil (compaction, erosion, and local contamination) if not implemented properly (Hopwood et al. 2021). These methods can disrupt soil structure, deplete nutrients, and increase soil vulnerability to erosion. Additionally, using chemical fertilizers and pesticides in large quantities further deteriorates soil health, harming beneficial microorganisms and reducing long-term fertility. Cultivating fragile or marginal lands, which are less suitable for intensive farming, can lead to soil erosion, a decline in vegetation cover, and, eventually, desertification. Sustainable land management and conservation practices are therefore crucial for maintaining soil health and mitigating land degradation.

Overexploitation for fossil fuels ranks among the most environmentally damaging industries worldwide. The extraction, processing, and transportation of fossil fuels significantly damage land resources. Mining processes, a key part of fossil fuel extraction, generate various byproducts that can contaminate nearby soil and water, posing substantial risks to ecosystems. Disposing of these byproducts often leads to soil pollution, rendering the land less fertile and unsuitable for agricultural or ecological use. In some instances, the damage can be so severe that the land becomes completely barren when mining operations conclude, leading to irreversible land degradation. Additionally, mining activities disrupt the physical structure of the land, causing increased erosion, habitat destruction, and loss of biodiversity. Restoring these areas can be a lengthy and expensive process, and in many cases, the land remains unusable, contributing to further degradation and loss of valuable land resources. Sustainable resource management, combined with efforts to transition away from fossil fuels, is critical for mitigating the impact of overexploitation on land.

Pollution is a significant factor contributing to land degradation, primarily through the disposal of hazardous substances and contaminants such as chemicals, industrial waste, and radioactive materials from nuclear power plants. These pollutants can severely damage soil quality, often remaining in the environment for many years and making recovery difficult. In addition to industrial waste, using chemical fertilizers and pesticides to boost agricultural production driven by population growth also contributes to soil pollution. These chemicals infiltrate the soil and nearby surface water bodies, leading to broader ecosystem contamination. The accumulation of pollutants in the soil reduces its fertility by harming the biological activity of essential microbial communities that play a key role in maintaining soil health. Over time, this results in the depletion of vital nutrients, diminishing the soil's productivity and its ability to support plant growth.

The above discussion highlighted the human-driven causes of land degradation. In addition to these, natural factors also play a significant role in degrading land quality. Natural causes of land degradation include extreme weather events such as heavy rainfall, flooding, droughts, strong winds, earthquakes, and volcanic eruptions. These natural processes can lead to soil erosion, nutrient depletion, and a decline in overall land productivity. It is also important to note that many of these natural processes are influenced and intensified by human activities. To address the issue of land degradation, the United Nations has established the Sustainable Development Goals (SDGs), which include a specific target, SDG 15.3. This target focuses on "combating desertification, restoring degraded land and soil, including land affected by desertification, drought, and floods, and striving to achieve a land degradation-neutral world by 2030." This goal underscores the global commitment to reversing the impacts of land degradation as part of a broader effort to ensure environmental sustainability and ecological health.

19.3 Contaminated Land Remediation Approach

When considering financial and timing implications, a systematic approach is necessary for the characterization and remediation of contaminated sites. The most important tasks of such a systematic approach include (i) site characterization, (ii) risk assessment, and (iii) the selection of an effective remedial action. Figure 19.2 outlines such a systematic approach. In addition, innovative integration of various tasks of remediation can often lead to a faster, cost-effective remedial program.

To develop a remediation strategy, site characterization or development of a conceptual site model (CSM) is usually the first step. It consists of the collection and assessment of data representing contaminant type, distribution, and transmission/migration pathways at a site under investigation. The results of a site characterization form the basis for risk assessment and decisions concerning the requirements of remedial goals and action. Additionally, the results serve as a guide for the design, implementation, and monitoring of the remedial system. An effective site characterization includes the collection of data pertaining to (i) site geologic data, including site stratigraphy and important geologic formations, (ii) hydrogeological data, including major water-bearing formations and their hydraulic properties, such as groundwater depth, direction of flow, and

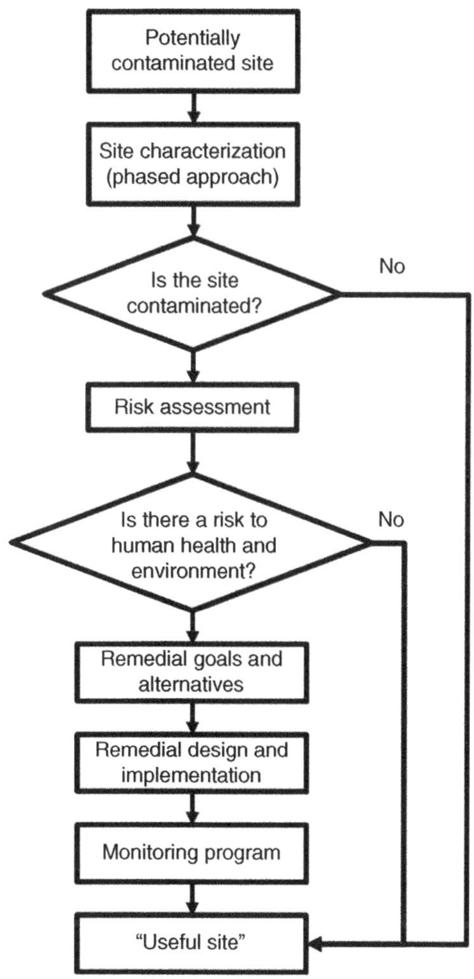

Figure 19.2 Systematic approach for contaminated site remediation.

gradient, and (iii) site contamination data, including type, concentration, phase, and distribution, and lateral and vertical extent of the contamination. Additionally, surface conditions both at and around the site must be taken into consideration.

Once characterization activities have confirmed the presence of site contamination, a risk assessment may be performed. The USEPA has developed comprehensive risk assessment procedures. The USEPA procedure was originally developed by the United States Academy of Sciences in 1983. It was adopted with modifications by the USEPA for use in Superfund feasibility studies and RCRA corrective measure studies (USEPA 1989). This procedure provides a general, comprehensive approach for performing risk assessments at contaminated sites. It consists of four steps: (i) hazard identification, (ii) exposure assessment, (iii) toxicity assessment, and (iv) risk characterization. Detailed information on the four steps is provided in Chapter 12.

Remediation is required when risks posed by the contamination at the site are deemed unacceptable. Generally, remediation methods are divided into two categories: in-situ and ex-situ remediation methods. In-situ methods treat contaminated soils and/or groundwater in place, eliminating the need to excavate the contaminated soils and/or extract groundwater. In-situ methods are advantageous because they are often less expensive, less invasive, often allow site surface activities to continue, and commonly provide increased safety to both on-site workers and the public within the vicinity of the remedial project. Successful implementation of in-situ methods requires a thorough understanding of subsurface conditions. In-situ containment, which may include the use of bottom barriers, vertical walls, and caps, may be a feasible strategy to minimize the risk posed by the contamination at some sites. Ex-situ methods are used to treat excavated soils and/or extracted groundwater. Surface treatment may be performed either on-site or off-site, depending on site-specific conditions. Ex-situ treatment methods are attractive because they may lessen the need for extensive in-situ monitoring during the remediation of subsurface conditions in such cases. Ex-situ treatment is also often faster than in-situ treatment and offers easier control and monitoring during remedial activity implementation.

19.4 Green and Sustainable Remediation Technologies

Remedial technologies are classified into two groups based on their scope of application: (i) vadose zone or soil remediation technologies and (ii) saturated zone or groundwater remediation technologies. The reader should refer to Sharma and Reddy (2004) for more detailed information on these remediation technologies. The most common practice used to remediate vadose zone contamination is excavation. Impacted soil is typically characterized through testing, allowing for appropriate transportation and disposal measures, commonly involving landfill disposal. Following excavation activities, confirmation samples are often collected from the sidewalls and base of the resulting excavation to ensure remedial goals are achieved, and clean fill materials are used to backfill the resulting excavation. When the excavation of contaminated soils is not a feasible option, several conventional and innovative treatment methods may be utilized. These methods may either be in-situ or ex-situ methods (Figure 19.3). Common vadose zone soil remedial methods include the following:

- Soil vapor extraction (SVE) consists of three basic components: an extraction system, an airflow system, and an off-gas treatment system. By applying a vacuum to the subsurface within the contaminant zone, the extraction system induces the movement of volatile organics and facilitates their removal and collection. Collected vapors pass through the airflow system and are delivered

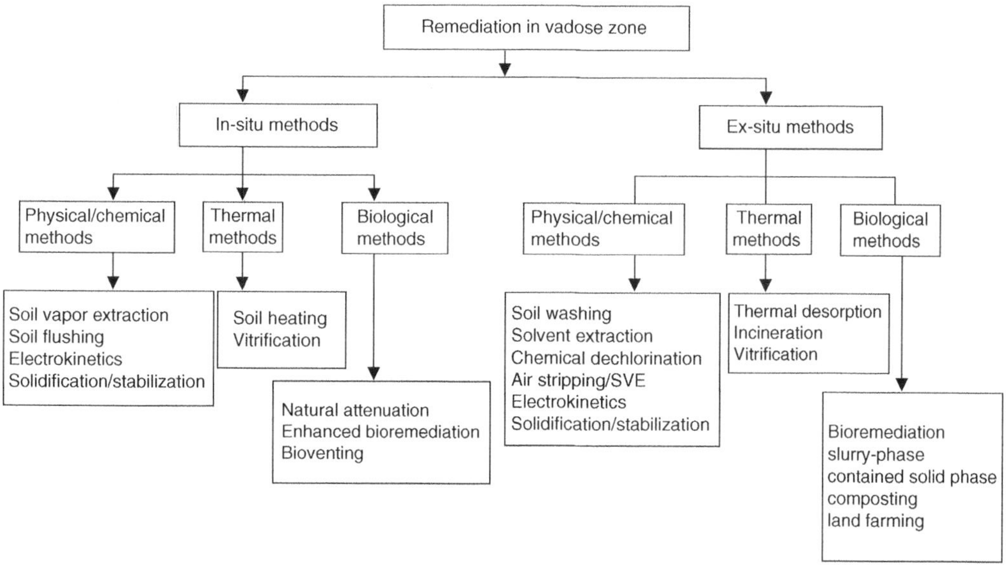

Figure 19.3 Vadose zone (soil) remediation technologies.

to the off-gas treatment system or, if regulatory limits permit, are emitted directly into the atmosphere.

- In-situ soil flushing involves the extraction of contaminants from the soil using water or other selected aqueous wash solutions. The flushing agent may be introduced into the subsurface in several ways, and once introduced, the flushing agent moves downward through the contaminated zone. Once the migrating agent/contaminant solution encounters the water table, it will mix with the groundwater and flow downgradient to a withdrawal point where it may be extracted, often via conventional extraction wells.
- Chemical oxidation technologies have also evolved as a preferred remedial alternative for in-situ or ex-situ remediation of soils and groundwater. With this technology, an oxidizing agent is introduced and mixed into the subsurface. Chemical oxidation typically involves reduction/oxidation (redox) reactions that chemically convert hazardous contaminants to nonhazardous or less toxic compounds that are more stable, less mobile, or inert.
- Soil stabilization and solidification involve applying additives or processes to contaminated soil to chemically bind and immobilize contaminants, preventing mobility.
- Electrokinetics involves the application of a low electric potential gradient across a contaminated soil zone to induce contaminant movement toward electrodes for the subsequent removal.
- Bioremediation utilizes microorganisms to biologically degrade contaminants into harmless end-products.
- In-situ vitrification (ISV) employs electrical power to heat and melt contaminated soil to destroy organic contaminants and stabilize inorganic contaminants. Alternatively, low-temperature soil heating is used to decontaminate soils through vaporization, steam distillation, and stripping.

If groundwater remediation is required, some of the aforementioned remedial technologies may be applied to saturated soils, including soil flushing, electrokinetics, and bioremediation. In addition, the following remedial methods may also be used:

- Groundwater may be extracted for treatment ("pump-and-treat") to remove any free-phase contaminants and/or contaminated groundwater with dissolved contaminants. While pump-and-treat may be successful during the initial stages of implementation, performance drastically decreases at later times. As a result, significant amounts of residual contamination can remain unaffected by continued treatment. Due to these limitations, the pump-and-treat method is now primarily used for free product recovery and to control contaminant plume migration.
- Air sparging involves injecting a gas, usually air, into the saturated soil zone below the lowest known level of contamination. Due to the effect of buoyancy, the injected air will rise toward the surface. As the air encounters contamination, it will, through a variety of mass transfer and transport mechanisms, strip the contaminant away or assist in in-situ degradation.
- Dual-phase extraction, also known as vacuum-enhanced recovery, is a hybrid remediation technique that combines technology from pump-and-treat and SVE. During implementation, groundwater is extracted through the application of a vacuum, allowing for the removal of the dissolved contaminants within the extracted groundwater as well as the contaminant vapors due to the applied vacuum.
- Permeable reactive barriers (PRBs) incorporate a permeable, reactive media to adsorb, degrade, or destroy contamination within groundwater as it passes through the permeable barrier. Common reactants include zero-valent iron, zeolites, organobentonites, and hydroxyapatite.

In some cases, it may be impractical or undesirable to actively remediate contamination in soils and/or groundwater via in-situ methods. This can be due to the presence of surface obstructions, such as structures or utilities, or the lateral/vertical position of contamination such that it cannot be readily addressed. In such situations, containment systems may be considered (Figure 19.4). Surface capping involves the installation of a surface barrier that prevents or limits the ability of underlying contaminated subsurface media to be encountered. Vertical cutoff barriers, vertical cutoff walls, or barrier walls are embedded or keyed into a low permeability formation, preventing migration of contamination. Horizontal configuration can be circumferential, downgradient, or upgradient (Figure 19.4a).

In some cases, groundwater pumping well systems are used as active containment systems to manipulate and manage groundwater for removing, diverting, and containing a contaminated plume or for adjusting groundwater levels to prevent plume movement (Figure 19.4b). Subsurface drains are an alternative to pumping wells for the containment of contaminated groundwater (Figure 19.4c). They consist of a drainpipe surrounded by filter and backfill to intercept a plume hydraulically downgradient and then divert for collection and discharge to a treatment plant.

Using just one technology may not be adequate to remediate some contaminated sites where different types of contaminants exist (e.g. heavy metals combined with volatile organic compounds (VOCs)) and/or when the contaminants are present within a complex geological environment, such as a heterogeneous soil profile consisting of lenses or layers of low permeability zones surrounded by high permeability soils. Under these situations, different remediation technologies can be used sequentially to achieve remedial goals. The use of such multiple remediation technologies is often referred to as "treatment trains." Typical treatment trains used at contaminated sites include soil flushing followed by bioremediation, SVE followed by soil flushing, SVE followed by stabilization/solidification, and thermal desorption followed by solidification/stabilization, which is then followed by soil flushing. Alternatively, different remediation technologies can be used concurrently, such as SVE and air sparging, electrokinetics and bioremediation, and soil flushing and bioremediation. Further, technologies not typically considered sustainable may be combined with other technologies to develop multicomponent remedial programs that are sustainable.

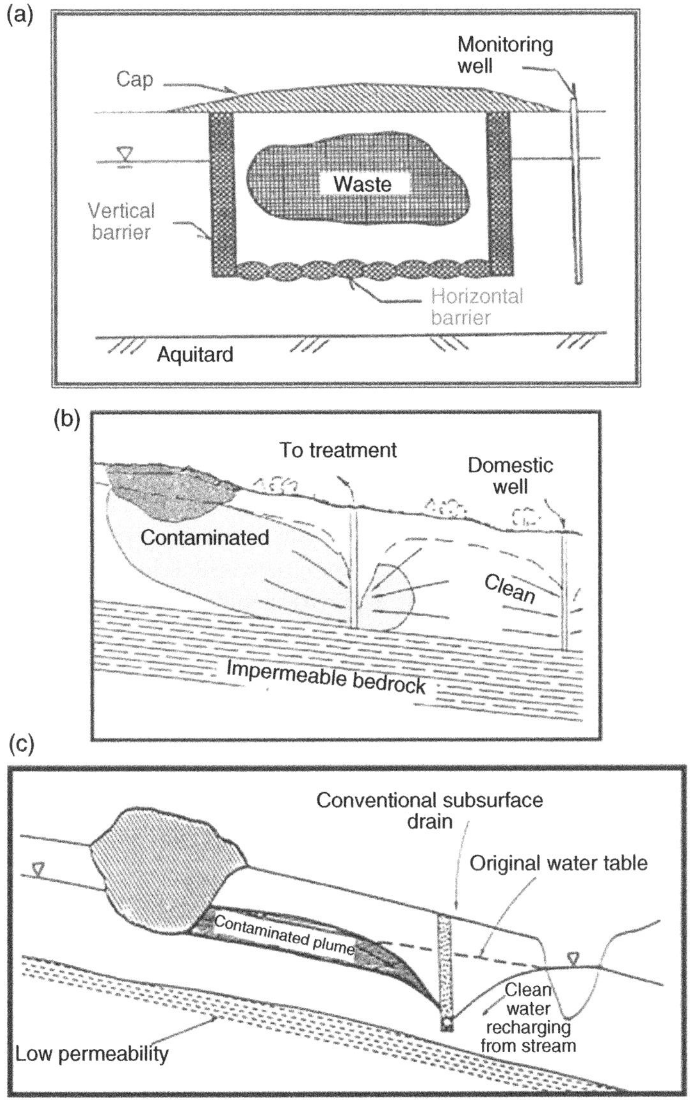

Figure 19.4 Containment technologies: (a) cap, vertical barrier, and bottom barrier; (b) pumping well systems; and (c) subsurface drain system.

It can be challenging to incorporate sustainability parameters into the process of selecting remedial technologies. Some technologies, such as pump-and-treat operations and incineration, are known to be energy-intensive and may not meet GSR criteria. An ideal remediation technology (and all associated on-site or off-site actions) should aim to

- Minimize the risk to public health and the environment in a cost-effective manner within a reasonable time;
- Minimize the potential for secondary waste and prevent uncontrolled contaminant mass transfer from one phase to another;
- Provide an effective, long-term solution;

- Minimize the impacts on land and ecosystem;
- Facilitate appropriate and beneficial land use;
- Minimize or eliminate energy input. If required, renewable energy sources (e.g. solar and wind) should be used;
- Minimize the emissions of air pollutants and GHGs;
- Eliminate freshwater usage while encouraging the use of recycled, reclaimed, and stormwater. Further, the remedial action should minimize the impact on water bodies;
- Minimize material use while facilitating recycling and/or the use of recycled materials.

Technologies that encourage uncontrolled contaminant partitioning between media (i.e. from soil to liquid or from liquid to air) or those that generate significant secondary wastes/effluents are not sustainable. Rather, technologies that destroy contaminants, such as bioremediation or chemical oxidation/reduction, minimize energy input, and minimize air emissions and wastes, are preferred. In-situ systems are often attractive, as they typically minimize GHG emissions and limit disturbance to the ground surface and the overlying soils.

A variety of remedial technologies satisfy core GSR criteria; however, the project life cycle for a specific technology should be considered to determine if it is appropriate for use at a given site. For example, ex-situ biological soil treatment may be considered an appropriate GSR technology; however, the impacts of transporting soil (if off-site treatment is required) should be evaluated. Similarly, enhanced in-situ bioremediation is also considered an attractive GSR technology, but the cumulative impacts that occur during its characteristically long treatment duration should be compared to those of other active remediation technologies that require less time. In general, passive containment systems such as phytoremediation and PRBs utilize little mechanical equipment and minimize energy input while resulting in minimal waste/effluent.

Some popular technologies used to treat residual contaminant concentrations are not considered effective in treating source remediation. Groundwater plumes with moderate to high dissolved contaminant concentrations may require a brief implementation of active remediation technologies to expedite contaminant mass reduction. Alternatively, many technologies appropriate for source removal are often ineffective in treating residual or lower concentrations that result from reduced contaminant diffusion and dissolution. Under such conditions, GHG emissions and energy usage associated with aggressive technologies may outweigh further contaminant mass removal/destruction. Further, a technology with lower energy requirements and emissions may be used to treat residual contamination. Large, dilute groundwater plumes may be treated using lower-energy passive technologies; this may extend the duration of the remediation program, but it will reduce overall net impacts on the environment.

The duration of the remediation program can itself be a major governing factor in remediation system selection. Remediation technologies such as bioremediation may require lower energy input, but they require longer treatment time. Further, given the duration of the remediation, cumulative energy use can often be greater as compared to a shorter but energy-intensive remediation program. Other anticipated or unanticipated side effects, such as incomplete mineralization, can render these as ineffective alternatives. Further, even energy-intensive aggressive technologies, such as thermally enhanced remediation, may become attractive from a sustainability standpoint if renewable energy sources are used.

Opportunities exist for reducing energy and carbon footprints from existing remediation systems. In particular, energy efficiency can be maximized by optimizing existing treatment systems, by critically evaluating the remedial design, and by upgrading the equipment. In addition, alternative sources of energy, including solar, wind, landfill gas, biomass, geothermal, tidal/wave, and cogeneration can be incorporated into existing systems. A growing number of existing projects have started to use solar or wind energy sources.

19.5 Sustainable Remediation Framework

A sustainable remediation framework is a systematic basis or means by which the sustainability of a remediation project may be assessed with respect to environmental, social, and economic factors. This assists in decision-making to increase the sustainability of a remediation project. Although a universally acceptable standardized framework has not yet been developed, several agencies and organizations in the United States and other countries have been active in developing frameworks for measuring and facilitating sustainability in the remediation of contaminated sites. The frameworks are developed by the USEPA, the Sustainable Remediation Forum (SURF), the Interstate Technology and Regulatory Council (ITRC), and the American Society of Testing and Materials (ASTM) and are explained in detail by Reddy and Adams (2015). A brief explanation of these frameworks is provided:

- *USEPA framework* – The framework emphasizes green remediation concepts and techniques that take into consideration a range of environmental effects. In emphasizing green remediation, the goal of the framework is to evaluate and select remediation alternatives and options that achieve maximum net environmental benefit during all phases of site characterization, remediation system implementation and operation, and post-remediation monitoring. The green remediation framework has incorporated five core elements to minimize environmental impacts, as shown in Figure 19.5.
- *Sustainable remediation forum framework* – The SURF framework consists of a tiered decision-making process that considers each phase of a remediation project: site characterization, remediation alternative analysis and selection, remediation system design and construction, operations and maintenance, post-monitoring, and closure. Tier 1 consists of a standardized, nonproject-specific, qualitative evaluation that utilizes checklists, lookup tables, guidelines, results from past project experience, rating systems, and matrices to identify best management practices (BMPs) that maximize positive sustainability impacts. Tier 2 consists of a semiquantitative approach using project-specific and nonproject-specific information as well as greater stakeholder involvement. The project-specific information can be evaluated using various assessment tools such as emission calculations, exposure calculations, scoring and weighing systems, spreadsheet-based tools, and simple cost-benefit analyses. This Tier 2 evaluation is best suited for sites that are moderately complex and require greater involvement of stakeholders. Tier 3 is the most comprehensive, detailed, quantitative evaluation for sustainability based on project-specific information. A Tier 3 evaluation requires a large quantity of project-specific data and utilizes sophisticated tools such as life-cycle assessment (LCA) for the sustainability evaluation of the project.

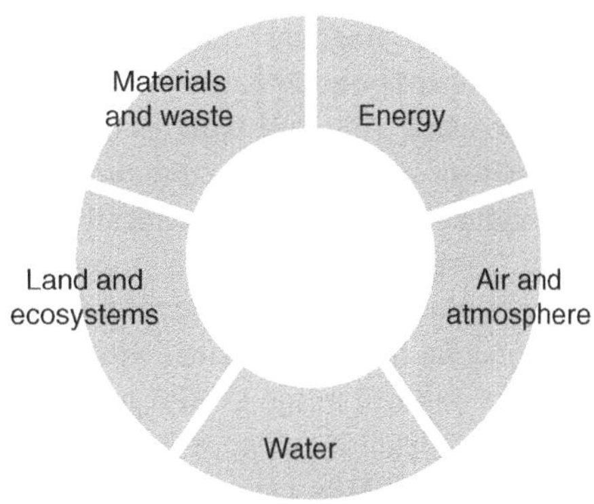

Figure 19.5 Core elements of the USEPA green remediation framework. *Source:* Green Remediation: Incorporating Sustainable Environmental Practices into Remediation of Contaminated Sites/United States Environmental Protection Agency/Public domain.

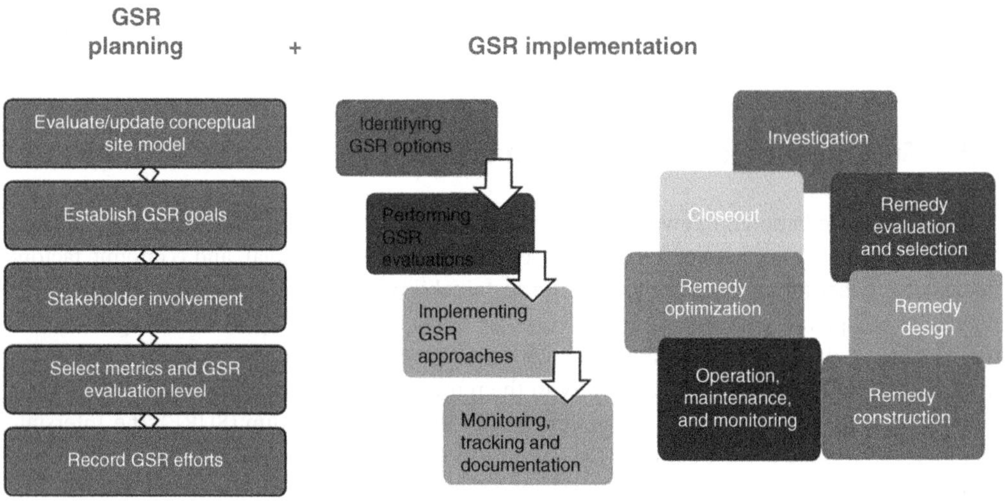

Figure 19.6 ITRC green and sustainable remediation framework. *Source:* ITRC (2011a,b).

- *Interstate technology and regulatory council framework* – Figure 19.6 depicts five generalized steps of this framework that can be performed to user-desired detail during each phase of the project. These steps include the following: (i) evaluation/update of a CSM; (ii) establishment of GSR goals for the project; (iii) project stakeholder involvement; (iv) selection of GSR metrics, evaluation level, and boundaries; and (v) documentation of GSR efforts. A three-level approach is recommended for evaluating and selecting GSR metrics: Level 1 consists of "common-sense"-based BMPs. These are selected to promote resource conservation and process efficiency. The net impact on the environment, community, or economics is not evaluated with this approach. Level 2 consists of the selection and implementation of BMPs with some degree of qualitative and/or semiquantitative evaluation. Qualitative evaluations may reflect trade-offs associated with different remedial strategies or use value judgments for different GSR goals to determine the best way to proceed. Semiquantitative evaluations are those that can be completed by using simple mathematical calculations or intuitive tools (e.g. conversion factors, online calculators, and spreadsheet-based programs). Level 3 consists of the selection and implementation of BMPs using a comprehensive quantitative evaluation. The evaluation often relies on LCA or footprint analysis approaches. This level requires more time, expense, and expertise.
- *American society of testing and materials framework* – ASTM has developed a standard guide for sustainable remediation specifically focused on "greener" cleanups in their ASTM E2893 standard. The standard describes a process for identifying, evaluating, and incorporating BMPs and, as appropriate, integrating a quantitative evaluation that facilitates an overall net reduction in environmental impact associated with remediation projects. This guide addresses the five core elements outlined in the USEPA green remediation framework as described above. The standard provides detailed guidance on planning and scoping a remediation project, implementation of appropriate BMPs, employing a quantitative evaluation when appropriate, and documentation and reporting of sustainability-related performance. ASTM has developed another standard for sustainable remediation projects (Figure 19.7), ASTM E2876, which provides a framework for integrating environmental, economic, and social aspects into remediation projects. BMPs implemented under this guide can incorporate all three aspects of sustainability (environmental, economic, and social) into remediation projects that are designed to address human health, public safety, and ecological risks (Figure 19.8).

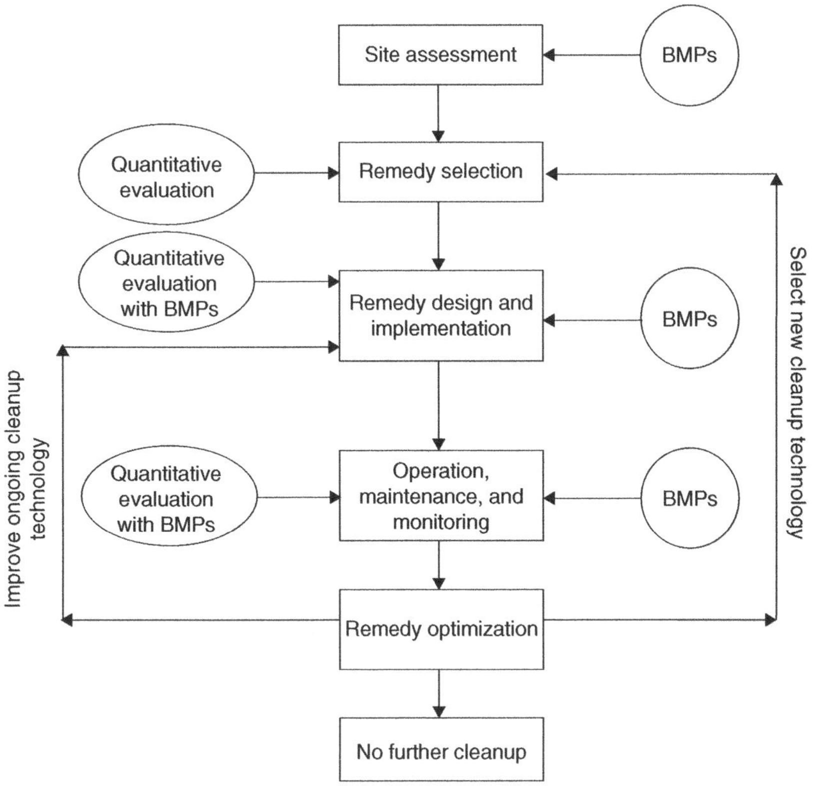

Figure 19.7 ASTM greener cleanup overview. *Source:* ASTM (2014)/ASTM International.

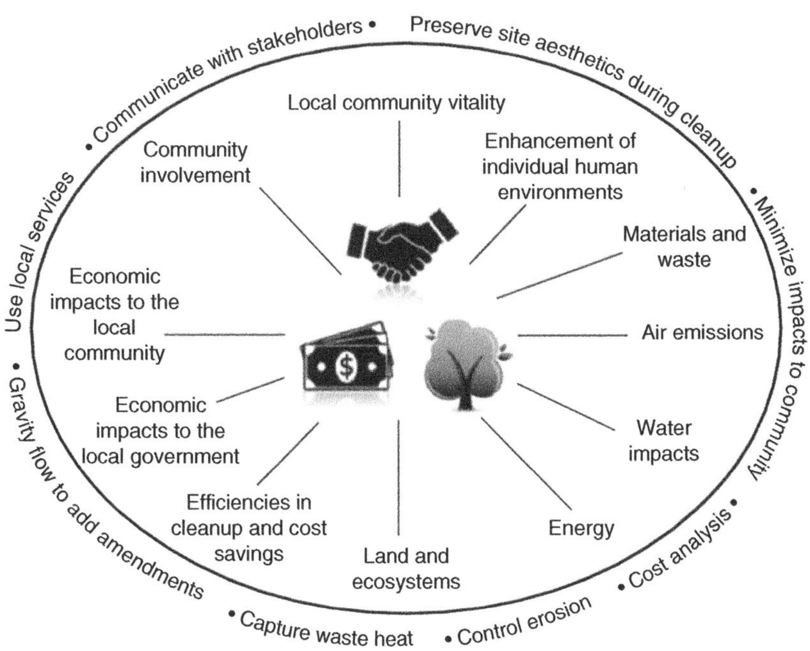

Figure 19.8 ASTM sustainability framework: relationship between the sustainable aspects (center), core elements (spokes), and some example BMPs (outer rim of wheel). *Source:* ASTM (2014)/ASTM International.

19.6 Sustainable Remediation Indicators, Metrics, and Tools

As stated in Chapter 6, sustainability indicators must have attributes of "SMART" and should be conducive to express the environmental impact in quantitative terms. Environmental indicators may include the following:

- GHG and other air emissions,
- Contributions to climate change,
- Use of freshwater resources,
- Impacts to soil,
- Utilization of natural resources,
- Impacts on surface water or groundwater,
- Use of recycled/repurposed materials,
- Overall waste generation,
- Diversion of waste materials from or to landfill facilities.

Economic sustainability indicators that may be considered for the remediation project include the following:

- Direct and indirect job creation within the community,
- Direct and indirect investment within the community,
- Facilitation of government grants for the project and community as a whole,
- Long-term tax and revenue generation within the enhanced community,
- Degree of highest and best use (HBU) achieved by the remediated property,
- Potential to "upzone" the property and nearby properties due to remediation activity.

Some key indicators of social sustainability include:

- Enhancement of community aesthetics,
- Enhancement of quality-of-life features (e.g. improved transportation opportunities or recreational facilities),
- Public participation in decision-making,
- Educational and job training opportunities,
- Interaction between community groups,
- Emotional ownership of the community in a remediation project,
- Improved physical and mental health and well-being of members of the community,
- Enhanced social opportunities for members of the community,
- Strengthening or enhancement of existing community institutions (e.g. recreational organizations, charitable foundations, houses of worship).

Once key indicators and related metrics have been devised for sustainability analyses, they may be formally evaluated using a sustainability assessment tool. These tools can range from simple decision trees or spreadsheets to full LCAs. Several qualitative, semiquantitative tools, and quantitative tools have been developed (Reddy and Adams 2015) and are briefly summarized as follows.

The purpose of qualitative assessment tools is to screen remediation technology and/or identify best management practices (BMP) for alternatives based on anticipated impacts across the environmental, economic, and societal dimensions of sustainability. These commonly consist of guidance documents or advisory manuals that outline an appropriate selection process, including relevant criteria. Two examples of qualitative tools have been developed by public regulatory agencies, including the Illinois Environmental Protection Agency (EPA) Greener Cleanup Matrix and the Minnesota Pollution Control Agency Toolkit for Greener Practices.

While qualitative tools offer a screening tool for BMPs or other remediation-related factors, semiquantitative tools offer more rigor in the analysis of sustainability. Typically, these tools will offer a "scorecard"-like approach in which potential quantitative factors may be ranked and scored, resulting in a weighted average or cumulative score that allows for a direct numerical comparison among several potential remediation alternatives. These semiquantitative tools are typically straightforward and do not incorporate advanced numerical modeling; rather, they may be used for screening or feasibility assessment when considering remediation alternatives for a project as well as alternative applications or design of a particular remediation technology that may have been selected for a project. Examples of semiquantitative assessment tools are the California Green Remediation Evaluation Matrix (GREM) and the Social Sustainability Evaluation Matrix (SSEM) (Reddy and Adams 2015).

For many projects, the use of a qualitative or semiquantitative analysis tool will prove to be useful for analyzing the sustainability aspects of one or more remediation alternatives. This is especially the case when a project is relatively simple or straightforward, or when the tool is applied as a screening tool to assess the feasibility of a remediation project. In many instances, however, the results of a qualitative or semiquantitative analysis may be too limited to be of much use for sustainability analysis. This is especially the case for more complex remediation projects where a wide range of parameters must be carefully and thoroughly assessed.

When warranted by the degree of complexity of a project, quantitative analysis tools should be incorporated for sustainability analysis. These advanced tools for sustainability evaluations typically offer a far more detailed and rigorous assessment of the environmental, social, and economic impacts of remediation. Some common tools developed and used in practice include SiteWise™, USEPA Environmental Footprint Analysis Tool (SEFA), and general ISO LCA. For details, see Reddy and Adams (2015).

No single tool can cover every type of project. Rather, it is important to assess several key aspects of a project, which can then be used to select the most appropriate tool for analysis.

19.7 Case Studies

Several examples have been reported where sustainability evaluation was performed to select the most sustainable remediation based on the site-specific conditions (Reddy and Adams 2015; Sadasivam and Reddy 2017). Three case studies that use different sustainability assessment tools are presented:

- *Case study #1* – An industrialized site contaminated with VOCs, polycyclic aromatic hydrocarbons (PAHs), pesticides, metals, total organic content, and abnormal pH levels. Several potential soil and groundwater contamination remediation technologies have been considered for the site based on applicability, cost range, limitations, and commercial availability. For soils, excavation/disposal, phytoremediation, in-situ chemical oxidation (ISCO), and solidification/stabilization have been identified as potential remediation alternatives. For groundwater, pump-and-treat, in-situ flushing, PRB, and monitored natural attenuation (MNA) have been identified as potential remediation alternatives. A comparative assessment of potential remedial technologies was performed based on the BMPs as well as qualitative and quantitative assessments. The general BMPs for the selected technologies have been assessed based on the BMPs listed in the Greener Cleanup Matrix developed by the Illinois EPA and the Toolkit for Greener Practices developed by the Minnesota Pollution Control Agency (ITRC 2011a,b). In addition to BMPs, the GREM tool was used to perform a qualitative comparison of remediation technologies for sustainability and to evaluate adverse environmental impacts. A quantitative assessment was also performed based on the sustainability metrics. The sustainability metrics for the selected potential technologies were calculated using two tools: The sustainable remediation tool (SRT) and SiteWise. A combination of remediation methods was identified as the best alternative for the site. Solidification/stabilization (to be applied in areas of high contaminant concentrations) and phytoremediation (to be applied in other contaminated areas) have been recommended for the remediation of soils with PAHs and heavy metals, while MNA and phytoremediation have been recommended for the treatment of impacted groundwater.

- *Case study #2* – A degraded wetland site known as Indian Ridge Marsh (IRM) was found to be contaminated by heavy metals, pesticides, VOCs, PAHs, pesticides, and one observed instance of an Light Non-Aqueous Phase Liquid (LNAPL) plume containing petroleum hydrocarbons. Qualitative and quantitative analyses were conducted to evaluate potential environmental impacts associated with each remedial option using tools such as the GREM, SiteWise, and the SRT. Following a qualitative evaluation of sustainability metrics using GREM (i.e. noise, worker safety, aesthetics), a quantitative evaluation of energy/resource consumption was conducted using both SRT and SiteWise considering several project phases, including the remedial investigation, remedial action construction, operations and maintenance, and long-term monitoring. Additionally, the SSEM tool was applied to the IRM project to evaluate the social impacts of both remedial alternatives. The recommended strategy for remediation of IRM consists of the phyto-EB option. This alternative will act to stimulate existing soil microorganisms to enhance the degradation of organic contaminants at all identified areas of concern (AOCs). Native tree species with high growth and transpiration rates, deep rooting depths, and the ability to accumulate and/or sequester contaminants of concern will be employed.
- *Case study #3* – At a former zinc smelting site, soils were found to be contaminated with lead, arsenic, and copper. The SimaPro software was used to evaluate the life-cycle impact of two common methods of treatment: landfilling (excavation and hauling) and in-situ treatment by solidification and stabilization. The results of this analysis showed that excavation and hauling result in greater impacts in every category except human health – cancer, as compared to solidification/stabilization. Ultimately, this analysis indicated that given the large costs and disturbances associated with excavation and hauling, the solidification/stabilization option is the more attractive and sustainable remediation option.

19.8 Land Remediation Challenges and Opportunities

While there is growing interest in incorporating sustainability into environmental remediation, there are several challenges and opportunities that have to be addressed to promote sustainable remediation in practice. Some of these challenges are as follows:

- *"Greenwashing"* – Greenwashing refers to situations where sustainable remediation options have not been evaluated, and backup documentation is lacking but claims exist that say sustainable remediation approaches have been implemented. Greenwashing in a larger sense is commonly associated with a wide range of approaches, from consumer product marketing to legislative initiatives in which dubious "green" or "sustainable" claims are made. These claims may be misleading or outright false.
- *Lack of financial incentives for sustainable remediation* – A perception exists among many taxpaying citizens that only publicly funded projects are capable of absorbing the additional cost burden

associated with incorporating sustainability measures, which can hinder motivation to incorporate sustainable engineering approaches.

- *Lack of a regulatory mandate* – While many regulatory oversight agencies encourage sustainability-focused activities and efforts, no clear mechanism requiring the incorporation of such measures exists as applied to many remediation projects. Incorporation of BMPs and other activities that enhance the dimensions of sustainability in many cases is optional. With many agencies, the use of such measures may be encouraged, but in many cases, these activities are not required.
- *Lack of public awareness* – Despite the virtues of government-based incentives or mandates (the proverbial "carrot" or "stick") that would encourage the application of sustainability-focused activities or practices for site remediation, the government will not act in such a manner if it is not the will of the public.
- *Lack of specialty training on LCA, carbon balance, and other assessment tools for professionals* – Many remediation professionals do have the desire to incorporate sustainable measures into remediation projects; however, in many cases, they lack the skill set or knowledge of assessment tools to demonstrate the related benefits.
- *Greater academic focus* – This academic/practitioner model has worked for countless technological advances and will continue to advance the applications of environmental remediation. With a greater emphasis of sustainability as a common interest between academia and practitioners, sustainability-related applications would also be expected to evolve at an accelerated pace.
- *Further refinement and development of assessment tools and frameworks* – A move toward standardization, at least among similar assessment tools and frameworks, would be beneficial to the environmental remediation practice. Even if differences among computational processes within different tools remained, increased standardization in terms of reported output units, indicators, and metrics considered and greater agreement on the range of remediation activities that could be handled by different remediation tools would eliminate confusion.
- *Improved assessment of indirect consequences* – The difficultly in accounting for the indirect benefits exists for two primary reasons. First, system boundary selection will invariably affect the number of indirect consequences that are accounted for in an analysis. The system boundary expansion leads to a significant increase in the complexity and difficulty of an analysis in terms of both time and cost associated with the analysis. The second reason is that indirect consequences are not often properly accounted for by existing assessment tools, regardless of the choice of system boundary. Benefits such as reduced emissions, resource utilization, or construction jobs created by a remediation project would be considered typical and easily handled by a comprehensive analysis tool. However, benefits such as increased neighborhood tax receipts, increased life expectancy for residents near a project site, or increased number of species present in a rehabilitated natural habitat can be difficult to quantify with existing tools. As is the case with enhanced assessment tools, practitioners and academia could successfully collaborate to identify key indicators and metrics of indirect benefits resulting from sustainability-focused remediation projects as well as ways to incorporate and quantify into existing and future assessment tools.
- *Improved metrics/tools to address social issues* – Assessment frameworks and tools evolved, and the focus was primarily placed on environmental and economic dimensions because metrics associated with these dimensions were relatively easy to quantify and analyze. Further, whether associated with costs or physical units, economic and environmental metrics are relatively easy to objectively compare among remediation project alternatives. While there has been a general interest in measuring social-related sustainability impacts, tools and frameworks other than those cited in this book have been lagging the development of other more economically and environmentally focused tools.

19.9 Resiliency Considerations

As discussed in Chapter 5, the intensifying impacts of climate change are expected to increase the frequency and severity of extreme weather events. Hence, it is essential that sustainable land management practices incorporate resilience considerations. This is necessary not only to avoid further environmental degradation such as that caused by deforestation or fossil fuel extraction- but also to ensure long-term stability and effectiveness of agricultural practices and remedial activities. For instance, sustainable agroforestry practices offer a wide range of co-benefits compared to conventional agriculture: they sequester carbon, reduce soil erosion, enhance land's resilience to floods, landslides, and droughts, enhance soil organic matter, reduce the financial risks associated with crop failure, and promote greater biodiversity (Smith et al. 2017).

Sustainable and Resilient Remediation (SRR) activities should incorporate site-specific vulnerability assessments, climate-adaptive design practices, and continuous monitoring and restoration activities for improved climate resiliency (Kumar and Reddy 2020). Furthermore, considering the significant uncertainties with the extent and nature of data associated with climate change impacts, decision-making should be supported by structured approaches such as the evidence support logic methodology (Kumar and Reddy 2020). Organizations such as ITRC and SURF have expanded their guidance frameworks to explicitly address climate resilience in the design and implementation of site remediation (ITRC 2021). The SRR guidance developed by ITRC builds upon the earlier GSR framework by integrating a resilience component to address growing threats from extreme weather events and wildfires (ITRC 2021). Additionally, generalized frameworks such as the TQUALICSR framework, introduced in Chapter 13, offer a robust approach for the integrated quantification of resilience and sustainability aspects in remediation projects, facilitating their incorporation into design and decision-making processes. This integration is demonstrated through a case study presented in Chapter 21, which focuses on a contaminated site located near Lincoln, Nebraska, involving both soil and groundwater pollution. The case study evaluates the sustainability and resilience of various remediation technologies including emulsified zerovalent iron (EZVI) treatment, excavation and disposal, and electrokinetic remediation for soil remediation, and EZVI and pump-and-treat methods for groundwater remediation. The analysis results showed that EZVI treatment is the most sustainable and resilient option for both soil and groundwater remediation among the selected alternatives, highlighting the usefulness of such analysis in supporting informed decisions for SRR.

19.10 Summary

Land degradation, driven by both natural and human-induced factors, poses a critical threat to global ecosystems, livelihoods, and economies. Key drivers include overgrazing, deforestation, unsustainable agricultural practices, overexploitation of fossil fuels, and pollution, which collectively reduce soil fertility, disrupt ecosystems, and threaten biodiversity. Soil and groundwater pollution, caused by hazardous chemicals, pesticides, and industrial waste, exacerbates the issue, diminishing agricultural productivity and contaminating water resources essential for human and ecological health. Natural events such as droughts and floods, often intensified by human activities, further contribute to the problem. With over 40% of the Earth's land area degraded, affecting half of the global population and threatening $44 trillion in GDP; therefore, urgent action is needed in this regard. The United Nations Sustainable Development Goal 15.3 emphasizes combating desertification, restoring degraded land, and achieving land degradation neutrality by 2030, highlighting the importance of sustainable management to mitigate these impacts.

Environmental pollution, including soil and groundwater contamination, is one of the major problems faced by developed countries across the globe. Thousands of contaminated sites necessitating urgent remedial action have been identified. Environmental policies have evolved with respect to the characterization and remediation of contaminated sites. Systematic frameworks and approaches to contaminated site characterization and remediation have also evolved, such as the well-established procedures developed by the USEPA to appropriately characterize, assess, and remediate contaminated sites.

Several in-situ and ex-situ remediation technologies have been developed, and new and innovative technologies continue to be developed to address various contaminants and contaminated media. However, the net overall (environmental, economic, and social) benefit associated with using one remediation technology over other potential remedial alternatives for remediation of a contaminated site requires ongoing study. If not carefully assessed, the choice of a remediation technology and an overall remediation strategy itself may cause more harm than good when considered on a holistic or a triple-bottom-line basis.

GSR is emerging as a new paradigm for remediation of contaminated sites. Several national and international organizations have developed qualitative, semiquantitative, and quantitative sustainability assessment tools/frameworks that aid in adopting sustainable practices and alternatives in remediation strategy while ensuring minimizing risk to human health and the environment. Despite these efforts, there is still a lack of appropriate indicators, metrics, and assessment tools that can comprehensively analyze the environmental, economic, and social impacts of the remediation activities/strategies. Several challenges exist, which pose opportunities for academia and practicing professionals to develop innovative and reliable tools to incorporate sustainability principles into contaminated site remediation. Furthermore, incorporation of resiliency considerations for sustainable and resilient land management practices is paramount - as climate change exacerbates the impacts of extreme weather events such as floods and wildfires.

19.11 Questions

19.1 What is green remediation?

19.2 What is the definition of GSR as per ITRC?

19.3 What is the difference between green remediation and GSR?

19.4 Explain the systematic approach developed by USEPA for the remediation of contaminated sites.

19.5 What are the steps of the approach necessary for the characterization and remediation of contaminated sites?

19.6 What are the major tasks involved in the risk assessment procedure of contaminated sites developed by USEPA.

19.7 What are the two major categories of remediation technologies? Mention any two major differences between the two categories.

19.8 Discuss the advantages and disadvantages of both in-situ and ex-situ remediation.

19.9 List any five remediation technologies used for the remediation of contaminated soil in the vadose zone. Explain the technologies briefly.

19.10 List any five remediation technologies used for the remediation of contaminated groundwater. Explain the technologies briefly.

19.11 Describe the trade-offs of cost, timing, and environmental side effects (both positive and negative) associated with three remediation technologies. Also, describe a situation(s) where some of these factors may be significant and another situation where they may be negligible.

19.12 What are the attributes of an ideal remediation technology?

19.13 Explain the sustainability assessment framework developed by SURF.

19.14 Explain the sustainability assessment framework developed by ITRC.

19.15 Mention any five environmental, economic, and social sustainability indicators for sustainable remediation.

19.16 What are the challenges in implementing sustainable remediation of contaminated sites?

19.12 Problems

19.1 The environmental and economic impact indicator values for three different remediation technologies for remediating a contaminated site: in-situ air-sparging accompanied with bioremediation (IASB), ISCO, and pump-and-treat system (PT) are provided in the table below.

Indicator	Unit	IASB	ISCO	PT
GHG emissions	Metric ton CO_2e	36 396	246	402 689
Total energy used	MJ	2×10^8	1.4×10^6	4.74×10^9
Water used	m^3	5356	17.3	750 360
Electricity used	MWh	2763	9	388 674
Total NO_x	Metric ton	4.8	0.5	424
Total SO_x	Metric ton	11	0.4	1375
Total PM10	Metric ton	6.1	0.2	826
Accident risk – fatality	–	0.0004	0.00013	0.00051
Accident risk – injury	–	0.061	0.019	0.043
Nonhazardous waste landfill space	Tons	91.6	10	20
Lost hours from injury	Hours	0.5	0.2	0.3
Direct cost	USD	10 320 000	8 084 000	14 839 000
Social cost of CO_2	USD	1 310 270	8837	14 496 784
Social cost of NO_x	USD	1014	109	89 079

- Categorize the indicators into appropriate criteria similar to that of Problem 13.4
- Similar to Problem 13.4, calculate the integrated sustainability indexes for the three alternatives.
- Discuss your understanding of the sustainability of these technologies based on the sustainability indicator values.

19.2 A contaminated site requires remediation, and three methods are considered: electrokinetic remediation (EKR), excavation/disposal, and phytoremediation. Using the QUALICS (Quantitative Assessment of Life Cycle Sustainability) framework, calculate the overall sustainability index for each remediation option. The environmental, economic, and social weights are as follows: environmental (50%), economic (30%), and social (20%).

Criteria	EKR	Excavation/disposal	Phytoremediation
Environmental impact score	0.6	0.7	0.8
Economic impact score	0.5	0.6	0.9
Social impact score	0.8	0.5	0.7

- What is the sustainability index for each remediation alternative?
- If the weights are adjusted to prioritize economic requirements over the other two, what is the resulting impact on overall sustainability?

19.3 In implementing electrokinetic remediation (EKR) at a clay-rich contaminated site, different configurations are considered to reduce sustainability impacts. Each configuration varies by electrode material, energy source, and electrolyte type, influencing the sustainability outcomes as shown below.

Configuration	Electrode material	Energy source	Electrolyte	Environmental score	Cost (USD)	Social score
Config A	Steel	Diesel	10% H_2O_2	0.65	1 200 000	0.7
Config B	Titanium	Solar	30% $Na_2S_2O_8$	0.75	950 000	0.8
Config C	Graphite	Diesel	10% H_2O_2	0.6	1 400 000	0.6

Calculate the weighted sustainability score for each configuration based on the following criteria: environmental (60%), economic (20%), and social (20%). Note: You need to normalize the costs to a similar scale as environmental and social scores before calculating the sustainability score.

References

ASTM (2014). *ASTM E2893-16e1, Standard Guide for Greener Cleanups*. West Conshohocken, PA: ASTM International.

Baker, C.B., Smith, L.M., and Woodward, D.S. (2009). *The Sustainable Remediation Forum*. Wiley Periodicals, Inc.

EWG (2024). PFAS contamination in the U.S. Environmental Working Group. https://www.ewg.org/interactive-maps/pfas_contamination/ (accessed November 2024).

Hilker, T., Natsagdorj, E., Waring, R.H. et al. (2014). Satellite observed widespread decline in Mongolian grasslands largely due to overgrazing. *Global Change Biology* 20 (2): 418–428.

Hopwood, J., Frischie, S., May, E., and Lee-Mäder, E. (2021). *Farming with Soil Life: A Handbook for Supporting Soil Invertebrates and Soil Health on Farms*, 128 pp. Portland, OR: The Xerces Society for Invertebrate Conservation.

ITRC (Interstate Technology & Regulatory Council) (2011a). *Green and Sustainable Remediation: State of the Science and Practice. GSR-1.* Washington, DC: Interstate Technology & Regulatory Council, Green and Sustainable Remediation Team. www.itrcweb.org.

ITRC (Interstate Technology & Regulatory Council) (2011b). *Green and Sustainable Remediation: A Practical Framework, GSR-2.* Washington, DC: ITRC (Interstate Technology & Regulatory Council).

ITRC (Interstate Technology & Regulatory Council). 2021. *Sustainable Resilient Remediation SRR-1.* Washington, D.C.: Interstate Technology & Regulatory Council, SRR Team. https://srr-1.itrcweb.org/. (accessed 31 May 2025)

Kumar, G., and Reddy, K. R. (2020). Addressing climate change impacts and resiliency in contaminated site remediation. *Journal of Hazardous, Toxic, and Radioactive Waste*, 24(4), 04020026.

Reddy, K.R. and Adams, J.A. (2015). *Sustainable Remediation of Contaminated Sites.* New York: Momentum Press.

Sadasivam, B.Y. and Reddy, K.R. (2017). Approaches to selecting sustainable technologies for remediation of contaminated sites: case studies. In: *Sustainability Issues in Civil Engineering* (ed. G.L.S. Babu, S. Saride, and B.M. Basha), 271–306. Singapore: Springer.

Sharma, H.D. and Reddy, K.R. (2004). *Geoenvironmental Engineering: Site Remediation, Waste Containment, and Emerging Waste Management Technologies.* Hoboken, NJ: Wiley.

Smith, P., et al. (2017). *Special report on climate change, desertification, land degradation, sustainable land management, food security, and greenhouse gas fluxes in terrestrial ecosystems (SR2).* IPCC. https://www.ipcc.ch/site/assets/uploads/2018/07/sr2_background_report_final.pdf. (accessed 31 May 2025)

United Nations Convention to Combat Desertification (UNCCD) (1994). United Nations Convention to Combat Desertification in Countries Experiencing Serious Drought and/or Desertification, particularly in Africa. Retrieved from https://www.unccd.int/resource/convention-text (accessed 14 March 2025).

United Nations Convention to Combat Desertification (UNCCD) (2017). Land Degradation Neutrality TRANSFORMATIVE ACTION, tapping opportunities. Retrieved from https://www.unccd.int/sites/default/files/documents/2017-10/171006_LDN_TP_web.pdf (accessed 14 March 2025).

United Nations Convention to Combat Desertification (UNCCD) (2022). Global Land Outlook: 2nd Edition. Retrieved from https://www.unccd.int/resources/global-land-outlook/global-land-outlook-2nd-edition (accessed 14 March 2025).

USEPA (1989). *Risk Assessment Guidance for Superfund. Volume 1. Human Health Evaluation Manual (Part A).* U.S. Environmental Protection Agency. Report number EPA/540/1-89/002).

USEPA (2008). *Green Remediation: Incorporating Sustainable Environmental Practices into Remediation of Contaminated Sites, EPA 542-R-08-002.* Washington, DC: Office of Solid Waste and Emergency Response.

USEPA (2024). Cleaning Up Superfund Sites: Highlights of Bipartisan Infrastructure Law Funding. US Environmental Protection Agency. https://www.epa.gov/infrastructure/cleaning-superfund-sites-highlights-bipartisan-infrastructure-law-funding (accessed November 2024).

20

Climate Geoengineering

Chapter Goal and Objectives
The goal of this chapter is to explore the potential of climate geoengineering methods, including carbon dioxide removal (CDR) and solar radiation management (SRM), as strategies to mitigate the impacts of greenhouse gases and climate change, while addressing their applicability, theoretical foundations, and associated risks and challenges. The major objectives of the chapter are to (i) discuss the significance of greenhouse gases in driving global warming and climate change impacts; (ii) introduce and explain various CDR methods, such as subsurface, surface, marine organism sequestration, and direct engineered capture; (iii) present SRM techniques, including sulfur injection, reflectors, and mirrors; (iv) analyze the applicability and limitations of CDR and SRM methods in addressing climate change; (v) provide a theoretical framework for understanding the principles of climate geoengineering; and (vi) highlight the risks and challenges associated with the deployment of climate geoengineering techniques.

20.1 Introduction

As discussed in previous chapters, increased levels of atmospheric carbon dioxide and other GHGs from anthropogenic sources have contributed to an increase in global temperatures. This effect, in turn, is contributing to changes in global climate and to adverse side effects that result from climate change. As depicted in Figure 20.1, reflected solar radiation and radiation emitted from the Earth's surface can be trapped by these increased GHGs, thus maintaining a steady temperature on the Earth's surface. Figure 20.2 depicts a model of sources and sinks associated with the global carbon cycle. This figure demonstrates the flux that occurred during the preindustrial age as well as the increases that have occurred as a result of human industrial activities. Figures 20.3–20.7 further highlight the increases in carbon dioxide emissions and how the accumulation of carbon dioxide and other GHGs in the atmosphere may affect the measured global surface temperature change, which is captured by a range of climate models (IPCC 2021, 2023). Figure 20.3 shows the shared socioeconomic pathway (SSP) scenarios and their projected global temperature outcomes by 2100 relative to pre-industrial levels. The SSP scenarios are grouped into five families: SSP1: sustainability, SSP2: middle of the road, SSP3: regional rivalry, SSP4: inequality, and SSP5: fossil-fuel development. These scenarios emphasize the variability in future climate outcomes depending on

Sustainable and Resilient Engineering: Drivers, Metrics, Tools, and Applications, Second Edition. Krishna R. Reddy, Claudio Cameselle, and Jeffrey A. Adams.

Companion website: www.wiley.com/go/reddy/sustainableresilientengineering2e

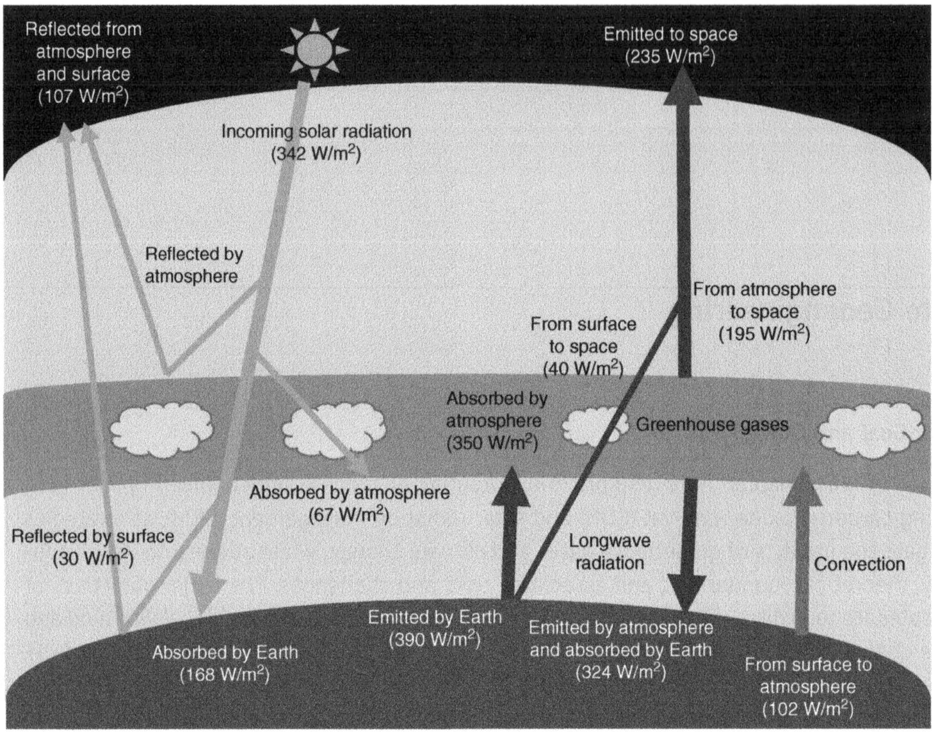

Figure 20.1 Global average energy budget in the atmosphere. *Source:* Adapted from Kiehl and Trenberth (1997). © American Meteorological Society.

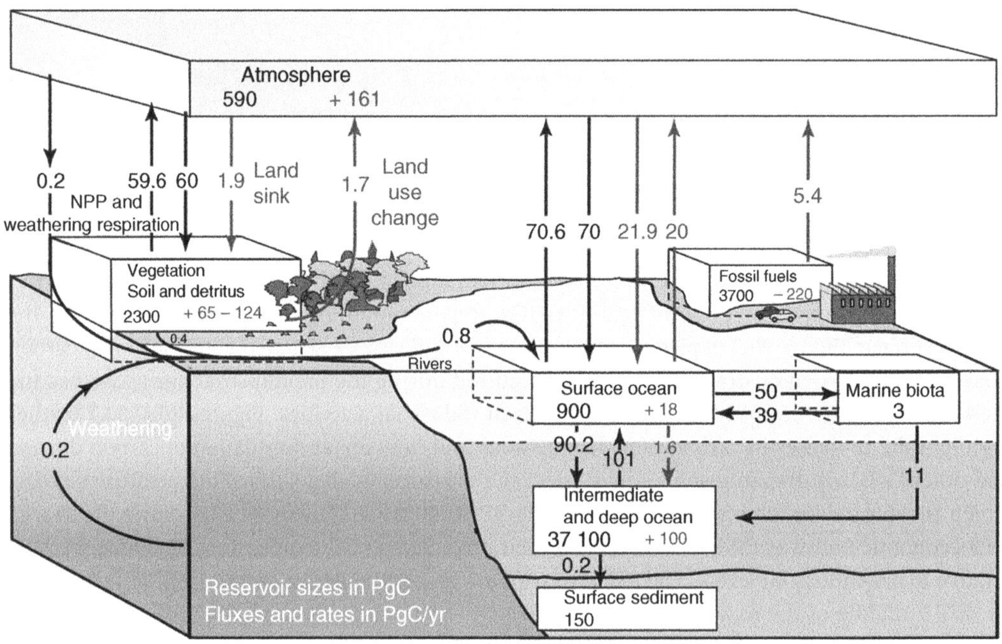

Figure 20.2 Global carbon cycle. *Source:* Sarmiento and Gruber (2002)/with permission of American Institute of Physics.

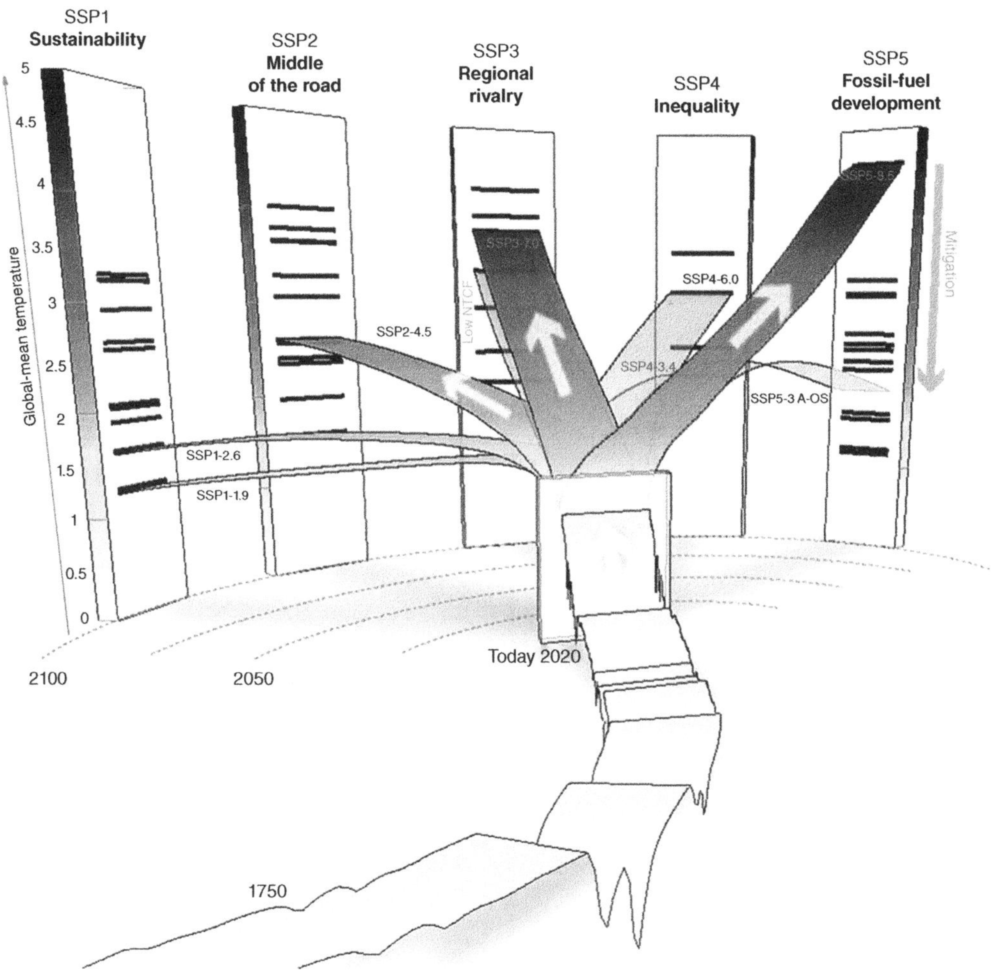

Figure 20.3 The SSP scenarios and their five socio-economic SSP families. *Source:* Meinshausen (2020)/ European Geosciences Union (EGU)/CC BY 4.0.

socioeconomic choices and policy actions. Figure 20.4 shows the prediction of GHG (CO_2, CH_4, and N_2O) emissions and concentrations in the atmosphere that constitute the driving factors in the IPCC scenario models. The figure compares the GHG concentration trajectories of SSP scenarios to those of representative concentration pathways (RCPs). RCPs are scenarios developed in climate modeling and assessment by the IPCC as part of the Fifth Assessment Report (AR5). In the latest IPCC report (IPCC 2023), RCPs have been complemented or replaced by SSPs, which combine socioeconomic assumptions with emissions and concentration trajectories to provide more detailed and integrated scenarios. Figure 20.5 depicts the variation of CO_2 in the atmosphere and the average surface temperature over millions of years and the projections for the future under different scenarios. The GHG emissions from human activity show an average increase of 0.5 °C during the twentieth century, but the possible scenarios predict a range of 1.5–4.5 °C increase in global mean temperatures by 2100, much above 1990 levels (Figure 20.6). Figure 20.7 shows projected global temperature changes for mid-term (2041–2060) and long term (2081–2100) under various SSP scenarios that confirm that most of the global land and ocean areas will be warmer by the middle and

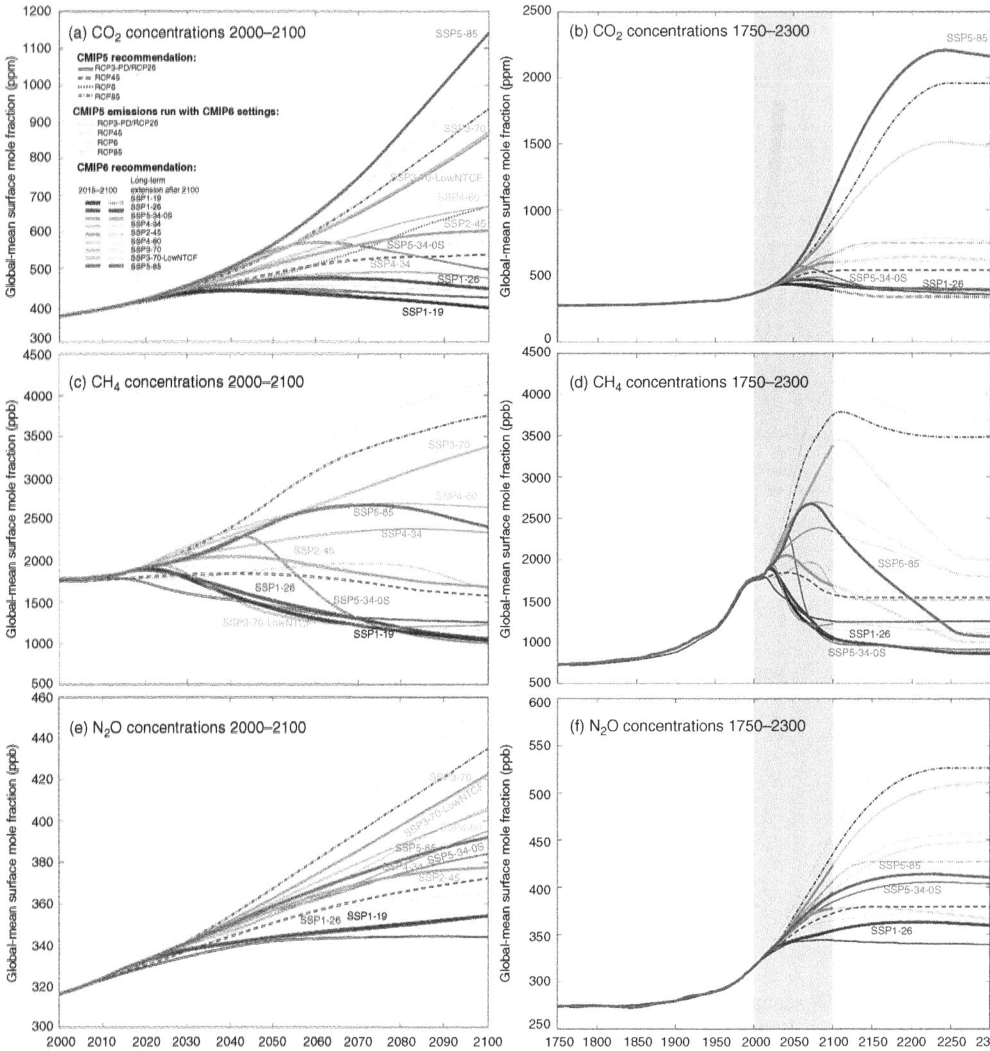

Figure 20.4 Greenhouse gas concentrations under different scenarios: (a) CO_2 concentrations 2000–2100, (b) CO_2 concentrations 1750–2300, (c) CH_4 concentrations 2000–2100, (d) CH_4 concentrations 1750–2300 (e) N_2O concentrations 2000–2100, (f) N_2O concentrations 1750–2300. *Source:* Meinshausen et al. (2020)/ European Geosciences Union (EGU)/CC BY 4.0.

the end of the twenty-first century than in 1995–2014. These results emphasize the importance of mitigation for reducing future climate impacts.

As climate change evolves, several public and private entities ranging from small local groups to large international organizations have proposed and sought means to mitigate climate change and its effects. Many of these entities have pointed to a general solution – a drastic reduction in GHG emissions to approximate levels that were present before the Industrial Revolution. A range of sustainable engineering approaches have been implemented, including renewable and zero-emission energy sources, the use of green and sustainable buildings and infrastructure, and other strategies. However, all the efforts in reducing GHG emissions have not been enough to reduce the emissions

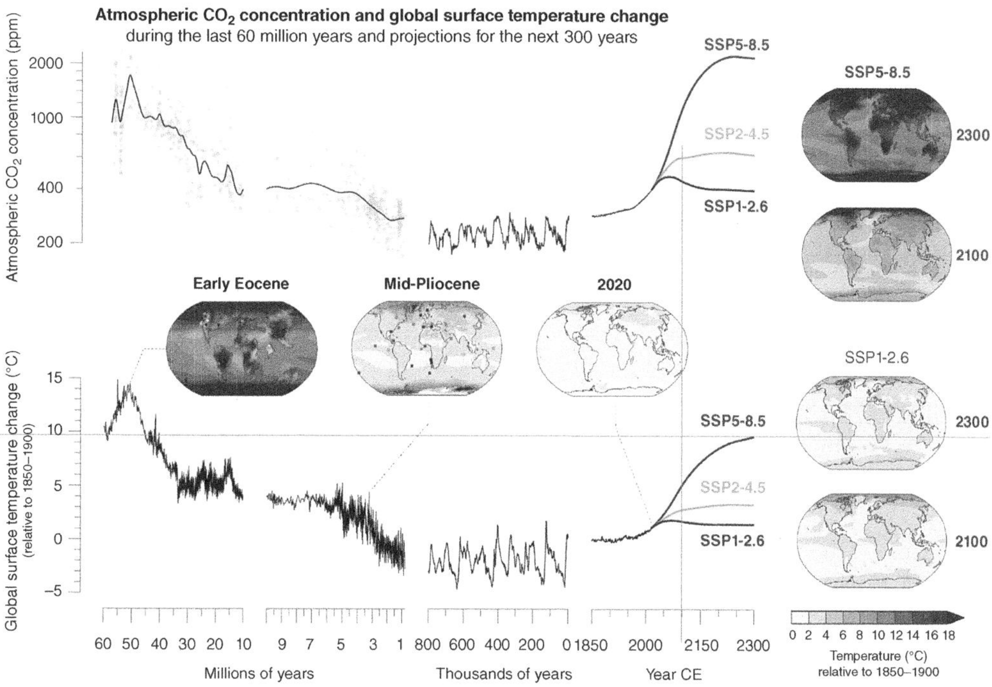

Figure 20.5 Changes in atmospheric CO_2 and global surface temperature (relative to 1850–1900) from the deep past to the next 300 years. *Source:* IPCC (2021)/Intergovernmental Panel on Climate Change/CC BY 4.0.

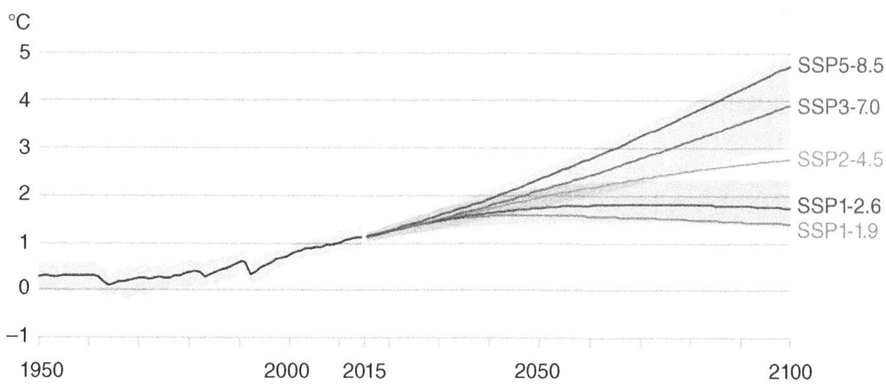

Figure 20.6 Projections in global surface temperature changes relative to 1850–1900 under different scenarios. *Source:* IPCC (2021)/Intergovernmental Panel on Climate Change/CC BY 4.0.

below a confidence level to avoid dangerous impacts on Earth's temperature and climate. As more stringent emission targets are proposed and adopted, it is hoped that greater actions for emission reduction will occur, including the identification and implementation of emission-curbing strategies. However, there is still a risk that those actions will not be implemented in time to avoid irreversible changes and effects on climate (IPCC 2023).

As presented in Figure 20.6, it is likely that surface temperatures will increase by 2 °C or greater during this century unless the global GHG emissions are reduced by at least 50% of 1990 levels by

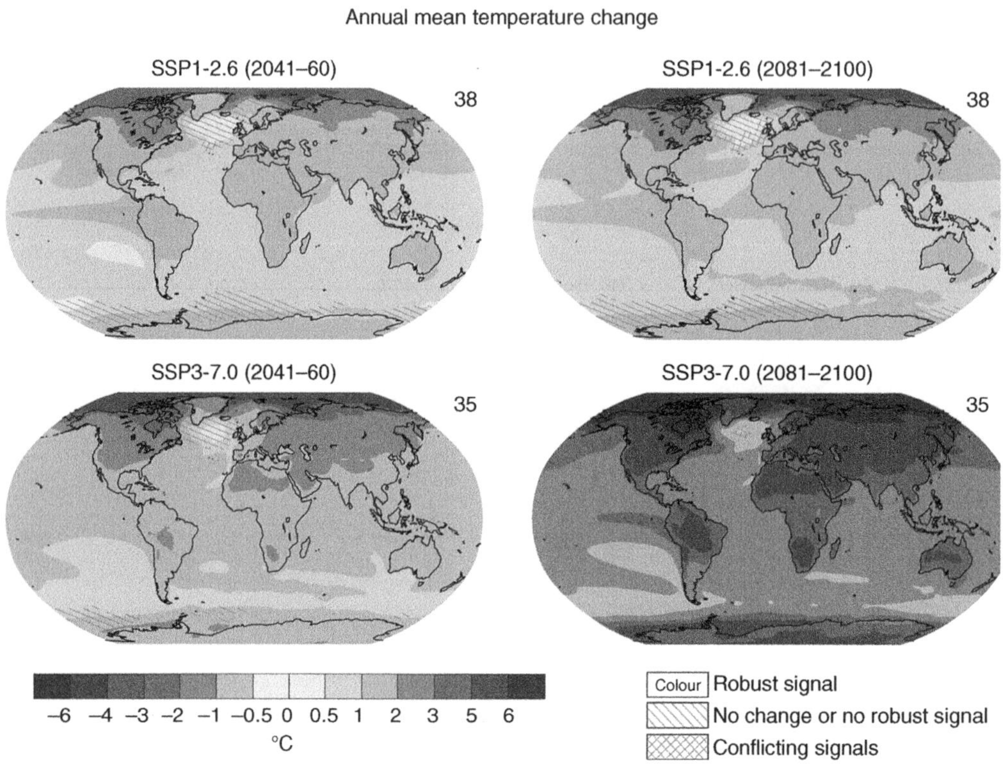

Figure 20.7 Mid- and long-term change on annual mean surface temperature. *Source:* IPCC (2021)/ Intergovernmental Panel on Climate Change/CC BY 4.0.

2050 and even more thereafter (IPCC 2023). Further, there are no credible predictive models in which the global mean temperatures show a peak by the year 2100 and then start to decline. Unless future reductions of GHG are much more successful than they have been so far, additional action will be necessary to cool the Earth this century. This chapter discusses such actions (also known as climate geoengineering) briefly.

20.2 Climate Geoengineering

Climate geoengineering is an evolving field that seeks large-scale intervention strategies to mitigate climate change. Such strategies are deliberate interventions that need to be of a large enough scale to have a measurable effect on the Earth's climate (Shepherd 2012). These strategies, while evolving and highly innovative, generally fall into one of the two categories (Zhang et al. 2015) below:

- Carbon dioxide removal (CDR) methods address the original cause of climate change. CDR methods are designed to remove GHGs from the atmosphere, focusing on the sequestration of carbon dioxide, which allows more outgoing longwave radiation to escape back into space. These methods may remove GHGs directly or seek to influence natural processes to remove gases indirectly. These methods are comprehensive but may take years to fully work (Leung et al. 2014).

- Solar radiation management (SRM) methods attempt to offset the effects of increased GHG concentrations in the atmosphere reducing the absorption of solar radiation by the Earth. Various methods have been designed to deflect the incoming solar radiation back toward space (Ming et al. 2014).

20.3 Carbon Dioxide Removal (CDR) Methods

Four CDR methods include subsurface sequestration, surface sequestration, marine organism sequestration, and direct engineered capture.

20.3.1 Subsurface Sequestration

Subsurface sequestration involves the capture, liquefaction, and injection of carbon dioxide into underground geologic formations (Holloway 2005). Figure 20.8a shows underground geologic carbon sequestration. Geologic carbon sequestration primarily involves the separation and capture of CO_2 at the point sources of emissions, followed by its storage in deep underground geologic formations. This is sometimes also referred to as carbon capture and storage (CCS). The physical means of geologic sequestration involves capturing CO_2 and trapping it in some natural or man-made cavities (e.g. caverns, mines) or in underground rock formations (e.g. depleted oil and gas reservoirs, and aquifers). The injection of CO_2 and oil extraction efforts may be combined for simultaneous enhanced oil recovery and CO_2 sequestration (Pham and Halland 2017). The United States is the leading nation in enhanced oil recovery technology. The United States as of August 2024 has injected about 63.16 million metric tons of CO_2 in the United States as a part of DOE's Clean Coal Research, Development, and Demonstration programs (USEPA 2024). This figure reflects the increasing implementation of carbon capture and storage (CCS) technology over the years, with annual sequestration rates peaking at 16.3 million metric tons in 2023 alone. Figure 20.8b shows the depiction of sequestration implementation on a platform similar to an oil drilling platform in the North Sea. This system has been tested to evaluate the long-term stability of CO_2 storage (Heinemann et al. 2013). The chemical means of carbon sequestration involves transforming the CO_2 into another stable substance via chemical reactions (e.g. dissolving CO_2 in underground water, forced/accelerated mineral carbonation forming stable precipitates, and adsorption).

The cost of subsurface sequestration can vary widely depending on the specific geological conditions and technologies used. On average, the cost ranges from \$40 to \$90 per metric ton of CO_2. This includes capturing, compressing, transporting, and CO_2 injection into suitable underground formations. For full-scale deployment, the subsurface sequestration methods have been estimated to cost around \$500 billion per year between 2024 and 2050, to meet the necessary capture and storage of 7–9 billion metric tons of CO_2 per year by 2050 to induce meaningful reductions in CO_2 in the atmosphere to achieve the global warming targets set by the Paris Agreement (Kali et al. 2022; Smith et al. 2024). Successful implementation is dependent on cooperation between corporate and governmental entities. From a technical standpoint, there is uncertainty regarding the stability of sequestered carbon, and suitable repositories must be present near the carbon source that needs to be sequestered. Some of the other major concerns regarding geological sequestration include the longevity of the captured and sequestered CO_2 to remain trapped, the hazardous risks of underground pressurized explosion and triggering of faults in rocks, the net energy use and carbon emission reduction from building and operating the carbon sequestration facilities, and the cost-effectiveness of the sequestration technologies.

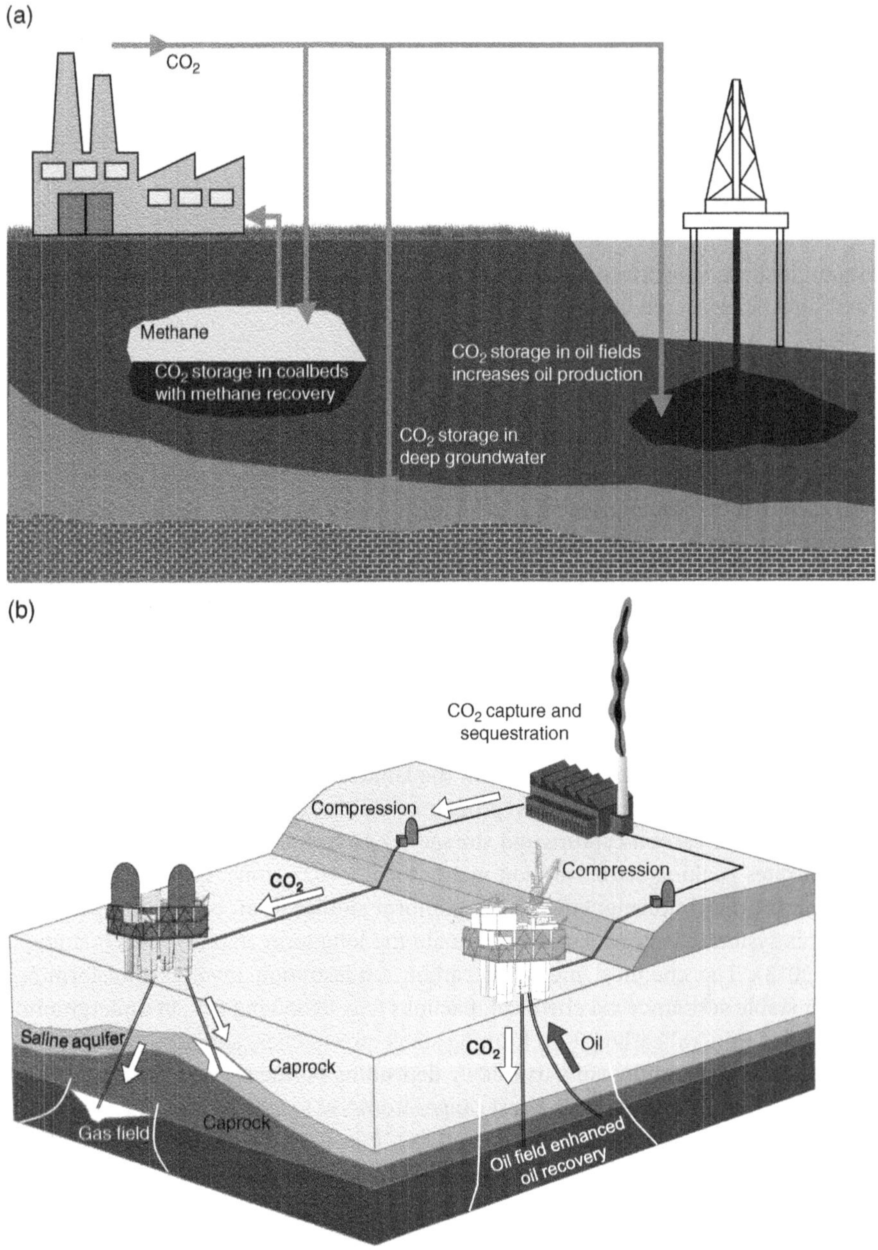

Figure 20.8 (a) Underground geologic carbon sequestration. (b) Carbon dioxide sequestration on a platform in the sea, similar to an oil drilling platform.

20.3.2 Surface Sequestration

Surface sequestration involves planting of fast-growing trees and vegetation to incorporate carbon dioxide into cellulosic plant matter (Hui et al. 2017). As with subsurface sequestration, surface sequestration requires international cooperation and agreement between numerous corporate and governmental entities. Sequestration rates remain uncertain, and such strategies have substantial implications with respect to widespread forest stewardship. Further, there are implications for the nitrogen budget and other ecosystem factors. The development of urban vegetation may also help to mitigate the CO_2 emissions to the atmosphere (Fares et al. 2017).

20.3.3 Marine Organism Sequestration

Marine organism sequestration involves the stimulation of marine organism growth, during which CO_2 is incorporated into the shells and bodies of organisms. One approach that has garnered much attention (and at least an equal amount of controversy) is the use of iron fertilization (Chisholm 2000). The concept is fairly straightforward – an iron-based compound would be released into the ocean as a "seed" – it would act to stimulate plankton growth, which would result in the adsorption of CO_2 from the atmosphere due to the increased metabolic processes (Figure 20.9). Upon death, the plankton settles out of the water column, resulting in the sequestration of the atmospheric carbon. Figures 20.10 and 20.11 show the results of an experiment performed in the southern hemisphere and reported by Chisholm (2000). As demonstrated, the seeding did result in an increase in algal growth and a measurable reduction in atmospheric CO_2. However, the full extent of the effects on marine ecosystems is not understood, and this technique is highly controversial. If it were to be performed on a wide scale, it would be estimated to cost about \$10–\$20 billion, which represents an annual per capita cost of \$10–\$20 per person in the developed world. International agreements would be required and even advocates acknowledge that the ocean would be acidified, affecting marine life and marine shell production. Further, such a large-scale plan would require a large fleet of ships and an enormous amount of iron.

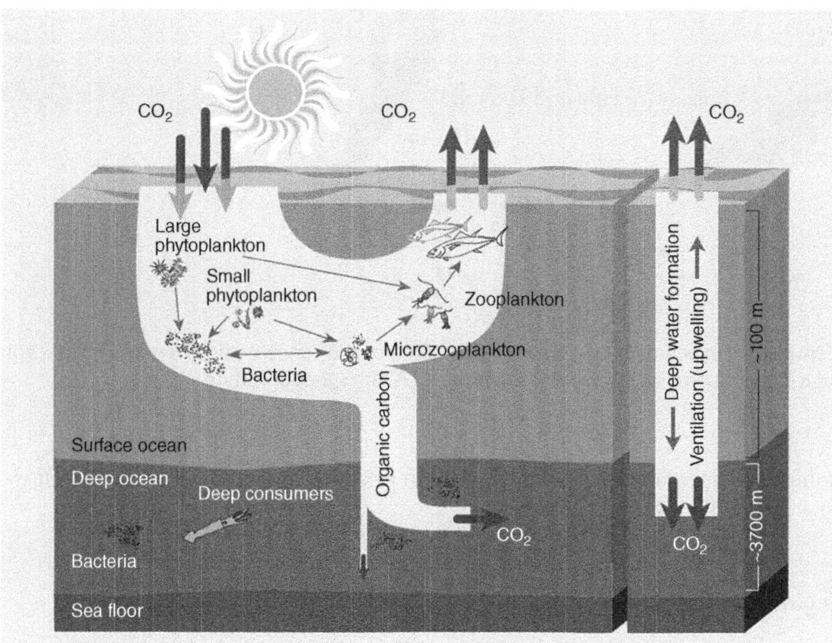

Figure 20.9 Iron fertilization of the oceans to CO_2 storage. *Source:* Chisholm (2000). Reproduced with permission of Springer Nature.

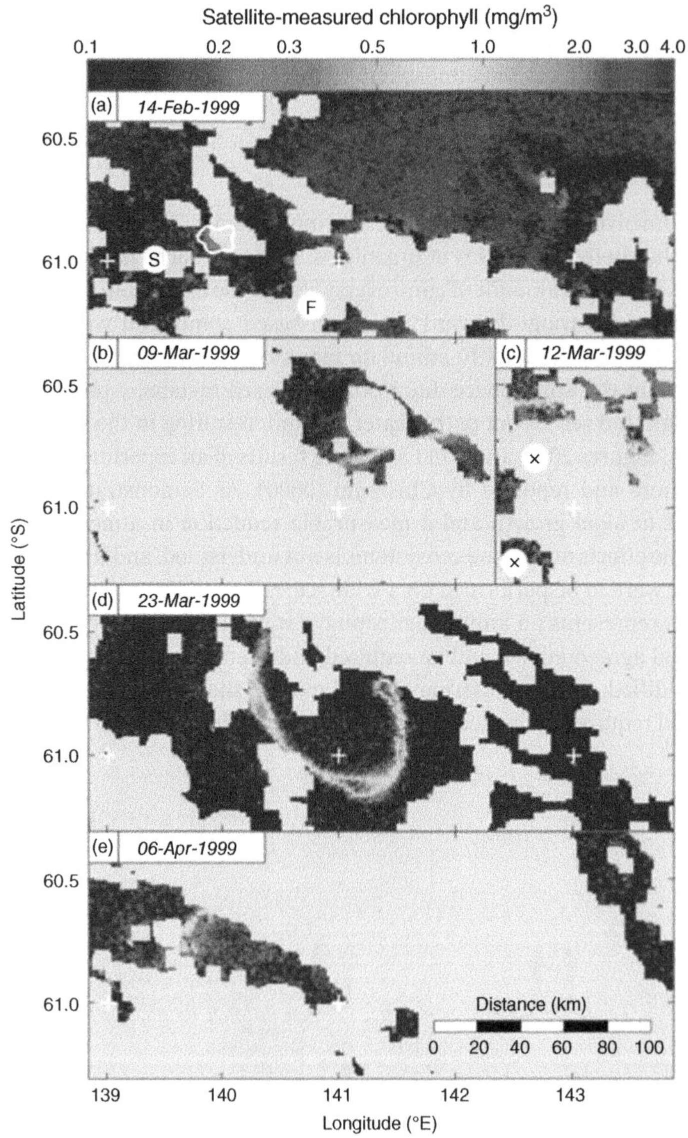

Figure 20.10 Iron fertilization experiment in Southern Ocean. Chlorophyll concentration from 14 Feb to 6 Apr 1999. *Source:* Reproduced with permission from Abraham et al. (2000)/Springer Nature.

Finally, it is debatable as to how much of the ocean is in an iron-limited state, which would allow this technique to be successful (Hauck et al. 2016).

20.3.4 Direct Engineered Capture

Direct engineered capture is also a straightforward method for carbon sequestration. Sorbent materials are deployed, capturing and sequestering carbon dioxide flowing through the adsorbent medium (Figure 20.12). Common adsorption materials include calcium hydroxide or calcium oxide (Socolow et al. 2011).

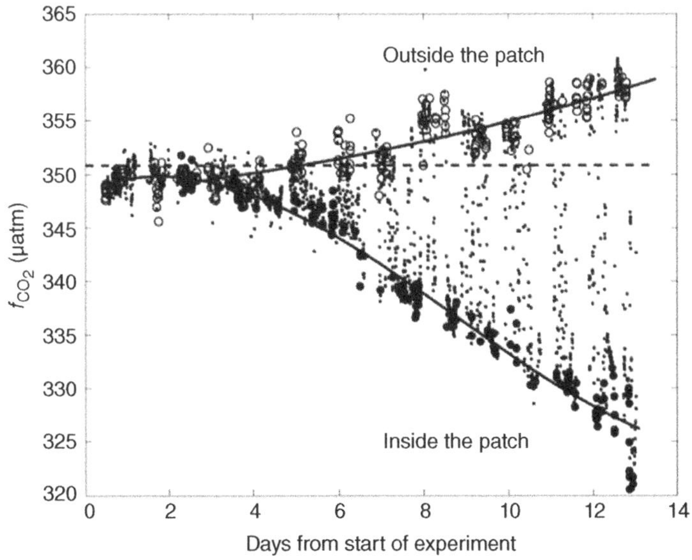

Figure 20.11 Iron fertilization experiment in Southern Ocean. Carbon dioxide reduction as a result of the enhanced plankton growth. *Source:* Watson et al. (2000). Reproduced with permission of Springer Nature.

Figure 20.12 Direct engineered capture of CO_2 using adsorbent materials. *Source:* Climeworks, Switzerland.

20.4 Solar Radiation Management (SRM) Methods

SRM methods act to increase the amount of radiation reflected away from the Earth, thereby reducing the amount of solar radiation that reaches the surface and is absorbed by the Earth. On a small scale, some thought has been given to increase the surface reflectivity, or "albedo," of the Earth through a range of methods, including white structural roofs ("cool roofs"), planting crops with a high relative reflectivity, or covering large areas of the desert with reflective material (Ming et al. 2014). Methods have also been proposed to enhance the reflectivity of marine clouds. On a grander scale, methods have been developed to inject sulfate aerosols into the lower stratosphere, mimicking the effect of large-scale volcanic eruptions (Kalidindi et al. 2015). On an even larger scale, consideration has been given to deploy large-scale shields or reflectors in space to deflect a fraction of the solar energy that reaches the Earth.

SRM technologies are designed to directly modify the inflow and outflow of solar energy to the Earth. Large-scale deployment of these methods would likely have an effect on climate within a few years. These methods do not deal with the original cause of the climate change, the increasing concentration of GHG in the atmosphere. However, the quick response offered could be useful in a time of emergency, such as when a climate "tipping point" is approached.

20.4.1 Sulfur Injection

Sulfur injection involves the injection of sulfate aerosol particles into the stratosphere to reflect sunlight away from the Earth, reducing the amount of solar radiation reaching the Earth's surface and acting to counteract climate change resulting from the greenhouse effect. Past evidence from large-scale volcanic eruptions, such as the eruption of Mt. Pinatubo in 1991, indicates that the addition of aerosols to the stratosphere can cool the planet (McCormick et al. 1995). Mt. Pinatubo's massive eruption resulted in a global scale cooling of 0.5 °C in its aftermath (Figure 20.13). As per de IPCC (2023) report, volcanoes have an influence on global temperatures due to the dispersion of aerosols that block the sunlight, causing a temporary decrease in the temperature, but the effect in the long term is minor compared to human activities (Figure 20.14). Given the proper size and large-scale deployment of aerosol particles, it is theoretically possible to cool the Earth, as demonstrated in Figure 20.15, which shows modeling of reduced aerosol temperatures resulting from deployed aerosols (Rasch et al. 2008), while sounding somewhat farcical, such an approach has proponents. For instance, Paul Crutzen, a 1995 Nobel Laureate in chemistry, opined that research on the potential of

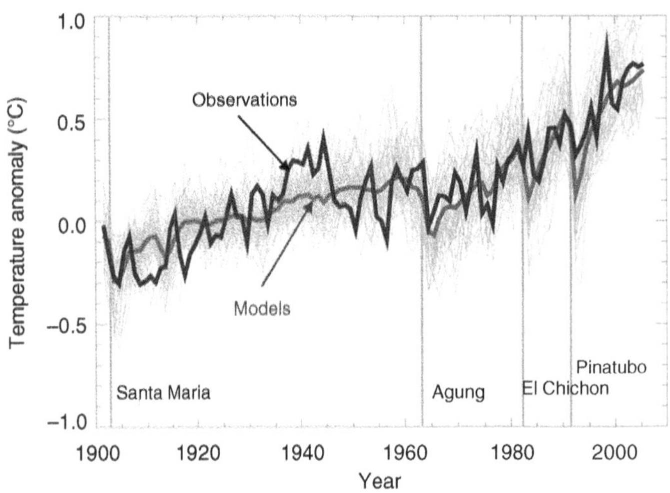

Figure 20.13 Effect of volcanic eruptions in the Earth's surface temperature due to the emission of sulfate aerosols. *Source:* IPCC (2007)/Intergovernmental Panel on Climate Change/CC BY 4.0.

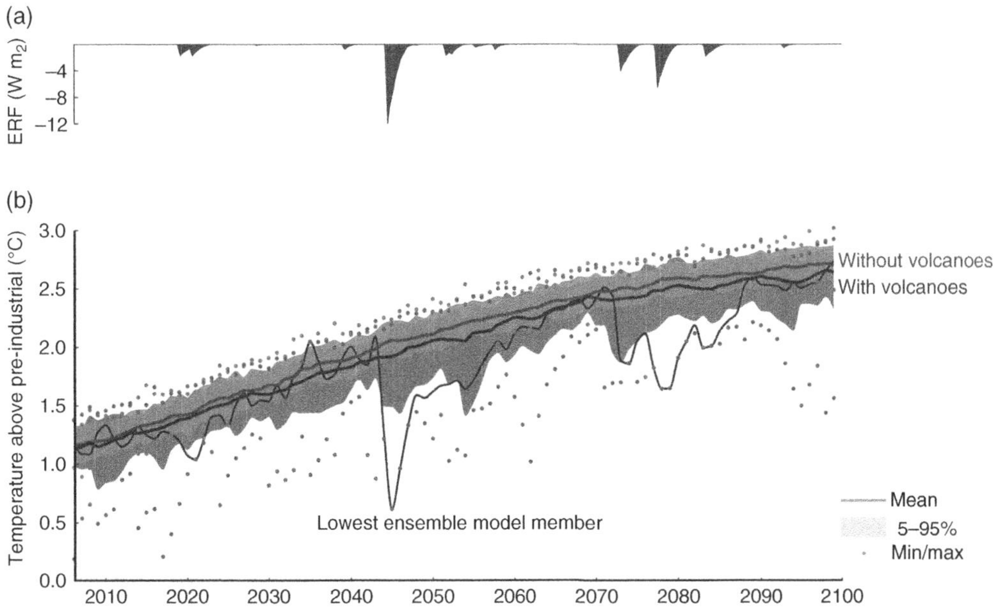

Figure 20.14 Potential impact of volcanic eruption on future global temperature change. (a) potential low likelihood high impact twenty-first century volcanic future. (b) impact of eruptions upon twenty-first century GSAT projections. *Source:* IPCC (2021)/Intergovernmental Panel on Climate Change/CC BY 4.0.

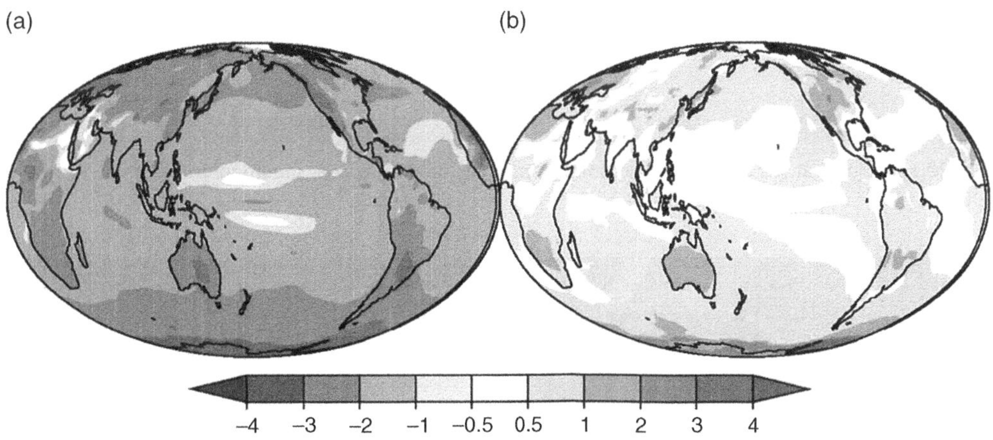

Figure 20.15 Surface temperature change from (a) present-day to (b) a doubling of current carbon dioxide levels scenario with the dispersion of 2 Tg/yr sulfur particles.

increasing the Earth's albedo by large-scale injection of sulfur aerosols into the stratosphere should be performed (Crutzen 2006).

If such an approach were to be employed, in order to be successful, between 1 and 2 Tg of aerosols would need to be injected annually. The large-scale injection could be performed in one of three ways. In the first alternative, ballistic shells or missiles (likely on the order of several thousand per day) could be fired into the stratosphere. As a second alternative, several thousand daily injections could be made from large aircraft delivery in the stratosphere. As a third alternative, ground releases could be made in the tropics. Using one of the three alternatives would require an annual expenditure of between $25 and $50 billion dollars, which represents an annual per capita cost of $25–$50 per person in the developed world. The need for international agreements for this practice should be explored. While never tested on any appreciable scale, volcanic eruptions do provide evidence that the mechanics of this technique would have an effect on the climate. Additionally, since its effects are temporary, a long-term commitment would not be necessary.

20.4.2 Reflectors and Mirrors

Equally grand or revolutionary, if not unconventional, is the potential use of reflectors or mirrors in space. The goal of such an application would be to deploy mirrors in space to deflect incoming solar radiation (Govindasamy et al. 2003). Figure 20.16 demonstrates the deployment location at the Lagrangian Point, located at an intermediate point between the Earth and the Sun. The system could be opaque reflectors or mirrors that would totally divert some fraction of sunlight away, or they could consist of a system that would diffract sunlight to some lesser degree, allowing a fraction of impacted sunlight to get through to the earth (Figure 20.17). In either case, a space vehicle or satellite would be needed to deploy the reflection media to a proper location. To maintain a proper distance from the Earth during its revolution around the Sun, an active positioning system, likely consisting of compressed gases, would need to be used on an ongoing basis. Because these gases would eventually be exhausted, it would be necessary to maintain or replace reflective instruments

Figure 20.16 Mirror at the L1 Lagrangian point to partially block the solar radiation on Earth.

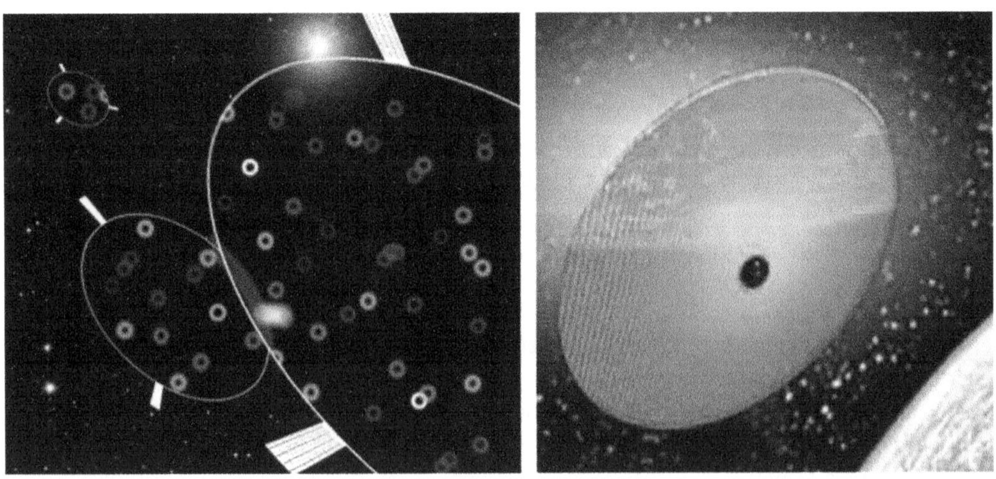

Figure 20.17 Reflectors of diffraction gratings in space to reduce the solar energy on Earth.

on a regular basis. The cost of deployment would likely be between \$50 and \$500 billion on an annual basis, which represents an annual per capita cost of \$50–\$500 per person in the developed world. International agreements would be required for deployment, and a long-term commitment would be required due to ongoing maintenance or replacement.

20.5 Applicability of CDR and SRM

The CDR and SRM climate geoengineering methods are very ambitious in reducing the effects of global warming and climate change. They could be considered for implementation but only as part of a wider set of options to reduce the GHG emissions, global warming, and climate change (Blackstock and Long 2010). When comparing these two groups of methods, CDR methods are preferable over SRM methods because they mitigate the negative effects of GHG in the long term. CDR methods show various characteristics that make them preferable over other methods: they are safe, effective, sustainable, and affordable. Furthermore, they can be developed and applied alongside other conventional mitigation methods as soon as the technological applications are developed and the cost is affordable. SRM methods may be applied only when there is an urgent need to limit or reduce the average global temperatures. SRM methods may be an effective tool in case of the necessity of rapid reduction of temperatures. However, there is limited information about the effects of SRM in the medium and long term. Thus, these methods must only be considered for application for a short period, in conjunction with other GHG removal methods, CDR, or conventional CO_2 mitigation methods. Thus, the SRM methods can be scaled back or discontinued at an appropriate time. As depicted in Figure 20.18, techniques from these two classes could be applied together in an integrated application.

When considering both CDR and SRM technologies, their deployment costs and their effects on the environment are of a very large scale. Arguably, these would be considered drastic, "last resort" measures. However, some argue that climate change has already reached, or may soon reach, a "tipping point" or a "point of no return" in which climate effects and the effects imparted to the environment may be irreversibly damaging. There is a fear that these dynamics may result in a

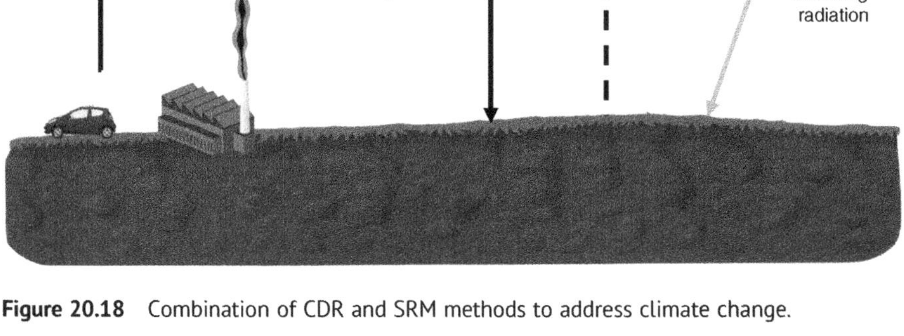

Figure 20.18 Combination of CDR and SRM methods to address climate change.

new stable state in which accelerated positive feedback effects occur. One example would be the collapse of Arctic ice, which could trigger the release of methane on a large scale from permafrost in Siberia. A "nightmare" scenario exists where a "domino effect" would occur, with successive parts of the climate system collapsing in succession, facilitating additional strain on subsequent parts and leading to a rapid, sudden climate change.

20.6 Climate Geoengineering – A Theoretical Framework

Because of the grand scale of deployment, and the exorbitant costs that would be associated with such use, as well as the international agreement and cooperation that would be necessary, it is critical to consider a framework in which such technologies would be analyzed, selected, and implemented with strong international consensus. First, there is general agreement that such technologies should only be considered when large-scale intervention is necessary and then only to the extent that is needed. Further, a thoughtful investigation is needed to integrate economic and cultural dimensions, as well as a focus on systems as opposed to subsystems. Such an approach should also be inclusive and transparent with clear accountability. Metrics have to be developed and implemented that will allow for the tracking of progress and the identification of problems or undesirable side effects. Further, there is a need to realize that any approach would involve high levels of uncertainty and should allow for mid-course intervention and correction. Systems should be designed to allow for incremental adjustment or reversibility as well as resiliency and redundancy.

20.7 Risks and Challenges

Such a large-scale approach with such widely ranging uncertainty inherently carries risks and has resulted in criticisms. First, the deployment costs would dwarf any large-scale engineering project contemplated to date in history. However, as demonstrated in Figures 20.19 and 20.20, once deployed, the incremental cost of mitigation with respect to a unit reduction of emissions mitigation declines drastically and may be attractive in the long run as compared to a comparable large-scale deployment of conventional approaches. Due to technical and implementation concerns, some have called for a moratorium on climate geoengineering research and consideration. Some of these reasons include concerns over effectiveness, control, predictability, unintended side effects, reliability, the potential of weaponization, effects on sunlight and cloud cover, moral objections, a lack of global control, and termination shock. Further, from an ethical standpoint, there is a concern that humans should not view the Earth as their property or artifact and that to engineer systems that could strongly affect every global ecosystem is not considered ethical. Additionally, there is a concern that such an approach does not address the cause of the problem (increasing GHG emissions) but only mitigates the effects. Finally, there is a strong concern regarding the unpredictability of these systems and that their implementation cannot be adequately or accurately modeled.

All of the proposed climate geoengineering systems require large-scale implementation to have any appreciable effect on climate change. Even the least expensive technologies would require expenditures in billions of dollars, and some of the more complex technologies could be significantly more. Further, for some of the more complex systems, a robust technological development process is necessary even before a prototype or pilot study could be performed, let alone operational

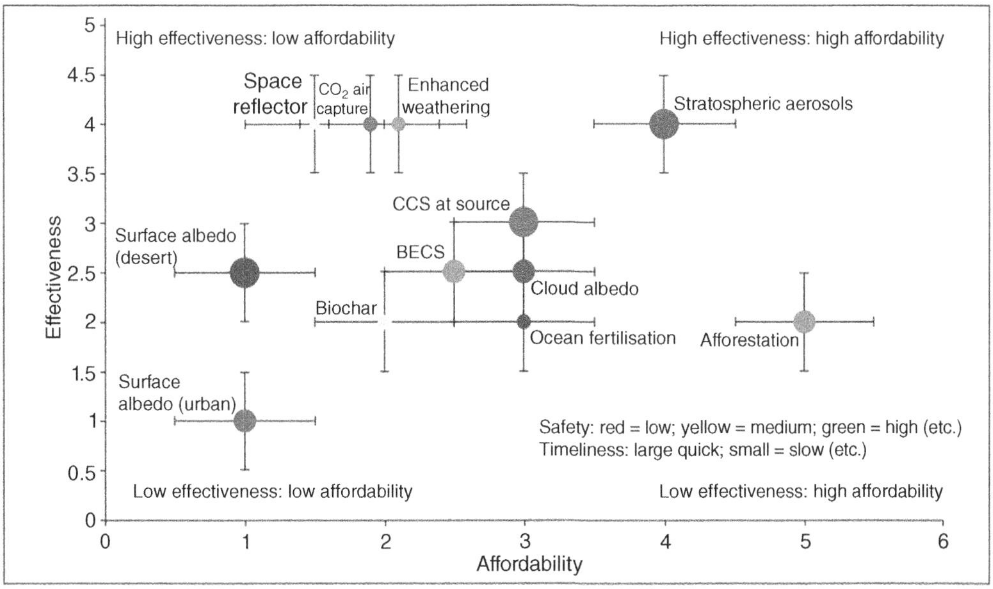

Figure 20.19 Evaluation of geoengineering technologies for mitigation of climate change. *Source:* Shepherd (2009).

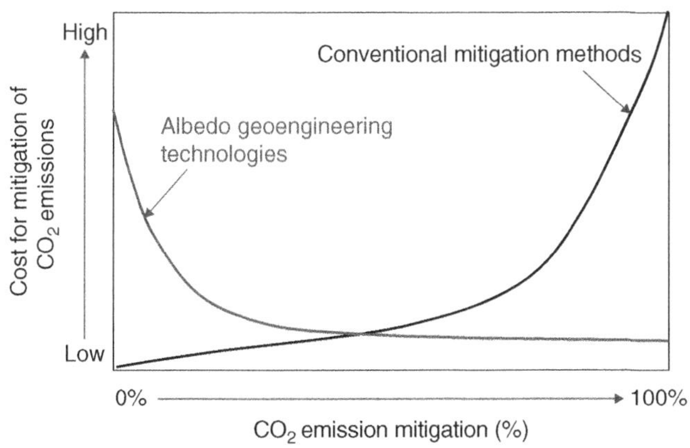

Figure 20.20 Comparative cost analysis of the conventional mitigation of climate change and geoengineering methods. *Source:* Keith (2000).

deployment. At this time, no clear institutional framework is in place to handle the related research and development program. As a result, many potentially feasible technologies do not have adequate development or experimental data to demonstrate their efficacy.

Even if a promising technology has been developed, it will be at least as difficult to implement. It is unclear who would have the authority or the ability to implement a global-scale technology. It is equally unclear how to allocate responsibility for implementation and operation as well as how to fund its use. Further, use could result in significant ecosystem changes on a global scale, resulting in billions of people being adversely affected while billions of others benefit. Any deployment would surely result in effects that cross many, if not all, international borders. The related legal issues and liability are even difficult to fathom. It is critical to develop these regulatory, legal, and economic frameworks before deployment could be contemplated.

While many challenges and concerns exist within the scientific and regulatory community, these concerns are dwarfed by the concerns that will likely develop within the general public. Public attitudes toward climate geoengineering as well as transparent public engagement in development will play a critical role in the potential implementation of climate geoengineering projects in the future. It is important to note that perceptions of risks involved, the level of trust in ongoing research, the transparency of all actions taken, and a clear delineation of vested interest will ultimately decide the political will and feasibility for future use. If there is to be even a marginal deployment of climate geoengineering methods in the future, an active international societal dialogue will be important to satisfy the demand for information for the possible environmental, social, and economic impacts and unintended consequences.

From a regulatory standpoint, no governing policies currently exist with respect to climate geoengineering. Parties to the United Nations Framework Convention on Climate Change (UNFCCC) should make increased efforts to adapt to and mitigate the effects of climate change, in particular by pursuing ongoing reductions of global GHG emissions, regardless of the potential use of climate geoengineering in the future. The UNFCCC has acknowledged that stabilization of emissions and the mitigation of climate change effects should be pursued.

Because of all these challenges with regard to climate change, the ongoing research and development should continue to advance. Such work can better inform decisions about implementation, and climate change effects can be more accurately modeled and predicted. Additionally, ongoing research and development may result in the development of lower-risk technologies that could be more attractively deployed, either on a regional or a global scale. Further, a greater data set could be collected in the meantime with respect to climatic models, including the development of carefully planned and executed experiments.

20.8 Summary

Because of the costs involved and the likely dramatic effects that may result from implementation, climate geoengineering, while offering the promise of being a significant mitigation approach to climate change, remains highly controversial. Many even argue that to pursue such an approach would be unethical, both from a professional and a moral standpoint. However, if faced with an imminent and dire climate crisis, it would be difficult to justify not acting with such an approach if a reasonable chance of success existed. Further, there are questions as to how to proceed if significant uncertainty of effectiveness and unintended consequences exist.

Several key organizations have at least implicitly agreed that climate geoengineering approaches should at least be considered and studied. These agencies include the Parliament of the United Kingdom, the Royal Society, the Institute of Mechanical Engineers, and NASA. On the other hand, some groups have been vocally opposed, including Greenpeace and Friends of the Earth. Controversy exists because it is generally believed that the most effective way to moderate climate change is to reduce the GHG emissions before irreversible changes take place. No climate geoengineering technology can provide an easy or readily acceptable alternative solution to mitigate climate change. However, all the methods described in the chapter could be useful in the future to mitigate, in part, climate change, and therefore, a more detailed research and analysis will be necessary before their practical implementation.

Geoengineering of the Earth's climate could very likely be technically feasible. However, the potential technical methodologies are in their beginning stages, and there are major uncertainties regarding effectiveness, cost, and environmental impact, both intended and unintended. Further international collaboration is critical in defining a research and development approach on promising technologies as well as a critical study of the feasibility, costs, benefits, risks, and opportunities presented by these technologies. Further, there is a pressing need to develop a governance and implementation framework to guide research, development, funding, and deployment, and engagement of all stakeholders through a transparent dialogue.

20.9 Questions

20.1 What is climate geoengineering? Explain the two major categories of climate geoengineering strategies.

20.2 Discuss climatic or environmental scenarios where geoengineering interventions could be necessary.

20.3 Name a few carbon dioxide removal methods.

20.4 Describe specific drawbacks associated with oceanic or marine carbon sequestration.

20.5 What are subsurface and surface carbon dioxide sequestration methods?

20.6 Name a few solar radiation management methods.

20.7 What are the disadvantages of the sulfur injection method for solar radiation management?

20.8 Assess in detail the financial and political considerations of CDR and SRM approaches.

20.9 What are the challenges associated with implementing CDR and SRM methods for climate change mitigation? Explain briefly.

20.10 Develop and discuss a financial incentive program that could be used to encourage developed and developing countries to collaborate on CDR and SRM approaches.

20.10 Problems

20.1 A potential CO_2 storage site is identified in a deep saline aquifer with a thickness of 200 m, a porosity of 20%, an area of 10 km^2, and a depth of 1500 m. Calculate the maximum volume of CO_2 (in tons) that can be stored in this aquifer, assuming a maximum allowable CO_2 saturation of 80%. The formation is at a pressure of 50 MPa and a temperature of 50 °C. (Hint: use ideal gas law)

20.2 The primary products of burning a dry plant are ash and CO_2; hence, an approximate estimate of the carbon sequestered by a plant can be obtained by controlled burning of a plant that is dried in a controlled environment. Suppose such tests on five different types of plant species of similar age, which are capable of rapid carbon removal, revealed the following results (Table P.20.1).

Table P.20.1 Results of carbon sequestration analysis on different plant species.

Plant type	Weight of an average adult plant after drying (g)	Average weight of ash after burning (g)
Type 1	25	23
Type 2	80	13
Type 3	300	120
Type 4	250	50
Type 5	100	20

A garden of area 2 hectares had a total of 2500 plants (type 1 – 800, type 2 – 450, type 3 – 200, type 4 – 400, and type 5 – 650). Calculate the following assuming that the plants of each type in the garden are of approximately the same size and age as the plants tested.

- Carbon sequestration efficiency of each type of plant (kg CO_2/kg mass of plant)
- Total carbon sequestered by each plant species in the garden
- Total carbon sequestered by all the plants in the garden

References

Abraham, E.R., Law, C.S., Boyd, P.W. et al. (2000). Importance of stirring in the development of an iron-fertilized phytoplankton bloom. *Nature* 407 (6805): 727.

Blackstock, J.J. and Long, J.C. (2010). The politics of geoengineering. *Science* 327 (5965): 527–527.

Chisholm, S.W. (2000). Oceanography: stirring times in the Southern Ocean. *Nature* 407 (6805): 685–687.

Crutzen, P.J. (2006). Albedo enhancement by stratospheric sulfur injections: a contribution to resolve a policy dilemma? *Climate Change* 77 (3): 211–220.

Fares, S., Paoletti, E., Calfapietra, C. et al. (2017). Carbon sequestration by urban trees. In: *The Urban Forest* (ed. D. Pearlmutter, C. Calfapietra, R. Samson, et al.), 31–39. Springer International Publishing.

Govindasamy, B., Caldeira, K., and Duffy, P.B. (2003). Geoengineering Earth's radiation balance to mitigate climate change from a quadrupling of CO_2. *Global and Planetary Change* 37 (1): 157–168.

Hauck, J., Köhler, P., Wolf-Gladrow, D., and Völker, C. (2016). Iron fertilisation and century-scale effects of open ocean dissolution of olivine in a simulated CO_2 removal experiment. *Environmental Research Letters* 11 (2): 024007.

Heinemann, N., Wilkinson, M., Haszeldine, R.S. et al. (2013). CO_2 sequestration in a UK North Sea analogue for geological carbon storage. *Andean Geology* 41 (4): 411–414.

Holloway, S. (2005). Underground sequestration of carbon dioxide—a viable greenhouse gas mitigation option. *Energy* 30 (11): 2318–2333.

Hui, D., Deng, Q., Tian, H., and Luo, Y. (2017). Climate change and carbon sequestration in forest ecosystems. In: *Handbook of Climate Change Mitigation and Adaptation*, 2e (ed. W.-Y. Chen, M. Lackner, and T. Suzuki), 555–594. New York: Springer. ISBN: 978-3-319-14408-5.

IPCC (2007). Climate change 2007 – the physical science basis. In: *Contribution of Working Group I to the Fourth Assessment Report of the IPCC*. New York, NY: Cambridge University Press. ISBN: 978-0521-88009-1.

IPCC (2021). *Climate Change 2021: The Physical Science Basis. Contribution of Working Group I to the Sixth Assessment Report of the Intergovernmental Panel on Climate Change*, 2391. Cambridge and New York, NY: Cambridge University Press https://doi.org/10.1017/9781009157896.

IPCC (2023). Climate change 2023: synthesis report. Intergovernmental Panel on Climate Change. https://www.ipcc.ch/report/ar6/syr/ (accessed September 2024).

Kali, R.F.S., Fan, D., Hazel, N., and Striolo, A. (2022). Techno-socio-economic analysis of geological carbon sequestration opportunities. *Environmental Science: Advances* 1 (2): 138–155.

Kalidindi, S., Bala, G., Modak, A., and Caldeira, K. (2015). Modeling of solar radiation management: a comparison of simulations using reduced solar constant and stratospheric sulphate aerosols. *Climate Dynamics* 44 (9–10): 2909–2925.

Keith, D.W. (2000). Geoengineering the climate: history and prospect. *Annual Review of Energy and the Environment* 25 (1): 245–284.

Kiehl, J.T. and Trenberth, K.E. (1997). Earth's annual global mean energy budget. *Bulletin of the American Meteorological Society* 78 (2): 197–208.

Leung, D.Y., Caramanna, G., and Maroto-Valer, M.M. (2014). An overview of current status of carbon dioxide capture and storage technologies. *Renewable and Sustainable Energy Reviews* 39: 426–443.

McCormick, M.P., Thomason, L.W., and Trepte, C.R. (1995). Atmospheric effects of the Mt Pinatubo eruption. *Nature* 373 (6513): 399.

Meinshausen, M., Nicholls, Z.R.J., Lewis, J. et al. (2020). The shared socio-economic pathway (SSP) greenhouse gas concentrations and their extensions to 2500. *Geoscientific Model Development* 13: 3571–3605. https://doi.org/10.5194/gmd-13-3571-2020.

Ming, T., de Richter, R., Liu, W., and Caillol, S. (2014). Fighting global warming by climate engineering: is the Earth radiation management and the solar radiation management any option for fighting climate change? *Renewable and Sustainable Energy Reviews* 31: 792–834.

Pham, V. and Halland, E. (2017). Perspective of CO_2 for storage and enhanced oil recovery (EOR) in Norwegian North Sea. *Energy Procedia* 114: 7042–7046. https://doi.org/10.1016/j.egypro.2017.03.1845.

Rasch, P.J., Tilmes, S., Turco, R.P. et al. (2008). An overview of geoengineering of climate using stratospheric sulphate aerosols. *Philosophical Transactions of the Royal Society A: Mathematical, Physical and Engineering Sciences* 366 (1882): 4007–4037. https://doi.org/10.1098/rsta.2008.0131.

Sarmiento, J.L. and Gruber, N. (2002). Sinks for anthropogenic carbon. *Physics Today* 55 (8): 30–36.

Shepherd, J.G. (2009). *Geoengineering the Climate: Science, Governance and Uncertainty*. Royal Society.

Shepherd, J.G. (2012). Geoengineering the climate: an overview and update. *Philosophical Transactions of the Royal Society A: Mathematical, Physical and Engineering Sciences* 370 (1974): 4166–4175.

Smith, S.M., Geden, O., Gidden, M.J. et al. (ed.) (2024). *The State of Carbon Dioxide Removal*, 2e. https://doi.org/10.17605/OSF.IO/F85QJ.

Socolow, R., Desmond, M., Aines, R. et al. (2011). *Direct Air Capture of CO_2 with Chemicals: A Technology Assessment for the APS Panel on Public Affairs (No. EPFL-BOOK-200555). American Physical Society.*

USEPA (2024). *Supply, underground injection, and geologic sequestration of carbon dioxide*. US Environmental Protection Agency. Greenhouse Gas Reporting Program (GHGRP). https://www.epa.gov/ghgreporting/supply-underground-injection-and-geologic-sequestration-carbon-dioxide (accessed November 2024).

Watson, A.J., Bakker, D.C.E., Ridgwell, A.J. et al. (2000). Effect of iron supply on Southern Ocean CO_2 uptake and implications for glacial atmospheric CO_2. *Nature* 407 (6805): 730.

Zhang, Z., Moore, J.C., Huisingh, D., and Zhao, Y. (2015). Review of geoengineering approaches to mitigating climate change. *Journal of Cleaner Production* 103: 898–907.

Section IV

Sustainable and Resilient Engineering Applications

21

Sustainable and Resiliency Assessment in Engineering Projects

Chapter Goal and Objectives
As discussed in previous chapters, comprehensive sustainability and resiliency assessments that integrate environmental, economic, and social dimensions can be applied to a wide range of projects and products. This chapter presents case studies demonstrating how to conduct such assessments using a general framework and various tools introduced earlier in this book. The main objectives are to (i) highlight the importance of life cycle resiliency and sustainability assessments in the preliminary design stage of an engineering project; (ii) provide a generic framework with a detailed methodology for conducting these assessments; (iii) understand the applicability of different tools introduced in this book by applying them to a broad range of engineering projects; (iv) interpret the results of such comprehensive assessments; and (v) understand how such assessments can aid in sustainable and resilient decision-making.

21.1 Introduction

The United Nations has outlined 17 Sustainable Development Goals (SDGs) to be achieved by 2030, with climate action serving as a central focus (UN DESA 2016). Sustainable development, which emphasizes balancing environmental conservation, social equity, and economic growth for the benefit of future generations, has gained critical significance (Reddy et al. 2024). Within this context, embedding sustainability into engineering practices is not just desirable but essential. Resiliency, a core attribute of sustainable systems, pertains to the ability of engineered systems to adapt to and recover from unforeseen challenges, stressors, or uncertainties. Recognizing resiliency as a prerequisite for sustainability is crucial (Reddy et al. 2022).

Professional engineering organizations, such as ASCE and ITRC, have issued guidelines for incorporating sustainability and resiliency into engineering design and implementation (ASCE 2023a,b; ITRC 2021). However, achieving true integration requires the ability to quantify both sustainability and resiliency from the earliest stages of project development. While frameworks for quantifying these aspects have been developed, most address sustainability and resiliency independently, with few efforts to integrate the two holistically. The interdependence of these concepts cannot be overstated. Resilient systems that lack sustainability may worsen climate change and other environmental impacts, undermining their resiliency over time. Conversely, sustainable systems that fail to withstand extreme events are inherently unsustainable. Thus, the separation of

Sustainable and Resilient Engineering: Drivers, Metrics, Tools, and Applications, Second Edition. Krishna R. Reddy, Claudio Cameselle, and Jeffrey A. Adams.

Companion website: www.wiley.com/go/reddy/sustainableresilientengineering2e

these concepts is counterproductive to creating systems capable of supporting human survival and well-being (Reddy et al. 2024).

A substantial body of literature exists on sustainability quantification, with frameworks continually evolving. However, much of such work focuses narrowly on environmental impact assessments, often overlooking the broader social and economic dimensions of sustainability, known collectively as the triple bottom line (TBL). Literature on resiliency quantification in engineering systems is also gaining momentum with growing impacts of climate change. Yet, there remains a scarcity of unified frameworks that integrate the TBL aspects of sustainability with resiliency assessments, treating them as interconnected rather than separate concepts.

In view of this, Reddy et al. (2024) have developed a comprehensive framework to perform an integrated resiliency and sustainability assessment for any engineering project. This framework called the TQUALICSR (*T*iered *Qua*ntification of *L*ife *C*ycle *S*ustainability and *R*esiliency) framework, as introduced in Chapter 13, helps streamline the sustainable and resilient design process through a comprehensive assessment of sustainability and resiliency using various tools introduced in the earlier chapters of this book. Hence, this chapter aims to summarize the methodology specific to the aforementioned framework and provide case studies of such comprehensive assessments through the application of the same for a broad range of engineering projects.

21.2 Integrated Sustainability and Resiliency Assessment Framework

A general methodology framework for performing integrated sustainability and resiliency assessments, as proposed by Reddy et al. (2024), is presented in Figure 21.1. Brief descriptions of the different stages of this methodology, as applied in the case studies presented in this chapter, are provided in this section. A comprehensive discussion of the methodology can be found in the original paper by Reddy et al. (2024).

21.2.1 Resiliency Assessment

A general methodology for performing a comprehensive resiliency assessment on engineering projects is presented in Figure 21.2. The first step is to identify climate or other disruptions and stressors specific to the project location. To quantify the resiliency of the engineering system to these identified shocks or stressors, resiliency indicators must be established for each disruption. An indicator-based approach to quantifying resiliency to various climate change impacts can effectively highlight both positive and negative changes in resiliency in a more concrete manner. This approach enables stakeholders to conduct practical and contextual analyses, supporting informed and efficient decision-making.

Functionality curves are commonly used to assess engineering resiliency by illustrating performance changes over time during disruptions. These curves demonstrate the resiliency of an engineered system both in terms of performance loss and time loss. Some of the case studies in this chapter utilize technical resiliency indicators corresponding to various components of a functionality curve. For instance, a technical resiliency indicator is quantified both in terms of performance loss, time loss, and the time and extent of recovery, i.e. the loss in the performance of a specific technical function of the engineered system and the time needed to restore its functionality to its original state or better. Such indicators provide a comprehensive assessment of the true

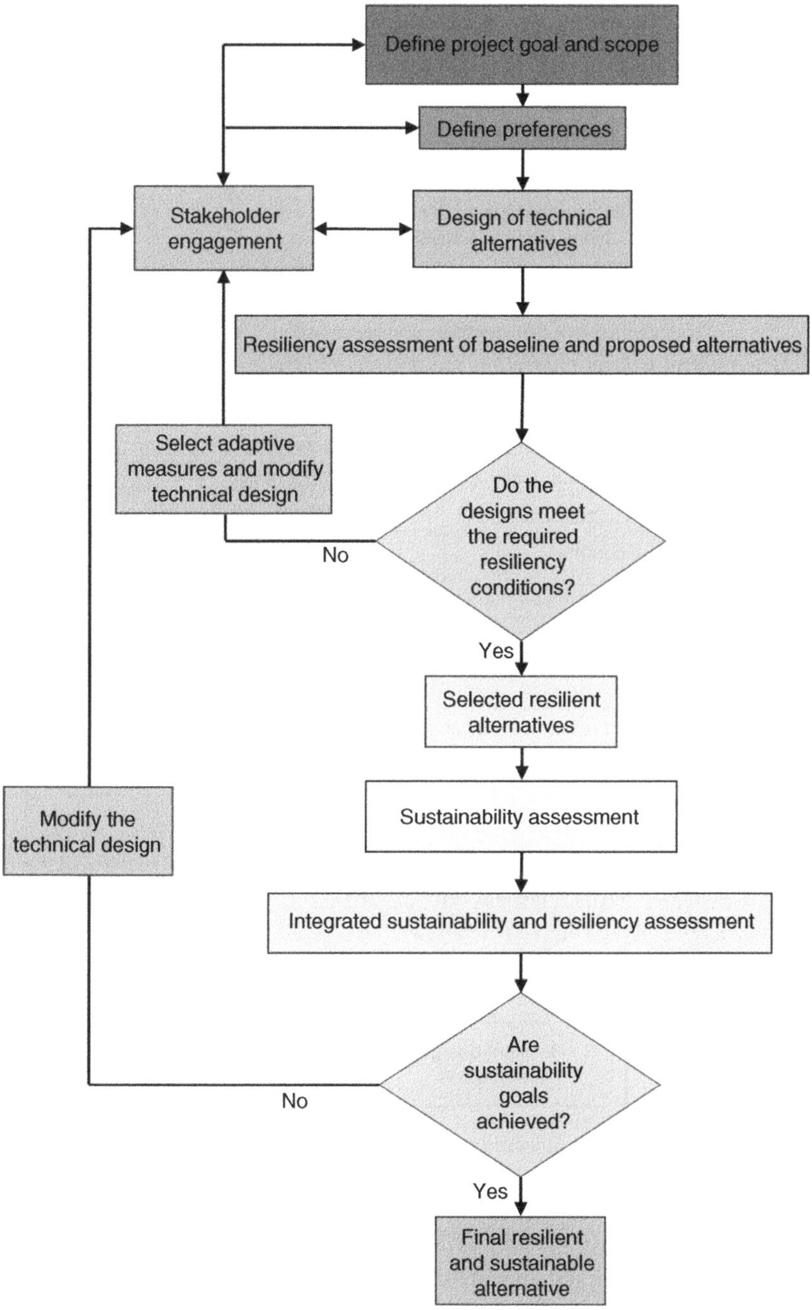

Figure 21.1 Integrated resiliency and sustainability assessment framework. *Source:* Reddy et al. (2024)/with permission of Springer Nature.

resiliency of the engineered system under identified shocks or stressors. Further discussion on the different shapes of functionality curves and their characteristics can be found in Reddy et al. (2024).

In addition to using resiliency indicators derived from the functionality curve for technical resiliency quantification, it is crucial to consider the cascading impacts across TBL dimensions.

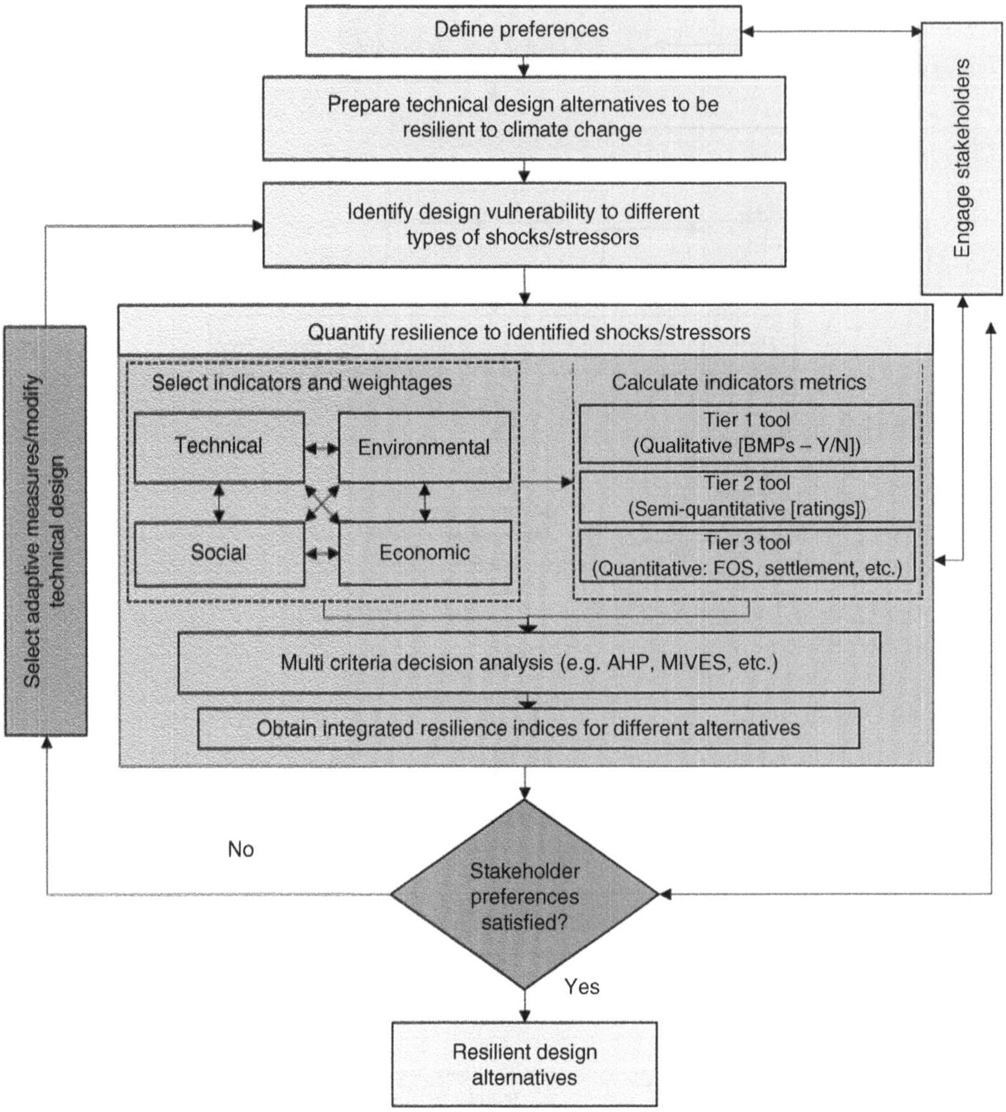

Figure 21.2 Resiliency assessment framework. *Source:* Reddy et al. (2024)/with permission of Springer Nature. Abbreviations: BMPs – Best Management Practices, FOS – Factor of Safety, AHP – Analytical Hierarchy Process, MIVES – Integrated Value Model for Sustainability Assessment.

For instance, the failure of an engineering system and its subsequent recovery can affect people, the environment, and the economy of surrounding communities. These cascading effects must be considered to capture the full scope of resiliency. If the system is designed to withstand extreme, low-probability events (e.g. a 100- or 500-year flood), the environmental, economic, and social costs of such a design should also be calculated. These cascading impacts can be broken down into specific criteria and indicators.

Established tools for quantifying resiliency indicators are rare and are often data-intensive to be able to quantify in preliminary design stages. Hence, the case studies presented in this chapter use ratings with a scale ranging from 0 to 4 (least resiliency to highest resiliency) to quantify the identified resiliency indicators. Resiliency assessment specific to each disruption should be performed, and the comprehensive resiliency index obtained should include resiliency with respect to all identified shocks and stressors. The integration of the resiliency indicators can be performed as per the methodology provided in Section 21.2.3.

21.2.2 Sustainability Assessment

Performing a comprehensive TBL sustainability assessment poses challenges in indicator selection, metric quantification, and integration of diverse indicators. As a systematic approach to address these challenges, Figure 21.3 provides the general methodology for the sustainability assessment of engineering projects. The assessment begins with defining boundaries and a functional unit,

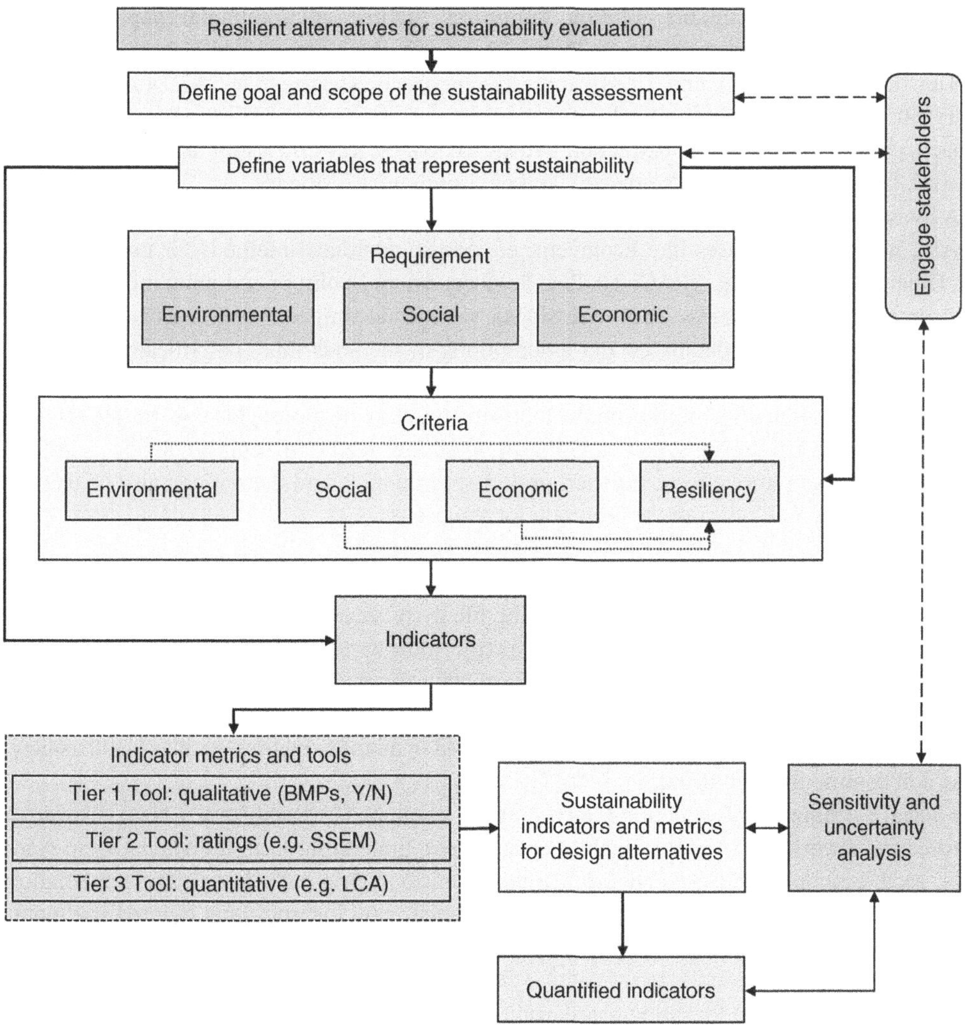

Figure 21.3 Sustainability assessment framework. *Source:* Reddy et al. (2024)/with permission of Springer Nature. Abbreviations: BMPs – Best Management Practices, SSEM – Social Sustainability Evaluation Matrix, LCA – Life Cycle Analysis.

ensuring neither extremely narrow nor broad scopes. Variables are identified hierarchically as requirements, criteria, and indicators. For example, environmental sustainability (requirement) encompasses criteria like air, water, ecosystems, and human health, each with specific indicators. Indicators should be SMART (Specific, Measurable, Achievable, Relevant, Time-bound) and SPICED (Subjective, Participatory, Interpreted, Cross-checked, Empowering, Diverse). Relevant indicators can be extracted from prior research or refined using the Delphi method, which iteratively gathers expert consensus through surveys (Dalkey and Helmer 1963).

Selected indicators can then be categorized into tiers based on available metrics and tools. Quantification involves evaluating these indicators for resilient design alternatives while accounting for uncertainties and assumptions. The following sections briefly describe the quantification process for environmental, economic, and social sustainability.

21.2.2.1 Environmental Sustainability

An environmentally sustainable solution minimizes lifetime environmental impacts. These impacts can be quantified as shown in Figure 21.3 using tools categorized into Tier-I (Yes or No), Tier-II (rating systems), and Tier-III tools like life cycle assessment (LCA). LCA, guided by ISO 14040, evaluates environmental impacts using a "cradle-to-gate" or "cradle-to-grave" approach, covering raw material extraction to disposal stages. Key criteria for assessment include air, water, land, energy, materials, toxicity, and environmental resiliency.

LCA tools like SimaPro, GaBi, OpenLCA, and SEFA (Spreadsheets for Environmental Footprint Analysis), alongside databases like Ecoinvent, provide standardized methods for impact assessment. Different methods (e.g. TRACI 2.0, ReCiPe) focus on mid-point or end-point indicators, balancing intermediate measures (e.g. emissions) with final impacts (e.g. biodiversity loss). Geographical relevance of databases and assumptions in methods must be critically evaluated. In general, the case studies presented in this chapter use TRACI 2.1 V1.01/US 2008 method which quantifies the environmental impacts on the following impact categories: global warming, ecotoxicity, eutrophication, carcinogens, non-carcinogens, fossil fuel depletion, ozone depletion, acidification, smog, and respiratory effects. Further, project-specific tools and metrics can also be used to quantify any additional indicators appropriate for the project.

21.2.2.2 Social Sustainability

Assessing social sustainability requires identifying life cycle stages (e.g. raw material acquisition, construction, operation, end-of-life) and stage-specific indicators to holistically quantify impacts on individuals and communities. Indicator selection often depends on data availability and project type. For example, Reddy et al. (2014) proposed tailored indicators for contaminated site remediation. Stakeholder surveys and expert opinions can also be used to quantify metrics, such as health, safety, justice, and community well-being.

The Social Sustainability Evaluation Matrix (SSEM), developed by Reddy et al. (2014), provides a structured framework for social sustainability assessment, integrating socio-individual, socio-community, socio-economic, and socio-environmental dimensions. Originally designed for remediation projects, its flexibility allows adaptation to diverse domains by incorporating tailored mid-point indicators. Some of the case studies presented in this chapter use an SSEM-based approach to assess the social sustainability.

The scoring systems used for rating-based evaluations or surveys should be simple and transparent. The social sustainability evaluations in most of the case studies presented in this chapter use a rating scale between −2 and +2 for negative to positive impacts across indicators, enabling aggregation of scores within and across dimensions. Supplementary tools like EJScreen, introduced in

Chapter 13, can also be used to integrate environmental justice considerations to enhance the social sustainability of an engineering project.

21.2.2.3 Economic Sustainability

To achieve a holistic assessment, economic sustainability should ideally be evaluated across the entire life cycle of a project, considering direct and indirect costs and benefits at each stage. Key tools for this purpose include life cycle costing (LCC) and cost-benefit analysis (CBA), which are well-established methods for quantifying economic impacts. However, these tools often focus on direct monetary costs and benefits, overlooking indirect costs and benefits, such as monetary costs of environmental impacts or interruption to business activities.

LCC focuses on evaluating the total costs of a project or product across its life span, from inception to end-of-life, emphasizing comprehensive cost considerations. Furthermore, direct benefits such as the revenue generated by the project under consideration during its lifetime can be quantified for comprehensive CBA. Indirect economic impacts, such as environmental costs, are less frequently quantified but are critical to a complete assessment. Tools like SimaPro and frameworks, such as Ecovalue 2008 and Environmental Prices, can estimate these impacts in monetary terms, addressing costs related to environmental degradation or resource depletion. For simplicity, the case studies presented in this chapter use the social cost of carbon provided by United States Environmental Protection Agency (USEPA) as the estimate for indirect costs of environmental impacts. In addition, other indirect costs and benefits – such as employment generation and impacts on local business – are often subjective and can be quantified through surveys, simple ratings, or binary (yes/no) indicators.

21.2.3 Integrated Resiliency and Sustainability Assessment

A flowchart showing the methodology to integrate various indicators quantified in the resiliency and sustainability assessments to develop an integrated resilient sustainability index (RSI) is presented in Figure 21.4.

Indicators quantified during the resiliency and sustainability assessments are of various magnitudes and units and hence pose challenges for direct integration and comparison. Therefore, first, all the indicators need to be normalized to a predefined scale. Various methods that can be used for the normalization of indicators are listed in Reddy et al. (2024); however, the case studies presented in this chapter use a value function, defined by Eq. (21.1), to normalize the values of each indicator on a scale of 0 to 1, where 0 indicates the least sustainable or resilient and 1 indicates the most sustainable or resilient option.

$$V_i = \left[B \times \left[1 - e^{-K_i \times \left(\frac{|S - S_{min}|}{C_i} \right)^{P_i}} \right] \right] \tag{21.1}$$

$$B = \left[\frac{1}{1 - e^{-K_i \times \left(\frac{|S_{max} - S_{min}|}{C_i} \right)^{P_i}}} \right] \tag{21.2}$$

where V_i is the normalized value of an indicator and B is a factor that keeps the value between 0 and 1. S is the value of the indicator quantified from the sustainability and resiliency analyses, and S_{max} and S_{min} are the terms defining the ideal and worst case (i.e. values of indicator corresponding to values 1 and 0, respectively). P_i is a shape factor that defines the shape of the curve (concave – $P_i < 1$;

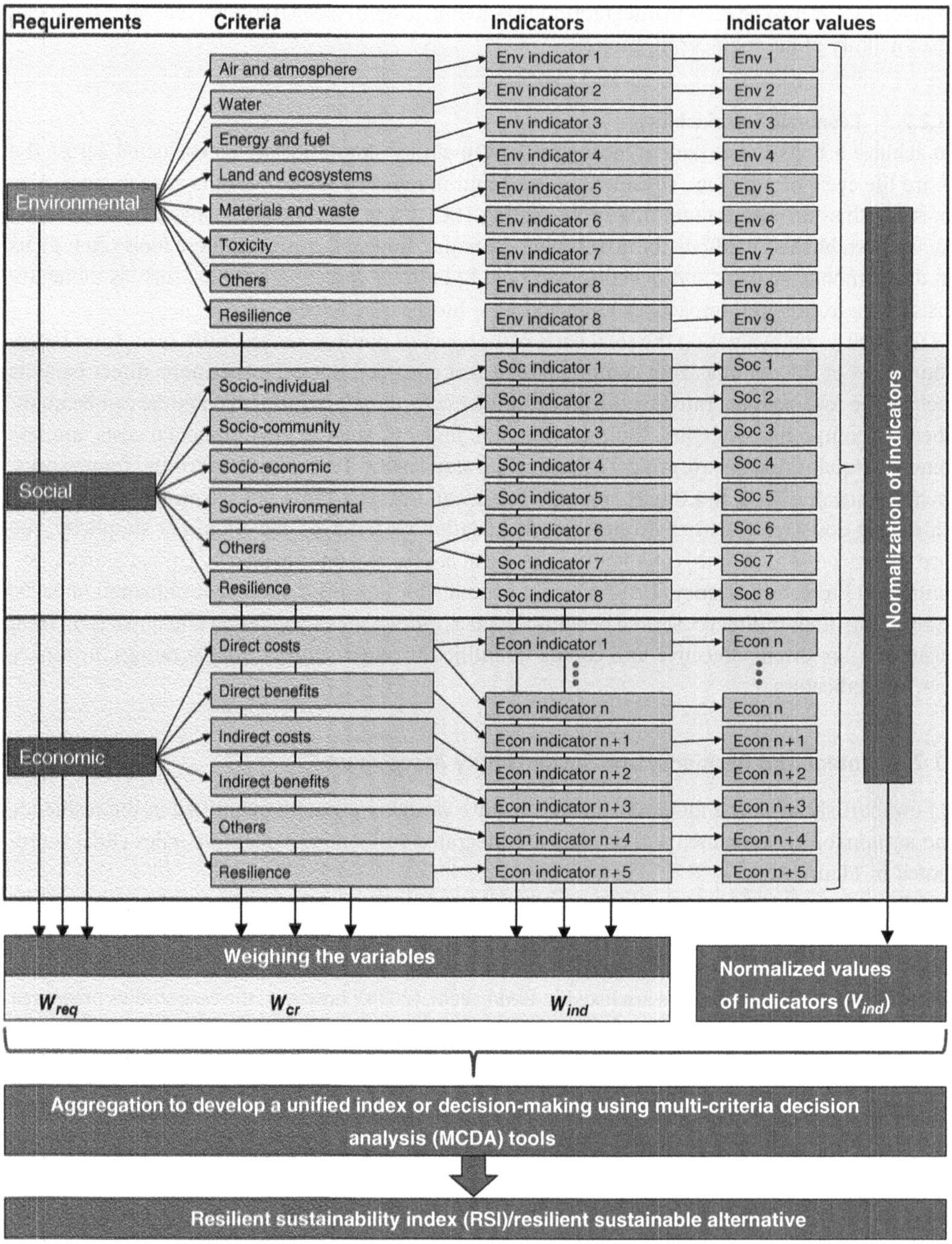

Figure 21.4 Integration of resiliency and sustainability indicators to develop a resilient sustainability index.

convex or S-shaped – $P_i > 1$; and linear – $P_i \approx 1$), C_i is the abscissa of the point of inflection in curves with $P_i > 1$, and K_i decides the ordinate of point of inflection. Depending on the variable type and definition of value curves, all the above-mentioned parameters can be defined and used to normalize the value of each indicator to lie between 0 and 1.

The variables identified during the assessment need to be weighted according to their relative importance for integration. Various multi-criteria decision analysis (MCDA) tools, such as the Analytical Hierarchy Process (AHP), are available for this purpose. However, the case studies presented in this chapter either use equal weights or manually assign weights based on expert judgment regarding the importance of each variable. For more information on methods to determine the weights of variables in such assessments, refer to Reddy et al. (2024).

The normalized values of the indicators, along with the weights of the variables, can then be used to develop a RSI or to rank alternatives based on their relative performance. Several MCDA tools can be used for this purpose, and the most common methods used for this purpose are discussed in Reddy et al. (2024). However, the case studies in this chapter use a simple weighted additive aggregation function, as defined by Eqs. 21.3, 21.4, and 21.5), to calculate the RSI for each alternative.

$$V_{cr} = \sum V_{ind} \times W_{ind} \tag{21.3}$$

$$V_{req} = \sum V_{cr} \times W_{cr} \tag{21.4}$$

$$V_{final} = \sum V_{req} \times W_{req} \tag{21.5}$$

where V_{ind}, V_{cr}, and V_{req} are the values of each indicator, criteria, and requirement, respectively, while W_{ind}, W_{cr}, and W_{req} are their corresponding weights. V_{final} will give the final integrated RSI based on the entire analysis. Weights of various requirements, criteria, and indicators can be varied based on stakeholder preference to perform a sensitivity analysis to determine the controlling variables. Values of RSI can aid practitioners, stakeholders, and decision makers in making decisions regarding the most viable option among the alternatives analyzed.

21.3 Resiliency and Sustainability Assessment of Water Disinfection Technologies at Stickney Water Reclamation Plant

21.3.1 Problem Statement

In the face of a growing world population, declining freshwater resources, and, above all, poor water management, recycling and reusing water from wastewater treatment plants is emerging as a critical solution to address both environmental problems and the growing water demand. A crucial aspect that warrants scrutiny is the historical patterns of water allocation, wherein only 12% of the water extracted from the freshwater supply is earmarked for public needs, and a mere 1% is set aside for domestic use. The remaining 87% of the freshwater supply is consumed by six endeavors: thermoelectric power, irrigation, industrial applications, aquaculture, livestock, and mining (Dieter et al. 2018). Mismanagement of existing water supplies and the increasing incidence of extreme weather events undermine water security, creating the need to address the sustainability of water supplies in terms of both quantity and quality (Garcia and Alamanos 2023), as well as the resiliency of their infrastructures.

Once considered a burden to be disposed of, wastewater is now recognized as a valuable resource. Through advanced treatment processes, it is transformed, stripped of its former contaminants, and made viable for various non-potable and even potable applications. From irrigating agriculture to industrial processes and augmenting drinking water supplies, the potential use of reclaimed water is vast and versatile. However, the transition from wastewater to a reusable resource is challenging. Sound regulatory frameworks must accompany technological advances to ensure safety and quality. Public perception and acceptance also play a crucial role in realizing the full potential of water reuse initiatives. These aspects can be addressed through resiliency and sustainability assessments of the technologies being considered for wastewater treatment.

To address these global challenges, this study investigates the potential of wastewater reuse through a case study of the Stickney Water Reclamation Plant (WRP). By evaluating tertiary disinfection technologies and their applications, this research aims to provide a sustainable and resilient approach to transforming wastewater into a valuable resource for reuse.

21.3.2 Project Background

The project focuses on the Stickney WRP as a case study for implementing tertiary effluent treatment to facilitate wastewater reuse. The Stickney WRP, located in Cicero, Illinois, is one of the world's largest wastewater treatment facilities. Operated by the Metropolitan Water Reclamation District of Greater Chicago, Stickney plays a vital role in managing wastewater from the Chicago metropolitan area. The Stickney WRP serves residents in the central part of Chicago and 46 other communities within a 260-square-mile area. This WRP consists of two plants: the original West Side Plant, which was placed in service in 1930, and the Southwest Plant, which began operation in 1939. The Stickney WRP currently serves over 2.3 million people, treats an average of 700 million gallons of wastewater per day, and has the capacity to treat up to 1.4 billion gallons per day (MWRD 2023).

Stickney employs advanced treatment processes to ensure that the wastewater it receives is thoroughly cleaned before being discharged. These processes include primary treatment to remove large solids and secondary treatment to break down organic matter using bacteria. The treated water is then released into the Chicago Sanitary and Ship Canal, meeting stringent regulatory standards to protect public health and the ecosystem. However, the quality parameters of the water discharged from the secondary treatment currently do not meet the requirements defined by USEPA for reuse applications. The primary concern among these parameters is the presence of a high concentration of fecal coliforms, approximately 7790 mg/l (MWRD 2024). Therefore, tertiary disinfection treatment is required to reduce the high concentration of fecal coliforms (Metcalf and Eddy 2003). Thus, the two technologies chlorination/dechlorination and UV treatment, were evaluated. For the assessment, the current scenario at Stickney WRP was referred to as "Untreated Water" and the proposed scenario as "Treated Water."

The main objectives of this project are to propose the design of a disinfection system for the Stickney WRP and to evaluate the feasibility of reusing the final effluent. Specific non-potable water applications considered for this project include irrigation in nearby open spaces and gray water use in nearby institutions. Gray water reuse encompasses applications, such as residential, commercial, and industrial bathroom sinks, bathtub shower drains, and clothes-washing equipment drains (USEPA 2017a). Along with designing disinfection systems, distribution pipelines for the effluent were also designed and included in the assessment as part of the proposed system.

To ensure a holistic evaluation, the project assesses the resiliency of the two disinfection technologies (chlorination/dechlorination and UV treatment) through a semi-quantitative approach and evaluates their sustainability using the TBL framework. Finally, an RSI will be obtained to determine the most beneficial technology to implement. Based on these results, detailed comparisons and recommendations can be made to inform decision-making for wastewater reuse at the Stickney WRP.

21.3.2.1 Technical Design of Alternatives

A technical design was developed for the two suggested systems – UV and chlorination/dechlorination – with a design life of 20 years. The design was based on the system boundaries and functional units, including magnitude, duration, and qualitative parameters. The magnitude of the disinfection system was defined as the amount of water required to be reused within a 2-mile radius from the Stickney WRP per day. Lastly, the qualitative parameter considered was fecal coliform concentration in the effluent at Stickney WRP. The proposed scenario for chlorination/dechlorination and UV systems can be seen in Figure 21.5.

Water Requirement

Selecting the water reuse application regions was the initial stage in the technical design process. Using Google Maps, the project's qualifying application regions were chosen within a 2-mile radius (Figure 21.6). Institutional areas where the effluent can be used as gray water (indicated in blue) and open spaces (not used for agricultural purposes) were identified as eligible for water reuse application in this study. The open spaces, consisting of parks, golf courses, and cemeteries, are shown with green markers in Figure 21.6. Based on the application areas, the number of pipelines and pumps required for water distribution was estimated, and the results were used as the inputs in the sustainability assessment for this study. The water needed per capita for the public supply, obtained from the database of United States Geological Survey (USGS) Water Resources, was multiplied by the population at each site to determine the total amount of water needed for the institutional sites. According to the calculations, the schools and colleges within the project's boundary would require about 3.15 million gallons per day (MGD). The US Department of Energy's (USDOE 2023) guidelines were used to calculate the water needed for irrigation at open-space locations, yielding a result of about 0.1 MGD.

Chlorine/Dechlorination Disinfection

The chlorination/dechlorination contact chamber was designed based on the required water flow of 3.25 MGD for reuse. A peak factor of 1.5 was also considered in the design to account for seasonal variations. Assuming a hydraulic retention time of 15 minutes and a calculated peak flow of 4.875 MGD, the volume of the chamber was calculated. The dimensions of the chlorination/dechlorination system's channels were 90 feet long, 9 feet wide, and 10 feet deep. Three channels were required to provide sufficient contact time between chlorine and untreated water. Baffle walls were provided to ensure adequate mixing of chlorine with the untreated wastewater. Reinforced concrete was used to construct the contact chamber. Based on previous literature, the estimated volume, dimensions (length, width, and thickness) of the side walls, baffle walls, end walls, and base slab were assumed (Gilmore 2016). The quantities of reinforced steel and concrete were determined during the building stage. Additionally, the amounts of chlorine for disinfection and sulfur dioxide for dechlorination were determined annually as part of the operation phase.

UV Disinfection

The UV system was designed considering several parameters, including the designated water flow for reuse applications of 4.875 MGD, UV transmittance, UV dose, fouling factor, and the concentration of fecal coliform in the Stickney WRP effluent, as discussed in the previous section. The UV transmittance for secondary treated effluent approaching the disinfection system without filtration was 65%. The UV dose required to treat 10^4 to 10^6 fecal coliform is approximately 80–100 mW/cm^2 (Metcalf and Eddy 2003). A dose of 100 mW/cm^2 was selected to ensure a significant fecal coliform reduction. The dimensions of the UV channels were 45 feet long, 4 feet wide, and 6 feet deep. The length and depth were estimated based on the previous literature (Gilmore 2016), and the width

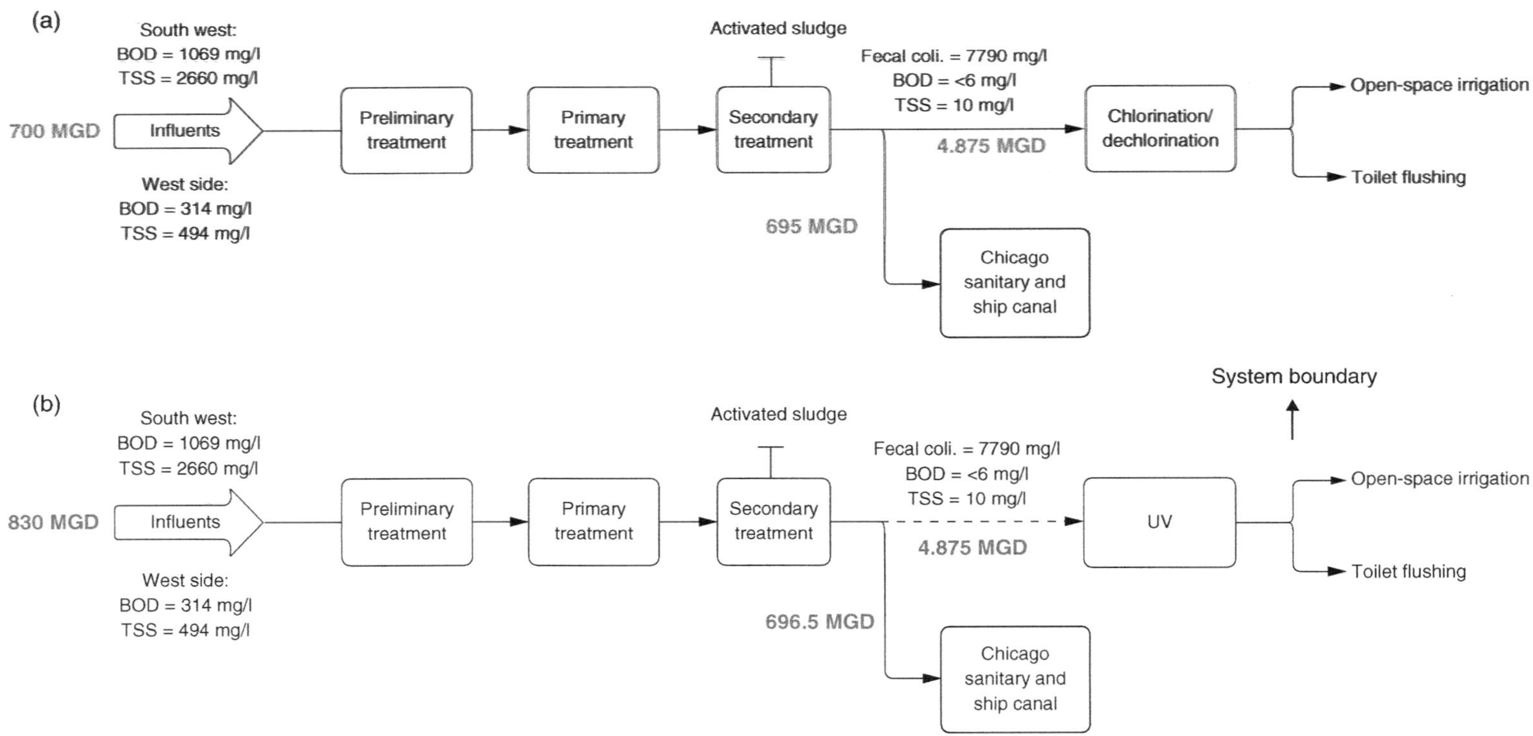

Figure 21.5 Proposed (a) chlorination/dechlorination system and (b) UV system for water disinfection.

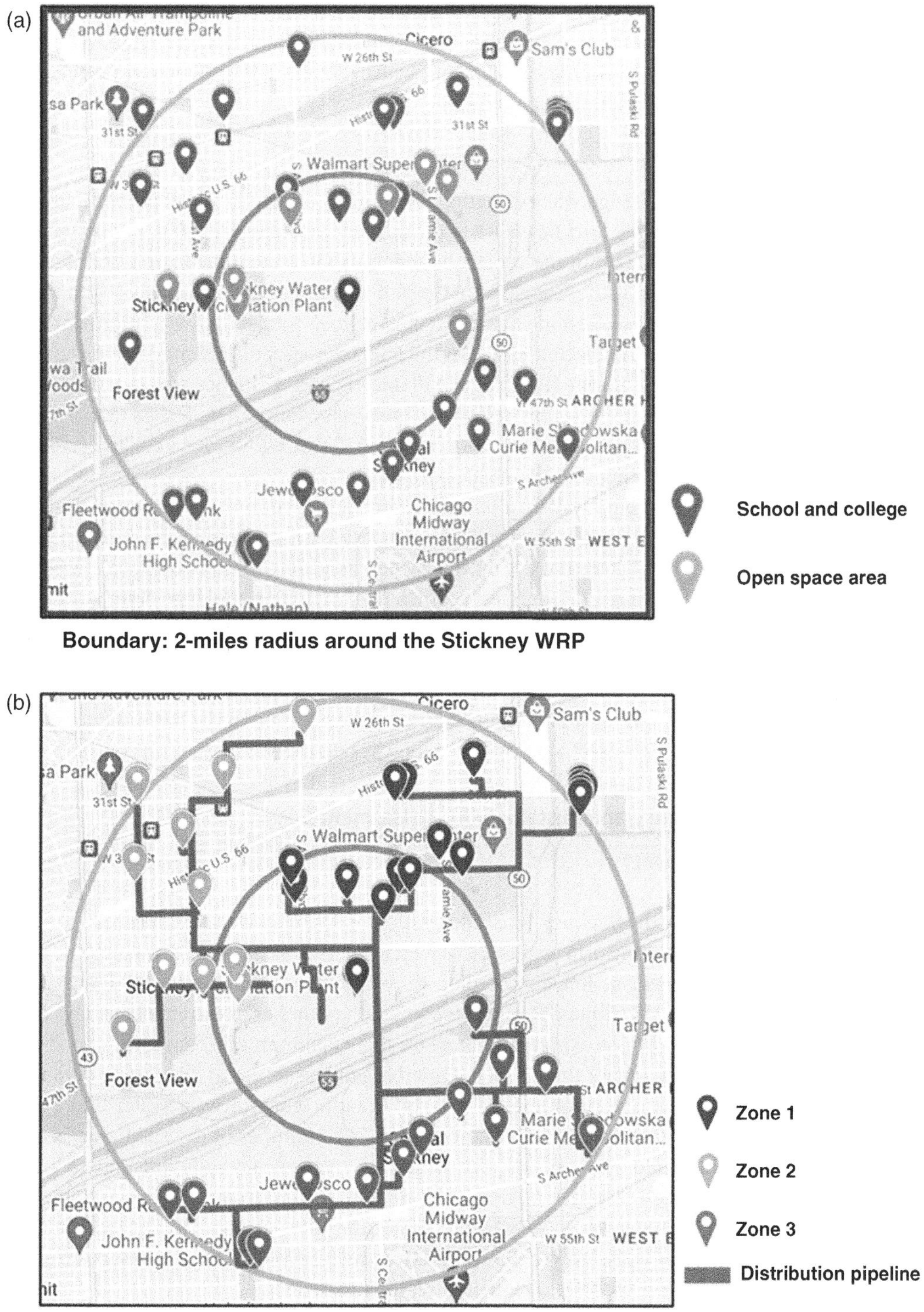

Figure 21.6 (a) Water reuse applications sites and (b) Stickney water distribution pipes layout.

was designed to accommodate 12 horizontal modules, each with 8 lamps. These modules were calculated based on the UV transmittance, UV dose, and designated water flow for reuse. As previously discussed, the amounts of concrete and steel needed for the UV system were calculated and will be incorporated into sustainability analysis.

Effluent Distribution Pipelines

The final step in the technical design was estimating the number of pipelines and pumps required for the water distribution from the Stickney WRP to the application areas. The effluent distribution to open spaces and institutions within the project boundary is illustrated in Figure 21.6. The distribution pipeline was designed using Google Maps to plot routes along the roadways connecting the application areas with the disinfection chamber. The pipeline's total length was approximately 112 000 ft. Connection parts were selected based on the characteristics of each node, and the pipe material chosen was polyvinyl chloride (PVC) in two diameter sizes: 10 inches and 14 inches. Two pipeline sizes were selected because the flow velocity in the pipe system should be less than 5 ft/s to avoid damage and tear to the pipes and connection parts. PVC was chosen for its high corrosion resistance and lightweight, making it easy to transport and install.

Due to the energy loss in the pipe system, a pump was implemented in the distribution pipeline design. The required energy in terms of the head was computed using EPANET 2.2. The pipeline, nodes, and reservoir were first plotted to set up the model, assuming all network nodes were at elevation 0. The pipe length, diameter, roughness coefficient of each pipe, and base demand at each node were then input into the model. As a result, a total head of approximately 87 ft was computed to meet the water pressure requirements, which states that the water pressure in the system must be higher than 17 psi to prevent contaminated water from entering the pipe system (MECC 2023). Consequently, the resulting head (H) and design flow rate (Q) were used in the performance curve provided by the manufacturer to select a pump. For energy input, the pump was run continuously.

21.3.3 Integrated Resiliency and Sustainability Assessment

21.3.3.1 Resiliency Assessment

Climate Concerns

The resiliency assessment for the Stickney WRP addresses three primary climate concerns: extreme temperatures, dry spells, and rainstorm events, all of which could impact the performance and efficiency of wastewater treatment systems. According to the Climate Mapping for Resilience and Adaptation (CMRA) and the US Climate Resilience Toolkit, the number of extreme heat days over the next 20 years (the design life) is expected to increase to a maximum of 52 days per year. Temperatures exceeding 95 °F can impair the activity of microorganisms essential for wastewater treatment, thereby reducing system performance. Similarly, the efficiency of wastewater and water treatment technologies may decrease at high temperatures. The number of dry spell days is also projected to rise, reaching up to 208 days per year, which can concentrate pollutants due to lower influent flow rates, increasing the burden on treatment systems. Furthermore, rainstorm events are projected to increase in frequency by 4%, with the maximum intensity potentially reaching 50 inches over the next 20 years. These events can strain infrastructure and, in some cases, dilute chemicals used in the disinfection processes, reducing treatment efficiency.

Technical/Engineering Resiliency Indicators

This project has defined three climate hazards, and for each, we have established corresponding technical resiliency indicators based on their potential impact on the wastewater treatment plant. Table 21.1 displays these technical resiliency indicators along with cascading environmental, social,

Table 21.1 Resiliency indicators, metrics, and tools to assess different disinfection technologies.

Resiliency requirement	Indicator	Metric	Tool
Technical	Increase bacteria in wastewater (extreme temperature)	Rating on a scale of 0 to 4	Expert rating
	Increase concentration of suspended sol (extreme temperature)		
	More pathogens (rainstorm)		
	Overflow/more capacity (rainstorm)		
	Dilutes the concentration (rainstorm)		
	Increase in pollution concentration (dry spell)		
	Less water to dilute contaminants/less capacity (dry spell)		
Cascading environmental	Resistance development of microorganisms	Rating on a scale of 0 to 4	Expert rating
	Alteration of microbial communities		
	Disinfection by-product discharge		
Cascading social	Individual health	Rating on a scale of 0 to 4	Expert rating
	Employment opportunities		
	Worker safety and training		
	Public health		
	Community engagement and perception		
	Loss of community trust		
	Recreational use of water bodies		
	Employment rate		
	Consumption rate		
	Consumption cost		
	Algal blooms		
	Spread of antibiotic resistance		
	Impact pathogen discharge into water bodies		
Cascading economic	Cost of technology	Rating on a scale of 0 to 4	Expert rating
	Maintenance cost		
	Operation cost		
	Labor cost		
	Utilities cost		
	Waste management		
	Productivity		
	Long term cost saving		
	Public health benefits		
	Quality-of-life improvements		
	Resource conservation		

and economic indicators selected for the current study for each climatic event. Details regarding the cascading impacts are discussed in the subsequent section. The primary resiliency indicators considered include reduction in treatment efficacy, decrease in performance (based on the process involved, environmental impacts, economic costs to maintain the process over extended periods), and social costs if contamination reaches water sources. For example, we considered the reduction of treatment efficiency as an indicator of resiliency in the case of extreme heat because over the next 20 years, temperatures may exceed 90 °F while the bacteria used in wastewater treatment plants function optimally at temperatures between 68 °F and 95 °F. These microorganisms can still function at a slower rate at lower temperatures, but significant thermal elevation impairs bacterial activity and viability, rendering the process ineffective. Similarly, in the case of rainstorms, increases in flow rate to the wastewater treatment facility can put pressure on system performance, potentially exceeding the facility's treatment capacity. Finally, during dry spells, the performance of the WRP may decrease as reduced influent flow under severe drought conditions can lead to an increase in downstream pollutants.

A semi-quantitative rating is used to evaluate technical resiliency indicators, where a score of zero indicates no resiliency due to the extreme level of negative impact and a score of 4 indicates a very high resiliency with a low level of negative impact. In this project, the resiliency was evaluated across three scenarios. The first is the baseline scenario, where no treatment is implemented (current facilities). The second scenario evaluates the UV technology, and the third examines the chlorination/dechlorination technology. The three climate concerns and their corresponding indicators are analyzed for each scenario.

First scenario – no treatment: In the first scenario, an event of extreme heat can cause extreme negative impacts on the treatment efficiency and decrease the performance of the WRP because temperatures are expected to exceed 90 °F in the next 20 years, and the reproduction rates of bacteria in wastewater increase rapidly when temperatures exceed 95 °F. If the plant effluent does not meet the reuse standards now, it will likely perform worse in the future. Therefore, we have assigned a score of 0 in this case. Dry spell periods can also decrease the performance of the current infrastructures. As the influent flow decreases under severe drought conditions, the downstream effluent (into the Chicago Sanitary and Ship Canal) experiences a marked increase in pollutant concentration, rendering the current facilities less efficient. Therefore, a score of 1 (low resiliency) is assigned. During rainstorms, the events can strain system performance. Heavy storms increase the influent flow to the wastewater treatment facility, potentially exceeding the facility's treatment capacity. Although the Stickney WRP does not receive rainwater through the sewer system, most of the infrastructure is outdoors, which can lead to hydraulic failures. Thus, the current scenarios demonstrate low resiliency, and the corresponding score is 1.

Second Scenario – UV Technology: In the second scenario, extreme temperatures, such as extreme heat, can reduce the treatment efficiency. Pathogenic microorganisms in water are more sensitive to UV radiation at higher temperatures, which can lead to a higher inactivation rate. However, when temperatures exceed 95 °F, bacteria present in wastewater reproduce more rapidly. Therefore, the effluent from the secondary treatment would have a higher pollutant load than currently recorded, and the UV treatment's effectiveness could be slightly reduced. In this case, the UV technology demonstrates high resiliency, and the respective score is 3. On the other hand, a dry spell scenario can decrease the performance of the technology. This reduction in the flow rate may affect the efficiency of the UV treatment, as the microorganisms in the water may receive less exposure to UV radiation. As the influent flow decreases, the downstream effluent (into the Chicago Sanitary and Ship Canal) experiences a marked increase in pollutant concentration. Therefore, even with a tertiary UV treatment, the final effluent may still have higher pollutant concentrations due to the lack of dilution water in the discharge channel flow. Thus, the score assigned is 2, indicating medium resiliency. Additionally, rainstorm events can reduce treatment efficiency. As explained before, the Stickney WRP does not receive rainwater through the sewer system, much of the infrastructure, including the UV treatment, is outdoors, leading to accumulated rainwater that can increase pathogen levels and decrease treatment performance. However, the advantage of the UV treatment is that it requires minimal contact time to be effective. Consequently, this translates to the medium negative impact but high resiliency, and the corresponding score is 3.

Third Scenario – Chlorination/Dechlorination Technology: In this scenario, extreme temperatures can significantly pressure the system's performance. High temperatures may affect the activity of microorganisms in wastewater. Although chlorination aims to disinfect and control microbial populations, temperature variations can alter the overall microbial response to chlorine, especially when the secondary treatment effluent carries a higher contaminant load. Consequently, this can imply a very high negative impact, indicating that this technology exhibits a low resiliency under these conditions; thus, a score of 1 is assigned. Similarly, dry spells can stress the technology. With decreasing wastewater flow, the concentration of contaminants may increase. This could intensify the chlorination processes due to an increased presence of organic matter, suspended solids, or other substances that react with chlorine, leading to a score of 1. Additionally, rainstorm events can reduce the treatment efficiency of the chlorination/dechlorination system. Heavy rain can significantly increase wastewater flow, which could dilute the chemicals used in the chlorination/dechlorination processes. This dilution can affect the final concentration of microbes and substances in the wastewater, including diluting the residual chlorine, thereby impacting the effectiveness of disinfection. Therefore, with this climate concern, the technology can experience a very high negative impact, and the corresponding score is 1 (indicating low resiliency).

Cascading Resiliency Impacts

The cascading environmental, social, and economic consequences are established based on the technical indicators of resiliency previously determined. Cascading environmental implications refer to the series of reactions from adverse effects that occur when one environmental change triggers a series of subsequent impacts across interconnected ecosystems or ecological processes. Economic cascading consequences describe the chain reaction of financial effects initiated by a single economic event or policy change, affecting various sectors due to interconnected dependencies and feedback loops within the economic system. Lastly, cascading social consequences refer to the chain reaction of impacts within societies when a single social change or event triggers a series of subsequent effects across various aspects of social life. These consequences often result from the interconnectedness of social systems and the complex relationships between individuals, communities, and institutions.

Cascading environmental resiliency indicators – The primary cascading environmental indicators are related to the effects on the vulnerability of each scenario and the consequences if the systems fail. These include the impact on local aquatic ecosystems, alteration of microbial communities, byproduct discharge into water bodies (such as pathogens and chlorine), development of UV-resistant strains, risk to biodiversity, and more. We are providing the justifications for all the indicators considered in our study. For instance, it impacts the local aquatic ecosystem because the effluent may contain pathogens and other harmful organisms, posing a severe environmental risk and indicating no resiliency. Similarly, this indicator is considered a risk to biodiversity due to the extremely high risk to biodiversity from untreated wastewater discharges, reflecting no resiliency.

In the first scenario, extreme heat can impact local aquatic ecosystems. Higher temperatures can increase pathogen survivability and metabolic rates, exacerbating their impact on ecosystems. Therefore, the current facilities show no resiliency, and the corresponding score is 0. Similarly, the microbial communities can be altered in seasons of prolonged dry spells. Dry conditions can reduce water flow and increase the concentration of microbes, disrupting natural microbial balance. Thus, extreme negative impacts can occur if the tertiary treatment is not implemented. In other words, the current facilities have no resiliency, and the score is 0. On the other hand, rainstorm events can dilute disinfectant concentrations, potentially reducing their efficacy and allowing for the survival and proliferation of resistant strains. In this case, the current infrastructure will face a shallow negative impact, showing a very high resiliency; hence, the score is 4.

In the second scenario, extreme heat can alter microbial communities. Extreme temperatures can alter the rate at which UV light affects different microbes, as well as microbial metabolism, and growth rates, which could impact the efficacy of UV treatment. Therefore, the score assigned is 1 (low resiliency) because of the probability of a very high negative impact. On the other hand, ecological impact due to UV disinfection can occur in dry spell periods. Reduced water flows can lead to higher concentrations of UV disinfection by-products, if any, and more direct exposure of aquatic life to UV light. However, since UV disinfection typically has minimal by-products, the primary concern would be ensuring the system adjusts to changes. Consequently, UV technology will face only medium negative impacts and show high resiliency (score 3). Similarly, changes in microbial diversity due to UV disinfection can occur in rainstorm events. These events can introduce a variety of new microorganisms into water bodies. The effectiveness of the UV system may vary depending on the influx of different microbial types and concentrations brought in by heavy rains. Thus, UV technology is highly resilient and will face only a medium negative impact (score 3).

In the third scenario, extreme heat can impact local aquatic ecosystems. Higher temperatures can exacerbate chlorine's toxic effects on marine organisms. Thus, chlorination/dechlorination technology can cause very high negative consequences, and it is considered to have low resiliency for this scenario, with a score of 1. In dry spell events, chemicals from the chlorination/dechlorination system can run off into local ecosystems. Lower water levels can concentrate chlorine by-products during dry spells, increasing toxicity. Therefore, it is assigned a score of 1, indicating low resiliency. Alteration of microbial communities can occur in rainstorm events. Rain can dilute chlorine's effects but may also spread altered microbial communities. Thus, it is considered a medium resiliency (score 2) due to the high negative impacts that this technology can cause after implementation.

Cascading economic consequences – The primary cascading economic indicators relate to the effects on the vulnerability of each scenario and the consequences if the systems fail. These include the impact on water pricing, cost of technology maintenance and operation, healthcare costs from waterborne diseases, cost of outbreak management, among others. We are providing justifications for all the indicators considered in our study. For instance, it is regarded as the cost of technology maintenance and operation because the absence of disinfection processes reduces operational costs, indicating very high economic resiliency in this narrow aspect. Costs associated with the operations and maintenance (O&M) of UV technology – owing to its energy efficiency and lack of chemical use – can lead to long-term savings, indicating high economic resiliency. Similarly, this indicator accounts for healthcare costs due to the potential increase in the prevalence of waterborne diseases, which can lead to significant healthcare costs, representing no economic resiliency.

In the first scenario, extreme heat can cause economic impacts from environmental degradation. Higher temperatures can worsen the effects of pollution, degrading ecosystems and reducing their monetary value. As a result, the current facilities cannot prevent extreme economic impacts. Thus, a score of 0 is assigned, indicating no resiliency. Additionally, extended dry spells can affect fisheries and agriculture. Dry conditions can concentrate pollutants, harming aquatic life and agriculture. Therefore, considering these factors, the current facilities demonstrate no resiliency (score 0) and can result in an extreme economic impact. On the other hand, the cost of technology maintenance and operation in the current facilities can support any heavy rain. However, over the next 20 years, increased rainfall may lead to greater damage and higher maintenance and operation costs. Therefore, the resiliency in this case is low, and a score of 1 is assigned.

In the second scenario, UV technology's maintenance and operation costs at high temperatures can be affected. Extreme temperatures may increase the cooling requirements for UV disinfection systems, potentially raising energy use and operational costs. However, the overall resiliency is high as these systems are generally energy efficient and can cope with temperature fluctuations. Therefore, this technology exhibits high resiliency and is assigned a score of 3. Similarly, high resiliency is noted for impacts on water pricing. Dry spells may reduce the overall water supply, which can increase the value of treated water. However, the high economic resiliency of UV systems, due to reasonable operational costs, can mitigate the impact on water pricing. Similarly, runoff may carry various contaminants into the water treatment system after rainstorm events. The UV treatment reduces the need for environmental mitigation related to disinfection by-products, which is why the resiliency is rated medium, as the primary concerns are handling increased water flow and turbidity rather than chemical contamination. Therefore, the cost of environmental mitigation is scored with 3.

In the third scenario, extreme heat can impact the liability and remediation costs. Higher temperatures can amplify chlorine's adverse effects, increasing potential liabilities. Therefore, this technology is assigned a score of 1, indicating low resiliency. Dry spell seasons can affect the cost of chemical procurement and handling. The concentration of chemicals may need adjustment during dry spells, impacting costs. Therefore, it is assigned a score of 2, representing medium resiliency. Similarly, the cost of regulatory compliance and monitoring during rainstorm events implies more frequent monitoring for by-products and can increase costs. Therefore, this scenario demonstrates a medium resiliency with a high negative impact, and a score of 2 is assigned.

Cascading social resiliency indicators – The cascading social indicators are related to the effects on the vulnerability of each scenario and the consequences if the systems fail, such as public health crises, impact on local recreation and lifestyle, employment opportunities, and others. For example, the potential for widespread public health issues due to pathogen spread is high, indicating no social resiliency. Continuous vigilance is required. Additionally, employment rates are unlikely to be affected by the absence of water disinfection, and implementation of UV technology may create specialized jobs. In contrast, chlorination may not significantly change employment opportunities. Thus, it is considered an indicator of economic resiliency.

In the first scenario, extreme heat can trigger a public health crisis. Heat may exacerbate the spread and virulence of waterborne diseases and, if additional treatment is not implemented, it can negatively impact human health. Therefore, a score of 0 is assigned, indicating no resiliency. Additionally, dry spell seasons can erode community trust in water management. Drought conditions may underscore the consequences of inadequate water treatment. Therefore, the current facilities do not demonstrate resiliency for future events, and a score of 0 is assigned.

Public health implications can be affected by extreme temperatures. Heat may increase pathogen activity; UV resiliency ensures effective inactivation, maintaining public health. Therefore, this technology demonstrates medium resiliency and receives a score of 2. On the other hand, dry conditions necessitate reliable water treatment, securing jobs in UV system maintenance. Thus, in this case, UV systems exhibit high resiliency with a score of 3. Similarly, rainstorms increase the demand for clean water; effective UV treatment enhances public trust. Therefore, community engagement and perception will not be adversely affected, indicating high resiliency for this scenario (score 3).

In the third scenario, extreme heat can impact public health, as it may increase the formation of harmful by-products, raising health concerns. Therefore, this technology exhibits medium resiliency and is assigned a score of 2. Similarly, reduced water flow during dry spells can increase by-product concentrations, affecting recreation. In other words, this climate hazard can impact the recreational use of water bodies; consequently, the system has medium resiliency (score 2) and represents a high negative impact. On the other hand, rainstorms may wash more by-products into water bodies, heightening community concerns. Thus, community engagement and perception will be negatively affected, representing a low resiliency for the chlorination/dechlorination technology.

Overall Resiliency Index

After completing all resiliency assessments (technical and cascading environmental, economic, and social consequences), the results were standardized and normalized into a single comparable unit using the Integrated Value Model for Sustainability Assessment (MIVES) model. The MIVES model assigns weights to each category and scales the result from 0 to 1, with the higher score indicating the better option. The four categories were analyzed with equal weight (25%). This means that technical (25%), environmental (25%), economic (25%), and social (25%) aspects are equally important in this analysis. Similarly, the three climate hazards were assigned equal weight, reflecting their equal importance: extreme temperatures (33%), rainstorms (33%), and dry spells (33%). The main results are presented as follows.

In Figure 21.7a, the weighted results are presented based on the probability of occurrence and severity of impact, among other factors, for the second scenario, when a chlorination/dechlorination system is implemented. Storms have a significant impact on the technical resiliency of this system. For environmental resiliency, the three climatic hazards must be considered to avoid future environmental impacts. The extreme heat highly impacts the social resiliency of this technology, while economic resiliency was most affected by dry spell periods. Figure 21.7b presents the weighted results for the third scenario, when a UV system is implemented. Rainstorms significantly impact the technical resiliency of this system.

For environmental resiliency, extreme heat and dry spells are critical factors to be considered for avoiding future environmental impacts. Extreme heat and rainstorms strongly affect the social and economic resiliency of this technology. Finally, in Figure 21.8, the overall resiliency indices for decision-making are presented. The overall resiliency index shows that both proposed technologies achieve high resiliency; however, UV technology is more resilient (0.91) compared to chlorination/dechlorination (0.84). Additionally, the scenario where no disinfection treatment is implemented is the least resilient, showing zero technical, environmental, and social resiliency. The current facilities show an economic resiliency score of 0.13 as no initial investment is required, and the facilities are already self-sufficient in terms of operating and maintenance costs. UV technology and chlorination/dechlorination have lower scores in the environmental dimension, 0.20 and 0.17, respectively.

Another scenario was considered for the MIVES resiliency index analysis by assigning different weights to the requirements. The assigned weights are technical (50%), environmental (16.66%), economic (16.66%), and social (16.66%), and the resiliency index results in this case differed slightly from those in the case where equal weights were used. However, UV technology still exhibits a

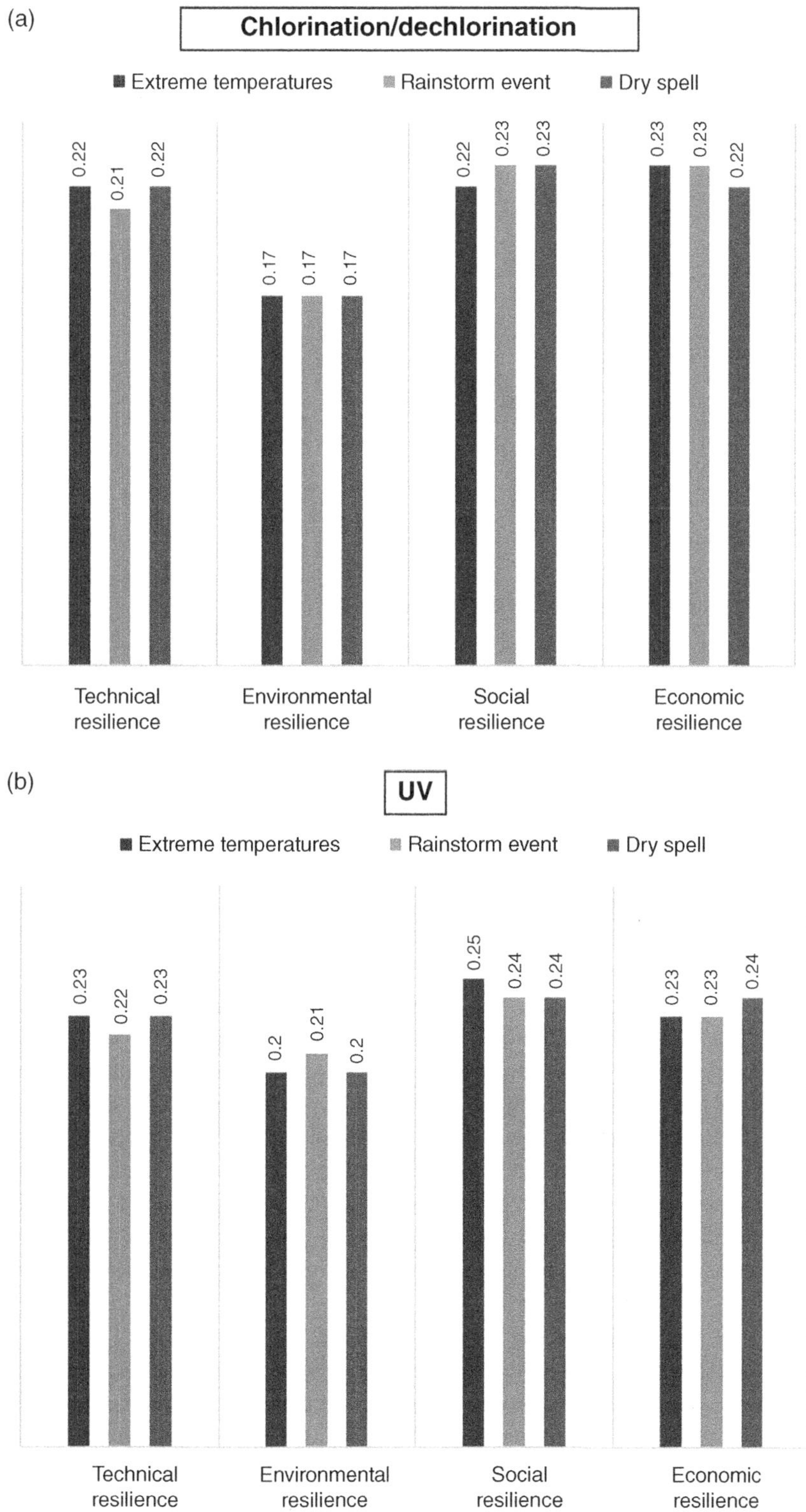

Figure 21.7 Resiliency index values for (a) chlorination/dichlorination and (b) UV technologies under different climate events.

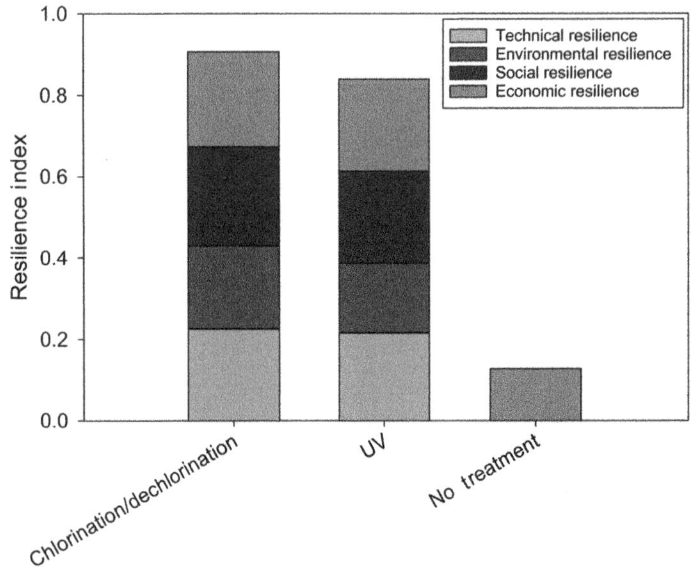

Figure 21.8 Overall Resiliency Index for the three scenarios.

higher resiliency (0.91) compared to chlorination/dechlorination (0.85), and the scenario without any disinfection treatment implemented was the least resilient (0.08).

In this project, we developed a resiliency assessment based on a semi-quantitative analysis. Although resiliency assessment is still a subject of ongoing studies, we aim to take the first step toward understanding and quantifying the resiliency of water disinfection technologies implemented at Stickney WRP. A common gap in resiliency assessment is the tendency to focus strictly on specific components, such as infrastructure or technology, while overlooking broader systemic factors. This narrow scope can miss interdependence between system components and external factors, such as regulatory frameworks, institutional capacity, and community participation, all of which are critical determinants of resiliency. Resiliency assessment often relies heavily on quantitative metrics and indicators, such as infrastructure performance metrics or economic cost-benefit analyses, which can overlook qualitative aspects of resiliency, like social cohesion, adaptive capacity, and community empowerment. Incorporating qualitative data and participatory approaches can provide a more holistic understanding of resiliency and inform more effective strategies for building resiliency. Therefore, the analysis performed in our resiliency assessment includes technical, economic, environmental, and social aspects, each with their respective qualitative analysis and rating. Additionally, while many resiliency assessments focus on short-term or immediate impacts and responses, we have considered both short-term and long-term climate concerns for our analysis.

UV technology is widely recognized for its effectiveness in disinfecting wastewater by inactivating pathogens, such as bacteria, viruses, and protozoa. Its reliability in achieving high levels of disinfection is well established, contributing to improved public health protection and compliance with regulatory standards for effluent water quality. Additionally, this system demonstrates its adaptability to various environmental factors, such as changes in flow rate and water quality characteristics. In summary, after analyzing technical, environmental, economic, and social parameters, this technology shows reliable results on resiliency, coping well with extreme heat, rainstorms, and periods of drought. Compared to chlorination/dechlorination systems, the UV system is still more resilient. However, chlorination/dechlorination technology also exhibits high resiliency. It is widely recognized for its effectiveness in achieving significant microbial reduction, providing additional protection against waterborne diseases. Nevertheless, the reliability of chlorination/dechlorination systems can be influenced by water quality, contact time, and chlorine dosage, which may vary depending on site-specific conditions and operational practices. Therefore, as the results indicate, this technology has somewhat less resilient than UV technology in social and environmental aspects. After evaluating these two scenarios, maintaining only the current infrastructure without adding disinfection treatment is the least resilient scenario if the objective is water reuse.

21.3.3.2 Sustainability Assessment

The first step of an integrated sustainability assessment involves the identification of environmental, social, and economic indicators, as mentioned before. The indicators identified for this case study, along with the corresponding metrics and tools used to quantify them, are presented in Table 21.2. A detailed description of quantification methodology followed and the results of each pillar of sustainability assessment along with the overall RSI are presented in this section.

Table 21.2 Sustainability indicators, metrics, and tools to assess different disinfection technologies.

Sustainability requirement	Indicator	Metric	Tool
Environmental	Ozone depletion	kg CFC-11 eq	LCA
	Global warming	kg CO_2 eq	
	Smog	kg O_3 eq	
	Acidification	kg SO_2 eq	
	Eutrophication	kg N eq	
	Carcinogenic	CTUh	
	Non-carcinogenic	CTUh	
	Respiratory effects	kg PM2.5 eq	
	Ecotoxicity	CTUe	
	Fossil fuel depletion	MJ surplus	
	Environmental resiliency index	Scores from resiliency assessment	Resiliency assessment
Social	Quality of life of an individual	Rating on a scale of −2 to +2	Public Survey
	Public health of an individual associated with the use of water discharged from treatment system		
	Degree of comfort of an individual associated with the use of water discharged from the treatment system		
	Degree of concerns of an individual due to odor problems and presence of pathogens in the water discharged from the treatment system		

(*Continued*)

Table 21.2 (Continued)

Sustainability requirement	Indicator	Metric	Tool
	Potential risks and safety hazard associated with the operation of the treatment system on the employees		
	Degree of disruption to the community for the construction of the water distribution pipelines		
	Incorporation of sustainability in the construction of infrastructure required for water reuse		
	Participation of research institutes for sustainable projects		
	Cost required for the construction of the treatment technology		
	Long-term cost required for the operation and maintenance of the treatment system		
	Economic benefits of reusing the discharged water from the treatment system		
	New employment opportunities for the construction and operation of the treatment technology		
	Employee training needed for the treatment technology		
	Energy consumption required by the treatment system		
	Impact of greenhouse gas emissions generated from the treatment system on global warming		
	Degree of impact with which water treatment system will affect other media (air and water)		
	Social resiliency index	Scores from resiliency assessment	Resiliency assessment
Economic	Building costs	USD	Life cycle costing
	Operation costs		
	Distribution pipes and pumps		
	Labor burden		
	Contingency		
	Distribution pipes and pumps		
	Energy		
	Cleaning		
	Miscellaneous allowance		
	Distribution pipes and pumps		
	Economic resiliency index	Scores from resiliency assessment	Resiliency assessment

System boundaries – The construction, transportation, and operation phases are considered for the sustainability assessment (Figure 21.9). The system boundaries of this project include all material and energy inputs that could affect environmental impacts during the operation of the systems. Electricity consumption for operating the chlorination/dechlorination system was assumed to be negligible; hence, it was not included. However, UV technology essentially requires electricity for UV lights and the equipment necessary for their functioning. The chlorination system requires the

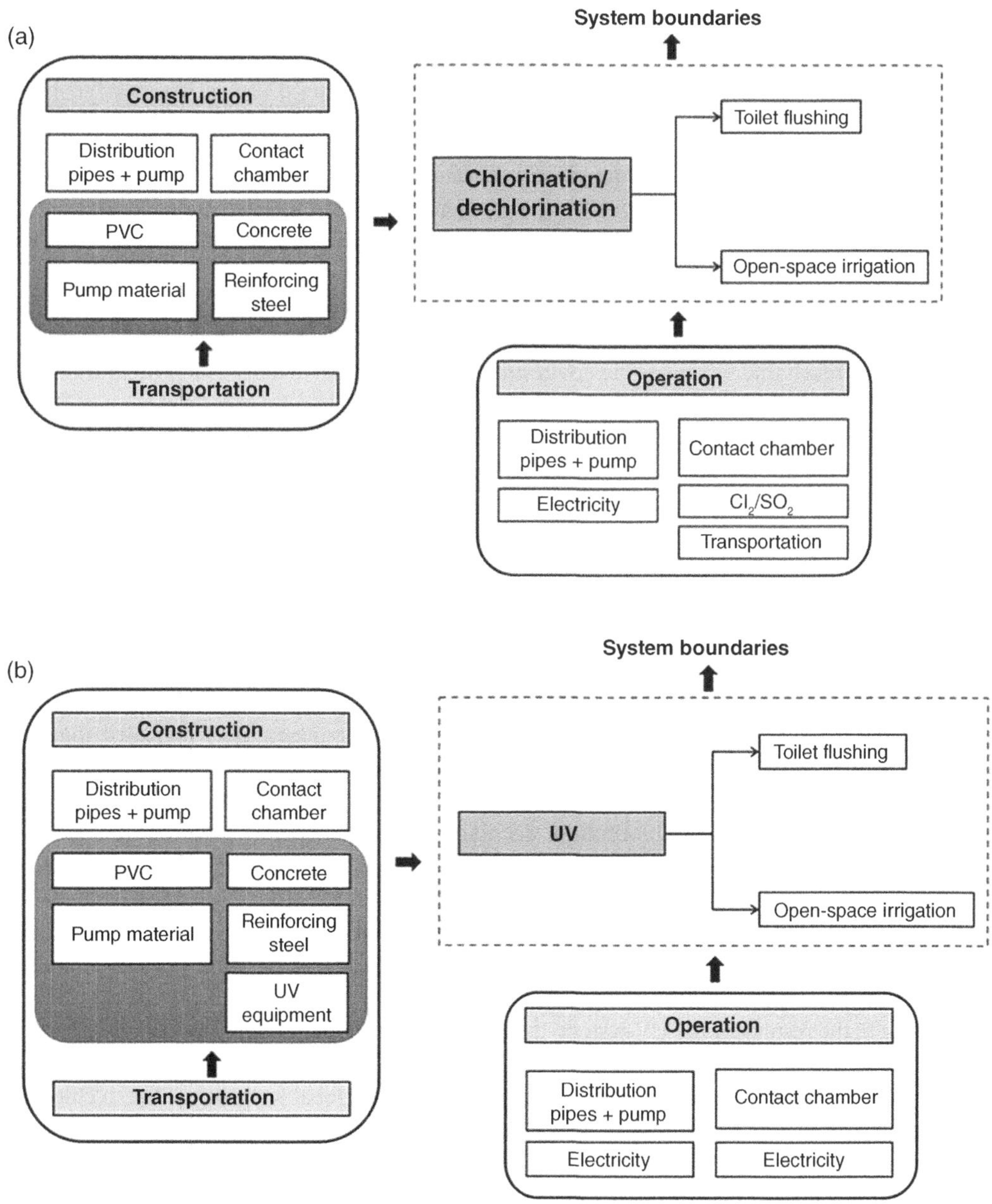

Figure 21.9 System boundaries for environmental assessment of (a) chlorination/dechlorination, and (b) UV technologies.

transportation of chemicals such as chlorine. Both systems are considered from two aspects. The main aspect involves analyzing each technology's construction, transport, and operation, including the requirements for water distribution for reuse purposes, such as irrigation and graywater. The second aspect includes only the operations of technology.

Environmental Sustainability

An environmental sustainability analysis was conducted to assess the environmental impact of incorporating disinfection systems into the existing infrastructure of the Stickney WRP. After establishing the system's boundaries, the preliminary inputs for SimaPro were based on the quantities of materials, energy, and transportation required for the technologies. Additionally, for input data, TRACI 2.1 V1.01/US 2008 and the materials and transportation inventory built into the software were used (Reddy et al. 2020). The leading environmental and human health indicators evaluated for this project included ozone depletion, global warming, smog, acidification, eutrophication, carcinogens, non-carcinogens, respiratory effects, ecotoxicity, and fossil fuel depletion. In addition to these environmental indicators, Table 21.2 displays the social and economic indicators selected for conducting the TBL assessment in the current study.

Life cycle assessment (LCA) – A LCA was performed to quantify the environmental impact of the proposed disinfection systems (chlorination/dechlorination versus UV) at Stickney WRP. It is a cradle-to-grave approach that evaluates the environmental effects of an activity or a process throughout its entire life cycle (from material acquisition to waste disposal and reuse) (Reddy et al. 2020). The aims of the LCA are to quantify and evaluate the environmental burdens associated with a product or process, including resource depletion, energy consumption, greenhouse gas emissions, air and water pollution, and waste generation. By analyzing these impacts across the life cycle, LCA helps identify opportunities for improvement and optimization to minimize environmental impacts and enhance sustainability. For this project, SimaPro v9.0 software was used to evaluate the environmental sustainability of both UV and chlorination/dechlorination technologies.

Figure 21.10 shows the results of the main scenario, which considers construction materials, transportation, the energy consumption of the different operations, and the water distribution system for reuse. According to the results, UV technology has the least impact on the environment and human health. The results from the environmental sustainability assessment indicated that the chlorination/dechlorination system had higher adverse environmental impacts in almost all of the impact categories, except for fossil fuel depletion, compared to the UV system. One of the possible reasons for this was that the chlorination/dechlorination system involved the substantial daily use of chlorine gas, which had a higher negative environmental impact. Additionally, the amount of concrete required to build the contact chamber for the chlorination/dechlorination system was greater than the UV system. All the factors discussed above concerning chlorination/dechlorination outweigh the negative impacts caused by the concrete used in building the UV chamber and the electricity consumed during its operation.

According to the results for the UV system, the leading environmental indicators that will have a significant negative impact include global warming, smog, and fossil fuel depletion. In the case of the chlorination system, for example, the substantial impact on global warming is due to chlorine production, which can indirectly contribute to greenhouse gas emissions if fossil fuels are used in its manufacturing. On the other hand, UV technology can negatively impact global warming because electricity generation is closely linked to global warming due to the greenhouse gas emissions associated with many methods of generating electricity, such as fossil fuel combustion and carbon intensity.

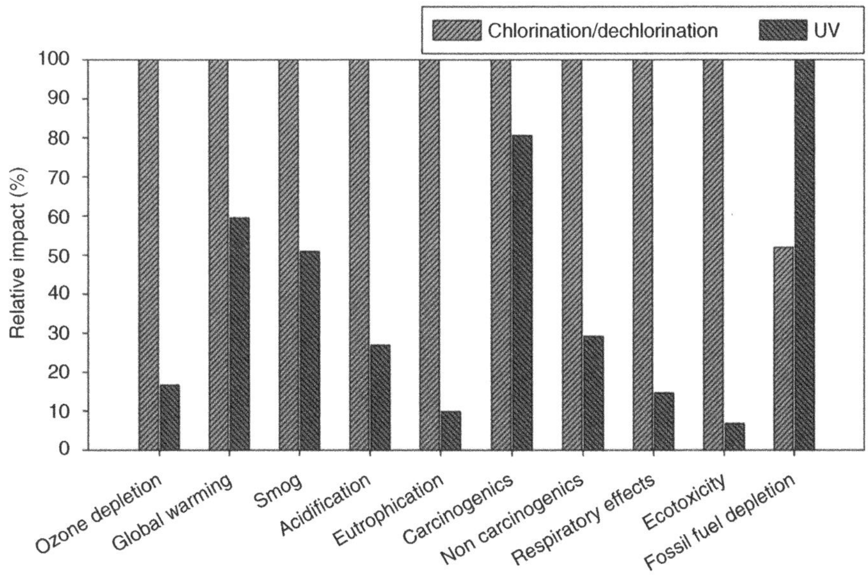

Figure 21.10 Characterization impacts of chlorination/dechlorination versus UV system including pumps and distribution pipes.

Social Sustainability

A social sustainability assessment was performed to evaluate the social impacts that the disinfection technologies and the current untreated wastewater scenario at Stickney WRP may have on the community. Social sustainability is an integral part of the TBL assessment and is essential for ensuring stakeholders' contribution. A stakeholder is defined as any individual, group, or organization that can affect or be affected by business or project actions, objectives, and policies. Stakeholder participation offers many benefits, including fostering complete project transparency with the local community. Therefore, we have identified the main stakeholders in this project as follows:

Direct stakeholders:

1) Stickney WRP management and operations team: They oversee the treatment processes and are responsible for implementing and managing the new disinfection technology.
2) Local regulatory authorities: They ensure compliance with environmental and public health regulations for water treatment and reuse.
3) Residents within the 2-mile radius: They are the immediate recipients of the water reuse benefits and are potentially affected by the changes in water quality.
4) Businesses using treated water: Local industries and commercial entities that may use the treated water.

Indirect stakeholders:

1) Environmental NGOs: They are interested in the environmental impact of wastewater treatment and its effects on the Chicago Sanitary and Ship Canal.
2) Media: They report information about the project and influence public perception and awareness.

3) Research and development organizations: Institutions that might use project data to develop better treatment solutions or may have contributed to technology.
4) Tourism and recreation sector: Entities dependent on the water bodies for their business are affected by the quality and reputation of the water in the canal.
5) Downstream communities and users: Communities and users further along the Chicago Sanitary and Ship Canal who would be affected by the changes in water quality due to the upstream treatment processes.

Public survey – Once the main stakeholders were identified, we surveyed them to establish their preferences and concerns. The survey was broken down into two main sections. The first section of the survey consisted of 4 general questions designed to determine the knowledge and experience of the respondents. The next section consisted of 20 questions for each proposed disinfection technology and the existing scenario at Stickney WRP. These 20 questions were organized into 5 categories: individual, community, economic, environmental, and resiliency assessment. This last category is included in the survey because one of our objectives is to determine an RSI. Therefore, we need to include social resiliency indicators in the sustainability assessment (Table 21.2). Thus, in this last category, the questions pertain to resiliency to extreme temperatures, public health implications, worker safety, community engagement, and perception of disaster. Respondents were asked to rate each question on a scale of −2 to 2. The rating system was as follows: −2 (unacceptable), −1 (diminished), 0 (neutral), 1 (improved), and 2 (ideal).

Figure 21.11 shows the final tabulated results from the social survey. It presents the results for each category and the overall results. After evaluating the five categories, the results show that UV technology is the highest-rated option. Stakeholder preferences are favorable to the implementation of this technology. Additionally, it is evident that no treatment was the least preferred option. Also, the chlorine system shows lower scores than UV; this may be because the population is concerned about the disinfection by-products of chlorine. Finally, in the category of resiliency assessment, we can see that according to stakeholders' perception, UV is more resilient than chlorine systems. The results also reveal that the lowest scores are in the socio-environmental category. This

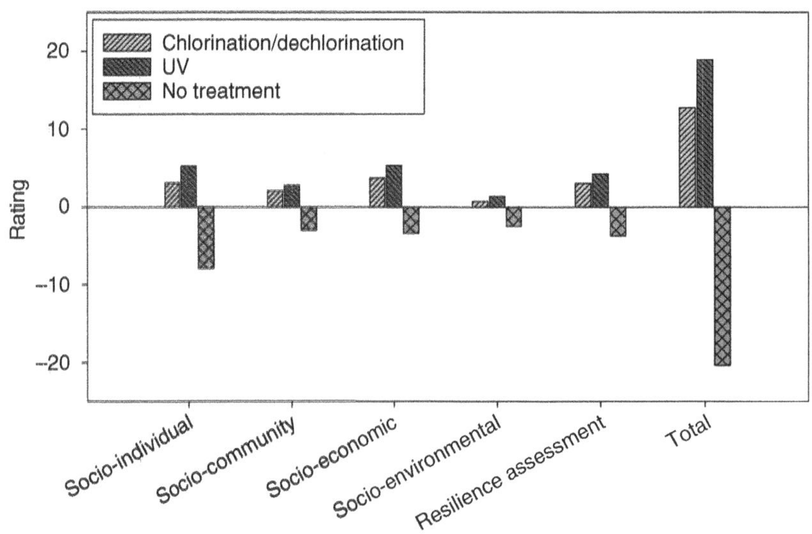

Figure 21.11 Final social survey results.

means that stakeholders consider all three scenarios to potentially impact the environment, yet UV technology remains the preferred technology.

Economic Sustainability

The economic sustainability assessment was conducted after calculating the cost for the two proposed systems, distribution pipelines and pumps. The economic assessment categories were divided into direct costs, O&M cost, and indirect costs. Direct costs consist of all construction and material costs required for disinfection systems, distribution piping, and pumps. For instance, the costs for the chlorination/dechlorination system include concrete, steel, and chemicals (chlorine). Similarly, for the UV technology, the costs considered are for concrete, steel, and UV lamps. The direct costs of the water distribution pipes and pumps were also included. Indirect costs include the cost of transportation of materials for each process. Finally, the operation and maintenance costs of each of the technology and the pump system were considered for a service life of 20 years, which corresponds to our design life (Table 21.2).

Figure 21.12 shows the results of the economic assessment. The total direct cost for chlorination/dechlorination, including distribution pipelines and pumps, was estimated at around $14 270 500 compared to $13 981 500 for UV disinfection. Transportation costs were also taken into consideration based on the distance from the chemical and material suppliers to the Stickney WRP for construction and operation. The life cycle inventories for raw materials and transportation were then input into LCA for both systems. The transportation cost for the chlorination/dechlorination system was about $500 000 for the first year, while the transportation cost for the UV system was calculated to be about $14 000. The total annual O&M cost for UV disinfection, at $220 400, was about 50% higher than for chlorination/dechlorination, which was approximately $106 900. UV systems require periodic maintenance, such as cleaning and replacing UV lamps and sleeves, and they consume electricity to power the UV lamps and associated equipment. The total indirect costs were quite comparable between chlorination/dechlorination and UV, at roughly around $4 261 000 and $4 198 600, respectively. UV technology generally involves higher upfront capital costs compared to chlorination. UV systems require the installation of UV lamps, reactors, and associated

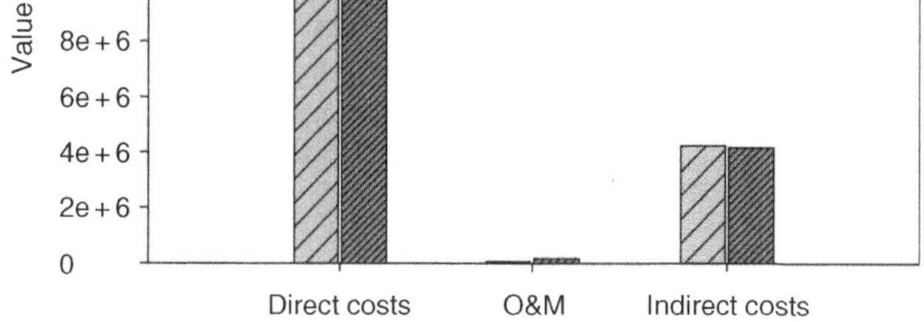

Figure 21.12 Cost comparison between chlorination/dechlorination versus UV systems.

control equipment, which can be expensive to purchase and install. In contrast, chlorination systems have lower initial capital costs.

21.3.3.3 Integrated Resilient Sustainability Index

In the integrated resiliency and sustainability analysis, we unified the resiliency indicator defined in Section 21.3.3.1 in the sustainability assessment. The cascading environmental and economic resiliency indicators were incorporated into the environmental and economic sustainability assessments. The social resiliency assessment is already included in the public survey results developed for the social sustainability assessment. Thus, the results were standardized and normalized into single comparable units. For this purpose, the MIVES model was adopted, and the corresponding resiliency indicators were included in each category. The MIVES model assigns weight to each category and scales the result from 0 to 1 (with the higher score indicating a better option). The three categories were analyzed with different weights, indicating that environmental, economic, and social aspects vary in importance in this analysis.

The integrated RSI is presented in Figure 21.13. This figure graphically represents the cumulative scores based on the MIVES model, accounting for varying environmental, economic, and social sustainability weights. The specific weightings for each sustainability category vary across the plots, emphasizing the importance of each category under different scenarios.

Figure 21.13a indicates the integrated index scores when all three sustainability categories are weighted equally, with environmental, social, and economic sustainability each contributing 33.33% to the total score. Chlorination/dechlorination scores 0.49, UV treatment scores 0.64, and no treatment scores 0.33. These values illustrate that under a balanced weighting system, UV disinfection provides a superior combination of resiliency and sustainability compared to chlorination/dechlorination, with the no-treatment option lagging significantly behind. Figure 21.13b changes the weighting to favor environmental sustainability (60%), with social and economic sustainability each reduced to 20%. This alteration in weighting still results in UV disinfection being the more sustainable option, scoring 0.75 and demonstrating its robustness in environmental terms. Meanwhile, no treatment scores poorly at 0.20.

Figure 21.13c significantly shifts the emphasis to social sustainability (60%), with environmental and economic sustainability each accounting for 20%. This shift still results in a trend where UV treatment scores higher at 0.59, reflecting its strong social acceptability, while the score for chlorination/dechlorination slightly decreases to 0.46. The score for no treatment remains the same at 0.21. Finally, Figure 21.13d presents a scenario where economic sustainability is most heavily weighted (60%), with the other two categories at 20%. Here, "no treatment" shows the highest score of 0.59, slightly higher than UV treatment's score of 0.58. This high score for the "no treatment" option could be due to zero expenses associated with this option compared to the other two alternatives, which involve certain costs with their implementation.

Through these figures, the results illustrate that UV treatment consistently ranks the highest across the majority of the scenarios except the last one, asserting its position as the most resilient and sustainable option for wastewater disinfection. The scores and weightings reflect a thorough evaluation, considering the complex interplay among technical, environmental, social, and economic aspects for each technology.

21.3.4 Conclusions and Recommendations

Implementing UV technology as a disinfection treatment at Stickney WRP is a more robust and sustainable alternative to incorporating a chlorination/dechlorination system. The current design

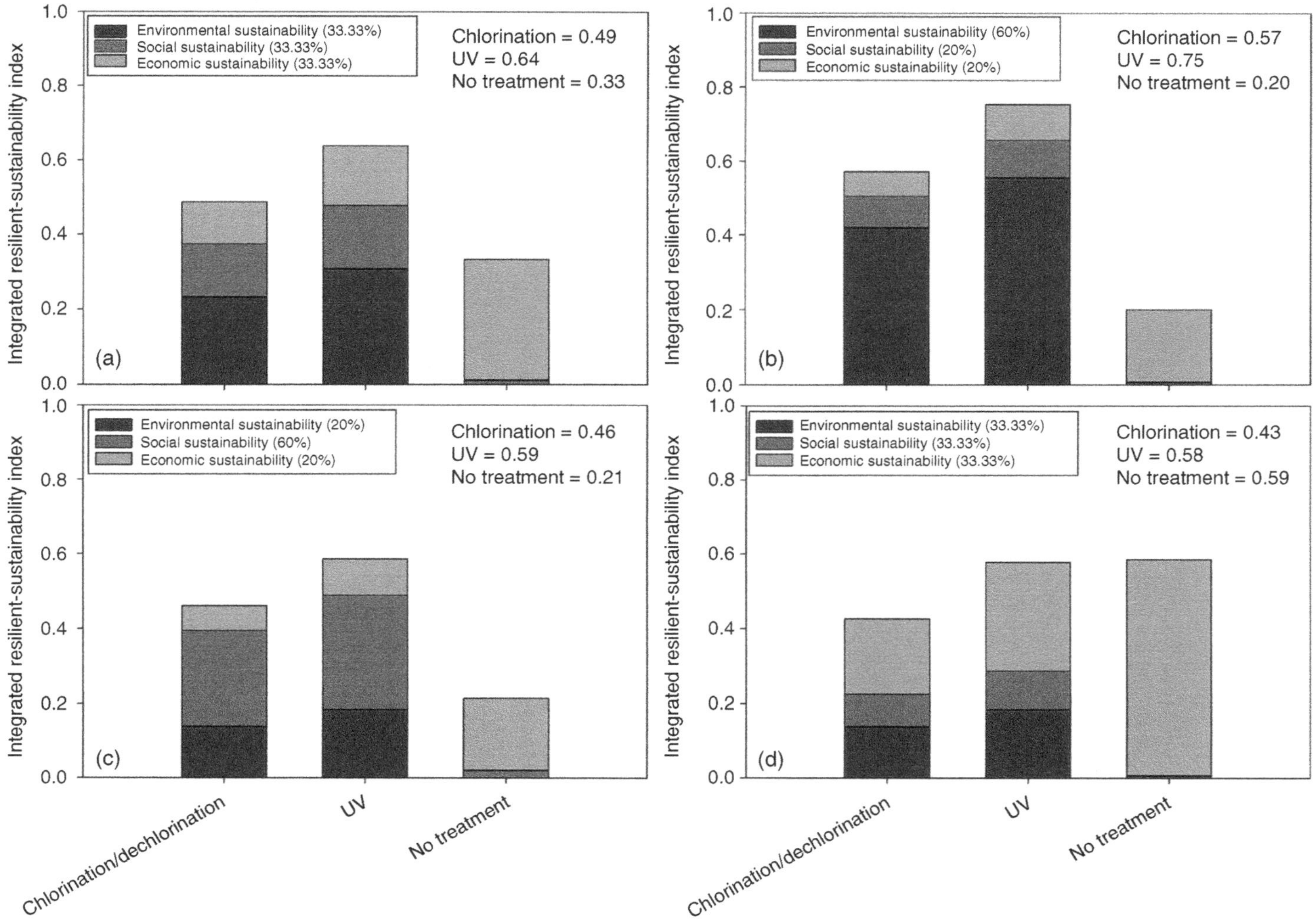

Figure 21.13 Integrated RSI for the three scenarios: chlorination/dechlorination, UV, and no disinfection, with different weights for environmental, social, and economic sustainability.

of UV technology offers high resiliency and sustainability; therefore, it is recommended to add this technology to reuse the effluent in irrigation safely and as graywater for toilet flushing.

Chlorination/dechlorination also shows positive results in terms of resilience and sustainability. However, some of this system's primary environmental and human health challenges are related to disinfection by-products, some of which are carcinogenic. These by-products can be produced when residual chlorine contacts organic matter and ammonia present in the wastewater. However, this technology can be applied with a strategy to prevent the formation of disinfection by-products, such as removing excess natural organic matter before chlorination, for example, by using granular activated carbon. While this strategy can enhance the resiliency and sustainability of the system, a more comprehensive analysis considering the additional costs is needed.

UV disinfection is not affected by temperature or humidity, unlike chemical disinfection techniques, and does not produce hazardous residues in the environment. This makes UV technology particularly suitable and resilient for wastewater treatment facilities operating in various climates. However, this technology requires energy, which entails environmental impacts. Hence, implementing renewable energy becomes essential to increase the sustainability of this technology. In addition, the high operating and maintenance costs are areas that could be improved, considering that implementing this system can also increase employment rates, we can argue that it is comparable to the cost disadvantages.

Not applying any technology is the least acceptable option. It implies that the effluent cannot be reused and that there is a possible risk of system failure due to future extreme weather events. Since water is a limited resource, priority must be given to its conscious and proper management. Therefore, the opportunity to implement improvements to safeguard natural resources cannot be overlooked.

Resiliency assessment is still under development, and thus, the initial steps taken in this project could shed light on developing tools for an integrated assessment of resiliency and sustainability. Filling the gaps in resiliency assessment requires a more holistic, dynamic, and participatory approach that considers the full spectrum of risks, uncertainties, and complexities inherent in wastewater treatment systems. By incorporating diverse perspectives, embracing complexity and uncertainty, and fostering adaptive capacity at all levels, resiliency assessment can evolve into a more effective tool for building resilient and sustainable infrastructure.

21.4 Resiliency and Sustainability Assessment to Design and Remediate a Contaminated Site

21.4.1 Problem Statement

Contaminated sites can severely affect human and ecosystem health based on the extent of contamination and the contaminant type. Soil and groundwater contamination can easily reach the communities nearby, increasing the risk of contaminated sites. Remediation of the site is a way forward if the contamination levels harm the ecosystem and the human life in and around the site (Sharma and Reddy 2004). Further, there have been increasing concerns over the changing climate and its impacts on engineering projects in recent years. It is necessary to ensure that the remediation projects are resilient to withstand the stressors caused by these climate impacts. Ensuring any engineering project is sustainable and does not add to the existing problem through environmental, social, and environmental impacts is also essential. To address the concerns above, a systematic assessment is needed to evaluate the vulnerability of remediation technologies to external stressors and disruptions. Additionally, a sustainability assessment should focus on the long-term effects of remediation technology on the environment and its social and economic aspects. This project involves a case study of contaminated soil and groundwater at an oil refinery in Lincoln, Nebraska. The study presents a comprehensive assessment of the resiliency of the remediation alternatives against changing climate conditions and the sustainability of the alternatives concerning their long-term environmental, social, and economic impacts. Among the feasible remediation technologies, in situ electrokinetic remediation (EK), excavation and off-site disposal, and emulsified zero-valent iron nanoparticles (EZVI) were selected as alternatives for soil remediation at the source area of the contamination. Pump-and-treat and EZVI were chosen as alternatives for groundwater remediation in the migrated contaminated plume.

21.4.2 Project Background

This case study aims to provide a comprehensive resiliency and sustainability assessment for designing remediation methods at a contaminated site, using a remediation case study at an oil refinery in Lincoln, Nebraska. The site considered in this study is a facility used for soybean extraction, processing, and oil refinery located northeast of Lincoln. The facility, constructed in 1950 for soybean grain elevators, is owned by Archer Daniels Midland (ADM). The site was investigated for potential contamination by Huff & Huff, Inc., in 2008, following the discovery of possible carbon tetrachloride contamination in the water wells downstream from the facility by the USEPA. The contaminants were primarily found in the soil near the bean silos and in groundwater in the contaminant plume that extends up to 0.75 miles northeast of the facility. The contaminants of concern included carbon tetrachloride, carbon disulfide, chloroform, methylene chloride, and 1,2-dibromoethane. All contaminants were found in soil and groundwater, except for carbon disulfide, which was only detected in groundwater. The contamination was understood to have occurred in two stages: first, an accidental release of petroleum in 1999 involving the discharge of 750 gallons of fuel from broken underground pipes; second, a continuous release of carbon tetrachloride over several years, which was used as a fumigant before ADM took ownership of the site.

The project site and its surrounding area are underlain by a 19-foot clay layer, with a 2-foot layer of sandy clay beneath it, followed by a 37-foot layer of sand. Beneath these layers lies a 40-foot Dakota sandstone formation. The terrain on the surface slopes to the north, directing water flow from south to north and northeast of the facility into the nearby Steve Creek. Consequently, the contaminants released from the source area disperse slowly through the silty layer until they reach the sand layer, where the flow accelerates, eventually reaching the groundwater.

The contaminants were mainly discovered in the source area (soil contamination) and the deep zone (groundwater) (Figure 21.14a,b). The area near the bean silos' southwest quadrant had the highest quantities of each contaminant in the soil. Additionally, as shown in the figure, the groundwater contaminated plume from the source location crosses Steve Creek, extending 0.75 miles northeast. Since none of the contaminants have a viscosity lower than water and all have a specific gravity greater than water, they will all sink in and flow downward more quickly than water (Davidson et al. 2005). The study focused on bringing down the contaminant concentrations in the soil and the groundwater to the remedial goals.

21.4.2.1 Technical Design of Alternatives

Several remediation technologies are available to remediate groundwater and the soil, such as monitored natural attenuation, excavation, air sparging, EK remediation, soil flushing, bioremediation, soil vapor extraction (SVE), and the EZVI method. However, not all these methods are appropriate

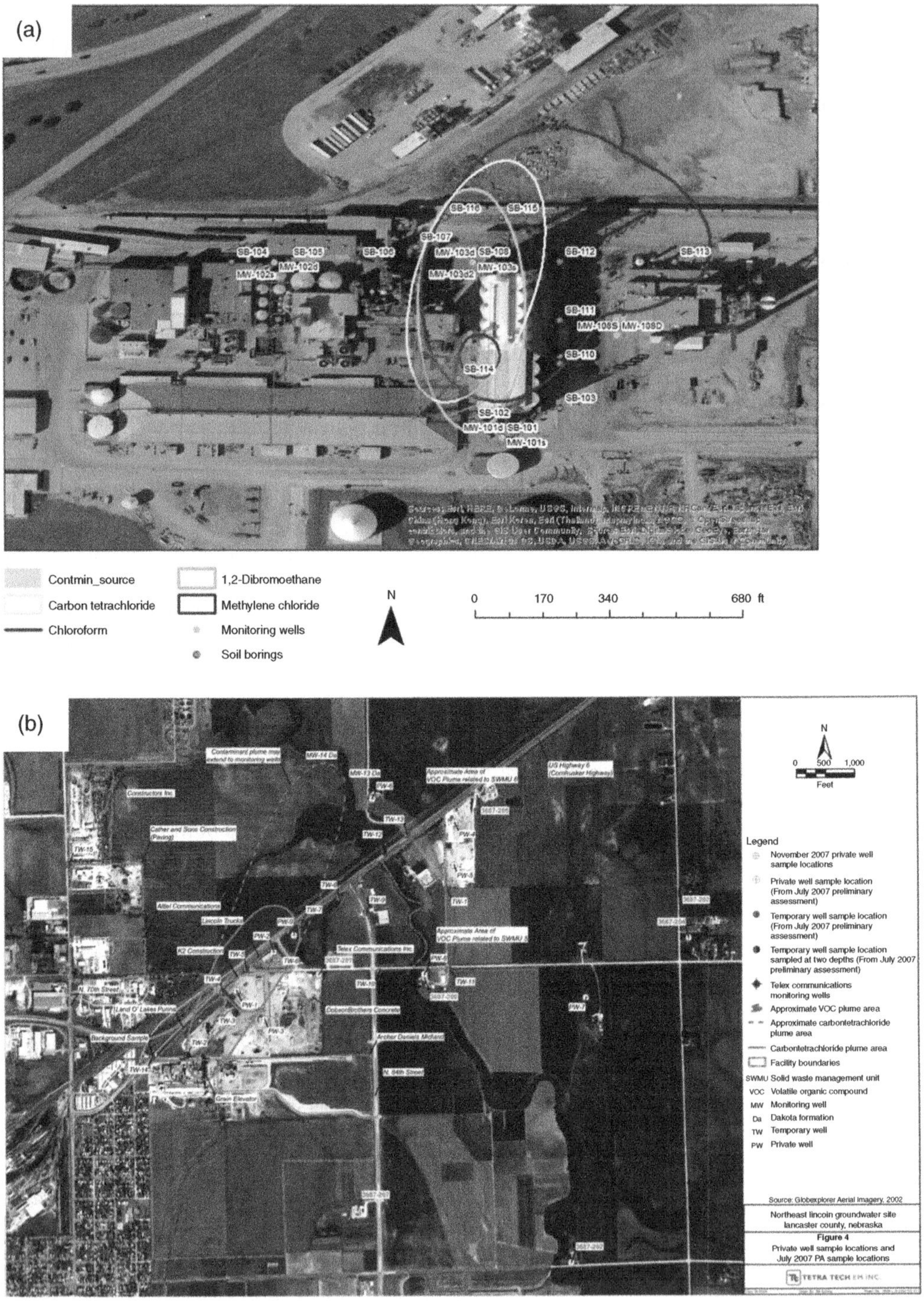

Figure 21.14 Contaminated area: (a) soil contamination, (b) groundwater plume.

for application at this contaminated site. A technical screening of various remediation techniques was performed using the federal remediation technologies roundtable (FRTR) and the USEPA guidelines as a screening matrix method. The remediation technologies were rated as average, above average, and below average based on different criteria. Technologies such as excavation, air sparging, EK remediation, soil flushing, bioremediation, SVE, and the EZVI method were considered for soil remediation. Pump-and-treat, air sparging, EK remediation, permeable reactive barrier, bioremediation, SVE, and EZVI methods were assessed for groundwater remediation. Based on the scores after the technical screening, excavation, EK remediation, and EZVI were selected as technical alternatives for soil remediation, and pump-and-treat and EZVI were selected for groundwater remediation.

Soil Remediation

Excavation – It is a commonly used method in environmental cleanup that involves excavating the contaminated soil or material from the site. This process often addresses immediate risks to human health or the environment. The excavated hazardous material is then safely transported for disposal or to a treatment facility for processing if necessary (USEPA 2001).

In the scope of the study, excavation is to be done within a contaminated area of 100 ft × 100 ft. The initial phase will involve the complete clearance of the site to facilitate the excavation process. Subsequently, the soil will be excavated, reaching a substantial depth of 35 ft to ensure complete contaminant removal. Post-excavation, the contaminated soil will be securely transported for disposal to an off-site landfill. Refilling material will be sourced to the site after successfully disposing of the contaminated soil. This systematic process of excavation, disposal, and refilling ensures the effective remediation of the site, rendering it safe and suitable for subsequent use.

Electrokinetic remediation – This method involves applying an electrical potential difference to the sub-surface to create movement of ions, facilitating the removal of contaminants through various methods, such as electro-migration and electro-osmosis.

In the contaminated area measuring 100 ft × 100 ft, an array of 18 anodes and 18 cathodes (Figure 21.15a) will be installed to a maximum depth of 15 ft. The anodes consist of steel bars encased in PVC, while the cathodes are hollow stainless-steel bars. The electrolytes used in this setup are sodium hydroxide (NaOH) for the anodes and ethylenediaminetetraacetic acid (EDTA) for the cathodes. Pumps are to be installed to ensure the adequate circulation of these electrolytes. Additionally, separate tanks will be used to store electrolytes for the anodes and cathodes. The quantiles of electrolytes necessary and the time of remediation for operation were calculated based on the initial concentrations of the contaminants found in the soil and the remedial goals.

Emulsified zero valent iron – EZVI is a remediation method that mixes food-grade vegetable oil with a surfactant, elemental iron, and water to obtain an emulsified mixture, which is then injected into the ground. It creates a specific structure that enables direct destruction of dense non-aqueous phase liquids (DNAPLs) using a combination of abiotic and biotic processes (Quinn et al. 2005). It is a viable option considering long-term biodegradation from oil, minimal labor and waste generation, and the fact that it can treat soil and groundwater.

In the remediation plan, 27 locations with a radius of 10 ft and 5 locations with a radius of 5 ft were proposed for EZVI injection (Figure 21.15b). The injection will be carried out at intervals of 2.5 ft in depth, starting from the top of the impacted soils (estimated to be at 10 ft) until a depth of 35 ft to reach the groundwater. At each interval, the injection consisted of 102 l of an emulsified mixture (10% nano-scale zero valent iron, 51% water, and 39% vegetable oil). This is immediately followed by injecting 342 l per interval of a 2% emulsified vegetable oil solution. This plan ensures the effective application of EZVI, tailored to the site's specific conditions.

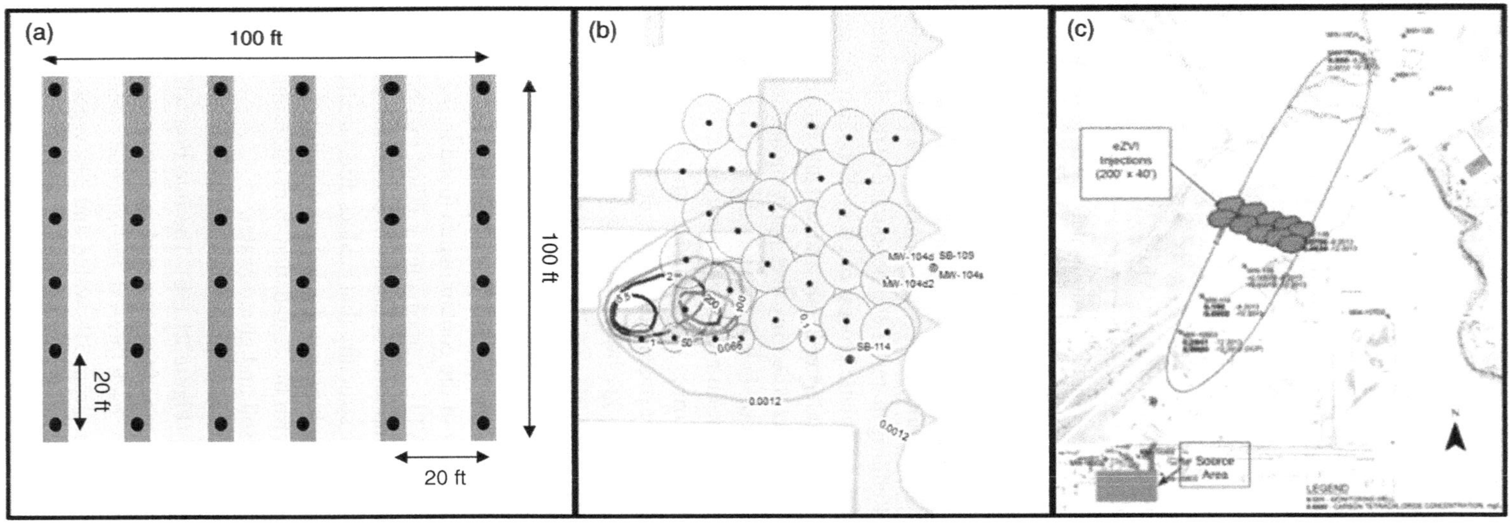

Figure 21.15 Proposed plan of remediation for (a) electrokinetic remediation of soil; (b) EZVI remediation of soil; and (c) EZVI remediation of groundwater.

Groundwater Remediation

Pump-and-treat – It is a standard method for remediating groundwater contaminated with dissolved chemicals, including industrial solvents, metals, and fuel oil. Groundwater is pumped from wells to an above-ground treatment system to remove the contaminants. Pump-and-treat systems are also used to "contain" the contaminant plume (Mackay and Cherry 1989; USEPA 2017b). Containment of the plume prevents it from spreading to nearby drinking water wells, wetlands, streams, and other natural resources.

In the northeast region of the refinery, spanning an area of 200 ft × 40 ft, the proposed plan involves installing 8 extraction wells and 8 injection wells to control the migration of contaminants effectively. The wells are drilled to a depth of 60 ft to address contamination in the deep plume. Our system is designed to extract water at a rate of 100 gallons per minute. Since the contaminants used are VOCs (Volatile Organic Compounds), tray air stripping was considered as the treatment method, with pumped water flowing through trays equipped with a blower to remove dissolved contaminants during the treatment process (Heron et al. 1998). The operation period will be calculated based on the remedial goals and the initial concentrations found in the groundwater. Further, extensive monitoring will allow for tracking the progress of the remediation efforts and making necessary adjustments to optimize the effectiveness of the treatment.

Emulsified zero valent iron – Similar to the soil remediation plan, the proposed plan involves injecting EZVI into 20 locations, each with a radius of 10 ft, within an area of 200 ft by 40 ft (Figure 21.15c). The EZVI mixture (10% nano-scale Zero Valent Iron, 51% water, and 39% vegetable oil), followed by the 2% emulsified vegetable oil injection at 2.5-foot intervals, remains the same as the soil remediation process. Injection will start at the base of the sandy clay–sand interface and continue down to an estimated depth of 60 ft.

Monitoring

A total of 12 monitoring wells are proposed for installation, three of which will be within the soil remediation area for surface monitoring, and one will be installed as a background reference point during the upgrade. Further, two wells will be positioned downgradient to measure the remediation impacts effectively. Six groundwater monitoring wells will be located at three distinct locations within the middle and deep zones.

The primary function of these wells is to measure the effects of the remediation process, both downgradient and cross-gradient. This monitoring strategy ensures that we have a thorough understanding of the remediation plan's effectiveness and the migration of contaminants. The data collected from these wells will provide valuable insights into the progress of the remediation efforts and guide any necessary adjustments to the approach.

21.4.3 Integrated Resiliency and Sustainability Assessment

21.4.3.1 Resiliency Assessment

Using different online tools, such as Climate Check, Federal Emergency Management Agency (FEMA), Climate Risk and Resiliency Portal (ClimaRR), and CMRA, potential climate change events at the site were identified. These events include flooding, frost days and ice storms, tornadoes, and droughts. Each disaster was assessed based on different risk indices, including Climate Mapping, National Risk Index, US Vulnerability Index, and Climate Vulnerability Index. The risks were rated from "relatively high" to "very high," indicating the severity and likelihood of the event occurring. This comprehensive risk assessment allows us to understand the potential threats and incorporate appropriate mitigation strategies in the remediation efforts.

The Resiliency Index (RI) of each alternative was obtained following a thorough resiliency assessment. The indicators used, metrics, and the scale adopted for rating are presented in Table 21.3 The

Table 21.3 Resiliency indicators, metrics, and tools to assess different remediation technologies.

Resiliency requirements	Criteria	Indicators	Metric	Tool
Technical	Performance deterioration	Performance loss	Rating on a scale of 0 to 4	Expert rating
		Monitoring wells		
		Contaminants dispersion		
	Time for deterioration	Performance loss		
		Monitoring wells		
		Contaminant dispersion		
	Time in deteriorated state	Monitoring wells		
		Remediation process		
	Performance recovery	Monitoring wells performance		
		Remediation efficiency		
	Time for recovery	Monitoring wells		
		Redesign		
Cascading environmental	Air	Release of VOCs	Rating on a scale of 0 to 4	Expert rating
	Water	Impact on private wells		
		Impact on nearby creek		
		Impact on groundwater		
	Human health	Exposure to workers		
		Exposure to the public		
	Land	Impact on topsoil		
Cascading social	Social individual	Exposure to remediation workers	Rating on a scale of 0 to 4	Expert rating
		Impact on jobs of remediation workers		
		Exposure to refinery workers		
		Impact on jobs of refinery workers		
	Socio-community	Impact on nearby residents		
		Impact on agricultural area		
		Impact on trust of community		
	Socio-economic	Effect on agricultural yield		
		Effect on local businesses		
	Socio-environmental	Effect of air quality		
		Effect of water quality		
		Effect on soil quality		
Cascading economic	Direct costs	Cost of redesign	Rating on a scale of 0 to 4	Expert rating
		Increased maintenance costs		

Table 21.3 (Continued)

Resiliency requirements	Criteria	Indicators	Metric	Tool
	Indirect costs	Operations of refinery		
		Decrease in property values		
		Refinery business		
	Direct benefits	Redesign employment		
	Indirect benefits	Improved water management post redesign		
		Improved soil quality management post redesign		

Note: Rating based on level of negative impact: 0-extreme; 1-very high; 2-high; 3-medium; 4-low.

RI of three alternatives for the source area are presented in Figure 21.16a. The results indicate that soil remediation excavation is the most resilient option, with an RI of 0.975, compared to 0.80 for EZVI and 0.65 for EK remediation methods. As excavation involves removing all the contaminated soil from the site, the process does not pose any future threat due to climate change impacts. Hence, the excavation has a perfect score of 0.25 in the technical resiliency, cascading environmental, and social dimensions. However, the economic resiliency score for excavation is slightly below 0.25, as the other two methods provide employment opportunities during maintenance and redesign. EZVI performs better than the EK method as the EK involves above-ground structures that can be affected by external stressors. Of the two alternatives considered for groundwater remediation, EZVI, with an RI of 0.92, was more resilient than the pump-and-treat method (RI = 0.72) (Figure 21.16b). EZVI performs better in all dimensions as the pump-and-treat involves continuous operation and has structures above ground that could be affected by climate change impacts and disruptions. These results affirm the project's resiliency and ability to deliver reliable outcomes in changing environmental conditions and extreme events.

21.4.3.2 Sustainability Assessment

The first step of an integrated sustainability assessment involves the identification of environmental, social, and economic indicators, as mentioned before. Indicators identified specific to this case study, and the corresponding metrics and tools used for the quantification of the indicators are presented in Table 21.4. A detailed description of quantification methodology followed and the results of each pillar of sustainability assessment along with the overall integrated RSI are presented in this section.

Environmental Sustainability

Environmental sustainability involves evaluating the environmental effects of remediation methods. This dimension includes analyzing emissions, resource consumption, waste management, and the overall ecological footprint of each technology. The environmental impacts of the alternatives were evaluated using SimaPro software, which assesses impacts across various categories based on the materials and processes involved. Similarly, SiteWise, a tool designed for evaluating the environmental impacts of remediation alternatives, was used to quantify environmental indicators.

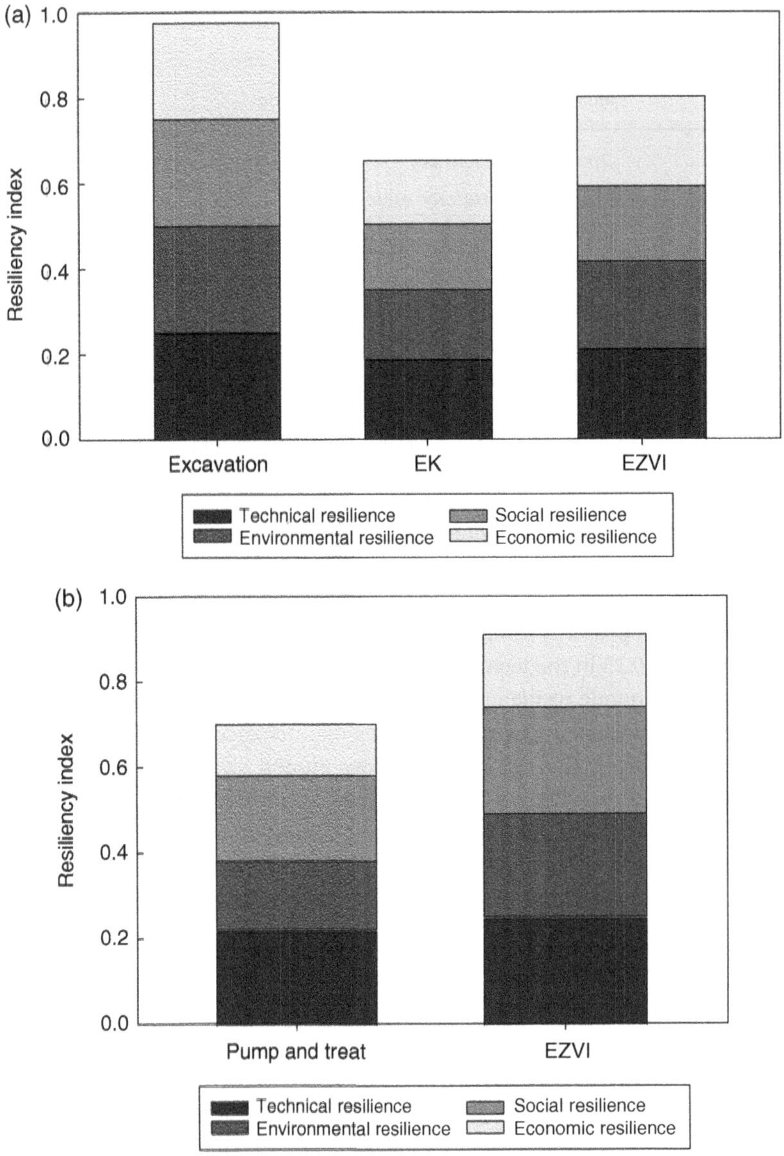

Figure 21.16 Resiliency Index of alternatives (a) soil remediation; (b) groundwater remediation.

The steps in the remediation methods and system boundaries for the environmental impact analysis are shown in Figure 21.17, while the SiteWise results for soil and groundwater remediation are presented in Figure 21.18. The impact categories from SiteWise include GHG (Greenhouse gas) emissions, energy use, water consumption, NO_x, SO_x, and particulate matter emissions. SiteWise analysis shows that the EZVI method has a significantly lower environmental impact than other methods for both soil and groundwater remediation. The lower environmental impact of EZVI is attributed to its lower energy and water consumption, making it a more environmentally friendly

Table 21.4 Sustainability indicators, metrics, and tools to assess different remediation technologies.

Sustainability requirements	Criteria	Indicators	Metric	Tool
Environmental sustainability	Air	Ozone depletion	kg CFC-11 eq	LCA
		Global warming	kg CO_2 eq	
		Smog	kg O_3 eq	
		Total particulate matter emissions	Metric ton	
		Total NO_x emissions	Metric ton	
		Total SO_x emissions	Metric ton	
	Water	Total water usage	Gallons	
		Acidification	kg SO_2 eq	
		Eutrophication	kg N eq	
	Human-health	Carcinogenic	CTUh	
		Non-carcinogenic	CTUh	
		Respiratory effects	kg PM2.5 eq	
	Land and ecosystem	Ecotoxicity	CTUe	
		Fossil fuel depletion	MJ surplus	
	Resiliency	Environmental resiliency index	Scores from resiliency assessment	Resiliency assessment
Social sustainability	Social-individual	Effect on quality of life	Rating on a scale of 0 to 4	Expert Rating and Survey
		Effect on public health		
		Contribution to learning opportunities and skill development		
		Effect on community/civic pride		
		Effect on perception of refinery		
		Effect on recreational activities		
	Socio-community	Changes in land use pattern	Rating on a scale of 0 to 4	
		Effect on housing conditions		
		Effect on recreational facilities		
		Effect on transportation		
	Socio-economic	Effect on local businesses	Rating on a scale of 0 to 4	
		Employment opportunities		
		Effect on new investment		
		Additional tax revenue		

(*Continued*)

Table 21.4 (Continued)

Sustainability requirements	Criteria	Indicators	Metric	Tool
		Effect on new business creation		
		Effect on collaboration with outside investors		
	Socio-environmental	Effectiveness of the remediation method	Rating on a scale of 0 to 4	
		Degree of disruption (Noise, Truck traffic)		
		Increase in energy consumption		
		Increase in water consumption		
		Impact on other environmental media (emissions, air pollution)		
		Impact on environment through operation		
	Resiliency	Social resiliency index	Score from resiliency assessment	Resiliency assessment
Economic sustainability	Direct costs	Design	USD	Cost-analysis
		Equipment		
		Operations		
		Waste treatment/disposal		
		Maintenance		
		SiteWise operations cost	USD	SiteWise
	Indirect costs	Property value reduction	Rating on a scale of 0 to 4	Expert rating
		Land use		
	Direct benefits	Revenue		
		Employment		
		Land property value increase post remediation		
		Resource conservation		
	Indirect benefits	Awareness		
		Competency		
		Welfare		
	Resiliency	Economic resiliency index	Score from resiliency assessment	Resiliency assessment

Note: Rating based on level impact: 0-unacceptable; 1-diminished; 2-no impact; 3-improved; 4-ideal.

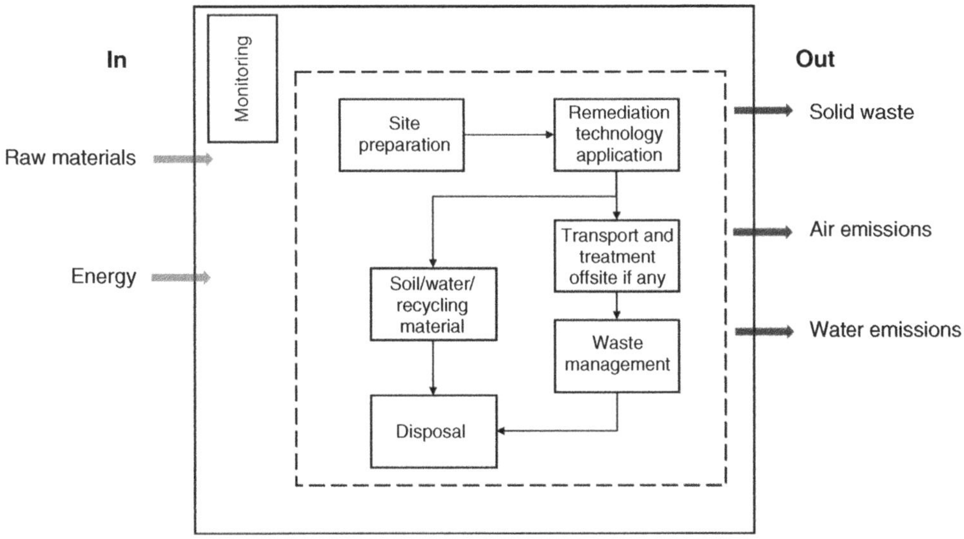

Figure 21.17 System boundaries of remediation methods.

option than other methods considered in the study. In contrast, excavation requires heavy machinery for soil removal and disposal, while EK remediation and pump treatment involve prolonged operation and high energy consumption. These benefits are critical for minimizing the overall environmental footprint of remediation activities.

SimaPro results comparing the alternatives in soil remediation are presented in Figure 21.19a. While excavation is effective at removing pollutants, it contributes significantly to global warming, acidification, and fossil fuel depletion compared to other methods. These impacts stem from heavy machinery use, high energy demands for material transport and processing, and disruption of terrestrial ecosystems. On the other hand, EK remediation had higher environmental impacts in several categories, as the energy-intensive nature of the electrokinetic process and the associated chemical treatments add significant emissions and resource use. EZVI in the source area has a lower impact in most categories, primarily due to its lower energy requirements. However, it has the highest impact in eutrophication, as the EZVI material can oxidize and react with organic matter in the soil, forming potentially harmful compounds. Figure 21.19b, showing the SimaPro results for groundwater remediation alternatives, demonstrates that EZVI has lower environmental impacts than the pump-and-treat method. This is primarily because EZVI treats contaminants on-site, reducing the need for extensive water pumping, processing, and treatment associated with pump-and-treat systems, which require constant energy input and water management.

The results highlight the importance of choosing remediation techniques compatible with sustainability goals. EZVI's lower impact on critical environmental metrics in both soil and groundwater remediation scenarios indicates that it is a better alternative for projects that aim to minimize ecological disturbance. In contrast, methods such as EK remediation and excavation cause significant environmental impacts despite their effectiveness in pollutant removal due to their higher resource and energy requirements. These assessments highlight the need to consider a wide range of environmental impacts. By choosing methods such as EZVI, it is possible to achieve restoration goals by adhering to the principles of environmental sustainability.

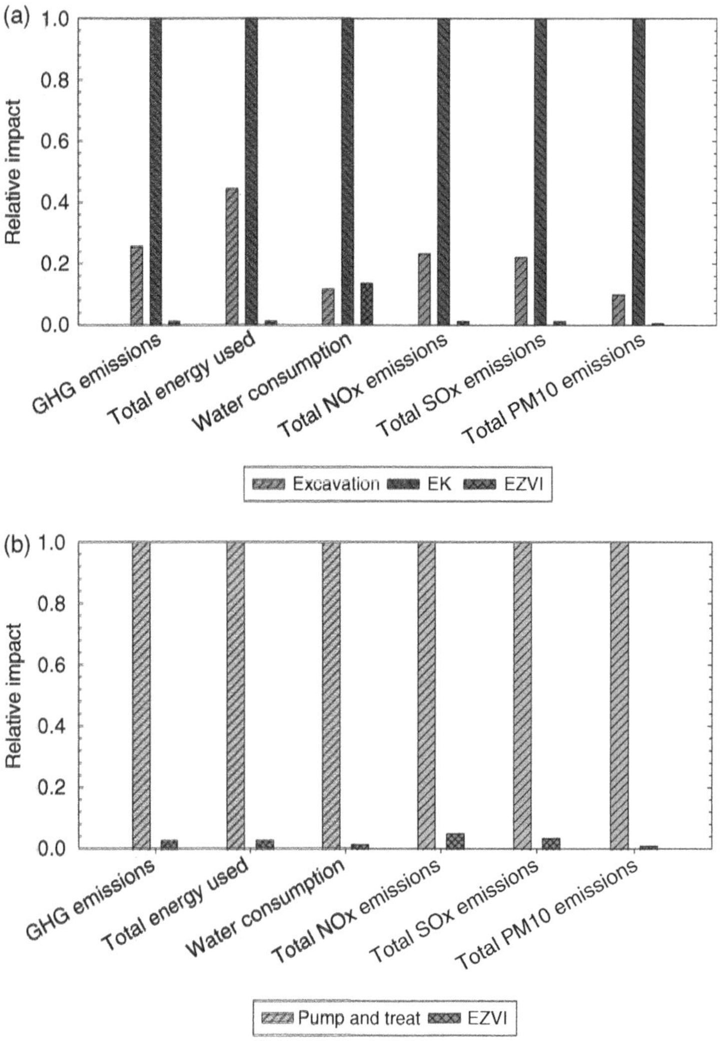

Figure 21.18 SiteWise environmental impact comparison among alternative remediation technologies for (a) soil remediation, and (b) groundwater remediation.

Social Sustainability

Essential factors to consider for social sustainability include the effects on public health, the community's well-being, and on the general standard of living for people who live close to remediation sites. The social sustainability indicator approach is structured to assess the impact of a remediation technology from the socio-individual, socio-community, socioeconomic, and socio-environmental perspectives (Li et al. 2021). EJScreen provided a detailed demographic and socioeconomic profile of the community surrounding the contaminated site. This assessment is critical to ensure that remediation efforts do not disproportionately affect vulnerable populations and that community engagement strategies are appropriately tailored. To assess social sustainability using the information collected from the EJScreen, the assessment is divided into two

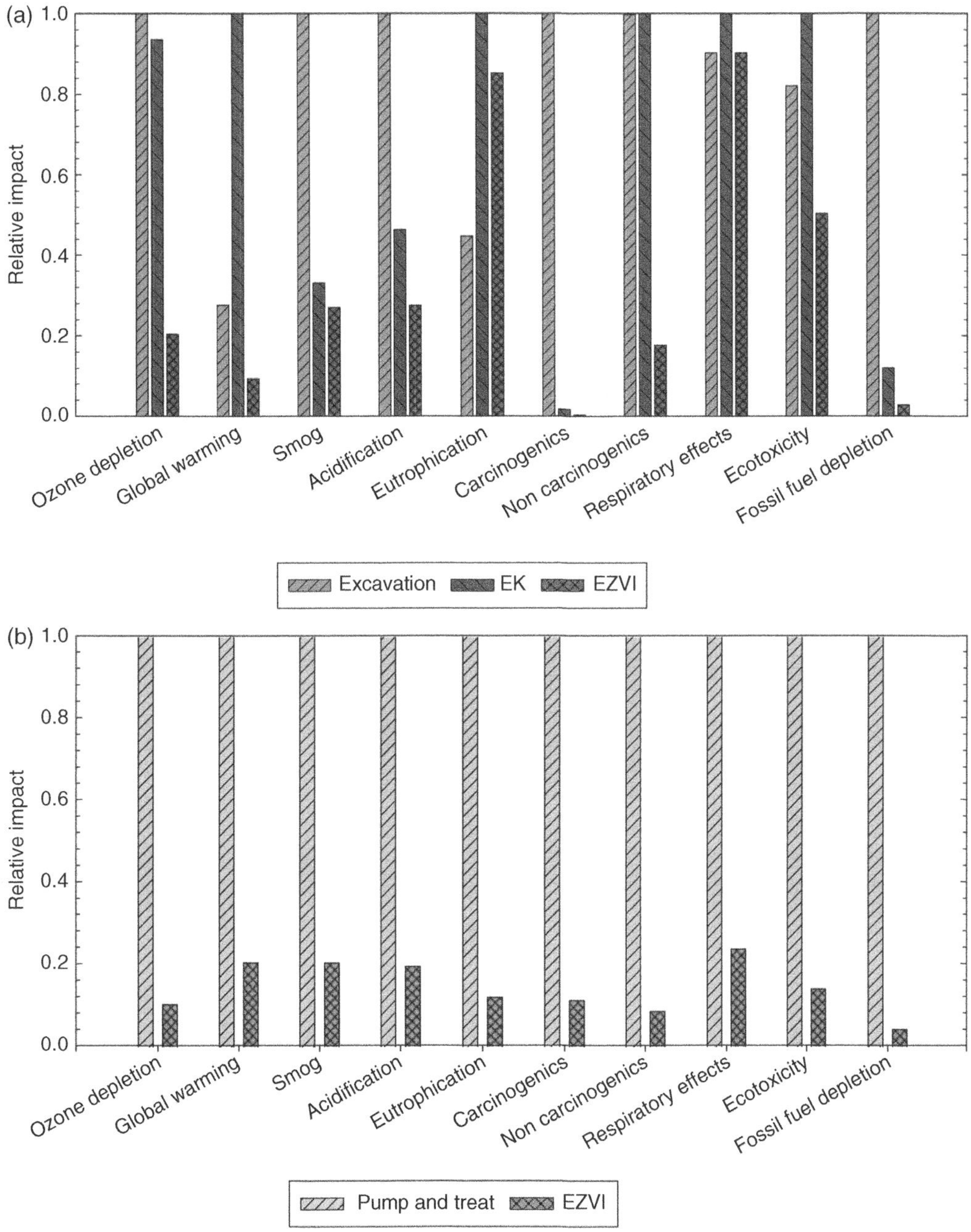

Figure 21.19 SimaPro environmental impact comparison among alternative remediation technologies for (a) soil remediation, and (b) groundwater remediation.

categories: a research group and a survey, each evaluating impacts across five critical social sustainability criteria: Social Individual, Social Community, Social Economic, Social Environmental, and Social Resiliency. Under each of these criteria, various indicators were considered based on the SSEM.

The results from both research group scores and the survey show that excavation was least beneficial socially while the EZVI had the most beneficial scores. These scores reflect the fact that continuous operation of the EK method and extensive initial excavation work significantly affect the community and the businesses around the contaminated site. Similarly, pump-and-treat involves continuous operation for a prolonged period, indicating less beneficial social impacts than the EZVI method in groundwater remediation. However, the scores from the research group show that it is beneficial to use EZVI over excavation and EK for soil remediation and EZVI over pump-and-treat in groundwater remediation. The integrated results from the survey and the group scaled to 1 are presented in Figure 21.20. Although the benefits are reflected in the survey scores, the differences are not as substantial, highlighting the importance of stakeholders' involvement in the decision-making process.

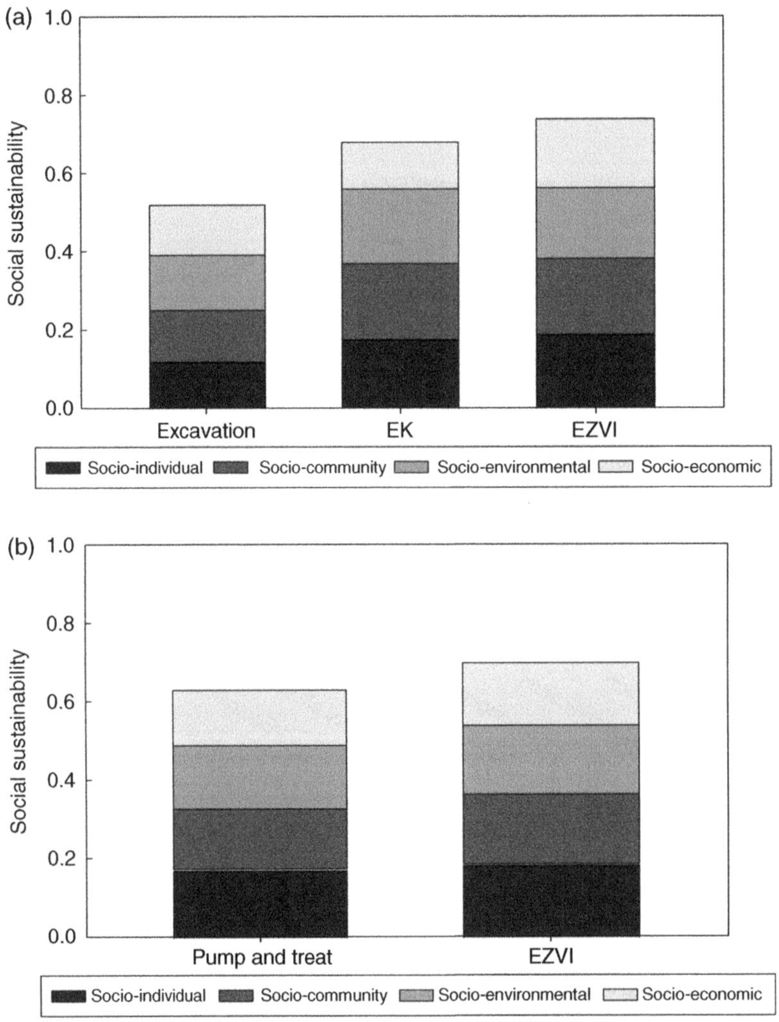

Figure 21.20 Social sustainability comparison among alternative remediation technologies for (a) soil remediation, and (b) groundwater remediation.

Economic Sustainability

Economic sustainability involves the economic benefits and cost-effectiveness of remediation alternatives as the main components. This covers the direct expenses of installing the remediation technology and maintaining the methods and the indirect costs and benefits, including property value changes, effects on nearby businesses, and possible job creation. The goal is to strike a balance that will support the long-term economic stability of the contaminated area. The costs of the technologies were estimated based on the literature review and FRTR reference guide (Goldstein and Ritterling 2001). In soil remediation, the excavation method incurred direct costs of $1 100 664, with significant expenditures directed toward excavation, transport, landfilling, backfilling, compaction, and planning and design. This reliance on physical labor and logistics reflects the method's significant operational demands. The EK remediation method costs $1 086 200, with planning and design constituting the largest share, underscoring a substantial investment in preparation and strategy. Operational costs accounted for 20% of the expenditure, highlighting the ongoing expenses required for the remediation process. Meanwhile, the EZVI remediation method, with a total cost of $817 524, allocated most of its budget to planning, design, and the injection of EZVI, focusing resources on initial setup and direct treatment activities. Among the three methods, EZVI demonstrated the lowest total cost, emphasizing efficiency in planning and intervention. Regarding indirect costs, EK remediation was higher than excavation, while EZVI had the lowest. These differences reflect each method's varying long-term maintenance costs or secondary expenses. Economic analysis showed that all three methods shared similar indirect benefits, including impacts on ADM refinery revenue, employment generation during remediation, property value enhancements post-remediation, and resource conservation. However, excavation and EK remediation offered notable additional benefits, such as raising community awareness about remediation technologies and showcasing immediate competency in achieving cleanup goals. The sustainability evaluation concluded that EK and EZVI methods provided comparable economic advantages to excavation.

In groundwater remediation, the pump-and-treat method, with a total cost of $979 500, devoted most of its budget to pumping equipment, making it the primary cost driver. By contrast, EZVI remediation for groundwater costs $684 466 and focuses expenditures on planning, design, and EZVI injection while incurring significantly lower indirect costs. The results upon integration are scaled and presented in Figure 21.21. Comparative scoring of direct and indirect benefits revealed that EZVI is better than the pump-and-treat method, offering a more economically sustainable alternative.

21.4.3.3 Integrated Resilient Sustainability Index

To find RSI of the alternatives, resiliency scores in each dimension were incorporated into the TBL sustainability evaluation. During the integration, equal weights were used for all the indicators under each criterion. In social sustainability, to incorporate results from both the group assessment and survey, each criterion was assigned 10%, except for social resiliency, which was assigned 20%. The list of indicators, metrics, and tools utilized in the assessment is presented in Table 21.4, and the RSI values of different alternatives are presented in Figure 21.22. Environmental sustainability considers how each alternative impacts the environment and reduces contamination in the region. Social sustainability assesses the remediation's impact on the community, considering public health, safety, and community disruption. On the other hand, economic sustainability evaluates the technology's cost-effectiveness, considering long-term financial costs and benefits. Integrating resiliency into each of these TBLs and formulating an RSI helps in understanding a holistic view of

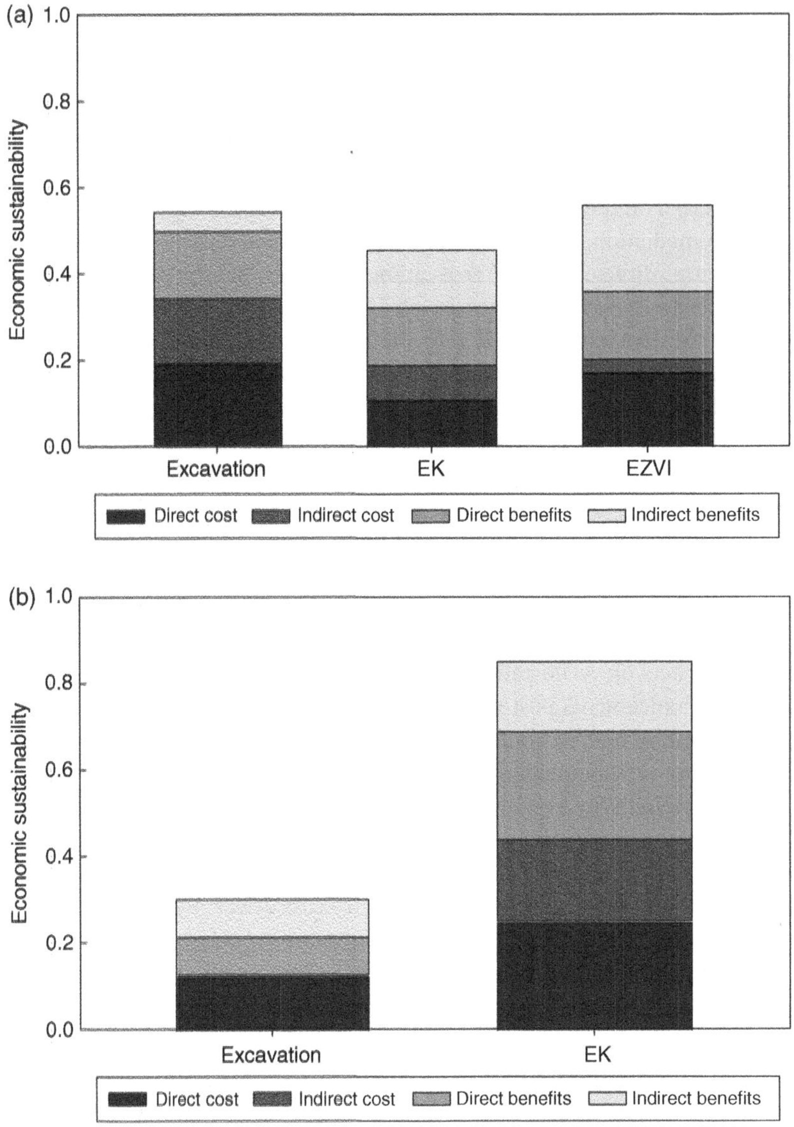

Figure 21.21 Economic sustainability comparison among alternative remediation technologies for (a) soil remediation, and (b) groundwater remediation.

the long-term impacts of the remediation method. It also assesses the short-term resiliency of the system to climate change and other external stressors.

Among the soil remediation methods, excavation showed moderate environmental sustainability, with EK remediation having the highest impact and lowest sustainability and EZVI showing the lowest impact and highest environmental sustainability. Social sustainability was low for excavation, whereas EZVI and EK displayed comparable social sustainability. Despite its higher direct costs, the excavation method exhibited economic sustainability comparable to EZVI and significantly higher than EK, attributed to greater economic resiliency in excavation and lower resiliency in EK. However, the overall RSI score for excavation was slightly lower than that of EK remediation, with EZVI emerging as the most resilient and sustainable alternative.

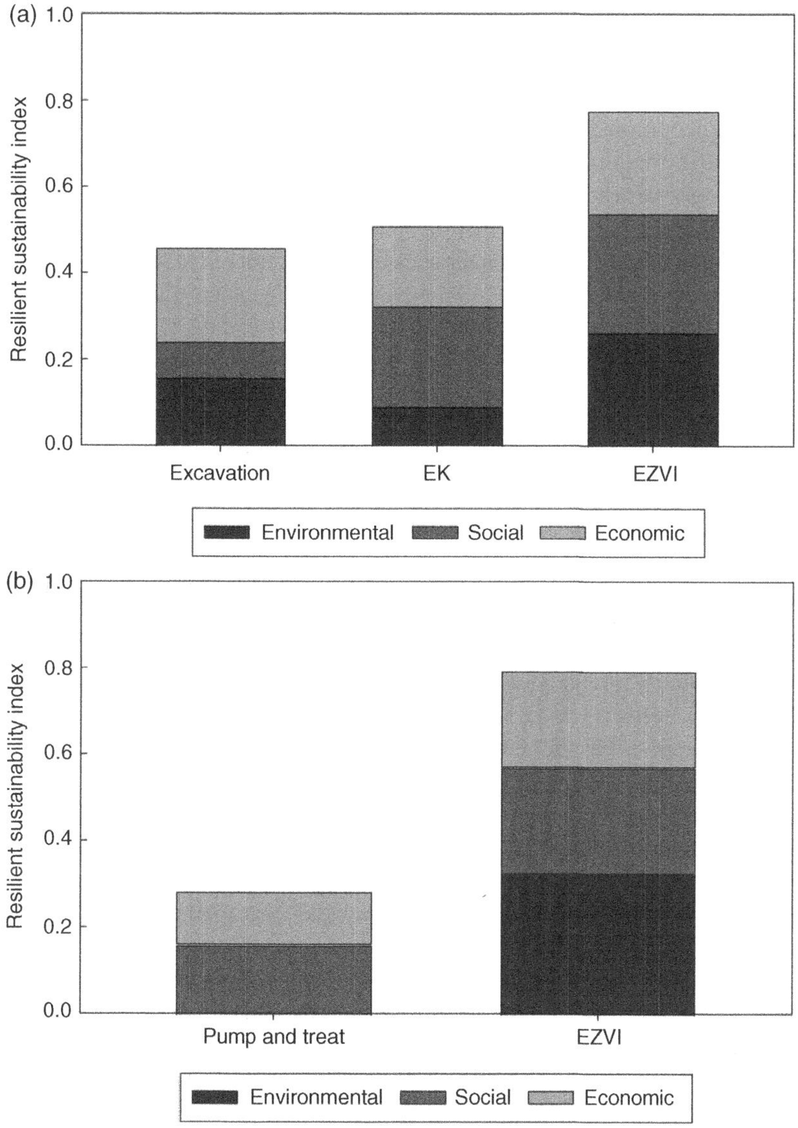

Figure 21.22 RSI of remediation technologies for (a) soil remediation, and (b) groundwater remediation.

EZVI demonstrated high environmental sustainability in groundwater remediation compared to the pump-and-treat method, reflecting its lower environmental impact. Its high social sustainability is attributed to reduced disruption to communities and surrounding areas. EZVI achieved higher economic sustainability due to its lower costs and associated benefits. In contrast, the pump-and-treat method showed lower environmental sustainability, primarily due to higher environmental impacts from its prolonged operational period. Social sustainability was also low, indicating more significant disruptions to the community. Similarly, its economic sustainability lagged behind EZVI, driven by higher operational expenses. These findings highlight EZVI as the most resilient and sustainable alternative, outperforming the pump-and-treat method across all sustainability dimensions in groundwater remediation.

21.4.4 Lessons Learned

Combining resiliency and sustainability assessments into a comprehensive RSI provides valuable insights into sustainable environmental management, particularly for determining the most effective approaches to remediate contaminated sites. Evaluating methods such as excavation, EZVI, and pump-and-treat highlights the importance of balancing immediate effectiveness in contamination removal, resiliency to extreme weather events and external stressors, and long-term impacts on the TBL dimensions of sustainability. The study further emphasizes the necessity of selecting remediation methods based on each site's specific characteristics and requirements.

Trade-offs are an inherent part of environmental management. For instance, a highly resilient method like excavation may underperform on sustainability metrics, leading to lower overall RSI scores compared to methods like EZVI. Recognizing these trade-offs is crucial for making informed decisions addressing immediate needs and long-term goals. Additionally, the successful implementation of remediation technologies requires input from diverse stakeholders, including residents, environmental professionals, and regulatory bodies. Their feedback is essential to ensure that the chosen methods are socially equitable and aligned with broader community and environmental objectives.

21.4.5 Conclusions

Remediating contaminated sites presents unique challenges that vary significantly depending on site-specific conditions. With numerous remediation technologies available, selecting the best alternative that aligns with the site's needs is crucial. In this study, a technical screening process narrowed the options to three alternatives for soil remediation and two for groundwater remediation. These alternatives were then evaluated for their resiliency and sustainability.

The soil remediation resiliency assessment identified excavation as the most resilient option compared to EZVI and EK remediation. However, EZVI emerged as the more resilient and sustainable choice when sustainability factors were considered. Similarly, EZVI was found to be resilient and sustainable for groundwater remediation compared to the pump-and-treat method.

Combining resiliency and sustainability into a single RSI offers a comprehensive framework for selecting remediation technologies in environmental management. This approach evaluates each method's long-term viability under changing climate conditions and external stressors while ensuring that remediation efforts align with environmental, social, and economic objectives. By leveraging the RSI, decision-makers can make informed choices and effectively communicate the benefits of technologies that address contamination and promote the long-term health and viability of ecosystems and neighboring communities.

21.5 Resiliency and Sustainability Assessment of Different Alternatives for Typical Highway Pavement

21.5.1 Problem Statement

The United States has over four million miles of roads, with about 47 000 miles of these being interstate highways. Looking at the paved roadways, each roadway has multiple underlying layers which play a pivotal role in how the roadway functions and dramatically affect the sustainability and resiliency of the roadway. While different mixtures of materials, including aggregate, concrete, asphalt, etc., are used in different layers of pavements, most of the paved roads in the United States use hot mixed asphalt (HMA) owing to its price, reliability, and excellent drivability. Although one of the cheaper options available, asphalt has significant environmental sustainability concerns owing to the petroleum-based raw materials and other resource-intensive material use in base and subbase layers. Jointed concrete pavements have become a popular alternative to asphalt, owing to their excellent durability. Although better than asphalt in some aspects of sustainability, concrete pavements still demand significant use of resources which includes the use of natural aggregate in the sub-base and subgrade courses. Recently, the use of recycled concrete as a substitute to virgin aggregate for pavement applications has gained popularity, owing to the technical robustness along with sustainability benefits.

As roadways are a critical part of the transportation system, they serve as a main artery of any country's economy and can be a driving force of a booming economy; therefore, engineers have a responsibility to construct roadways that are not only long-lasting and resilient but also reduce negative environmental impacts, are low cost, and help increase the well-being of roadway users and members of the community, all while meeting all the structural and loading needs.

21.5.2 Project Background

The Illinois Tollway Authority recently completed reconstructing a 14-mile stretch of Interstate 90 from Schaumburg to Elgin, Illinois (Figure 21.23). Many innovative sustainable practices were implemented in this project, including the use ofjointed concrete pavement with a recycled aggregate subgrade. This method used existing milled and crushed on-site pavement to be repurposed as an aggregate subgrade, exemplifying a commitment to eco-conscious infrastructure development. In this case study, this method will be compared to two other traditional alternatives, jointed concrete with virgin aggregate and full-depth asphalt pavement with recycled aggregate subgrade. Sustainability and resiliency will be investigated for the three alternatives to find the most resilient and sustainable option. Determining an equivalent unit of functional units is crucial for sustainability assessments since it allows all the data to be referred to a baseline unit. The functional unit selected for the assessment in this case study was 1 mile of a 12-foot-wide lane highway pavement.

21.5.2.1 Technical Design of Alternatives

The design of the 14-mile section, following the guidelines of the 1993 AASHTO Design Guide and the IDOT Design Guide, for each alternative are presented in Figure 21.24 and are briefly described below:

Alternative 1: This methodology used in the 2016 reconstruction uses a 13-inch-thick jointed concrete pavement, a 3-inch-thick asphalt subbase, and a 9-inch-thick recycled aggregate subbase. The existing asphalt layer was milled and stored on-site, where the existing concrete pavement was broken and crushed and used as an aggregate in the subgrade for the new pavement. Re-using the existing pavement reduces the transport of material off-site for processing, re-delivering the material back to the site, reduces congestion in the local roads, reduces emissions from trucking, and saves time and money. This alternative has a 50-year design life.

Alternative 2: Consists of 13-inch-thick jointed concrete pavement, 3-inch-thick of asphalt subbase, and 9-inch-thick of virgin aggregate. This alternative was compared with alternative 1, which used recycled aggregate as a subbase, to see how the recycled aggregate would perform compared to the virgin aggregate. This alternative will require removing existing pavement,

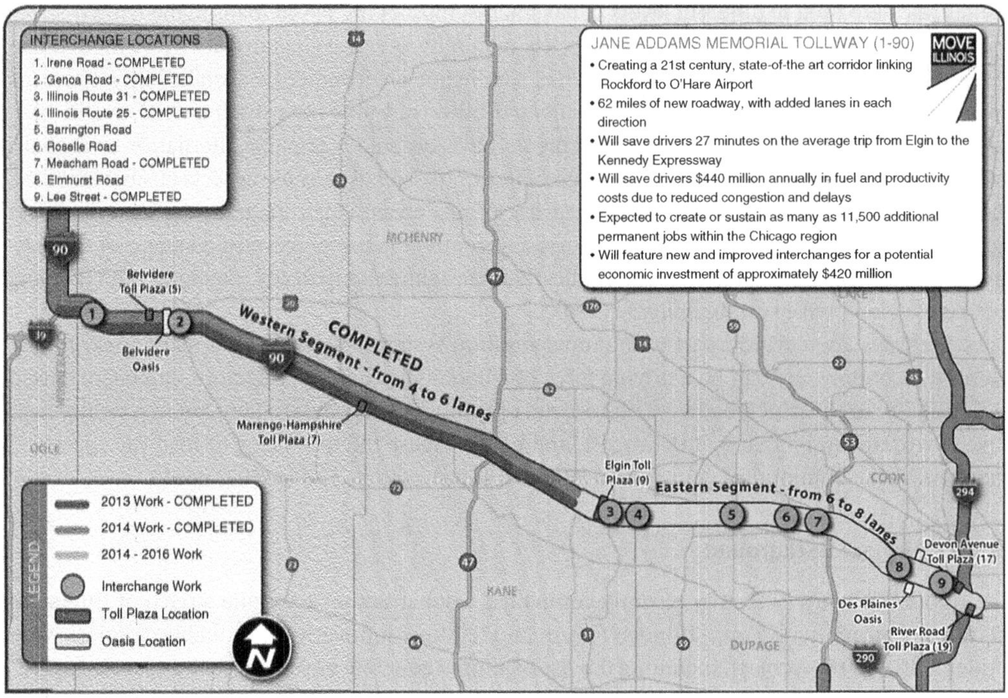

Figure 21.23 Location of the project from Schaumburg to Elgin (Illinois, US).

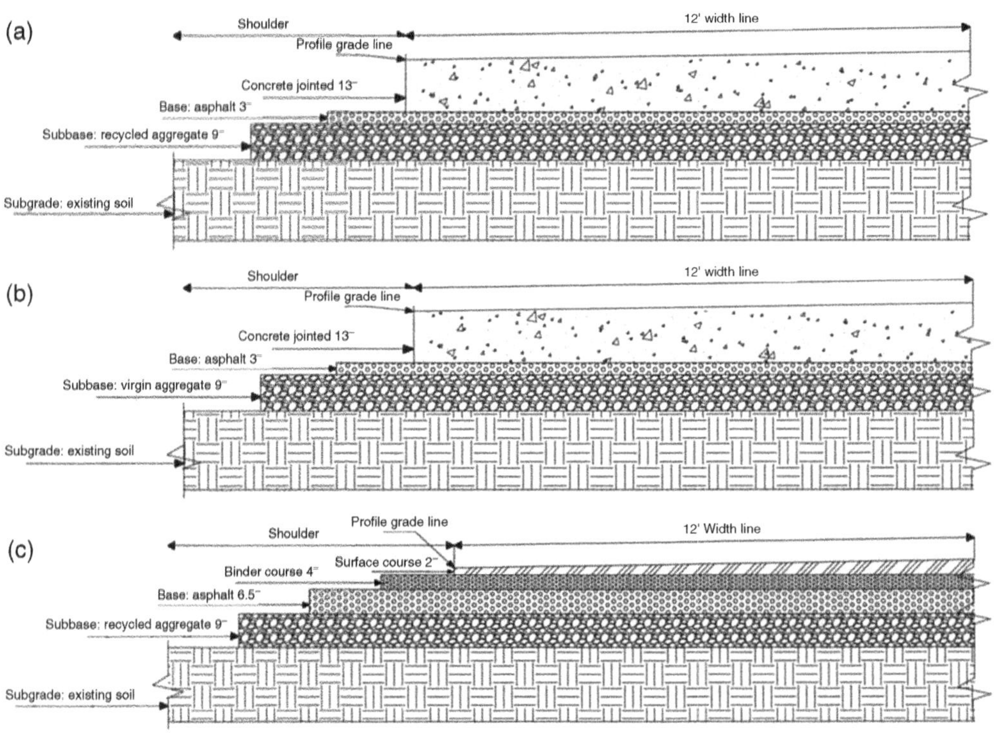

Figure 21.24 Pavement sections for (a) alternative 1, (b) alternative 2, and (c) alternative 3. *Source:* Seyrfar et al. (2021). with permission of The National Academies of Sciences, Engineering, and Medicine.

transporting and disposing off-site to a landfill, and using virgin aggregate for subgrade, which has a 50-year design life.

Alternative 3: Consists of 2-inch-thick asphalt as surface course, 4-inch-thick asphalt as binder course, and 6.5-inch-thick asphalt as subbase course. This alternative provides a contrast between asphalt and concrete pavements in this study. Generally, asphalt pavement is cheaper, has a shorter duration of installation, and provides smoother rideability. This alternative has a 20-year design life, with maintenance to extend to a 50-year life.

21.5.3 Integrated Resiliency and Sustainability Assessment

21.5.3.1 Resiliency Assessment

In this case study, tools such as the climate check website and general information on weather patterns were used to choose short-term and long-term climate effects that can impact pavement performance specific to the location. Floods and changes in temperature were the most significant climate events that would affect the longevity of the pavements in the presented area and thus were selected. A selected set of indicators, as shown in Table 21.5, were used to rate the resiliency of the pavement alternative against these climate impacts. The cascading environmental, social, and economic impacts of resiliency were also quantified using the indicators shown in Table 21.5.

Long-term or gradual climate change effect – Change in temperature: Chicago experiences significant temperature variations throughout the year, with hot summers and cold winters, leading to thermal expansion and contraction of pavement materials. These temperature

Table 21.5 Indicators and corresponding metrics and tools used to assess the resiliency of alternative pavements.

Resiliency requirement	Indicator	Metric	Tool
Technical	Resiliency to higher temperatures Cracks due to freeze-thaw cycles Fatigue resistance	Rating on a scale of 0 to 4	Expert rating
Cascading environmental	Resource use for restoration measures Energy use for restoration measures Emissions from restoration measures Urban heat island effects due to temperature changes	Rating on a scale of 0 to 4	Expert rating
Cascading economic	Costs of repair Traffic delays Environmental degradation costs	Rating on a scale of 0 to 4	Expert rating
Cascading social	Life and safety Hazard exposure Mobility impact	Rating on a scale of 0 to 4	Expert rating

(Rating scale definition: 0 – Extremely high impact/Very low resiliency, 1 – High impact/Low resiliency, 2 – Moderate impact/Medium resiliency, 3 – Low impact/High resiliency, 4 – No impact/Very high resiliency).

fluctuations exacerbate the aging process of asphalt pavements, accelerating the development of cracks, rutting, and surface distresses. Asphalt surfaces have a strong ability to absorb heat. During winter, freeze–thaw cycles further contribute to pavement deterioration as moisture penetrates the pavement, freezes, and expands, causing cracking and pothole formation. Over time, repeated exposure to extreme temperatures weakens the structural integrity of roadways, necessitating frequent maintenance and rehabilitation efforts to address pavement deficiencies and ensure safe and reliable transportation infrastructure for the city's residents and commuters.

Short-term or immediate climate change effect – Flooding: Flooding has many significant adverse effects on paved roads and can lead to erosion and washouts, surface deterioration, and weakening of subsurface layers. In areas of cold temperatures such as Chicago, as snow melts and refreezes, it can weaken the pavement and accelerate surface deterioration; salt and other chemicals used for melting ice and snow can seep into the pavement and cause corrosion, accelerating the decline of the roadways. Chemicals can also weaken the structural integrity of roadways, making them more susceptible to damage during flooding, making flooding a significant effect that needs to be investigated. Over the next 30 years, this area has a significant chance of flood water reaching the structure. Flood hazard zones may be prevalent in this area. Excessive flooding or ponding in the future would mean designing the pavement for higher design loads and proper drainage systems to avoid erosion of the aggregate base and subgrade.

For the resiliency assessment, equal weights were given to both disasters based on an equal probability of impacting the pavement in the next 30 years. The indicators were quantified using expert rating by employing a scale between 0 and 4 (0 indicating the least resiliency and 4 indicating the highest resiliency). The technical resiliency to both disasters was comparable in both the concrete alternatives, as studies showed that recycled concrete aggregate performs similar or better than virgin aggregate in most cases (Kim et al. 2011). Asphalt pavements are more vulnerable to rutting and fatigue due to temperature changes, while concrete pavements can be significantly impacted by the cracks induced by freeze and thaw cycles. Each alternative was rated for resiliency against the identified climate impacts following these general principles, and the resulting technical resiliency was similar for both the concrete alternatives and low for asphalt pavement with recycled aggregate subgrade (alternative 3) (Figure 21.25).

The cascading environmental impacts of resiliency were quantified using indicators, such as resource consumption, energy use, and emissions, that can result from repair or restoration activities. Generally, virgin aggregate subgrade and asphalt pavement require more resources and result in higher emissions compared to their equivalent counterparts, while concrete pavements demand higher energy usage. Furthermore, due to high heat absorption by asphalt pavements, an urban heat island phenomenon will be more significant when using asphalt pavements. Using these general principles, the corresponding indicators were rated and the resulting environmental resiliency showed high resiliency for alternative 1 followed by alternative 3, and alternative 2.

Similarly, such general principles were used to rate the cascading social and economic impacts of resiliency. The integrated resiliency index obtained after integration using the MIVES methodology described before is shown in Figure 21.25. Similar resiliency indices were obtained for alternatives 1 and 3, while alternative 2 where concrete pavement with virgin aggregate was used showed lower resiliency.

21.5.3.2 Sustainability Assessment

The first step of an integrated sustainability assessment involves the identification of environmental, social, and economic indicators, as mentioned before. Indicators identified specific to this

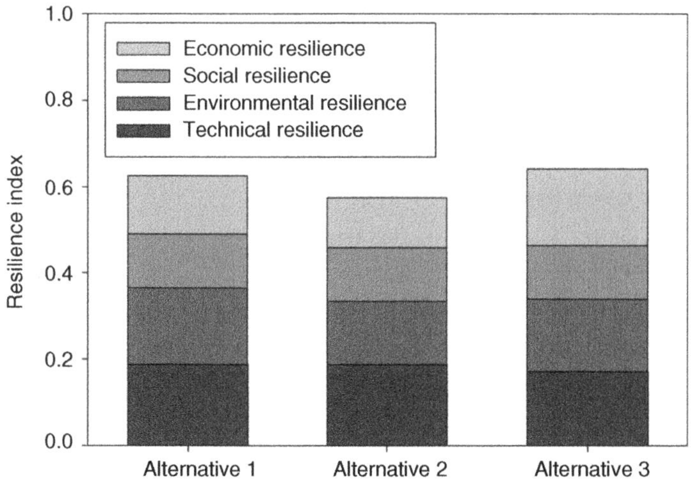

Figure 21.25 Results of the resiliency assessment for the three alternative pavements.

case study, and the corresponding metrics and tools used for quantification of the indicators are presented in Table 21.6. A detailed description of quantification methodology followed and the results of each pillar of sustainability assessment along with the overall integrated RSI are presented in this section.

Environmental Sustainability

SimaPro was used to assess the life cycle environmental impacts of all the three alternatives. Developed by the Dutch company PRé Sustainability, SimaPro allows users to analyze and report on the sustainability of their products, from raw material extraction through manufacturing, distribution, use, and disposal. In this case study, equipment consumption, materials, and transportation during the life cycle stages of the project were considered (Figure 21.26) for the inventory analysis of the three alternatives, and the resulting inventory was used as inputs in SimaPro for the environmental sustainability assessment.

The inventory analysis involved estimates of the diesel consumption by various equipment, including truck-mounted demolition systems, crushers, excavators, vibro-compactors, dozers, loaders, millers, concrete/asphalt pavers, and trucks for transport of materials to and from the construction site. In addition, the quantities of materials used in the construction of pavements were estimated, as shown in Table 21.7. The resulting inventory was then used as input to quantify the environmental impacts quantified using TRACI 2.1 V1.01/US 2008. The environmental impact categories quantified using the TRACI methodology include global warming, ecotoxicity, eutrophication, carcinogens, non-carcinogens, fossil fuel depletion, ozone depletion, acidification, smog, and respiratory effects. The resulting relative environmental impacts are presented in Figure 21.27.

The characterization of the environmental impact assessment categories showed that the environmental impact of all three alternatives mainly comes from materials – most of the effects from the construction phase. The overall comparison of the three alternatives is shown in Figure 21.27. Alternative 3 showed the highest environmental impact in 6 of the 10 categories, resulting from the

Table 21.6 Indicators along with metrics and tools used for assessing the sustainability of the three alternatives.

Requirement	Criteria	Indicator	Metric	Tool
Environmental sustainability	Air	Ozone depletion	kg CFC-11 eq	LCA
		Global warming	kg CO_2 eq	
		Smog	kg O_3 eq	
	Water	Eutrophication	kg N eq	
	Human health	Carcinogenic	CTUh	
		Non-carcinogenic	CTUh	
		Respiratory effects	kg PM2.5 eq	
	Land	Ecotoxicity	CTUe	
		Fossil fuel depletion	MJ surplus	
	Resiliency	Environmental resiliency index	Scores from resiliency assessment	Resiliency assessment
Social sustainability	Social-individual	Quality of life	Rating on a scale of −2 to +2	SSEM
		Public health and happiness		
		Construction safety		
		Inclusivity		
		Driving quality		
	Socio-community	Construction duration		
		Aesthetic appeal		
		Road safety		
		Maintenance frequency		
		Green practices		
	Socio-economic	Disruption of business activity		
		New business creation		
		Employment		
		Tax revenue		
	Socio-environmental	Noise pollution		
		Sustainable practices		
		Environmental degradation		
	Resiliency	Social resiliency index	Score from resiliency assessment	Resiliency assessment
Economic sustainability	Direct costs	Equipment	USD	Life cycle costing
		Infrastructure		
		Maintenance		
		Energy		

Table 21.6 (Continued)

Requirement	Criteria	Indicator	Metric	Tool
	Indirect costs	Environmental cost		
	Direct benefits	Employment	Rating	Expert rating
		Resource conservation		
	Indirect benefits	Awareness		
		Competency		
		Welfare		
	Resiliency	Economic resiliency index	Score from resiliency assessment	Resiliency assessment

Note: LCA – Life cycle assessment, SSEM – Social sustainability evaluation matrix.

Figure 21.26 Life cycle stages of the pavement considered for the environmental impact assessment of the three alternatives.

Table 21.7 Quantities of materials used in each alternative pavement.

Material	Alternative 1	Alternative 2	Alternative 3
Concrete (yd^3)	3997	3997	–
Asphalt (tons)	1148	1148	7828
Virgin aggregate (kg)	–	2560	–

excessive amount of asphalt material used compared to alternatives 1 and 2. The use of recycled aggregate slightly lowered the environmental impacts of alternative 1 compared to alternative 2.

Social Sustainability

Social assessment was divided into five criteria, as seen in Table 21.6. The SSEM (Reddy et al. 2014) was used to rate each alternative using a scale between −2 and +2. The results of the social sustainability assessment are presented in Figure 21.28. Recycled aggregate in the subgrade significantly impacted the social sustainability scores, as alternatives 1 and 3 scored better compared to alternative 2. Further, considering similar aggregates, concrete pavement exhibited higher acceptability compared to asphalt pavement, as alternative 1 scored better than alternative 3.

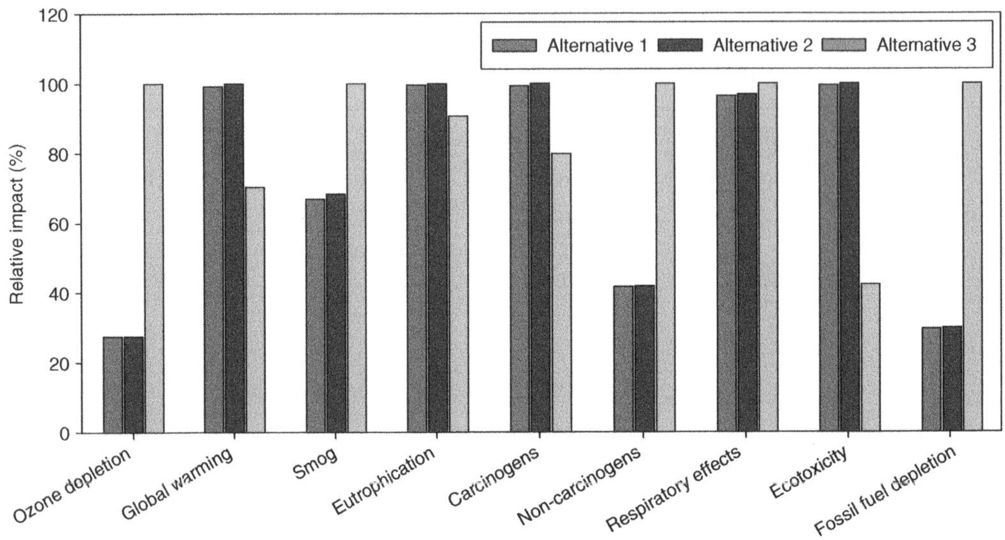

Figure 21.27 Comparison of environmental impacts of the three alternative pavements.

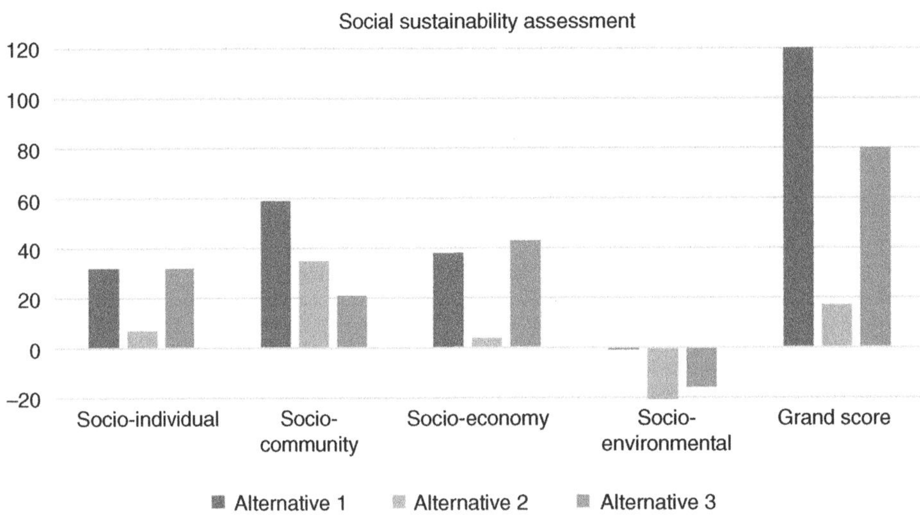

Figure 21.28 Social sustainability scores of the three alternative pavements (Seyrfar et al. 2021).

Economic Sustainability

Following the quantities estimated for each activity during the inventory analysis, total costs were calculated using cost per unit, which was based on bid tabs for construction, maintenance, and disposal. The total direct and indirect costs were estimated, as presented in Figure 21.29.

Direct costs: Costs were analyzed for aggregate, paving, maintenance, and demolition stages. Maintenance is the highest cost across the board for all three alternatives. Alternative 3 has the lowest direct cost of $1 797 952.43. Alternative 1 has the highest direct cost of $2 657 190.92.

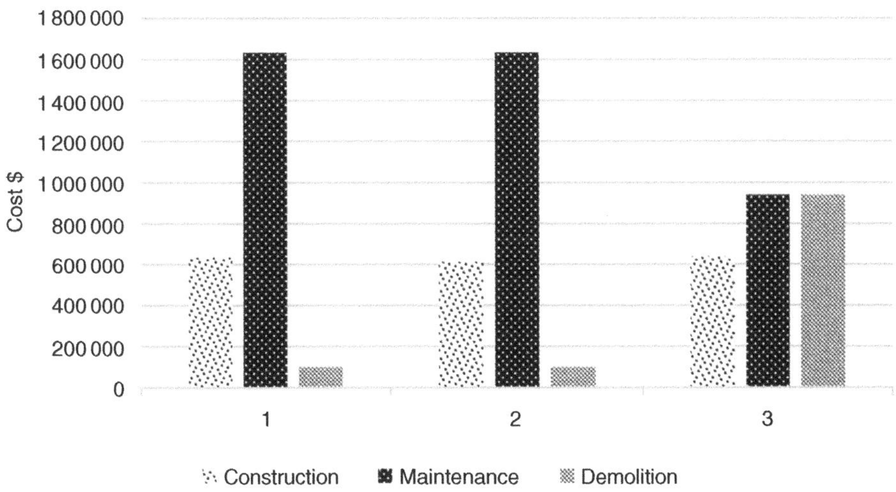

Figure 21.29 Direct and indirect costs of the three alternatives. *Source:* Seyrfar et al. (2021). with permission of The National Academies of Sciences, Engineering, and Medicine.

Indirect costs were calculated using CO_2 emission data from LCA and CO_2 social cost based on a 3% discount rate determined by the USEPA (2017c). Alternative 3, while having the lowest direct costs, has the highest indirect costs due to the negative environmental impact of asphalt. Alternatives 1 and 2 have similar indirect costs, but alternative 2 has the lowest (alternative 1 indirect cost = \$134 820.00; alternative 2 indirect cost = \$128 940.00; alternative 3 indirect cost = \$516 600.00). Overall, alternative 3 is the cheapest, despite the high indirect costs, which total \$2.3 million. Alternatives 1 and 2 had comparable total costs, but alternative 1 is the costliest option at \$ 2.79 million (Figure 21.29).

Direct and indirect benefits were quantified using ratings for the three alternatives, and the resulting indicator values were included in the calculation of the final integrated sustainability index. Results are not shown here for brevity purpose.

21.5.3.3 Integrated Resilient Sustainability Index

The integrated RSI is determined by aggregating the indicator values as described in Section 21.2. Environmental, social, and economic resiliency indices obtained from the resiliency assessment were included as separate criteria in the corresponding requirements to truly integrate the resiliency and sustainability aspects to develop an integrated RSI.

Integrated resilient sustainability indices calculated using different weights for environmental, social, and economic requirements are presented in Figure 21.30. When equal weights were assigned to all requirements, alternative 1 was identified as the most sustainable, followed by alternative 3 and alternative 2. However, the rankings changed when the weights were adjusted, as the overall sustainability indices varied significantly with changes in weight distribution. For instance, when economic sustainability was assigned the highest weight, alternative 3 emerged as the most sustainable. Conversely, this same alternative ranked as the least sustainable when the environmental requirement was assigned the highest weight. These variations in sustainability indices underscore the critical importance of stakeholder engagement to account for their preferences and priorities in the decision-making process.

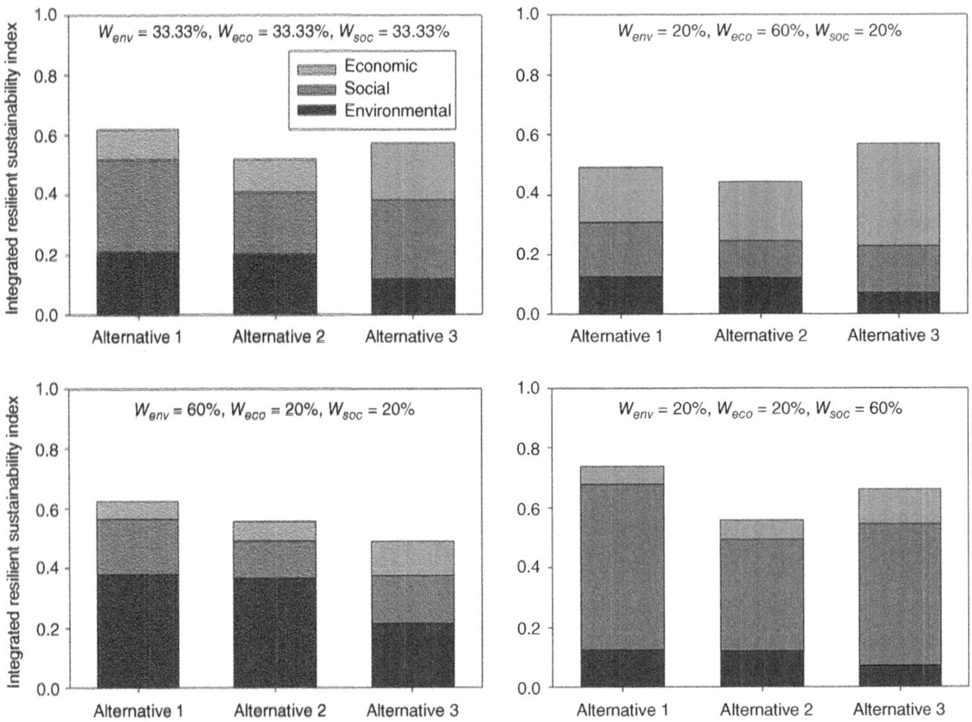

Figure 21.30 Integrated resilient sustainability indices for the three alternative pavements.

21.5.4 Conclusions

The resiliency and sustainability of pavements are crucial, as a significant amount of global resources are consumed in pavement construction and maintenance. Additionally, roadways play a vital role in the economic sustainability of global, national, and local economies. Sustainable practices in pavement construction, such as using recycled aggregate materials in subbase and subgrade courses, are increasingly being adopted as the push for sustainability gains momentum. In this context, quantifying the resiliency and sustainability of various alternatives can assist in the decision-making process by identifying the most resilient and sustainable options. This case study assessed a 14-mile section of the Interstate 90 highway in Illinois, considering three alternative pavements: (i) concrete pavement with a recycled aggregate subgrade, (ii) concrete pavement with a virgin aggregate subgrade, and (iii) asphalt pavement with a recycled aggregate subgrade. The resiliency assessment revealed that all three alternatives demonstrated similar resiliency to climate extremes, although alternative 3 showed slightly higher vulnerability. Furthermore, the integrated resiliency and sustainability assessment indicated that the most resilient and sustainable option depends on the weight assigned to each pillar of sustainability. When greater importance was given to economic sustainability, the asphalt pavement option emerged as the most sustainable. Conversely, when greater importance was given to environmental sustainability, the asphalt pavement option ranked as the least sustainable. These findings highlight the complexity of evaluating pavement alternatives and emphasize the need for a balanced and stakeholder-driven approach to decision-making for sustainable and resilient pavements.

21.6 Resiliency and Sustainability Assessment of Community Gardens Versus Hydroponics: Chicago's Sustainable Food Systems

21.6.1 Problem Statement

Chicago's Back of the Yards neighborhood exemplifies the challenges urban communities face. Historically known for its industrial roots and the iconic Union Stock Yards, this area has undergone significant transformations over the years, yet some issues persist. While the area once thrived as a hub of meatpacking and industrial activity, the decline of these industries left behind vacant lots and economic disparities. Among these challenges is the prevalence of food insecurity and the designation of food deserts, where access to fresh and healthy food is limited.

21.6.2 Project Background

Unsustainable farming methods compound these challenges, as conventional agricultural practices prioritize efficiency and profit over environmental sustainability and community well-being (Russo et al. 2017). Large-scale monoculture farming, reliant on chemical inputs and intensive land use, has significant environmental consequences, depleting soil health, polluting waterways, and contributing to climate change. These methods often prioritize the production of commodity crops rather than diverse, nutrient-rich foods essential for a balanced diet. In urban environments like the Back of the Yards, limited access to fresh produce exacerbates health disparities, contributing to higher rates of diet-related illnesses, such as obesity, diabetes, and cardiovascular disease. Without access to affordable, nutritious food options, residents rely on convenience stores and fast-food outlets, perpetuating a cycle of poor health outcomes and economic hardship. Amid these challenges, initiatives like the Plant Chicago Hydroponics Garden and community gardens offer alternative models prioritizing resiliency and sustainability. Hydroponics and community gardens are one of many urban agriculture systems that are defined as farming activities that take place within the boundaries of a city (Wilhelm and Smith 2018). By embracing innovative farming techniques, such as hydroponics and community-led gardening, these projects provide access to fresh, locally grown produce and promote environmental stewardship, economic empowerment, and community resiliency (Martin et al. 2019).

21.6.2.1 Technical Design of Alternatives

This project comprehensively analyzes the environmental, social, and economic impacts of two distinct gardening systems: a vertical hydroponic garden and a traditional community garden. Both versions of these gardening systems are designed to accommodate the same number of users, allowing for a comparative assessment of their sustainability. The study focuses on the hydroponics garden at Plant Chicago and the raised community garden in the Back of the Yards neighborhood on the Southwest Side of Chicago. Over 10 years of environmental data have been gathered and analyzed for seasonality and infrastructure considerations for the hydroponics garden. The number of plants per season was considered the LCA functional unit. Plant Chicago grows various crops, including winter kale, lettuce, arugula, tomatoes, melons, dill, and chard. It aims to equip city residents with the space, knowledge, and tools to grow their food.

The size and design of the scale-up were informed by the specifications of the single-module system described in the literature (Sullivan et al. 2015) and the dimensions of the room housing the single-module system at Plant Chicago. The 480-square-foot room can theoretically fit up to six vertical hydroponics modules, with 2 to 3-foot-wide walking paths in between, resulting in 12 growing

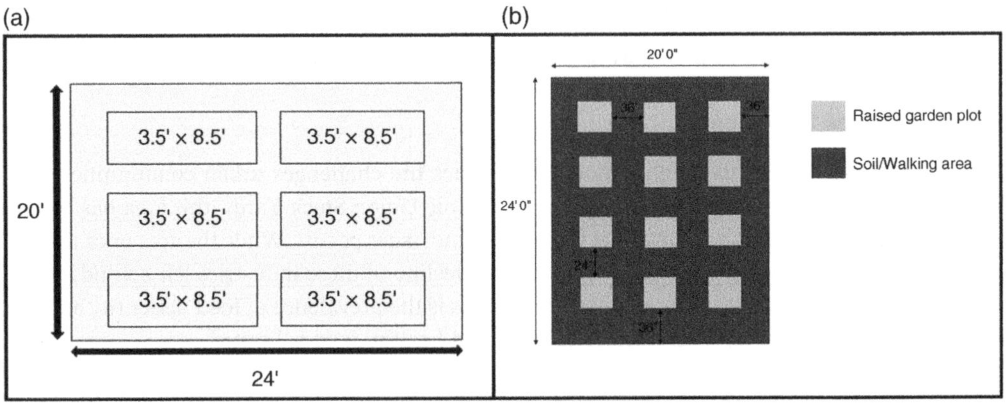

Figure 21.31 Technical design layout of scaled (a) hydroponics gardening; (b) community gardening plot (drawing not to scale).

beds shared between 24 growers. The layout of the scale-up design considered in this study is shown in Figure 21.31a. LED lights were considered the light source for plants. To assess the scale-up hydroponics system, all data and details from the case study were scaled and adjusted to account for the expansion from one single two-tier module to six two-tier module systems. An example of the hydroponic garden technical design is shown in Figure 21.32a.

The community garden was designed with 3 ft (L) × 3 ft (W) × 1.5 ft (H) raised beds made of wood, with a capacity of growing seven plants per bed. For comparison with the scaled hydroponics system, it was considered an outdoor area of 480 square feet for the garden, which can fit up to 12 raised beds, with 2- to 3-foot-wide walking paths in between, keeping the same total number of 24 growers, with 2 growers splitting one bed (Figure 21.31b). An example of the scaled community garden technical design is shown in Figure 21.32b. Three inventory categories were defined for the operational phase of community gardening: infrastructure materials, plant-growing materials, and transportation. The infrastructure included wood and nails for constructing the growing beds. The plant-growing materials included water, nutrients/fertilizer, compost, and topsoil, and the transportation accounts for the plant-growing materials (except for water), and the users commute to the community garden on average three days/week. No plant-growing energy inputs were accounted for in this model.

Figure 21.32 shows the system boundaries for the two alternative scenarios: hydroponic and community-based raised gardens. This information helps identify the inputs and outputs for the inventory analysis in each life cycle stage for the resiliency and sustainability assessment.

21.6.3 Integrated Resiliency and Sustainability Assessment

21.6.3.1 Resiliency Assessment

The resiliency assessment was performed based on the climate impacts that could potentially occur in the Back of the Yards neighborhood. The vulnerability to extreme climate impacts and changing climate conditions were identified using the Risk Factor software. The resiliency of the systems was assessed against strong winds, dangerously hot days, and pollution release, which represent poor air quality conditions. Further, the indicators were identified based on the impact the stressors could have on the two alternative systems and their cascading environmental, social, and economic effects (Table 21.8). The evaluation was conducted to identify the vulnerabilities in both alternatives. The resiliency indices obtained from the analysis are presented in Figure 21.33. The community garden

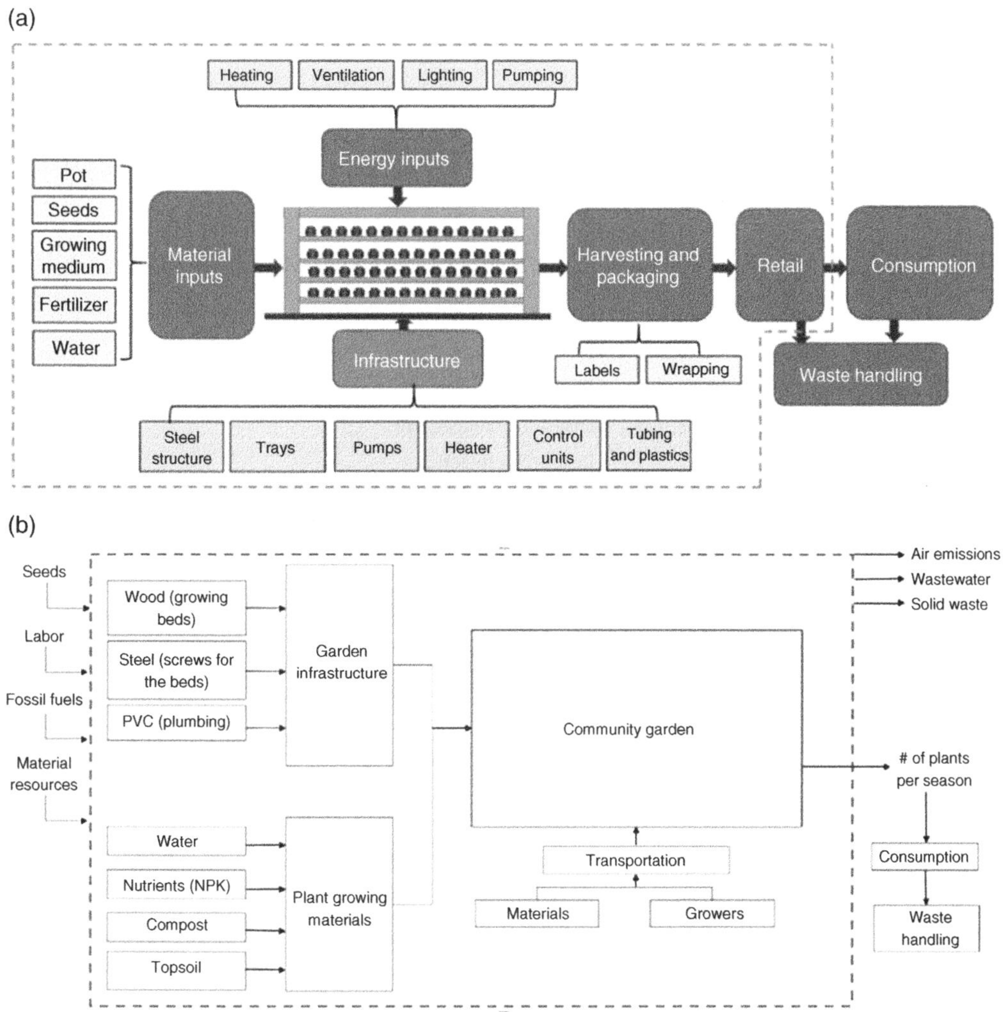

Figure 21.32 A diagram showing the boundaries of (a) hydroponic garden; (b) community garden.

was found to have an overall resiliency index of 0.84, while the hydroponic garden had a resiliency of 0.89. Being in the open, the community garden is inherently vulnerable to external stresses, thus resulting in a lower score across all the four requirements of resiliency assessment. Furthermore, the hydroponics garden is easily manageable and less vulnerable to external factors as it is a controlled system. However, the closed system of hydroponics gardens also means that the plants are less resistant to pests and can spread rapidly and become more challenging to manage.

21.6.3.2 Sustainability Assessment

The first step of an integrated sustainability assessment involves the identification of environmental, social, and economic indicators, as mentioned before. Indicators identified specific to this case study, and the corresponding metrics and tools used for the quantification of the indicators are presented in Table 21.9. A detailed description of quantification methodology followed and the results of each pillar of sustainability assessment along with the overall integrated RSI are presented in this section.

Table 21.8 Resiliency indicators, metrics, and tools to assess hydroponics and community gardens.

Resiliency requirements	Criteria	Indicators	Metric	Tool
Technical rResiliency	Performance deterioration	System reliability	Rating on a scale of 0 to 4	Expert rating
		Disaster management		
		Resource efficiency		
	Time for deterioration	System reliability		
		Disaster management		
		Resource efficiency		
	Time in deteriorated state	System reliability		
		Disaster management		
		Resource efficiency		
	Performance recovery	System reliability		
		Disaster management		
		Resource efficiency		
	Time for recovery	System reliability		
		Disaster management		
		Resource efficiency		
Cascading environmental impacts	Air	Greenhouse effect	Rating on a scale of 0 to 4	Expert rating
		Quality of air (consistency)		
	Water	Impact on local aquatic ecosystems		
		Nutritional value of water		
	Human health	Indoor air quality (gases)		
		Contacts with chemical compounds		
	Land	Alteration of microbial communities		
		Resistance development		
Cascading social impacts	Social-individual	Mental health	Rating on a scale of 0 to 4	Expert rating
		Respiratory issues/disease		
		Weakening of immune systems		
	Socio-community	Public health hazards		
		Quality of life		
		Stress levels due to medical burdens		
	Socio-economic	Medical bill costs		
		Govt expenditure		
		Medical insurance premium increase		
		Economic disparities		
	Socio-environmental	Environmental stewardship		
		Greenhouse gas emissions		

Table 21.8 (Continued)

Resiliency requirements	Criteria	Indicators	Metric	Tool
		Public health hazards		
		Climate change contributions		
Cascading economic impacts	Direct costs	Labor cost changes	Rating on a scale of 0 to 4	Expert rating
		Input cost variability		
		Expensive PPE kits with filters		
		Equipment and machinery costs		
	Indirect costs	Transportation costs		
		Overheads costs for pollution levels monitoring		
	Direct benefits	Revenue generation from food production		
		Source of income for skilled individuals		
		Less use of fertilizers and preservatives		
	Indirect benefits	Reduction in grocery bills		
		Fuel savings due to less transportation		
		Healthy and organic food in plates		

Note: Rating based on level of negative impact: 0-extreme; 1-very high; 2-high; 3-medium; 4-low.

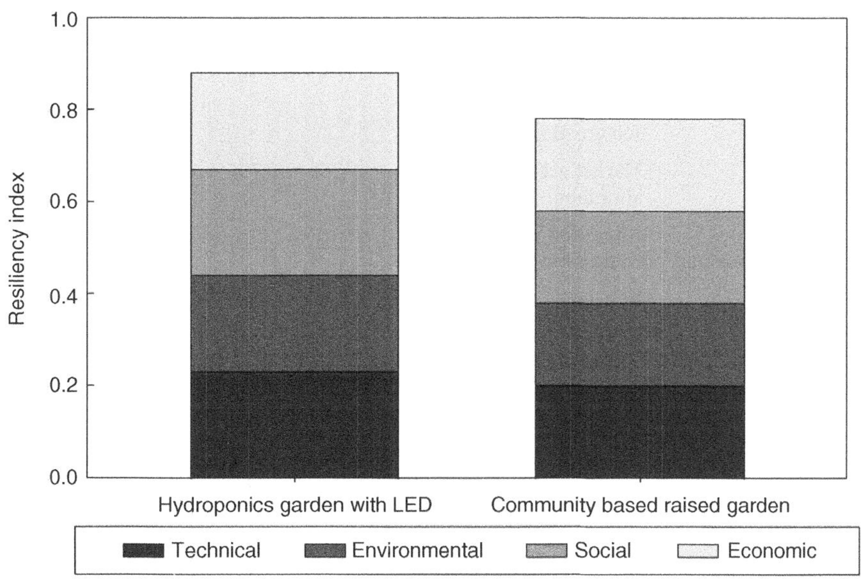

Figure 21.33 Resiliency Index of hydroponics and community gardens.

Table 21.9 Sustainability indicators, metrics, and tools to assess hydroponics and community gardens.

Sustainability requirements	Criteria	Indicators	Metric	Tool
Environmental sustainability	Air	Ozone depletion	kg CFC-11 eq	LCA
		Global warming	kg CO_2 eq	
		Smog	kg O_3 eq	
	Water	Acidification	kg SO_2 eq	
		Eutrophication	kg N eq	
	Human Health	Carcinogenic	CTUh	
		Non-carcinogenic	CTUh	
		Respiratory effects	kg PM2.5 eq	
	Land	Ecotoxicity	CTUe	
		Fossil fuel depletion	MJ surplus	
	Resiliency	Environmental resiliency index	Scores from resiliency assessment	Resiliency assessment
Social sustainability	Social-individual	Access to fresh produce	Rating on a scale of 0 to 4	Survey
		Quality of life		
		Social cohesion		
		Skill development		
		Improved health/ nutrition		
		Education		
	Socio-community	Green spaces and mental health		
		Environmental stewardship		
		Youth engagement		
		Food security		
	Socio-economic	Input cost variability		
		Labor cost changes		
		Job creation		
		Market access and affordability		
	Socio-environmental	Alteration of microbial communities		
		Resistance development		
		Ecological impact due to chemical use		
		Impact on local aquatic ecosystems		
	Resiliency	Social resiliency index	Scores from resiliency assessment	Resiliency assessment
Economic sustainability	Direct costs	Infrastructure	USD	Cost analysis

Table 21.9 (Continued)

Sustainability requirements	Criteria	Indicators	Metric	Tool
		Transportation		
		Plants growing material		
		Maintenance		
	Indirect costs	Environmental cost		
	Direct benefits	Revenue	USD	Benefits analysis
		Employment	Rating on a scale of 0 to 4	Expert rating
		Resource conservation		
	Indirect benefits	Awareness		
		Competency		
		Welfare		
	Resiliency	Economic resiliency index	Scores from resiliency assessment	Resiliency assessment

Note: Rating based on level impact: 0-unacceptable; 1-diminished; 2-no impact; 3-improved; 4-ideal.

Environmental Sustainability

The environmental impact assessment was done using the software tool SimaPro. It helps in evaluating the ecological footprint of their offerings throughout their entire life cycle, from raw material extraction to production, use, and disposal. The inventory information was sourced from a collaborator at Plant Chicago, available online vendors, and literature studies (Sullivan et al. 2015). The normalized values (Quantities) of inputs required for the construction and operation of the gardening facilities were identified based on the information gathered from Plant Chicago, available online vendors, and literature reviews (Weingaertner and Moberg 2014; Peruzzini et al. 2017; Samangooei et al. 2016). The inventory for the hydroponics garden included nutrients their transportation, along with racks, decks, growing beds, growing medium, lights and their support, floating boards, water reservoir, and PVC pipes, while the community garden inventory included raised bed planks, compost, topsoil, and fertilizers. Additionally, both models included transportation of growers.

Hydroponic systems stand out for their efficient use of resources, particularly in saving water through recycling, a stark comparison to traditional gardening methods. Both gardening methods are efficient in terms of space, making them suitable for urban environments. Hydroponics gardens require tiny land and can be integrated vertically. Additionally, community gardens support local biodiversity and provide habitats for wildlife, while hydroponics minimizes land disruption and helps prevent soil erosion, reducing its impact on ecosystems. The environmental impacts from the SimaPro analysis of the two alternative scenarios are presented in Figure 21.34. The results show that hydroponic gardens have more environmental impact in 6 out of 10 categories. The categories of significance for the hydroponics garden included carcinogenic, acidification, eutrophication, non-carcinogenic, and ozone depletion. The main contributor to the environmental impacts is the heater used to regulate the temperature in the room. Similarly, the community garden has higher impacts in the ecotoxicity category, primarily due to the high-water input used to grow

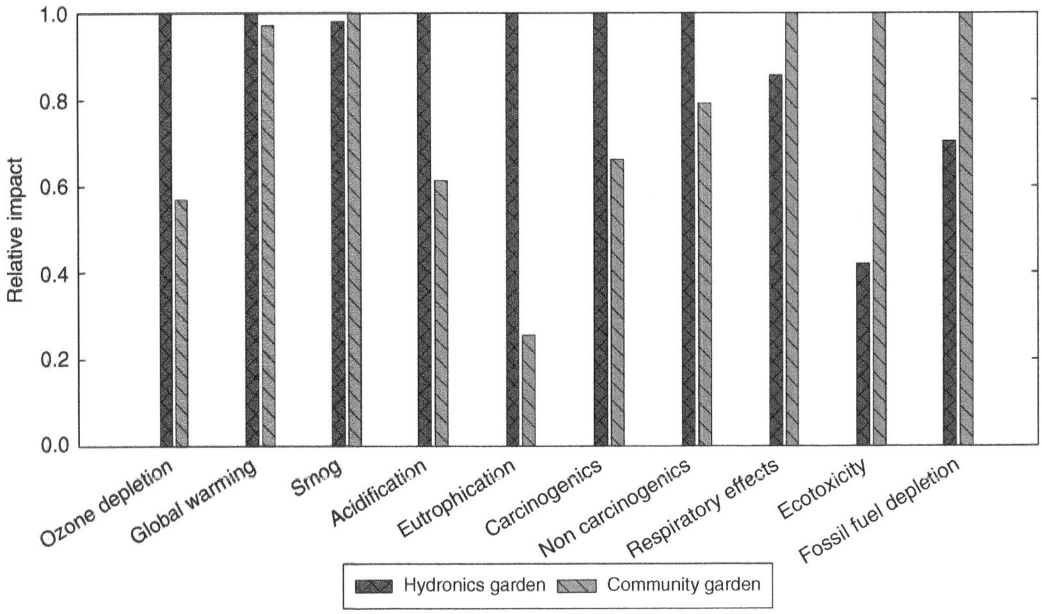

Figure 21.34 Environmental impacts of hydroponic garden and community garden.

the plants. Other categories of significance for the community gardens included fossil fuel depletion and respiratory effects due to higher transportation impacts.

Social Sustainability

The social sustainability assessment aimed to compare the various social impacts on the neighborhood residents involved in hydroponic gardening versus those participating in community garden volunteering. A survey questionnaire focusing on five socio-categories: social-community, socio-economic, socio-individual, socio-resiliency, and socio-environmental was sent out to understand the stakeholder preferences. The survey's objective was to gauge the social implications within the community following the implementation of the two alternative projects.

Both alternatives are hubs for social interaction, collective labor, and collaborative problem-solving. Moreover, these gardens serve as valuable educational spaces, offering practical lessons in sustainable agricultural practices and raising awareness about nutrition to a diverse range of community groups. Through workshops, demonstrations, and hands-on experience, individuals of all ages and backgrounds can learn valuable skills that empower them to grow their food and make healthier choices. Beyond education, community gardens significantly contribute to the health and wellness of individuals (Menconi et al. 2020). They promote physical fitness, mental well-being, and a sense of vitality by encouraging active lifestyles and providing access to fresh, nutritious produce.

Additionally, gardening can be therapeutic, offering stress relief and relaxation amidst the hustle and bustle of daily life. Ultimately, community gardens serve as vibrant, inclusive spaces that nourish the body and nurture a sense of community spirit and resiliency, making them essential assets to the social fabric of any neighborhood. In the survey, the respondents agreed with the assumptions and perceived both gardening techniques as highly effective. They recognized the potential of both methods in addressing food insecurity and food deserts, although many lacked familiarities with

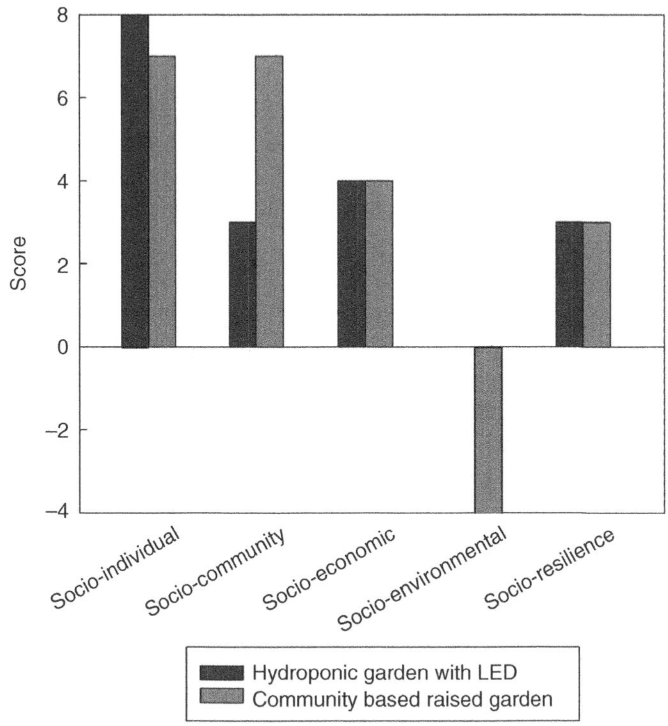

Figure 21.35 Social sustainability evaluation matrix (SSEM) results.

hydroponic gardening. The survey results were used to assess the positive and negative impacts on the community using the SSEM, and the results are presented in Figure 21.35.

Maintaining high levels of sanitation in hydroponic gardening necessitates strict adherence to cleanliness protocols to prevent disease and contamination, making it a challenge to prioritize social engagement. Meanwhile, community gardening fosters social interaction among residents through frequent volunteering opportunities, promoting broader social engagement. Additionally, environmental considerations were factored in, such as the energy consumption required for hydroponic gardens to operate effectively (Barthel and Isendahl 2013). The study revealed that Hydroponic Gardening scored slightly higher than community gardening.

Economic Sustainability

A LCC analysis was conducted to compare hydroponics with community gardens. Both alternatives were evaluated within the same system boundary, with identical lifetime and functional unit assumptions used in the LCA. The capital costs included expenses related to purchasing the necessary infrastructural materials. The economic comparison was refined by calculating the equivalent annual annuity (EAA) based on a discount rate of 2.6%. This calculation allows the capital costs to be spread evenly over the operational lifetime, providing a clear annual cost comparison. The annual operating costs considered in this analysis included expenses for energy, water, and plant-growing materials, such as fertilizer, soil, and compost. These costs are critical as they recur throughout the life of the gardens, and the operational period for both systems was set at 10 years. Further, indirect costs and direct and indirect benefits were considered other economic sustainability criteria. Indirect costs included environmental costs, while the benefits included revenue, employment, awareness regarding sustainable development practices in the community, and welfare, which corresponded to the two alternatives.

The difference in operating expenses is critical, especially when considering gardening systems' sustainability and economic feasibility over a 10-year operational period. It was found that there were significant differences in the total normalized costs associated with each gardening method. Specifically, the hydroponics garden exhibits a lower cost of \$7.017 per plant per season, while the community garden presents a higher cost of \$12.514 per plant per season. These operating costs for the community garden include higher expenditures for traditional gardening necessities, such as water, soil, and compost. In contrast, the hydroponics system, which utilizes a more resource-efficient setup, incurs lower operational costs. Figure 21.36. shows the economic sustainability assessment for the two alternatives, including the direct and indirect benefits, direct and indirect costs, and economic resiliency. The hydroponics garden design would incur lower costs and higher benefits than the raised community garden. Both the community-based and hydroponic urban gardens help create jobs within the community. These gardens generate employment opportunities ranging from managing the garden itself to roles in educational outreach, further stimulating economic activity in the community. Further, hydroponic systems stand out for their economic efficiency. They produce high yields in a limited space and require fewer pesticides, leading to lower operational costs over time. This combination of benefits makes urban gardening a compelling economic proposition.

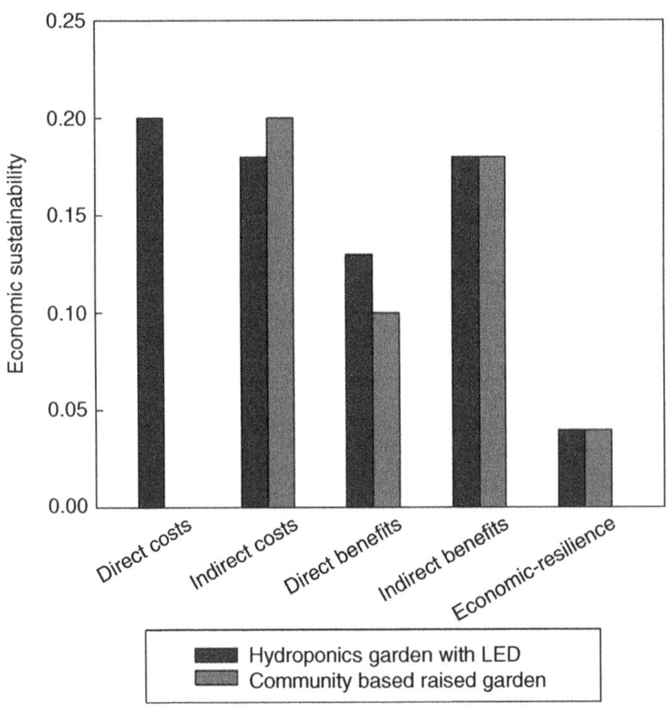

Figure 21.36 Economic sustainability of hydroponics garden and community-based raised garden.

21.6.3.3 Integrated Resilient Sustainability Index

A resilient RSI was calculated for the two alternatives, the hydroponics garden and the community-based raised garden. This methodology calculates a non-dimensional RSI by calculating a weighted sum of normalized individual environmental, social, and economic indicators based on each indicator's minimum and maximum values and distribution shape. The indicators, metrics, and the tools utilized for calculating the RSI are listed in Table 21.9, while the resulting RSI values are presented in Figure 21.37. The hydroponic garden was the most sustainable and resilient option compared to the community-based raised garden. Both alternatives' environmental and social impacts were similar, giving them similar social sustainability results. However, the hydroponics garden has significant economic benefits and lower costs, making it a resilient and sustainable alternative gardening method. Hydroponic gardens were found to have an RSI of 0.65 compared to 0.57 for the community-based raised garden.

21.6.4 Lessons Learned

The lessons learned reveal the potential of both hydroponic and community gardens to support urban environments. Hydroponic gardens excel in resource efficiency, particularly water usage, making them highly effective for producing abundant food in resource-limited urban areas. Meanwhile, community gardens foster social connections, promote gardening education, and strengthen community bonds. The study suggests that incorporating both types of gardens in cities can yield optimal results. However, hydroponic gardens demonstrate greater resiliency and offer higher economic benefits while incurring lower costs. Continuous advancements can further enhance urban farming practices, including research and experimentation with innovations, such as vertical gardens, aquaponics, and green roofs.

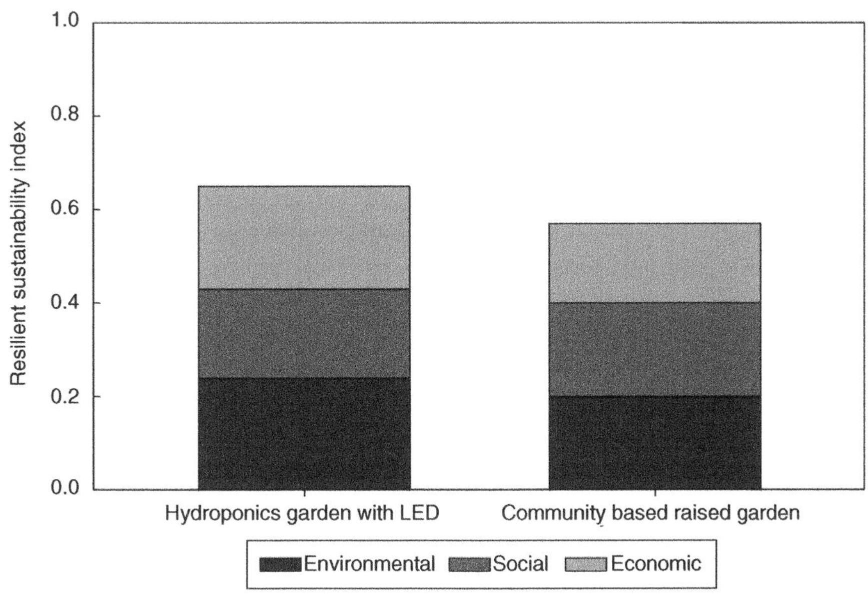

Figure 21.37 Integrated resilient-sustainability index of hydroponics garden and community-based raised garden.

21.6.5 Conclusions

The results from the comprehensive comparative resilient sustainability assessment of hydroponic gardens and community-based raised gardens suggest that hydroponic gardens are more resilient and sustainable. The hydroponics garden had an RSI of 0.65 compared to 0.57 for the community garden. Hydroponics offers significant economic benefits due to high yield per area and reduced reliance on pesticides, contributing to lower long-term costs. One major limitation of the study included the difficulty of fully incorporating all relevant aspects when comparing alternatives for food production that share many social similarities. For example, hydroponics can serve as a social community connection spot year-round versus seasonally in community gardens. However, this benefit may not be immediately apparent to the stakeholders. Hydroponics offers unique, sustainable benefits and resiliency, but there are areas of opportunity for improvement. The system can be further optimized to reduce the impacts associated with energy consumption, such as improving the room insulation, using a more efficient heating system, or adopting a renewable energy source.

21.7 Resiliency and Sustainability Evaluation of Green Infrastructure in a Chicago Neighborhood

21.7.1 Problem Statement

For over 100 years, the city of Chicago has undertaken significant civil engineering endeavors, including reversing the flow of the Chicago and Calumet Rivers and establishing various drainage districts, stormwater systems, and levee districts in the counties surrounding the Chicagoland area to manage its water quantity and quality (Barker 1986; Cain 1978). Despite these measures, urban flooding remains a persistent challenge, making it imperative to consider strategies to mitigate and safeguard the city's infrastructure and inhabitants. When the local storm sewer systems cannot effectively handle the runoff, they are discharged into rivers, streams, and, ultimately, into Lake Michigan. This discharge is called the combined sewer overflow (CSO) (MWRDGC 2023). CSO events in the city of Chicago have become more and more frequent due to an increase in storm events, which contribute to runoff excess.

Additionally, the outdated infrastructure of sewer systems exacerbates the problem. Moreover, the lack of evapotranspiration on flat concrete streets and insufficient green spaces for stormwater infiltration further contribute to the issue. Furthermore, these CSO events pose a significant public health hazard. They result in untreated sewage backing up, blocking roadways, polluting waterways, and often causing flooding in residential areas. The Tunnel and Reservoir Plan (TARP) was introduced in the 1960s to address these issues, which included three large reservoirs, four deep tunnels, and over 100 miles of underground tunnels to capture overflow (Dept. of Water Management, City of Chicago 2016). TARP's primary objectives were to reduce flooding, improve city water quality, and provide protection during large storms. Despite TARP's implementation, the Chicagoland area has experienced increased flood damage (Hawthorne 2020). The plan has incurred \$4 billion for Chicago; however, specific neighborhoods continue to suffer annual damages exceeding \$2 million.

Further, data analysis reveals that urban flooding problems in Chicago are not distributed evenly. The aging sewer network in the city, coupled with the intensifying storms resulting from climate change, has made urban flooding a pressing issue, specifically as it disproportionately affects the underprivileged communities on the south and west sides of Chicago. As the percentage of households of color in an area increases, there is an exponential rise in the number of flood insurance claims. Thirteen zip codes account for 75% of the flood damage claims paid in Chicago, with 93% of the residents being people of color and 25% living below the poverty line.

Green infrastructure, including bioswales and permeable pavement, offers resilient solutions that can provide immediate social and environmental advantages to these communities while maintaining economic viability (Flynn and Traver 2013; Weiss et al. 2019). Additionally, green infrastructure provides many health benefits, reduces the urban heat island effect, enhances air quality (Deshmukh et al. 2019), increases rainwater infiltration, and can even positively influence the mental health of the residents in the community. This study systematically evaluates the resiliency and sustainability of the baseline model of conventional pavements and green infrastructure stormwater control measures in Humboldt Park, Chicago. The main goal was to analyze the eco-friendly infrastructures compared to conventional roadways and concrete, focusing on their resiliency to extreme weather events, long-term changing climate, and overall sustainability.

21.7.2 Project Background

The impact of urban flooding can vary within a city, with different neighborhoods experiencing varying levels of rainfall during storms. To address this issue, green infrastructure implementation in specific neighborhoods provides a localized solution for flood-prone areas. In this case, we will analyze the potential reduction of volumetric stormwater flows in Chicago by evaluating the resiliency and sustainability of two green infrastructure designs implemented in Humboldt Park. Humboldt Park, located in the western part of Chicago, has been selected for this study due to its history of frequent flooding. The inadequate water infiltration in Humboldt Park is primarily attributed to its soil composition. Approximately 50% of the soil in this area consists of clay and silty soil, which hinders the natural infiltration of water (USDA 2019). Urban flooding has caused significant consequences, including property damage and negative effects on residents' lives, mental well-being, and work. As Humboldt Park is mainly a residential area, the study focused on six blocks, totaling 31.6 acres (Figure 21.38). The results of this study can potentially be applied to adjacent blocks in the future as they face similar circumstances.

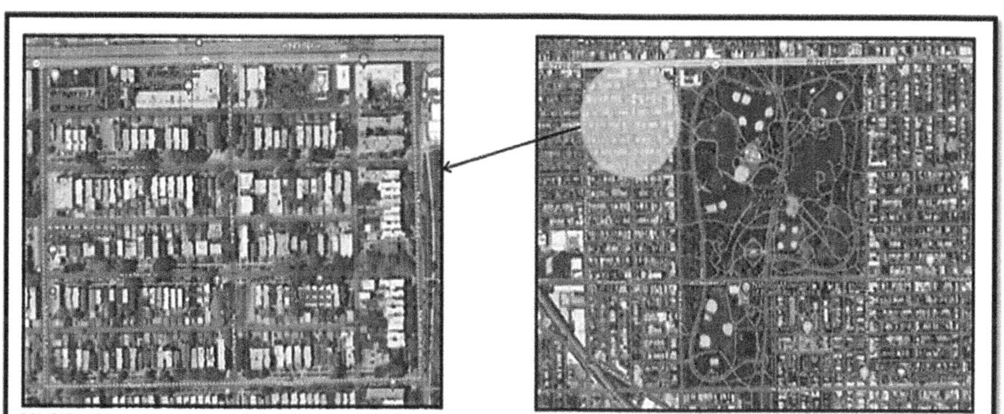

Figure 21.38 Northwest 6-Block Area of Humboldt Park (Chicago, US).

21.7.2.1 Technical Design

Existing Condition

Land use, soil type, roads, and other geographical features data were gathered from the City of Chicago, the USA Elevation data, and USGS. The sub-catchment areas were determined to facilitate hydraulic computations. The curve number, a widely used parameter ranging from 0 to 100, indicates the runoff potential of an area, with higher values signifying lower infiltration (Beven 1986). An average curve number was computed for each sub-catchment, with an overall average curve number of approximately 92, indicating a highly impermeable surface.

Data collection for the hydraulic design in the study area involved gathering information on rainfall events from the Chicago Stormwater Management Department. Traditional stormwater systems are typically designed for 5- to 25-year storm events, but this project considers a 50-year, 24-hour storm. This analysis will offer valuable insights into the resiliency of the system. The rainfall pattern follows the SCS Type II method, specific to the Chicago region. The square footage of each catchment area was determined, and a runoff coefficient was assigned and averaged for each sub-catchment. By utilizing Personal Computer Storm Water Management Model (PCSWMM), a software developed by the USEPA, calculations were performed to determine the total runoff volume and peak discharge for each sub-catchment and the combined runoff volume and peak discharge.

Proposed Green Infrastructure (GI)

The technical design for the proposed green infrastructure involved exploring various best management practices, such as green roofs, dry wells, and wet ponds, to determine the most suitable option for the community. Due to the limited space for detention ponds and the predominance of residential buildings, the study focused on public right-of-way areas like sidewalks, alleys, roads, and parkways. Bioswales and permeable pavement were identified as the green infrastructure for parkways and sidewalks/alleys, respectively. These measures aimed to reduce the overloading of the local storm sewer system and CSO. The decision to implement permeable pavement was based on its ability to infiltrate a significant portion of annual rainfall, reduce peak discharge rates, and divert stormwater from the system. However, bioswales were chosen for their effectiveness in slowing runoff velocity, purifying water, and replenishing the groundwater table. These choices were validated through a thorough evaluation involving literature and volumetric hydraulic calculations. Implementing this green infrastructure not only addresses the issue of excess runoff but also enhances the overall design by increasing the time of concentration for runoff, a critical design constraint.

Professionally designed and consistently maintained permeable pavement diminishes peak runoff flow rates and absorbs a substantial portion of runoff volume. In terms of its effectiveness during the harsh winter conditions in Chicago, it maintains its infiltration capacity (Weiss et al. 2019). Permeable pavement necessitates less salt for equivalent safety in snow melting procedures, thereby lessening water runoff pollution (Weiss et al. 2019). The enhancements outlined in Figure 21.39 involve the replacement of current surfaces.

The total storage capacity of the proposed green infrastructure was calculated considering the voids in the engineered soils, subgrade aggregate, and silty clay soils. This calculation aimed to demonstrate the structure's ability to store or infiltrate water and divert it from the storm sewer system. To maximize stormwater retention, the City of New York Department of Environmental Protection provided two designs from their Green Infrastructure Design and Construction Standard Drawings, aligning with our goals. These designs included deeper aggregate subgrades and stone columns of varying sizes beneath them. An active Excel sheet, developed by the city of New York in

Figure 21.39 Typical catchment design of proposed bioswales and proposed permeable pavers.

conjunction with their standard drawings, was utilized to assess the storage capacity of the proposed design. The bioswales exhibited a volumetric water storage of 157 984 cubic feet, while the permeable pavement showed a storage capacity of 196 986 cubic feet. Thus, implementing the green infrastructure improvements could divert 354 972 cubic feet of water. The calculations suggest that the total runoff under existing conditions amounts to 581 580 cubic feet. However, by implementing the proposed green infrastructure, it is possible to achieve a significant reduction of 61% in water runoff. This reduction will significantly minimize the strain on local sewers and reduce the risk of flooding. Figure 21.40 illustrates the storage volume hydrograph, representing the decrease in the overall inflow of cubic feet per second (cfs) into the designated area. Implementing these improvements can yield social, economic, and environmental benefits by reducing the intensity and frequency of flooding events. Native vegetation in the bioswales will attract wildlife and enhance the community's aesthetics. Further, numerous studies have indicated that the inclusion of vegetation leads to a reduction in air pollution (Deshmukh et al. 2019).

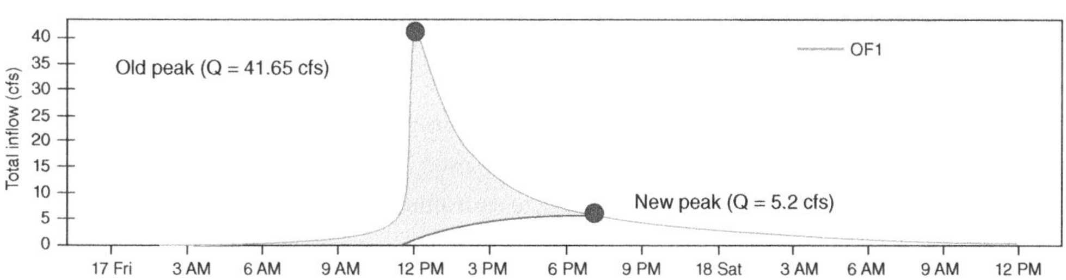

Figure 21.40 Storage volume hydrograph comparison of peak cubic feet per second (cfs) between existing conditions and proposed improvements.

21.7.3 Integrated Resiliency and Sustainability Assessment

21.7.3.1 Resiliency Assessment

The resiliency assessment indicators were selected by considering factors related to climate change impacts, including temperature, heat index, wind speed, precipitation, wildfire, and flooding, as well as stakeholder preferences and concerns. Table 21.10 presents the indicators under each resiliency requirement. The assessment identified three primary shocks that could adversely affect the project: outdated storm sewer systems, temperature fluctuations (due to climate change and urban

Table 21.10 Resiliency indicators, metrics, and tools to assess conventional and green infrastructure.

Resiliency requirements	Criteria	Indicators	Metric	Tool
Technical resiliency	Performance deterioration	Structural integrity	Rating on a scale of 0 to 4	Expert rating
		Death of vegetation		
		Material degradation		
	Time for deterioration	Structural integrity		
		Death of vegetation		
		Material degradation		
	Time in deteriorated state	Structural integrity		
		Death of vegetation		
		Material degradation		
	Performance recovery	Structural integrity		
		Death of vegetation		
		Material degradation		
	Time for recovery	Structural integrity		
		Death of vegetation		
		Material degradation		
Cascading environmental impacts	Air	Increase in urban heat island effect	Rating on a scale of 0 to 4	Expert rating
	Land	Drought		
		Soil fertility		
Cascading social impacts	Social individual	Impacts of individual due to failure	Rating on a scale of 0 to 4	Expert rating
	Socio-community	Loss of community assets		
		Residents move out rate increase		
Cascading economic impacts	Direct cost	Cost of repair damage	Rating on a scale of 0 to 4	Expert rating
		Maintenance cost		
		Damage to agricultural land		
	Indirect cost	Agriculture product demand increase		

Note: Rating based on level of negative impact: 0-extreme; 1-very high; 2-high; 3-medium; 4-low.

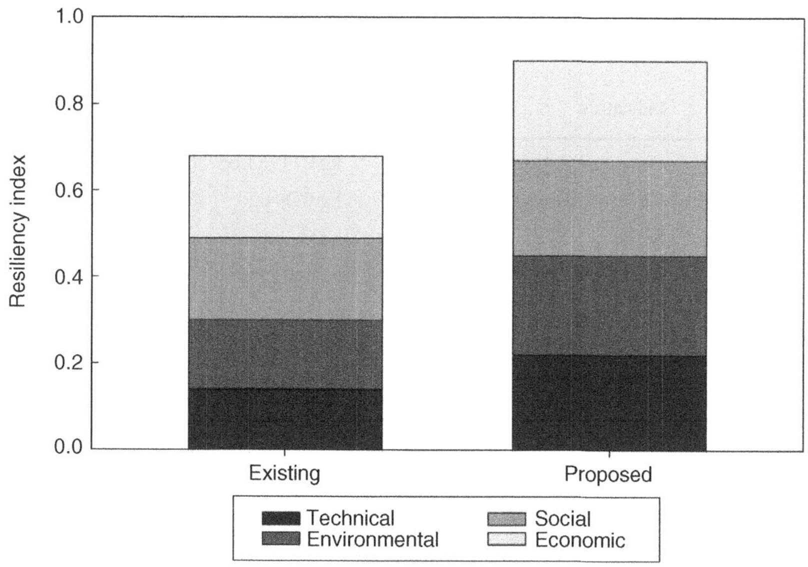

Figure 21.41 Comparison of resiliency indices of existing conditions (baseline model) versus proposed green infrastructure.

heat islands), urban flooding (attributed to climate change), and sudden economic changes caused by events, such as war, funding challenges, or cost uncertainties.

Figure 21.41 compares the resiliency scores for the baseline model and the proposed green infrastructure. Results demonstrate a significant resiliency improvement, with the baseline model scoring 0.68 compared to 0.9 for the proposed green infrastructure. This enhancement is primarily due to stormwater management features, such as bioswales and permeable pavements, which reduce runoff volume, control peak flow rates, improve water quality, and support biodiversity. Additionally, these features provide socio-economic benefits and exhibit superior durability, longevity, and adaptability to changing environmental conditions, ensuring sustained performance over time.

21.7.3.2 Sustainability Assessment

The first step of an integrated sustainability assessment involves the identification of environmental, social, and economic indicators, as mentioned before. Specific indicators identified in this case study, and the corresponding metrics and tools used for the quantification of the indicators are presented in Table 21.11. A detailed description of quantification methodology followed and the results of each pillar of sustainability assessment along with the overall integrated RSI are presented in this section.

Environmental Sustainability

The environmental sustainability assessment of alternative options was performed using SimaPro software. The environmental impacts obtained from SimaPro were considered indicators under four criteria: air, water, human health, and land. The assessment compares the construction of green infrastructure to current conventional conditions in residential neighborhoods, focusing on Humboldt Park in Chicago. The focus is on replacing cast-in-place concrete alley slabs and sidewalks with precast permeable pavers. Additionally, bioswales are being implemented in the parkway areas situated between sidewalks and curb/gutter.

Table 21.11 Sustainability indicators, metrics, and tools for conventional and green infrastructure systems.

Sustainability requirement	Criteria	Indicators	Metric	Tool
Environmental sustainability	Air	Ozone depletion	kg CFC-11 eq	LCA
		Global warming	kg CO_2 eq	
		Smog	kg O_3 eq	
		Air quality improvement	Scaled from 0 to 1	Expert rating
	Water	Acidification	kg SO_2 eq	LCA
		Eutrophication	kg N eq	
		Hydrological cycle volume infiltrated	Scaled from 0 to 1	NYC GI guide
	Human health	Carcinogenic	CTUh	LCA
		Non-carcinogenic	CTUh	
		Respiratory effects	kg PM2.5 eq	
	Land	Ecotoxicity	CTUe	
		Energy use	Scaled from 0 to 1	NYC GI guide
		Fossil fuel depletion	MJ surplus	LCA
	Resiliency	Soil fertility	Rating on a scale of 0 to 4	Scores from resiliency assessment
		Erosion		
		Impact to endangered species		
Social sustainability	Social individual	Impact of flooding	Rating on a scale of 0 to 4	Survey
		Degree of comfort		
		Mental and physical health		
		Property damage		
		Safety hazards		
		Impact on education/business		
	Socio-community	Degree of disruption		
		Infrastructure development (streets and water system)		
	Socio-economic	Lost tourism revenues		
		Maintenance cost		
		Damage to roadways		
		Damage to agricultural land		
	Socio-environmental	Energy consumption (construction and maintenance)		
		Impact of greenhouse gas emissions		
	Resiliency	Pedestrian walkways and their connectivity	Rating on a scale of 0 to 4	Scores from resiliency assessment
		Population with disrupted transportation access		

Table 21.11 (Continued)

Sustainability requirement	Criteria	Indicators	Metric	Tool
Economic sustainability	Direct costs	Materials and equipment	USD	Cost analysis
		Labor		
		Transportation and disposal		
		Fuel (Hrs burned)		
		Operation and maintenance		
	Indirect costs	Risk of injury	Rating on a scale of 0 to 4	Expert rating
		Reducing costs of using vehicles by improving walkability		
		Air pollutant removal		
		Social cost of CO_2	USD	USEPA cost of carbon
	Direct benefits	Water quality cost	USD	Benefits analysis
		Reduced insurance flooding claims		
		Increased property values		
		Future investment	Rating on a scale of 0 to 4	Expert rating
	Indirect benefits	Reduced energy cost (pumping)	USD	Benefits analysis
		Enhanced employment opportunities	Rating on a scale of 0 to 4	Expert rating
	Resiliency	Damage to properties	Rating on a scale of 0 to 4	Scores from resiliency assessment
		Maintenance cost		
		Tourism revenue		

Note: Rating based on level impact: 0-unacceptable; 1-diminished; 2-no impact; 3-improved; 4-ideal.

The functional unit considers six sub-catchments over a 30-year design life, focusing on stormwater bioretention benefits. System boundaries for the LCA encompassed manufacturing, transportation, construction, and material disposal. The life cycle inventory data for the alternatives are presented in Table 21.12.

The LCA using SimaPro (Figure 21.42) reveals a significant environmental impact from the suggested green infrastructure. This impact arises from the need for additional engineered soil layers and extensive subgrade excavation to meet stormwater design specifications. However, this effect is temporary, as subsequent maintenance activities will have a significantly smaller footprint. Manufacturing significantly contributed to the environmental impacts of the proposed permeable pavers. Moreover, constructing permeable pavers involves more excavation for the subgrade layers, resulting in larger quantities of disposal and longer transportation hours. Figure 21.42 shows a comparison of the environmental impact for various scenarios, including those without bioswales, with bioswales, existing pavers, and permeable pavers. The results indicate that overhauling existing conditions to implement green infrastructure would significantly affect most impact categories in the study.

Table 21.12 Life Cycle Inventory for the green infrastructure: bioswales and permeable pavers.

		Quantity				
		Bioswales		Permeable pavers		
	Type	Conventional	Proposed	Conventional	Proposed	Unit
Manufacturing	Concrete	0.00	991.01	3033.45	5983.88	ton
	Gravel round	0.00	4959.90	3296.70	6503.20	ton
	Sand	1539.30	4734.45	0.00	900.00	ton
	Asphalt	0.00	138.60	0.00	0.00	ton
	Cardboard	0.00	0.15	0.00	0.15	ton
	Limestone crushed	0.00	3218.00	0.00	3700.00	ton
Transportation	Truck	24 772.41	213 378.57	345 963.97	756 488.60	ton.km
On-site work	Excavation	2932.00	10 087.00	4495.00	7279.00	cu.yd
	Backfilling	1466.00	10 627.00	2997.00	8512.00	cu.yd

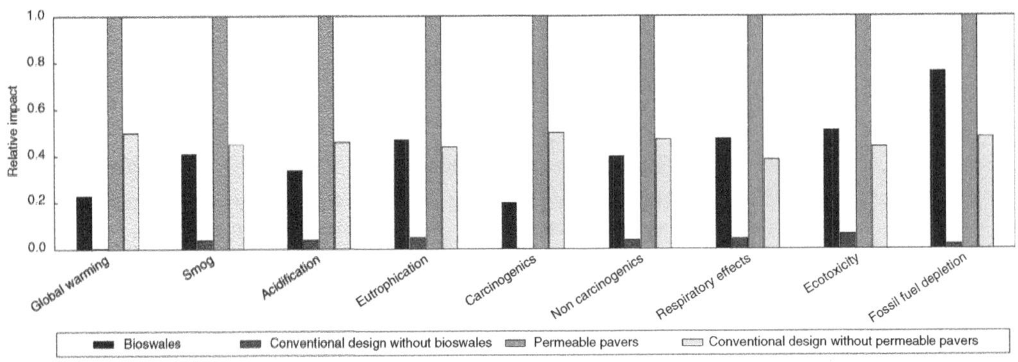

Figure 21.42 Environmental impact for various scenarios: no bioswales, bioswales, existing pavers, and permeable pavers.

Social Sustainability

A public opinion survey was conducted to better understand the social impacts experienced by Chicago residents and to actively engage them in the project. The survey utilized the SSEM methodology. The survey was structured around five key indicators: individual social impact, community social impact, socio-economic factors, socio-environmental factors, and resiliency. It comprised 36 questions, and the results were analyzed and standardized to quantify social sustainability. Of the 23 participants, 85% were Chicago residents, with 5% being green infrastructure experts, whose input was emphasized over non-expert responses. Over half of the respondents reported that flooding had impacted their mental, social, or physical well-being. Approximately 30% of participants reported property damage requiring repairs due to flooding. Additionally, 85% of participants supported allocating tax funds to implement green infrastructure in Chicago. The primary concerns identified by participants regarding sustainability and social justice included walkable communities, infrastructure development (streets and water systems), and social equity, health, safety, and wellness.

The SSEM results demonstrate a more positive social sustainability for the proposed green infrastructure than the alternative to the existing conditions (no intervention). The EJScreen tool and the public survey reveal that flooding significantly affects individual and community well-being, emphasizing the need for efficient stormwater management solutions. The widespread approval for dedicating tax revenues to implementing green infrastructure in Chicago, mainly in Humboldt Park, highlights the public acknowledgment of its social advantages. The SSEM results show that the proposed green infrastructure positively affects individual, community, economic, and environmental dimensions of social sustainability, along with resiliency. The key concerns raised by participants, such as pedestrian-friendly neighborhoods and infrastructure enhancement, highlight the importance of integrating social justice and sustainability principles into urban planning. The results presented in Figure 21.43 advocate for green infrastructure as a vital component of urban development, promoting resiliency, equity, and community well-being.

Economic Sustainability

The economic sustainability assessment was based on the direct cost estimate, indirect cost estimates, and direct/indirect benefit estimates. Some indicators, including the social cost of CO_2, were sourced from published literature and organizations like the USEPA. The economic sustainability results are presented in Figure 21.44, comparing the baseline model (existing infrastructure) and the proposed GI. According to the city of Chicago's best management practices guidelines, permeable pavement costs twice as much as conventional options. Even though the direct costs for the proposed GI are higher in all indicators compared to the existing conditions, the indirect costs were

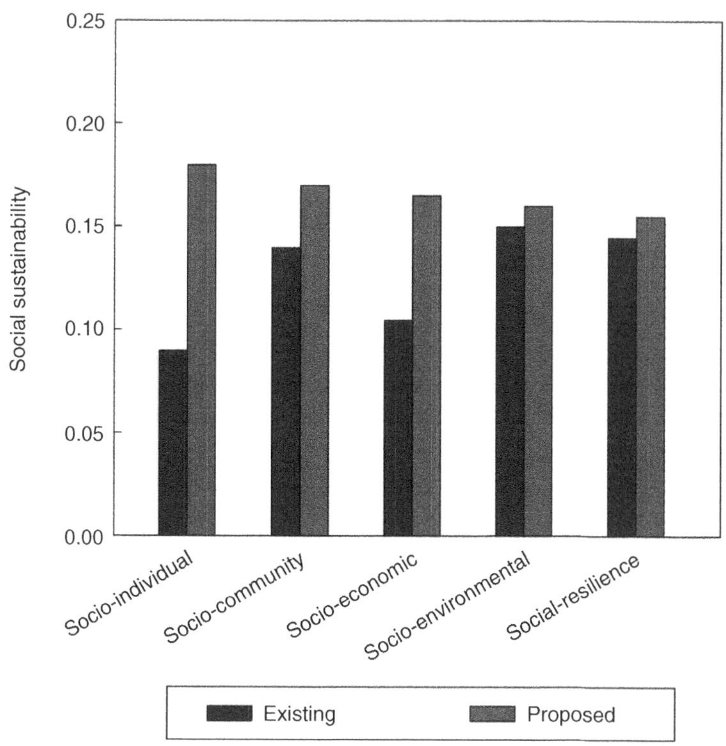

Figure 21.43 Social sustainability assessment results.

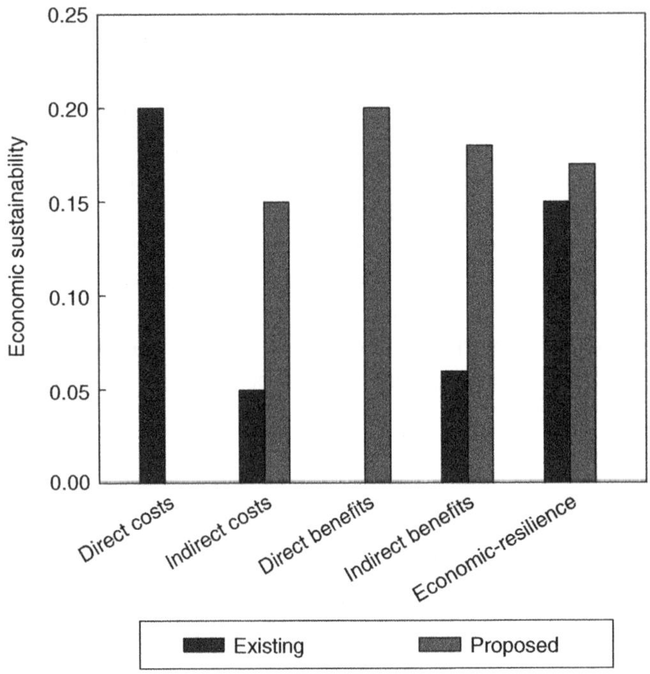

Figure 21.44 Economic sustainability assessment results.

lower. Moreover, the proposed GI's direct and indirect benefits are significantly greater and outweigh its costs. Additionally, green infrastructure offers economic benefits by mitigating infrastructure damages from flooding.

Comparing the baseline model to the proposed GI provides compelling evidence for its economic sustainability and resiliency. Despite higher material and labor costs, the benefits of GI, such as increased property values and investment prospects, establish it as a cost-efficient solution with substantial long-term advantages.

21.7.3.3 Integrated Resilient Sustainability Index

After evaluating social, environmental, and economic sustainability, along with resiliency aspects, the integrated RSI was calculated as a weighted sum. The results were normalized and presented in Figure 21.45, and the indicators, metrics, and tools used are presented in Table 21.11. The assessment determined that the proposed green infrastructure is more resilient and sustainable than conventional pavements, walkways, and alleys. The results show that the resiliency of the proposed design to stresses and climate impacts is significantly higher than the existing conditions. Moreover, despite higher initial costs and associated environmental impacts, the social and economic benefits outweigh these drawbacks, making the proposed green infrastructure both resilient and sustainable.

21.7.4 Lessons Learned

While green infrastructure may require higher initial investments, it provides long-term benefits by mitigating environmental risks, improving social well-being, and delivering economic advantages. This stresses the importance of forward-thinking in infrastructure planning and investment. This study emphasizes the critical role of engaging stakeholders and incorporating community feedback

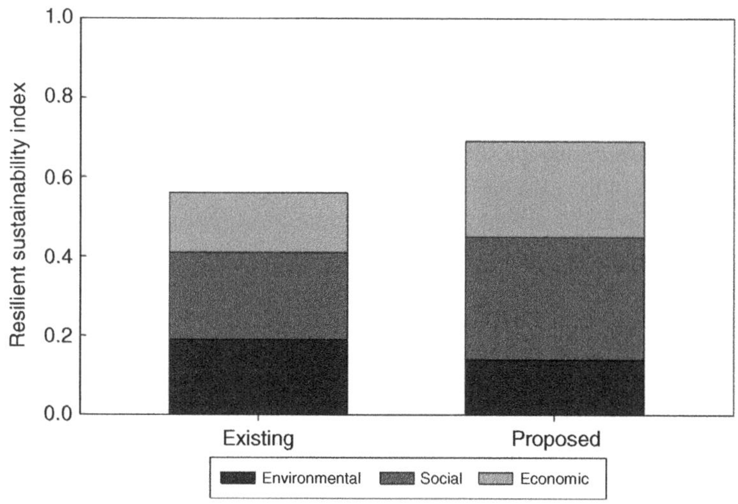

Figure 21.45 Resilient sustainability index.

in developing sustainable infrastructure projects. We gain valuable insights into local priorities, concerns, and preferences by actively involving residents, experts, and policymakers through surveys, consultations, and public discussions. This participatory approach promotes ownership, fosters trust, and ensures that infrastructure solutions align with the community's needs and values. Additionally, our analysis has underscored the importance of resiliency in infrastructure design and planning. Our analysis underscores the critical need to integrate sustainability and resiliency into infrastructure planning. Addressing disruptions such as outdated storm sewer systems, climate change, and economic uncertainties enhances the ability of infrastructure projects to withstand future challenges.

21.7.5 Conclusions

GI has emerged as an effective solution to mitigate the impacts of flooding in urban areas. GI encompasses sustainable techniques like bioswales and permeable pavement, which are resilient to climate change impacts and other risks. Green infrastructure reduces surface runoff and minimizes flooding risks compared to traditional pavement. Bioswales, for instance, are vegetated channels designed to intercept, collect, and transport stormwater runoff while simultaneously removing pollutants. Permeable pavement allows water to infiltrate the ground, recharging groundwater and reducing runoff. The incorporation of GI practices has shown that they can significantly reduce runoff volume by 61% and decrease peak discharge by 88%.

This study's sustainability and resiliency assessment suggests that GI practices are more resilient than conventional pavers and green sidewalks against climate impacts and stresses. Although implementing green infrastructure requires more materials and maintenance and may have temporary environmental consequences, its long-term ecological benefits make it more sustainable. A case study in Humboldt Park, Chicago, demonstrated that implementing green infrastructure, such as bioswales and permeable pavement, involves high initial costs and temporary environmental consequences. However, they play a crucial role in mitigating the impacts of flooding and creating more resilient and sustainable communities. This study encourages investment in developing and implementing GI in underprivileged communities with low median income, facing higher flood risk and social vulnerability.

References

ASCE (2023a). *Standard Practice for Sustainable Infrastructure.* Reston, VA: ASCE. https://doi.org/10.1061/9780784415733.

ASCE (2023b). *Integrating Resilience and Sustainability into Civil Engineering Projects* (ed. C. Field and C. Zawislak). Reston, VA: ASCE. https://doi.org/10.1061/9780784484814.

Barker, B. (1986). Lake Diversion at Chicago. *Case Western Reserve Journal of International Law* 18: 203. https://scholarlycommons.law.case.edu/jil/vol18/iss1/10.

Barthel, S. and Isendahl, C. (2013). Urban gardens, agriculture, and water management: sources of resilience for long-term food security in cities. *Ecological Economics* 86: 224–234.

Beven, K. (1986). Runoff production and flood frequency in catchments of order n: an alternative approach. In: *Scale Problems in Hydrology: Runoff Generation and Basin Response* (ed. V.K. Gupta et al.), 107–131. Dordrecht: Springer Netherlands.

Cain, L.P. (1978). *Sanitation Strategy for a Lakefront Metropolis: The Case of Chicago.* Northern Illinois University Press. ISBN: 9780875800646.

Dalkey, N. and Helmer, O. (1963). An experimental application of the Delphi method to the use of experts. *Management Science* 9 (3): 458–467.

Davidson, C.I., Phalen, R.F., and Solomon, P.A. (2005). Airborne particulate matter and human health: a review. *Aerosol Science and Technology* 39 (8): 737–749.

Dept. of Water Management, City of Chicago (2016). *Stormwater Management Ordinance Manual.* City of Chicago. Retrieved 13 December 2024 https://www.chicago.gov/content/dam/city/depts/water/general/Engineering/SewerConstStormReq/2016StormwaterManual.pdf.

Deshmukh, P., Isakov, V., Venkatram, A. et al. (2019). The effects of roadside vegetation characteristics on local, near-road air quality. *Air Quality, Atmosphere & Health* 12: 259–270.

Dieter, C.A., Maupin, M.A., Caldwell, R.R. et al. (2018). *Estimated Use of Water in the United States in 2015.* U.S. Geological Survey Circular.

Flynn, K.M. and Traver, R.G. (2013). Green infrastructure life cycle assessment: a bio-infiltration case study. *Ecological Engineering* 55: 9–22.

Garcia, J.A. and Alamanos, A. (2023). A Multi-Objective Optimization Framework for Water Resources Allocation Considering Stakeholder Input. ECWS-7 2023, 32. https://doi.org/10.3390/ECWS-7-14227.

Gilmore, K.R. (2016). Teaching life cycle assessment in environmental engineering: a disinfection case study for students. *The International Journal of Life Cycle Assessment* 21 (12): 1706–1718. https://doi.org/10.1007/s11367-016-1138-1.

Goldstein, M.L. and Ritterling, J.M. (2001). A practical guide to estimating cleanup costs. *Remediation Journal: The Journal of Environmental Cleanup Costs, Technologies & Techniques* 11 (4): 103–121.

Hawthorne, M. (2020). Heavy rain overwhelms $3 billion deep tunnel again, forcing sewage into Lake Michigan. Chicago Tribune. https://www.chicagotribune.com/news/environment/ct-heavy-rains-sewage-chicago-20200515-4rxpjn7hargw3avja6igpvwgxu-story.html.

Heron, G., Van Zutphen, M., Christensen, T.H., and Enfield, C.G. (1998). Soil heating for enhanced remediation of chlorinated solvents: a laboratory study on resistive heating and vapor extraction in a silty, low-permeable soil contaminated with trichloroethylene. *Environmental Science & Technology* 32 (10): 1474–1481.

ITRC (Interstate Technology & Regulatory Council) (2021). *Sustainable Resilient Remediation SRR-1.* SRR team. Washington, DC, USA: ITRC. www.itrcweb.org.

Kim, S., Ceylan, H., Gopalakrishnan, K. et al. (2011). Comparative performance of concrete pavements with recycled concrete aggregate (RCA) and virgin aggregate subbases. In: *Transportation and*

Development Institute Congress 2011: Integrated Transportation and Development for a Better Tomorrow, pp. 710–719. https://doi.org/10.1061/41167(398)68.

Li, X., Cundy, A.B., Chen, W., and Lyu, S. (2021). Systematic and bibliographic review of sustainability indicators for contaminated site remediation: comparison between China and western nations. *Environmental Research* 200: 111490.

Mackay, D.M. and Cherry, J.A. (1989). Groundwater contamination: pump-and-treat remediation. *Environmental Science & Technology* 23 (6): 630–636.

Martin, M., Poulikidou, S., and Molin, E. (2019). Exploring the environmental performance of urban symbiosis for vertical hydroponic farming. *Sustainability* 11 (23): 6724.

MECC (2023). *Pressure in the Distribution System*. Mountain Empire Communty College. https://water.mecc.edu/concepts/pressuredistr.html.

Menconi, M.E., Heland, L., and Grohmann, D. (2020). Learning from the gardeners of the oldest community garden in Seattle: resilience explained through ecosystem services analysis. *Urban Forestry & Urban Greening* 56: 126878.

Metcalf, L. and Eddy, H.P. (2003). *Wastewater Engineering*, 4e. McGraw-Hill.

MWRD (2023). *Stickney Water Reclamation Plant*. Metropolitan Water Reclamation District of Greater Chicago. https://mwrd.org/locations/stickney-water-reclamation-plant.

MWRD (2024). *Metropolitan Water Reclamation District of Greater Chicago*. https://mwrd.org/what-we-do/documents-and-data/water-reclamation-plant-data-and-reports.

Peruzzini, M., Gregori, F., Luzi, A. et al. (2017). A social life cycle assessment methodology for smart manufacturing: the case of study of a kitchen sink. *Journal of Industrial Information Integration* 7: 24–32.

Quinn, J., O'Hara, S., Krug, T. et al. (2005). Emulsified zero-valent iron-treatment of chlorinated solvent DNAPL source areas. In: *US EPA Workshop on Nanotechnology for Site Remediation*.

Reddy, K.R., Sadasivam, B.Y., and Adams, J.A. (2014). Social sustainability evaluation matrix (SSEM) to quantify social aspects of sustainable remediation. In: *ICSI 2014*, pp. 831–841. https://doi.org/10.1061/9780784478745.078.

Reddy, K.R., Ghimire, S.N., Wemeyi, E. et al. (2020). Life cycle sustainability assessment of geothermal heating and cooling system: UIC case study. In: *E3S Web of Conferences*, Volume 205, 07003. https://doi.org/10.1051/e3sconf/202020507003.

Reddy, K.R., Robles, J.R., Carneiro, S.A.V., and Chetri, J.K. (2022). Tiered quantitative assessment of life cycle sustainability and resilience (TQUALICSR): framework for design of engineering projects. In: *Advances in Sustainable Materials and Resilient Infrastructure* (ed. K.R. Reddy, R.K. Pancharathi, N.G. Reddy, and S.R. Arukala), 1–19. Springer Singapore. https://doi.org/10.1007/978-981-16-9744-9_1.

Reddy, K.R., Janga, J.K., Verma, G., and Nagaraja, B. (2024). Integration and quantification of resilience and sustainability in engineering projects. *Journal of the Indian Institute of Science* 104 (2): 435–488.

Russo, A., Escobedo, F.J., Cirella, G.T., and Zerbe, S. (2017). Edible green infrastructure: an approach and review of provisioning ecosystem services and disservices in urban environments. *Agriculture, Ecosystems & Environment* 242: 53–66.

Samangooei, M., Sassi, P., and Lack, A. (2016). Soil-less systems vs. soil-based systems for cultivating edible plants on buildings in relation to the contribution towards sustainable cities. https://urn.fi/urn:nbn:de:hebis:34-2016050350228 (accessed 14 December 2024).

Seyrfar, A., Osman, I., Yarveicy, V. et al. (2021). Life cycle sustainability assessment of flexible and rigid pavement designs: a case study of Illinois Interstate Highway. In: *Transportation Research Board 100th Annual Meeting*, Washington, DC, USA, TRBAM-21-03544.

Sharma, H.D. and Reddy, K.R. (2004). *Geoenvironmental Engineering: Site Remediation, Waste Containment, and Emerging Waste Management Technologies*. Wiley.

Sullivan, C., Hallaran, T., Sogorka, G., and Weinkle, K. (2015). An evaluation of conventional and subirrigated planters for urban agriculture: supporting evidence. *Renewable Agriculture and Food Systems* 30 (1): 55–63.

UN DESA (2016). *The Sustainable Development Goals Report 2016 – July 2016.*New York, USA: United Nations Department of Economic and Social Affairs (UN DESA).

USDA (2019). *Web Soil Survey*. US Department of Agriculture. https://websoilsurvey.nrcs.usda.gov/app/ (accessed 14 December 2024).

USDOE (2023). *Guidelines for Estimating Unmetered Landscaping Water Use*. U.S. Department of Energy. https://www.energy.gov/femp/articles/guidelines-estimating-unmetered-landscaping-water-use.

USEPA (2001). *A Citizen's Guide to Excavation of Contaminated Soil*. U.S. Environmental Protection Agency (USEPA). https://semspub.epa.gov/work/05/953560.pdf (accessed 14 December 2024).

USEPA (2017a). *Water Recycling and Reuse: The Environmental Benefits*. United States Environmental Protection Agency. https://19january2017snapshot.epa.gov/www3/region9/water/recycling/ (accessed 14 December 2024).

USEPA (2017b). Air sparging. In: *How to Evaluate Alternative Cleanup Technologies for Underground Storage Tank Sites: A Guide for Corrective Action Plan Reviewers*. https://www.epa.gov/ust/how-evaluate-alternative-cleanup-technologies-underground-storage-tank-sites-guide-corrective (accessed 14 December 2024)

USEPA (2017c). *The Social Cost of Carbon. Estimating the Benefits of Reducing Greenhouse Gas Emissions*. US Environmental Protection Agency. https://19january2017snapshot.epa.gov/climatechange/social-cost-carbon_.html (accessed 14 December 2024).

Weingaertner, C. and Moberg, Å. (2014). Exploring social sustainability: learning from perspectives on urban development and companies and products. *Sustainable Development* 22 (2): 122–133.

Weiss, P.T., Kayhanian, M., Gulliver, J.S., and Khazanovich, L. (2019). Permeable pavement in northern North American urban areas: research review and knowledge gaps. *International Journal of Pavement Engineering* 20 (2): 143–162.

Wilhelm, J.A. and Smith, R.G. (2018). Ecosystem services and land sparing potential of urban and peri-urban agriculture: a review. *Renewable Agriculture and Food Systems* 33 (5): 481–494.

Index

a

acidification impact category 218
acid rain 65–68
aesthetic degradation 85
AI. *see* artificial intelligence (AI)
air emissions 397–400
air pollution 63–65, 429
air sparging 506, 579
albedo 531
American Council of Engineering Companies (ACEC) 476
American Public Works Association (APWA) 476
American Society of Civil Engineers (ASCE) 34, 175–176, 466, 475, 476
  ASCE 7-22 176
  ASCE Policy Statement 27, 517
  infrastructure resilience division (IRD) 176
American Society of Testing and Materials (ASTM) framework 509, 510
ammonium ingestion 245
analytic hierarchy process (AHP) 364, 553
anthropogenic carbon dioxide emissions 397
anthropogenic/natural flows for elements 235–236
artificial intelligence (AI) 375–378
ASCE. *see* American Society of Civil Engineers (ASCE)
ash 416
Asphalt pavement embodied carbon tool (asPECT) 356
atom economy 432, 433
Attest Engagements on GHG Information 261

b

background consumption 146
bay area rapid transit (BART) 477
Beijing, population growth rate 8–9
benzene 63
Berryessa Transit Center 477–481
best management practices (BMPs) 476, 509
Big Data 376–378
bio-based materials 387–388
biochar 388
biodegradable wastes 434
biodiversity 60–61, 274
  loss of 17
biogas 77
biological oxygen demand (BOD) 70
biomass energy 413–416
biomimicry 375, 388–389
bioremediation 505, 579
bioswales 491, 492
BMPs. *see* best management practices (BMPs)
BREEAM tool 356
budget
  construction 238
  definition 236–237
  national 328
  of natural resources 236–237
Building Research Establishment Environmental Assessment Method (BREEAM) 230, 457
built-infrastructure projects
  stormwater treatment/green infrastructure 486–493
  transportation infrastructure 484
  water and wastewater treatment 485

Sustainable and Resilient Engineering: Drivers, Metrics, Tools, and Applications, Second Edition. Krishna R. Reddy, Claudio Cameselle, and Jeffrey A. Adams.

Companion website: www.wiley.com/go/reddy/sustainableresilientengineering2e

c

California Green Remediation Evaluation Matrix (GREM) 513
carbon calculator by EFFC (European Federation of Foundation Contractors), and DFI (Deep Foundations Institute) 356
carbon capture and sequestration 403
carbon capture and storage (CCS) 527
carbon capture and underground storage 405
carbon dioxide emissions 521, 523
carbon dioxide removal (CDR) methods 271
  direct engineered capture 530, 531
  marine organism sequestration 529–530
  subsurface sequestration 527–528
  surface sequestration 528–529
carbon disclosure project (CDP) 117
carbon footprint 220, 356
  case study 264–270
  emission measurement 257–261
  GHG emission mitigation 270–272
  GHG inventory 262
  global warming potential 256–257
  independent verification 261
  offset unavoidable emissions 258
  scope of 257
  standards for calculating 261
  targets and strategies 258
  tools for GHG inventory 262–263
  University of Vigo 264–270
carbon markets 274
carbon monoxide 63, 64
carbon nanotubes (CNTs) 384
carbon neutral approaches 454–455
CDR methods. *see* carbon dioxide removal (CDR) methods
chemical oxidation technology 505
chemicals of potential concerns (COPCs) 345
Chicago neighborhood, green infrastructure in
  problem statement 616–617
  resiliency assessment 620–621
  sustainability assessment 621–626
  technical design 618–619
Chicago's sustainable food systems
  community gardens *vs.* hydroponics 605–616
  problem statement 605
  resiliency assessment 606–607
  sustainability assessment 607–615
  technical design of alternatives 605–606
Chile's wildfires by drought and record heat 2016 165–166
chlorination 555, 556, 576
chlorofluorocarbons (CFCs) 61
chromium cycle 238
circular economy 442
  engineered materials for 383–384
  resilient 444
City of Chicago, pedestrian infrastructure 316–318
civil infrastructure
  "C" grade 475
  Envision rating system 476–481
  report card 475
Clean Air Act (1970) 64
Clean Air and Clean Water Acts 79
clean energy 403–405
climate action plans 272–274
climate change 11, 48–57, 158–163
  and extreme weather impacts 15–6
Climate Change Action Plan 258
climate geoengineering
  CDR methods
    applicability of 535–536
    direct engineered capture 530, 531
    marine organism sequestration 529–530
    subsurface sequestration 527–528
    surface sequestration 528–529
  risks and challenges 536–538
  SRM methods
    applicability of 535–536
    reflectors and mirrors 534–535
    sulfur injection 532–534
  theoretical framework 536
climate leadership in parks tool (CLIP) 263
Climate Mapping for Resilience and Adaptation (CMRA) 558
climate resilience evaluation and awareness tool (CREAT) 3.0 178, 229
climate "tipping point" 532
CO_2 emissions 54
Committee of Institutional Means for Assessment of Risks to Public Health 344
community and institutional structures 103
community resources 103

comparable emissions database (CEDB) 263
Comprehensive Assessment System for Built Environment Efficiency (CASBEE) 230
Comprehensive Environmental Response, Compensation, and Liability Act (CERCLA) 80, 81, 430
concentrated animal feed operations (CAFOs) 107
conceptual site model (CSM) 503
constructed wetlands 491, 492
containment systems 506, 507
contaminated land remediation 503–504
contaminated site remediation 576–594
  groundwater remediation 581
  monitoring 581
  problem statement 576–577
  resiliency assessment 583
  soil remediation 579
  sustainability assessment 583–594
  technical design of alternatives 577–581
conventional LCA 335–337
copper use 238
cost–benefit analysis 290, 359, 551
"cradle to grave" 429
criteria maximum concentration 72
cycle 237

d

2024 DANA 165
data gap analysis 286
dechlorination 555, 556, 576
decline of ecosystems 16
deforestation 59–60, 502
democracy and governance 100
desertification 57–59
Design for Environment (DfE) approach 426
developed and developing societies 99–100
Development, and Demonstration programs (DOE, 2018) 527
digital twinning 380–382
direct engineered capture 530, 531
disaster debris management 443
disaster resiliency
  challenges 195
  2016 Chile's wildfires by drought and record heat 165–166
  climate change and extreme events 158–163
  2024 DANA 165
  2024 Dubai flooding 165
  2024 Hurricane Helene 164
  2024 Hurricane Milton 164
  2012 Hurricane Sandy in New York City 163
  managed retreat and cost-benefits 194
  overview of 157–158
  resiliency (*see* resiliency)
  soft infrastructure 194–195
  2017 worst South Asian monsoon floods 166
distributional justice 108
diversity 100
Division for Sustainable Development Goals (DSDG) 26
DOE's Clean Coal Research 527
domestic hidden flows (DHF) 247
dose–response assessment 346–347
droughts 161–162
dry lands 57
dual-phase extraction 506
2024 Dubai flooding 165
durable goods calculator (DGC) 263

e

Earth observation (EO) 378
Easter Island 4–5
Eaton Fire in Altadena 166
eco-design and green materials 426–427
ecological resiliency 166
ecological risk assessment 350–352
economic assessment framework 111–113
economic assessment tools 358–359
  cost–benefit analysis 359
  life-cycle costing 358
economic input–output (EIO) analysis 220–221
economic input–output life-cycle analysis (EIO-LCA)
  available models 330
  boundary definition and circularity effects 330
  vs. conventional LCA 335–337
  environment sector 330
  industry transactions 330
  interpretation of results 331
  online tool 330
  projects using 332–334
  uncertainty 331–332

economic input–output (EIO) model
  cause-and-effect relationship 329
  constant technical coefficients 329
  GDPs 328
  input–output transaction table 328–330
  limitations 330
  mass balance markets 329
  national budgets 328
  NFAs 328
  vs. physical input–output (PIO) analysis 337–339
  two-sector economy 328, 329
economic issues 110–115
  economic assessment framework 111–113
  life cycle costing 113–114
  true-cost accounting 114–115
economic resiliency 167
economic sustainability 551
  Chicago neighborhood, green infrastructure in 625–626
  Chicago's sustainable food systems 613–614
  contaminated site remediation 590–591
  highway pavement 602–603
  water disinfection technologies 573–574
ecosystem capital 16
ecosystems, decline of 16
ecosystem services valuation tools 360
eco-toxicity impact category 218
effective concentration (EC) 346
electrokinetic remediation (EK) 577, 579
electrokinetics 505
electronic waste (e-waste) 79
elemental cycles 136–141
"elephants and scorpions" analogy 247
embodied carbon 454–455
Embodied Carbon in Construction Calculator (EC3) 356
employee commute emissions calculator 263
emulsified zerovalent iron nanoparticles (EZVI) 577, 579, 587, 590
endpoint analysis 291
end-use sector emissions 398–399
energy budget, atmosphere 547, 548
energy efficiency and reduced consumption 403
energy efficient buildings 403
energy infrastructure 473
energy-intensive aggressive technologies 508
energy recovery 435
engineering resiliency 167
environmental accounting 111
Environmental and Social Impact Assessment (ESIA) 102
environmental assessment tools/indicators 356–358
environmental CBA (eCBA) 358
environmental concerns 48
  acid rain 65–68
  aesthetic degradation 85
  air pollution 63–65
  deforestation 59–60
  desertification 57–59
  eutrophication 74
  global warming and climate change 48–57
  land contamination 83–84
  land use patterns 85
  loss of habitat and biodiversity 60–61
  noise pollution 86
  odors 84–85
  ozone layer depletion 61–62
  salinity 74–75
  smog 65
  thermal pollution 85–86
  visibility 84
  waste generation and disposal 75–83
  water usage and pollution 68–73
environmental health risk assessment
  ecological risk assessment 350–352
  emergence of the risk era 344
  risk assessment and management 344–350
environmental impact assessment (EIA) 102, 290
environmental impacts
  current products and activities 285
  of energy generation
    air emissions 397–400
    land resource use 401
    solid waste generation 400
    water resource use 401
  matrix for accounting the 307
  relative 307–309
  when selecting between several product and activity options 286
environmentalism 22, 97
environmental justice (EJ) 107–110

Environmental Justice Screening and Mapping Tool (EJSCREEN) 361
Environmental Justice Strategic Enforcement Assessment Tool (EJSEAT) 361
environmental justice tools 361–362
environmental LCA (eLCA) 358
environmental LCC (eLCC) 358
environmentally responsible product matrix (ERP matrix) 307
environmental pollution 6–7
environmental product declarations (EPDs) 427
environmental remediation techniques 10
Environmental, Social and Governance (ESG) Reporting 116–117
environmental sustainability 550
 Chicago neighborhood, green infrastructure in 621–624
 Chicago's sustainable food systems 611–612
 contaminated site remediation 583–589
 highway pavement 599–601
 water disinfection technologies 570
Envision 476–481
EPA's electronic greenhouse gas reporting tool (e-GGRT) 263
equity 100, 107
euryhalines 75
eutrophication 74
eutrophication impact category 218
events emissions calculator 263
excavation 579
excessive nutrient loading 71
expanded SLCA 310–312
exposure assessment 347–349
ex-situ remediation method 504
extended producer responsibility (EPR) 427–428
extreme events 158–163
extreme weather events 160

f

Federal Highway Administration (FHWA) 356
federal remediation technologies roundtable (FRTR) 579
Federal Water Pollution Control Act of 1948 71
financial CBA (fCBA) 358
financial LCC (fLCC) 358
fleet emissions calculator 263
flooding/drought 57
floods 161–162
foreign hidden flows (FHF) 247
fossil fuels
 coal 126
 combustion 398–400
 natural gas 127–129
 oil shales and tar sands 127
 petroleum 126–127
Fourth National Climate Assessment (NCA4) 160
fracking 127
fresh water 134–135
full environmental LCC (feLCC) 358
Fuzzy Evaluation for Life Cycle Integrated Sustainability Assessment (FELICITA) 362–363
fuzzy logic 362
FY2008 Consolidated Appropriations Act 262

g

GaBi 550
geographic information systems (GIS) 379
geothermal energy 410–413
GHG inventory 220
GHG protocol calculation tools and guidance 263
global carbon cycle 53–54, 136, 521, 522
Global Change Research Act of 1990 160
global climate change impact category 218
global reporting initiative (GRI) standards 117
global warming 48–57
global warming potential (GWP) 49–52, 256–257
governance issues 115–116
graphene 384
gray infrastructure 194
green accounting 111
green and sustainable remediation (GSR) 501
 technologies 504–509
green building
 average savings of 453
 components of
  features of 455–458
  geothermal energy system 457
  hot water generation with solar energy and circulation system 456
  solar power 455, 456
  stormwater harvesting 455

green building (*cont'd*)
concepts 454–455
design and operation 453
history 452
impact of 453
LEED 457–466
resiliency considerations 466–468
Green Building Certification Institute (GBCI) 462
green chemistry 432–433
green engineering (GE) 29
Greener Cleanup Matrix 513
greenhouse effect 12, 48–49
greenhouse gases (GHGs) 11–14, 218–219, 255, 384, 430, 500
CDR 526
climate change and extreme weather impacts 15–6
developments in the United States 262
global warming potential 256–257
mitigation 270–272
SRM methods 527
UNFCCC 538
greenhouse gas reporting program (GHGRP) 262
green infrastructure 389, 486
GreenLITES 356
green remediation 500
Greenroads 356
green roofs 487, 488
green stormwater infrastructure
bioswales 491, 492
constructed wetlands 491, 492
green roofs 487, 488
permeable pavements 487–489
rain gardens and planter boxes 491
rainwater harvesting 488, 490
tree canopies 491
green vehicles 404
greenwashing 514–515
GREET model 263
gross domestic products (GDPs) 328, 501
groundwater plumes 508, 578
groundwater remediation 505, 581

h

hard infrastructure 473
hazard identification 345–346
hazard index (HI) 350
hazardous waste
integrated waste management 437–440
landfill 438
RCRA regulations 429–430
HBS Tata Hall Executive Education Center 463–466
high-level waste (HLW) 132
highway pavement
problem statement 595
resiliency assessment 597–599
sustainability assessment 599–604
technical design of alternatives 595–597
horizontal drilling 127
hot mixed asphalt (HMA) 595
human health cancer impact category 218
human health noncancer impact category 218
human health particulate matter (PM) impact category 218
human population 7–9
2024 Hurricane Helene 164
2024 Hurricane Milton 164
Hurricane Sandy in New York City 2012 15–16, 163
hydrofluorocarbons (HFCs) 256
hypoxia 74

i

Illinois EPA 514
Indian Ridge Marsh (IRM) 514
Industrial Emissions, Acid Rain Impact from 66–68
industrial smog 65
industrial solid wastes 429
industrial waste 242
infrastructure. *see* civil infrastructure; sustainable infrastructure
infrastructure resilience division (IRD) 176
innovative energy storage technologies 386–387
in-situ remediation method 504
in-situ soil flushing 505
in-situ vitrification (ISV) 505
institute for sustainable infrastructure (ISI) 476
integrated resilience assessment tools 230, 364
integrated resiliency and sustainability assessment 551–553
Chicago neighborhood, green infrastructure in 620–626

Chicago's sustainable food systems 606–615
contaminated site remediation 576–594
typical highway pavement 595–604
water disinfection technologies 558–574
integrated resilient engineering 35
integrated resilient sustainability index
Chicago neighborhood, green infrastructure in 626
Chicago's sustainable food systems 615
contaminated site remediation 591–594
highway pavement 603–604
water disinfection technologies 574
integrated sustainability assessment tools 230, 362–364
Integrated Value Model for Sustainability Assessment (MIVES) 364, 564, 574
integrated waste management 437–440
Intergovernmental Forum on Forests (IFF) 60
Intergovernmental Panel on Climate Change (IPCC) 49, 175
Assessment Report 56–57, 523
Intergovernmental Panel on Forests (IPF) 60
International Auditing and Assurance Standards Board (IAASB) 261
international financial reporting standards (IFRS) sustainability disclosure standards 117
International Geosynthetics Society (IGS) sustainability calculator 356
International standard on Assurance Engagements 261
Internet of Things (IoT) 382
Interstate Technology and Regulatory Council (ITRC) framework 509–510
INVEST (Infrastructure Voluntary Evaluation Sustainability Tool) 356
IPAT equation 20–21
iron fertilization 529, 531
ISO 14040:2006 283
ISO 14044:2006 283

k

Kyoto Protocol (1997) 274

l

land contamination 83–84
land degradation 57–59, 499
landfill gases 416
landfill-induced groundwater contamination 11
landfills 7, 10, 11, 40, 53, 76–80, 82–84, 92, 188, 190–193, 209, 238, 239, 244, 256, 259, 263, 287, 306, 383, 385, 397, 400, 401, 405, 413–416, 429, 435, 436, 439–442, 444, 464, 473, 477, 500, 504, 508, 512, 514, 579, 590, 597
hazardous waste 438
nonhazardous waste 437, 438
land management, sustainable and resilient
challenges and opportunities 514–516
contaminated land remediation approach 503–504
green and sustainable remediation technologies 504–509
problems 501–503
sustainable remediation indicators, metrics, and tools 512–513
sustainable remediation framework 509–511
land resource use 401
land use patterns 85
large-scale sulfur dioxide emissions 63
LCA. *see* life cycle analysis (LCA)
leachate 76–77
leachate-based groundwater pollution 76
leached solid wastes 10
lead 63
Leadership in Energy and Environmental Design (LEED) 221, 356
certification an attractive benefit 462
credentials 462
energy and atmosphere 459
GBCI 462
HBS Tata Hall Executive Education Center 463–466
indoor environmental quality 459
innovation 460
integrative process 459
life-cycle aspects 459
locations and transportation 459
materials and resources 459
regional priority 460
scoring system
certification level based 460, 462
for new constructions 460, 461
sustainable sites 459
water efficiency 459
LEED. *see* Leadership in Energy and Environmental Design (LEED)

legal issues 141–142
less/least developed countries (LDCs) 99–100
lethal concentration (LC) 347
levelized cost of energy (LCOE) 219, 416
life cycle analysis (LCA) 32, 113, 441, 509
  definition and objective 280–281
  framework and guidance 284–285
  goal and scope definition 285–286
  history 282–284
  interpretation 294
  ISO standards 281–282
  LCI 286–289
  LCIA 289–294
  overview 279–280
  procedure 281–282
  tools and applications 296–299
life-cycle assessment (LCA) 220, 550, 570
life-cycle costing (LCC) 113–114, 358–359, 551
life cycle impact assessment (LCIA) 289–294
life cycle inventory (LCI)
  feedback loop 288
  format 289
  material flows within systems 288
  outputs from raw materials 288
  specific stages for the life cycle 288
  system boundary 286–289
  treatment of data collection 288
  unit of function 288
  unresolved problems 289
light detection and ranging (LiDAR) sensors 378
London Fog 65
loss of habitat 60–61
lowest observable effect concentration (LOEC) 347
low impact development (LID) 486
low/intermediate level radioactive waste (LILW) 132

m

management of waste
  energy recovery 435
  green chemistry 432–433
  hierarchy of 429
  integrated 437–440
  landfills 437
  minimize and/or reduce waste generation 433–434
  pollution prevention 431–432
  reuse/recycling 434–435
March Point Landfill site 188
marine organism sequestration 529–530
Master Equation 20–21
material flow analysis (MFA)
  chromium cycle 238
  copper use 238
  wastes 242–245
material flow and budget analysis 220
material recovery facility (MRF) 434
material recycling 243
maturity 100
maximum allowable toxicant concentration (MATC) 347
measured mineral resource 132–133
medical waste 79
Medical Waste Tracking Act (MWTA) 79
mental and physical health 101
metallic ores consumption 5–6
metals 435
methane (CH_4) 536
midpoint effect 291
Millennium Ecosystem Assessment 16, 60
mineral resource 132–133
Minnesota Pollution Control Agency 514
misconsumption 146
money economy 110
monitored natural attenuation (MNA) 513
monsoon floods 166
more/most developed countries (MDCs) 99–100
MOtor Vehicle Emission Simulator (MOVES) 263
multi-criteria decision analysis (MCDA) 364, 553
municipal solid waste (MSW) 10–11, 75, 416, 429

n

nanomaterials 383–384
National Ambient Air Quality Standards (NAAQS) 64
national budgets 328
National Emission Standards for Hazardous Air Pollutants (NESHAPs) 64

National Financial Accounts (NFAs) 328
national material account (NMA) 245–250
National Pollution Discharge Elimination System (NPDES) 72, 486
National Priority List (NPL) 430
National recycling coalition's (NRC) environmental benefits calculator 263
national stormwater calculator with climate assessment tool 177–178
natural eutrophication 74
natural habitats 487
natural resources
  budget of 236–237
  Easter Island example 4–5
  elemental cycles 136–141
  fossil fuels 126–132
  metallic ores consumption example 5–6
  mineral resource 132–133
  radioactive fuels 132
  resource depletion 141–147
  water resources 134–136
nature-based solutions 387–389
net present value (NPV) 416
nitrogen oxide emissions 397
nitrous oxide (N_2O) 64, 256
NOAA sea level rise viewer 178
NOAA Surging Seas 178
noise pollution 86
nonhazardous waste
  landfill for 437
  municipal solid waste 429
  RCRA regulations 429
no observed adverse effects level (NOAEL) 347
northeast recycling council's environmental benefits calculator 263
noxious by-product chemicals 7
nuclear energy 401–403

o

odors 84–85
office footprint calculator™ 263
off-set unavoidable emissions 258
OpenLCA 550
optimization model for reducing emissions of greenhouse gases from automobiles (OMEGA) 263
organic wastes 70
overconsumption 146
overexploitation for fossil fuels 502
overgrazing 501
overshooting 146
ozone-depleting substances (ODS) 62
ozone depletion 61–62
  impact category 218

p

Palisades Fire in Pacific Palisades 166
paper 435
paper use emissions calculator 263
Paris Agreement (2015) 274
particulate matter 63
passive solar heating 405
Pavement Life-cycle Assessment Tool (PaLATE) 356
perfluorocarbons (PFCs) 256
permeable pavements 487–489
permeable reactive barriers (PRBs) 506
persistent organic pollutants (POPs) 83
personal GHG emissions calculator 263
photochemical smog 65
  formation impact category 218
photovoltaic cells 405
physical economy 110
physical input–output (PIO) analysis 337–339
phytoremediation technology 439
plasma arc torch methods 439
plastics 435
political and social resources 103
pollution 502
pollution controls 403
pollution prevention (P2) 431–432
polylactic acid (PLA) 387
polynuclear land use approach 19
polyvinyl chloride (PVC) 558
population characteristics 103
poverty 101
precipitation 56
primary pollutants 63
procedural justice 108
product life cycle 279, 307–309
projections in global surface temperatures 521, 525

public policies and legislative actions 286
public water system (PWS) 485
pump-and-treat systems 581

q
quality of life 100
Quantitative Assessment of Life Cycle Sustainability (QUALICS) 363
quantitative environmental health risk assessment 221

r
radioactive fuels 132
radioactive waste 79
rainwater harvesting 488, 490
RCRA corrective measure study 504
RCRA Subtitle C regulations (USEPA 2017b) 429
RCRA Subtitle D regulations (USEPA 2017b) 429
recognition justice 108
recycled content (ReCon) tool 263
recycled waste materials 383
recycling 9, 434–435, 438, 439
reference dose (RfD) 347–349
reinforced concrete 555
remedial technology
  bioremediation 508
  cap, vertical barrier and bottom barrier 506, 507
  GSR criteria 507–508
  incineration 507
  pump-and-treat operations 507
  pumping well systems 506, 507
  saturated zone/groundwater 504
  subsurface drain system 506, 507
  vadose zone soil remedial methods 504–505
remote sensing 378–380
renewable energy 273, 403
  biomass energy 413–416
  geothermal energy 410–413
  solar energy 405–406
  water energy 409–410
  wind energy 406–409
renewable feedstock 432
Renewable Portfolio Standards (RPSs) 417
representative concentration pathways (RCPs) 523
Resilience and Disaster Recovery (RDR) Tool Suite 229
Resilience-based Earthquake Design Initiative (REDi™) rating system 228
resiliency 33
  assessment framework 178–183
  assessment tools 228–229
  cascading impacts of 182–183
  considerations 418
  ecological 166
  economic 167
  engineering 167
  indicators 221–226
  infrastructure 176–178
  initiatives and policies on 172–176
  metrics 226–228
  resilient environmental remediation 188–190
  resilient materials 443
  San Francisco firehouse resilient design 184
  San Francisco resilient CSD design 184–188
  social 167
  vs. sustainability 34, 168–169
  technical 180–182
resiliency assessment 546–549
  Chicago neighborhood, green infrastructure in 620–621
  contaminated site remediation 583
  highway pavement 597–599
  water disinfection technologies
    cascading resiliency impacts 561–564
    climate concerns 558
    overall resiliency index 564–567
    technical/engineering resiliency indicators 558–561
Resiliency Index (RI) 581
resilient circular economy 444
resilient engineering 33–34
  integrated 35
resilient environmental remediation 188–190
Resilient Node Cluster Analysis Tool (ReNCAT) 229
resilient waste management 443–444
Resource Conservation and Recovery Act 1976 (RCRA) 10, 79, 429
resource conservation, in energy sector 384–387
resource depletion
  causes of 141–142

description 141
effects of 142–145
overshooting 146
urban metabolism 147
restorative justice 108
retrofitting 431
reuse 9
reuse of treated wastewater 486
rising temperatures 158–159
risk assessment 290
risk assessment and management
dose–response assessment 346–347
exposure assessment 347–349
hazard identification 345–346
risk characterization 349–350
schematic 344
risk characterization 349–350
robotics 380
Rockefeller Foundation's 100 Resilient Cities Initiative 173

S

Safe Drinking Water Act (SDWA) 72
salinity 74–75
San Francisco firehouse resilient design 184
San Francisco resilient CSD design 184–188
sanitary landfills 437
SBToolPT 356
SBToolPT-UP 356
SDGs. *see* Sustainable Development Goals (SDGs)
sea level change curve calculator tool 177
sea level rise 160
secondary and/or refined/synthesized products 16
secondary pollutants 63
sediment runoff 7
sensors 378–380
Sertraline 32
sewage-borne pathogens 71
Sewer System Improvement Program (SSIP) 184
shared socioeconomic pathway (SSP) 521
shipping emissions calculator 263
SimaPro 280, 550, 587, 599, 621, 623
SiteWise^{TM} 513
SLCA. *see* streamlined life-cycle assessment (SLCA)
"SMART" 512, 550
smart technologies 376–382
smog 65
snow cover 57
social CBA (sCBA) 358
social cohesion/interconnection 100
social impact assessment (SIA) 102–106
social indicators 100–102
social injustice 17–19
social issues
developed and developing societies 99–100
equity 107
environmental justice 107–110
social impact assessment 102–106
social indicators 100–102
social sustainability implementation 107
social vulnerability assessment 99–100
society 98–99
sustainability concept 100
social LCA (sLCA) 358
social resiliency 167
social sustainability 550–551
Chicago neighborhood, green infrastructure in 624–625
Chicago's sustainable food systems 612–613
concept 100
contaminated site remediation 589–590
highway pavement 601–602
water disinfection technologies 571–573
social sustainability evaluation matrix (SSEM) 263, 513, 550, 624, 625
social sustainability implementation 107
social vulnerability assessment 99–100
societal LCC (sLCC) 358
society 98–99
soft infrastructure 194–195, 473
soil contamination 578
soil flushing 579
soil remediation 579, 593
soil stabilization 505
soil vapor extraction (SVE) 504–505, 577, 579
solar energy 405–406
solar power 455, 456
solar radiation management (SRM) methods
reflectors and mirrors 534–535
sulfur injection 532–534
solar thermal technology 405

solidification 505
solid waste 416, 437
Solid Waste Disposal Act 1965 78
solid waste generation 400
South Asian monsoon floods 2017, 166
Southern California wildfires 166
spreadsheets for environmental footprint analysis (SEFA) 550
Stickney Water Reclamation Plant (WRP), water disinfection technologies at 553–576
stormwater harvesting 455
stormwater infrastructure 486–493. *see also* green stormwater infrastructure
stormwater runoff 486
streamlined life-cycle assessment (SLCA) 220
  alternative graphical representation 308–309
  applications of 318–323
  environmental concerns 306–307
  expanded 310–312
  graphical representation 308–309
  life cycle stages 306–307
  matrix for accounting the environmental impacts 307–308
  overview 305
  product-related 306
  relative environmental impact 307–309
  simple example 312–318
subsurface contamination 500
subsurface sequestration 527–528
sulfur dioxide emissions 397
sulfur hexafluoride (SF6) 256
sulfur injection 532–534
Superfund Amendments and Reauthorization Act (SARA) 80–81
Superfund feasibility study 504
Superstorm Sandy 163
supplementary cementitious materials (SCMs) 429
surface capping 506
surface reflectivity 531
surface temperature 521, 525
sustainability 21–32, 47
sustainability assessment 549–551
  Chicago neighborhood, green infrastructure in
    economic sustainability 625–626
    environmental sustainability 621–624
    integrated resilient sustainability index 626
    social sustainability 624–625
  Chicago's sustainable food systems
    economic sustainability 613–614
    environmental sustainability 611–612
    integrated resilient sustainability index 615
    social sustainability 612–613
  contaminated site remediation
    economic sustainability 590–591
    environmental sustainability 583–589
    integrated resilient sustainability index 591–594
    social sustainability 589–590
  highway pavement
    economic sustainability 602–603
    environmental sustainability 599–601
    integrated resilient sustainability index 603–604
    social sustainability 601–602
  water disinfection technologies
    economic sustainability 573–574
    environmental sustainability 570
    integrated resilient sustainability index 574
    social sustainability 571–573
sustainability and resiliency indicators, attributes of
  actionable/achievable 206
  measurable 206
  relevant 206
  specific 206
  timely 206
sustainability assessment tools 219–221
sustainability evaluation matrix (SSEM) 513, 550
sustainability indicators 206–217
  fiscal sustainability 215
  in healthcare buildings 212–213
  for industrial organizations 214
  list of 216
  in manufacturing 211
  for UN Sustainable Development Goals 207–209
sustainability metrics 217–219
sustainable airports 484
sustainable and resilient energy engineering
  energy consumption 396
  energy sources in the United States 395–397
  environmental impacts of energy generation 396–401

nuclear energy 401–403
renewable energy 405–416
strategies for clean energy 403–405
sustainable buildings 466
Sustainable Development Goals (SDGs) 23–27, 207–208, 545
sustainable engineering (SE) 27–32
sustainable infrastructure
greenhouse gas (GHG) emissions 482–483
leadership 482
natural world 482
principles of 474
quality of life 482
resiliency considerations 494
resource allocation 482
of water infrastructure
green stormwater infrastructure 487–491
groundwater 485
public drinking water system 485
stormwater runoff 486
sustainable materials
eco-design and green materials 426–427
environmental product declarations 427
extended producer responsibility 427–428
sustainable remediation, defined 500
Sustainable Remediation Forum (SURF) framework 509
sustainable remediation framework
ASTM framework 510–511
ITRC framework 509–510
SURF framework 509
USEPA framework 509
sustainable remediation tool (SRT) 514
sustainable waste management 440–442
sustainable water treatment 485–486
synthetic biology (SynBio) 375, 389

t

tailings 400
task force on climate-related financial disclosures (TCFD) 117
telecommunications infrastructure 474
thermally enhanced remediation 508
thermal pollution 85–86
3D printing 380, 485
Tiered Quantification of Life Cycle Sustainability and Resiliency (TQUALICSR) 364–366, 467, 494, 546
Tool for Reduction and Assessment of Chemicals and Other Environmental Impacts (TRACI) 218, 297
Toolkit for Greener Practices 514
transportation infrastructure 473, 484
treated water 554
tree canopy 491
tropical storms 57
true-cost accounting (TCA) 114–115
two-sector EIO model 334

u

uncontrolled urban sprawl 19–20
UN Convention to Combat Desertification (UNCCD) 59
underground geologic carbon sequestration 527
The United Nations (UN) 173–174
United Nations Convention on Biological Diversity (UNCBD) 61
United Nations Convention to Combat Desertification (UNCCD) 501
United Nations Department of Economic and Social Affairs (UNDESA) 26
United Nations Environmental Program (UNEP) 481
United Nations Framework Convention on Climate Change (UNFCCC) 538
United States Army Corps of Engineers (USACE) 481
United States Environmental Protection Agency (USEPA) 504, 551
United States Geological Survey (USGS) 555
University of Vigo (UIC) 264–270
unmanned aerial vehicles (UAVs) 378
unsustainable agricultural activities 502
untreated water 554
urban metabolism 147
urban sprawl 19–20
US Climate Resilience Toolkit 558
US Department of Defense (DOD) 175
USEPA 262
USEPA Environmental Footprint Analysis Tool (SEFA) 513
USEPA green remediation framework. 509

US Federal Government Initiatives 173
US nonfuel minerals 235–236
US Presidential Policy Directives (PPDs) 172–173
US Resource Conservation and Recovery Act (RCRA) 10, 79, 429, 444, 499
UV disinfection 555–558, 576
UV radiation 560

V

vacuum-enhanced recovery 506
vadose zone soil remediation technologies 504–505
Valley Transportation Authority (VTA) 477
vehicular GHG emissions 258
visibility 84
vitrification (ISV) 505
volatile organic compounds (VOCs) 7, 506, 581

W

waste 242–245
  circular economy 442
  definition 428
  effects and impacts of 430
  gasification 439
  generation 9–11, 428–430
  generation and disposal 75–83
  generation flow 242
  hazardous waste 429
  nonhazardous waste 429
  streams 453
  sustainable waste management 440–442
  waste management
    energy recovery 435
    green chemistry 432–433
    hierarchy of 429
    integrated 437–440
    landfills 435
    minimize and/or reduce waste generation 433–434
    pollution prevention 431–432
    resilient 443–444
    reuse/recycling 434–435
waste reduction model (WARM) 263
waste-to-energy incinerators 435
waste to energy (WTE) process 78
water and environment infrastructure 473
water and wastewater treatment 485
water conservation 273–274
water disinfection technologies, Stickney Water Reclamation Plant 553–576
  chlorination/dechlorination 555
  effluent distribution pipelines 558
  problem statement 579–580
  resiliency assessment 558–567
  sustainability assessment 567–574
  technical design of alternatives 555 558
  UV disinfection 555–558
  water requirement 555
water energy 409–410
water footprint 135, 136
water infrastructure
  green stormwater infrastructure 487–491
  groundwater 485
  public drinking water system 485
  stormwater runoff 486
water pollution 7, 71
water resources 134–136
water resource use 401
water supply wells 485
water treatment 485–486
water usage and pollution 68–73
Western Australia Council of Social Services (WACOSS) 100
wetlands, constructed 491, 492
Whitmarsh Landfill 188
wildfires 62–163, 165–166
wind energy 406–409
World Commission on Environment and Development 168

Z

Zofnass Program for Sustainable Infrastructure 476

Printed and bound by CPI Group (UK) Ltd, Croydon, CR0 4YY
17/09/2026
14964335-0008